Tumors of the Fetus and Infant

Springer
New York
Berlin
Heidelberg
Barcelona
Hong Kong
London
Milan
Paris
Singapore
Tokyo

Hart Isaacs, Jr., MD

Associate Pathologist, Department of Pathology
Children's Hospital San Diego, and
Associate Clinical Professor of Pathology
University of California, San Diego School of Medicine
San Diego, California

Tumors of the Fetus and Infant

An Atlas

With 208 Figures in 585 Parts

Springer

Hart Isaacs, Jr., MD
Associate Pathologist
Department of Pathology
Children's Hospital San Diego
Associate Clinical Professor of Pathology
University of California, San Diego School of Medicine
San Diego, CA 92123-4282, USA
hisaacs@uscd.edu

Library of Congress Cataloging-in-Publication Data
Isaacs, Hart.
Tumors of the fetus & infant : an atlas / Hart Isaacs, Jr.
p. ; cm.
Includes bibliographical references and index.
ISBN 0-387-95186-5 (alk. paper)
1. Tumors in children—Atlases. 2. Infants—Diseases—Atlases.
3. Fetus—Diseases—Atlases. I. Title: Tumors of the fetus and infant. II. Title.
[DNLM: 1. Neoplasms—Infant—Atlases. 2. Neoplasms—embryology—Atlases. QZ 17
I73t 2002]
RC281.C4 I727 2002
618.92'992—dc21 2001054922

Printed on acid-free paper.

Production coordinated by Chernow Editorial Services, Inc., and managed by Lesley Poliner; manufacturing supervised by Joe Quatela.
Typeset by Matrix Publishing Services, York, PA.
Printed and bound by Maple-Vail Book Manufacturing Group, York, PA.
Printed in the United States of America.

9 8 7 6 5 4 3 2 1

ISBN 0-387-95186-5 SPIN 10789371

Springer-Verlag New York Berlin Heidelberg
A member of BertelsmannSpringer Science+Business Media GmbH

To my family,
without whose support
this atlas would not
have been possible

Preface

This atlas attempts to bring together in a simple and clear fashion the wealth of information presently available concerning neoplasia in the young. The purpose is to present a concise account, which will serve as the basis for a useful teaching aid, of the clinical and pathological features of neoplastic disease and tumorlike conditions in the fetus and infant.

Throughout I have used my own material as fully as possible and have introduced accounts of personally studied cases supplemented by ones from other colleagues illustrative of the subjects under discussion. For the same reason, I make frequent reference to my own published articles and to earlier publications, *Tumors of the Fetus and Newborn*, and the chapter entitled Tumors in *Potter's Pathology of the Fetus and Infant* (Refs. 1 and 2 of Chapter 1), the contents of which in many ways supplement the present work. Most photographs and tables have been produced from personally studied material, while others were taken, by permission, from various articles appearing in the literature.

The atlas is addressed primarily to pathologists, but I hope that clinicians, residents, and medical students in pathology and other medical specialties may find it useful and beneficial as a concise, easy-to-use, quick ready reference source.

HART ISAACS, JR., MD

Acknowledgments

My grateful thanks are due to my colleagues and friends who have assisted me in various ways during the preparation of this book, especially to all who have allowed me to use illustrations from their published work and to the publishers of journals and books from which they have been produced. Particular thanks is due to Dr. Henry Krous of the Department of Pathology, Children's Hospital, San Diego, for his continual encouragement and support. The help given me with the electron photomicrographs by Ann Peters of the Department of Pathology, Children's Hospital, San Diego, has been invaluable. I want also to thank my publishers for their helpfulness in the preparation and production of this atlas.

HART ISAACS, JR., MD

Contents

1
General Survey

Although tumors and tumorlike conditions of the fetus and infant occur infrequently, they present challenging diagnostic and treatment problems. There is considerable drama and terrible parental apprehension when dealing with a tumor in an infant or unborn child. Unfamiliarity with these conditions may lead to an erroneous diagnosis and unnecessarily aggressive therapy. Neoplasms and tumorlike conditions occurring during the first year of life are not entirely the same as those observed in the adolescent or adult—the types, incidence, and clinical features differ, as do behavior and response to treatment [1–14] (Tables 1.1–1.3).

No tumor of the adult grows as rapidly as does the normal embryo. In the early stages of development, normal embryonic cells may have some of the characteristics of neoplastic cells. Generally it is accepted by pathologists that the diagnosis of malignancy depends on certain microscopic criteria, but these are not always helpful or valid in a young child. Normal developing organs and tissues display significant mitotic activity and contain immature or embryonic-appearing structures that may simulate malignancy. It is often only from knowledge of the postnatal course of a tumor that definite proof can be obtained of its malignancy [2].

Tumors that are not histologically malignant may cause death in utero or shortly after birth because of their location—for example, a large lymphangioma, mature teratoma, or fibromatosis involving vital structures in the neck or mediastinum [2].

Tumors occurring in the perinatal period differ markedly in their presenting clinical signs and symptoms from those found in older individuals (Tables 1.3 and 1.4) [1,10,11]. Maternal dystocia may be the first sign of a large-space-occupying congenital tumor such as a giant intracranial or sacrococcygeal teratoma. Fetal hydrops is another unique manifestation [15]. Rupture during delivery of a large tumor, such as a neuroblastoma or a hepatoblastoma, with fatal exsanguination of the fetus, is a dramatic illustration of another unusual presentation.

The antenatal diagnosis of fetal neoplasms is one aspect of fetal disease that only recently has been implemented, allowing an understanding of the prenatal natural history and pathophysiology of these tumors [12,16,17]. Advances in imaging techniques, particularly ultrasonography and rapid magnetic resonance imaging, are responsible for making an early diagnosis possible. Tumors most often diagnosed by antenatal ultrasonography are teratomas, mesoblastic nephroma, liver tumors, and neuroblastoma [16,17].

Tumors Occurring in Patients with Congenital Malformations and Syndromes

An important relationship exists between certain congenital malformations and syndromes and the development of neoplasms [1,10,11,18–24] (Table 1.4). In this setting the tumor occurs after birth or later in life in individuals with specific inherited diseases, or congenital anomalies or malformation syndromes, many of whom probably have an underlying chromosomal defect (Table 1.5).

Hereditary conditions such as the neurocutaneous syndromes, also known as "*phakomatoses,*" are characterized by a high incidence of neoplasia. For example, tuberous sclerosis and neurofibromatosis, which are autosomal dominant, predispose the individual to the development of gliomas and malignant peripheral nerve tumors [20,21] (Tables 1.4 and 1.5).

Incidence

Tumors of the newborn are uncommon, with an incidence of 7.2 in 100,000 live births in the United Kingdom [3]. Most fetal and newborn tumors are benign rather than malignant, with cancer estimated to occur in 3.65 of 100,000 live births in the United States [13]. Malignant tumors, although rare, are an important cause of perinatal mortality.

Several series indicate that the most common malignancies noted at birth are neuroblastoma, leukemia, brain tumors, and sarcomas in that order [3,4,7,8,11,13] (Table 1.1). Almost all

TABLE 1.1. Comparison of five surveys consisting of 607 perinatal tumors.

Tumor	Institutions[a,b] CCRG [10]	HSC [4]	CHB [3]	CHLA [1]	CHSD (work in 1983–2000)	Total tumors[b]
Neuroblastoma	27 (26.7)	48 (47.0)	31 (18)	36 (21.6)	16 (19.3)	158 (26.0)
Teratoma	16 (15.8)		49 (29)	62 (37.1)[c]	25 (30.1)[d]	152 (25.0)
Leukemia	17 (16.8)	8 (7.8)	21 (12)	18 (10.8)	10 (12.1)	74 (12.2)
Brain tumors	12 (11.9)	9 (8.8)	14 (8)	15 (9.0)	13 (15.7)	63 (10.4)
Sarcoma	13 (12.9)	12 (11.8)	17 (10)	8 (4.8)	8 (9.6)	58 (9.6)
Renal tumors	8 (7.9)	4 (3.9)	8 (5)	6 (3.6)	4 (4.8)	30 (4.9)
Liver tumors	3 (3.0)	1 (1)	8 (5)	18 (10.8)	7 (8.4)	37 (6.1)
Retinoblastoma		17 (6.7)	14 (8)	4 (2.4)		35 (5.8)
Total[e]	101	102	170	167	126	607 (100)

[a] CCRG, Childhood Cancer Research Group, Oxford; HSC, The Hospital for Sick Children, Toronto; CHB, Children's Hospital, Birmingham, U.K.; CHLA, Children's Hospital of Los Angeles; CHSD, Children's Hospital of San Diego.
[b] Numbers in parentheses are percentage of total cases in study for the respective institution or category.
[c] Two of 62 with yolk sac tumor.
[d] Two of 25 with yolk sac tumor.
[e] Total for cases recorded in the respective study, not the total of the columns in the table.

TABLE 1.2. Differential diagnosis and distribution of abdominal masses in the newborn.

Gastrointestinal (15%)
- Intestinal atresia
- Imperforate anus
- Meconium plug
- Meconium ileus (cystic fibrosis)
- Meconium cyst
- Intestinal duplication
- Mesenteric cyst
- Lymphangioma
- Intussusception
- Pyloric stenosis
- Gastric teratoma

Liver and bile ducts (5%)
- Hemangioma
- Mesenchymal hamartoma
- Hepatoblastoma
- Polycystic disease
- Choledochal cyst
- Simple cyst
- Subcapsular hematoma
- Metastatic neuroblastoma, leukemia

Urinary (55%)
- Megacystis
- Hydronephrosis
- Hydroureter
- Urachal cyst
- Urachal diverticulum
- Multicystic kidney
- Polycystic kidneys
- Renomegaly (renal vein thrombosis)
- Mesoblastic nephroma
- Wilms' tumor
- Rhabdoid tumor

Adrenal (5%)
- Neuroblastoma
- Hematoma
- Cyst

Genital (15%)
- Hydrometrocolpos
- Ovarian cyst(s)
- Granulosa cell tumor
- Ovarian torsion

Miscellaneous (5%)
- Sacrococcygeal or retroperitoneal teratoma
- Fetus-in-fetu
- Anterior meningomyelocele
- Splenic hematoma
- Splenic cyst
- Subdiaphragmatic extralobar pulmonary sequestration

Source: Koop, CE [25]; Kirks DR, et al [26]; Schwartz MZ, Shaul DB [27]; Isaacs H Jr [1].

TABLE 1.3. Examples of unusual clinical presentations of perinatal tumors.

Maternal dystocia	Rupture of tumor during delivery with fatal exsanguination
Teratoma	Teratoma
Neuroblastoma	Neuroblastoma
Congenital mesoblastic nephroma	Hepatoblastoma
Astrocytoma	Hepatic hemangioma
Ependymoma	Congestive heart failure shortly after birth
Sialoblastoma	Arteriovenous malformation
Fetal hydrops	Cavernous hemangioma
Teratoma	Teratoma
Neuroblastoma	Congenital mesoblastic nephroma
Congenital mesoblastic nephroma	Astrocytoma (glioblastoma multiforme)
Wilms tumor	Cardiac rhabdomyoma
Hepatoblastoma	Cardiac myxoma
Cardiac rhabdomyoma	Cardiac fibroma
Astrocytoma (glioblastoma multiforme)	Polyhydramnios
Primitive neuroectodermal tumor	Teratoma
Lymphangioma (cystic hygroma)	Hemangioma
Hemangioma	Congenital mesoblastic nephroma
Neurofibroma	Wilms tumor
	Astrocytoma
	Bronchopulmonary myofibroblastic tumor

Source: Arceci RJ, Weinstein HJ [10]; Isaacs H Jr [1].

TABLE 1.4. Examples of syndromes and congenital malformations associated with childhood tumors.

Malformation or syndrome	Neoplasm
Beckwith-Wiedemann syndrome and hemihypertrophy	Wilms tumor Adrenocortical adenoma Adenocortical carcinoma Hepatoblastoma Pancreatoblastoma
Aniridia	Wilms tumor
Genitourinary system anomalies and Perlman syndrome	Nephroblastomatosis Wilms tumor
Hirschsprung disease	Neuroblastoma
Poland syndrome	Leukemia
Drash syndrome	Wilms tumor
Tuberous sclerosis	Cardiac rhabdomyoma Angiomyolipoma Astrocytoma
Multiple endocrine neoplasia (MEN-2)	Thyroid medullary carcinoma Pheochromocytoma Submucosal neuromas
Nevoid basal cell carcinoma (Gorlin) syndrome	Basal cell carcinoma Medulloblastoma Ovarian fibroma Cardiac fibroma
Wiskott-Aldrich syndrome	Lymphoma
Bloom syndrome	Leukemia
Down syndrome	Leukemia

Source: Bolande RP [28]; Mulvihill JJ [20]; Isaacs H Jr [1].

TABLE 1.5. Examples of chromosomal anomalies associated with childhood tumors.

Chromosomal defect	Childhood tumor
1p del, p32p36del, double minutes	Neuroblastoma
11p13 del , trisomy 18	Wilms tumor
13q14 del	Retinoblastoma Osteosarcoma
Trisomy 21 (Down syndrome), monosomy 7, t(11q23), t(1;22), t(9;11)	Leukemia Gonadoblastoma
Gonadal dysgenesis (46,XY; 45X/46,XY)	Germinoma
Klinefelter syndrome (XXY)	Teratoma
t(11;22)(q24;q12), t(21;22)	Ewing sarcoma and primitive neuroectodermal tumor (PNET)
t (1;13) t(2;13)	Alveolar rhabdomyosarcoma
del 1q	Astrocytoma

Source: Mulvihill JJ [20]; Isaacs H Jr [1].

tumors occurring during the first year of life are essentially the same histologically as those found later on, but the order of frequency of the neoplasms in the two age groups is different. Leukemia-lymphoma, brain tumors, neuroblastoma, and soft tissue sarcomas, in decreasing order of occurrence, are the more common neoplasms found in older children and adolescents less than 15 years of age [14].

References

1. Isaacs H Jr. *Tumors of the Fetus and Newborn.* Philadelphia: WB Saunders; 1997. Major Problems in Pathology. Series; Vol 35.
2. Isaacs H Jr. Tumors. In: Gilbert-Barness E, ed. *Potter's Pathology of the Fetus and Infant.* Vol 2. St. Louis, MO: CV Mosby; 1997:1242.
3. Parkes SE, Muir KR, Southern L, et al. Neonatal tumours: a thirty-year population-based study. *Med Pediatr Oncol* 1994;22:309.
4. Campbell AN, Chan HSL, O'Brien A, et al. Malignant tumours in the neonate. *Arch Dis Child* 1987;62:19.
5. Dehner LP. Neoplasms of the fetus and neonate. In: Naeye RL, Kissane JM, Kaufman N, eds. *Perinatal Diseases.* Baltimore: Williams & Wilkins; 1981:286. International Academy of Pathology, Monograph No. 22.
6. Werb P, Scurry J, Ostor A, et al. Survey of congenital tumors in perinatal necropsies. Pathology 1992;24:247.
7. Isaacs H Jr. Neoplasms in infants: a report of 265 cases. *Pathol Annu* 1983;18(2):165.
8. Isaacs H Jr. Perinatal (congenital and neonatal) neoplasms: a report of 110 cases. *Pediatr Pathol* 1985;3:165.
9. Isaacs H Jr. Congenital and neonatal malignant tumors: a 28-year experience at Children's Hospital of Los Angeles. *Am J Pediatr Hematol Oncol* 1987;9:121.
10. Arceci RJ, Weinstein HJ. Neoplasia. In: Avery GB, Fletcher MA, MacDonald MG, eds. *Neonatology: Pathophysiology and Management in the Newborn.* 4th ed. Philadelphia: JB Lippincott; 1994:1211.
11. Broadbent VA. Malignant disease in the neonate. In Roberton NRC, ed. *Textbook of Neonatology.* 2nd ed. Edinburgh: Churchill Livingston; 1992:879.
12. Gallagher PG, Lampkin BC. Neonatal tumors. *Semin Perinatol* 1999;23:261.
13. Bader JL, Miller RW. US cancer incidence and mortality in the first year of life. *Am J Dis Child* 1979;133:157.
14. Young JL, Miller RW. Incidence of malignant tumors in US children. *J Pediatr* 1975;86:254.
15. Keeling JW. Fetal hydrops. In: Keeling JW, ed. *Fetal and Neonatal Pathology.* 2nd ed. Berlin. Springer-Verlag; 1993:253.
16. Garmel SH, Crombleholme TM, Semple JP, et al: Prenatal diagnosis and management of fetal tumors. *Semin Perinatol* 1994; 18:350.
17. Wienk MATP, Van Geijn HP, Copray FJA, et al. Prenatal diagnosis of fetal tumors by ultrasonography. *Obstet Gynecol Surv* 1990;45:639.
18. Narod SA, Hawkins MM, Robertson CM, et al. Congenital anomalies and childhood cancer in Great Britain. *Am J Hum Genet* 1997;60(3):474.
19. Koufos A, Grundy P, Morgan K, et al. Familial Wiedemann-Beckwith syndrome and a second Wilms tumor locus both map to 11p15. *Am J Genet* 1989;44:711.
20. Mulvihill JJ. Childhood cancer, environment and heredity. In: Pizzo PA, Poplack DG, eds. *Principles and Practice of Pediatric Oncology.* 2nd ed. Philadelphia: JB Lippincott; 1993:11.
21. Kousseff BG, Opitz JM, Gilbert-Barness E. Dysplasias. In: Gilbert-Barness E, ed. *Potter's Atlas of Fetal and Infant Pathology.* St. Louis, MO: CV Mosby; 1998:11.
22. Li FP. Genetic and familial cancer: opportunities for prevention and early detection. *Cancer Detect Prev* 1986;9:41.
23. Mulcahy GM, Goggins M, Willis D, et al. Pathology and genetic testing. *Cancer* 1997;80:636.
24. Gurney JG, Ross JA, Wall DA, et al. Infant cancer in the US: histology-specific incidence and trends, 1973–1992. *J Pediatr Hematol Oncol* 1997;19:428.
25. Koop CE. Abdominal mass in the newborn infant. *N Engl J Med* 1973;289:569.
26. Kirks DR, Merten DF, Grossman H, et al. Diagnostic imaging of pediatric abdominal masses; an overview. *Radiol Clin North Am* 1981;19:527.
27. Schwartz MZ, Shaul DB. Abdominal masses in the newborn. *Pediatr Rev* 1989;11:172.
28. Bolande RP. Neoplasia of early life and its relationships to teratogenesis. In: Rosenberg HS, Bolande RP, ed. *Perspectives in Pediatric Pathology.* Vol 3. Chicago: Year Book Medical Publishers; 1976:145.

2
Germ Cell Tumors

Germ cell tumors are a varied group of benign and malignant neoplasms derived from primordial germ cells [1–6] (Figure 2.1). They occur in several different sites, both gonadal and extragonadal, the latter in midline locations such as the sacrococcygeal, retroperitoneal, mediastinal, cervical, and pineal regions. Most germ cell tumors of the fetus and infant are histologically benign and are diagnosed as either mature or immature teratomas (Table 2.1). Yolk sac tumor (endodermal sinus tumor) occurs alone or in association with infant teratomas. Yolk sac tumors of the sacrococcygeal area and testis are far more prevalent than those arising from the cervix or vagina [1,2,5–8]. Neither germinoma nor choriocarcinoma has been described in association with infant teratomas [5,8]. Primary or metastatic choriocarcinoma may present in the newborn or infant with widespread metastases, elevated chorionic gonadotropin levels, and severe bleeding tendencies [6,9,10] (Table 2.1).

The classification of germ cell tumors by The World Health Organization (WHO) is the foundation for most contemporary classifications [4] (Table 2.2). According to this classification, germ cell neoplasms are divided into seven histological types: dysgerminoma (or seminoma), yolk sac tumor, embryonal carcinoma, polyembryoma, choriocarcinoma, teratoma, and mixed germ cell tumors [4]. Sometimes the all-inclusive term "germinoma" is applied to seminoma, dysgerminoma, and pineal germinoma, since all three are similar in microscopic appearance. Gonadoblastoma, a neoplasm typically found in dysgenetic gonads, is listed as a separate category of ovarian tumors, although germ cells are present in addition to Sertoli cells [1,4,11].

Teratomas

Teratoma is defined as a true tumor composed of multiple tissues foreign to, and capable of growth in excess of, those characteristic of the part from which it is derived [5]. However it is sometimes difficult to make a distinction between teratomas and structures that result from abortive attempts at twinning. An even progression can be traced from normal twins to conjoined twins, parasitic twins, and fetus-in-fetu [5]. Careful studies reveal a break in the progression from an oriented, longitudinal, partially symmetrical structure of a twin to the jumbled, disordered, irregular growth of a teratoma in which one, two, or three tissues predominate [5]. Despite the apparent progression from twins to fetus-in-fetu to teratomas, some investigators deny a relationship. The distinction is based primarily on the fact that teratomas are capable of independent growth, whereas structures included under malformations are limited in their potentiality for growth to a rate similar to the part of the body they resemble [5].

Teratomas are observed in several locations at birth, but the most common sites are the sacrococcygeal area and neck [2,5–7,12–14] (Figures 2.2–2.4). Other locations are the brain, pineal gland, retroperitoneum, anterior mediastinum, stomach, heart and pleura, pharynx, base of the skull, upper jaw, gonads, pelvis, liver, and subcutaneous tissue [5,12–19] (Figures 2.5–2.11). The types of tissues found in fetal and infant teratomas are practically the same regardless of the site of origin.

Teratomas may be discovered as unexpected or incidental findings on routine prenatal or neonatal imaging studies, in a

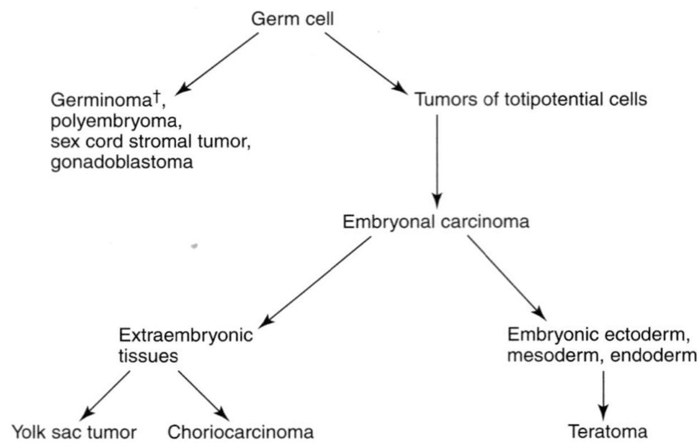

FIGURE 2.1. Origin of germ cell tumors; † indicates inclusion of dysgerminoma and seminoma. (From Teilum G [3], with permission.)

TABLE 2.1. Clinicopathological features of fetal and infant germ cell tumors.

Tumor	Clinical findings	Pathological findings[a]
Teratoma	Mass at site of origin (e.g., sacrococcygeal, cervical areas)	Tissues derived from all three germ layers Immature teratoma: immature neuroglial and neuroepithelial tissues
Fetus-in-fetu	Fetiform mass found most often in retroperitoneal area	Mature tissues and vertebral axis present
Yolk sac tumor	Gonadal or sacrococcygeal mass; associated with sacrococcgeal or other teratomas	Six histological patterns; reticular and papillary most common ones in infants; perivascular Schiller-Duval bodies; PAS + hyaline droplets; αFP + cytoplasms
Embryonal carcinoma	Mass alone or associated with a teratoma	Embryonal epithelial cells with large vesicular nuclei and big nucleoli; solid, papillary, and glandular patterns; tumor cells + for placental alkaline phosphatase and cytokeratin; αFP+; ± syncytiotrophoblastic hCG + giant cells
Polyembryoma	Gonadal or sacrococcgeal mass; found in combination with yolk sac tumor	Tiny embryoid bodies with two vesicles resembling amniotic and yolk sac cavities separated by a two- to three-cell-layer embryonic disk; αFP and hCG+
Choriocarcinoma	Abdominal mass, widespread metastases on imaging studies; severe bleeding tendencies; increased serum and urinary hCG	Soft, hemorrhagic, necrotic mass; cytotrophoblasts and giant syncytiotrophoblasts with cells intermediate between the two; hCG,[b] cytokeratin, and placental lactogen
Gonadoblastoma	Gonadal dysgenesis, 46,XY or 45,X0/46,XY karyotype	Malformed gonad(s) (e.g., streak ovary, with small, tan calcified nodules; large germ cells encircled by smaller, round, darkly staining Sertoli cells forming microfollicles with hyaline bodies and foci of calcification)

[a] αFP, α-fetoprotein; hCG, human chorionic gonadotropin; PAS, periodic acid–Schiff reaction.
[b] See text for exact details of immunohistochemical findings.
Source: Isaacs H Jr [6]; Scully RE, Young RH, Clement PB [4].

TABLE 2.2. WHO histological classification of germ cell tumors.

1. Dysgerminoma	5. Choriocarcinoma	3. Monodermal
Variant with syncytiotrophoblast cells	6. Teratomas	Struma ovarii
2. Yolk sac tumor (endodermal sinus tumor)	1. Immature	Carcinoid
Variants Polyvesicular vitelline tumor	2. Mature	Mucinous carcinoid
Hepatoid	Solid	Neuroectodermal tumors (specify type)
Glandular ("endometrioid")	Cystic (dermoid cyst)	Sebaceous tumors
3. Embryonal carcinoma	With secondary tumor (specify type)	Others
4. Polyembryoma	Fetiform (homunculus)	4. Mixed germ cell tumors (specify types)[a]

[a] For example, immature teratoma + yolk sac tumor.
Source: Scully RE, Young RH, Clement PB [4].

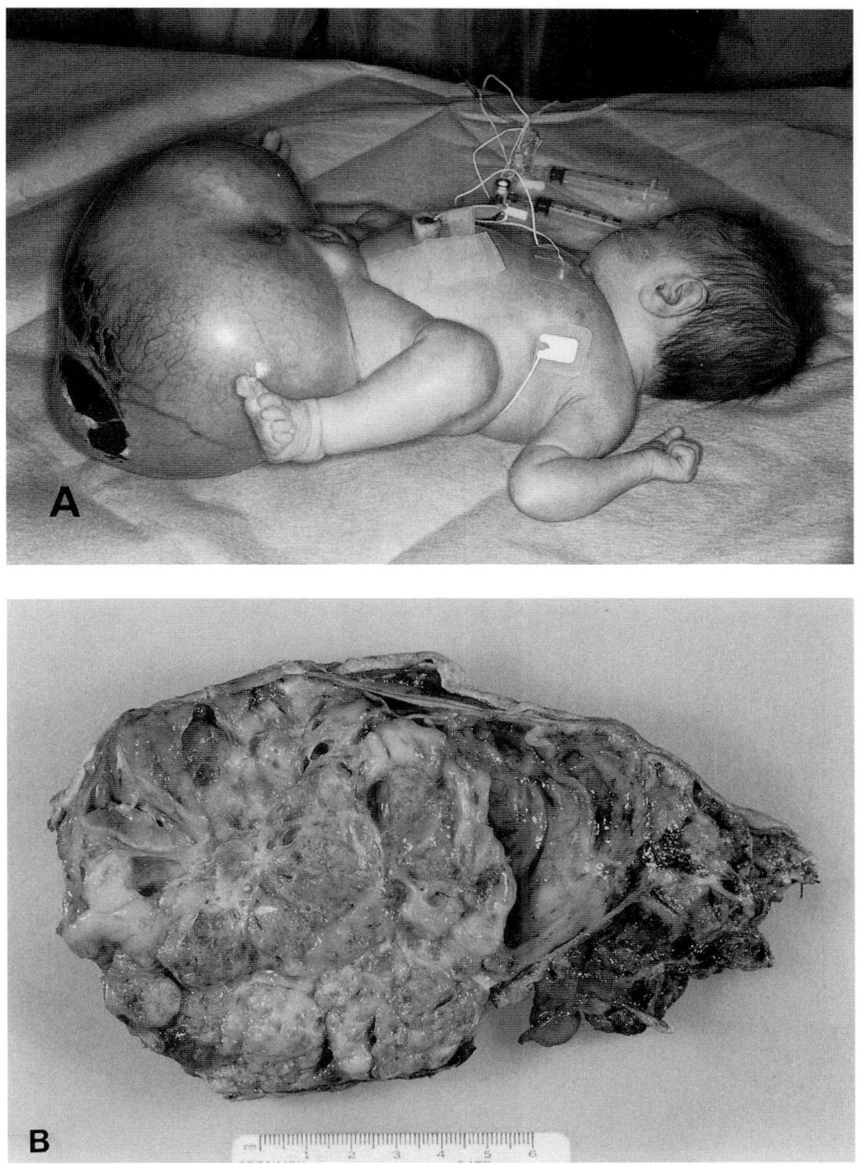

FIGURE 2.2. Solid sacrococcygeal teratoma. Sonography at 30 weeks gestation revealed an abnormality in the sacral region. Shortly before delivery fetal distress was noted and a cesarean section was performed. (A) The 4.4 kg female was born at 37 weeks gestation with a large sacrococcygeal tumor. (B) The tumor excised at 1 day of age weighed 1.5 kg (one third the baby's birth weight) and measured 15 × 13 × 10 cm^3. The bisected tumor reveals a cut surface with a variegated solid and cystic appearance. Skin partially covers the periphery of the specimen. Microscopically the tumor consists of immature neuroglial elements in addition to a variety of mature tissues characteristically found in these teratomas (see Figures 2.12–2.19).

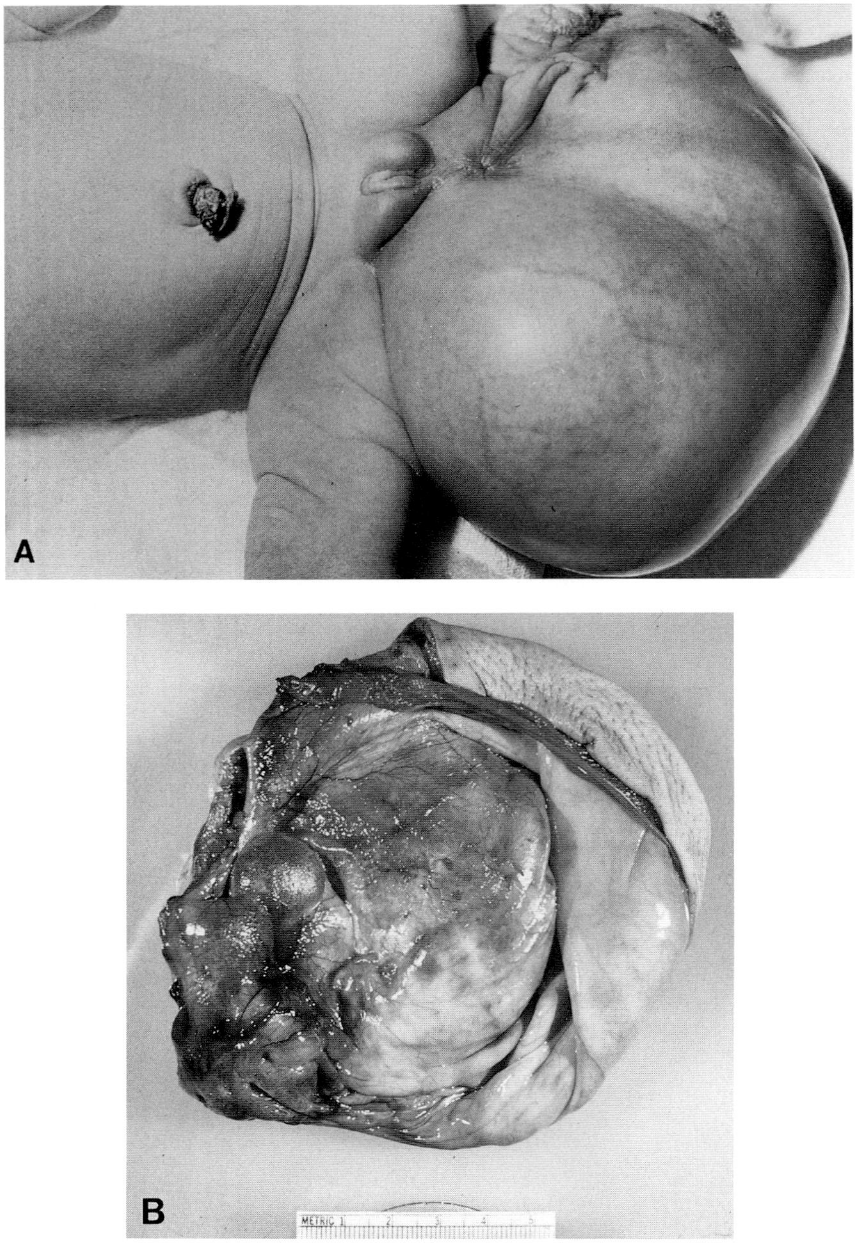

FIGURE 2.3. Cystic sacrococcygeal teratoma. (A) In this 12-day-old, 4.5 kg female, product of a full-term gestation, delivered by cesarian section because of dystocia, the external genitalia and anus are displaced anteriorly by the mass. (B) Gross specimen, 1090 g, 18.5 × 12.4 × 9.8 cm^3, consisted mostly of a large cyst partially covered by skin. The cyst was filled with 900 mL of clear fluid, which was drained prior to removal. Microscopically the teratoma consisted of both mature and immature tissues. (From Isaacs H Jr [7], with permission.)

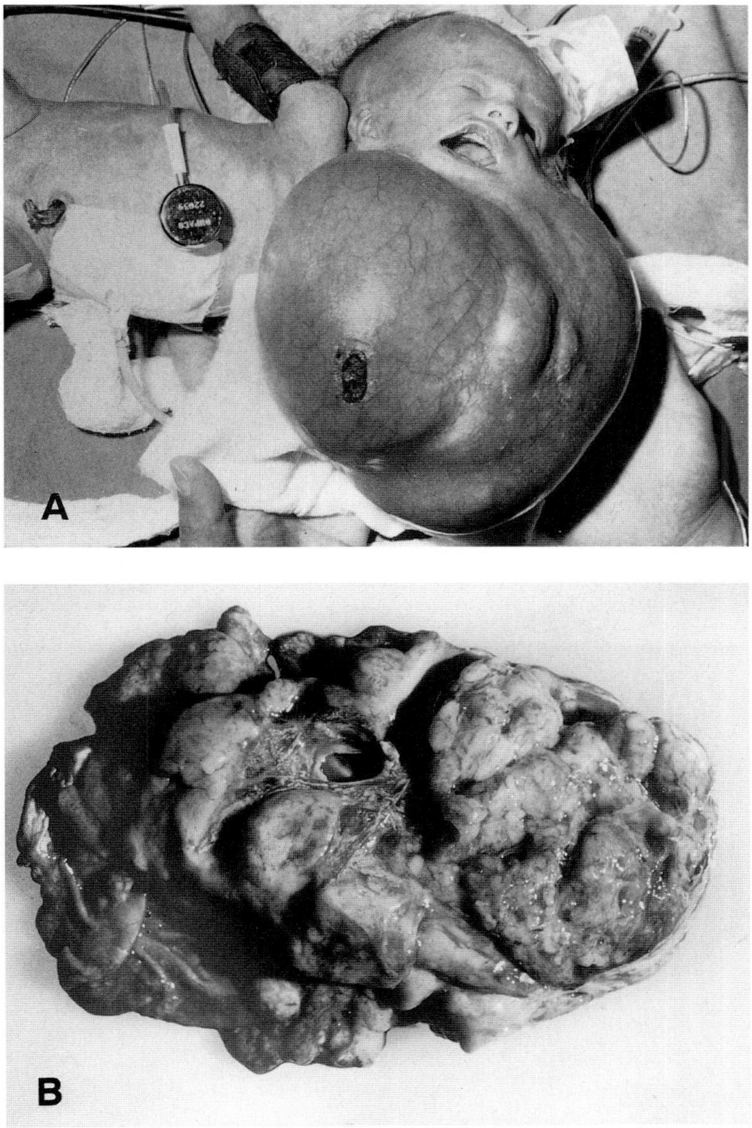

FIGURE 2.4. Thyrocervical teratoma. (A) Two-week-old male with a gigantic teratoma arising from the neck producing maternal dystocia. (B) The specimen weighed 1184 g and measured $24 \times 15 \times 8$ cm^3, consisting both of mature and immature neural tissue in addition to cartilage, respiratory, and gastrointestinal epithelium. (From Isaacs H Jr [7], with permission.)

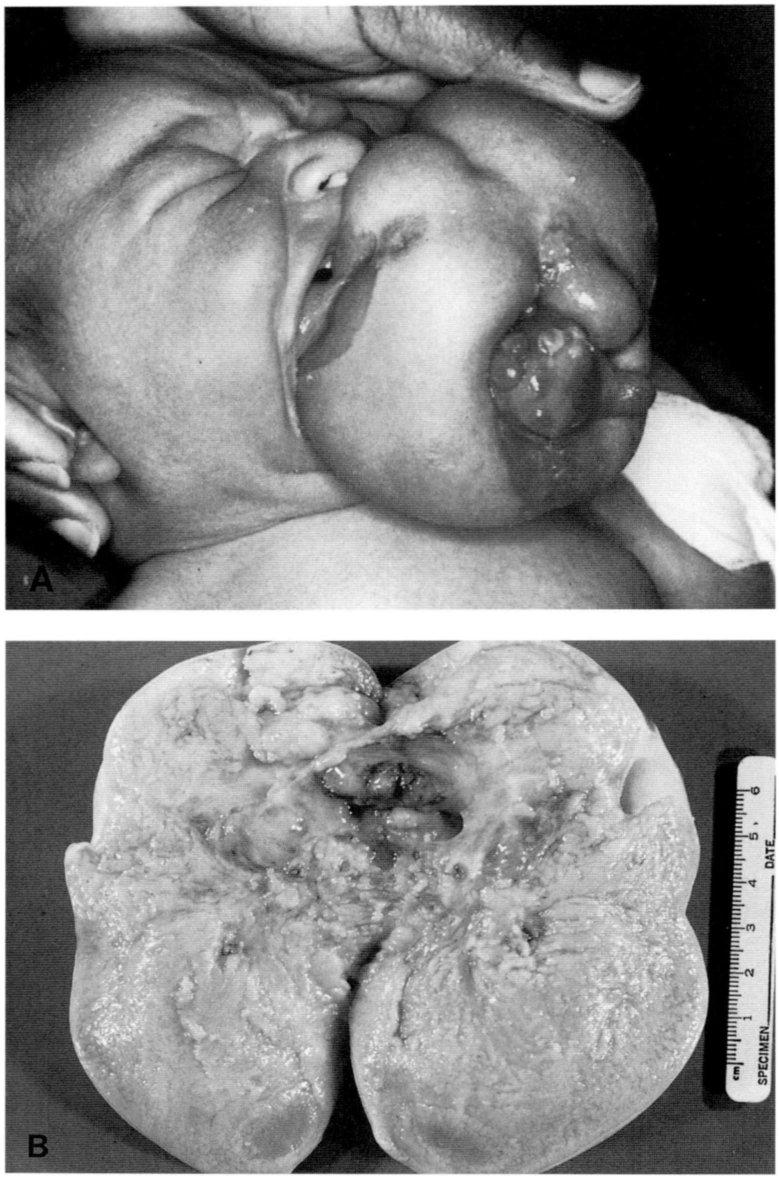

FIGURE 2.5. Epignathus. (A) A 4-day-old, full-term male infant with a solid teratoma attached to the right side of the hard palate. (B) The specimen consists mostly of fat. There is an appendage-like protrusion sticking out on the left side. The central nodular area is composed of mature tissues consisting of brain, gastrointestinal tract, and bone. Although the tumor is suggestive of "fetus-in-fetu" microscopically, neither a well-defined vertebra nor structural organization of the tissue components was found. (From Isaacs H Jr [35], with permission.)

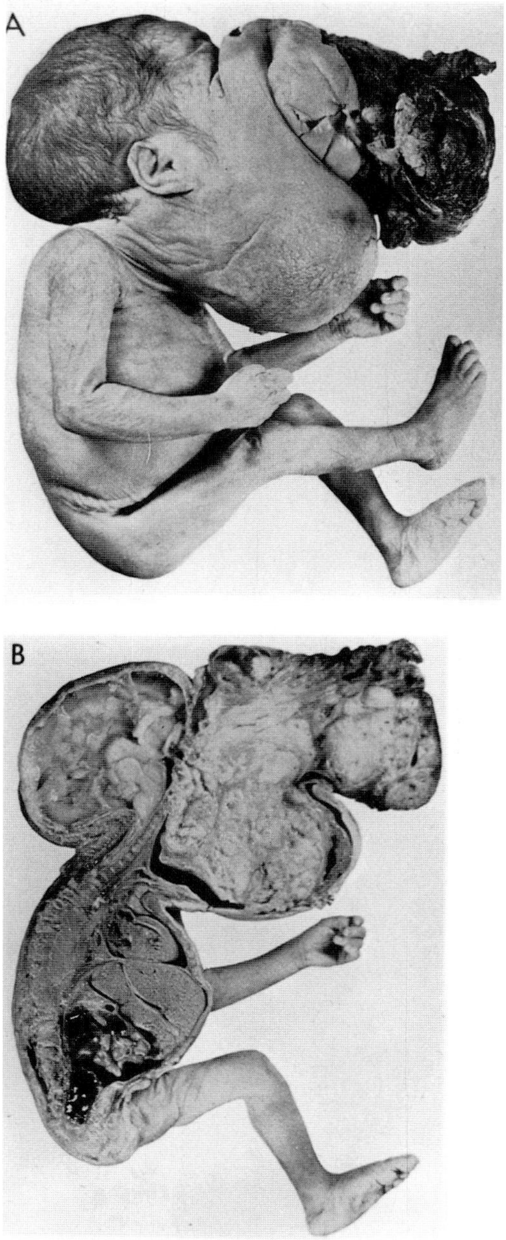

FIGURE 2.6. Pharyngeal teratoma with great distortion of the facial region. (A) Lateral view of the body surface. (B) Sagittal section throught the midline showing the extent of the tumor. (From Isaacs H Jr [5], with permission.)

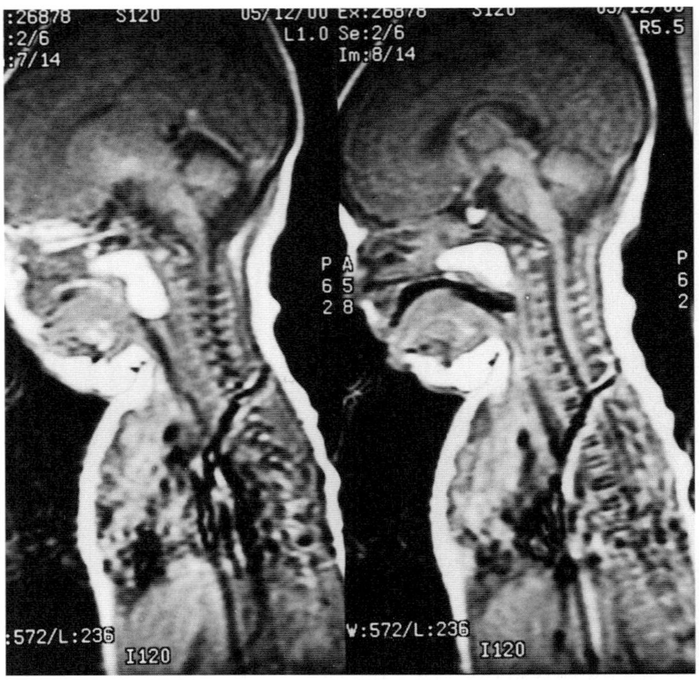

A

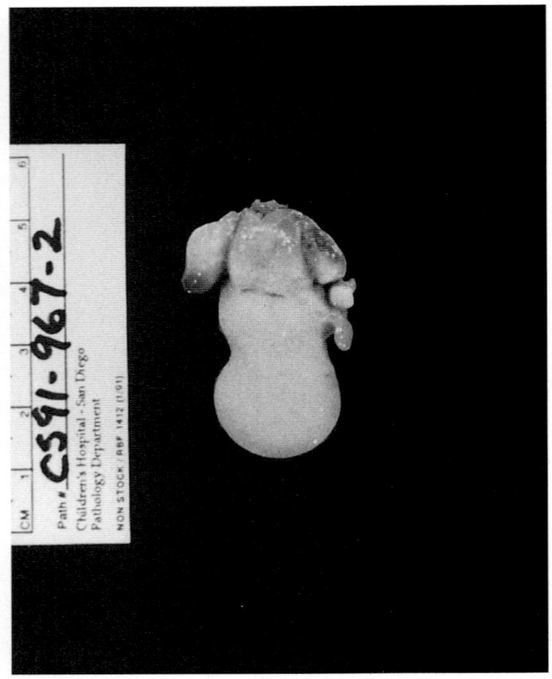

B

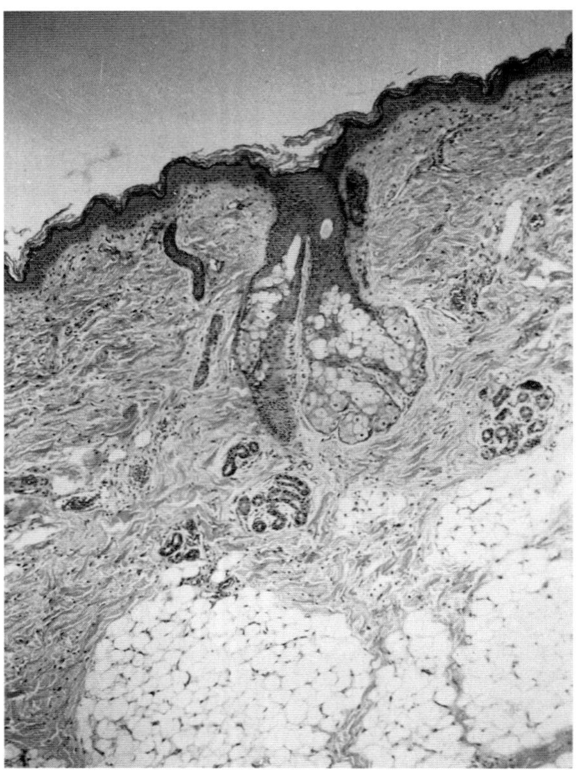

C

FIGURE 2.7. Nasopharyngeal dermoid ("hairy" polyp). Intermittent respiratory distress related to positioning of the baby was noted shortly after birth. Laryngoscopy revealed a fingerlike projection protruding from and obstructing the nasopharynx. (A) Radiograph shows a bright white mass within the nasopharynx and oropharynx with signal characteristics of fat. (B) Gross photograph showing the polypoid mass covered by skin with smaller, lateral projections. (C) Histologically the polyp consists only of skin with hair follicles, fat, and cartilage, not shown in this field. Hematoxylin and eosin (H&E); $\times 63$.

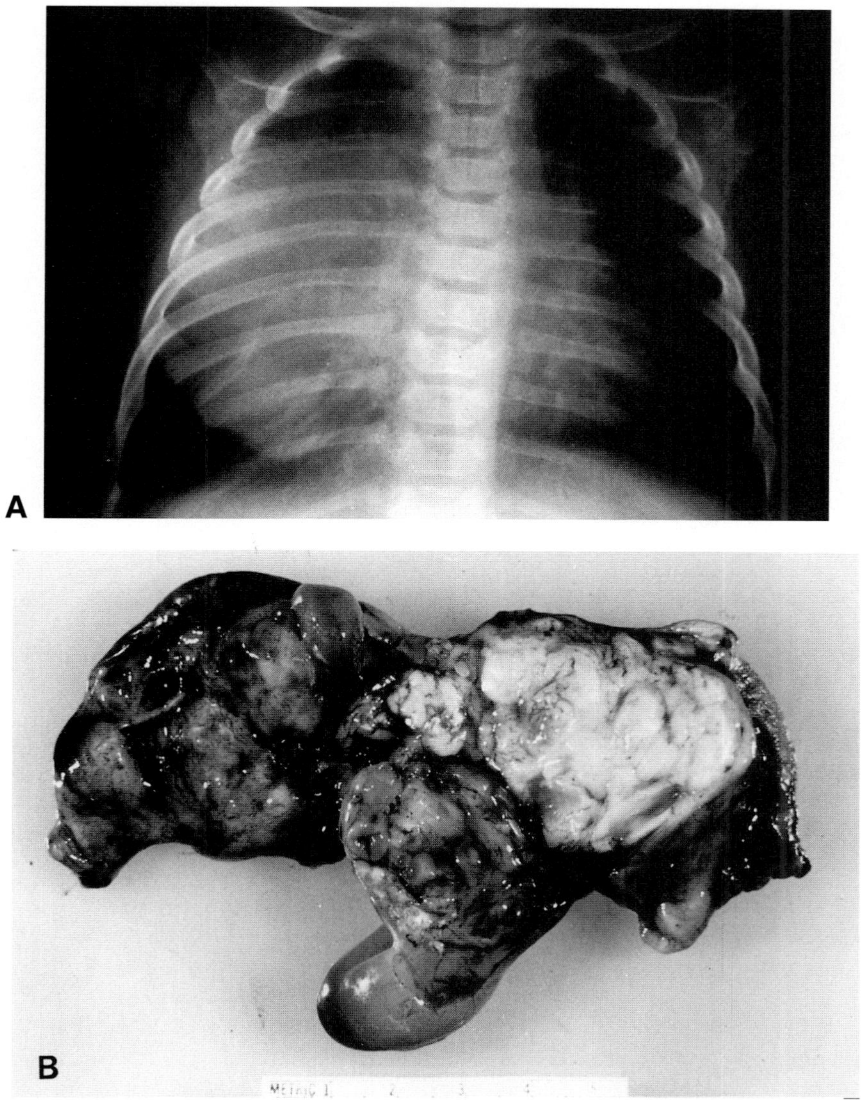

FIGURE 2.8. Mediastinal teratoma. A 4-month-old male was evaluated for an upper respiratory infection and a mediastinal mass was discovered on a chest radiograph. (A) Chest radiograph showing the mass. (B) The teratoma was attached to the thymus and consisted of both mature and immature tissues. (From Ablin A, Isaacs HI Jr [36], with permission.)

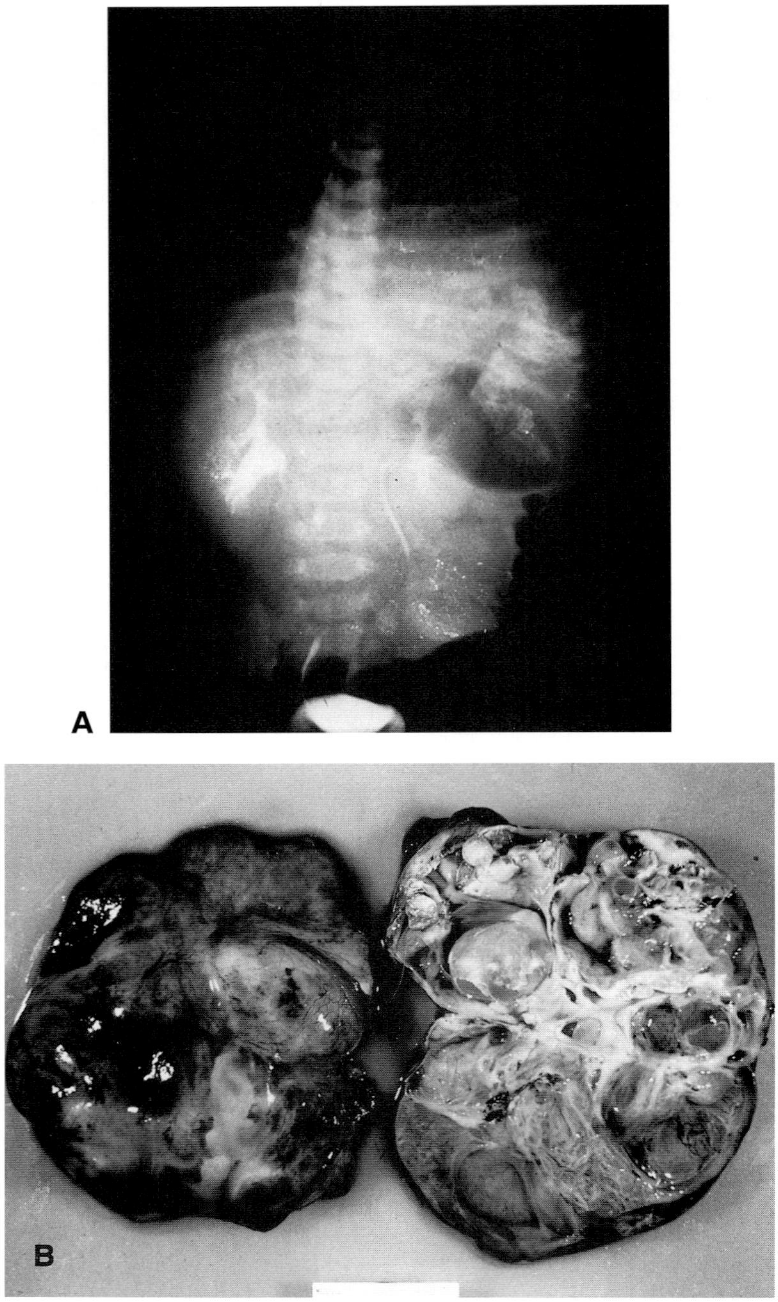

FIGURE 2.9. Gastric teratoma. A 4-month-old male with vomiting, feeding problems, and an abdominal mass. (A) Intravenous pyelogram depicts a tumor arising from the lesser curvature of the body of the stomach. (B) The cystic and solid tumor measured 16×14 cm^2 and contained immature neuroglial tissue, skin, choroid plexus, gastrointestinal and respiratory epithelium, and cartilage.

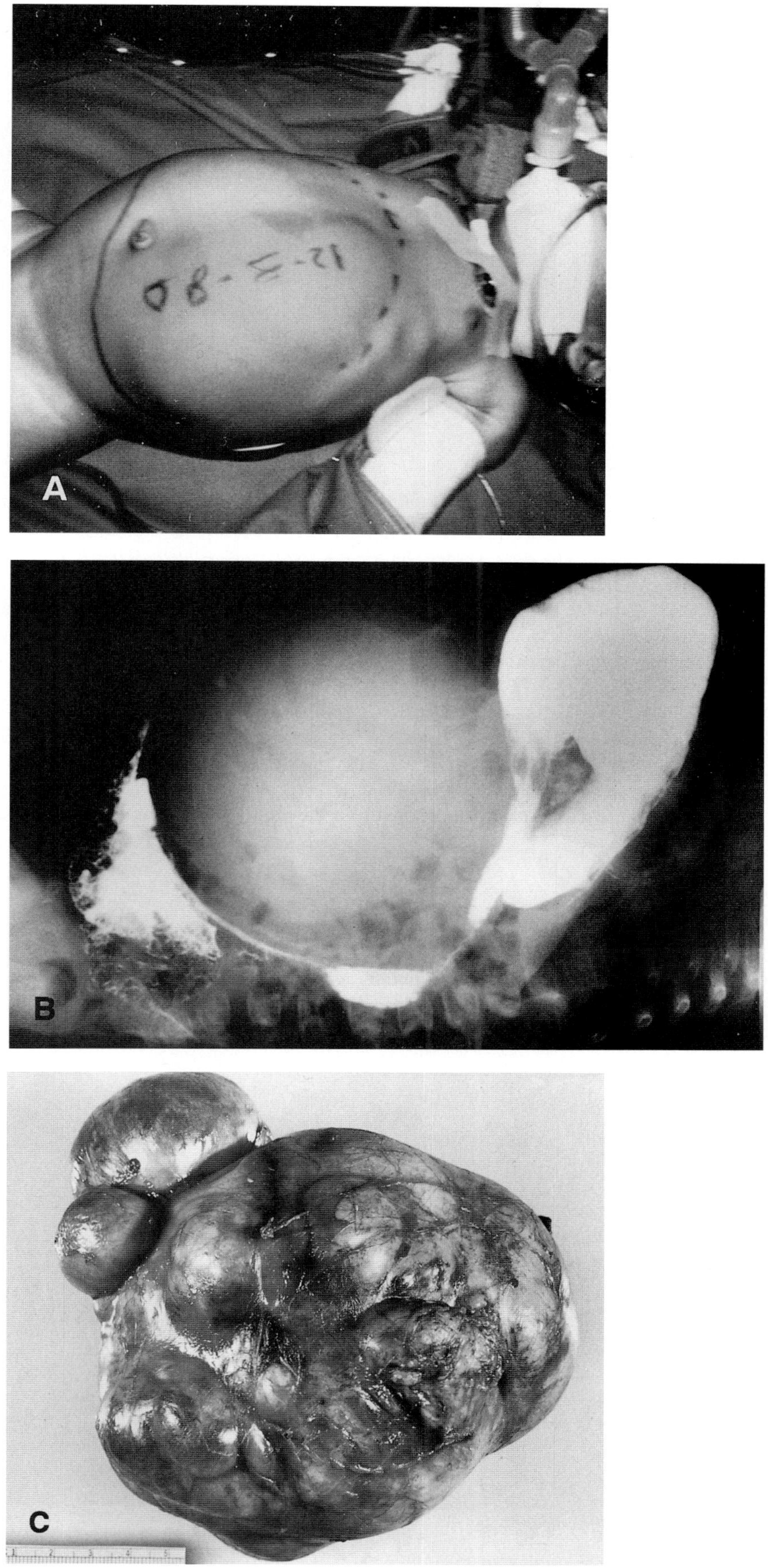

FIGURE 2.10. Retroperitoneal teratoma. The patient was a 2-month-old male with a history of constipation, progressive abdominal distension, and an enlarging abdominal mass since birth. (A) The infant's protuberant abdomen and tumor are outlined. (B) The abdominal imaging study reveals compression of the intestine by a large retroperitoneal mass. (C) The specimen, 799 g, $15 \times 15 \times 6$ cm^3, consists of solid and cystic areas. Both mature and immature tissue elements are present histologically.

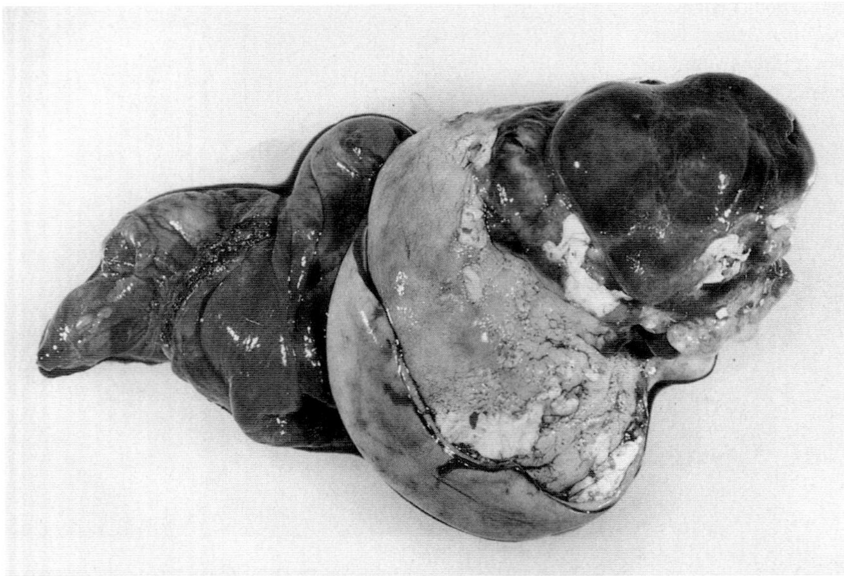

FIGURE 2.11. Teratoma of the testis. A testicular mass, 7.5 × 5 cm^2, was found in an infant during a routine bilateral inguinal hernia repair. The gross photograph shows a tumor composed of skin, white sebaceous material, and a variety of tissues. In addition to skin, bone, and gastrointestinal tract epithelium, immature neuroglial elements were present histologically.

stillborn, during a newborn physical examination, or later on during a routine follow-up visit or for some other unrelated clinical problem. Internal teratomas arising from the mediastinum, pericardium, and retroperitoneum are examples [6].

Histopathology

Teratomas are composed of tissues representing each of the three layers derived from the embryonic disk [1,2,4,20] (Figures 2.12–2.19). Endodermal components are the least common, but at times intestinal or gastric mucosa is remarkably well developed and may be surrounded by muscle layers. Ectodermal components, especially brain tissue, make up a large portion of most teratomas that are present at birth and are more prominent in these than in such tumors discovered later in life. This is particularly true of the sacral and coccygeal groups. The tissue of these tumors for the most part resembles neuroglia, although ganglion cells and cavities lined by cells simulating ependyma and choroid plexus are not uncommon [5] (Figures 2.12 and 2.13).

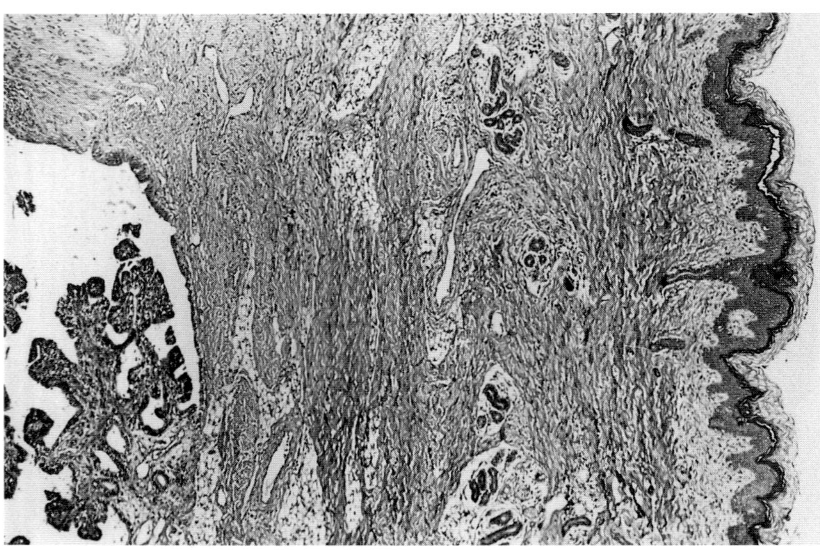

FIGURE 2.12. Teratoma with skin, neuroglial tissue, and choroid plexus. H&E; × 50. Most teratomas contain these three tissues, which were found in the tumors described in Figures 2.2 through 2.6 and 2.8 through 2.11.

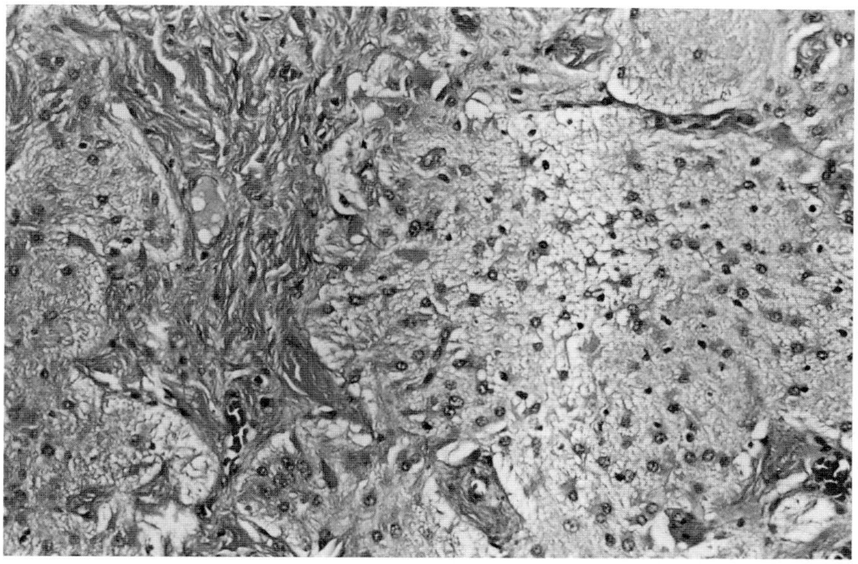

FIGURE 2.13. Neuroglial tissue, which is frequently noted in both mature and immature teratomas. Section was taken from a mature teratoma of the testis. H&E; × 150.

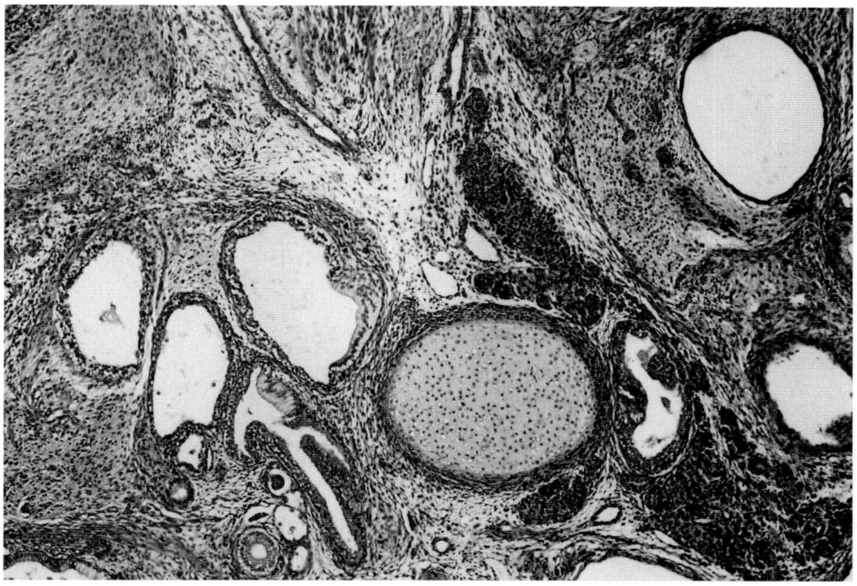

FIGURE 2.14. Photomicrograph of a mediastinal teratoma composed of mature neuroglial tissue with cysts lined by a variety of epithelia, in addition to cartilage and salivary gland. H&E; × 50.

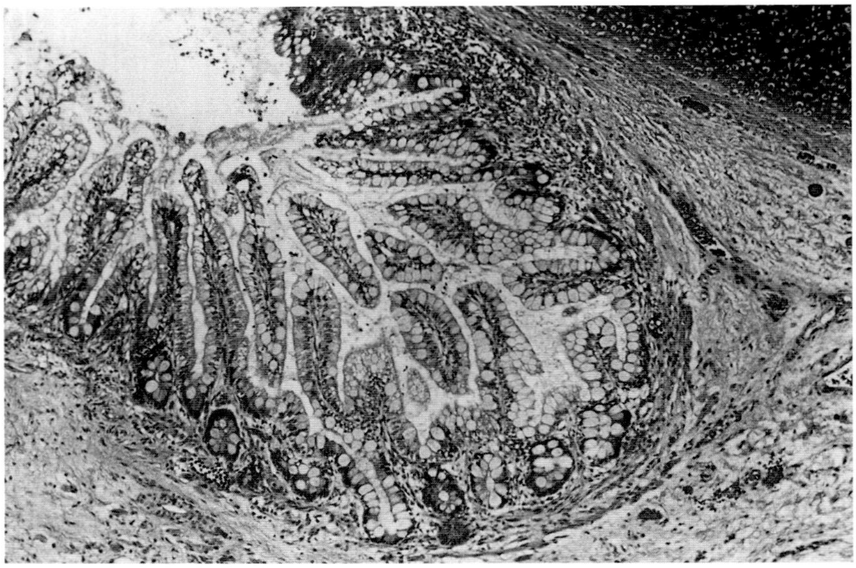

FIGURE 2.15. Mature sacrococcygeal teratoma with small bowel and cartilage components. H&E; × 50.

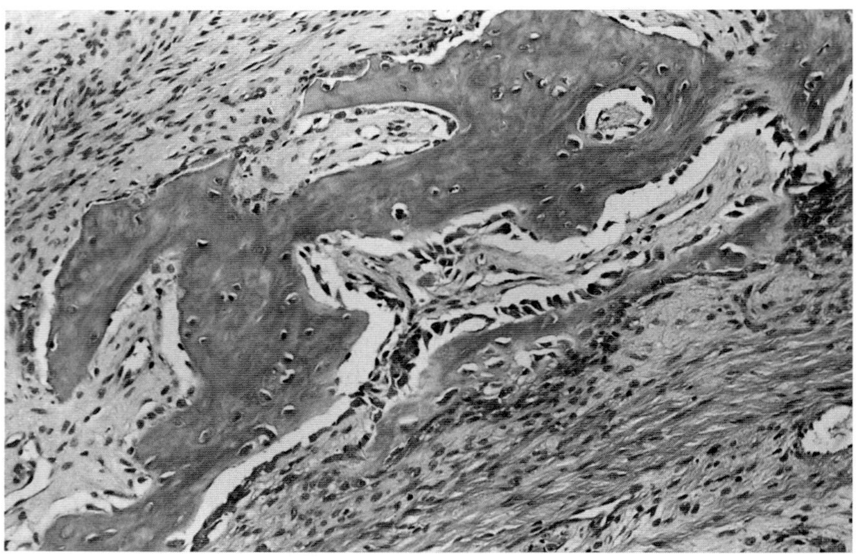

FIGURE 2.16. Trabeculae of bone from a testicular teratoma. Bone is often present in mature teratomas. H&E; × 120.

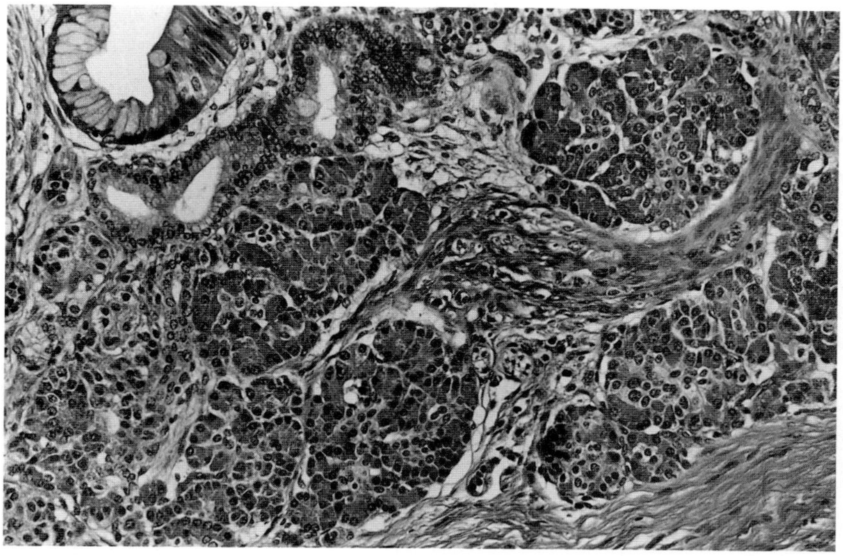

FIGURE 2.17. Mature sacrococcygeal teratoma with pancreas containing both acini and islets of Langerhans. H&E; × 120.

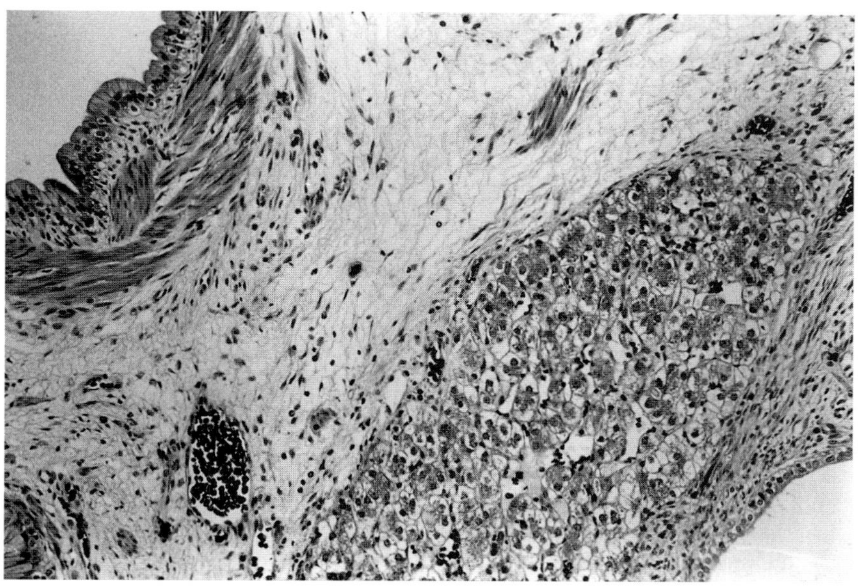

FIGURE 2.18. Mature sacrococcygeal teratoma with liver and primitive gut. H&E; × 50.

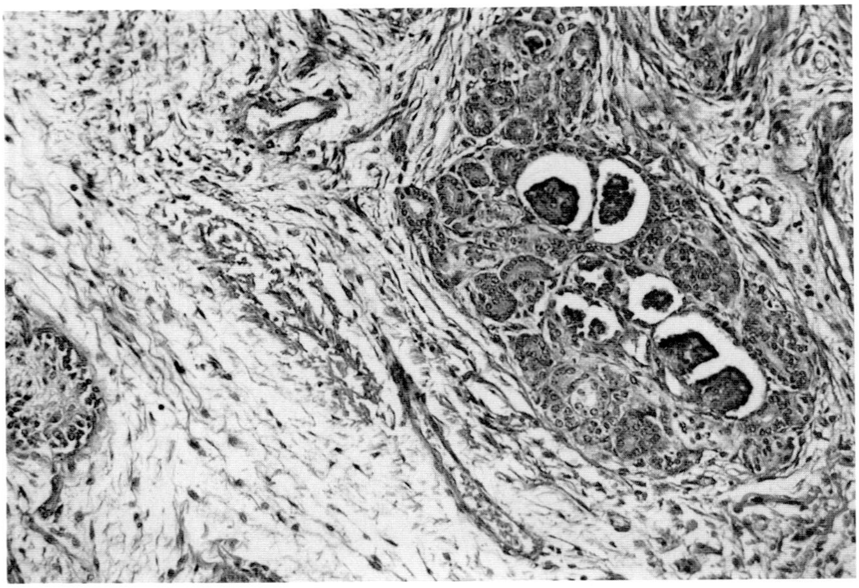

FIGURE 2.19. Nephrogenic tissue is an uncommon component of teratomas; section taken from an immature mediastinal teratoma. There is a small cluster of ganglion cells along the left lower margin. H&E; × 50.

The *immature teratoma* consists primarily of embryonic-appearing neuroglial or neuroepithelial components, which may coexist with mature tissues. They may display a worrisome histological appearance because of their hypercellularity, nuclear atypia, and increased mitotic activity (Figures 2.20–2.24). In most instances immature teratomas, high grade or otherwise, occurring in the fetus and infant are associated with a favorable prognosis [7,12,21]. The presence of microscopic foci of yolk sac tumor rather than the grade of immature teratoma per se is the only valid predictor of recurrence in pediatric immature teratomas at any site [21] (Figure 2.25). Therefore, grading of immature teratomas in the young is unnecessary at any site [6,21].

The neuroectodermal components of a teratoma are variably immunoreactive for one or more neural markers including glial fibrillary acidic protein (GFAP), neuron-specific enolase (NSE), S-100, neurofilament protein, synaptophysin, nerve growth factor receptor, glial filament protein, myelin basic protein, and polysialic acid [4]. α-Fetoprotein immunoreactivity in immature teratomas is usually confined to hepatic tissue, intestinal type epithelium, and yolk sac tumor, if present [4]. DNA ploidy, p53, and *ret* expression in teratomas are discussed by Herrmann et al. [20]. Occasionally tissues other than neural in origin, such as nodular renal blastema, are found in the immature teratoma [2,5].

Epidermis and dermal structures including hairs and sebaceous and sweat glands and fairly well developed teeth are generally present. Varieties of epithelium include columnar, pseudostratified, stratified, ciliated and nonciliated, secretory and nonsecretory. Glands in addition to those derived from the skin include salivary, thyroid, pancreas, and adrenal. Tissues resembling kidney, liver, and lung are rare. Mesodermal components such as fat, cartilage, bone, and muscle are present in almost all congenital teratomas [5] (Figures 2.12–2.19).

Fetus-in-Fetu

Fetus-in-fetu is defined as a pedunculated mass with vertebrae or notochord and structural organization exceeding that of a teratoma (but less than that of a twin) [2,5,12,22,23]. The retroperitoneum is the most common site (Figure 2.26). Both fetus-in-fetu and teratoma may occur together in the same individual, although this rarely happens, and a family history of twining is noted in some patients with teratomas [5,12]. Fetus-in-fetu has been detected by prenatal ultrasonography [23,24]. An abdominal retroperitoneal mass is the most common radiological and clinical finding [24] (Figure 2.26). Moreover, it is reported in association with an undescended testis, presenting as an abdominal mass and also within the liver [22].

Fetiform teratoma (homunculus) is a rare variant of teratoma resembling a malformed human fetus [4,5]. The caudal end of the homunculus including the lower extremities is typically more highly developed than in the cephalic portion. This tumor is distinguished from the more highly developed fetus-in-fetu [4]. Most cases of fetus-in-fetu are found in infants. In contrast to a homunculus, the fetus-in-fetu typically contains a highly developed, segmented vertebral column [4].

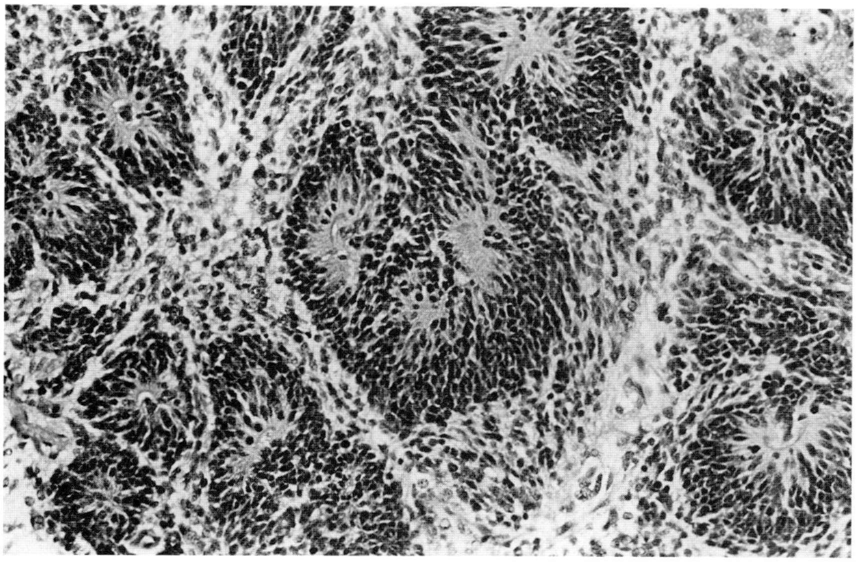

FIGURE 2.20. Immature cerebellar teratoma with ependymal rosette–like neuroglial elements. The main component of immature teratomas regardless of location are neuroglia elements, which may be derived from brain germinal matrix, ependyma, choroid plexus, or eye. In most instances the presence of immature neuroglial elements in a fetal or newborn teratoma is not associated with an unfavorable prognosis. H&E; × 50.

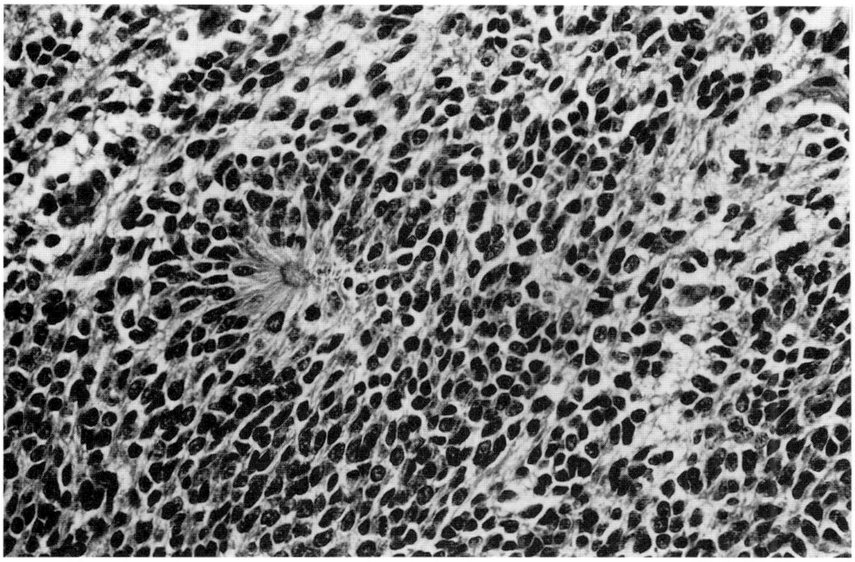

FIGURE 2.21. Higher power view of an ependymal rosette with foot processes. H&E; × 150.

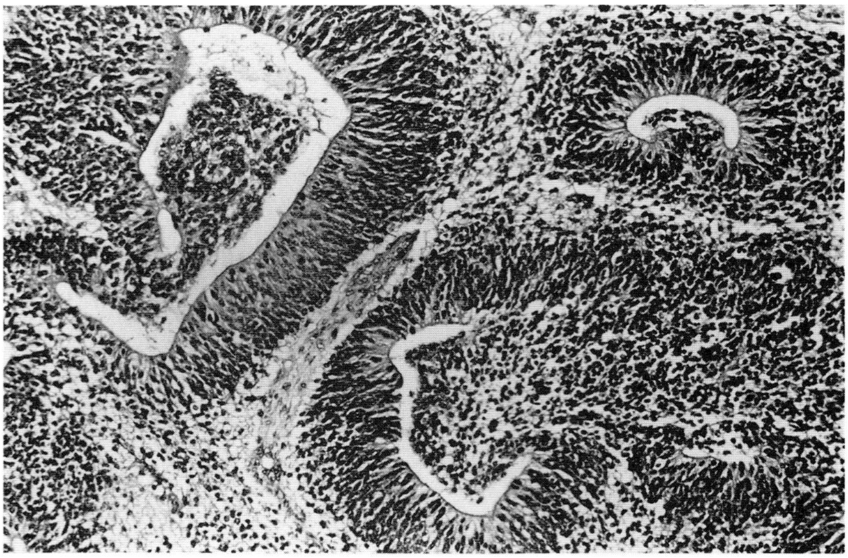

FIGURE 2.22. Immature cervical teratoma with neuroepithelial elements forming primitive ependymal-like canals. Similar structures are present in the immature mediastinal and gastric teratomas depicted in Figures 2.8 and 2.9, respectively. H&E; × 50.

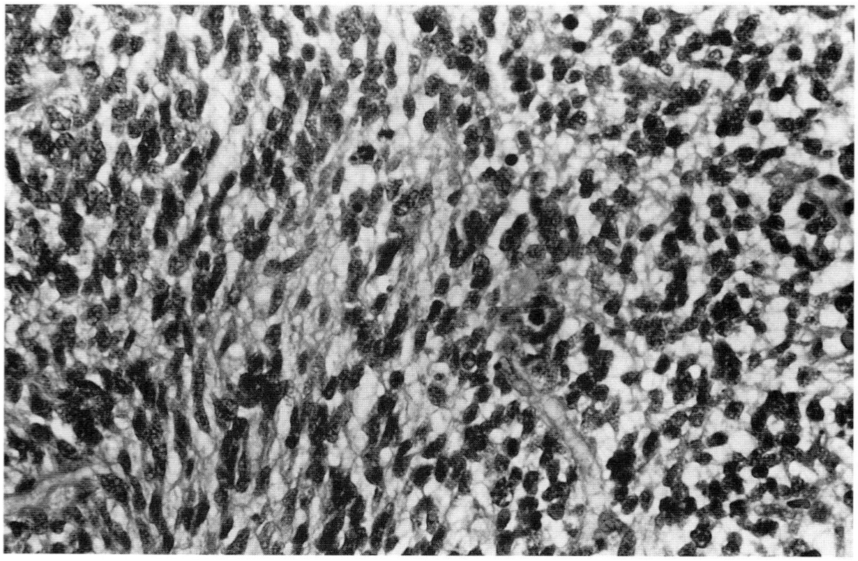

FIGURE 2.23. Immature cerebellar teratoma with germinal matrix–like cells having the appearance of a primitive neuroectodermal tumor. Note the numerous small dark round and spindle-shaped cells separated by fibrillar material (neuropil), which are histological features similar to those observed in both neuroblastoma and medulloblastoma. Foci of these immature elements blend in with more mature, recognizable neuroglial tissue, which is helpful in establishing the diagnosis of an immature teratoma. H&E; × 150.

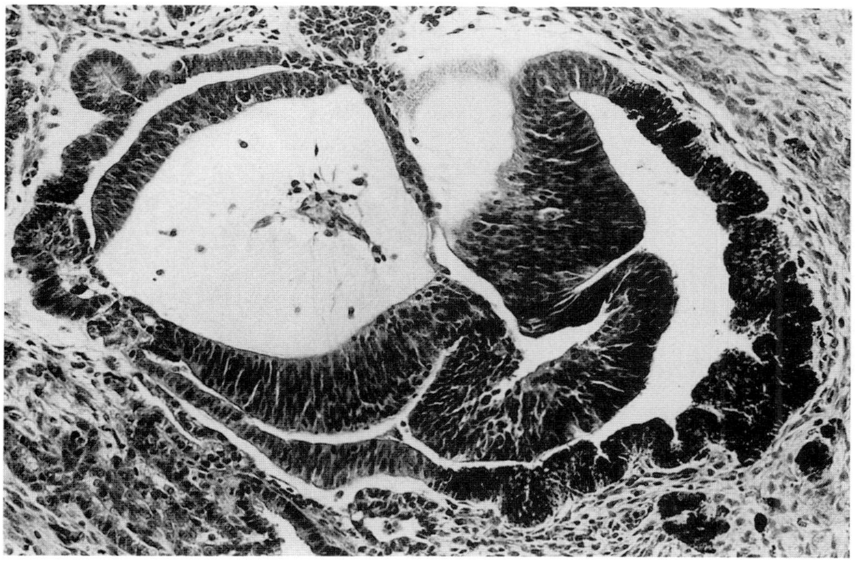

FIGURE 2.24. Immature cerebellar teratoma with pigmented optic cup–like structures. Heavily pigmented epithelium, consistent with retinal epithelium, is present along the right margin of the vesicle. H&E; × 150.

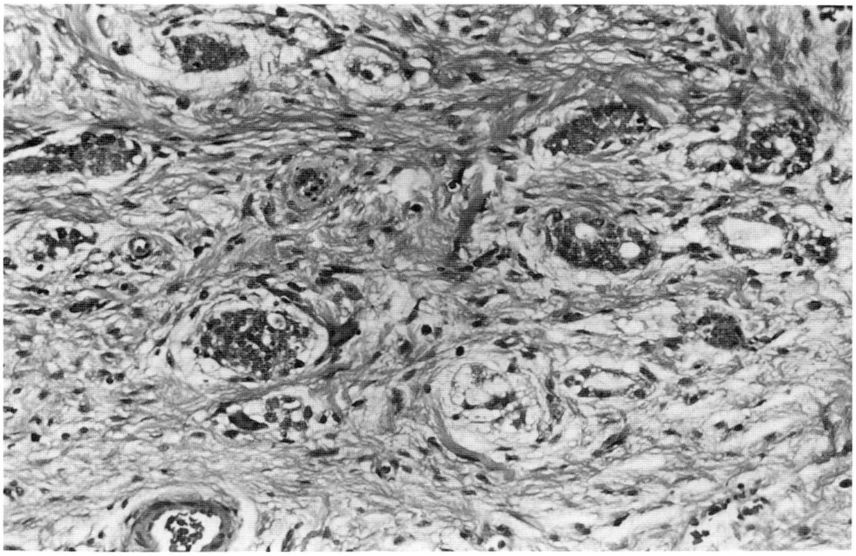

FIGURE 2.25. Subtle yolk sac tumor elements in a teratoma. Photomicrograph taken from an immature sacrococcygeal teratoma removed from a 6-day-old male. Small clusters of yolk sac tumor cells are sprinkled throughout the tumor mesenchyme. If not dilligently searched for, these tiny foci of malignant cells can easily be missed. The malignant tumor cells stain positively for α-fetoprotein, which is an aid in confirming the diagnosis. H&E; × 150.

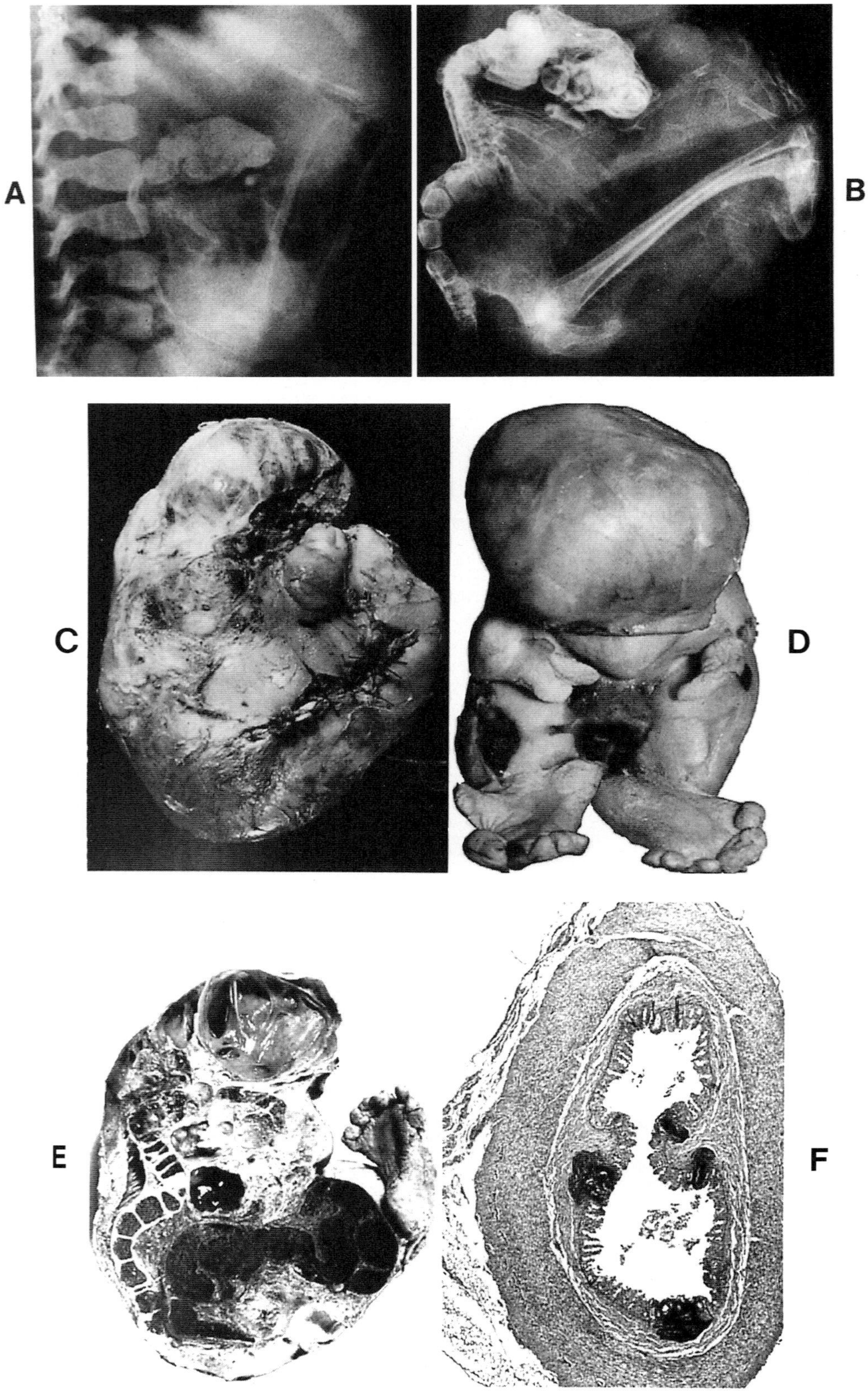

FIGURE 2.26. Fetus-in-fetu. Radiographs of a retroperitoneal mass resembling a fetus in a 1-year-old child. (A) Before removal from the abdomen. (B) After removal. (C) External surface of the mass shown in (A). (C) Before removal of the outer covering. (D) After removal of outer covering. (E) Sagittal section shows well-developed vertebra, spinal nerves, base of skull, and cystic brain. The large black convoluted structure is intestine filled with blood. (F) Cross section of intestine showing the presence of all muscle layers. Auerbach plexus appears identical to that of normal intestines. (From Isaacs H Jr [5], with permission.)

Yolk Sac Tumor

The yolk sac tumor (endodermal sinus tumor) is a malignant germ cell neoplasm characterized by a variety of distinctive morphological patterns, some recapitulating phases in the embryonic development of the normal yolk sac [3,4]. Enteric and hepatic differentiation suggests origin of the primitive gut and its derivatives from the yolk sac during embryogenesis [4]. It is the dominant malignant germ cell tumor of infants and children [5,8]. More arise from the sacrococcygeal area than from any other location during the first year of life. Less common primary sites in this age group are the testis, pelvic retroperitoneum, and vagina, but practically never the ovary.

Yolk sac tumors have a slimy, pale tan-yellow gross appearance with grayish-red foci of necrosis and small cyst formations. Generally they are very soft and mushy, falling apart upon removal. When the testis is involved, most of it is replaced by tumor, leaving a barely recognizable thin rim of light tan parenchyma (Figure 2.27).

Six or more histological patterns of yolk sac tumor have been described: papillary, reticular, solid, polyvesicular vitelline, hepatoid, and glandular including the endometrioid-like pattern [1,3,4,8]. The most common one is the reticular pattern, which consists of tumor cells with clear cytoplasms arranged in a network, as the name implies, distributed about spaces containing vacuolated pink-staining material (Figure 2.27B–E). The papillary form is composed of papillary projections with or without the classical perivascular endodermal sinus structures (Schiller-Duval body) (Figure 2.27C). Embryoid bodies are found in some yolk sac tumors (Figure 2.28).

Immunohistochemistry is helpful in establishing the diagnosis of yolk sac tumor [4]. Generally the tumor cell cytoplasms are immunoreactive, either focally or diffuse, with α-fetoprotein, α_{-1}-antitrypsin and cytokeratin but not with epithelial membrane antigen (Figure 2.27E). Most hyaline droplets, which are periodic acid–Schiff (PAS) positive and diastase resistant, stain variably with either α-fetoprotein or α_{-1}-antitrypsin. α-Fetoprotein, a useful biological marker present in the serum of these patients, can be used to monitor for the presence of recurrence and/or metastases and the effect of chemotherapy [1,4,25]. The lungs and liver are the sites of metastases (Figure 2.29). It is important to remember that normally the serum α-fetoprotein level is markedly elevated in the newborn, attaining normal levels at about 4 months of age [26] (Table 2.3).

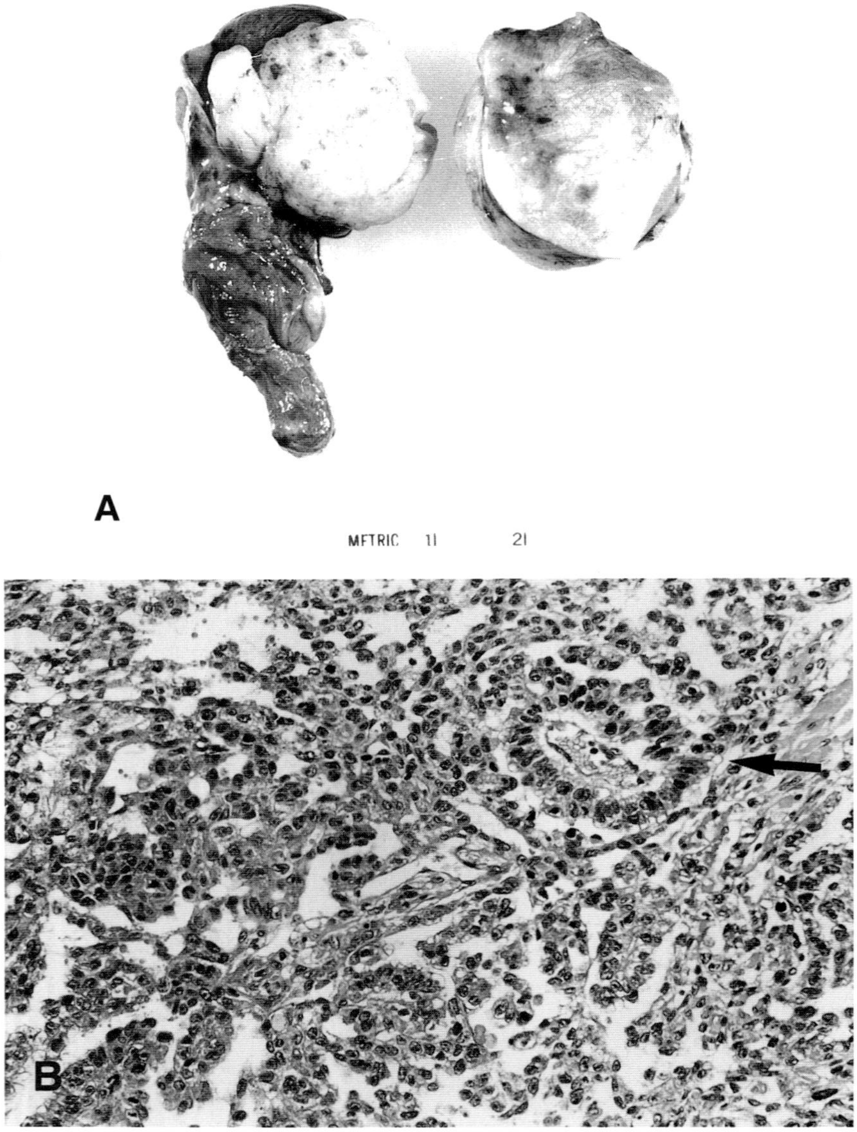

FIGURE 2.27. Yolk sac tumor of the testis: 11-month-old male with a scrotal mass. (A) The bisected testicular tumor and attached spermatic cord. (B) Photomicrograph reveals the characteristic papillary pattern of yolk sac tumor with perivascular endodermal sinus structures (Schiller-Duval body) (arrow). H&E; × 120. (C) Higher magnification of the papillary and endodermal sinus structures. Droplets of hyaline staining material (H&E) representing α-fetoprotein are present (arrows). × 380. (D) Histological section showing the reticular (networklike) pattern. Two small, immature testicular tubules are present near the lower magin of the photograph. H&E; × 120. (E) Tumor cells react with α-fetoprotein immunoperoxidase. × 300.

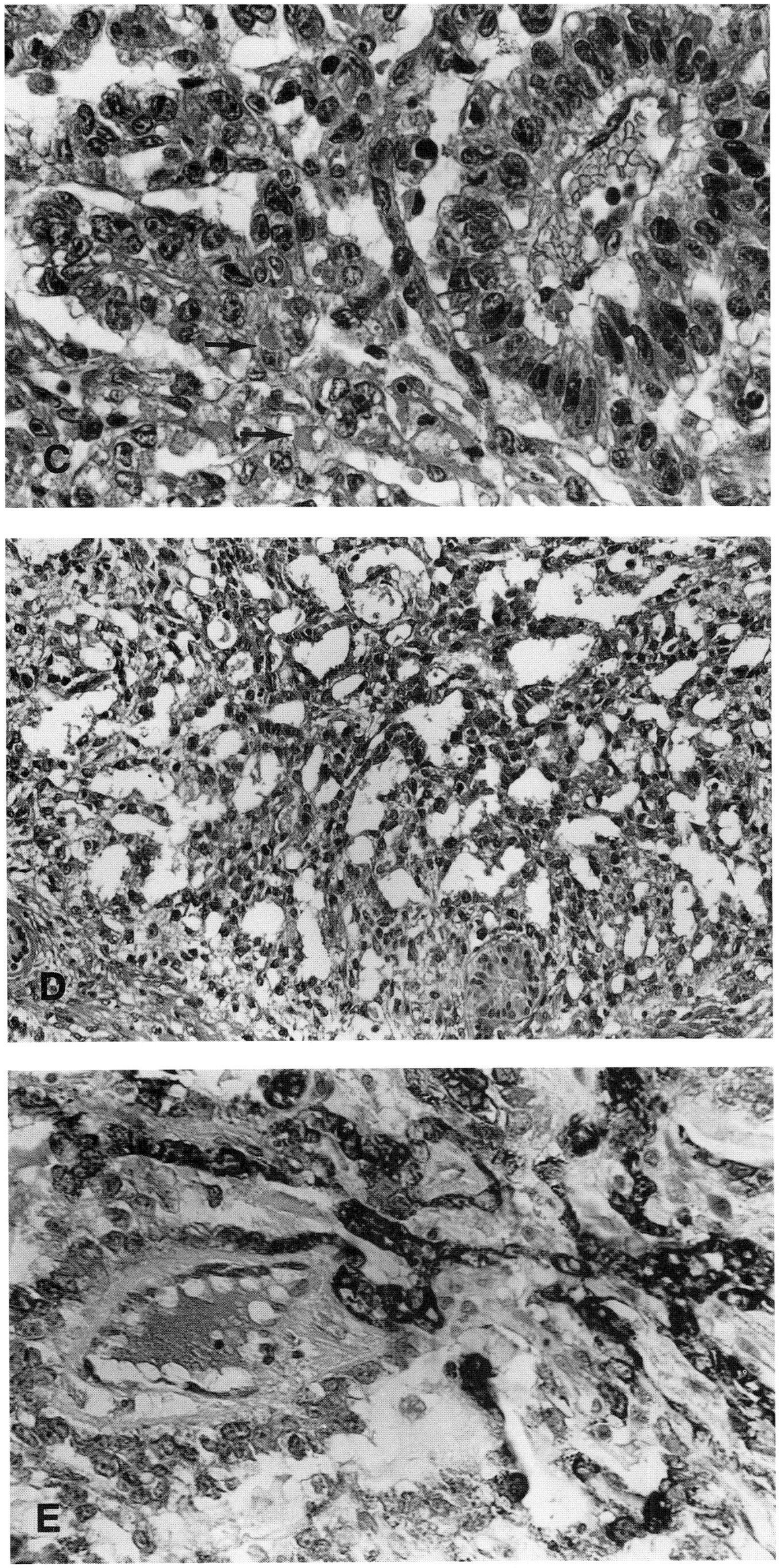

FIGURE 2.27. (*Continued*)

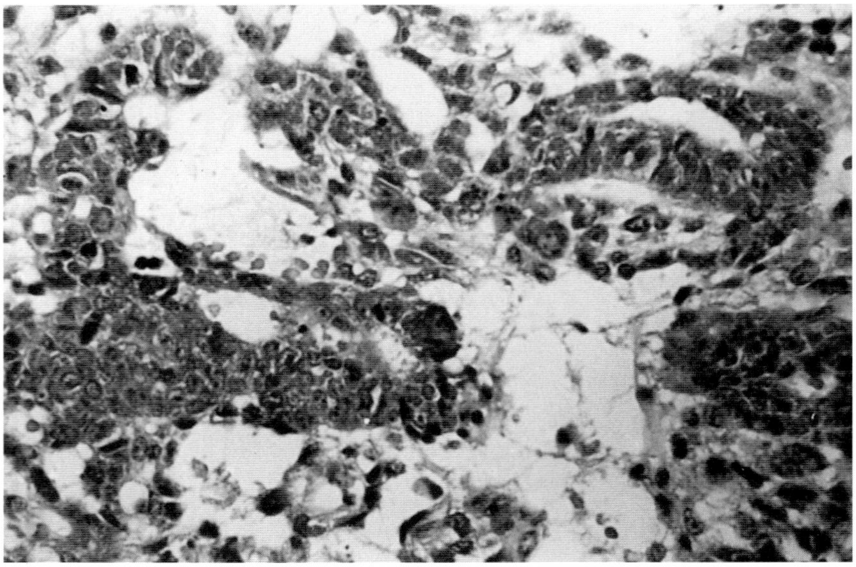

FIGURE 2.28. Embryoid bodies associated with yolk sac tumor: 1-year-old male with a yolk sac tumor of the testis. The embryoid body, upper right corner, resembles a tiny embryo consisting of two vesicles suggestive of yolk sac and amniotic cavity separated by a two- to three-cell-layer embryonic disk. H&E; × 300.

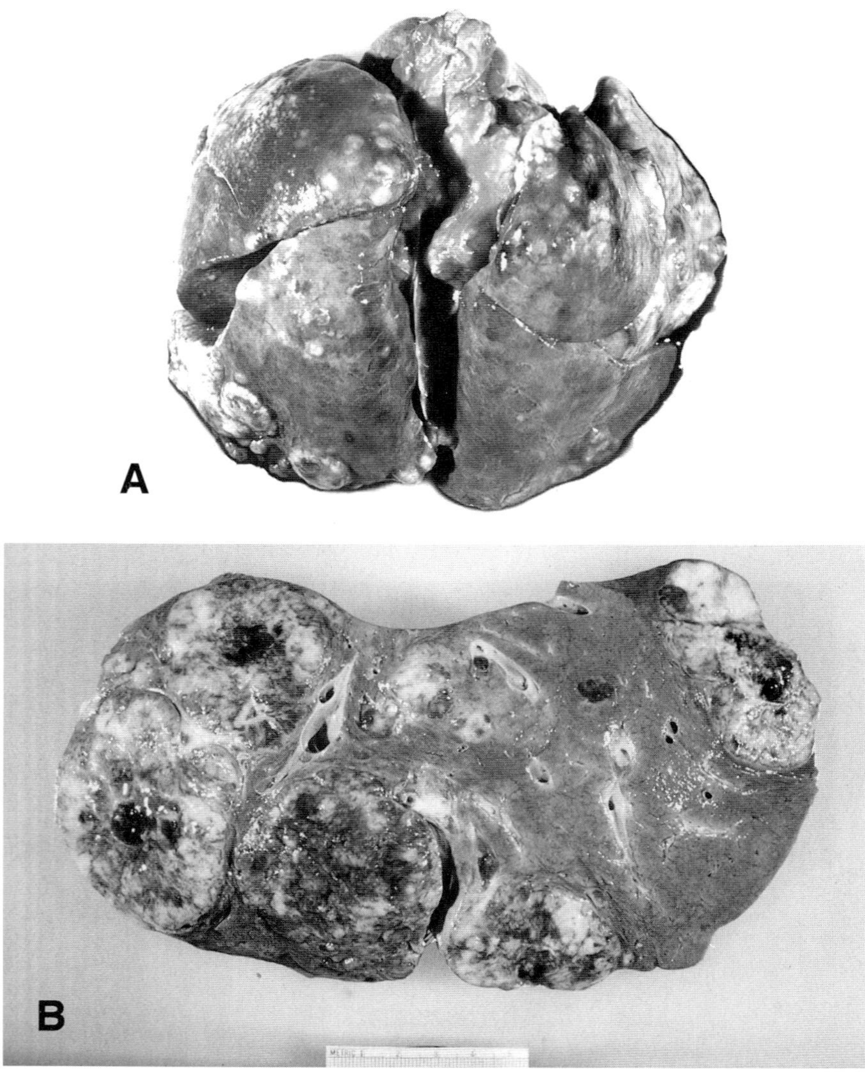

FIGURE 2.29. Sacrococcygeal teratoma with yolk sac tumor metastases. Yolk sac tumor is the most common malignant component of sacrococcygeal teratomas; the malignancy occurs in about 10% of these tumors at this location. (A) Lung metastases. (B) Liver metastases.

TABLE 2.3. Average normal serum α-fetoprotein levels in the newborn and infant.

Age	Mean ± SD (ng/mL)
Premature	134,734 ± 41,444
Newborn	48,406 ± 37,718
Newborn–2 weeks	33,113 ± 32,503
Newborn–1 month	9,452 ± 12,610
2 weeks–1 month	2,654 ± 3,080
2 months	323 ± 278
3 months	88 ± 87
4 months	74 ± 56

Source: Wu JT, Book L, Sudar K [26]. Used with permission.

Gonadoblastoma

Although gonadoblastoma typically arises from within a dysgenetic gonad, there are a few case reports of this tumor occurring in apparently normal testes [4,27,28]. The affected gonad may be of an indeterminate nature, a streak ovary, or a dysgenetic undescended testis [4,11,27]. Most patients with this tumor have a 46,XY or a 45,X0/46,XY mosaic karyotype.

Grossly the tumor consists of one or more small, firm, tan nodules characteristically containing flecks of calcification that can be appreciated on imaging studies. The distinctive histological features of gonadoblastoma are large germ cells surrounded by smaller, round, darkly staining Sertoli cells forming microfollicles with hyaline bodies and calcium deposits [3,4,6,11,27] (Figures 2.30 and 2.31).

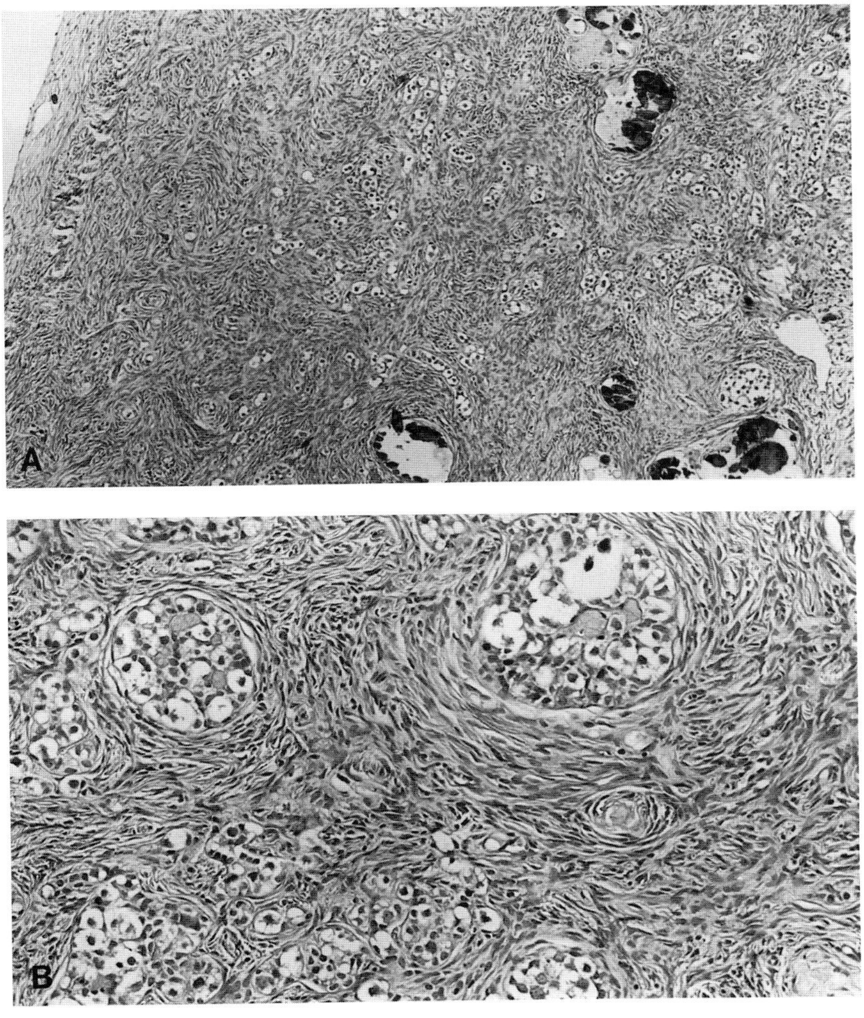

FIGURE 2.30. Gonadoblastoma arising from a dysgenetic gonad. The patient was a 7-year-old, 45,X0/46,XY child with ambiguous genitalia. (A) The left side of the field depicts the morphological features of a "streak gonad." H&E; × 75. Several microfollicles, some with calcification, are present. (B) Higher magnification reveals that the microfollicles are composed of nests of large germ cells and smaller, dark round to oval Sertoli cells. Some cell nests contain round spaces filled with hyaline-staining material. H&E; × 150. According to Teilum, gonadoblastoma recapitulates the structure of the fetal gonad [3]. Multiple foci of calcification form in the microfollicles, which can be appreciated grossly by a gritty sensation on section and also on imaging studies. (Courtesy of Rob Newbury, M.D., Department of Pathology, Children's Hospital, San Diego.)

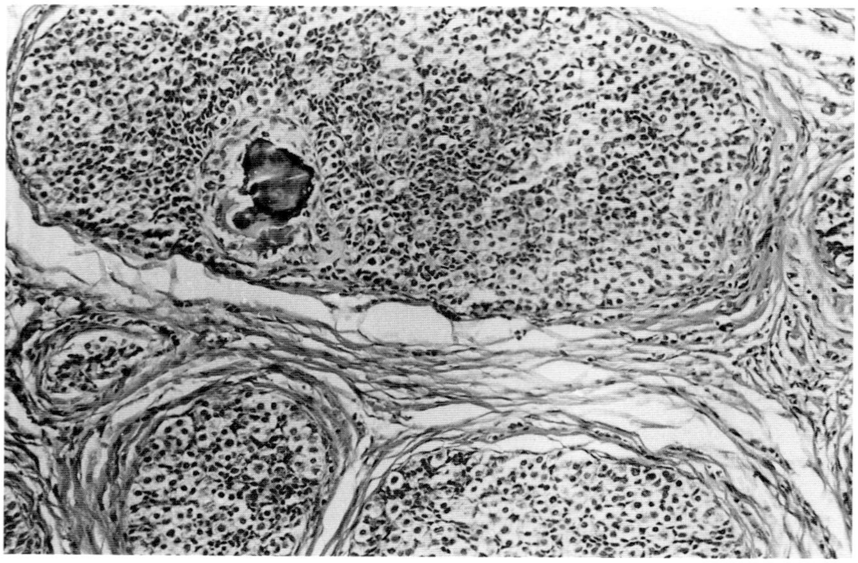

FIGURE 2.31. Gonadoblastoma arising from a dysgenetic gonad. The patient was a dysmorphic 4-month-old child with tetralogy of Fallot, ambiguous genitalia, and a karyotype of 45,XO/46,XY. During a bilateral herniorrhaphy, an undescended "streak" gonad, measuring 2×1.3 cm^2, was found on the right side and an undescended immature testis, 1.5×1.5 cm^2, was noted on the left. Tumor was present only in the streak gonad. The photomicrograph shows an early stage of gonadoblastoma consisting of nests of large germ cells and smaller, dark round to oval Sertoli cells situated in lobules surrounded by fibrous ovarian stroma. A focus of calcification is present. H&E; $\times$ 75.

Embryonal Carcinoma

Embryonal carcinoma is a malignant germ cell tumor characterized by large primitive epithelial cells resembling those of the embryonic germ disk [4,8]. It is far less common than yolk sac tumor and is seldom described in the first year of life. It occurs alone or in combination with a teratoma. Patients with this malignancy may have elevated serum human chorionic gonadotropin (hCG) and α-fetoprotein levels, which are useful in detecting recurrent tumor and/or metastases and in monitoring the effects of chemotherapy.

Grossly embryonal carcinoma has a variegated cut surface with light tan-gray to yellow soft areas with extensive hemorrhage and necrosis. Microscopically the tumor consists of embryonal-appearing epithelial cells with large vesicular nuclei, characteristic big nucleoli-forming solids, and papillary and glandular patterns [4,8] (Figure 2.32). Frequent mitoses, hyaline droplets similar to those found in yolk sac tumors, and syncytiotrophoblastic giant cells are variably present [4]. The tumor cells are immunoreactive with cytokeratin and placenta-like alkaline phosphatase and unreactive with epithelial membrane antigen [4]. Immunoreactivity with α-fetoprotein is variable and is present in less than half the cases. When present, the syncytiotrophoblastic cells are immunoreactive with hCG [4].

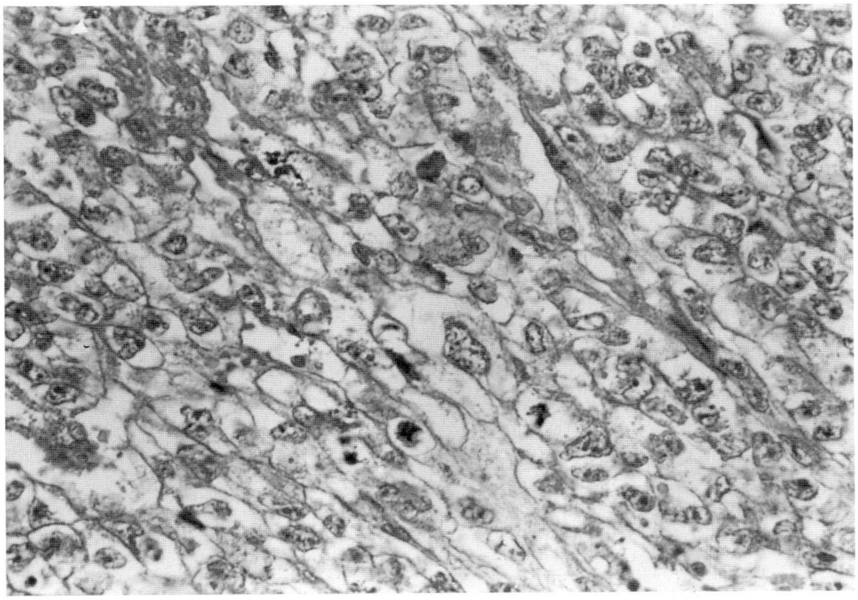

FIGURE 2.32. Presacral teratoma with embryonal carcinoma and yolk sac tumor. The patient was a 1-year-old female with a large presacral mass. Microscopically the tumor contains neuroglial tissue, cysts lined by squamous and respiratory epithelium, and nodules of cartilage. In addition there is a malignant germ cell tumor component composed of both yolk sac tumor (mostly) and focal embryonal carcinoma. The latter consists of sheets of large, pleomorphic, anaplastic epithelial cells with vesicular nuclei and prominent nucleoli. H&E; $\times$ 300.

Polyembryoma

Polyembryoma is a very uncommon, unusual germ cell tumor of the gonads characterized by the presence of embryoid bodies, which resemble presomite embryos [3,4]. Polyembryoma occurs alone or in combination with infant yolk sac tumor [4,6]. Patients may present with a sacrococcygeal or gonadal mass and elevated serum levels of hCG and/or α-fetoprotein [4]. Microscopically the embryoid bodies are similar to tiny embryos composed of two vesicles resembling yolk sac and amniotic cavity separated by a two- to three-cell-layer embryonic disk (Figure. 2.28). The polyembryoma stains positively for hCG and α-fetoprotein, suggesting both embryonic and extraembryonic differentiation [4].

Dysgerminoma and Choriocarcinoma

To my knowledge primary dysgerminoma (germinoma) has not been described in the first year of life, either alone or in combination with a teratoma [1,5]. Choriocarcinoma presents in the newborn or infant as either metastases secondary to a placental choriocarcinoma [9,29] or as a primary tumor arising from a variety of locations as the liver, lung, brain, kidney, and maxilla [4]. The most common site is the liver, followed by the brain and lung [10] (Figure 2.33). Approximately 25% of infants have symptoms at birth [10].

Grossly choriocarcinoma consists of friable, pale tan-gray hemorrhagic tissue (Figure 2.33). Histologically, it is composed of two main components, cytotrophoblasts and syncytiotrophoblasts, with cells intermediate between the two [4,8] (Figure 2.34). Cytotrophoblasts consist of large, round cuboidal cells with variable vesicular central nuclei and clear cytoplasm. Syncytiotrophoblasts form syncytia or knots similar to those of the placental villi; they are much larger cells, with abundant pink, vacuolated cytoplasm and large, hyperchromatic irregular nuclei. Cytotrophoblasts and syncytiotrophoblasts are found adjacent to one another and situated next to dilated vascular sinusoids (Figure 2.34C). Extensive necrosis and hemorrhage are characteristic both grossly and microscopically. Vascular invasion is a prominent feature [4,8,9,29] (Figure 2.34C). The syncytiotrophoblastic cells are immunoreactive for cytokeratin, hCG, human placental lactogen, and pregnancy-specific β_{1}-glycoprotein; the cytotrophoblast is reactive with cytokeratin; and the intermediate trophoblast with cytokeratin, human placental lactogen, and pregnancy-specific β_{1}-glycoprotein [4].

The diagnosis of choriocarcinoma should be considered in a young infant manifesting bleeding tendencies (e.g., hematemesis, hemoptysis, and anemia) accompanied by hepatomegaly, lung metastases, and elevated urinary and plasma chorionic gonadotropin levels. Infantile choriocarcinoma is an invariably rapidly progressive disease. There are only a few reports of successful treatment [10].

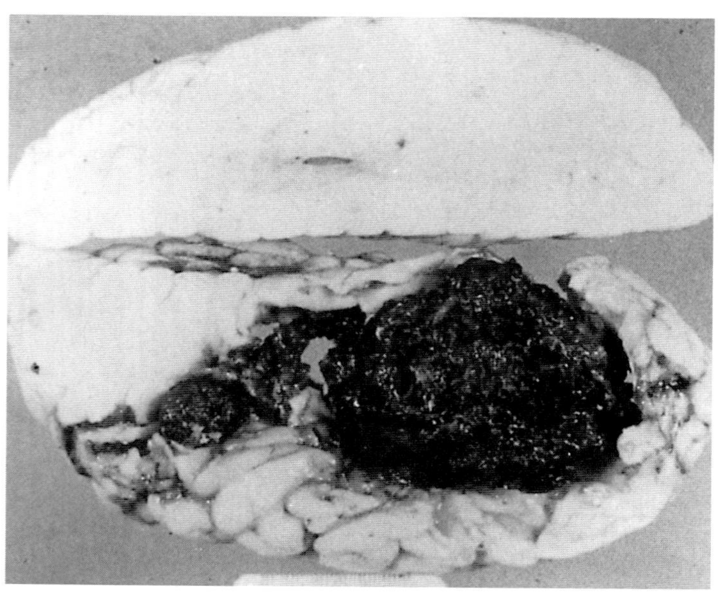

FIGURE 2.33. Choriocarcinoma in the brain of a newborn (metastases from a placental choriocarcinoma). A large hemorrhagic mass occupies the cerebral cortex. (From Isaacs H Jr [5], with permission.)

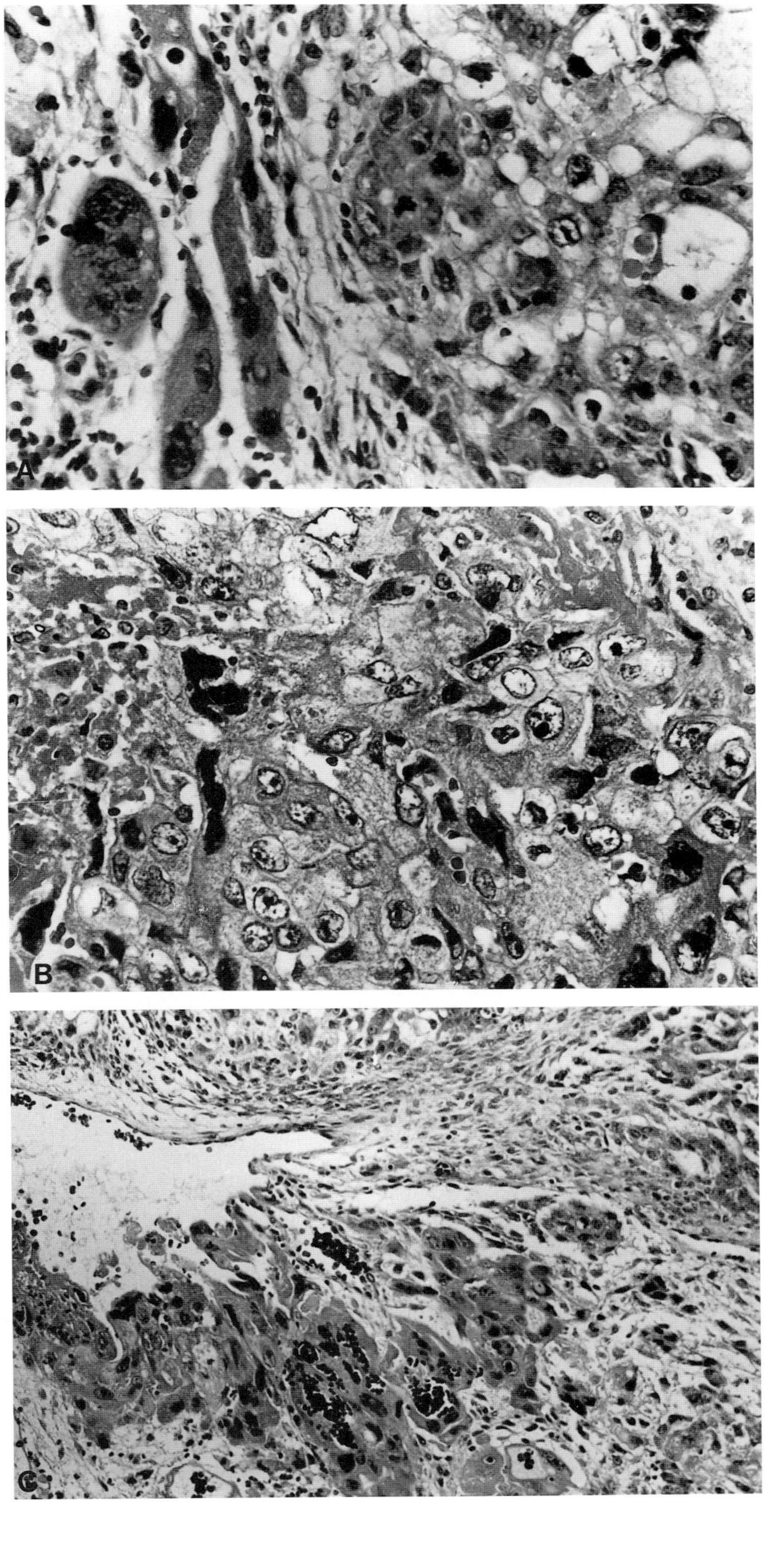

FIGURE 2.34. Choriocarcinoma. (A) Cytotrophoblasts occupy the right half of the field; giant multinucleated syncytial trophoblasts are noted on the left. H&E; × 300. (B) Cytotrophoblasts with large vesicular nuclei, prominent nucleoli and vacuolated to granular cytoplasms are shown. Larger cells intermediate between cytotophoblasts and syncytial trophoblasts are present. Featured prominently are foci of necrosis. H&E; × 300. (C) Vascular invasion is characteristic of this tumor. Tumor cells tend to line up along vascular channels. H&E; × 120.

Non–Germ Cell Gonadal Tumors

Tumors of the ovary and testes are rare in the fetus and infant [12,30]. Approximately 6% of childhood testicular tumors have been reported in the neonate [30]. Most primary tumors of the testis in older infants children are of germ cell origin, namely, teratoma and yolk sac tumor, and occasionally gonadoblastoma [6,30]. On the other hand, juvenile granulosa cell tumor has about the same incidence as germ cell tumors in the newborn testis [30] (Table 2.4). Ovarian cysts are far more common than ovarian tumors in the fetus and infant (Figure 2.35).

TABLE 2.4. Twenty-two neonatal testis tumors recorded in the prepubertal testis tumor registry.

Tumor	Number of Cases	Presentation	
		<24 h	1–30 days
Yolk sac tumor	6	1	5
Teratoma	1		1
Gonadal stromal tumor	6	4	2
Juvenile granulosa tumor	6	2	4
Gonadoblastoma	2		2
Hamartoma	1	—	1
Totals	22	7	15

Source: Levy DA, Kay R, Elder JS [30]. Used with permission.

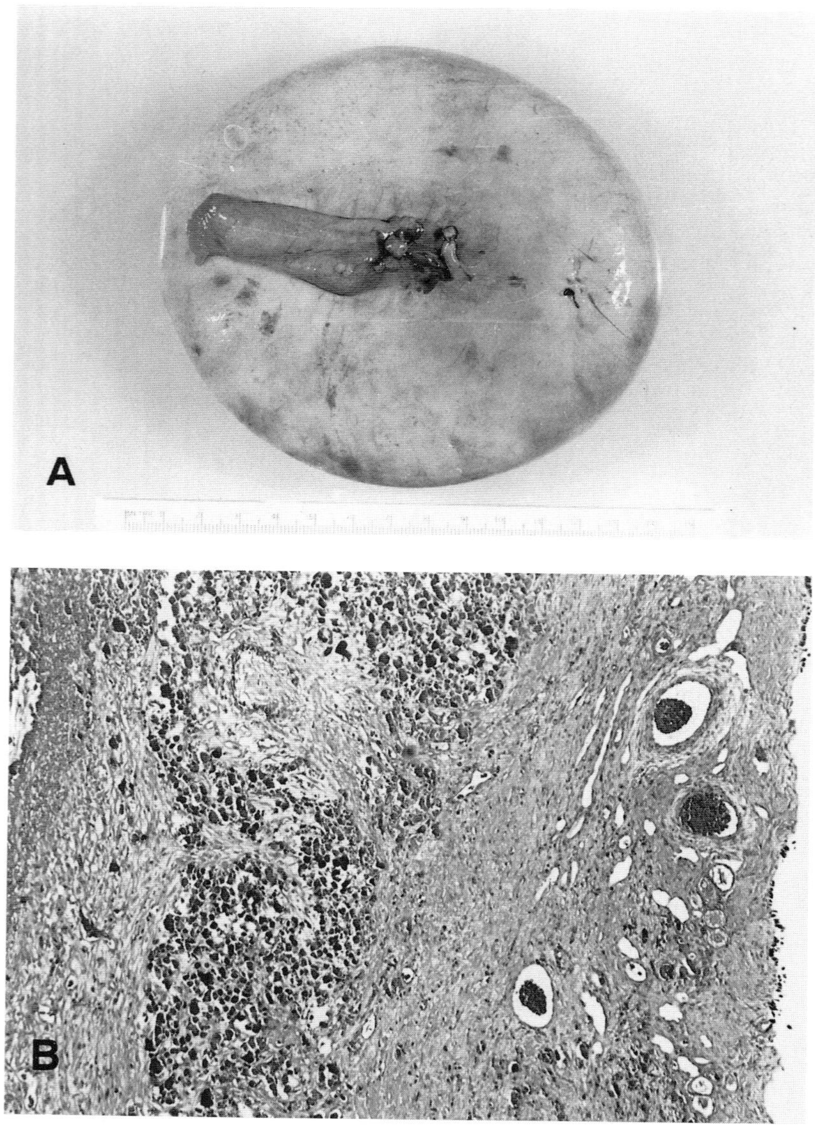

FIGURE 2.35. Follicular cyst of the ovary: 3-week-old female with an abdominal mass. (A) The left ovary was replaced by a 6 × 4.5 cm^2, 40 g cystic structure containing brown cloudy fluid and old blood clot. (B) Atrophic ovary and cyst wall containing foci of necrosis, calcification, old hemorrhage, and fibrosis. The histological findings are consistent with an old torsion (probably in utero) of a large follicular cyst followed by infarction. H&E; × 60.

Juvenile Granulosa Cell Tumor

Although the gonadal stromal tumors (juvenile granulosa cell, Sertoli cell, theca, and Leydig cell tumors) occur more frequently in adults than in infants and children, juvenile granulosa cell tumor is not only the foremost stromal tumor but also the leading gonadal tumor overall in the perinatal period [30–32].

The juvenile granulosa cell tumor of the testis is histologically analogous to its ovarian counterpart and similarly behaves clinically in a benign fashion. A favorable outcome for the infant can be anticipated with orchiectomy or oophorectomy alone as the sole treatment [32]. Testicular enlargement, which is the most frequent sign, abdominal swelling, torsion of the testis, an incidental finding in a hernia sac or at necropsy are the various presenting findings in the infant with this tumor [31,32].

Granulosa cell tumors of the ovary occur in the newborn as well [33,34]. Bilateral tumors are reported in this age group and may be associated with such endocrinological manifestations of sexual precocity as pigmented external genitalia and endometrial hemorrhage [6,33].

Grossly the juvenile granulosa cell tumor is composed of both solid and multiple cystic areas; some of the cysts are very large. Occasionally the involved gonad is replaced by numerous multilocular cysts filled with clear fluid [31,32,34]. His-

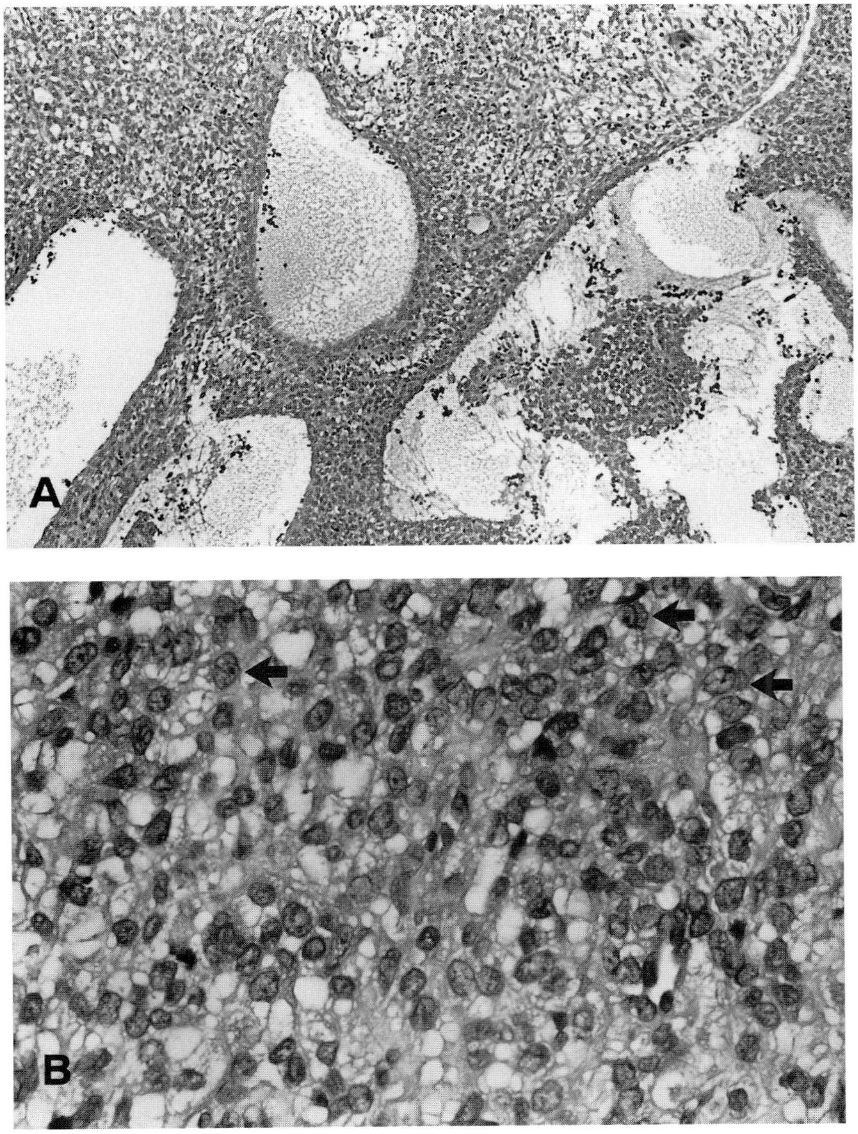

FIGURE 2.36. Juvenile granulosa cell tumor of the testis removed from a 5-month-old male. (A) Diffuse infiltrates of granulosa cells containing large cystic follicles filled with pale-staining secretion giving the tumor a "Swiss Cheese"–like appearance. H&E; × 60. (B) Higher magnification (× 400) showing a bubbly microcystic appearance and tumor cells with mild nuclear atypia and nuclear infolding. Some mitotic figures are present. H&E.

tologically the neoplasm shows diffuse sheets of granulosa cells and cystic follicles of various sizes with pale-staining secretion. The large cystic areas in some fields resemble "Swiss cheese" under low magnification (Figure 2.36A). The septa contain spindle cells resembling ovarian stroma, and the cysts are lined by granulosa cells with clear cytoplasm. The nuclei are round and slightly hyperchromatic, without the grooved appearance characteristic of the adult granulosa cell tumor (Figure 2.36B). Marked nuclear atypia is seen in some tumors accompanied by 5 to 7 mitoses per 10 high-power fields [31,32,34]. The tumor cells and normal Sertoli and granulosa cells stain positively for cytokeratin and vimentin. The spindle cells and normal peritubular myoid and theca externa cells react with vimentin, actin, and focally with desmin [31]. Leydig cells and theca interna cells from normal infant ovaries and testes express only vimentin [31]. These studies appear to confirm the concept that the granulosa tumor cells are related to ovarian and testicular stromal cells.

References

1. Dehner LP. Gonadal and extragonadal germ cell neoplasia in childhood. *Hum Pathol* 1983;14:493.
2. Gonzalez-Crussi F. Extragonadal teratomas. In: *Atlas of Tumor Pathology.* 2nd ser. Fascicle 18. Washington, DC: Armed Forces Institute of Pathology; 1982.
3. Teilum G. *Special Tumors of Ovary and Testis and Related Extragonadal Lesions: Comparative Pathology and Histological Identification.* 2nd ed. Philadelphia: JB Lippincott; 1976.
4. Scully RE, Young RH, Clement PB. Tumors of the ovary, maldeveloped gonads, fallopian tube, and broad ligament. In: *Atlas of Tumor Pathology.* 3rd ser. Fascicle 23. Washington, DC: Armed Forces Institute of Pathology; 1998.
5. Isaacs H Jr. Tumors. In: Gilbert-Barness E, ed. *Potter's Pathology of the Fetus and Infant.* Vol 2. St. Louis, MO: CV Mosby; 1997:1248.
6. Isaacs H Jr. Germ cell tumors. In: *Tumors of the Fetus and Newborn.* Philadelphia: WB Saunders; 1997:15. Major Problems in Pathology. Series; Vol. 35.
7. Isaacs H Jr. Perinatal (congenital and neonatal) neoplasms: a report of 110 cases. *Pediatr Pathol* 1985;3:165.
8. Hawkins EP. Pathology of germ cell tumors in children. *Crit Rev Hematol Oncol* 1990;10:165.
9. Witzleben CL, Bruninga G. Infantile choriocarcinoma: a characteristic syndrome. *J Pediatr* 1968;73:374.
10. Belchis DA, Mowry J, Davis JHSW. Infantile choriocarcinoma: re-examination of a potentially curable entity. *Cancer* 1993;72;2028.
11. Robboy SJ, Miller T, Donahoe PK, et al. Dysgenesis of testicular and streak gonads in syndrome of mixed gonadal dysgenesis: perspective derived from a clinico-pathologic analysis of twenty-one cases. *Hum Pathol* 13:700,1982.
12. Dehner LP. Neoplasms of the fetus and neonate. In: Naeye RL, Kissane JM, Kaufman N, eds. *Perinatal Diseases.* Baltimore: Williams & Wilkins; 1981:286. International Academy of Pathology, Monograph No. 22.
13. Dehner LP, Mills A, Talerman A, et al. Germ cell neoplasms of head and neck soft tissues: a pathologic spectrum of teratomatous and endodermal sinus tumors. *Hum Pathol* 1990;21:309.
14. Tapper D, Lack EE. Teratomas in infancy and childhood: a 54-year experience at the Children's Hospital Medical Center. *Ann Surg* 1983;198:398.
15. Berlin JA, Rich LS, Hahn JF. Congenital orbital teratoma. *Child's Brain* 1983;10:208.
16. Lack EE, Weinstein HJ, Welch KJ. Mediastinal germ cell tumors in childhood: a clinical and pathological study of 21 cases. *J Thorac Cardiovasc Surg* 1985;89:826.
17. Lakhoo K, Boyle M, Drake DP. Mediastinal teratomas: review of 15 pediatric cases. *J Pediatr Surg* 1993;28:1161.
18. Odell JM, Allen JK, Badura RJ, et al. Massive congenital intracranial teratoma. A report of two cases. *Pediatr Pathol* 7: 333,1987.
19. Byard RW, Jimenez CL, Carpenter BF, et al. Congenital teratomas of the neck and nasopharynx: a clinical and pathological study of 18 cases. *J Paediatr Child Health* 26:12,1990.
20. Herrmann M, Thompson K, Wojik EM, et al. Congenital sacrococcygeal teratomas: effect of gestational age on size, morphologic pattern, ploidy, p53 and *ret* expression. *Pediatr Dev Pathol* 2000;3:240.
21. Heifetz SA, Cushing B, Giller R, et al. Immature teratomas in children: pathologic considerations. *Am J Surg Pathol* 1998;22: 1115.
22. Alpers CE, Harrison MR. Fetus in fetu associated with an undescended testis. *Pediatr Pathol* 1985;4:37.
23. Heifetz SA, Alrabeeah A, St J Brown B, et al. Fetus in fetu: a fetiform teratoma. *Pediatr Pathol* 1988;8:215.
24. Fink AM, Cuckow PM, Scott R: Case report: fetus-in-fetu—imaging, surgical and pathological findings. *Clin Radiol* 1995; 50:274.
25. Wold LE, Kramer SA, Farrow GM. Testicular yolk sac and embryonal carcinomas in pediatric patients: comparative immunohistochemical and clinicopathologic study. *Am J Clin Pathol* 1984;81:427.
26. Wu JT, Book L, Sudar K. Serum alpha-fetoprotein levels in normal infants. *Pediatr Res* 1981;15:50.
27. Scully RE. Gonadoblastoma: a review of 74 cases. *Cancer* 1970; 25:1340.
28. Luisiri A, Vogler C, Steinhardt G, et al. Neonatal cystic testicular gonadoblastoma: sonographic and pathologic findings. *J Ultrasound Med* 1991;10:59.
29. Chandra SA, Gilbert EF, Viseskul C, et al. Neonatal intracranial choriocarcinoma. *Arch Pathol Lab Med* 1990;114:1079.
30. Levy DA, Kay R, Elder JS. Neonatal testis tumors: a review of the Prepubertal Testis Tumor Registry. *J Urol* 1994;151:715.
31. Groisman GM, Dische MR, Fine EM, et al. Juvenile granulosa cell tumor of the testis: a comparative immunohistochemical study with normal infantile gonads. *Pediatr Pathol* 1993;13:389.
32. Lawrence WD, Young RH, Scully RE. Juvenile granulosa cell tumor of the infantile testis: a report of 14 cases. *Am J Surg Pathol* 1985;9:87.
33. Zaloudek C, Norris HJ. Granulosa tumors of the ovary in children: a clinical and pathologic study of 32 cases. *Am J Surg Pathol* 1982;6:503.
34. Young RH, Lawrence WD, Scully RE. Juvenile granulosa cell tumor—another neoplasm associated with abnormal chromosomes and ambiguous genitalia: a report of 3 cases. *Am J Surg Pathol* 1985;9:737.
35. Isaacs H Jr. Neoplasms in infants: a report of 265 cases. *Pathol Annu* 1983;18(2):165.
36. Ablin A, Isaacs HI Jr. Germ cell tumors. In: Pizzo PA, Poplack DG, eds. *Principles and Practice of Pediatric Oncology.* 2nd ed. Philadelphia: JB Lippincott; 1993.

3 Soft Tissue Tumors

Soft tissue tumors are relatively common in the first year of life [1–17]. They are the source of a wide variety of neoplasms and tumorlike conditions often presenting as a palpable mass (Table 3.1). Soft tissue tumors in the fetus and infant are different in many respects from those of adults as far as their incidence, anatomical location, histopathology, and prognosis are concerned [3,8,12,13,16]. Excluding hemangiomas and lymphangiomas, they are the dominant group of tumors in infants overall and are second in frequency following teratoma in the newborn [11]. Neoplasms of fibrous connective tissue (myofibroblastic) origin are the most prevalent ones [7,9, 13,14]. Approximately a third of all neonatal and infant soft tissue tumors are classified histologically as malignant. The major malignancies are rhabdomyosarcoma, fibrosarcoma, primitive neuroectodermal tumor (PNET) and rhabdoid tumor [3,5,7–10,13].

In the normal developing fetus and newborn, cells appear immature to begin with, making the distinction between some benign and malignant tumors difficult, as exemplified particularly in the interpretation of soft tissue tumors, where the diagnosis of sarcoma must be made with caution [9,10,13,14, 17,18]. Congenital mesenchymal tumors are seldom malignant in spite of their large size or microscopic resemblance to soft tissue sarcomas in adults [8,14]. Although a soft tissue lesion such as a hemangioma or fibromatosis is called benign microscopically, it may cause death as the result of involvement of a vital structure in the neck or thorax, for example. In spite of this, most soft tissue neoplasms diagnosed in the first year of life, including some of the sarcomas, deserve a conservative therapeutic approach and generally have a favorable outcome [8–10,14,17].

Vascular Tumors

Hemangioma

Vascular lesions are responsible for over two thirds of all cases of soft tissue tumors and tumorlike conditions in infants and children [12,19,20]. Of the vascular lesions, hemangioma, lymphangioma and hemangiopericytoma comprise the majority, in that order, during the first year of life [13] (Figures 3.1–3.13).

Hemangiomas can proliferate in utero and present as fully developed tumors at birth. These congenital lesions can regress rapidly [21]. Hemangiomas in the newborn may be single or multiple and are extremely variable in extent (Figure 3.1). After birth they may remain the original size, grow with the body, or exhibit a more rapid growth. Many of those in the skin are not noticed until the child is a few weeks old and are typically located on the head and neck (60%), trunk (25%), or extremities (15%) [22]. Most regress after a limited period of growth and disappear by the end of the second year. Although at times they are very cellular (e.g., the solid, cellular capillary hemangioma) and show little elaboration into recognizable vascular structures, those present at birth have a favorable prognosis [14]. The skin is the most common site of hemangiomas, although they may be found in mucous membranes, deep connective tissue, or internal organs such as the liver, heart, and brain [13]. Cutaneous hemangiomas are occasionally associated with similar lesions in the brain, liver, and other organs, and the term "diffuse hemangiomatosis" is applied to this often fatal condition [13,23]. Hemangiomas in the region of the distribution of the trigeminal nerve may be associated with vascular anomalies of the choroid of the eye, the pia covering of the brain, and the brain itself in the area directly beneath the involved skin. Known as the Sturge-Weber syndrome, this condition is associated with other findings such as congenital glaucoma and buphthalmus, local thickening of the skull, atrophy of the body on the side opposite the cutaneous hemangioma, paralysis, aphasia, epilepsy, and mental retardation [14].

The hemangiomas associated with von Hippel-Lindau disease (cerebellar-retinal hemangiomatosis) may be present in early life, but symptoms are rarely recognizable until the second decade [14]. The vascular lesions may be single or multiple, cavernous, or composed of masses of solid capillary hemangiomata. They may involve the skin, cerebellum, retina, spinal cord, and liver. Adenomas or cystic disease of the

TABLE 3.1. Soft tissue tumors and tumorlike conditions of the fetus and infant: distinguishing features.

Tumor	Gross findings	Histological findings[a]
Fibromatosis	Firm, light tan-gray to white, whorled cut surface; no necrosis or hemorrhage	Regular, spindle-shaped fibroblastic cells with herringbone and storiform growth patterns; no nuclear atypia; few or no mitoses; vim +
Digital fibromatosis	Firm, light tan-gray nodule(s) on fingers or toes	Regular, spindle-shaped fibroblastic cells with herringbone and storiform growth patterns; no nuclear atypia; rare mitoses; ± red paranuclear inclusions
Torticollis	Firm, fibrous lump in sternocleidomastoid muscle; ± wry neck	Time dependent; early lesion resembles fibromatosis; late: hypocellular interstitial fibrosis with muscle fiber atrophy and loss
Myofibromatosis	Rubbery tan nodule with central congestion, necrosis; single or multiple lesions	Regular spindle cells resembling smooth muscle cells and fibroblasts; central blood vessels, necrosis and calcification; vim +, des +, actin +
Fibrous hamartoma of infancy	Firm, light tan-gray mass; blends in with surrounding fat	Myxoid areas with with small spindle cells; fibrous areas; tendonlike fibrous trabeculae; fat; vim +, des +
Calcifying aponeurotic fibroma	Poorly circumscribed, rubbery, gray-white nodule; ± gritty calcification	Bands of fibrous tissue with fibroblasts; hyalinized collagen; vacuolated chondrocytes ± calcification
Infantile fibrosarcoma	Firm to gelatinous tan-gray cut surface; foci of necrosis; larger than fibromatosis	Variable-sized, spindle-shaped cells with nuclear atypia; many mitoses; herringbone and storiform patterns; myxoid stroma; vim +; trisomy 11; EM: fibroblastic cells
Fibrous histiocytoma	Firm yellow-tan encapsulated	Spindle-shaped fibroblasts; round to oval histiocytes; herringbone and storiform patterns; nuclear atypia and mitotic rate variable; EM: immature histiocytes and fiboblasts; vim +; histiocyte markers ±
Juvenile xanthogranuloma	Firm tan-gray to tan-yellow nodule	Well-differentiated histiocytes with eosinophilic or foamy cytoplasms and vesicular nuclei; eosinophiles prominent; ± multinucleated Touton giant cells; EM: mature fibroblasts and histiocytes; vim +, actin ±; CD68 ±; CD1a +; factor VIIIa +
Angiomatoid fibrous histiocytoma	Red-brown mass with vascular, hemorrhagic center	Vascular channels bordered by atypical histiocytes and fibroblasts; mitoses; peripheral lymphoid nodules; vim +; des +; calponin +; actin ±
Capillary hemangioma	Tan-red nodules resembling lymph nodes; skin most common site	Lobules of proliferating capillaries
Cavernous hemangioma	Large blood-filled vascular channels in soft tissue	Vascular channels lined by regular endothelial cells and filled with blood; thrombi common
Lymphangioma	Hygroma: large cyst filled with lymph; cavernous: collections of dilated lymphatic channels filled with lymph	Single large cyst lined focally by endothelium or many cystically dilated lymphatics
Infantile hemangiopericytoma	Circumscribed, gray-tan to yellow, bulging; foci of necrosis and calcification	Lobular vascular pattern with peripheral slitlike vascular spaces; cells with round, regular open nuclei; capillary and pericyte reticulin patterns
Botryoid rhabdomyosarcoma	Grapelike masses protruding from a mucosal lined surface	Polypoid projection covered by mucosa of site of origin; submucosal mantle layer of tiny rhabdomyoblasts; pale myxoid stroma; vim +; des +
Embryonal rhabdomyosarcoma	Soft, pale tan-gray, gelatinous; foci of necrosis and hemorrhage	Spindle- and tadpole-shaped cells with cytoplasmic cross-striations; cellular areas alternating with pale myxoid areas; muscle markers +; EM: myofilaments; 11p del
Alveolar rhabdomyosarcoma	Soft, pale tan-gray, gelatinous; foci of necrosis and hemorrhage	Small-cell malignant tumor with an alveolar pattern; muscle markers +; EM: myofilaments; [t(2;13)(q35;q14)]
Spindle cell rhabdomyosarcoma	Soft, pale tan-gray, gelatinous; ± necrosis and hemorrhage	Uniform spindle-shaped rhabdomyoblasts; interdigitating growth pattern; muscle markers +; EM: myofilaments
Lipoma	Soft, light yellow circumscribed lobules of fat; single or multiple sites	Lobules of mature fat cells surrounded by delicate fibrous septa
Lipoblastoma	Light gray to yellow lobules of fat firmer than lipoma; single or multiple sites	Lobules of immature fat cells; vacuolated lipoblasts in gray myxoid areas; S-100 +; EM: developing white fat cells (lipoblasts)

(continued)

TABLE 3.1. (*Continued*)

Tumor	Gross findings	Histological findings[a]
Neurofibroma	Firm, pale, homogeneous, light gray-white tissue	Small round to oval Schwann cells, wavy neurites, pale myxoid stroma; S-100 +, Leu-7 +, myelin basic protein +
Plexiform neurofibroma	Shiny, pale, gray-white wormlike nodules	Tortuous, pale myxoid, enlarged nerve fibers with proliferation of nerve fibers and Schwann cells S-100 +, Leu-7 +, myelin basic protein +
Primitive neuroectodermal tumor (PNET)	Soft, friable, pale tan-gray tissue fragments; foci of necrosis and hemorrhage	Small blue-cell tumor with indistinct cytoplasm, round or oval nuclei, fine or clumped chromatin, small nucleoli; many mitoses; ± rosettes; vim +, HBA7 +, S-100 +, NSE +; EM: small primitive cells ± neurosecretory granules and neurofilaments; t(11;22) (q24; q11-12)
Melanotic neuroectodermal tumor	Tumor in craniofacial bones or soft tissues; tan, firm with black streaks or irregular pigmented areas	Irregular cystic spaces with small, round, dark neuroblasts and larger pigmented melanocytes; NSE +; synaptophysin +; HMB45 +; dopamine-β-hydroxylase +; EM: neuroblasts with neurosecretory granules and melanocytes with melanosomes
Ectomesenchymoma (gangliorhabdomyosarcoma)	Light tan-gray to white, bulging cut surface; gelatinous areas	Embryonal rhabdomyosarcoma cells; ganglioneuroma and/or neuroblastoma cells; Schwann cells; rhabdomyosarcoma cells: muscle markers +; Schwann cells S-100 +; ganglion cells and neuroblasts NSE +; EM: rhabdomyosarcoma: myofilaments; neuroblastoma: neurofilaments and neurosecretory granules
Polyphenotypic small-cell tumor	Soft, friable, pale gray, gelatinous	Nests and cords of small epithelial-like cells; neural, rhabdoid, skeletal muscle and PNET components; vim +; cytokeratin +; NSE +; actin +; EM: ± neurosecretory granules, intermediate filaments
Gingival granular cell tumor (epulis)	Mucosal covered tan-gray nodule(s)	Polygonal cells with prominent PAS + granules, regular, round, basophilic nuclei; vim +; S-100 –; muscle markers –; EM: membrane-bound pleomorphic cytoplasmic granules; cell membranes; intermediate filaments
Proliferative myositis and fasciitis	Soft to firm gray-white nodule	Fibroblasts; giant cells resembling ganglion cells; many mitoses; pale myxoid stroma; inflammatory cells; ganglion-like cells; vim +; actin +; CD68 +; EM: cells with fibroblastic, myofibroblastic and histiocytic differentiation

[a] Reactivity (+, –, or ±) with the following immunoperoxidoses is described: des, desmin; vim, vimentin; NSE, neuron-specific enolase; PAS, periodic acid–Schiff reagent. Electron microscopic studies indicated by EM.
Source: Isaacs H Jr [13]; Enzinger FM, Weiss SW [15]; Coffin CM [46].

pancreas, liver, and kidney may also be present. The disease is familial in approximately 20% of cases.

Patients with the Maffucci-Hast syndrome have enchondromas at birth; capillary, cavernous, or solid hemangiomas in the skin and subcutaneous tissue develop before puberty [14].

Capillary Hemangioma

A capillary hemangioma consists of anastomosing capillaries with lumina lined by a single layer of endothelial cells. A thin sheet of such anastomosing vessels lying in the dermis produces a pink to purple, noncompressible area on the skin surface and forms no tumor mass. Such is the nevus flammeus or port wine stain (nevus venosus) often seen on the face or nape of the neck (see Chapter 4: Figure 4.1). When such a proliferation involves a solid organ, the organ does not change its shape, although it may be enlarged. Characteristically capillary hemangiomas consist of lobules and masses of immature endothelial cells, some lining patent blood-containing spaces and some forming solid sheets (Figure 3.2), with little variation among the cells, although mitoses may be found. The cells and constituent vessels are distributed closely among normal muscle, subcutaneous tissue or parenchymal elements of the part involved. Most of the endothelial cells are intimately associated with reticulum fibers, although little or no connective tissue is present in most cellular portions of the tumor [14].

During the proliferative and early involuting phases, the capillary hemangioma becomes more lobular, with large

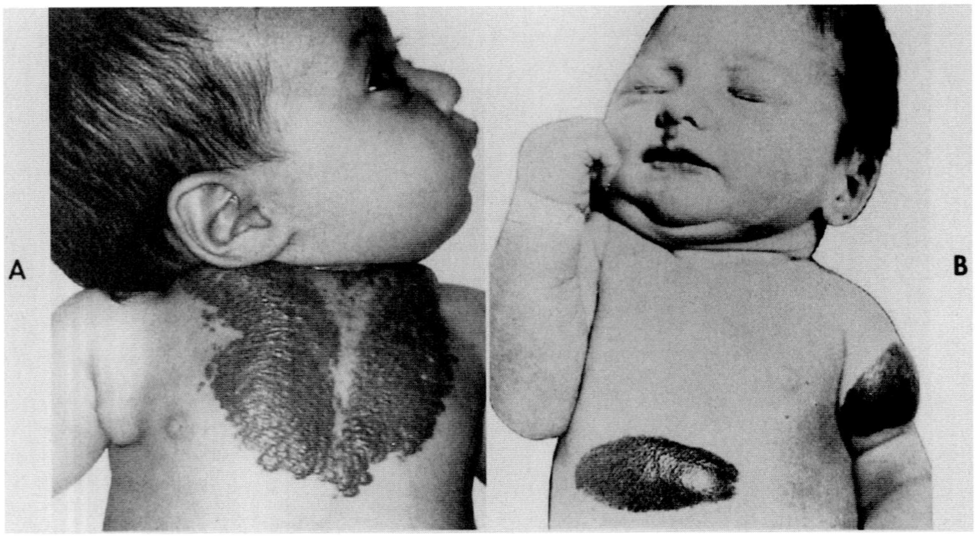

FIGURE 3.1. Hemangioma. (A) Raised hemangioma with multiple small excrescences involving the anterior chest wall. (B) Multiple hemangiomas having smooth, cushionlike surfaces. (From Isaacs H Jr [14], with permission.)

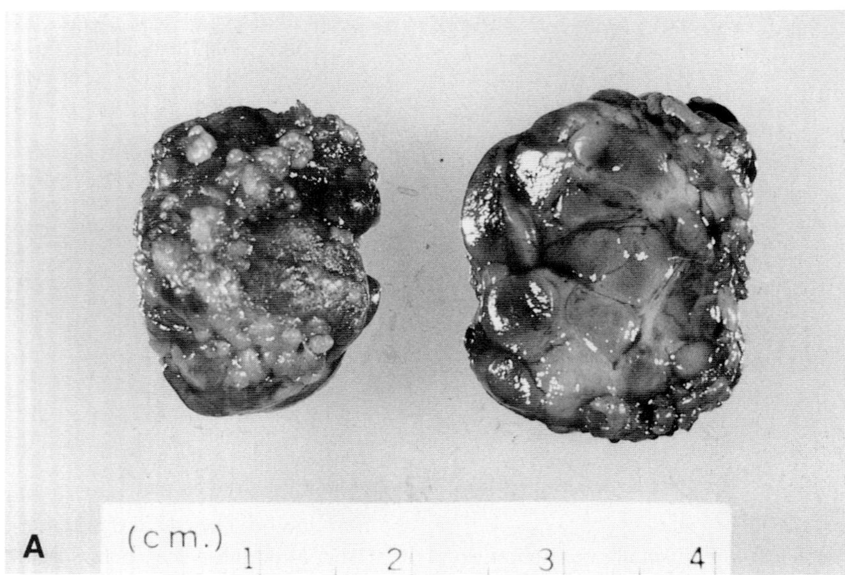

FIGURE 3.2. Capillary hemangioma. (A) A 2.5 × 1.5 cm^2 mass removed from the neck of a 3-month-old male. The lesion was brownish-red and nodular, resembling a mat of lymph nodes. (B) Low-power view (× 32) displaying the lobular architecture corresponding to the gross appearance. Hematoxylin and eosin (H&E). (C) The "open-and-closed hemangioma pattern" consisting of dense clusters of endothelial cells with a few capillary lumina. H&E; × 150. (D) An involuting capillary hemangioma taken from a 6-month-old female, showing abundant fat in the lobule separating residual nests of capillaries. Large "feeding" vessels situated along the left side of the picture are noted at the periphery of the lobules. H&E; × 32.

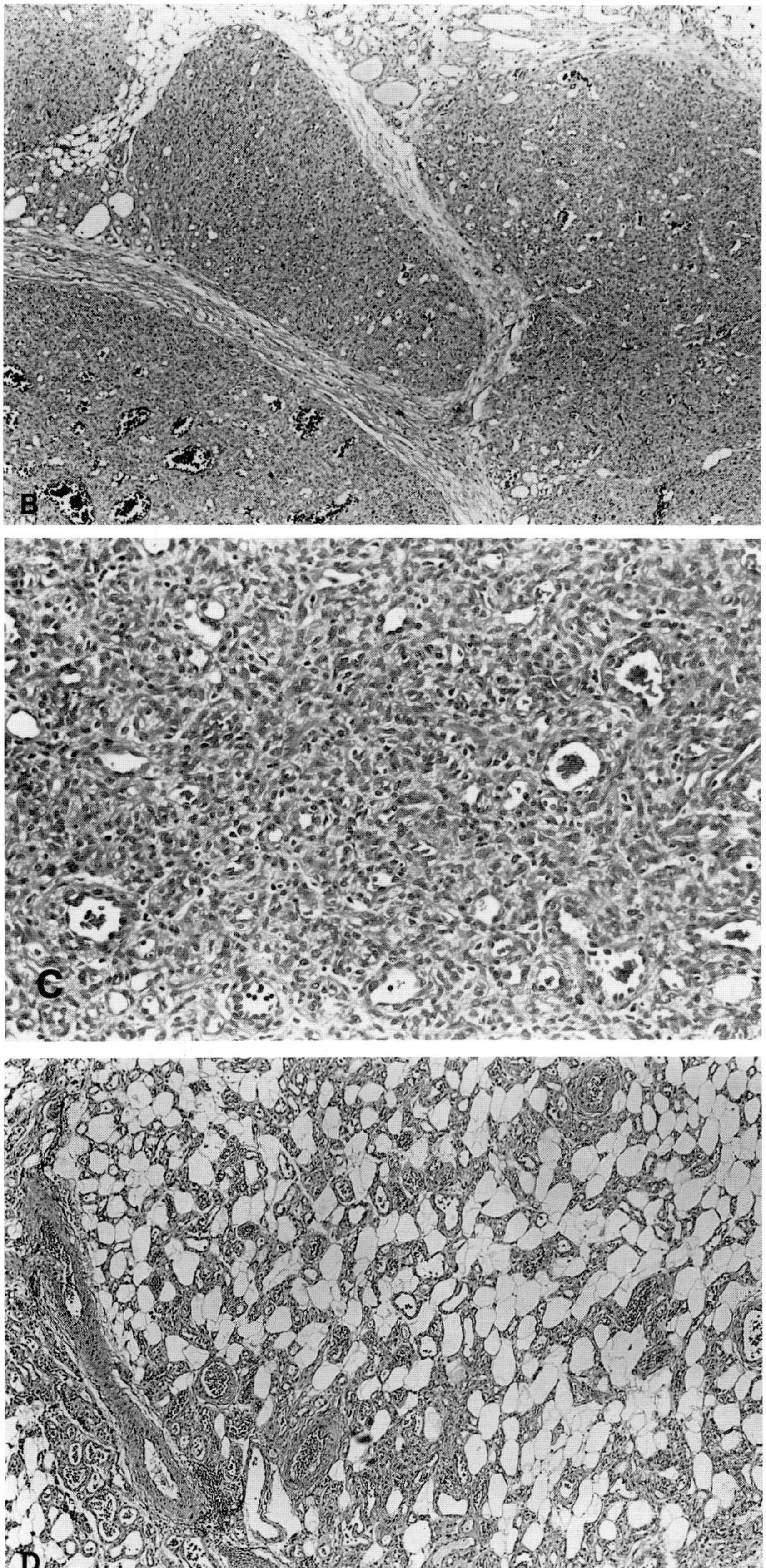

FIGURE 3.2. (*Continued*)

feeding and draining vessels located within fibrous septa adjacent to the lobules. The involuting phase, defined clinically as ages 1 to 7 years, is manifested histologically by decreasing cellular proliferation and a progressive increase in perivascular and interlobular fibrous tissue [19,21] (Figure 3.2D). Apoptosis is characteristic of the involuting phase, which begins before age 1 year, reaching its peak at 2 years, and gradually diminishing from then on [19,21]. With continued involution the endothelial cells flatten, the vascular channels become more dilated, and fibrofatty tissue increases in amount both within and between the lobules. At the end of the involuting phase, histological findings consist only of a few tiny vessels that are lined by flattened endothelial cells surrounded by fibrofatty tissue [14].

The clinical complications that arise from hemangiomas are the result of their size, location, and physiology [14]. They may impair vision by encroaching upon the eye; produce respiratory distress by tracheal compression; produce high-output congestive cardiac failure associated with nonimmune hydrops fetalis and polyhydramnios; cause central nervous system or gastrointestinal hemorrhage; or be responsible for a consumptive coagulopathy from platelet and fibrinogen trapping as in the Kasabach-Merritt syndrome [21,24,25].

Cavernous Hemangiomas

Cavernous hemangiomas have larger vascular spaces than the capillary hemangiomas and are compressible because of the large amounts of blood they contain. Those involving the skin are raised above the general skin level and often have a warty granular surface [14]. The raised vascular nevi such as strawberry nevi contain mixtures of cavernous and capillary nets. The vascular channels vary greatly in size and extent, and occasionally may involve large parts of the body (Figure 3.3).

The liver is the internal organ most frequently the site of a cavernous hemangioma. The extent of involvement may vary from a few millimeters to several centimeters. If superficial, trauma may cause rupture of some of the outermost channels, and fatal intra-abdominal hemorrhage may result. A tumor that is deep in the liver may never be discovered except as an incidental finding at necropsy [14]. Cavernous hemangioma of the lung in an infant is the subject of an isolated case report [26].

Microscopically the cavernous hemangioma consists of anastomosing thin-walled vascular channels lined by endothelial cells larger than those found in the capillary hemangioma (Figure 3.4). Vascular thrombi are common in these lesions. They are associated with large vascular malformations [14] (Figure 3.3B).

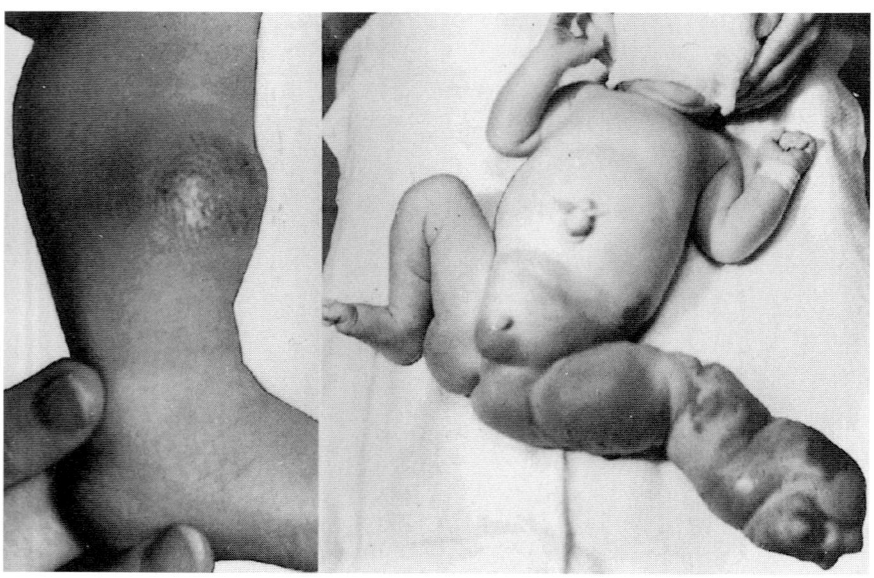

FIGURE 3.3. Deep-seated cavernous hemangioma. (A) Localized lesion on the right leg. (B) Generalized hemangioma involving the left leg, buttocks, and lower abdomen, composed of greatly dilated vascular spaces and responsible for local gigantism. The findings are consistent with the Klippel-Trenaunay-Weber syndrome. Death occurred at 10 days. (From Isaacs H Jr [14], with permission.)

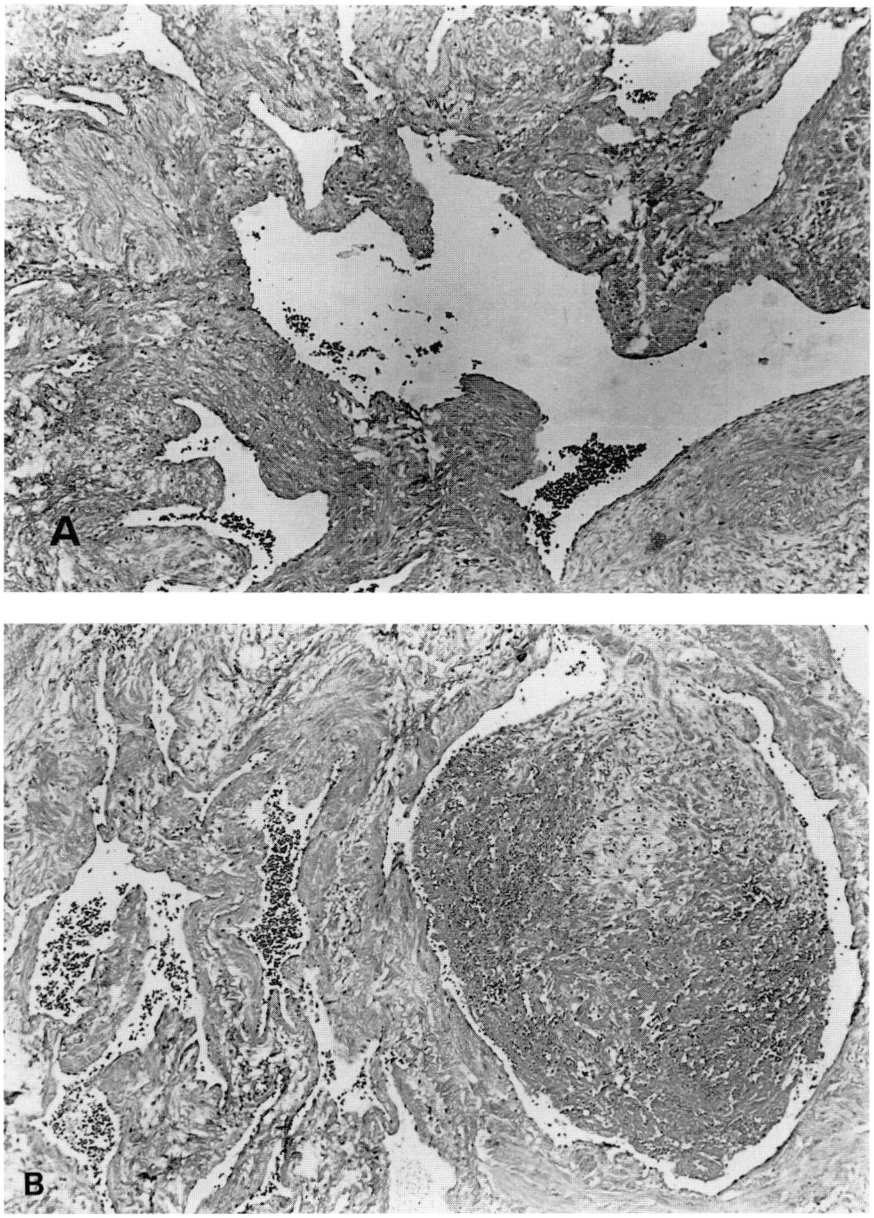

FIGURE 3.4. Cavernous hemangioma: 6×4 cm^2 red brown, nodular mass consisting of many dilated blood vessels was removed from the forearm of a male infant. (A) Large irregular vascular channels are lined by endothelium containing scattered erythrocytes. H&E; $\times$ 40. (B) Organizing thrombi are not uncommon in this lesion. H&E; $\times$ 40.

Vascular Malformations

Vascular malformations are developmental errors and consist of dysplastic vessels lined by inactive appearing endothelium. By definition they are present at birth, but they may become clinically evident any time during life (Figure 3.5). Vascular malformations consist of large endothelial-lined channels with varying amounts of smooth muscle within their walls. They may involve a large body area such as an extremity, producing gigantism of that part. Currently venous angiomas are considered to be vascular malformations. Generally arterial malformations, including arteriovenous anomalies, are diagnosed in late infancy or childhood [19].

Vascular malformations never regress but not infrequently undergo progressive expansion, producing clinical problems. In contrast to hemangiomas, vascular malformations do not express certain cellular markers, such as proliferating cell nuclear antigen, vascular endothelial growth factor, type IV collagenase, and basic fibroblast growth factor [23].

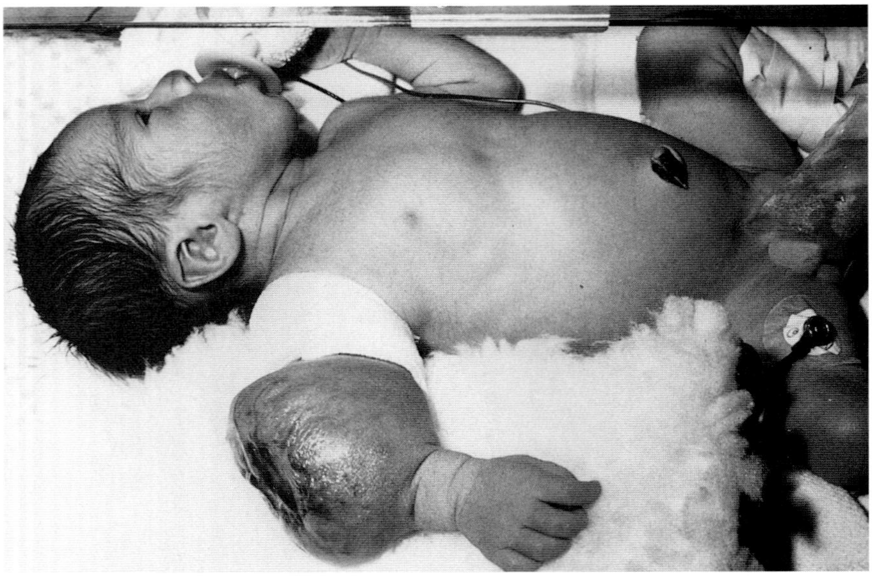

FIGURE 3.5. Arteriovenous malformation. Newborn with a large arteriovenous malformation of the right forearm that functioned as an arteriovenous fistula producing high-output congestive heart failure. Ligation of the artery leading into the malformation was required to control the life-threatening effect of the fistula.

Infantile Kaposiform Hemangioendothelioma

Most kaposiform hemangioendotheliomas are present at birth or develop shortly thereafter, typically as a red or purple cutaneous patch that rapidly enlarges, eventually becoming an indurated deep soft tissue mass [19]. This distinctive form of hemangioma often is associated with the Kasabach-Merritt syndrome [19,23]. Some investigators feel strongly that the Kasabach-Merritt phenomenon does not occur with the common hemangioma and that the previously reported cases of lesions associated with this phenomenon probably represented kaposiform hemangioendotheliomas rather than capillary or cavernous hemangiomas [27]. The author does not completely agree with this view.

Kaposiform hemangioendothelioma differs from the common childhood hemangioma in that the sexes are affected about equally and the former occurs more often on the trunk, in the retroperitoneum, and in the proximal portion of the extremities and less frequently in the cervicofacial region The most common locations are the retroperitoneum and upper extremities [19,27]. Their large size and infiltrative nature result in complications of hemorrhage, obstruction, or respiratory compromise, making complete surgical excision difficult [23]. Kaposiform hemangioendothelioma differs from the common hemangioma in that magnetic resonance imaging studies reveal diffuse enhancement with ill-defined margins, cutaneous thickening, stranding of subcutaneous fat, hemosiderin deposits, and small feeding and draining vessels [27].

Two histological patterns, angiomatous and kaposiform, are described [23]. The former consists of rounded or dilated vascular spaces lined by flattened endothelial cellls and the latter infiltrating fascicles, sheets, or nodules of spindled cells lining compressed (slitlike) or crescentic vascular channels. Mitoses are infrequently noted. Microthrombi, and hemosiderin deposits are variably present. It is the kaposiform pattern that differentiates this lesion from the common cellular hemangioma. Endothelial markers are present in the angiomatous areas but are variably expressed in the spindled cells [23].

Tufted Angioma

Tufted angioma, which occurs in the newborn and in early childhood, is a benign vascular lesion regarded as another form of hemangioma. The angioma is characterized by slow growth and the tendency to cover large areas of the trunk and neck. Other sites of involvement are the head and extremities. Patients with tufted angiomas may present with Kasabach-Merritt syndrome [28]. Histolologically the tumor consists of small lobules or vascular tufts scattered in a multiple "cannon ball-like" distribution in the dermis and subcutis (see Chapter 4). The endothelial cells are immunoreactive with *Ulex europaeus* lectin I and factor VIII, and the spindle cells between the vascular spaces are reactive with actin [23]. Endothelial cells and pericytes are noted with electron microscopy.

Lymphangioma

Lymphangioma is a benign vascular malformation composed of cystically dilated lymphatics. Many cystic lymphangiomas (hygromas) and diffuse or cavernous lymphangiomas are present at birth [10,13,14,29]. They are second to hemangioma

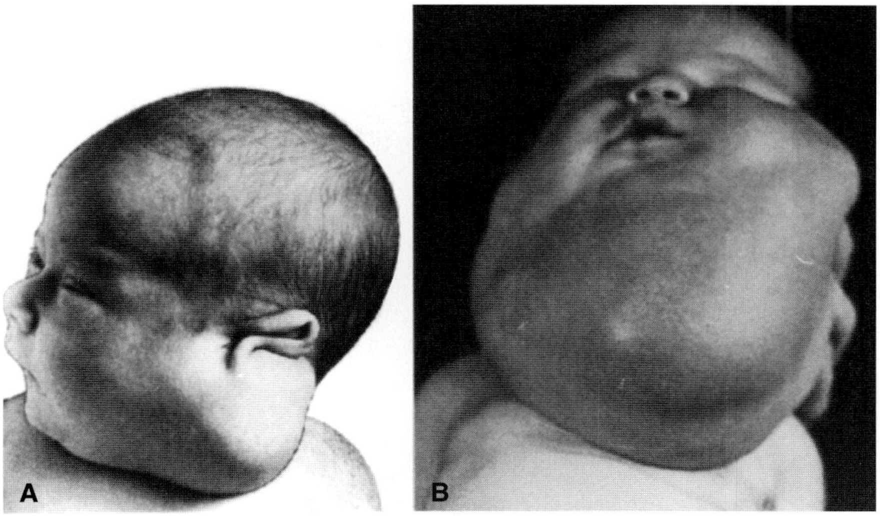

FIGURE 3.6. Cystic lymphangiomas (hygromas). (A) Unilocular tumor of the left side of the neck in an infant aged 4 days. (B) Multilocular tumor involving both sides and anterior part of the neck in an infant aged 4 weeks. (From Isaacs H Jr [14], with permission.)

as a cause of a soft tissue mass in the newborn and infant [10]. The incidence of cystic hygroma is estimated at 1 in 6000 pregnancies [29,30]. Lymphangiomas range in size from small, barely visible subepidermal skin or mucosal blebs (lymphangioma circumscriptum) to massively dilated cystic masses (hygroma), and they occur in almost any location, including the viscera and bones. However, the soft tissues of the neck, axilla, thorax, and the lower extremities are the major sites [14] (Figures 3.6–3.11).

Lymphangiomas result from failure of communication and obstruction of lymphatics. The term "cystic lymphangioma," or "hygroma," is applied to large single or slightly multilocular fluid-filled cavities, most often located on one side of the neck [14] (Figures 3.6–3.8). Occasionally they are multiple and may be found in other locations (Figures 3.9–3.11). Typically cystic hygromas develop late in the first trimester to early in the second trimester, becoming less common as pregnancy progresses. Fluid accumulation in dilated lymphatics and surrounding connective tissue leads to progressive lymphedema and in some instances hydrops fetalis [29,30].

The prognosis of a fetus with cervical cystic hygroma is poor when detected prenatally, especially before 30 weeks gestation, whereas when it is diagnosed after birth the prognosis is generally more favorable [29,31]. Bilateral hygromas of the neck with widespread lymphangiectasis are not uncommon in aborted fetuses (Figure 3.8). Lymphangiomas discovered in utero by sonography are associated with a variety of syndromes and birth defects [30,31]. Turner syndrome, which is noted in approximately 50 to 70% of cases [29], occurs in 1 of 5000 to 10,000 births and usually is the result a 45,X karyotype arising from chromosomal nondysjunction. Fetuses with cystic hygroma and Turner syndrome may have coexistent anomalies of cardiac development such as aortic valve defects, aortic arch hypoplasia, anomalous origin of the aortic arch vessels, and septal defects [29]. In utero fetal demise occurs frequently in 45,X fetuses [30]. Over 95% of nonmosaic 45,X fetuses are spontaneously aborted. Other abnormalities associated with cystic hygroma are hydrops fetalis, oligohydramnios, single-vessel umbilical cord, Noonan syndrome, fetal alcohol syndrome, diaphragmatic hernia (with Fryns syndrome), and trisomies 13,18, and 21; the malformation is described also in normal individuals [13,29–32].

Microscopically lymphangiomas are thin walled, filled with clear straw-colored fluid, lined by a single layer of flat endothelial cells, and surrounded by varying amounts of connective tissue and smooth muscle (Figures 3.7D, 3.10C, 3.11B). They are not compressible and are not visibly connected to lymphatic channels. Often the endothelial lining is missing. The endothelial-lined dilated lymphatic channels react positively with factor VIII, CD31, CD34, smooth muscle actin, and particularly CD34, which shows the most uniform pattern of endothelial reactivity [32,33]. Despite their propensity for recurrence lymphangiomas do not demonstrate the proliferative and involutional growth phases of hemangiomas [33]. They may become larger after birth because of increased secretion of fluid. The hygromas with Turner syndrome are composed of single cavities, whereas isolated cystic hygromas tend to be multiloculated [14].

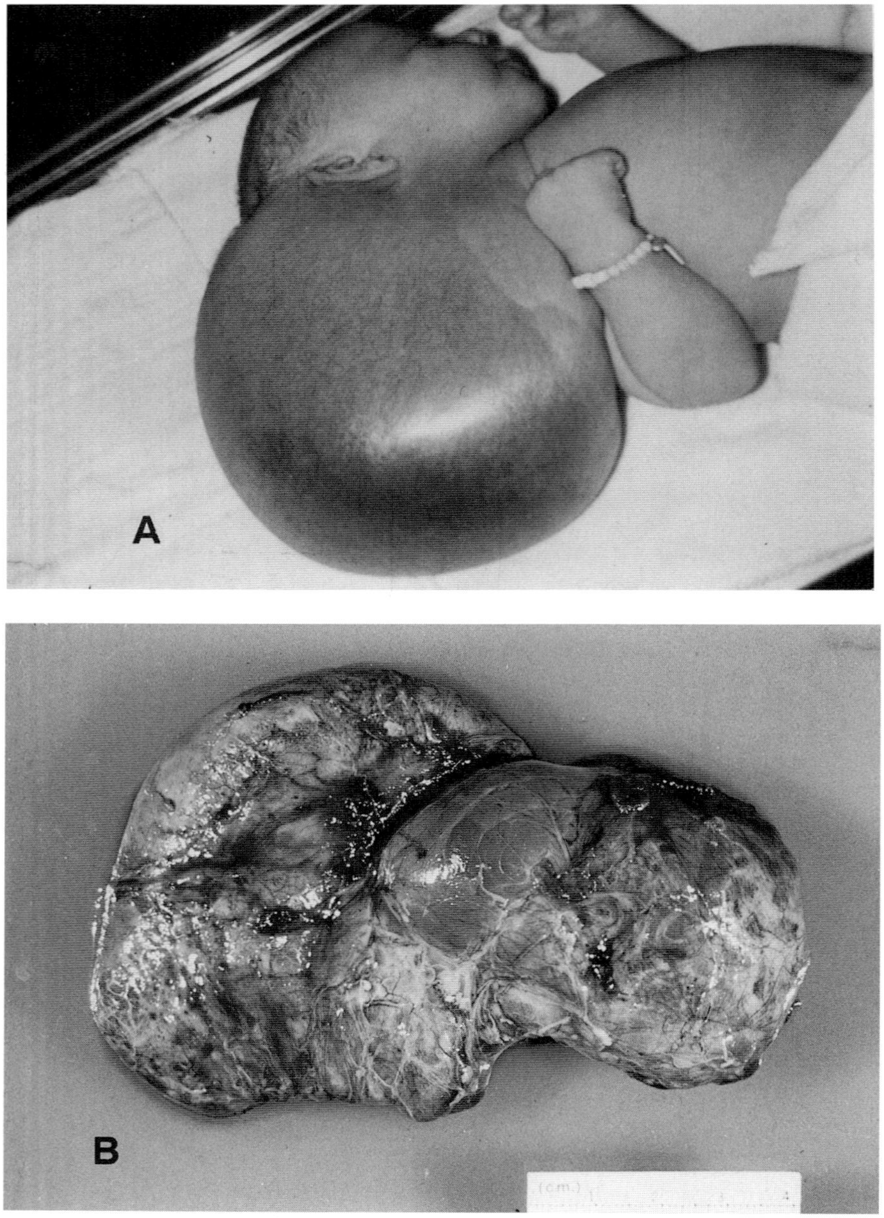

FIGURE 3.7. Cystic lymphangioma (hygroma) of the neck. (A) Newborn with a large, unilocular, cystic mass arising from the back of the neck and shoulder. (B) Unopened specimen taken from another infant with a similar but smaller hygroma, measuring 10×7 cm^2. (C) The opened specimen shows one large cyst and a smaller solid microcystic component situated at the top of the photograph. The cyst was filled with clear lymph fluid. (D) Several large vascular channels are lined by single rows of endothelial cells and filled with pale proteinaceous material. H&E; $\times$ 240.

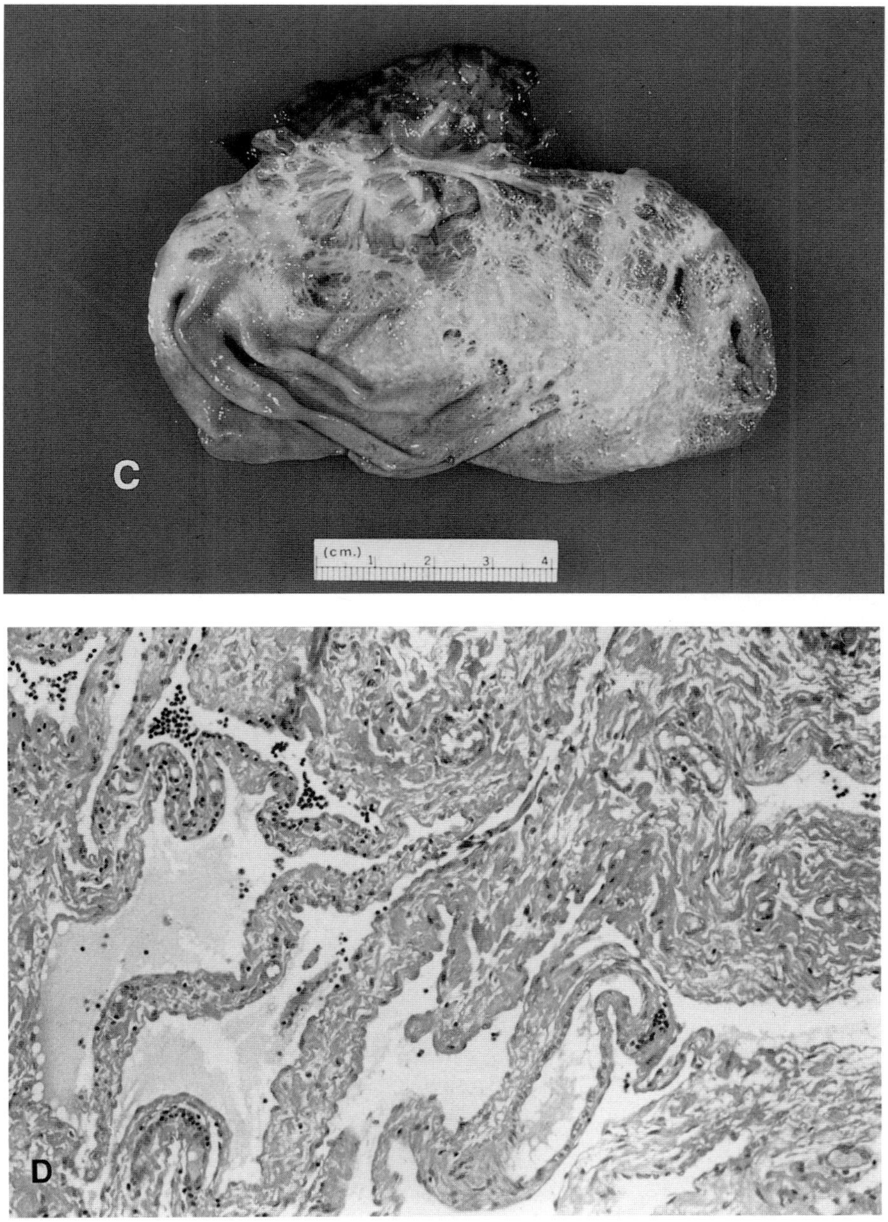

FIGURE 3.7. (*Continued*)

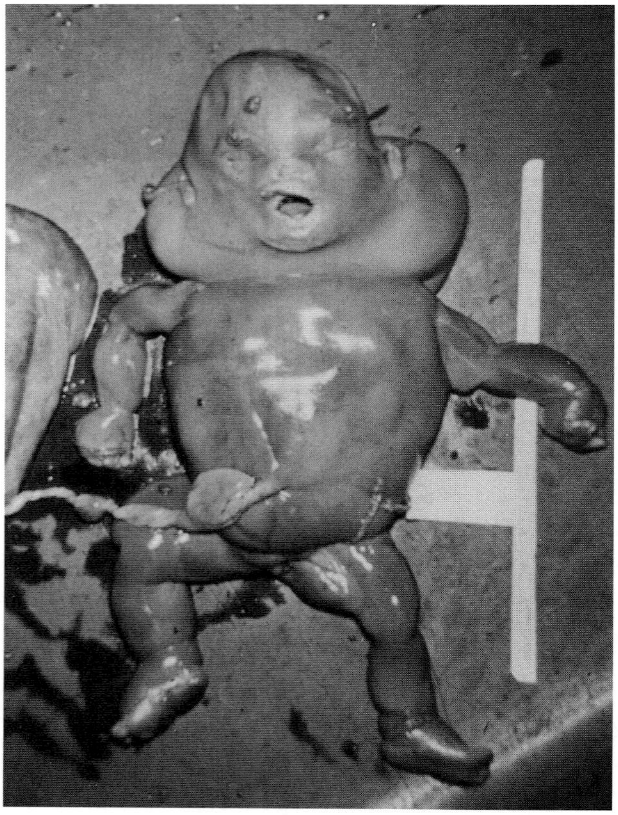

FIGURE 3.8. Fetus with bilateral cystic hygromas of the neck and generalized lymphangiectasis of subcutaneous tissue. There is a high incidence of Turners syndrome (45,X0) and other chromosomal anomalies (e.g., trisomy 21) associated with bilateral cervical hygromas.

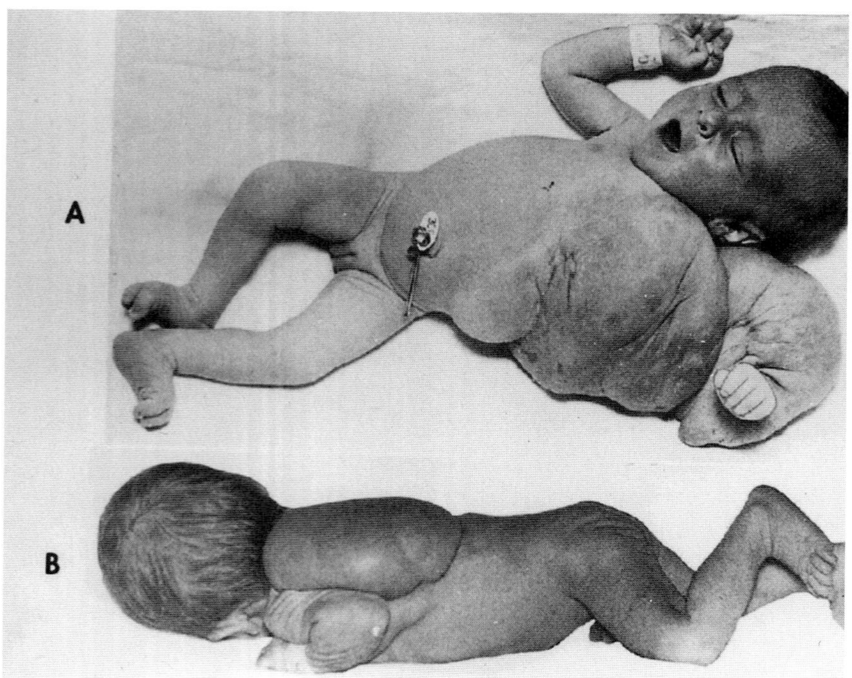

FIGURE 3.9. (A) Cavernous lymphangiomas involving the chest, back, and arm, the most common sites of such tumors, in an infant aged 5 days. Death followed an unsuccessful attempt at removal. (B) Cavernous lymphangioma of the left side of the back. (From Isaacs H Jr [14], with permission.)

FIGURE 3.10. Cavernous lymphangioma producing focal gigantism of an extremity. (A) One-year-old male was born with his left lower extremity much larger than the right. (B) The subcutaneous tissue is replaced by cystically dilated lymphatics of various sizes. Some are filled with blood (probably related to surgical artifact). (C) Numerous dilated lymphatics containing pale-staining lymph. H&E; × 50.

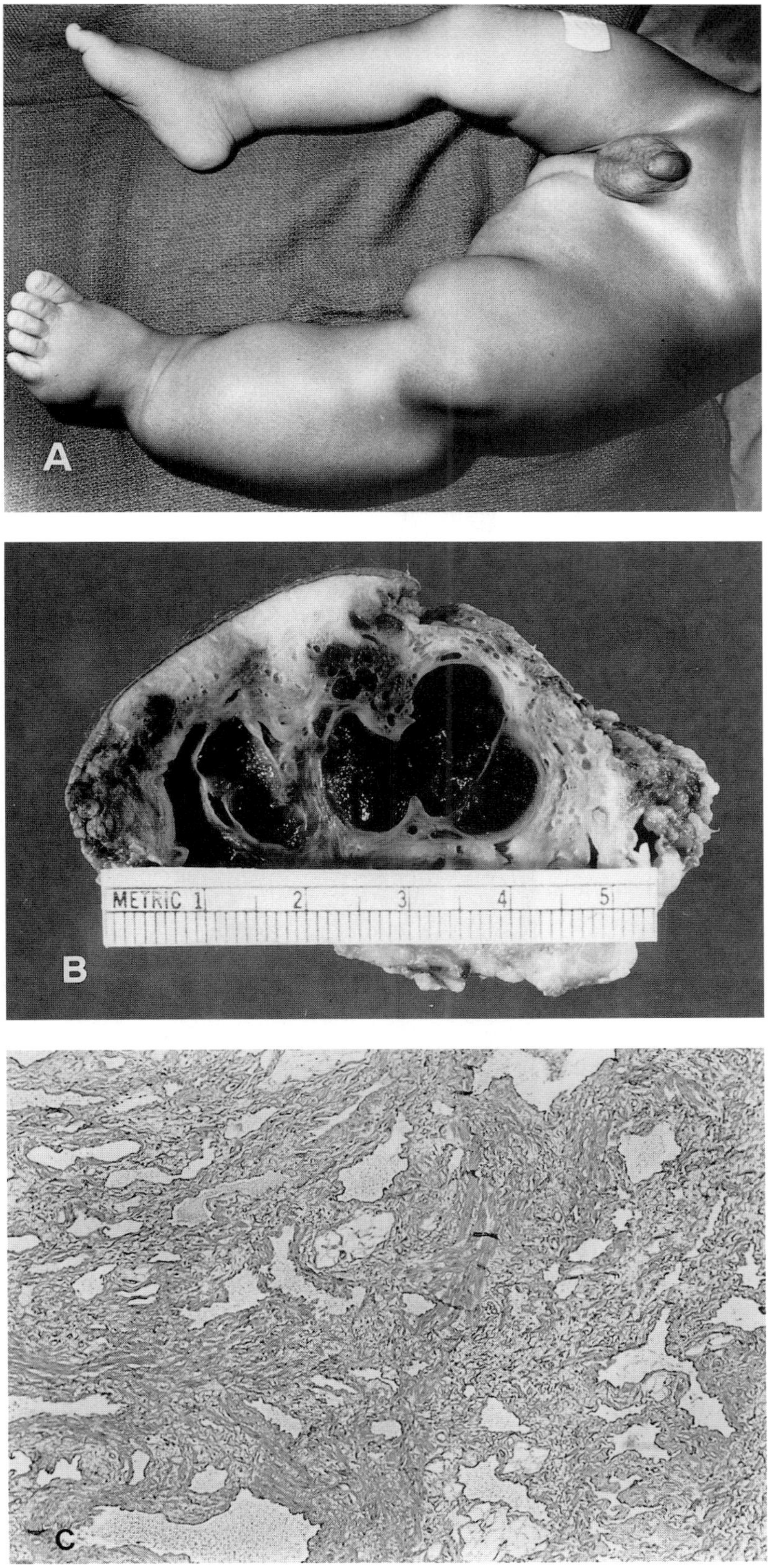

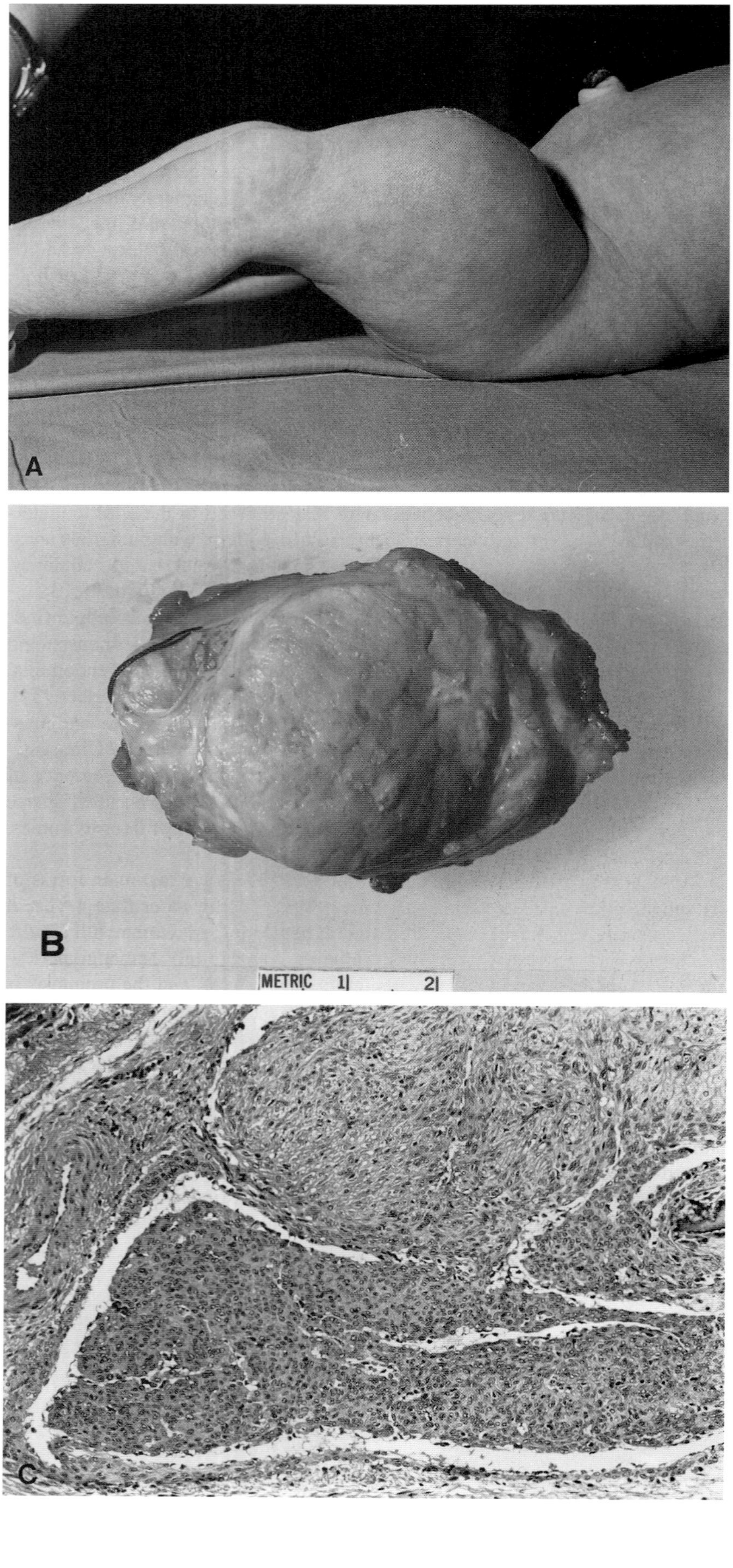

FIGURE 3.12. Infantile hemangiopericytoma. (A) A 1-month-old female with a left thigh mass present at birth. (B) The tumor weighed 43.6 g and measured 5.5×4.5 cm^2. The cut surface is lobular, bulging and tan-gray to yellow-brown with extensive necrosis. (C) Low power view ($\times$ 120) displaying the lobular architecture of the tumor. The lobules are rimmed peripherally by slitlike vascular spaces, which is a helpful diagnostic feature. H&E. (D) The tumor consists of cells with regular round to oval vesicular nuclei and inconspicuous nucleoli. A focus of calcification is present at the upper right corner. H&E; $\times$ 200. (E) Intravascular tumor thrombus within skeletal muscle. H&E; $\times$ 150. (F) Reticulin fibers surrounding nests of cells (hemangioma pattern) and individual cells (hemangiopericytoma pattern) with transition zones between the two. $\times$ 300. Verhoeff's elastic stain; $\times$ 300. (From Isaacs H Jr [10], with permission.)

FIGURE 3.10. Cavernous lymphangioma producing focal gigantism of an extremity. (A) One-year-old male was born with his left lower extremity much larger than the right. (B) The subcutaneous tissue is replaced by cystically dilated lymphatics of various sizes. Some are filled with blood (probably related to surgical artifact). (C) Numerous dilated lymphatics containing pale-staining lymph. H&E; $\times$ 50.

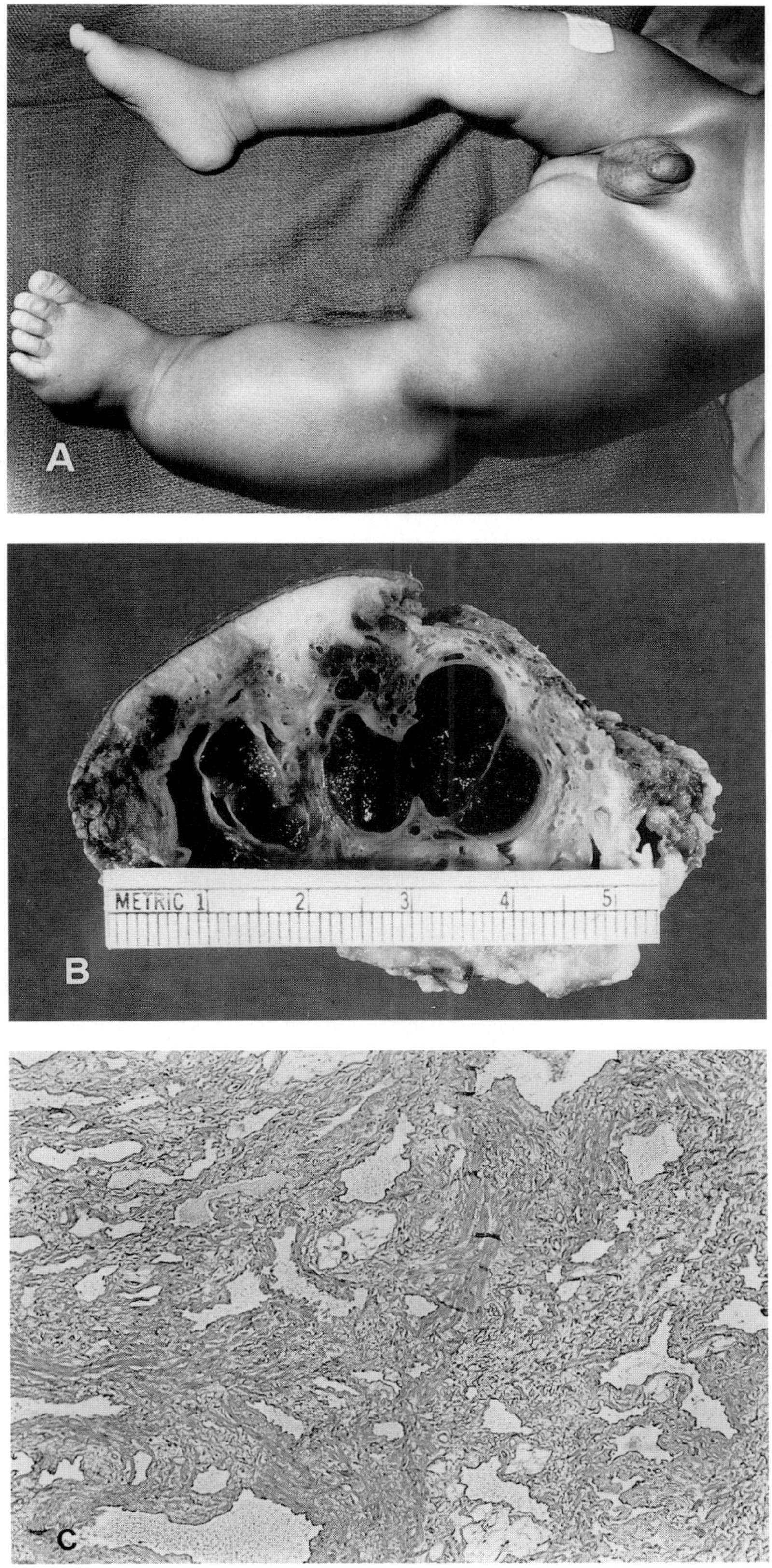

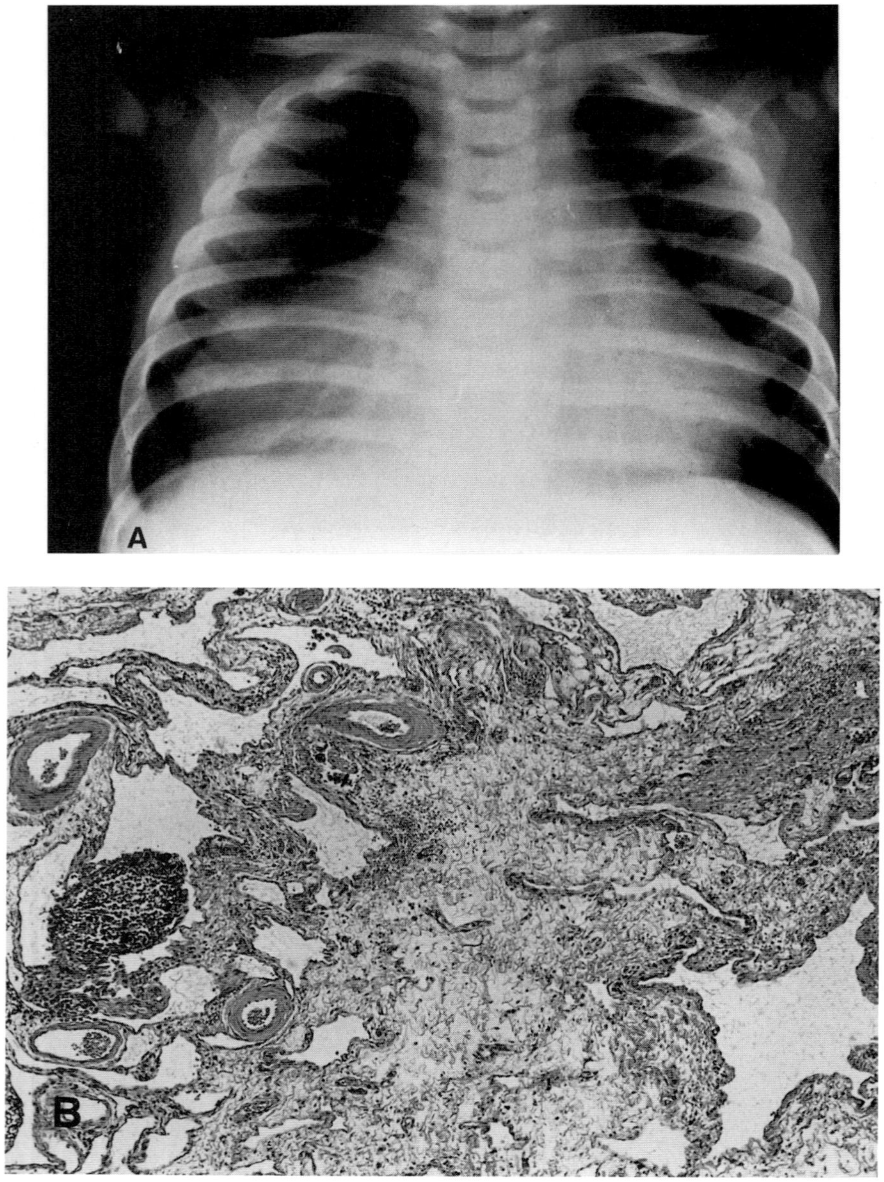

FIGURE 3.11. Lymphangioma of the mediastinum. (A) Chest x-ray of a large mediastinal mass with radiating vessels in a 5-month male with a history of increasing respiratory distress. (B) Biopsy reveals numerous cystically dilated lymphatics and a few blood vessels. Lymphoid nodules characteristically are present in lymphangiomas. H&E; × 120.

Lymphangiectasis

Usually visible at birth as a diffuse swelling of part or all of an extremity without other symptoms, lymphangiectasis is not progressive and remains limited to the area of original involvement. The skin and deep fascia are normal, but between them the space is unusually wide and is spongelike in appearance as a result of increase in connective tissue and size of lymph spaces. Histological examination reveals large lymphatic channels lined by a single layer of endothelial cells separated by normal connective tissue and blood vessels [14].

Hemangiopericytoma

Hemangiopericytoma is an uncommon, soft tissue, vascular neoplasm composed primarily of pericytes (pericapillary mesenchymal cells), which are intimately related to capillaries and share certain ultrastructural and immunohistochemical features in common with fibroblasts and smooth muscle cells [34–36]. Congenital (infantile) hemangiopericytoma is regarded as a distinct subtype of the classic hemangiopericytoma [12,37–39]. Typically lesions present as subcutaneous tissue masses in a variety of sites but they may occur also in other locations, such as the tongue, nasopharynx, stomach, and central nervous system [11,37–39]. The incidence of hemangiopericytoma is greater in adults than in children; less than 10% of cases are found in infants [10,11,15,34–36]. In addition, the neoplasm has been detected prenatally by ultrasound and noted in stillborns [13,37,39,40]. Perinatal hemangiopericytomas may mimic the clinical manifestations of hemangiomas. Life-threatening hemorrhage and hypovolemic shock from an ulcerated subcutaneous hemangiopericytoma, bleeding diathesis (anemia and hypofibrinogenemia) associated with a retroperitoneal primary, hematemesis from a gastric tumor, and high-output congestive heart failure are a few examples [13,37].

The subcutaneous tissues of the extremities and the head and neck are the most common locations of hemangiopericytoma [14,34,35,37,38]. They vary in size from 1 cm to 15 cm in diameter or more and display a variable gross appearance. Some are firm, tan-yellow to dark red-brown, nodular with bulging cut surfaces containing areas of necrosis and calcification (Figure 3.12) while others are friable, tan-gray, with foci of hemorrhage, necrosis, blood-filled cystic spaces similar to those observed in older patients [34] (Figure 3.13).

Histologically the tumors show a nodular vascular pattern consisting of plump to spindle-shaped pericytes with round or oval glassy nuclei. Tumors of infants typically display characteristic slitlike, crescent-shaped, endothelial-lined spaces situated about the periphery of the tumor nodules. In addition, there is a mixed histological picture composed of both hemangiopericytoma and capillary hemangioma components. The hemangioma component is composed of lobules of capillaries and endothelial cells situated inside the capillary basement membrane (Figure 3.12). Blend zones between the pericytoma and hemangioma components are present in infant tumors. Extensive necrosis and small foci of calcification are noted in most specimens [13,37] (Figure 3.12).

On the other hand some infant hemangiopericytomas have an appearance similar to that observed in the adult counterpart: a more prominent vascular pattern, with numerous anastomosing, dilated, thin-walled vessels, increased cellularity, nuclear atypia, and mitotic activity [12,37] (Figure 3.13).

The immunohistochemical and electron microscopic findings of congenital hemangiopericytoma are rather nonspecific [34,36,39]. Factor VIII related antigen staining of endothelial cells and vimentin positivity of tumor cells are described [34]. Desmin, S-100 protein, and keratin are nonreactive, and muscle actin (HHF35) stains focally positive [34,41]. Ultrastructural features supportive of the diagnosis are the presence of pericytes with interdigitating cytoplasmic processes, sparse cytoplasmic organelles, moderate numbers of free ribosomes, and few intercellular attachments [34,41].

Childhood hemangiopericytoma shows two main forms of clinical behavior. Generally in children older than 1 year it behaves simlar to the adult hemangiopericytoma with a more guarded prognosis. Infantile (congenital) hemangiopericytoma has a more benign clinical course, and the prognosis is generally favorable [10,11,15,37–39]. Moreover, spontaneous regression after partial resection occurs [35,39]. Reasons for these behavioral differences are unclear [35]. Occasionally metastases occur in infants with hemangiopericytoma [37]. The head and neck area is the site associated with the highest incidence of metastatic disease [37]. The treatment of choice is a conservative surgical approach with careful follow-up for an extended period of time because metastases may result one or more years after the initial surgery [37].

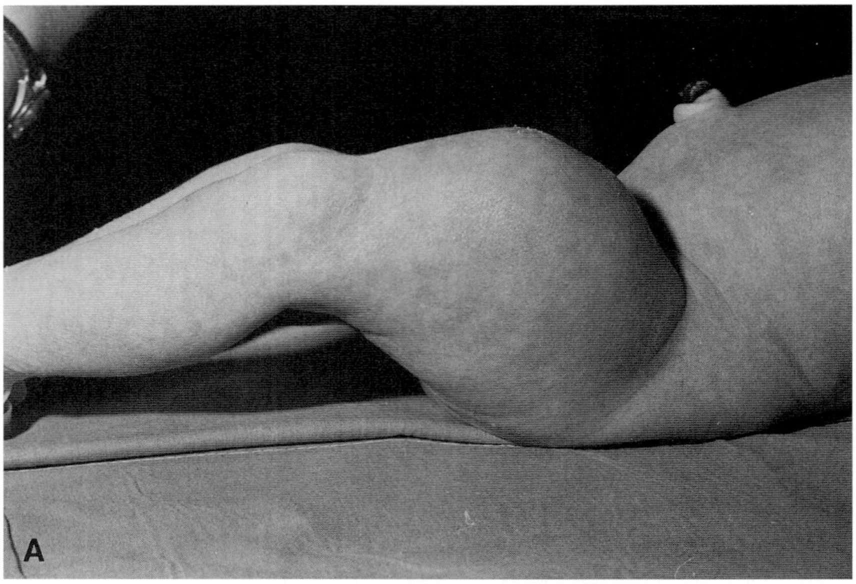

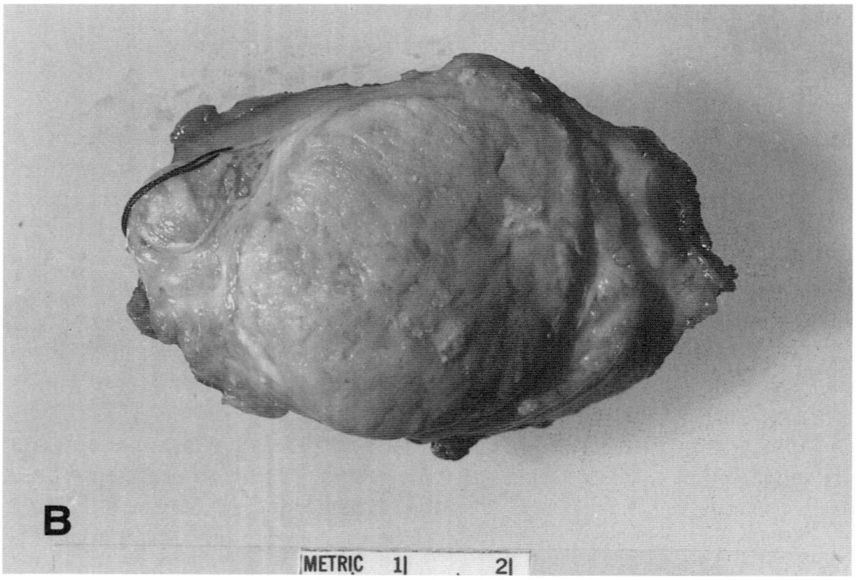

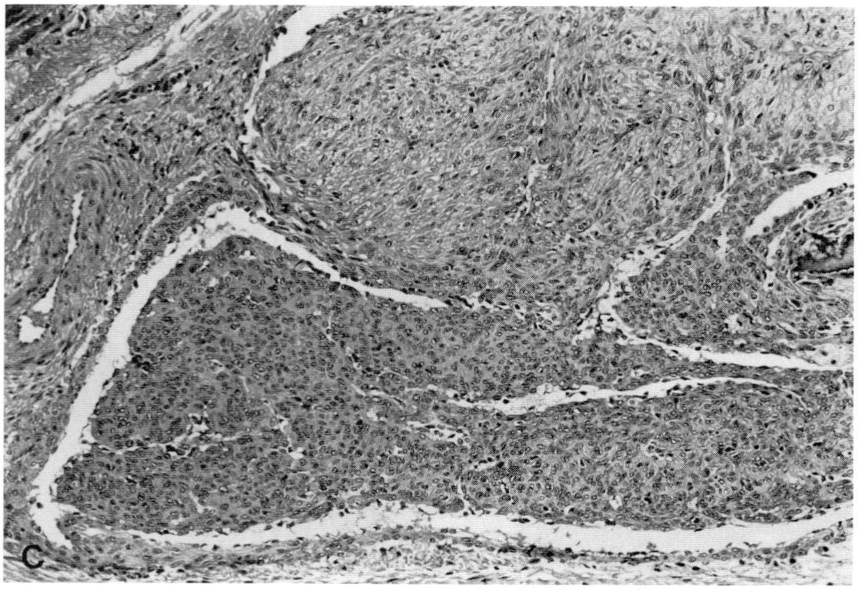

FIGURE 3.12. Infantile hemangiopericytoma. (A) A 1-month-old female with a left thigh mass present at birth. (B) The tumor weighed 43.6 g and measured 5.5×4.5 cm^2. The cut surface is lobular, bulging and tan-gray to yellow-brown with extensive necrosis. (C) Low power view ($\times$ 120) displaying the lobular architecture of the tumor. The lobules are rimmed peripherally by slitlike vascular spaces, which is a helpful diagnostic feature. H&E. (D) The tumor consists of cells with regular round to oval vesicular nuclei and inconspicuous nucleoli. A focus of calcification is present at the upper right corner. H&E; $\times$ 200. (E) Intravascular tumor thrombus within skeletal muscle. H&E; $\times$ 150. (F) Reticulin fibers surrounding nests of cells (hemangioma pattern) and individual cells (hemangiopericytoma pattern) with transition zones between the two. $\times$ 300. Verhoeff's elastic stain; $\times$ 300. (From Isaacs H Jr [10], with permission.)

FIGURE 3.12. (*Continued*)

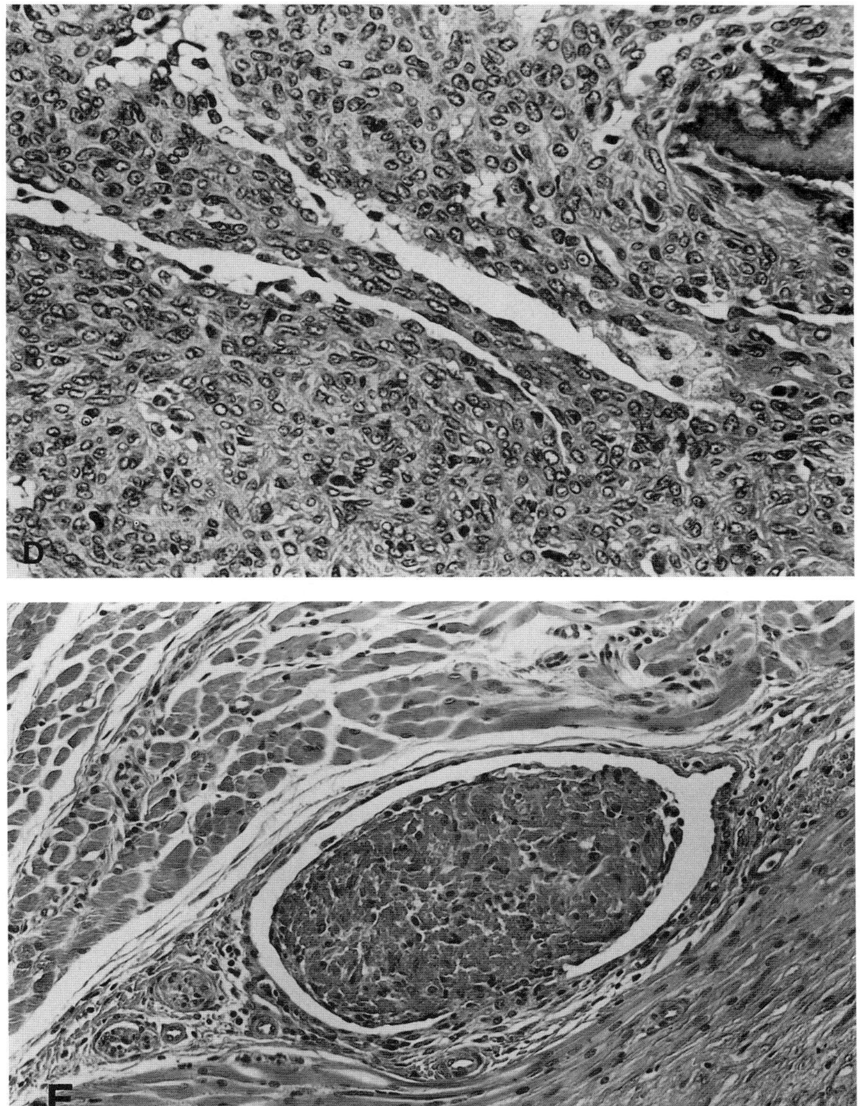

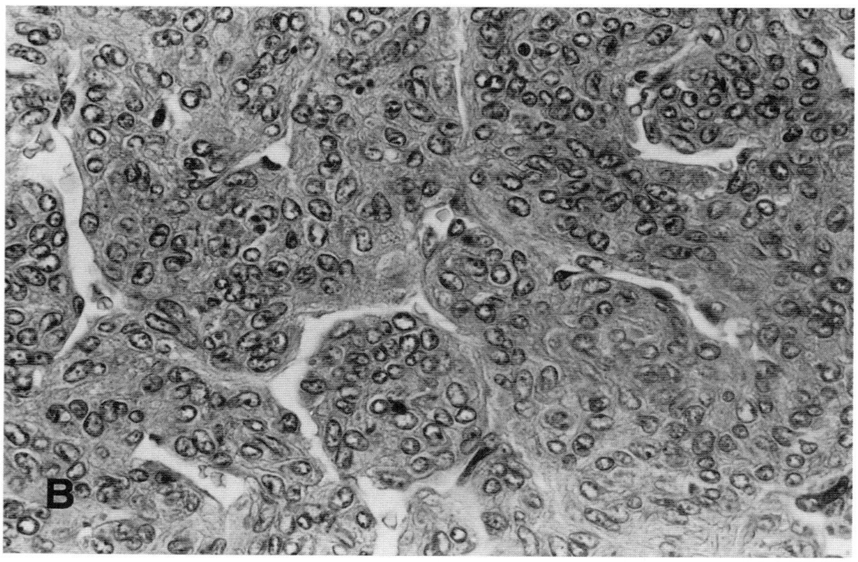

FIGURE 3.13. Infantile hemangiopericytoma. (A) External and cut surfaces of a 2 × 1.5 cm^2, hemorrhagic nodule removed from the posterior thigh of a 5-day-old male. (B) The tumor displays a lobular growth pattern and consists of cells with round to oval "glassy" vesicular nuclei. This tumor has some features more typical of the hemangiopericytomas found in older children and adults. H&E; × 400.

Fibrous Connective Tissue (Myofibroblastic) Tumors

The so-called fibrous connective tissue tumors are the largest category of soft tissue neoplasms found in the fetus and infant. They include the fibromatoses (desmoid type), the most common entity, myofibromatosis, infantile fibrosarcoma, digital fibromatosis, and fibrous hamartoma [9–11,14,42,43] (Figures 3.14–3.28). More than half involve the extremities, and approximately one third are malignant [3,8,17,44,45] (Table 3.1). Fibrosarcoma is the leading malignant soft tissue tumor in most fetal and infant series [3,8,10,11,14].

Electron microscopic and immunohistochemical studies show that the "fibrous connective tissue tumors," namely, fibromatosis, myofibromatosis, fibrous hamartoma, and infantile fibrosarcoma, are composed primarily of myofibroblasts and fibroblasts [13,15,46]. Although the fibrous connective tissue tumors share the same clinical presentation, namely, a palpable mass, and possibly a similar histogenesis, they show a variety of distinct gross and microscopic appearances. Generally members of this group are associated with a favorable outcome and deserve a conservative treatment approach.

Fibromatosis

Fibromatosis (juvenile or infantile desmoid fibromatosis) is a benign tumor presenting as a palpable mass in the fascia, skeletal muscle, or periosteum [9,13,15,42,46]. In the newborn and infant the most frequent sites are the extremities and the head and neck region, and it occurs more often in males than in females [10,11,15,43] (Figure 3.14). Sometimes it is difficult to distinguish histologically between a cellular fibromatosis and a low grade fibrosarcoma [10,11,15]. Although the two tumors are aggressive locally and often recur, generally they do not metastasize [10,11,15,46–48].

The appearance of fibromatosis is that of a firm light gray or white mass having a rubbery whorled cut surface (Figures 3.15 and 3.16). Often it appears encapsulated, but the tumor

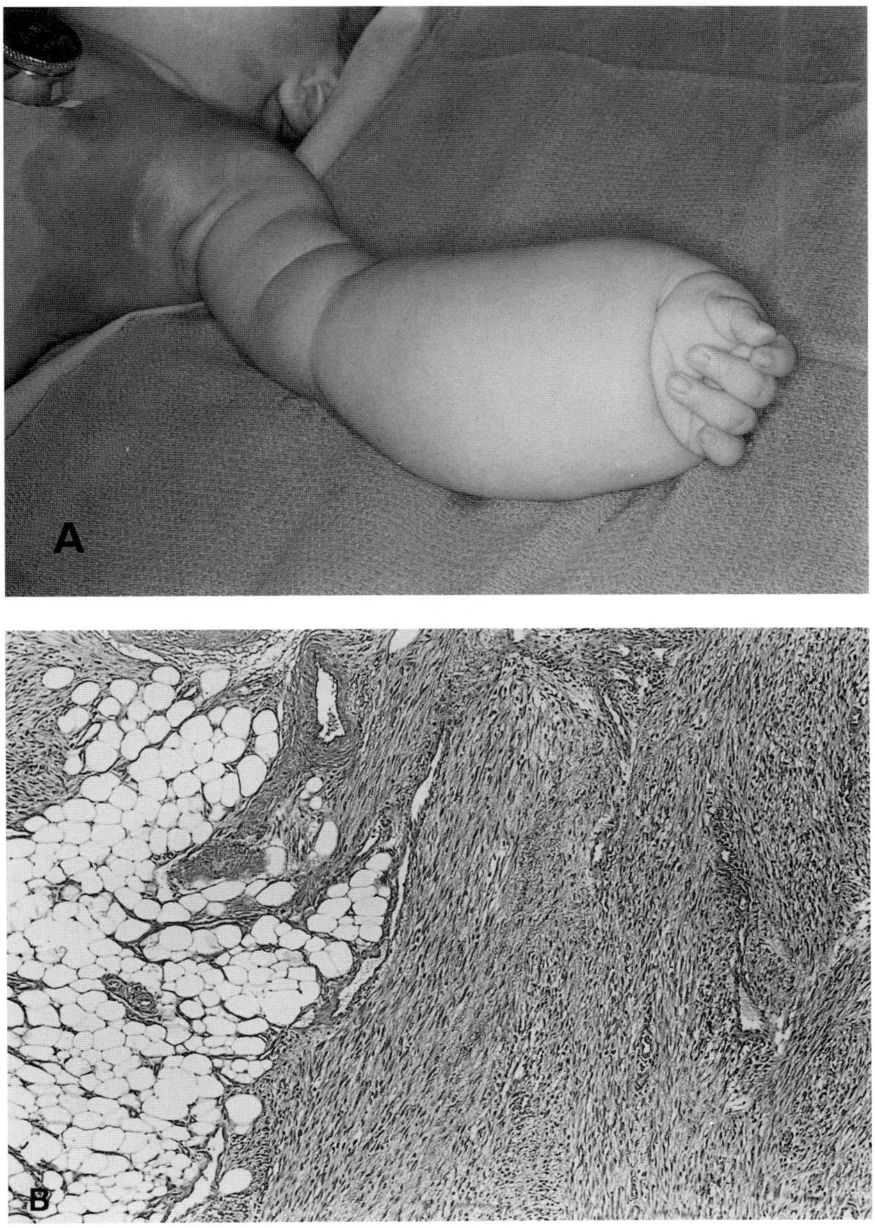

FIGURE 3.14. Congenital fibromatosis involving the forearm of a 3-month-old male. (A) Left forearm is diffusely enlarged by a firm mass. Note café-au-lait spots on the patiemt's chest and axilla. (B) Uniform, spindle-shaped cells invading fat and surrounding blood vessels. H&E; × 120.

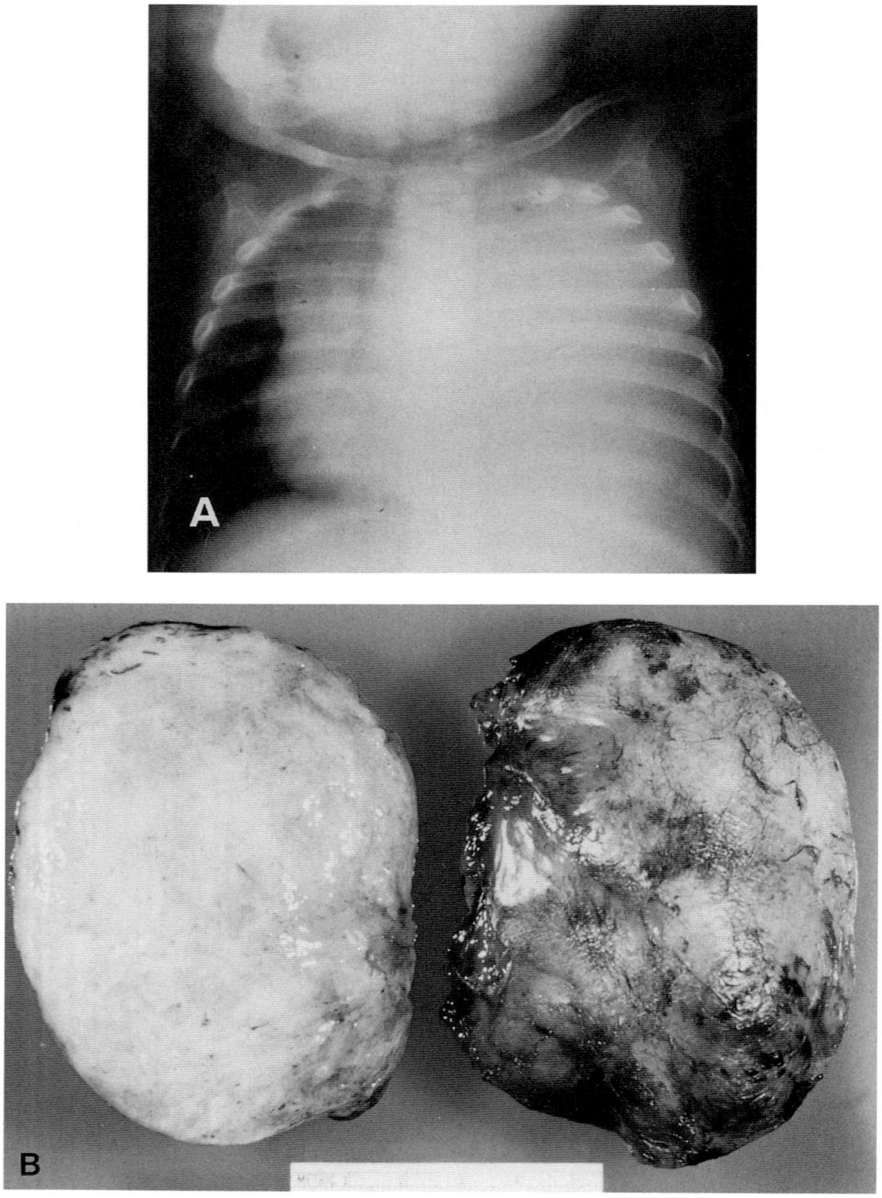

FIGURE 3.15. Fibromatosis of the mediastinum and supraclavicular area in a 7-month-old female who presented with a mass in the shoulder and chest. (A) Chest radiograph shows a large left mediastinal mass occupying much of the left thorax and displacing the mediastinal structures to the right. (B) Excised gross specimen, 10.5 × 8 cm^2, divided in half with the cut surface on the left and the external surface on the right. The cut surface has a pale gray appearance. (C) Interdigitating ("herringbone") and fanlike storiform (middle of field) growth patterns are typical for this group of tumors. H&E; × 300. (D) Higher power (× 480) view depicting the tumor cells in more detail. The cells resemble tissue culture–like fibroblasts. H&E. (From Isaacs H Jr [10], with permission.)

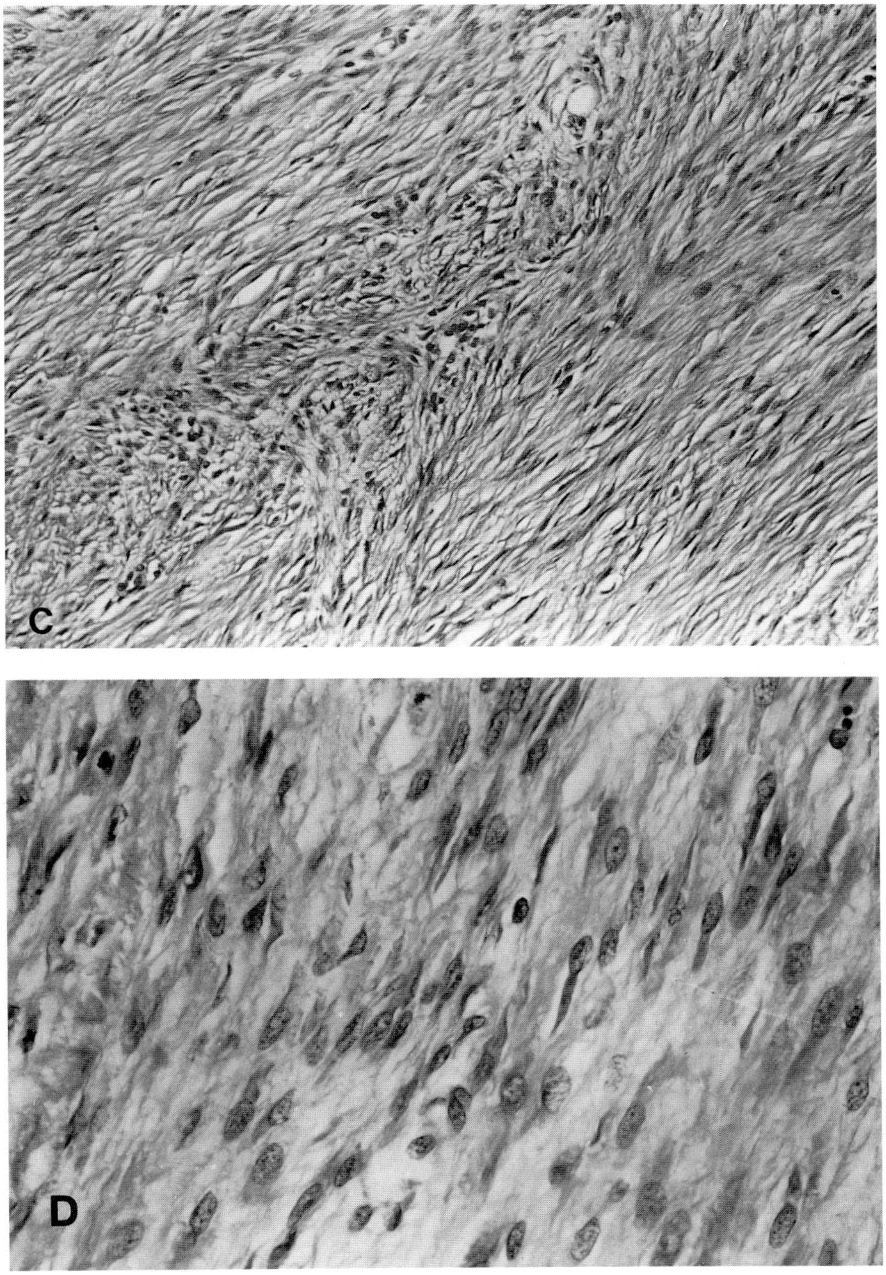

FIGURE 3.15. (*Continued*)

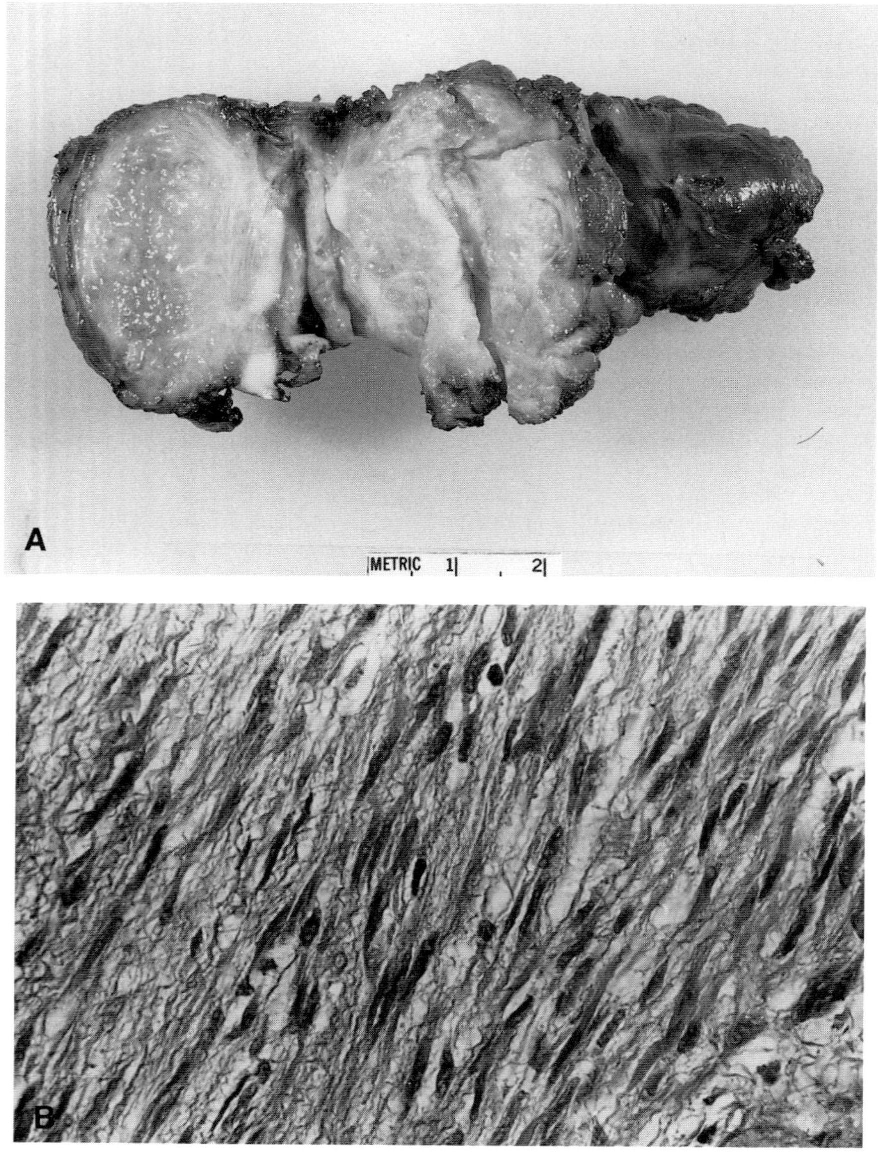

FIGURE 3.16. Congenital fibromatosis of the sternum in a 6-month-old male who had repeated resections and recurrences of his tumor. (A) The sternum is replaced by a firm, whitish-gray mass that extends into adjacent intercostal muscles. (B) The tumor consists of regular, small spindle-shaped cells separated by loose fibrous connective tissue. H&E; × 380.

invades imperceptibly into the adjacent soft tissues (Figure 3.16). Fibrosarcoma shows similar gross features but tends to be considerably softer, friable, and more myxoid or gelatinous appearing than a fibromatosis [10,11] (see later: Figures 3.25 and 3.26). Usually but not always, the fibromatoses are smaller than the fibrosarcomas, the former averaging 2 cm in greatest dimension and the latter 15 cm or more in diameter or more. Size is not always a reliable criterion [11,14].

The fibromatoses are basically spindle cell neoplasms with moderate variation in the cellularity and the amount of intercellular collagen (Figures 3.14–3.17). Some are hypocellular, with abundant collagen resembling scar tissue, while others are more cellular, with an occasional mitosis and a prominent pale-staining myxoid background [11,13,42,46,48]. Occasionally these differences in histological appearance can be appreciated in the same tumor, especially if it is large. Sampling error (e.g., resulting from tiny, barely visible, needle biopsies) can play a critical role in the diagnosis. The tumors stain positively with vimentin immunoperoxidase and variably with actin and desmin. Ultrastructural studies show that the tumor cells are fibroblasts and myofibroblasts.

The clinical features associated with recurrence are presentation at later than 5 years of age, extremity location of the tumor, and incomplete surgical removal. The histopathological

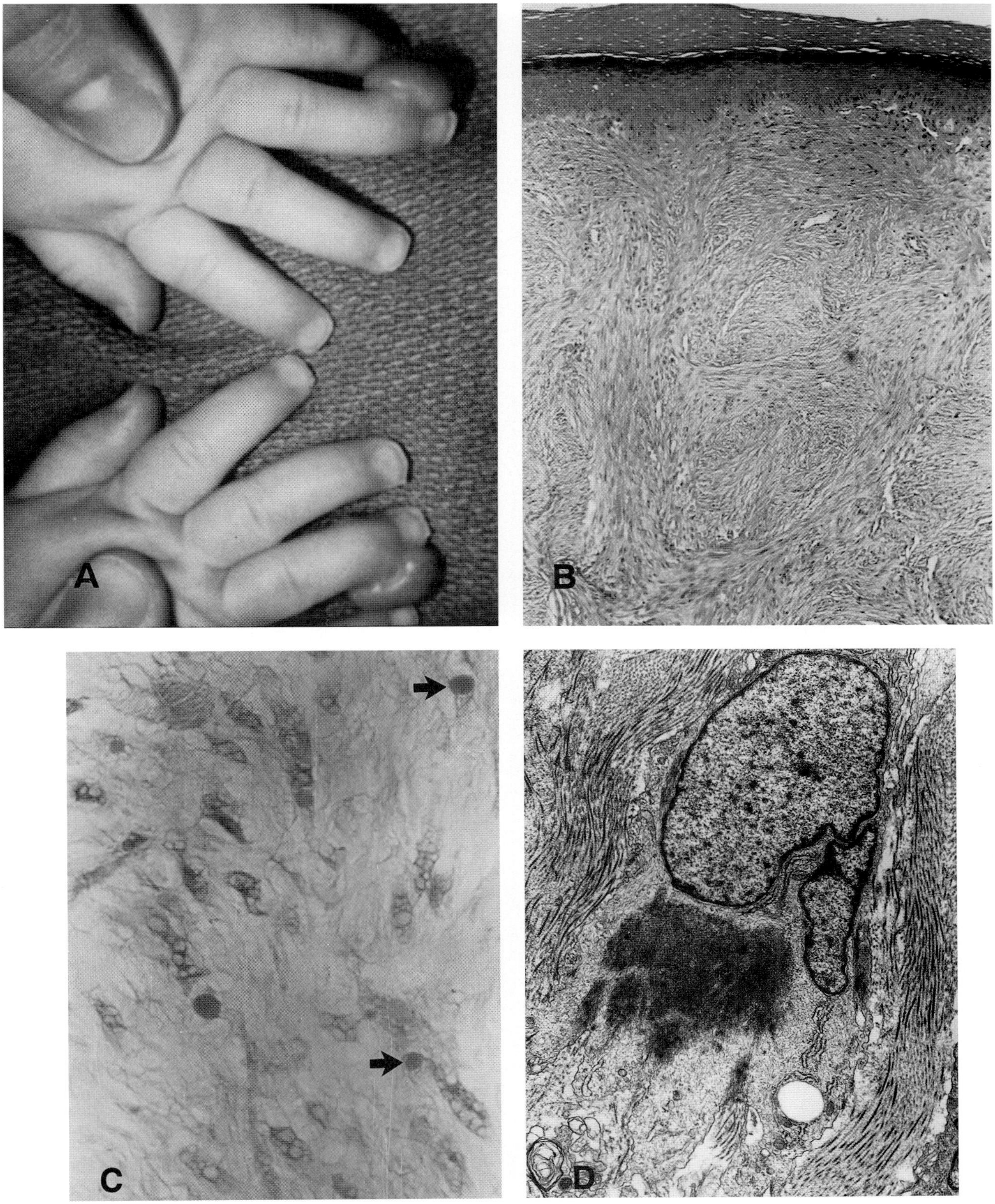

FIGURE 3.17. Digital fibromatosis in a 1-year-old male; nodules were present on the lateral aspects of the distal phalanges of the fourth fingers since birth. The nodules were excised but each time recurred; they eventually disappeared during adolescence. (A) Mirror image distribution of the nodular lesions. (B) The tumor is moderately cellular, consisting of small, regular spindle-shaped cells involving the dermis and subcutaneous tissue. The tumor displays an interdigiting or "herringbone" growth pattern. The epidermis is not involved. Trichrome; × 120. (C) Intracytoplasmic hyaline inclusions (arrows). H&E; × 615. (D) Electron micrograph (× 13,200) shows intracytoplasmic inclusions shows composed of actinlike filaments, which are situated beneath the nucleus. (A and B, from Isaacs H Jr [10], with permission. D, courtesy of Ann Peters, Department of Pathology, Children's Hospital, San Diego.)

findings seen with recurrence are microscopic evidence of tumor at resection margins, more than 5 or more mitoses per 10 high-power fields, and the presence within the tumor of necrosis and inflammation [43].

Surgical excision is the treatment of choice. The lesion recurs if not completely excised and will cause fatal complications if a vital structure is involved (Figure 3.16) [11].

Digital Fibromatosis (Recurring Digital Fibroma)

Infantile digital fibromatosis is a form of fibromatosis that characteristically involves the lateral aspects of the distal phalanges of the fingers and toes. Almost half are discovered within the neonatal period and the remainder by 1 year of life; most recur after simple excision [15,48]. The tumors may be single or multiple and are composed of white firm nodules with a broad base and shiny surfaces [14,15,46,48] (Figure 3.17).

Histological examination reveals the typical picture of a fibromatosis in this age group. The tumor consists of small, regular, spindle-shaped cells, showing an interlacing bundle or "herringbone" growth pattern with little or no mitotic activity, embedded in a collagenous matrix; it invades primarily the dermis and subcutaneous tissue [14,15,46,48] (Figure 3.17). Variably the trichrome stain displays red paranuclear intracytoplasmic inclusions, which by electron microscopy are seen to consist of packets of actin filaments. This finding is confirmed by immunohistochemistry [49]. Not unlike the other fibromatoses of infancy, ultrastructurally the cell of origin is considered to be the myofibroblast and fibroblast [42,49]. Digital fibromatoses invade locally; recurrence is frequent, and many eventually spontaneously regress and disappear [14,15,46,48]. Occasionally adjacent bone (phalanx) is involved. Conservative surgical excision is the recommended mode of therapy.

Torticollis

Torticollis is a tumorlike condition that produces a neck mass in an infant, presenting as a firm, palpable nodule in the sternocleidomastoid muscle. The lesion usually appears within the first two weeks of life and may be associated with a deformity called "wry neck." Torticollis is observed in less than 1% of newborns [15]. In about half the patients there is a history of complicated delivery (e.g., a breech birth or forceps extraction).

Although the other names used for this condition are fibromatosis colli and sternocleidomastoid tumor, recent studies indicate that the lesion probably represents a reactive, reparative process rather than a neoplasm [14,15,50].

The histological features of torticollis are time dependent. During the first few months of life the histological picture is similar to that of a fibromatosis: that is, the presence of cellular small, regular, spindle-shaped fibroblastic cells embedded in a collagenous matrix. However a few years later biopsy typically shows only bland, acellular fibrosis with skeletal muscle fiber loss and atrophy (Figure 3.18).

The recommended form of treatment in most the patients is physical therapy (i.e., early and prolonged stretching); surgery is reserved for patients with severe rotation limitation deformities, which are present in less than 10% of children with the disease [50,51].

Myofibromatosis

Myofibromatosis (congenital multiple fibromatosis, multiple vascular leiomyomas, mesenchymal hamartoma) is an unusual soft tissue neoplasm found primarily in the first year of life and presenting as one or more nodular lesions involving the soft tissues and occasionally the bones and the viscera [10,48,52–54]. Almost two thirds of the cases are discovered at birth.

Solitary myofibromatoses are more common than multiple ones [14,15,48,52] (Figures 3.19 and 3.20). The solitary myofibromatosis involves primarily the soft tissues of the head, neck, and trunk, whereas the multicentric tumors not only arise from the soft tissues situated over various parts of the body but also from the muscles, bones, and viscera [15,46, 48,521]. Congenital generalized fibromatosis is classified into two types: type I involves the skin, subcutaneous tissue, muscle and bone; type II involves primarily the viscera [48]. Tumors occur in the brain and spinal cord of patients with congenital visceral fibromatosis [15,48,52]. Moreover solitary congenital myofibromatosis of the small and large intestine has caused neonatal bowel obstruction and an abdominal mass [55]. Some of these lesions have been called sarcomas.

Although fibromatosis and myofibromatosis share several features in common such as the distribution of solitary lesions and the presence of fibroblasts and myofibroblasts as seen by light and electron microscopy, the findings of central necrosis and a prominent vascular pattern in the myofibromatosis are helpful diagnostic criteria used to distinguish the two [15,46,52]. Moreover, the tumor cells resemble smooth muscle cells (Figures 3.19 and 3.20). They are positively stained with phosphotungstic acid–hematoxylin (PTAH) and are immunoreactive with smooth muscle actin and vimentin but variably reactive with desmin [46,56].

Infantile hemangiopericytoma and and myofibromatosis may be closely related. In fact, some feel that they are essentially the same lesion [46,57]. The microscopic distinction between some fibromatoses and myofibromatoses is not always clear, particularly in tiny biopsy samples, for both lesions are composed of spindle cells with interdigitating growth patterns [10]. This is an important distinction because myofibromatosis has a greater tendency to spontaneously regress and resolve than the fibromatoses, which are more locally aggressive and have a greater recurrence rate [52]. Spontaneous regression of the former has been documented on several occasions [54,56,58]. However, when there is extensive involvement of multiple viscera, and especially the lungs, the

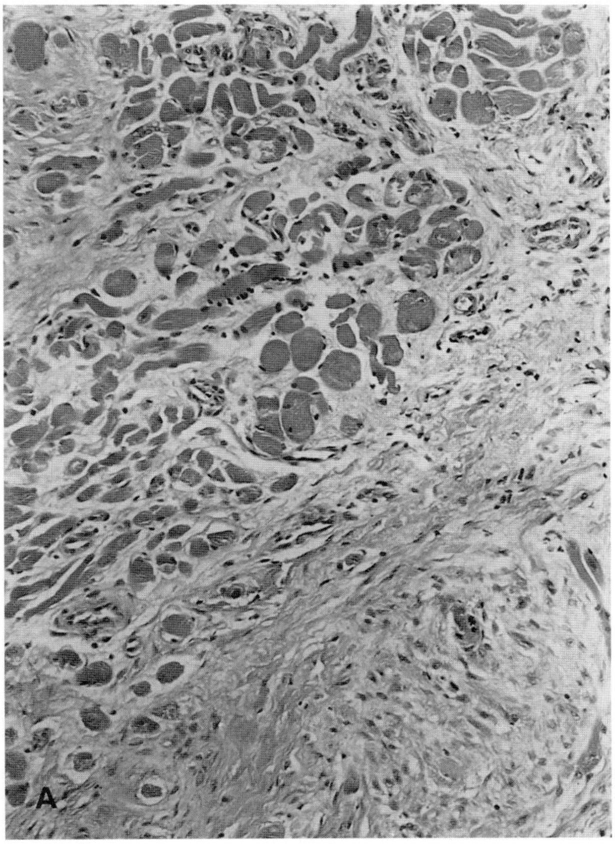

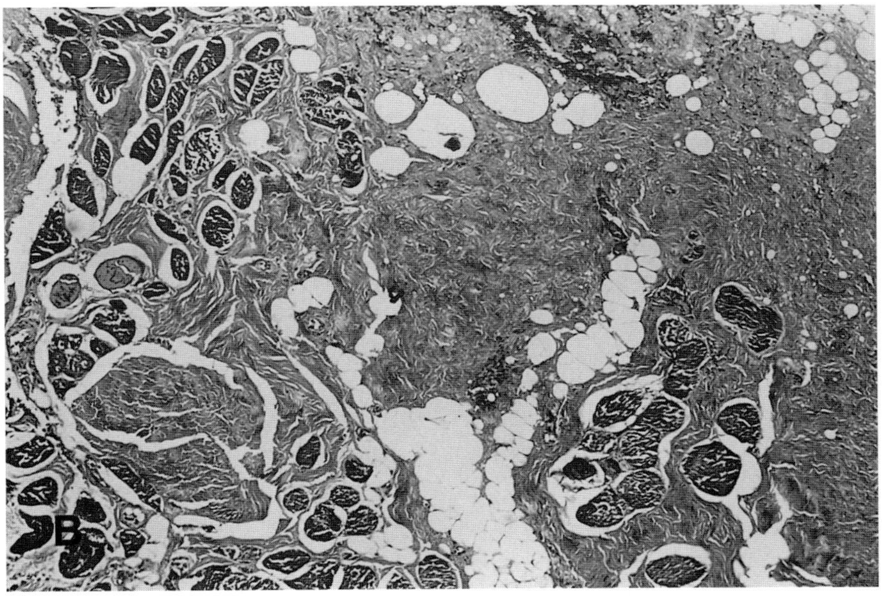

FIGURE 3.18. Torticollis (fibromatosis colli). The histological findings of this lesion are age dependent. (A) Photomicrograph of a portion of the sternocleidomastoid muscle removed from a 6-week-old male with torticollis. There is extensive skeletal muscle fibrosis and atrophy. At this early stage the appearance of interstitial fibrosis resembles fibromatosis. H&E; × 240. (B) Section taken from the sternocleidomastoid muscle of a 7-year-old boy with torticollis. In contrast to micrograph A, this one shows much more skeletal muscle fiber atrophy and loss; the interstital fibrosis appears bland and hypocellular resembling a scar. Trichrome; × 240.

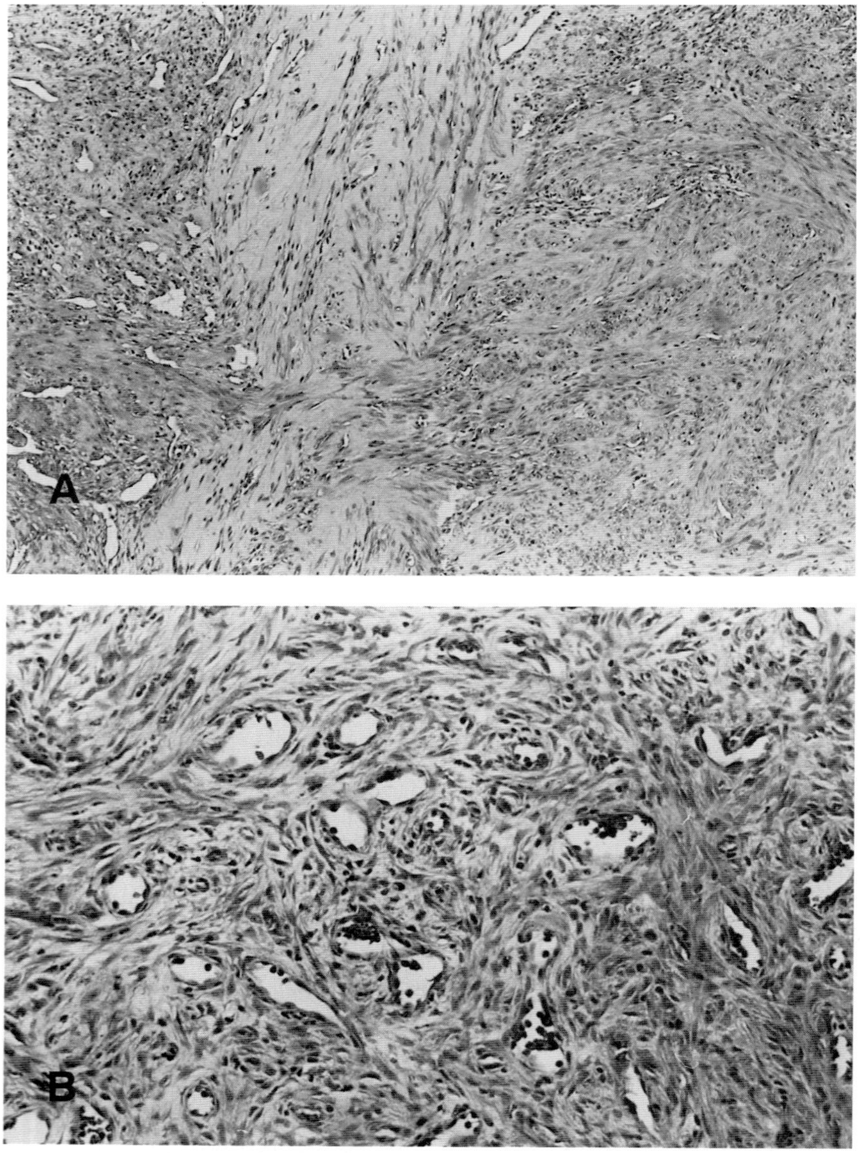

FIGURE 3.19. Infantile myofibromatosis. The lesions may be single or multiple, involving the subcutaneous tissue, muscle, viscera, or bones. (A) Photomicrograph of a solitary tannish-pink, 2×1.8 cm^2, subcutaneous nodule removed from a 2-month-old male.The tumor shows regular spindle-shaped myofibroblastic cells with a fibrous background and a focal vascular pattern. The cells appear more like smooth muscle cells than fibroblasts. Trichromes × 120. (B) The vascular pattern resembles a hemangioma or hemangiopericytoma. Trichromes × 120. (C) An area of necrosis. H&E; × 150. (D) Foci of calcification. H&E; × 150.

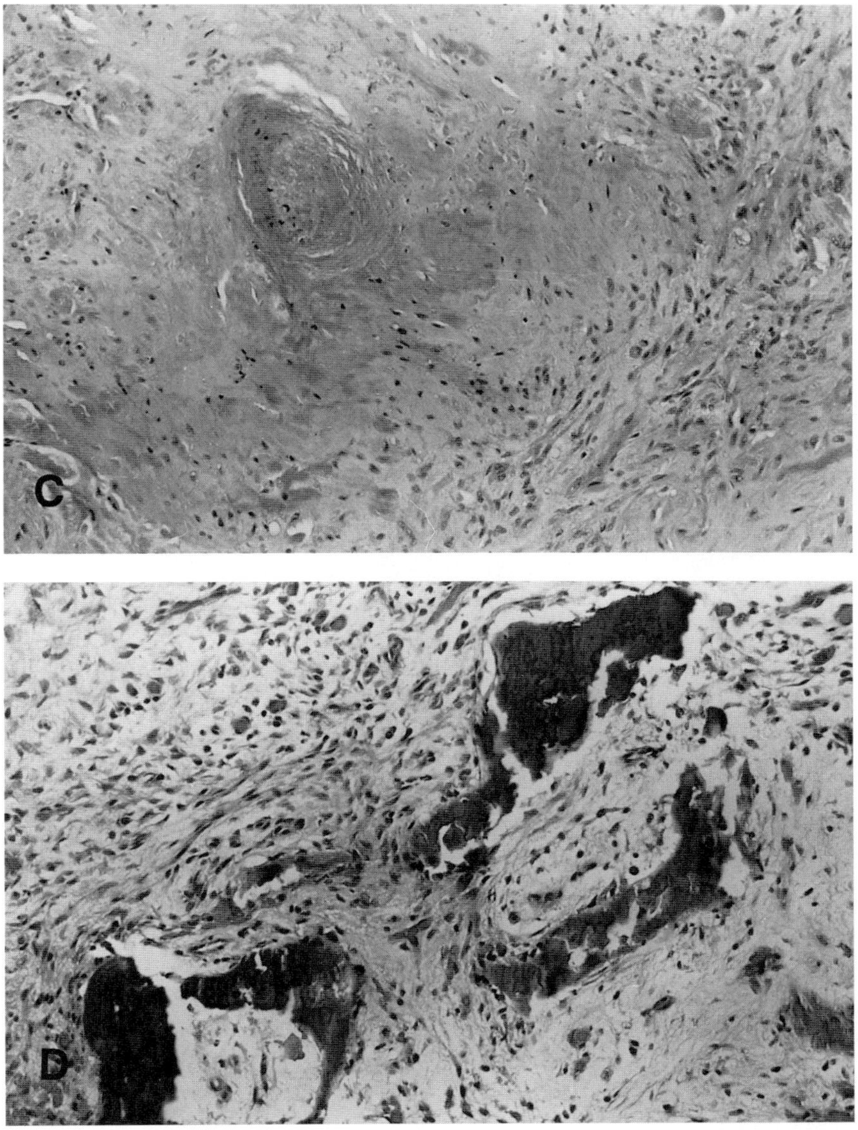

FIGURE 3.19. (*Continued*)

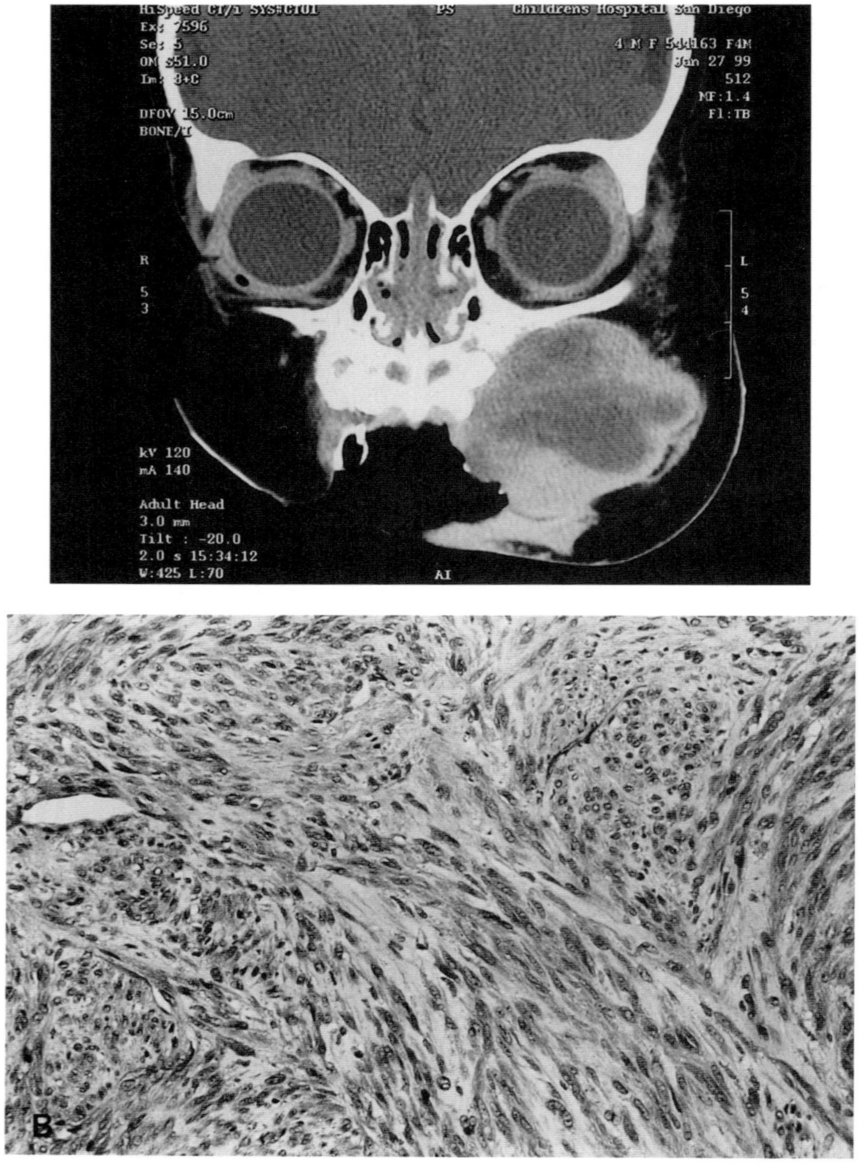

FIGURE 3.20. Infantile myofibromatosis. Soft tissue facial mass in a 4-month-old female. (A) Computerized tomography scan reveals a fairly well-circumscribed, nonhomogeneous, enhancing solid and cystic mass in the soft tissues of the left side of the face. The mass is adjacent to the lett maxilla and orbit, which show slight erosive bony changes. (B) Biopsy reveals bundles of regular spindle-shaped cells resembling smooth muscle cells situated in a pale myxoid background. Immunohistochemistry results were positive for vimentin and for actin and desmin focally. H&E; × 120.

patients with multicentric myofibromatosis have a poor prognosis [48,52,53]. The recurrence rate of single lesions is 7% after primary excision but cure is usually achieved after reexcision [52].

Fibrous Hamartoma of Infancy

Fibrous hamartoma is a relatively uncommon soft tissue neoplasm observed primarily within the first year of life; about 20% of cases are congenital [15,46,48,59,60]. The tumor presents as a palpable mass most often in the axilla, arm, thigh, shoulder, or back arising from the lower dermis or subcutaneous tissue [46,48,59,60]. Unusual primary sites reported in infants are the hand and genital areas [61]. The underlying skeletal muscle and nerves may be involved. Rarely, more than one lesion is found in the same patient. Generally the tumors do not recur after complete excision [15,62]. The lesion does not regress [63].

The fibrous hamartoma consists of a round, bulging mass with firm glistening, gray-white tissue and yellow nodules of fat (Figures 3.21 and 3.22). It rarely exceeds 8 cm in greatest diameter. Some appear well circumscribed while others blend imperceptively with the surrounding subcutaneous fat [11,46,48,59] (Figure 3.21).

Microscopic examination of the tumor shows three main components: nests of immature-looking, spindle-shaped cells embedded in a myxoid background; interlacing dense fibrous trabeculae resembling tendon; and lobules of mature adipose tissue situated between the other two components [14,46,48, 59–61] (Figure 3.23). Blood vessels are prominent in the myxoid areas. Tumor cells in the myxoid and fascicular fibroblastic areas react with vimentin and desmin [60]. The adipose tissue component is S-100 positive. By electron microscopy the spindle cells are found to consist of both fibroblasts and myofibroblasts, the latter having intracytoplasmic myofibrils with dense bodies, pinocytotic vesicles, and basement membrane [48,60].

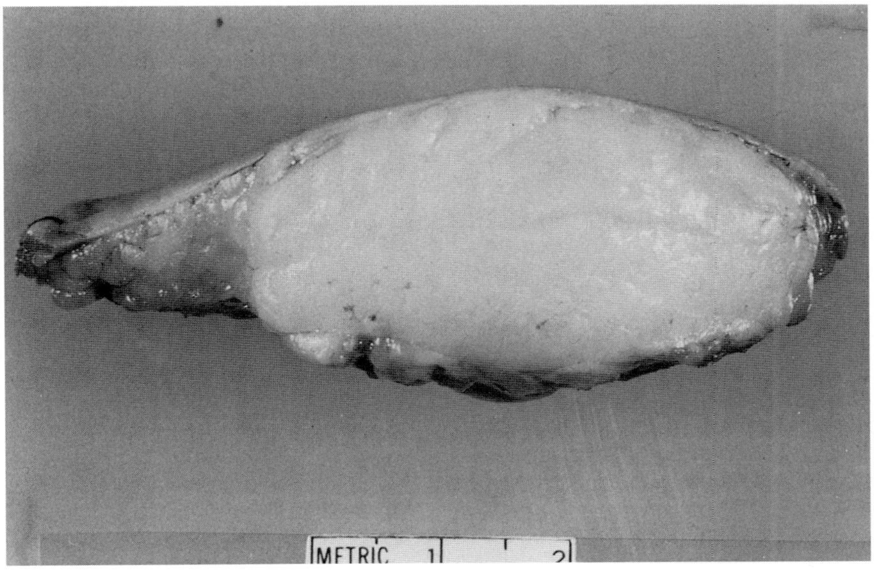

FIGURE 3.21. Fibrous hamartoma of infancy. Gross appearance of a 30 g, 7.4 × 3.6 cm^2 subcutaneous mass removed from the lateral aspect of the upper right arm of a 4-month-old female. The tumor was present since birth. It consists mostly of adipose tissue that blends imperceptibly with adjacent subcutaneous fat. The overlying skin is essentially normal.

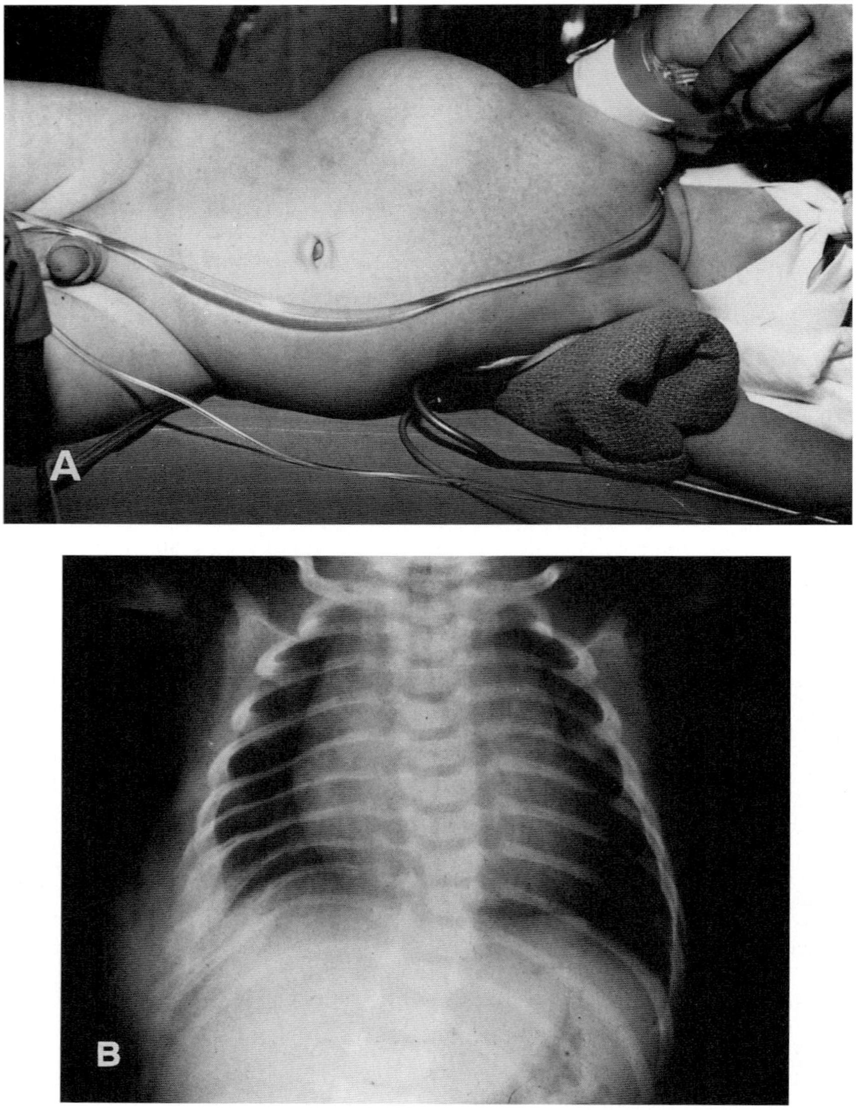

FIGURE 3.22. Fibrous hamartoma of infancy. (A) A 2-month old male with a right chest wall mass. (B) Chest radiograph reveals a soft tissue mass surrounding the ribs but without bone involvement. Scoliosis is present. (C) Cut surface shows a pale tan-gray infiltrative process within the soft tissues. (D) The tumor surrounds the ribs rather than arising from them, in comparison to, for example, a chest wall hamartoma of osseous origin (see Figures 17.2 and 17.3).

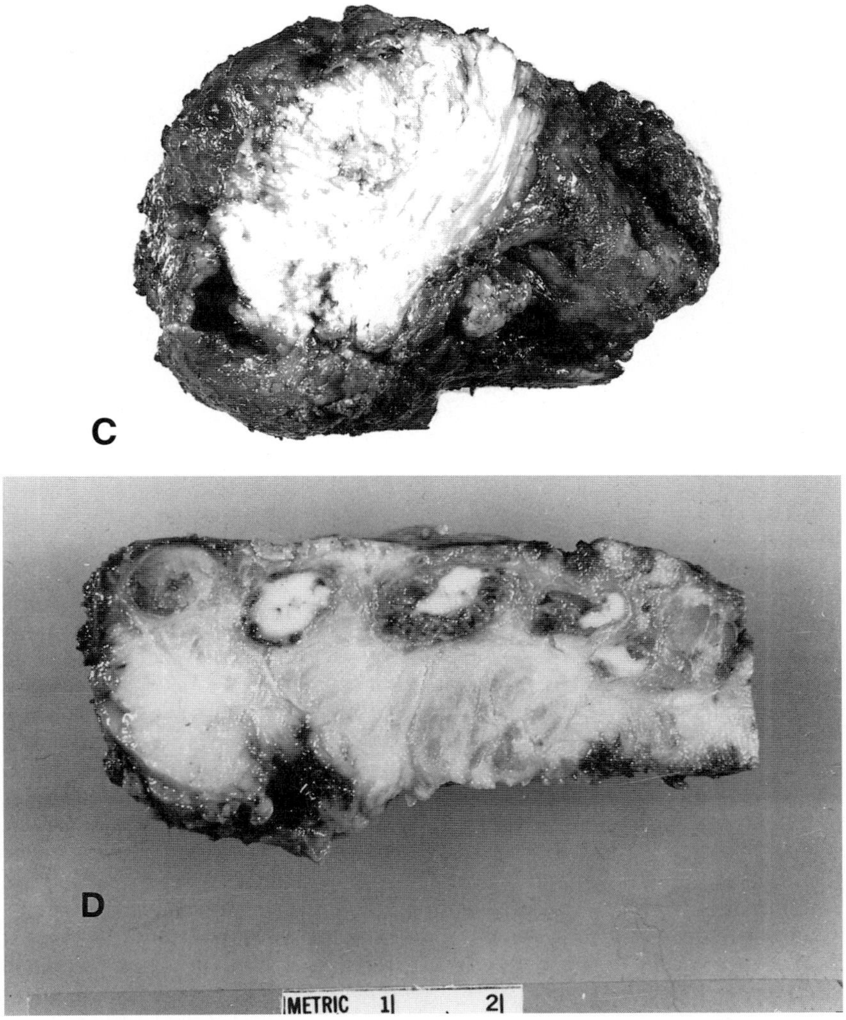

FIGURE 3.22. (*Continued*)

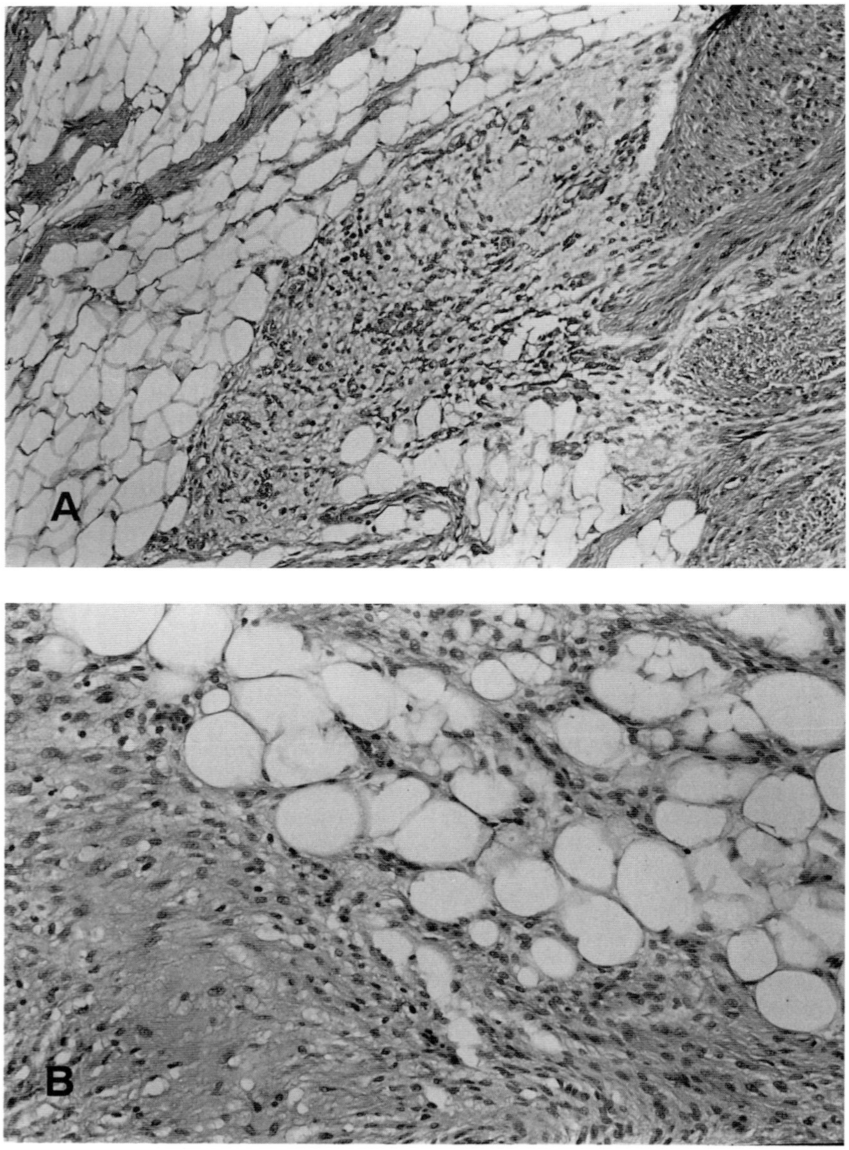

FIGURE 3.23. Fibrous hamartoma of infancy, histological findings. (A) The tumor consists of three components other than fat required to establish the diagnosis: fibroblastic areas on the right of the photograph, myxoid areas with small spindle-shaped cells in the center, and cordlike fibrous bands traversing the upper left hand corner. H&E; × 80. (B) Higher power view (× 150) of the immature spindle cells. H&E. (C) Tumor invading skeletal muscle. H&E; × 120. (D) Lymph node surrounded by tumor. H&E; × 120.

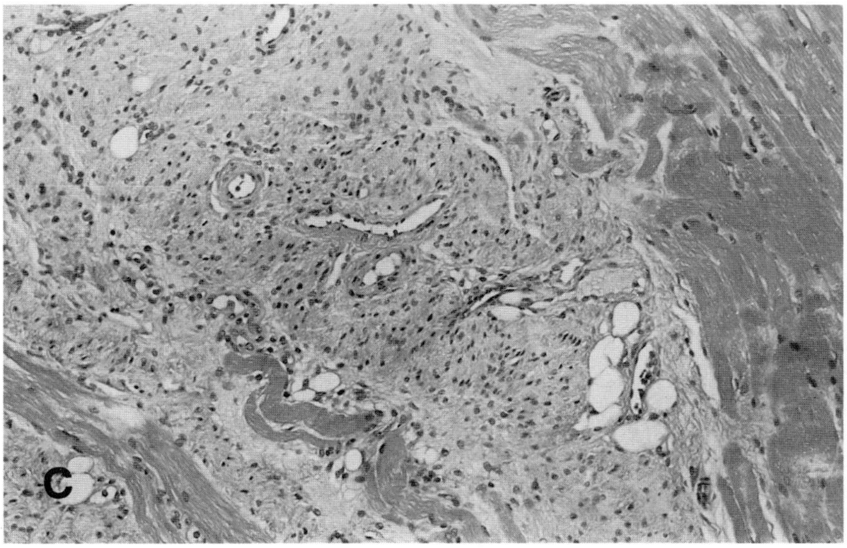

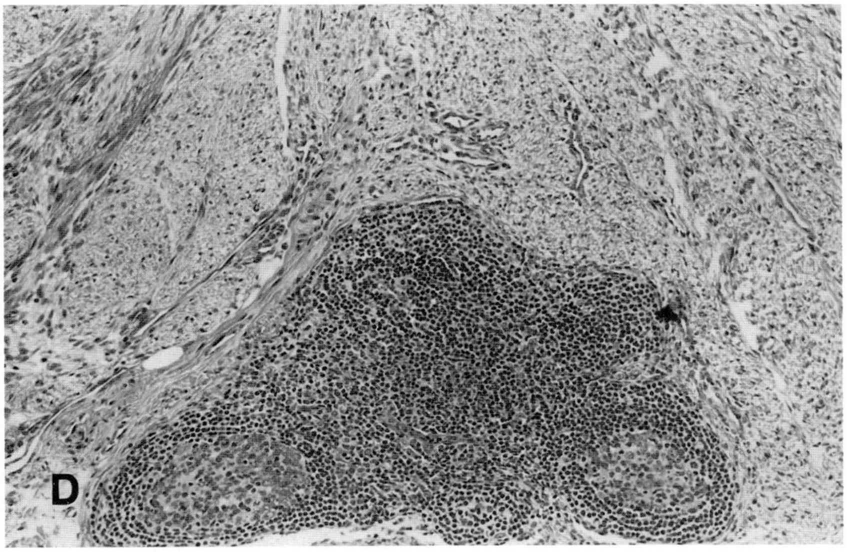

FIGURE 3.23. (*Continued*)

Calcifying Aponeurotic Fibroma

Calcifying aponeurotic fibroma is a fibrous connective tissue tumor characterized by the presence of focal calcifications [15,46]. Although there is a wide age range, It occurs primarily in the first two decades of life. The lesion is described rarely in the newborn or infant [16]. Typically it is found in the palms, soles, and digits of the hands and feet.

Grossly the tumor consists of a small poorly circumscribed, rubbery, gray-white nodule with or without gritty areas of calcification on section. The lesion may be attached to tendons, aponeuroses, or fascia in addition to fat or other soft tissues [15,46]. Histologically the calcifying aponeurotic fibroma consists of bands of fibrous tissue composed of fibroblasts with round or oval nuclei and ill-defined pink cytoplasms, areas of dense hyalinized collagen, and nodules of chondroid tissue. Foci of calcification are associated with vacuolated chondrocytes and are surrounded by pallisading fibroblasts showing little or no mitotic activity [15,46]. Clusters of multinucleated osteoclastic giant cell are present. Calcification may not be observed in tumors removed from young infants. Only fibrous areas with vacuolated chondrocytes are present.

Calcifying aponeurotic fibroma, particularly early examples without calcification, may be mistaken for fibrous hamartoma of infancy and other fibromatoses. However the presence of chondroid areas with vacuolated chondrocytes helps to establish the diagnosis of the former [46]. Conservative surgical excision is the recommended treatment; the lesion will recur if not completely excised and multiple recurrences are not unusual [15,46].

Granular Cell Tumor

Granular cell tumor (congenital epulis) is a neoplasm of controversial histogenesis occurring primarily in the gingiva of newborns and infants [15,64]. Two forms of granular cell tumor have been defined: one is the classic congenital epulis involving the gums, and the other occurs in the skin, the mucosa of the respiratory, gastrointestinal, and genitourinary tracts, and various sites throughout the body in older children and adults [15,62]. Immunohistochemical and ultrastructural studies suggest that the myofibroblast may be the cell of origin of the congenital form, but the debate regarding the histogenesis of this neoplasm continues [65].

In the newborn and infant the granular cell tumor arises as one or more mucosal covered nodules from the gingiva of the anterior alveolar ridge of either the mandible or the maxilla and occurs predominantly in females. Occasionally the tongue is involved. The tumor may present as a huge mass arising from the gingiva of the upper jaw, causing airway obstruction and feeding problems, thus mimicking a teratoma (epignathus) [14] (Figure 3.24).

Granular cell tumors are composed of polygonal cells with pink, PAS-positive granular cytoplasms and regular, round dark basophilic-staining nuclei (Figure 3.24). Immunostaining is reactive with vimentin and unreactive with S-100, muscle, and histiocytic markers; ultrastructural features include membrane-bound, pleomorphic cytoplasmic granules, cell membranes, and intermediate filaments with fusiform electron densities consistent with myofibroblasts [46,65]. The granular cell tumor is simply excised. However spontaneous regression of the lesion has been documented on several occasions even after incomplete removal or without surgery [64].

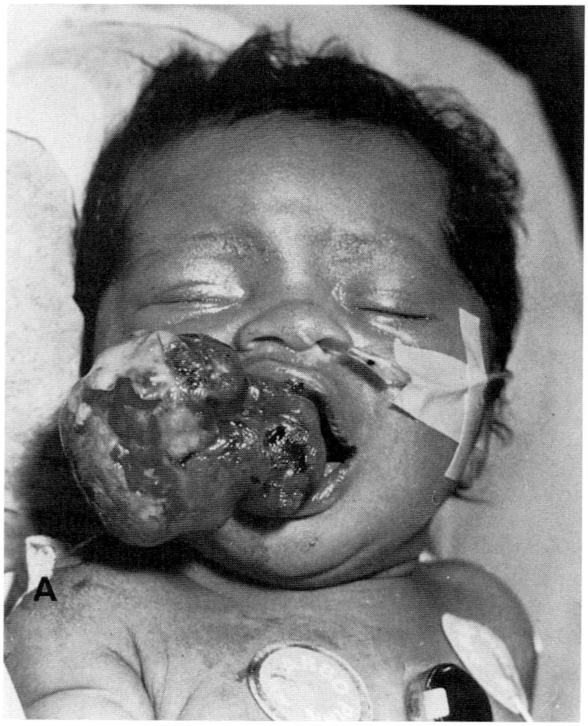

FIGURE 3.24. Congenital epulis (gingival granular cell tumor). This 5-day-old, full-term female, delivered by cesarean section, was born with a mass arising from the upper midline gingiva and alveolar ridge. The clinicians thought the tumor was a teratoma. (A) The mass originates from and obstructs the oral cavity. (B) The tumor, 6×5.5 cm^2, 39 g, has a uniform, gelatinous and gray-tan to yellow cut surface. (C) Closely packed, polygonal granular cells with prominent cell membranes are situated beneath the mucosa. H&E; $\times$ 120. (D) The tumor cells have characteristic cytoplasmic granules and regular round nuclei with inconspicuous nucleoli. The granules are positive for periodic acid–Schiff staining. Immunoperoxidase findings were not contributory. H&E; $\times$ 480.

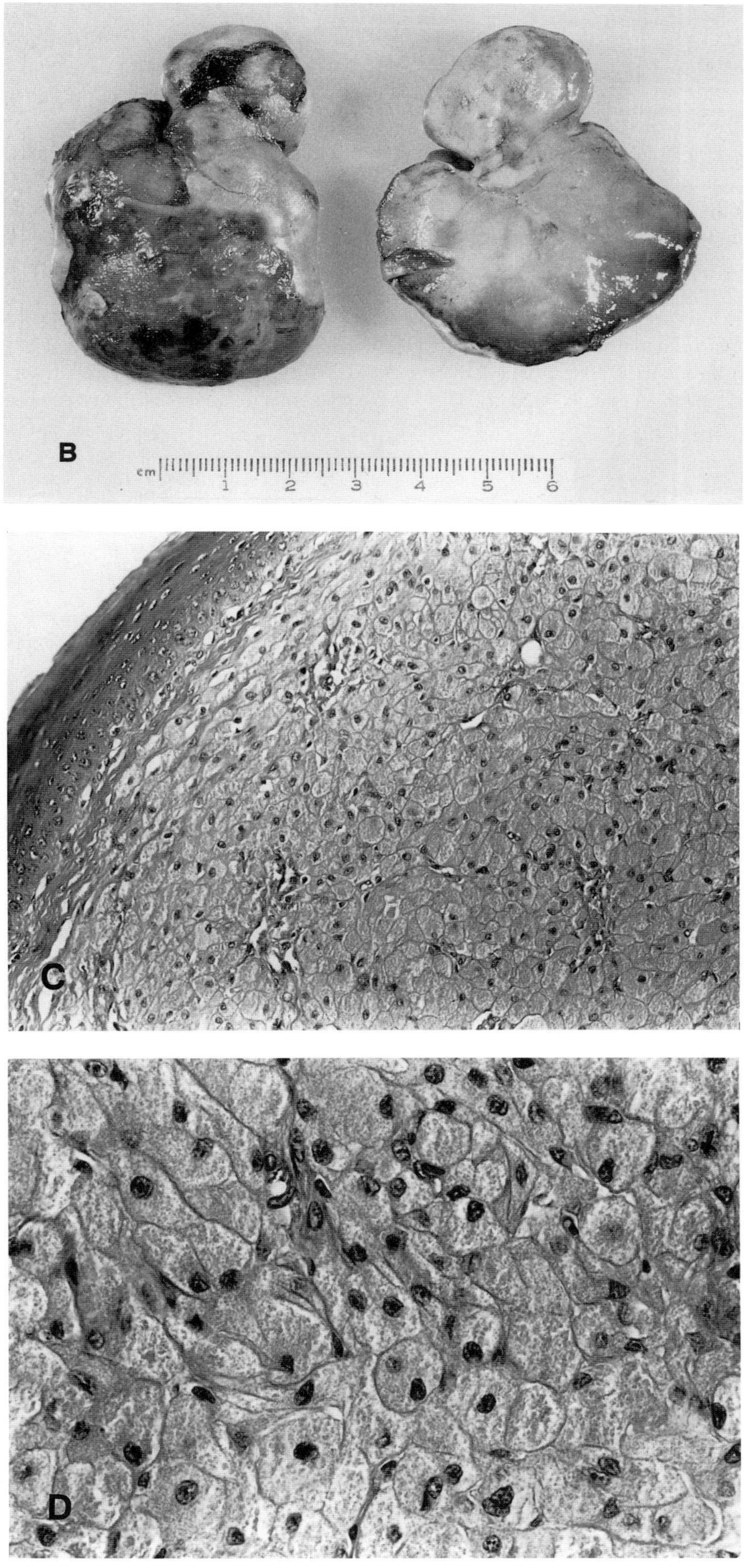

FIGURE 3.24. (*Continued*)

Infantile (Congenital) Fibrosarcoma

Congenital or infantile fibrosarcoma is relatively uncommon [9–11,44,48,66–69]. In a series of 110 cases of fibrosarcoma in infants and children reported from the Mayo Clinic, 40 (36%) were congenital [68]. Although the rate of local recurrence is significant, as high as 50% or more, metastases seldom occur [5,9–11,15,42,44,48,67,68,70]. Cytogenetic studies reveal an anomaly characterized by a nonrandom gain of chromosomes 11 and 20 [70]. Moreover, infantile fibrosarcoma and congenital cellular mesoblastic nephroma have a similar histological appearance and an identical trisomy 11 karyotype, suggesting that the two are histogenetically related [70].

Typically the newborn or infant with a fibrosarcoma presents with an impressive soft tissue mass varying in size from a few centimeters to as large as 20 cm or more often involving an extremity, the trunk, head, or neck [5,10,11,15,46, 66,67,69] (Figures 3.25, 3.26, and 3.27). The abdominal wall, retroperitoneum, inguinal area, and thoracic cavity are less common primary sites [69,71].

The tumor consists of firm, light gray, occasionally gelatinous tissue partially surrounded by a fibrous capsule. The larger tumors show areas of hemorrhage, necrosis, and small cystic formations (Figure 3.26). The gross appearance may be deceptive because extensions of tumor blend subtly with the adjacent soft tissues of the affected part so that just "shelling

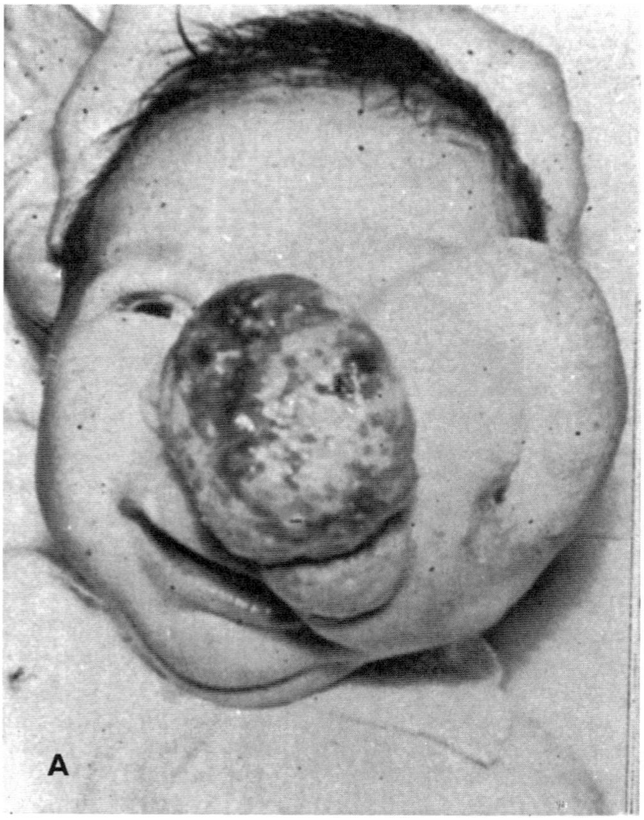

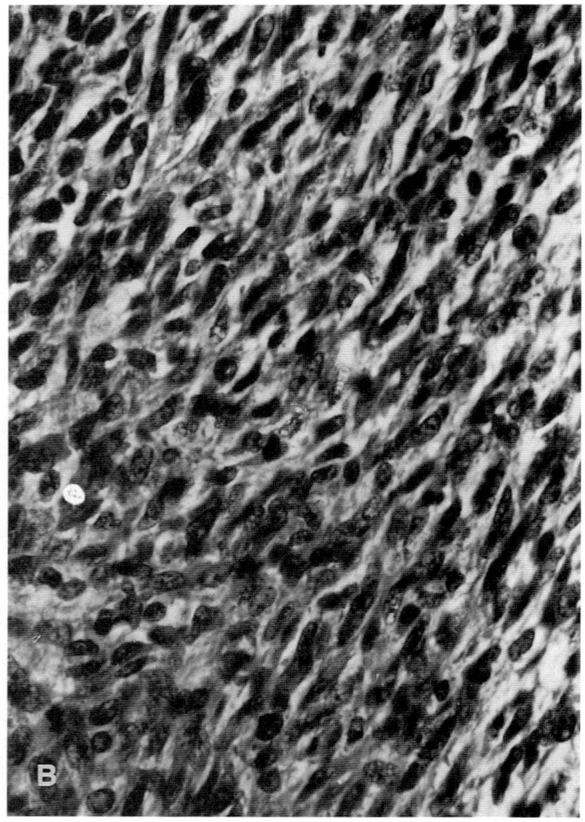

FIGURE 3.25. Congenital fibrosarcoma. (A) A 3-day-old female with a massive tumor, 16 × 16 × 10 cm^3, involving the face, nose, and maxillary sinus cured by wide local excision and grafting. (B) Hypercellular tumor composed of irregular, hyperchromatic spindle-shaped, fibroblastic cells with moderate nuclear atypia and mitotic acivity. H&E; × 380. (From Hays DM, Mirabal VQ, Karlan MS, et al. [67], with permission.)

FIGURE 3.26. Congenital fibrosarcoma. This 5-month-old male had a rapidly growing mass in the right thigh, discovered at 1 week of age. (A) Amputation specimen consists of a large soft, gelatinous tumor occupying most of the thigh. The head of the femur is visible on the medial aspect. (B) Photomicrograph showing the interdigitating, fasicular growth pattern of the tumor. H&E; × 120. (C) The cellular spindle cell neoplasm displays moderate nuclear atypia, 3 or 4 mitoses per high-power field, and the presence of scant interstitial collagen. H&E; × 600. (From Isaacs H Jr [10], with permission.)

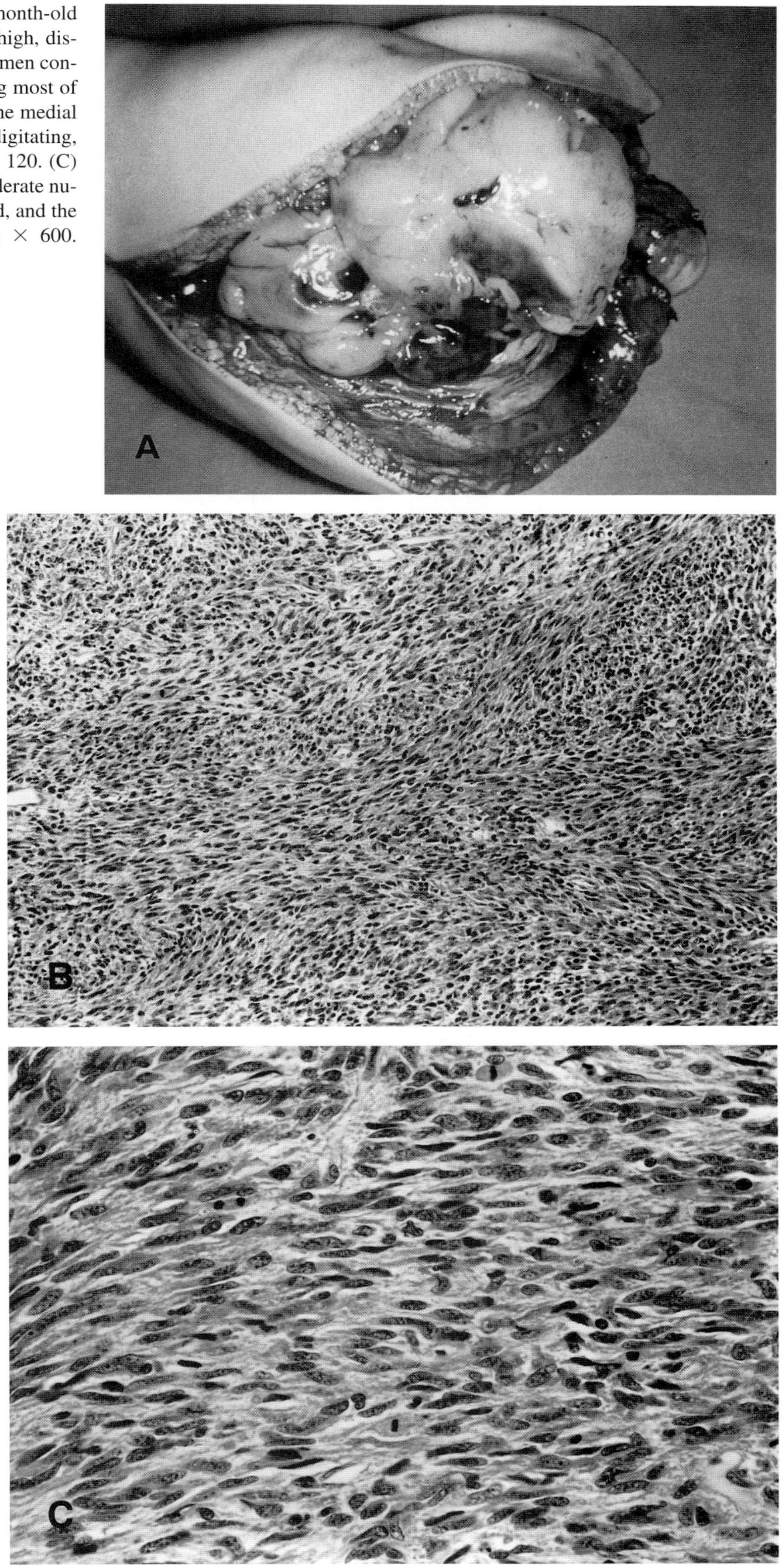

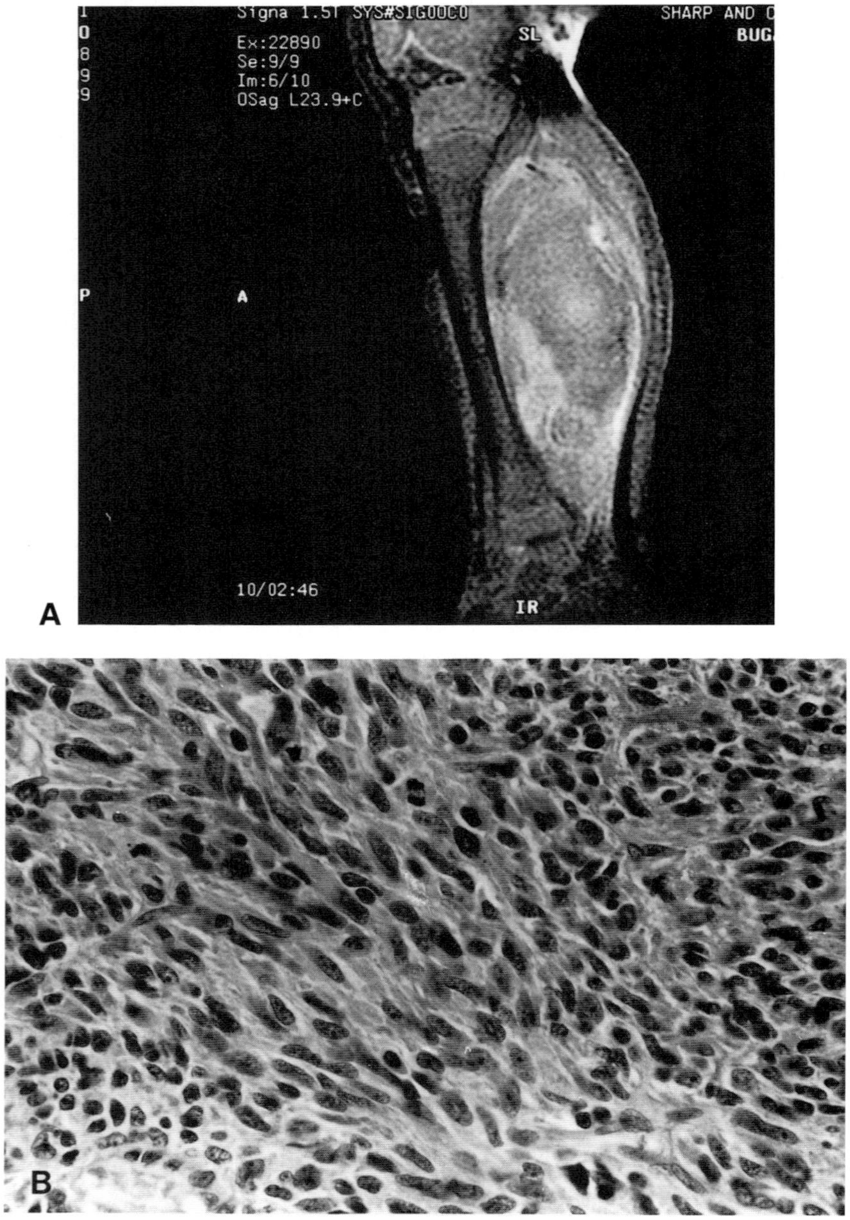

A

B

FIGURE 3.27. Congenital fibrosarcoma in a 2-month-old male born with a discrepancy in the size of his left lower leg, which was much larger than his right. (A) Magnetic resonance imaging. There is a large mass within the posterior compartment of the left lower leg. (B) The tumor biopsy material consists of fascicles of spindle-shaped cells with elongated irregular nuclei. Several mitoses are present. The cells are immunoreactive with vimentin and unreactive with actin and desmin. Karyotype perfomed on the tumor revealed a trisomy 11. Trichrome; × 300.

it out" leaves residual tumor behind. Therefore a wide local excision is recommended at the initial surgery, if this is technically feasible [14].

The tumor is composed of small, spindle-shaped cells separated by variable amounts of pale gray staining stroma having a gray, myxoid appearance (Figures 3.25–3.28). Differing amounts of eosinophilic-staining collagen depends on the degree of differentiation. Histologically fibrosarcoma may be almost identical to fibromatosis except that it displays more cellularity, as well as nuclear atypia, increased mitotic activity, and areas of necrosis than the latter [11]. The "herringbone" and storiform growth patterns are present in both tumors. The microscopic features of congenital fibrosarcoma are similar to those seen in the older child and adult, but the tumor cells tend to be smaller and less mature [15,46,48,66,68,69]. Some congenital fibrosarcomas consist of small round cells, rather than the customary spindle cells, embedded in an abundant pale-staining myxoid stroma displaying only moderate cellular atypia and a low mitotic rate.

Fibrosarcomas generally are immunoreactive with vimentin and type IV collagen but unreactive with cytokeratin, epithelial membrane antigen, neurofilament, S-100, actin, and desmin immunoperoxidase antibodies [46,69]. Ultrastructural studies reveal primarily immature fibroblasts and myofibroblasts [46,69] (Figure 3.28). There is no definite relation between the histological findings and either recurrence or metastases [11,44,46,66]. The incidence of metastases is low, ranging from 0 to 8% [11,42]. The presence of the trisomy 11 karyotype can be used to distinguish fibrosarcoma from other childhood sarcomas [46] (Figure 3.28E).

Management of this tumor is of prime concern because of its aggressive behavior and the functional impairment that may result from local treatment. Wide excision is the treatment of choice, with amputation as a last resort, if the tumor has caused a dysfunctional extremity. Adjuvant therapy is recommended for patients with axial and proximal extremity unresectable lesions [69,71]. Chemotherapy has been used following limb salvage surgery with some success [71]. The estimated 5-year survival for infantile fibrosarcoma is greater than 80% [14,15,46,69,72].

Fibrohistiocytic Tumors

Fibrohistiocytic tumors, which include a wide variety of both benign and malignant conditions, are composed of histiocytes and fibroblasts that involve the skin and the soft tissues. Not all members of this group are seen in the newborn and infant. The majority are found as subcutaneous nodules or soft tissue masses. During the first year of life the juvenile xanthogranuloma is the leading condition in this broad category. Both benign and malignant forms of fibrous histiocytoma occur in infants, but most are benign [14,15,73,74].

The principal fibrohistiocytic lesions found in infants are the juvenile xanthogranuloma, which is probably an inflammatory or reactive process rather than a neoplastic one, benign fibrous histiocytoma, and rarely, malignant fibrous histiocytoma [14,15,74].

Fibrous Histiocytoma

Like most soft tissue tumors in the young, the fibrous histiocytoma presents as a palpable mass. The benign fibrous histiocytoma consists of a firm, well-circumscribed or encapsulated tan-gray to yellow nodule, 2 to 3 cm in diameter. Microscopic examination reveals a moderately cellular tumor composed of spindle-shaped fibroblastic cells, and smaller numbers of round or oval vacuolated histiocytes (Figure 3.29). In addition, small numbers of multinucleated giant cells may be present (not the typical Touton giant cells of juvenile xanthofibroma). The interdigitating or "herringbone" and storiform growth patterns evident in some fields are similar to those observed in the broad category of fibromatoses. Generally the mitotic rate in the benign fibrous histiocytoma does not exceed 1 or 2 mitoses per high-power field. The vimentin immunoperoxidase stain is generally reactive, but the histiocytic markers, namely α_1-antitrypsin and α_1-antichymotrypsin, are variable in their degree of staining [14]. On the other hand, factor XIIIa is frequently expressed [15]. Electron microscopic studies show spindle-shaped fibroblasts with a prominent rough endoplasmic reticulum, larger, rounder histiocytic cells with variable numbers of lysosomes, and fibrohistiocytes composed of both endoplasmic reticulum and lysosomes [14,15,73,74]. Both the fibrous and histiocytic components are more immature than their mature soft tissue counterparts and show little or no evidence of phagocytosis or production of typical collagen fibers.

The main difficulty in establishing the diagnosis of fibrous histiocytoma is determining whether the tumor in question is benign or malignant [73]. Some helpful criteria used are the location of the tumor (deep-seated vs superficial), the mitotic rate (>3 mitotic figures per high-power field), and the degree of nuclear atypia (significant nuclear pleomorphism and bizarre giant cells) [15,73]. Most fibrous histiocytomas found in infants are benign but will recur if not completely removed. Malignant fibrous histiocytoma is similar biologically to the adult counterpart. Surgery is the main form of therapy, but the risk of local recurrence and pulmonary metastases indicates the need for adjuvant therapy in selected patients [75].

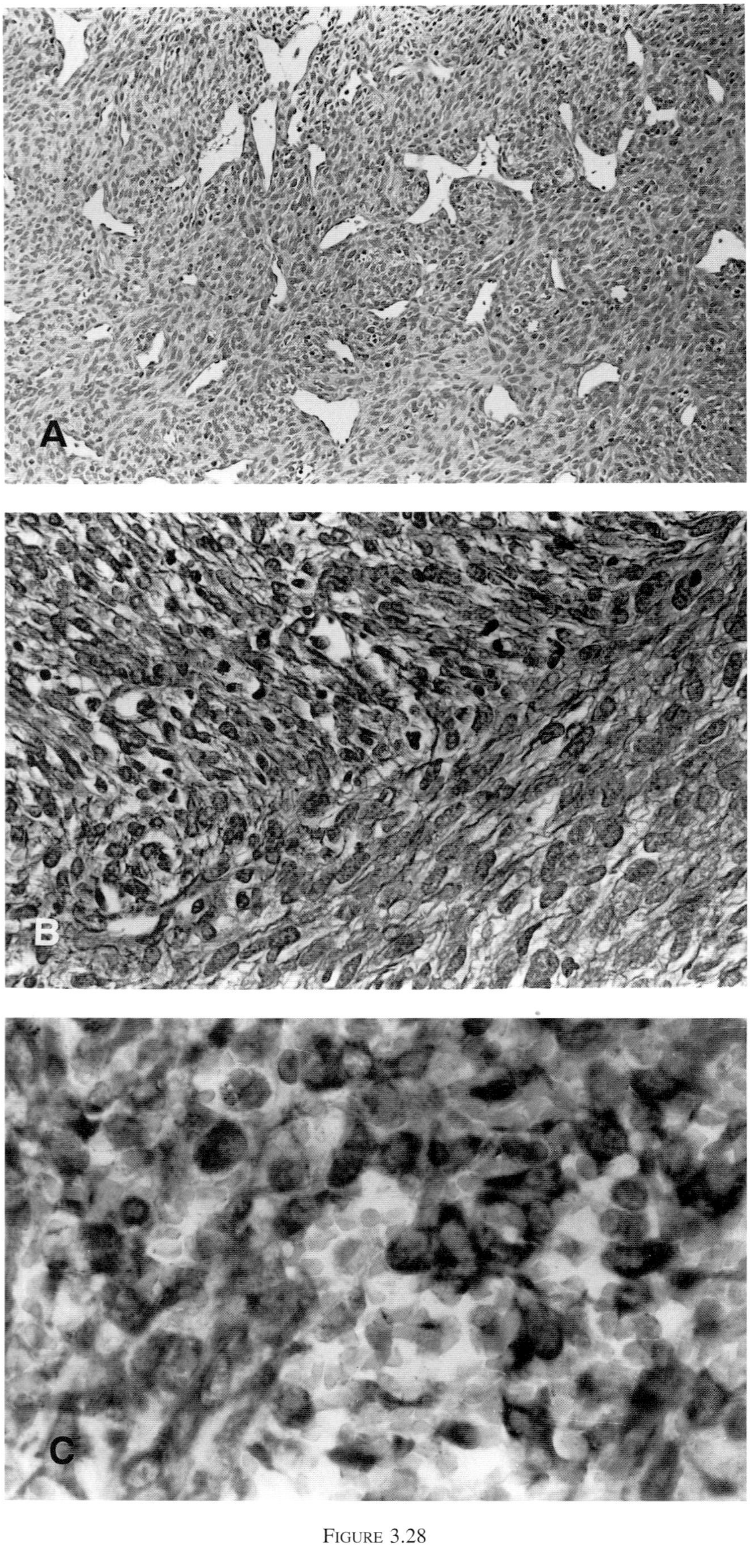

FIGURE 3.28

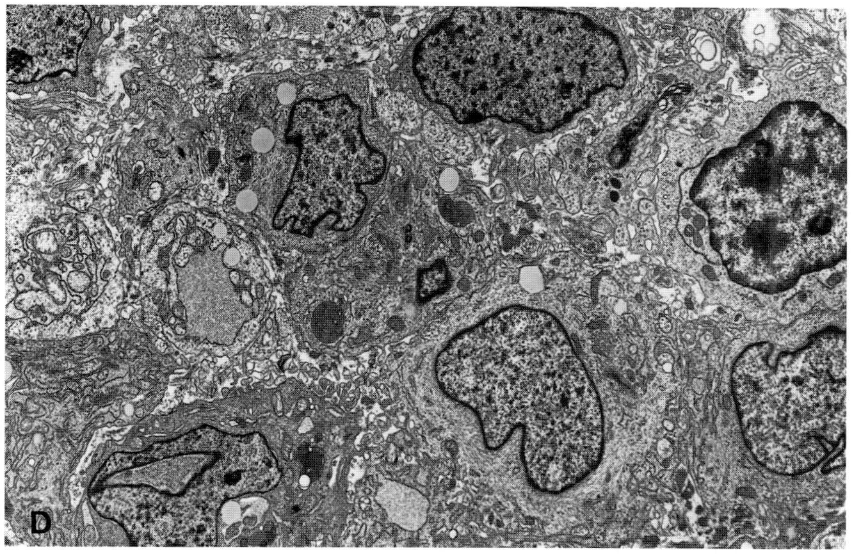

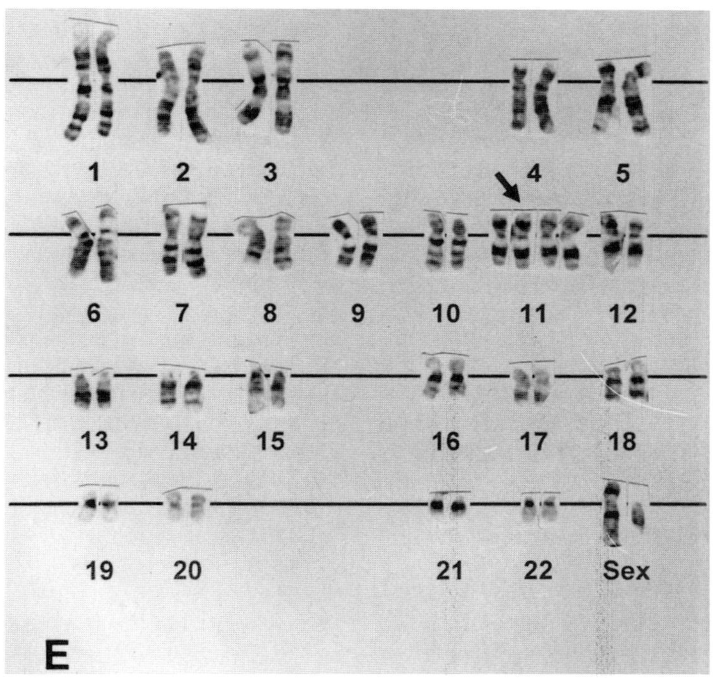

FIGURE 3.28. (*Continued*) Congenital fibrosarcoma in a 2-month-old male with a shoulder mass. (A) Prominent vascular pattern that might be confused with hemangiopericytoma. H&E; × 120. (B) Reticulin fibers separating tumor cells. Reticulin stain; × 400. (C) The tumor cells are reactive with vimentin but not with actin, desmin, or neuron-specific enolase. Vimentin immunoperoxidase; × 600. (D) Electron photomicrograph (× 5900) shows oval cells with slightly irregular, smooth cytoplasmic contours. The irregular nuclei have evenly spaced chromatin. Scattered small filaments are present within the cytoplasm. Focal intercellular collagen fibrils are noted. (Courtesy of Ann Peters, Department of Pathology, Children's Hospitals, San Diego.) (E) Karyotype of the tumor shows a trisomy 11. (Courtesy of James Mascarello, Ph.D., Department of Pathology, Children's Hospital, San Diego.)

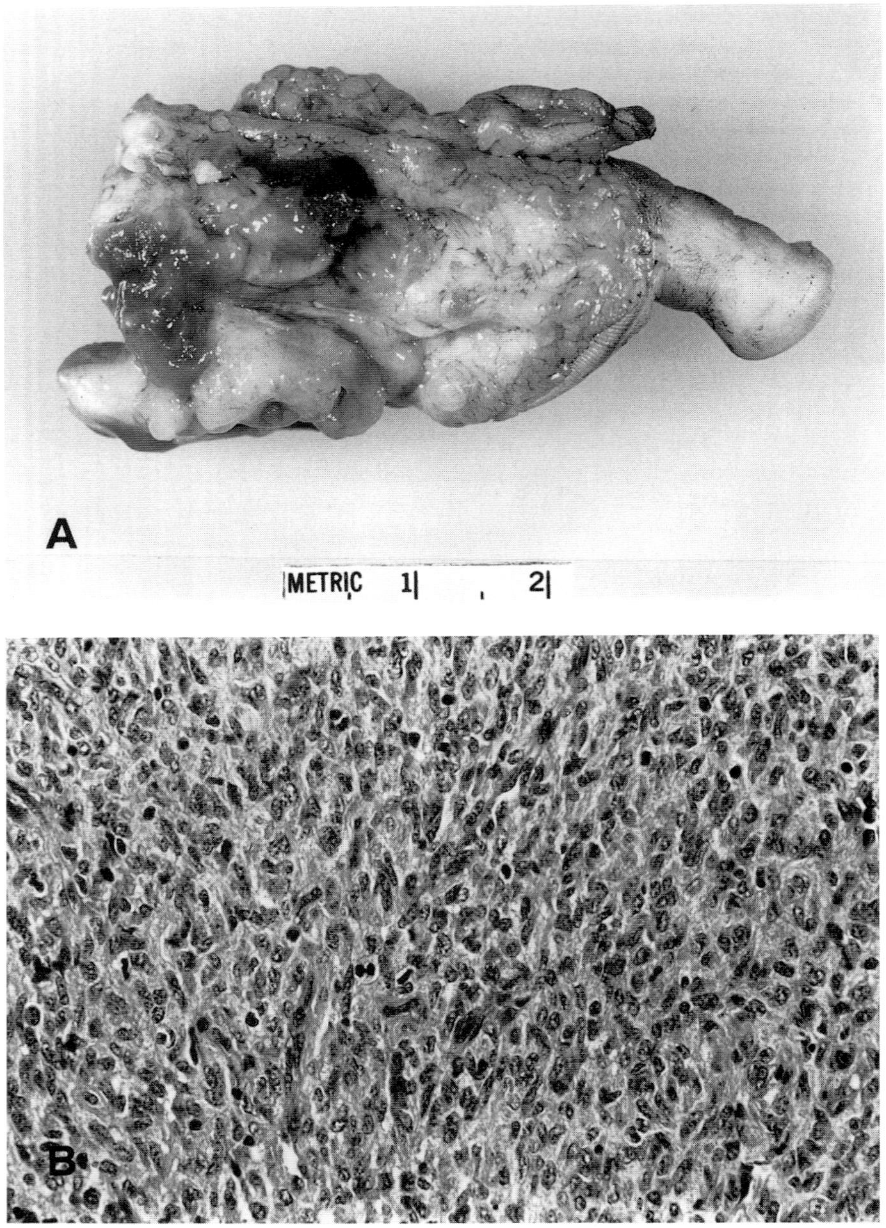

FIGURE 3.29. Fibrous histiocytoma: a 6-month-old female had an excisional biopsy of a mass on the dorsum of her left foot at 1 month of age. (A) Recurrent tumor consists of a brownish yellow mass situated in the interosseous muscles and fat. (B) The tumor consists of histiocytes and fibroblasts showing noticeable mitotic activity and mild to moderate nuclear atypia (Figure 3.31). H&E; × 600.

Grossly the malignant angiomatoid fibrous histiocytoma may be mistaken for a large soft tissue hematoma, a hemorrhagic hemangioma, or a lymph node (Figures 3.30 and 3.31) [14,15,74–77]. The characteristic light microscopic appearance consists of central cystically dilated vascular spaces lined by clumps of atypical-appearing fibroblasts and histiocytes with pleomorphic and hyperchromatic nuclei and moderate numbers of mitoses (instead of a single row of regular endothelial cells) (Figure 3.30). About the periphery, lymphoid follicles surround the central hemorrhagic zone, giving the tumor the appearance of a lymph node with a hemorrhagic center under low power (Figure 3.31) [15,74]. Because of these inflammatory changes and hemorrhage, the tumor is often misdiagnosed [78]. The myoifibroblastic phenotype of this tumor is supported by the reactive vimentin, desmin, and calponin staining and by the variable actin expression [75,77].

Angiomatoid fibrous histiocytoma has relatively low recurrence and metastatic rates [76,77]. It is regarded as a tumor of low to medium grade malignancy [77].

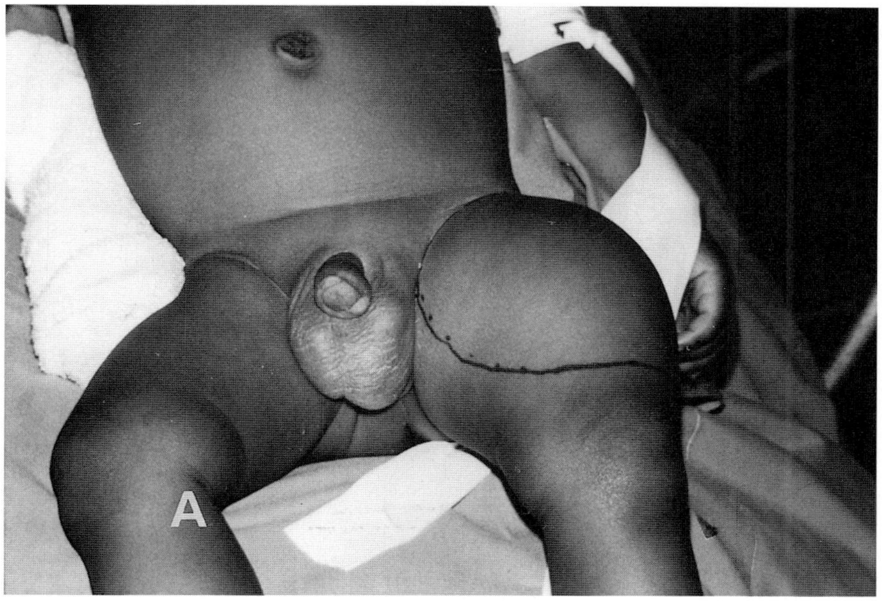

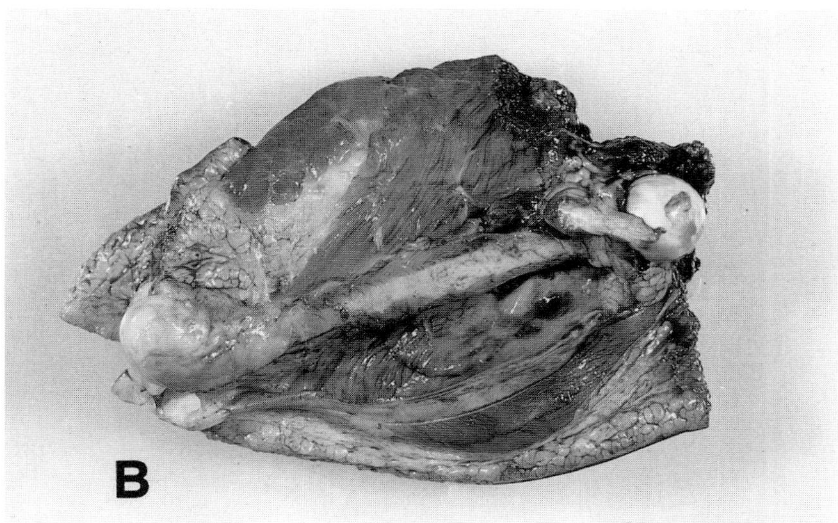

FIGURE 3.30. Angiomatoid fibrous histiocytoma. (A) A 3-month-old male with a left thigh mass, which gradually increased in size. (B) Amputation specimen showing the tumor (dark area) situated beneath the proximal femur in the belly of the vastus medialis muscle. The head of the femur is situated on the right-hand side of the photograph and the femoral condyles are on the left. (C) Close-up view of the involved muscle with tumor, which is composed of several, dark, hemorrhagic spongelike cavities. (D) Vascular channels are bordered by atypical histiocytes and fibroblastic cells. Trichrome; × 240. (E) Tumor cells rather than endothelial cells line the vascular channels. Trichrome; × 750.

(*Continued*)

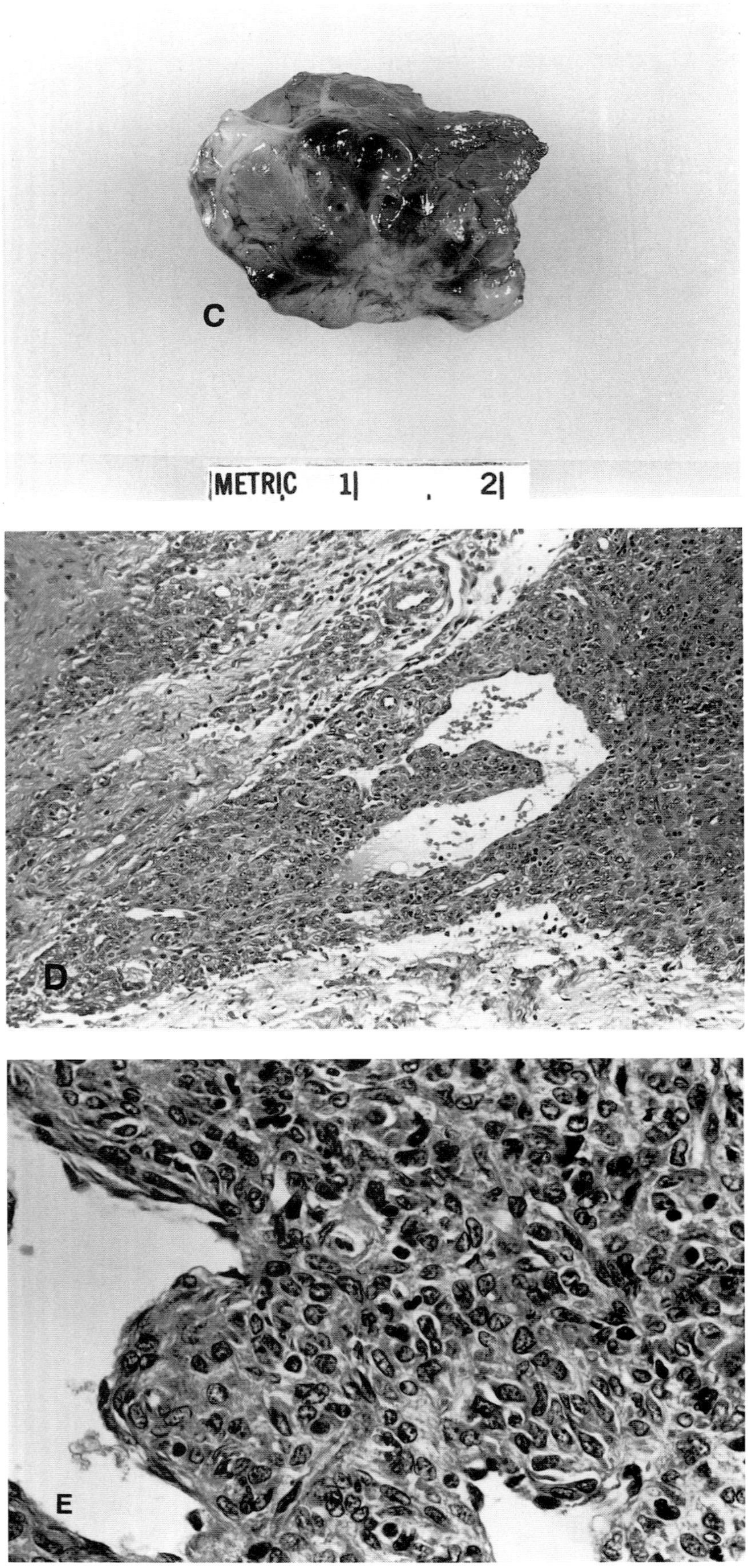

FIGURE 3.30. (*Continued*)

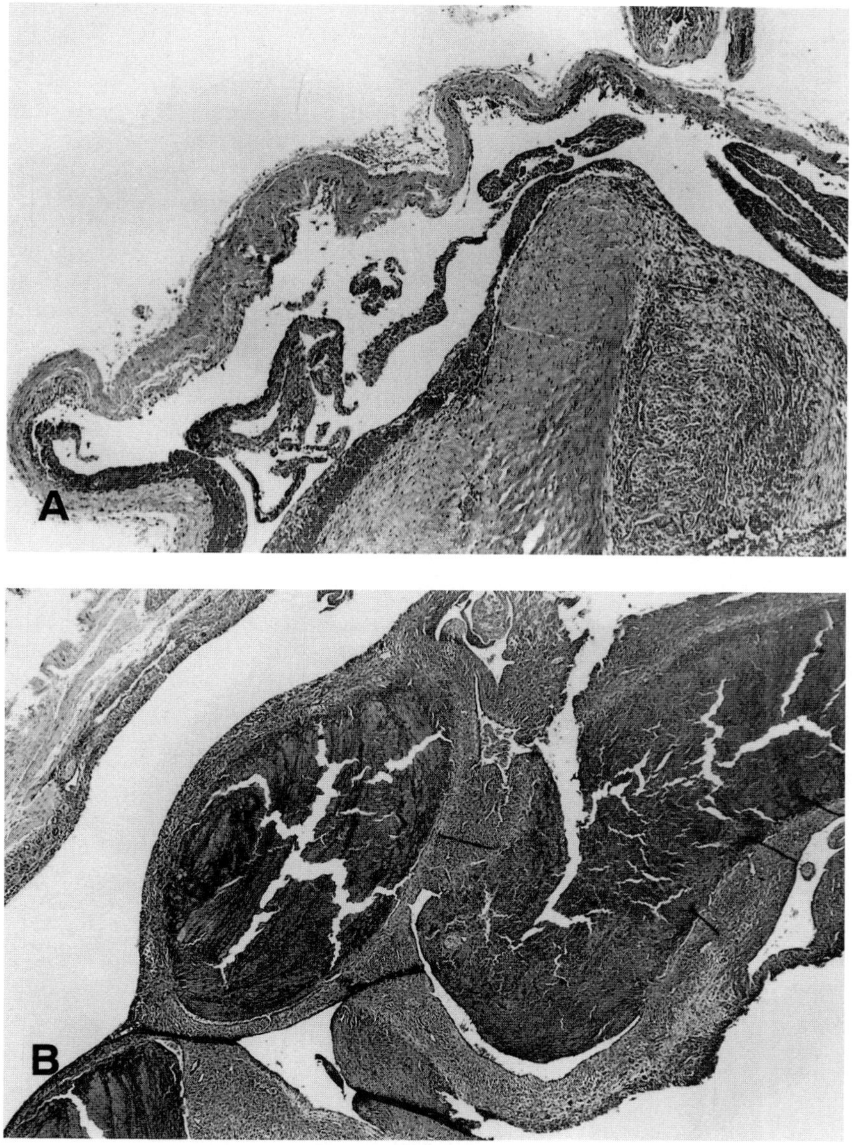

FIGURE 3.31. Angiomatoid fibrous histiocytoma from the neck of a 3-month-old male. (A) The partially cystic tumor mass surrounded by a dense fibrous pseudocapsule. Trichrome; × 30. (B) The prominent vascular component. Trichrome; × 30.

Juvenile Xanthogranuloma

Juvenile xanthogranuloma, a fibrohistiocytic proliferation found most often in newborns and infants, is characterized by one or more cutaneous nodules or, less frequently, by lesions in the deep soft issues or organs [15,46,72,79] (Figure 3.32). The cutaneous xanthogranulomas, which occur as red to yellow-brown papules distributed on the head, neck, and extremities, usually regress within two years of diagnosis [46,72,79,80]. The tumors are noted in other sites such as the eye, heart, lung, and deep soft tissues, which may or may not occur with cutaneous lesions [15,46,72,79,80].

Systemic (extracutaneous) involvement is rare but may be associated with significant morbidity and occasional deaths [80]. Symptoms are usually referable to bulky or infiltrative disease. The most frequent extracutaneous locations are the subcutaneous soft tissue, central nervous system, liver, spleen, lung, eye/orbit, oropharynx, and muscle—in that order [80]. Deep-seated xanthogranulomas occur as firm nodules averaging 2 to 4 cm in size and varying from tan-gray to tan-yellow on cross section [15,79]. The extracutaneous xanthogranulomas have essentially the same histological findings as the cutaneous ones. They show sheets of well-differentiated histiocytes with eosinophilic or foamy cytoplasms and vesicular nuclei. Typically there are multinucleated Touton giant cells characterized by peripheral wreaths of nuclei. Eosinophils about the periphery are a prominent feature. Variable numbers of lymphocytes and plasma cells are noted also.

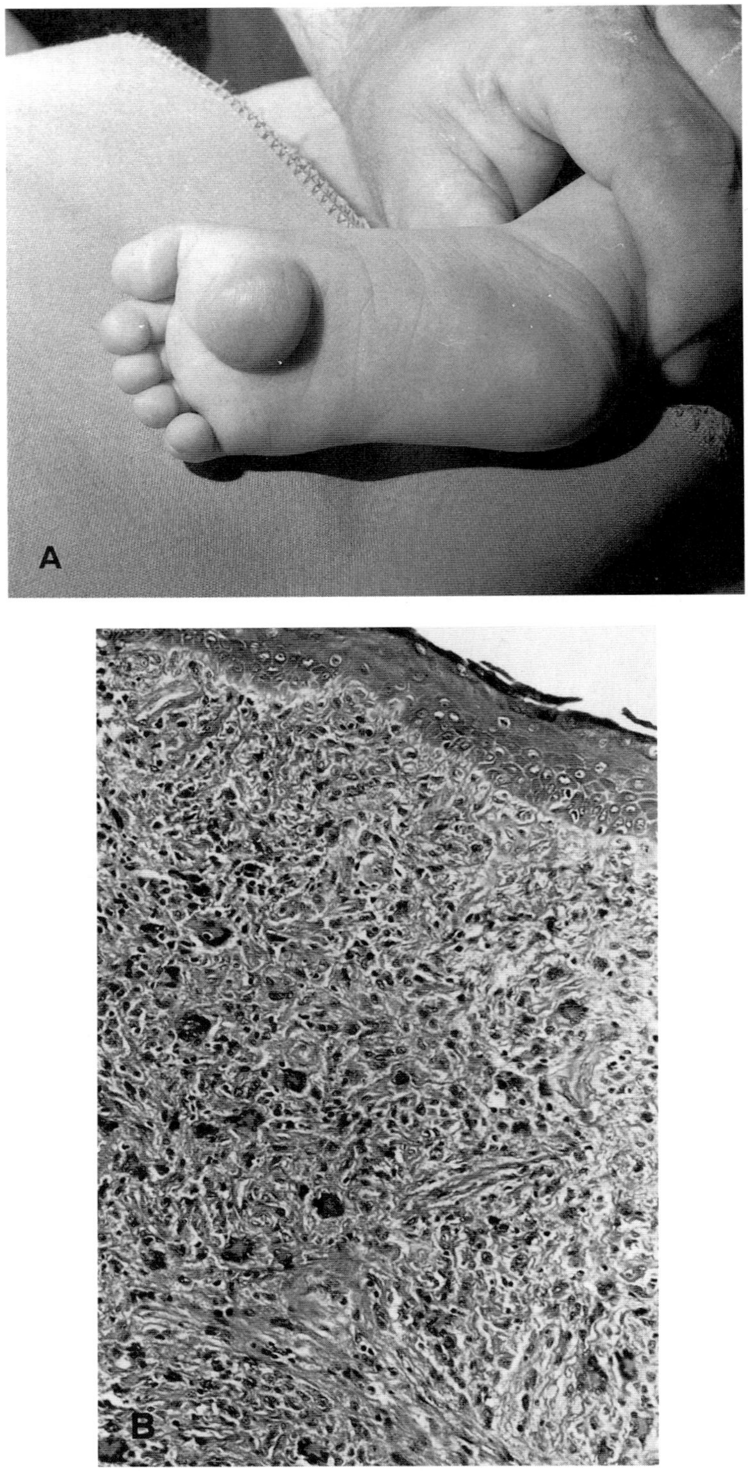

FIGURE 3.32. Juvenile xanthogranuloma. (A) A 5-month-old male with a mass on the plantar aspect of the right foot, present since birth. The specimen measured 3 cm in diameter and had a uniform, smooth tan-yellow cut surface. (B) There is a mixed cellular infiltrate involving the dermis and subcutaneous tissue composed of round to oval histiocytes, spindle-shaped fibroblasts, and scattered Touton multinucleated giant cells. Some histiocytes appear to be vacuolated. Fibrosis is present. The epidermis is not invaded. H&E; × 300.

Touton giant cells, which are helpful for establishing the diagnosis, may not be present always in the early active prolferative phase especially in extra-cutaneous locations. Lipid droplets that are oil red O positive are observed in the histiocyte cytoplasm [79]. In older lesions, foci of fibrosis may have the storiform and herringbone growth patterns seen in fibrous histiocytoma and the fibromatoses [15].

The histiocytic nature of the tumor cells is confirmed by electron microscopy. The cells are composed of pseudopodia, lysosomes, and lipid droplets [15,72]. However the racket-shaped Birbeck granules diagnostic of Langerhans cells are notably absent [15,72,79].

Immunohistochemical studies performed on this tumor show variable results. The histiocytic markers lysozyme, α_1-antitrypsin and α_1-antichymotrypsin are either nonreactive or sometimes focally or faintly positive and therefore not useful markers [46,79]. The macrophage or monocyte antigens CD68 (KP-1) and HAM-56 (HHF35) are expressed, as well as the mesenchymal cell markers vimentin and actin [46,72,79]. The cortical thymocyte antigen CD1a is reactive, suggesting a dermal dendrocytic origin; the dermal dendrocyte is believed to migrate from the skin to other locations [72]. Factor XIIIa, expressed principally by dermal dendrocytes, is expressed by both xanthogranuloma and fibrous histiocytoma cells [46,80].

Chromosomal analysis of juvenile xanthogranulomas show normal 46,XX or 46,XY karyotypes. DNA flow cytometry reveals diploid stem lines [72]. The prognosis is excellent for both cutaneous and extracutaneous tumors [15,79]. Most cutaneous and visceral lesions regress spontaneously without treatment.

Skeletal Muscle Tumors

Tumors of skeletal muscle are of two types, rhabdomyoma and rhabdomyosarcoma. The former is a rare, benign skeletal muscle lesion comprising less than 2% of all striated muscle tumors, whereas the latter is the most common soft tissue sarcoma in childhood [15,46,62]. Rhabdomyoma is subclassified further into fetal and adult types [62]. Generally *fetal rhabdomyomas* are found in infants and they more closely resemble developing skeletal muscle than the adult form, which is usually noted in later life [81]. Rhabdomyomas of the female genital tract are regarded as a special form of fetal rhabdomyoma [82]. Cardiac rhabdomyoma is a distinct lesion, probably harmartomatous, found primarily in cardiac muscle (see Chapter 15, Cardiac Tumors).

Rhabdomyoma

The most common site of the extracardiac fetal rhabdomyoma is the subcutaneous tissue of the head and neck region [72,81,82]. Rhabdomyomas present as slowly growing solitary masses within the soft tissue or submucosa, most often in the pre- and postauricular areas, followed in frequency by the neck, nasopharynx, and tongue [81]. The abdominal wall and the retroperitoneal area are other locations.

Grossly, fetal rhabdomyomas are circumscribed but not encapsulated, light gray to tan-pink nodules, averaging 3 cm in size. Microscopically they show haphazardly arranged, fetal-appearing, slender skeletal muscle cells with cross-striations separated by a pale-staining myxoid stroma containing small oval to spindle-shaped undifferentiated cells [62,81,82]. Mitoses are essentially absent, and nucleoli are small. The immunoperoxidase skeletal muscle markers are positive and the ultrastructural findings are consistent with skeletal myocytes in different stages of development [81,82]. Local recurrence rate is low, less than 1 in 10.

Rhabdomyosarcoma

Rhabdomyosarcoma is primarily a neoplasm of children and adolescents, comprising over half of all sarcomas in this age group [15,46]. Traditionally it is considered to be a sarcoma derived from the embryonic skeletal muscle cell, the rhabdomyoblast [15]. Rhabdomyosarcoma is a highly malignant neoplasm accounting for most soft tissue malignancies in infants, and it is second or third in frequency in the newborn, following fibrosarcoma [9–11,14,46].

There are associations between rhabdomyosarcoma and hereditary diseases or congenital malformations [13,83]. Postmortem studies performed on 115 children enrolled in the Intergroup Rhabdomyosarcoma Study revealed that 37 (32%) had congenital malformations, and over half of these involved the central nervous system [83]. An increased frequency of breast carcinoma and brain tumors is noted in families of patients with this sarcoma: the so-called family cancer syndrome [84]. Rhabdomyosarcoma is one of the malignancies occurring in patients with neurofibromatosis I.

Cytogenetic studies have shown karyotypic differences between embryonal and alveolar rhabdomyosarcomas [85]. The former exhibits a loss of heterozygosity of one of the short arms of chromosome 11, whereas the latter shows a translocation between chromosomes 2 and 13, {t(2;13) (q35;q14)}. The detection of the p-glycoprotein transmembrane protein, the product of the multiple drug resistance gene (*MDR*), is of diagnostic and prognostic significance, for its presence in a rhabdomyosarcoma is associated with a decreased survival rate [85].

Rhabdomyosarcoma is seldom noted at birth; approximately 3% of the pediatric age group with this neoplasm are diagnosed in the first month of life; 10% are found in patients less than 1 year of age [1,6,10,11,13,86]. In both the newborn and the infant, the head and neck region and the genitourinary tract are the most common primary sites.

Rhabdomyosarcomas can be divided into five histological types: *embryonal, botryoid, alveolar, spindle cell,* and *mixed.* As in the older child and adolescent, most rhabdomyosarcomas diagnosed in the first year of life are of the embryonal and botryoid types [10,13,15,87]. The alveolar form, associated with the worst prognosis, ranks third [5,10,13,15,87].

The *botryoid form of embryonal rhabdomyosarcoma* arises from mucosal-lined body cavities (e.g., vagina, urinary bladder, middle ear, biliary system) and has a distinctive appearance (Figure 3.33). Typically it forms lobulated, grapelike, polypoid projections with pale gray gelatinous cut surfaces (Figure 3.33). Tiny, round or spindle-shaped rhabdomyoblasts are situated directly beneath the epithelium located in a submucosal cellular rind or mantle (Figure 3.33). Well-defined cytoplasmic cross-striations are often very difficult to identify. The reactive desmin and actin immunoperoxidase stains are helpful in identifying these small cells as rhabdomyoblasts. The remainder of the polyp consists of pale-staining, myxoid-appearing connective tissue and blood vessels. The botryoid rhabdomyosarcoma is associated with the most favorable prognosis of all the histological types [14,87].

Embryonal and alveolar rhabdomyosarcomas consist of a gray-white, myxoid-appearing infiltrative mass, seldom encapsulated, with more extensive necrosis and hemorrhage than is observed in the botryoid type. *Embryonal rhabdomyosarcoma* shows considerable variation in its histological pattern, with cellular areas alternating with myxoid areas [15] (Figures 3.34 and 3.35). The more differentiated tumors show round to oval, sometimes tadpole- or racket-shaped cells with irregular nuclei and an eosinophilic-staining cytoplasm and occasionally, cytoplasmic cross-striations (Figure 3.36). The

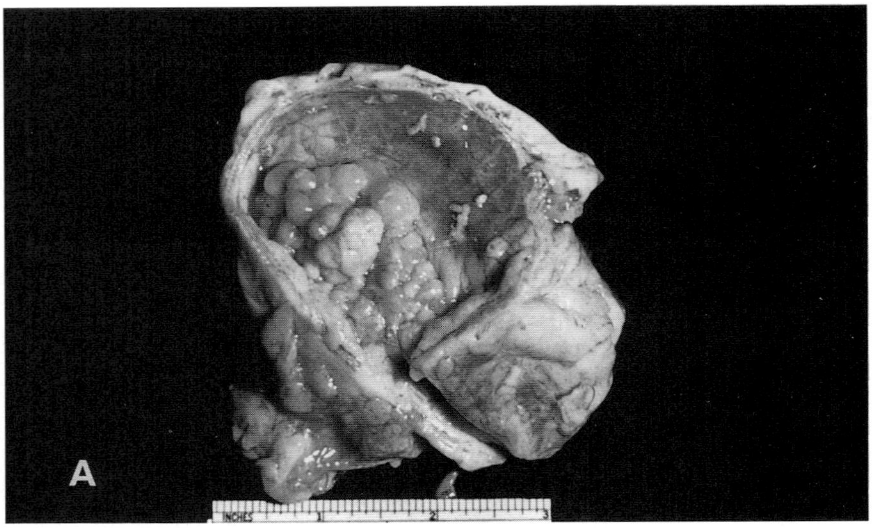

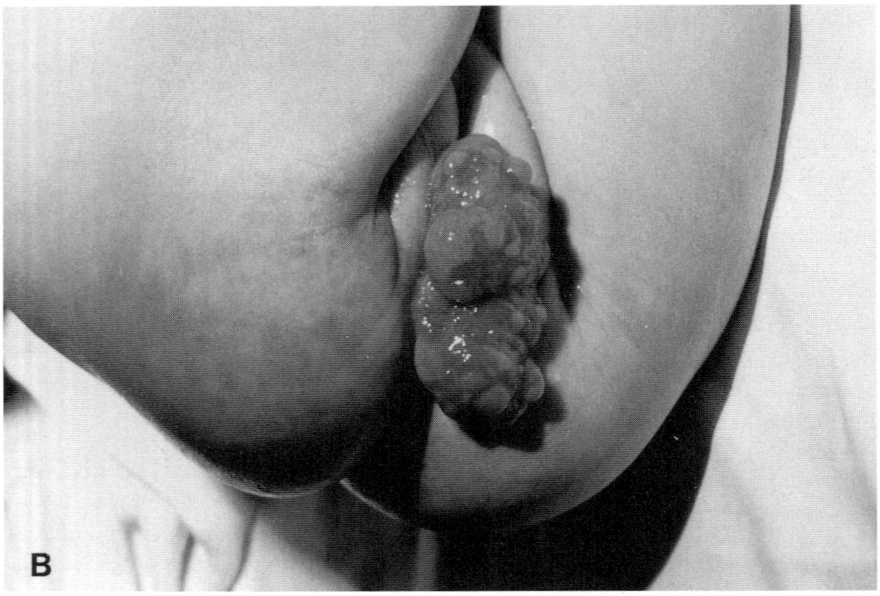

FIGURE 3.33. Botryoid embryonal rhabdomyosarcoma. (A) A 1-year-old male with grapelike tumor masses invading the trigone and posterior wall of the urinary bladder. (B) A 7-month-old female with similar appearing polypoid lesions protruding from the vagina. (C) Gross specimen of B, showing grapelike masses resting on a pale gray stroma. (D) Low-power view (× 120) of submucosal tumor nodules. Directly beneath the vaginal mucosa there is a cellular rind or mantle composed of rhabdomyoblasts. H&E. (E) Rhabdomyoblasts are shown in more detail. Some cells exhibit cytoplasmic cross-striations (arrowhead). H&E; × 480.

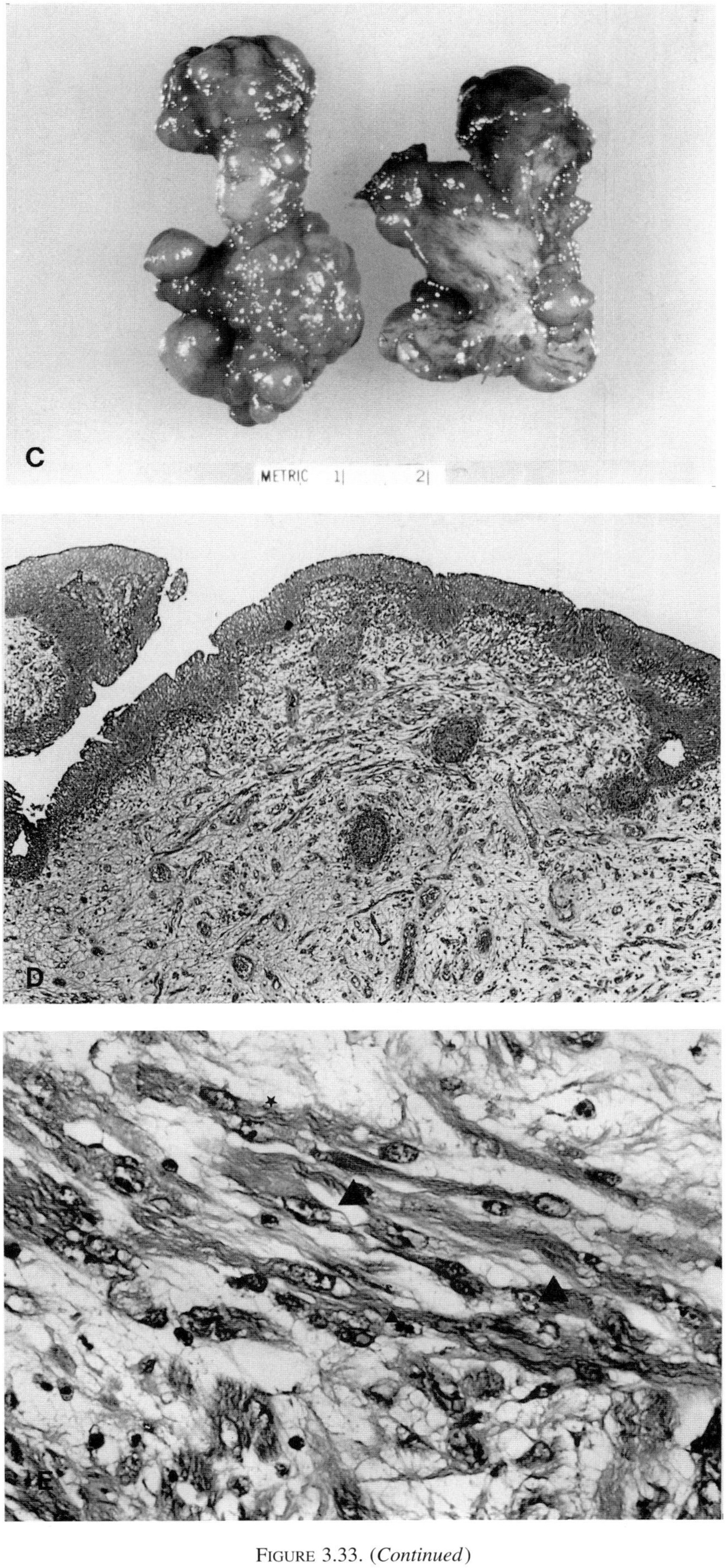

FIGURE 3.33. (*Continued*)

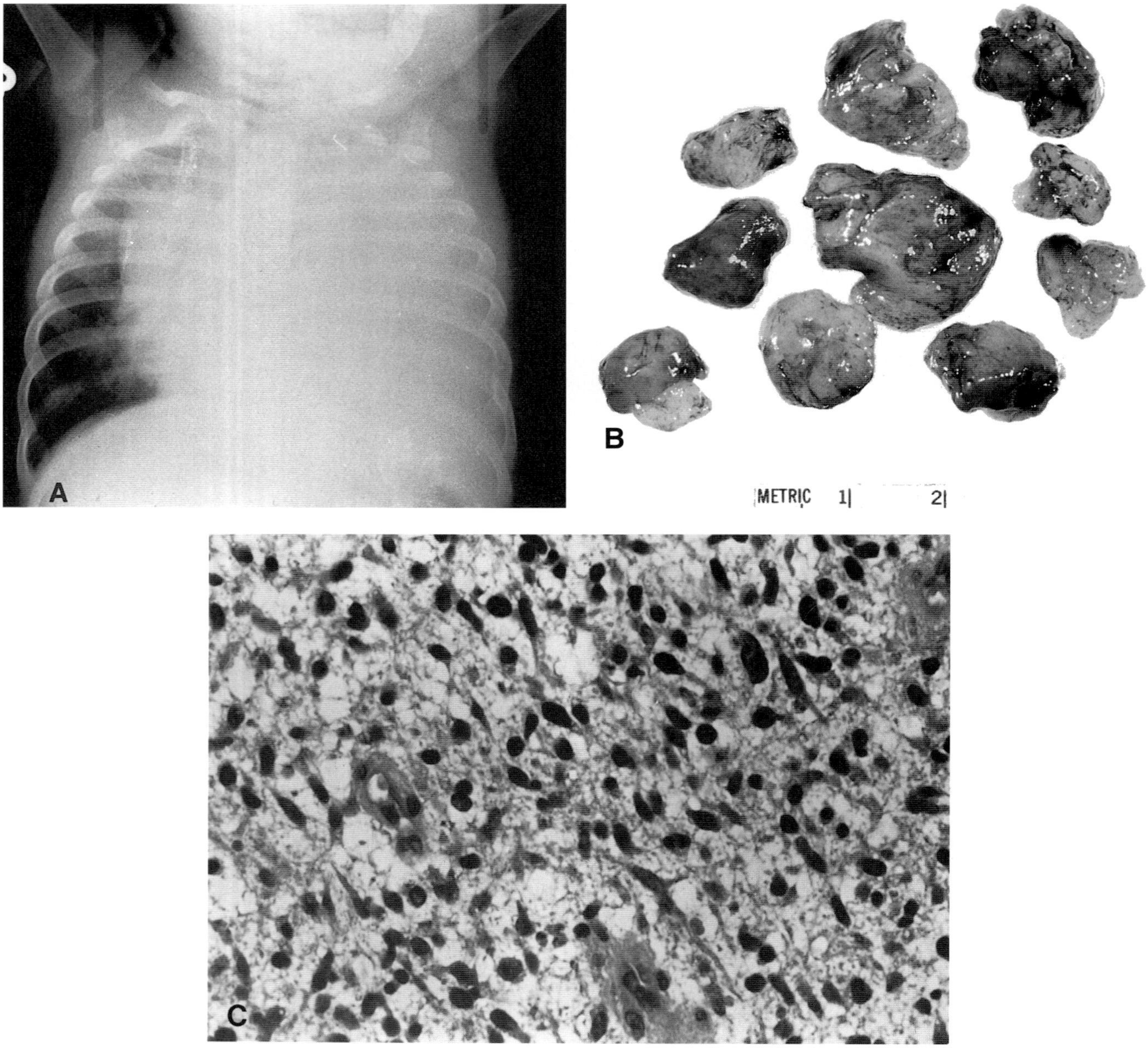

FIGURE 3.34. Embryonal rhabdomyosarcoma of the mediastinum. (A) Chest x-ray of an 8-month-old male with a 1-week history of a cough and progressive respiratory distress. The radiograph reveals consolidation of the left thorax with displacement of the mediastinum to the right. In the differential diagnosis of consolidation ("white out") of the thorax in an infant, the following neoplastic conditions should be considered: pleuropulmonary blastoma, pleuropulmonary PNET ("Askin tumor"), and rhabdomyosarcoma. Malignant lymphoma is seen more often in the older child. (B) Biopsy material obtained from the mediastinal mass, showing a soft gray gelatinous appearance. (C) The tumor consists of small, darkly staining rhabdomyoblasts in a myxoid background. The cells are round, tadpole shaped, or elogated. The cytoplasms are slightly vacuolated. Suggestive cross-striations are noted in some cells. H&E; × 320.

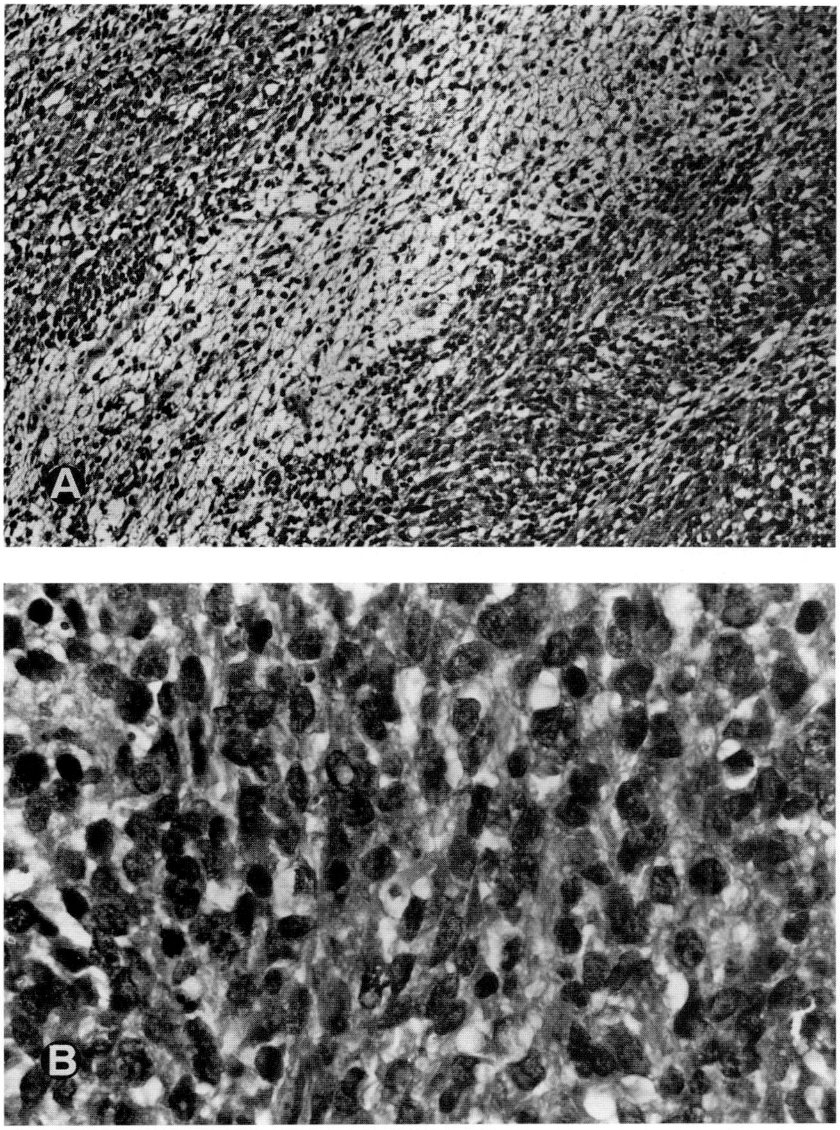

FIGURE 3.35. Embryonal rhabdomyosarcoma. (A) Myxoid areas alternating with hypercellular areas are characteristic of this tumor. H&E; × 150. (B) Cellular area composed small, poorly differentiated cells. Tumor cells were immunoreactive only with desmin, actin, and vimentin. H&E; × 480.

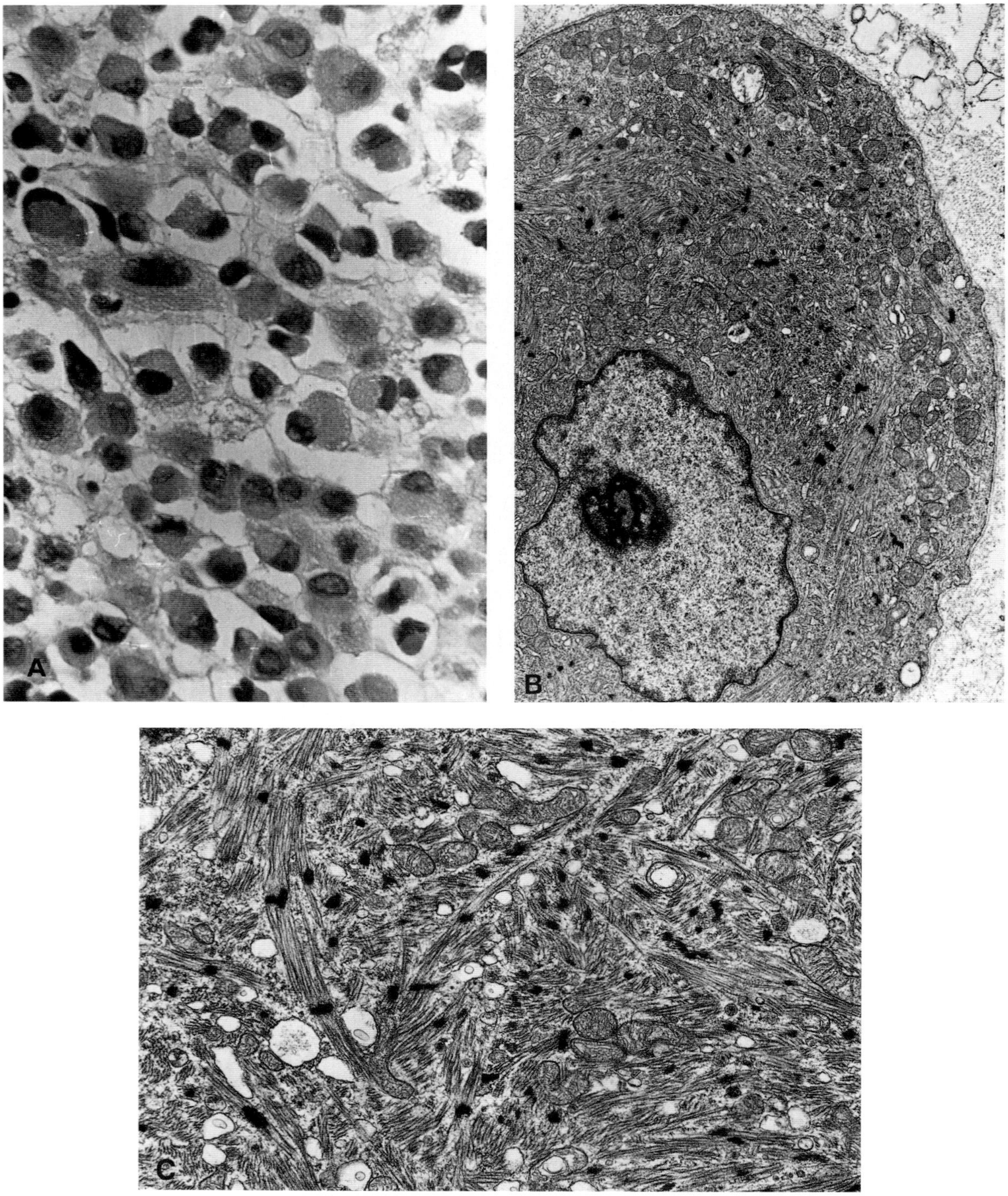

FIGURE 3.36. Embryonal rhabdomyosarcoma from the neck of a 2-year-old female. (A) Tumor consists mainly of well-differentiated rhabdomyoblasts. H&E; × 380. (B) Electron micrograph (× 5200) showing thick and thin filaments and a prominent basement membrane. (C) Higher magnification (× 17,800) of the thick and thin filaments with *Z* bands. (Electron photomicrographs courtesy of Ann Peters, Department of Pathology, Children's Hospital, San Diego.)

TABLE 3.2. Histopathological differential diagnosis of the small blue-cell malignant tumors.

Tumor	Light microscopy	Immunohistochemistry[a]	Ultrastructure
Neuroblastoma	Small round cells, rosettes, neurofilaments, nuclei with a peppery chromatin pattern	NSE +, NFP +, Vim +, des –, MSA –, LCA –, CK –, 013 –, B2MG –	Neurosecretretory granules, microtubules, neurofilaments
Rhabdomyosarcoma sarcoma	Spindle cells, eosinophilic cytoplasms, cross-striations, embryonal and alveolar patterns	des +, MSA +, vim +, NSE ±, NFP –, 013 –, CK –, LCA –	Spindle cells, primitive attachments, *Z* bands, thick and thin filaments, basement membranes; glycogen ±
Leukemia[b]	Diffuse infiltrates of small round cells, granules ±	LCA +, vim ±, MSA –, NFP –, CK –, 013 –	Round cells, no attachments, no matrix; granules ±
Primitive neuroectodermal tumor (PNET)[c]	Small round cells, rosettes ±, lobular pattern	NSE +, vim +, NFP ±, B2MG +, LCA –, des –, MSA –, CK –, HBA71 (013, CD99)+	Primitive cells, few organelles, rare neurosecretory granules, few intermediate filaments; primitive attachments ±
Rhabdoid tumor	Polygonal cells with pink cytoplasms, eosinophilic intermediate filament inclusions, round vesicular nucleus with one large nucleolus	vim +, CK +, des –, MSA –, NFP –, LCA –,	Bundles of cytoplasmic intermediate filaments

[a] Reactivity (+, –, or ±) with the following immunoperoxidases is described: NSE, neuron-specific enolase; NFP, neurofibrillary protein; vim, vimentin; des, desmin; LCA, leukocyte common antigen; MSA, muscle-specific actin; CK, cytokeratin; B2MG, anti-β_2-microglobulin.
[b] Myeloid leukemias may also present in extramedullary sites in the absence of diagnosable bone marrow involvement; this form of leukemia is called granulocytic sarcoma or chloroma. The granulocytic cell infiltrates often contain eosinophils and are reactive with myeloperoxidase and CD45. Megakaryoblastic leukemia (M7) is another example of a hematopoietic malignancy that may present with extramedullary bone and/or soft tissue masses (see Chapter 6: Leukemia).
[c] Includes the Askin thoracopulmonary tumor, Ewing sarcoma, and peripheral neuroepithelioma.
Source: Isaacs H Jr [14].

poorly differentiated embryonal rhabdomyosarcoma falls into the group of the "small-cell malignant tumors" of infancy and childhood, and ultrastructural and immunocytochemical studies are required to definitely establish the diagnosis [14,87]. Table 3.2 lists the histological differential diagnoses of the small blue-cell maligant tumors of infancy.

A "prognostically favorable variant" of embryonal rhabdomyosarcoma is called *spindle cell rhabdomyosarcoma* [87–89]. The paratesticular area and urinary bladder are the most common sites. Microscopically the tumor consists of uniform spindle-shaped rhabdomyoblasts displaying an interdigitating ("herringbone") growth pattern (Figure 3.37). The tumor cells react with vimentin and with the skeletal muscle markers actin and desmin [89].

Alveolar rhabdomyosarcoma has a characteristic microscopic picture showing small round to spindle-shaped cells forming irregular cystlike spaces mimicking pulmonary alveoli (Figure 3.38). In some instances with adequate preservation the rhabdomyoblasts attached to the lining of the fibrovascular septae appear to "spin off" into the lumen. Although well-defined cross-striations are difficult or sometimes impossible to find, the positive vimentin and desmin immunoperoxidase staining and the *Z*-band material and/or thick and thin filaments observed in the tumor cells by electron microscopy confirm the diagnosis (Figure 3.38). In addition there is a less well-differentiated form called *"solid" alveolar rhabdomyosarcoma* [90]. This variant is composed of sheets of small cells with round nuclei but without fibrous septa. The t(2;13)(q35;q14) chromosomal anomaly, characteristic of alveolar rhabdomyosarcoma, is present [91].

The classification currently used by the Intergroup Rhabdomyosarcoma Study for rhabdomyosarcoma and related sarcomas is the International Classification, which is based on tumor histology and prognosis [92] (Table 3.3). The classification consists of three groups: [favorable] intermediate, and [unfavorable] prognosis. Botryoid and spindle cell rhabdomyosarcomas are classified as favorable, embryonal rhabdomyosarcoma and extraosseous Ewing sarcoma (PNET) as intermediate, and alveolar rhabdomyosarcoma and undifferentiated sarcoma as unfavorable [92]. A fourth group has been added that includes subtypes whose prognosis is not presently evaluable such as the rhabdomyosarcoma with rhabdoid features [92].

The prognosis of patients with rhabdomyosarcoma depends on three main factors: the location, stage (extent of disease), and histological type [46,87,92–95] (Table 3.4). The reader is referred to the German Soft Tissue Sarcoma Study Group and the Intergroup Rhabdomyosarcoma Study findings for discussion of the outcome of children with this tumor [93,94].

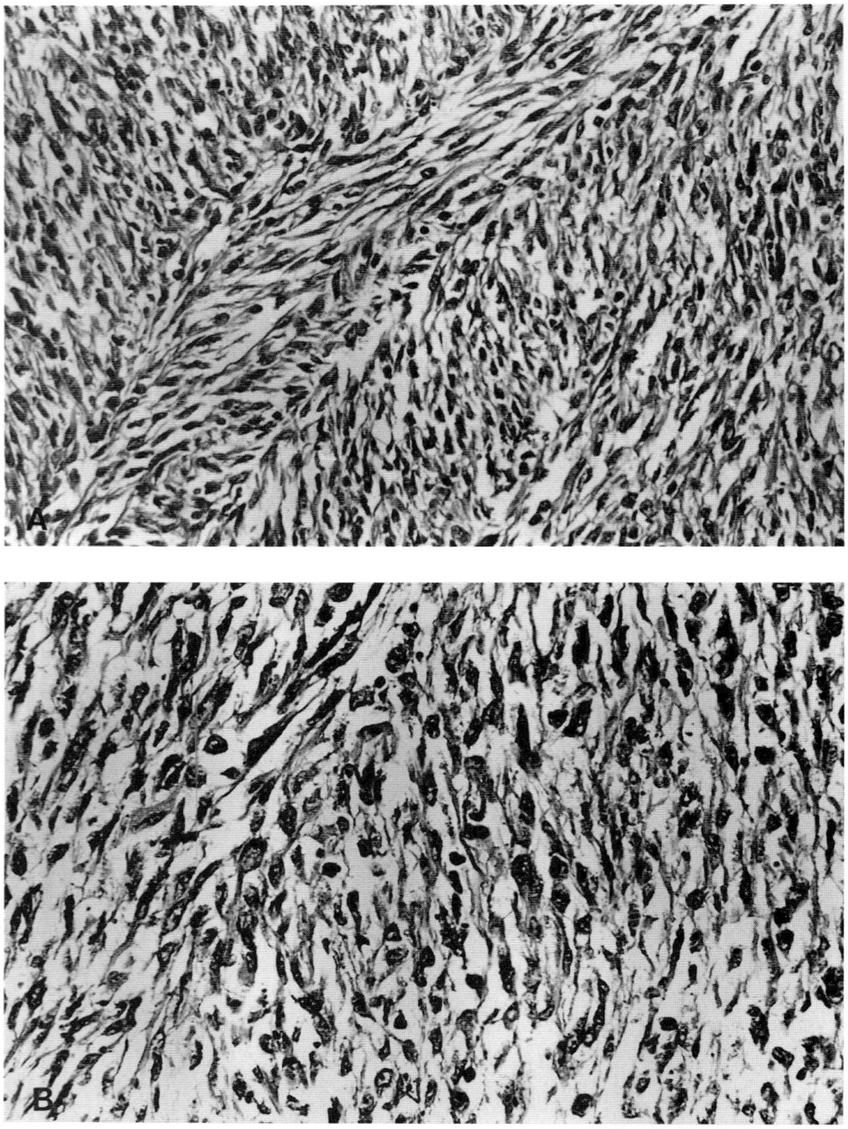

FIGURE 3.37. Embryonal rhabdomyosarcoma, spindle cell type. Biopsy material from a large, unresectable pelvic mass arising from the prostate and urinary bladder in a 6-month-old male. (A) The tumor consists of spindle-shaped rhabdomyoblasts displaying an interdigitating ("herringbone") growth pattern. H&E; × 150. (B) Higher power view (× 300) depicting elongated spindle-shaped rhabdomyoblasts with suggestive cytoplasmic cross-striations. Phosphotungstic acid–hematoxylin stain.

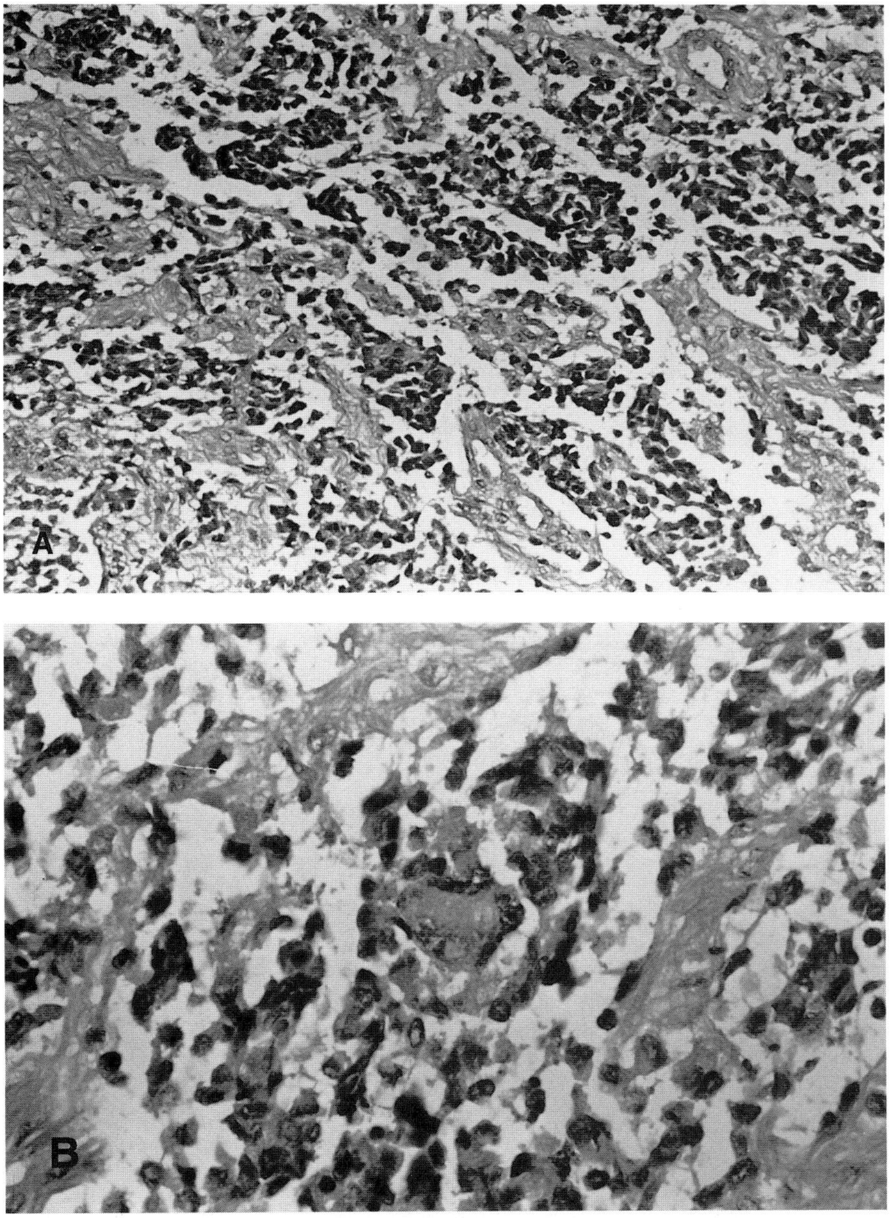

FIGURE 3.38. Alveolar rhabdomyosarcoma in a 7-month-old male with a thigh mass and metastases to the lip and cervical lymph nodes. (A) Small, darkly staining tumor cells are situated in alveolar-like spaces separated by slightly vascular fibrous septa. H&E; × 300. (B) Higher magnification (×750) depicting the variable morphology of the tumor cells. There is a tumor giant cell in the center of the field. H&E. (C) Tumor cells stain positively only for skeletal muscle markers desmin and for vimentin. Desmin immunoperoxidase; × 750. (D) Ultrastructure showing diagnostic cytoplasmic myofilaments. (Electron photomicrograph, × 18,000, courtesy of Darkin Chan, Children's Hospital, Los Angeles.)

(Continued)

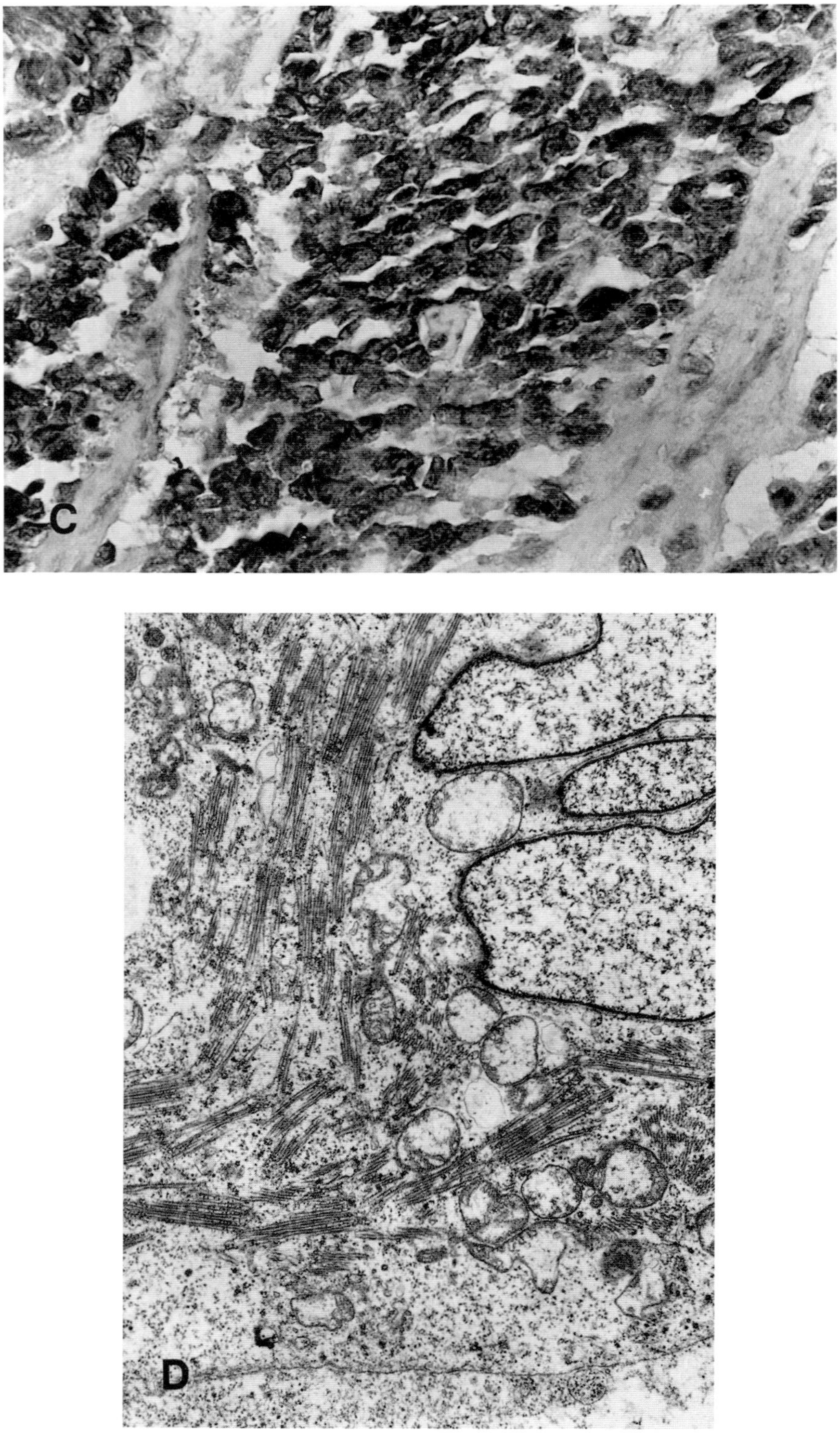

FIGURE 3.38. (*Continued*)

TABLE 3.3. International classification of rhabdomyosarcoma.

I.	Superior prognosis
	a. Botryoid embryonal rhabdomyosarcoma
	b. Spindle cell rhabdomyosarcoma
II.	Intermediate prognosis
	a. Embryonal rhabdomyosarcoma
III.	Poor prognosis
	a. Alveolar rhabdomyosarcoma
	b. Undifferentiated sarcoma
IV.	Subtypes whose prognosis is not presently evaluable
	a. Rhabdomyosarcoma with rhabdoid features

Source: Newton WA Jr, et al. [92].

TABLE 3.4. Clinical group stage system for rhabdomyosarcoma.

Clinical Stage	Group	Extent of disease/surgical result
I	A	Localized tumor, confined to site of origin, completely resected
	B	Localized tumor, infiltrating beyond site of origin, completely resected
II	A	Localized tumor, gross total resection, but with microscopic residual disease
	B	Locally extensive tumor (spread to regional lymph nodes), completely resected
	C	Locally extensive tumor (spread to regional lymph nodes), gross total resection, but with microscopic residual disease
III	A	Localized or locally extensive tumor, gross residual disease after biopsy only
	B	Localized or locally extensive tumor, gross residual disease after major resection (> 50% debulking)
IV		Any primary size tumor, with or without regional lymph node involvement, with distant metastases, without respect to surgical approach to primary tumor

Source: Qualman SJ, et al. [87].

Rhabdoid Tumor

In the 1970s rhabdoid tumor of the kidney was identified as a neoplasm different from that of Wilms tumor [96]. The tumor was given the name "rhabdoid" because microscopically it resembles a rhabdomyosarcoma, although it does not show skeletal muscle components either by immunoperoxidase studies or by electron microscopy. This highly malignant neoplasm, characterized clinically by early metastases and a high mortality rate, occurs in the newborn and during the first year of life and only rarely in the older age groups (see also Chapter 8, Brain Tumors, and Chapter 10, Renal Tumors). The malignancy occurs in the soft tissues, skin, central nervous system, heart, genitourinary tract, and other extrarenal sites [63,96–101]. It most frequently involves the soft tissues of the neck, proximal extremities, trunk, and retroperitoneal area [101]. Familial cases are reported in siblings.

Rhabdoid tumor may present as one or more cutaneous nodules prior to the discovery of the primary tumor [100,101]. Moreover, it may manifest in infants as widely disseminated disease without a definite renal or cerebral primary and prove to be rapidly fatal [99]. The tumor metastasizes to the placenta, skeleton, kidney, brain, and liver. Cytogenetic analysis shows deletions of chromosome 22q11 in most cases, confirming its relationship to rhabdoid tumors of the kidney and brain [99].

Histologically the rhabdoid tumors of soft tissue are similar to those of the kidney and brain. They consist of polygonal cells with large vesicular nuclei typically having one large round nucleolus. Eosinophilic staining cytoplasms contain inclusions composed of intermediate filaments, which are observed by electron microscopy and by immunohistochemistry [97,98,101] (see Chapter 10, Renal Tumors). The tumor cells are immunoreactive for both vimentin and keratin but are variably reactive for desmin, actin, neuron-specific enolase, and Mic2 [99]. Eleven cases studied in one series showed a diploid DNA content [96]. The histogenesis of this neoplasm is unknown [99]. The prognosis for rhabdoid tumor is dismal [97,100,101].

Adipose Tissue Tumors

Tumors of fatty tissue origin are not as common in infants and children as in adults, in whom they comprise most of the soft tissue neoplasms [15,46]. There are two main types of adipose tissue: white and brown fat [15]. Generally white fat is located in the subcutaneous tissue, abdomen, retroperitoneum, and mediastinum. The back, neck, mediastinum, axilla, and retroperitoneum are the principal locations for brown fat, which is found more commonly in infants and children [15] (Figure 3.39).

Benign adipose tissue tumors are classified into several categories [15]: lipoma (single, multiple, superficial, and deep); variants of lipoma, namely, lipoblastoma and angiomyolipoma; heterotopic lipoma (e.g., intra- and intermuscular lipoma, neural fibrolipoma); infiltrating proliferations of ma-

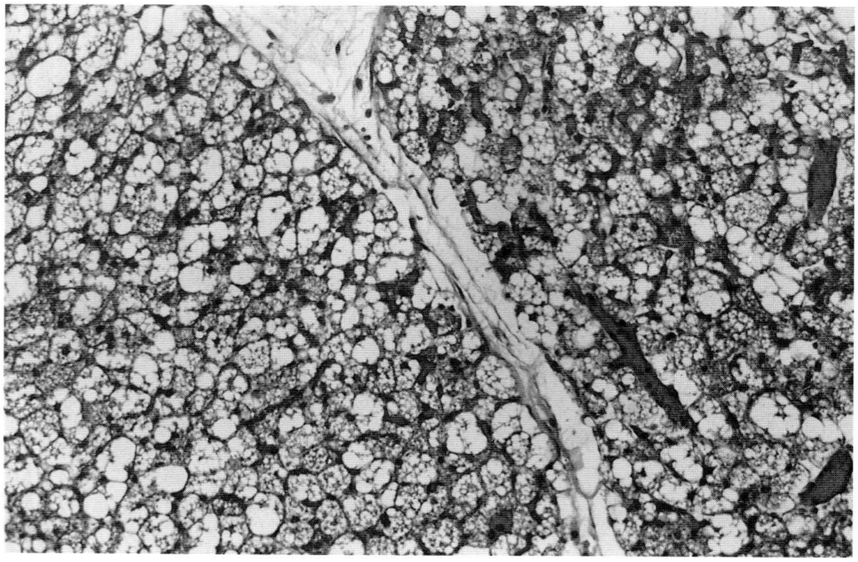

FIGURE 3.39. Brown fat. The specimen was removed incidentally from a 7-week-old male as part of the thymus during a thoracotomy for correction of a congenital heart defect. The adipocytes are arranged in lobules and are composed of finely vacuolated or granular cytoplasms. Nuclei are small and round and tend to be centrally located. Transitions between brown and clear white fat cells are evident. H&E; × 300.

ture fat cells (lipomatosis); and hibernoma, a benign neoplasm of brown fat [15]. Neural fibrolipomas are present not infrequently in babies as a component of their spinal dysraphic malformation (myelomeningocele). They may be simple lipomas, similar to those developing elsewhere in the body, or they may be more complex, including in addition to the lipomatous component a variety of unusual ectopic tissues of ectodermal, mesodermal, and/or endodermal origin. Liposarcoma in infants is the subject of an isolated case report [15,46].

Of a total of 102 soft tissue tumors occurring in infants, there were 23 lipomas and 2 lipoblastomas. The extremities with 7 and the spermatic cord with 14 were the main locations of the lipomas, and the shoulder and back were the sites of the lipoblastomas [12]. One of the largest series of lipoblastomas, consisting of 35 cases, came from the Armed Forces Institute of Pathology; the tumor most often involved the soft tissues of the lower and upper extremities [63]. Lipoblastoma was defined as the circumscribed form and lipoblastomatosis as the diffuse form of this tumor [63]. The diffuse form usually arises in deep soft tissue and has an infiltrative growth pattern and an increased tendency to recur. The neoplasm may compress or impair function of adjacent structures because of its size and location. Lipoblastoma occurs in the newborn [63,102,103].

Lipoma

Both the clinical features and the pathological appearance of a lipoma in an infant is essentially the same as in older children and adults [12,15,63]. Most present as a slowly growing, painless, soft subcutaneous mass (Figure 3.40). Less than a third are noted at birth.

Lipoblastoma

Lipoblastoma is a benign tumor related to embryonic white fat having the property for differentiation over a period of time [46,102]. It tends to grow slightly more rapidly than a lipoma and has a much firmer consistency. On cross section it is paler and more myxoid (gray) than a typical lipoma. These tumors range in size from 2 cm to over 14 cm in greatest diameter [14,63,103]. Microscopic examination shows immature fat cells with varying degrees of differentiation separated by connective tissue septa and loose myxoid areas. The presence of lipoblasts with a bubbly, vacuolated cytoplasm is requisite for the diagnosis. Mitoses are not observed. The underlying skeletal muscle may be involved. Over half the tumors in addition consist of mature fat cells and thus in some areas have the appearance of a lipoma (Figure 3.41). Lipid vacuoles stain positively with oil red O. Immunoperoxidase studies performed on lipoblastoma reveals that the tumor cells are S-100 positive and nonreactive with vimentin, actin, Leu-7, and factor VIII [102]. The ultrastructural findings of lipoblastoma resemble developing white fat and show lipoblasts in various stages of maturation [46,102].

Lipoblastomas and lipoma are cured by complete excision of the mass, and patients with these tumors usually have a favorable outcome [14,15,46,63,103]. Chung and Enzinger quoted a low recurrence rate of 14%, which they attributed to incomplete removal [63]. Sequential biopsies have demon-

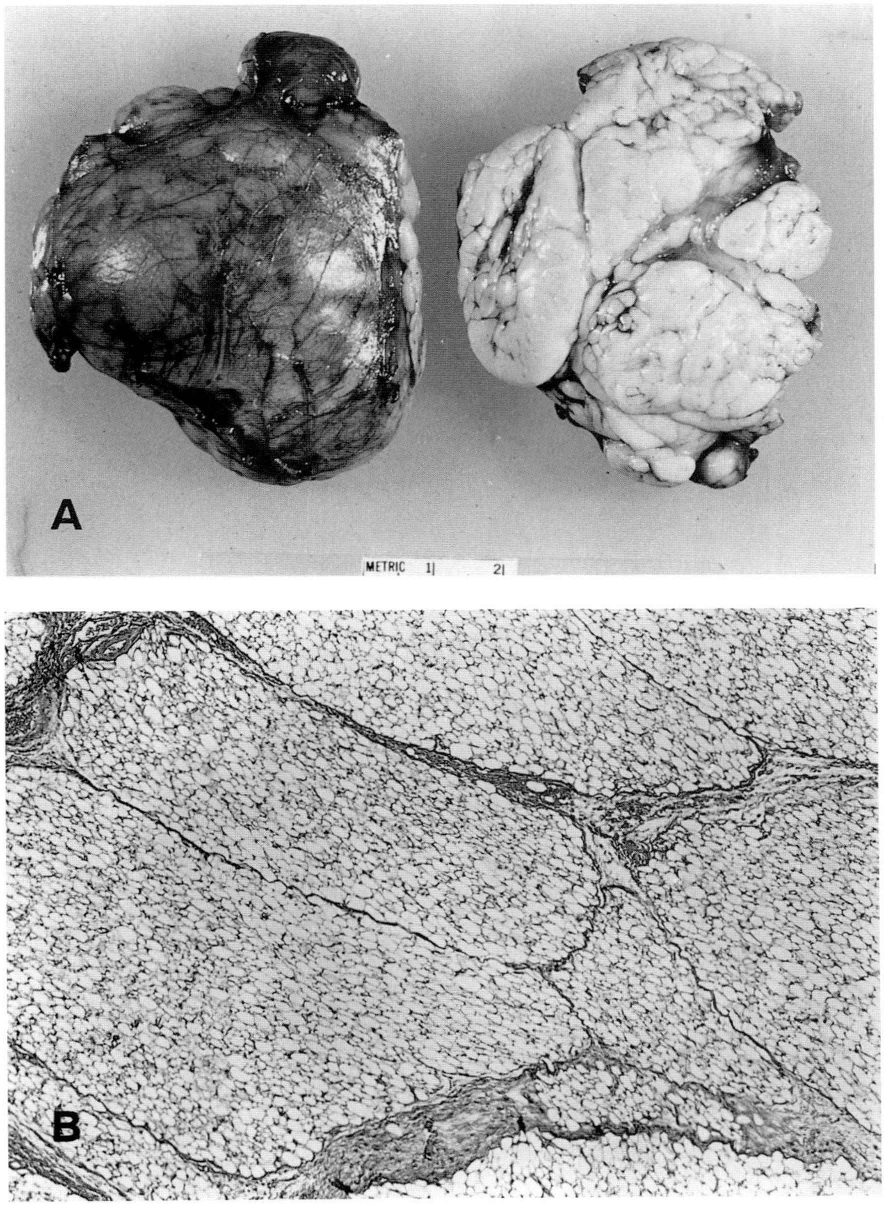

FIGURE 3.40. Lipoma. (A) A 6 × 5.5 cm^2 mass present for several months was removed from the right side of the neck of 6-month-old male. The cut surface demonstrates the typical lobular pattern seen in lipomas. (B) Low-power (× 48) view shows the lobular pattern and the uniform, mature fat cells. H&E.

strated that lipoblastomas tend to spontaneously mature to lipomas, which probably accounts for their good prognosis despite an inadequate excision [14,46,63,103].

Peripheral Nervous System Tumors

The peripheral nervous system tumors, which are considered to originate in the neural crest, include a wide variety of benign and malignant conditions [13,15,46,104–109]. Not all members of this category occur in the fetus and infant. However, the mundane neurofibroma and the more exotic entities primitive neuroectodermal tumor (PNET), melanotic PNET, ectomesenchymoma, medulloepithelioma, and malignant "triton" tumors are reported in the first year of life [106–110]. Schwannoma (neurilemoma) and malignant schwannoma (malignant peripheral nerve sheath tumor or "neurofibrosarcoma") are uncommon before adolescence [15,46,110]. However a few examples of congenital malignant peripheral nerve sheath tumors are mentioned [110].

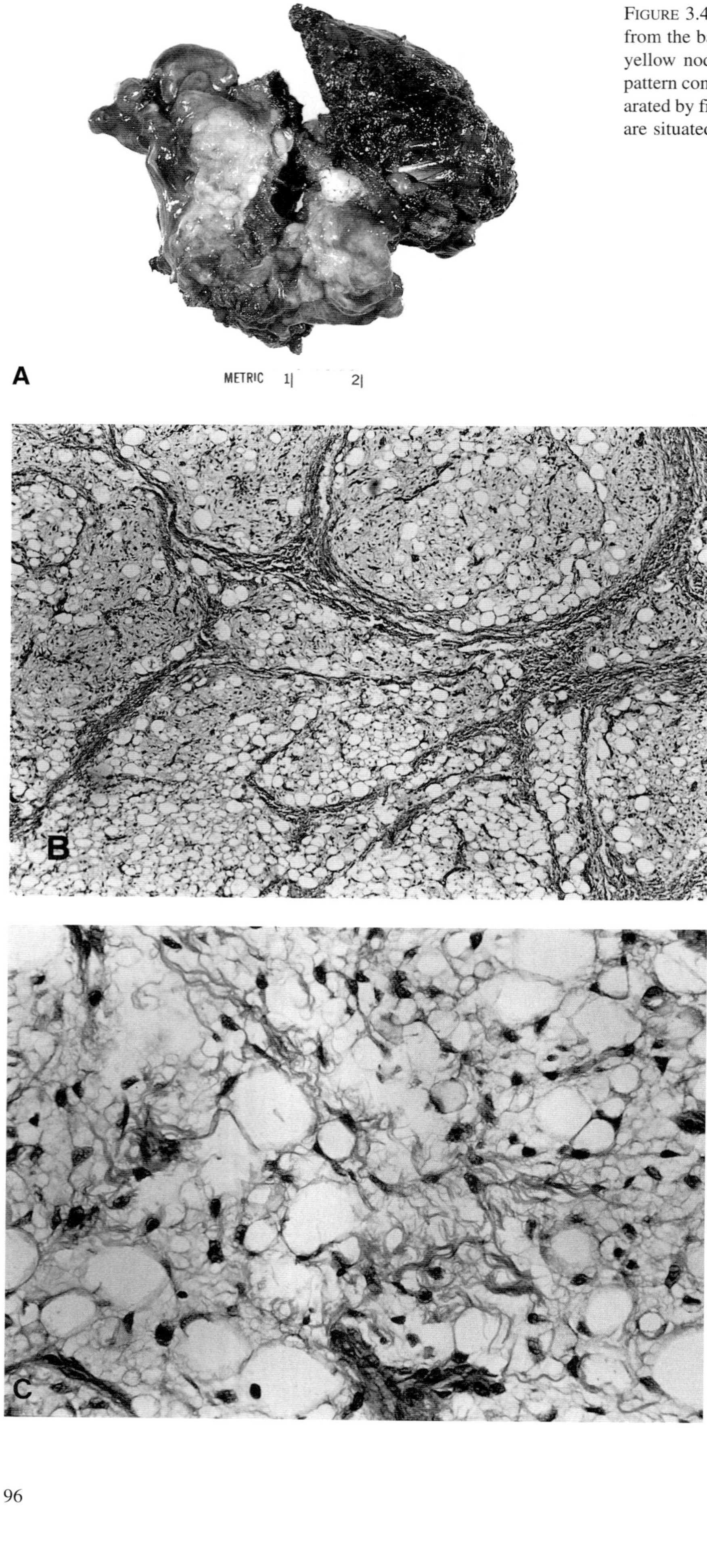

FIGURE 3.41. Lipoblastoma. (A) A 4 × 2 cm^2 mass removed from the back of a 1-year-old female. The tumor has a tan to yellow nodular cut surface. (B) The tumor shows a lobular pattern consisting of immature fat cells and myxoid areas separated by fibrous septa. H&E; × 120. (C) Vaculated lipoblasts are situated in myxoid areas. H&E; × 600.

Neurofibroma

Neurofibroma occurs either as a single dermal or soft tissue mass or as multiple tumors in association with café-au-lait spots and neurofibromatosis type I, an autosomal dominant disorder (Von Recklinghausen disease). Some neurofibromas are evident at birth, while others are not observed until later on in childhood, adolescence, or adulthood. They occur in any area in the body where nerve fibers are found [14]. Occasionally in the newborn neurofibromas present clinically in extraordinary ways [107,111].

Neurofibromas are grouped into three main histological types: solitary, plexiform, and diffuse [15]. Microscopically the solitary type consists of neurites, Schwann cells, and collagen bundles surrounded by a pale-staining myxoid stroma (Figure 3.42). The plexiform neurofibroma, the hallmark of neurofibromatosis type I, resembles a bunch of pale white worms and microscopically shows expansion of nerve fibers by irregular, tortuous, haphazardly arranged proliferations of Schwann cells, neurites, and collagen fibers with a pale-staining stroma (Figure 3.43). The diffuse form occurs in children with or without neurofibromatosis and in association with congenital melanocytic nevi with neural differentiation (see Chapter 4, Tumors and Tumorlike Conditions of the Skin). The diffuse neurofibroma subtly infiltrates the lower dermis and subcutaneous tissue as firm, homogeneous white

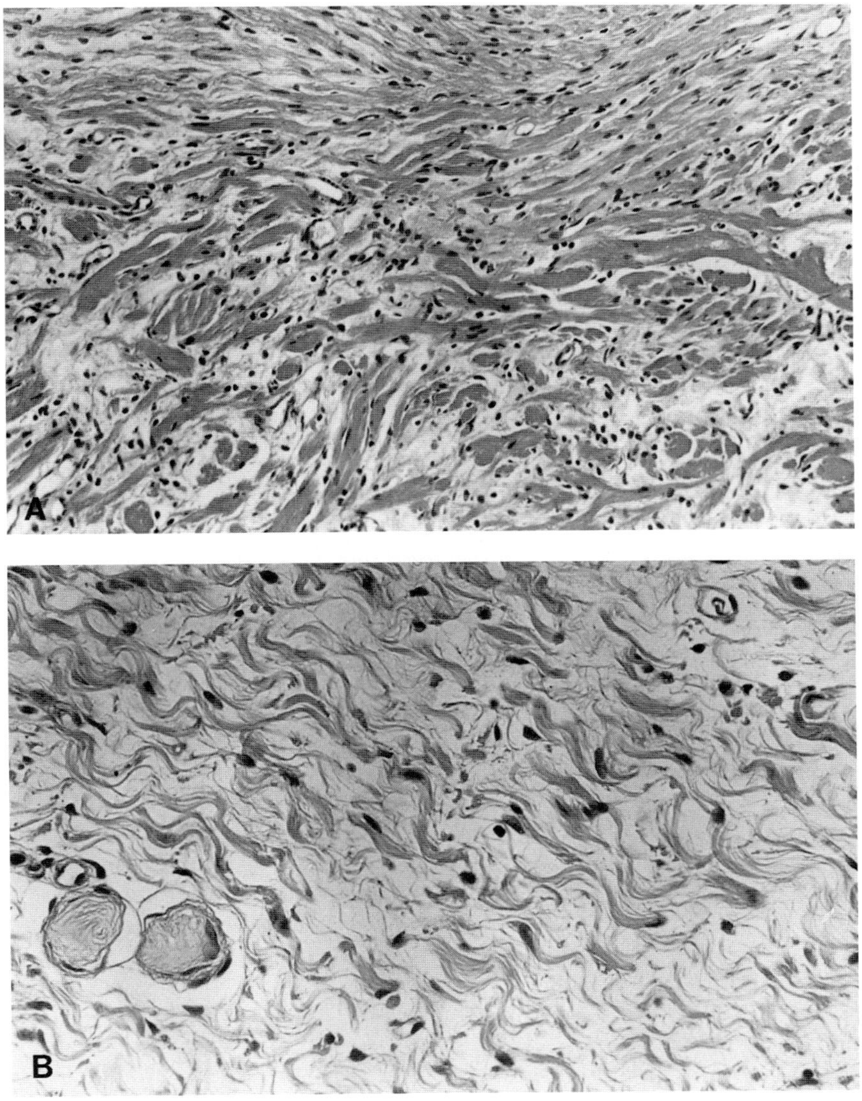

FIGURE 3.42. Neurofibroma in a 1-year-old male with a mass in the posterior fossa arising from the dura. (A) The tumor is composed of interlacing bundles of elongated cells with wavy, darkly staining nuclei. Wirelike strands of collagen and pale-staining mucoid material are present. H&E; ×150. (B) Another area of the tumor displays irregular, thick bundles of collagen and scattered Schwann cells separated by pale mucoid matrix. Structures consistent with Pacinian corpuscles are present in the lower left corner of the photograph. H&E; × 300.

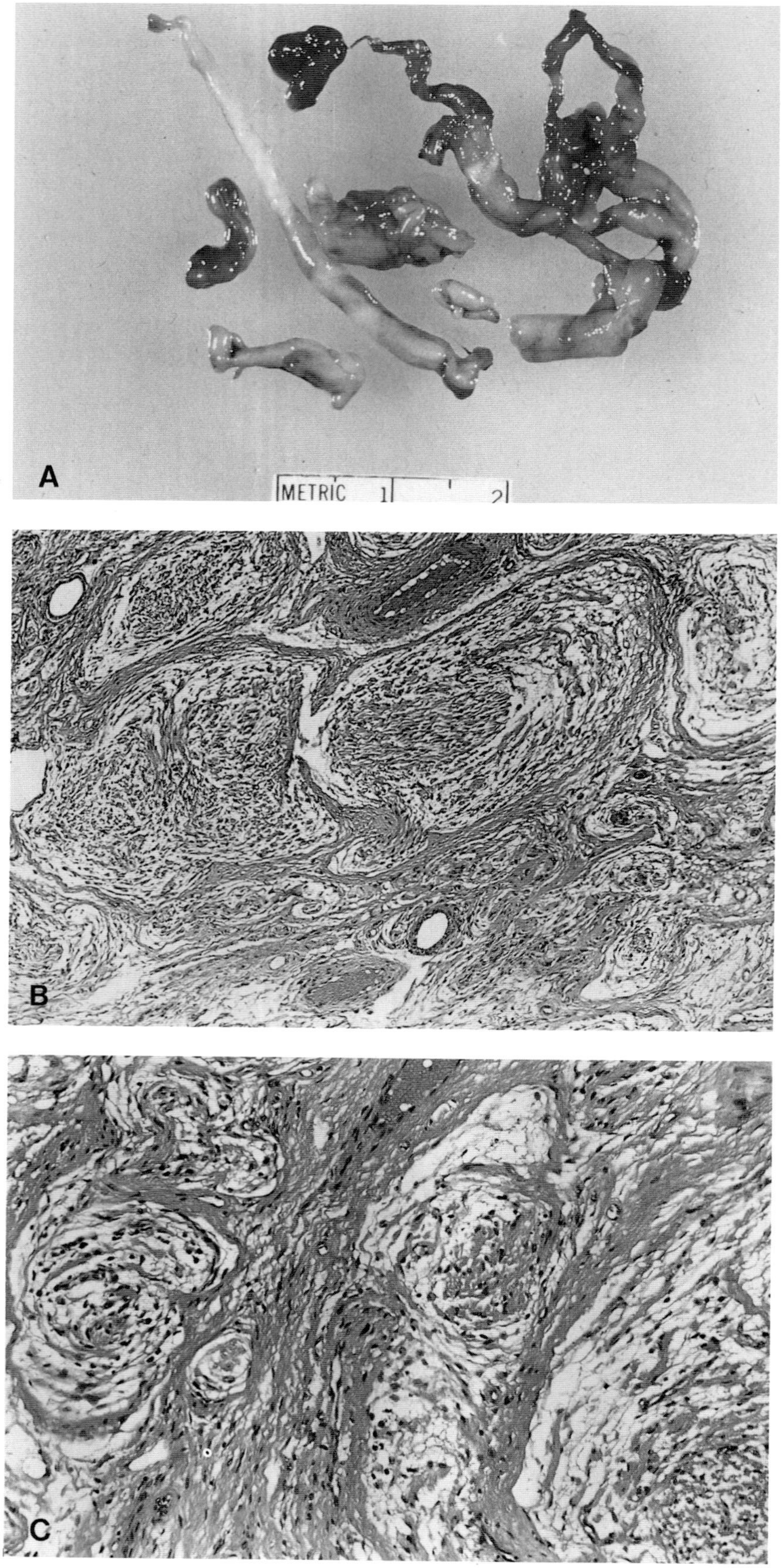

FIGURE 3.43. Plexiform neurofibroma in a 1-year-old male with neurofibromatosis I and a disfiguring facial mass present since birth. (A) The excised specimen consists of several, elongated, white and shiny worm like pieces of tissue. (B) Tumor is composed of enlarged, myxoid appearing nerve trunks with proliferations of neurites and Schwann cells. H&E; × 150. (C) Higher power view (× 300) showing the expanded nerve trunks with neurites and Schwann cells situated in a pale-staining stroma. The facial area is one of the most common sites for neurofibroma in young children. H&E.

tissue and histologically consists of small round to oval Schwann cells, wavy neurites, and a pale-staining connective tissue stroma [15]. Neurofibromas are immunoreactive with S-100, Leu-7, and myelin basic protein [46].

Skeletal malformations are noted in less than half of patients with neurofibromatosis type I [13,15]. Spinal deformities including vertebral dysplasia and kyphoscoliosis are the most frequent bone abnormalities, followed by congenital tibial pseudoarthrosis.

Several other neoplasms, both benign and malignant, have been described in association with neurofibromatosis. These include ganglioneuroma, pheochromocytoma, schwannoma, neurofibrosarcoma, rhabdomyosarcoma, Wilms tumor, and leukemia [13,15,46,107].

Primitive Neuroectodermal Tumor

The primitive neuroectodermal tumors are a group of small-cell malignant neoplasms arising from the central and peripheral nervous systems and soft tissues [15,46,104–109]. The PNET category contains a wide variety of neoplasms including medulloblastoma, retinoblastoma, neuroblastoma, Askin thoracopulmonary tumor, Ewing sarcoma, melanotic PNET of the jaw, peripheral neuroepithelioma of the soft tissues, and conceivably the neuroepithelial components found in immature teratomas [14,104,112].

Studies show that Ewing sarcoma (both osseous and extraosseous), PNET of bone, small-cell malignant tumor of the thoracopulmonary region (Askin tumor), and peripheral neuroepithelioma are probably histogenetically related. Neuronal differentiation (i.e., the presence of neurosecretory granules and neurofilaments) is observed by ultrastructural studies on these tumors. PAS stains are often positive. Immunohistochemical features (i.e., reactivity with vimentin, neuron-specific enolase, HBA71, and S-100 protein) are similar; and cytogenetic findings, namely, a t(11;22) (q24;q11–12) chromosomal translocation, are identical [14,15,46,112,113].

The chest wall, extremities, and face are the main primary sites [113–116]. PNETs show a variable gross appearance consisting of soft, friable, white to pale tan-gray tissue fragments with mucoid and cystic areas, foci of necrosis, and hemorrhage. Histologically they are small blue-cell tumors forming nests and lobules. Tumor cells have round or oval nuclei, coarse or clumped chromatin, small nucleoli, and indistinct or scant cytoplasms. Many mitoses are present (see Chapter 16, Tumors of the Lung). Homer-Wright rosettes similar to those observed in neuroblastomas are noted in some tumors [46,112,113]. The prognosis is poor. Tumors arising from all sites characteristically show early development of widespread metastases and a rapidly fatal clinical course.

Melanotic Neuroectodermal Tumor of Infancy

Several names have been applied to the peculiar neoplasm known as melanotic neuroectodermal tumor of infancy, including melanotic progonoma, retinal anlage tumor, melanotic adamantinoma, pigmented neuroectodermal tumor, and pigmented congenital epulis [15,117,118]. A neural crest origin has been proposed for this tumor; however its relationship to other small-cell tumors with neuroectodermal features such as neuroblastoma, Ewing tumor, and primitive neuroectodermal tumor remains unclear [6,116,119,120]. Studies suggest that melanotic neuroectodermal tumor is a dysembryogenetic neoplasm that recapitulates early embryonic retinal development [118,119]. Moreover the tumor cells are similar to those derived from retinal pigmented epithelium, and the pigment is essentially the same as that seen in the normal retina [118].

The melanotic neuroectodermal tumor is found at birth or during the first year of life, presenting as a rapidly growing mass most commonly in the anterior maxilla and less often in the mandible, brain, fontanelle, skull, and orbit [13,117]. Sometimes it occurs in the oropharynx or outside the head in the soft tissues and epididymis [118,121,122]. Levels of serum α-fetoprotein, and urinary vanillylmandelic acid (VMA) levels are inconsistently elevated in patients with this tumor [46,117,121].

The tumor ranges in color from gray to black, depending on the amount of melanin pigment present (Figure 3.44). One or more deciduous teeth may be found as part of the specimen when the tumor involves the alveolar process of either the maxilla or mandible [14].

Histologically the melanotic neuroectodermal tumor consists of two types of cell situated in irregular microcystic spaces: one a small, dark, round cell resembling a neuroblast and the other a slightly larger, paler cell containing melanin pigment (Figure 3.44). The tumor insiduously infiltrates bone and soft tissues and provokes an exuberant fibrotic and/or gliotic response depending on location. Ultrastructural studies reveal that the small dark round cells represent neuroblasts, since they contain neurosecretory granules and neurofilaments, and the larger paler cells melanocytes, since melanosomes are found in their cytoplasm [117,119].

Immunohistochemical studies reveal neural, melanocytic, and epithelial elements [46]. The small neuroblastic cells and the larger melanocytic cells are reactive with neuron-specific enolase, synaptophysin, HMB45, and dopamine β-hydroxylase. The larger melanocytic cells are immunoreactive with vimentin and cytokeratin [46,119,120,121,122]. Aneuploid DNA appears to be correlated with recurrence [119]. Flow cytometric studies are inconclusive and the usefulness of this procedure in predicting recurrence or metastases is uncertain [121,122].

The prognosis of melanotic neuroectodermal tumor is relatively good providing that it has been *completely* resected [6,14,119,121,122]. Currently, there is no effective adjuvant therapy for this tumor; radiation is precluded by the patients' young age, and chemotherapy trials have not shown long-term efficacy [121,122]. The recurrence rate is approximately 15%, and an overt malignant behavior is noted in about 5% of cases [6,117]. At present there are no reliable clinical or histological findings on which to predict outcome [46].

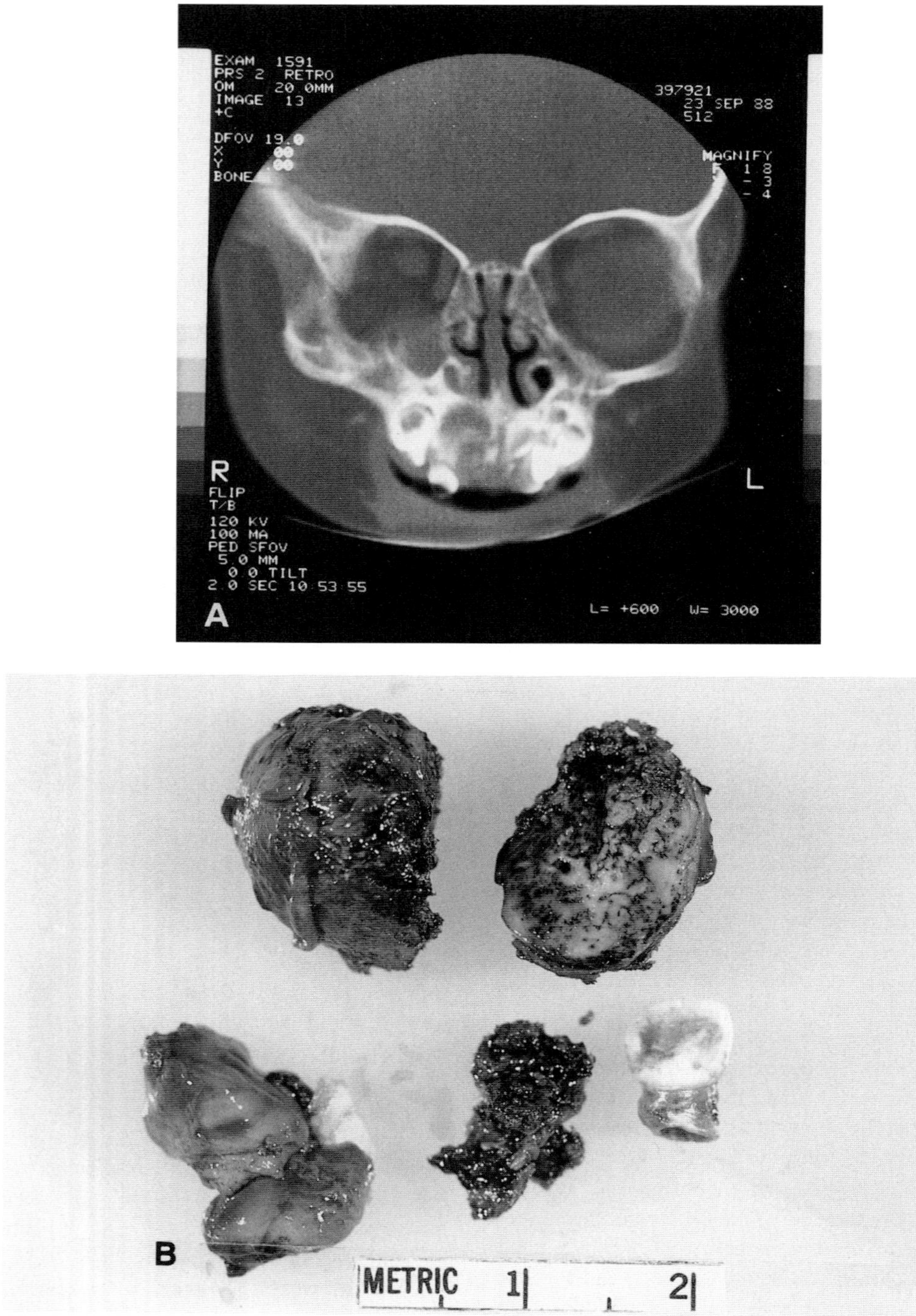

FIGURE 3.44. Pigmented (melanotic) primitive neuroectodermal tumor. (A) Magnetic resonance image of a 6-month-old female with a history of a mass in the hard palate for 2 months' duration. The imaging study shows a lytic lesion in the right maxilla and hard palate and erosion of the wall of the inferior orbit with displacement of teeth. (B) Bone fragments contain tumor that is tan, firm with irregular black pigmented areas. A tooth is part of the specimen. (C) Low-power view (× 120) showing irregular cystic structures containing nests of small darkly staining cells (neuroblasts) and larger pigmented melanocytes lining the microcysts. H&E. (D) Higher magnification (× 480) demonstrating the neuroblasts and melanocytes in more detail. H&E. (E) Neuroblastoma component consisting of a lobular pattern, collections of small, darkly staining cells and fine fibrillar material. H&E; × 120.

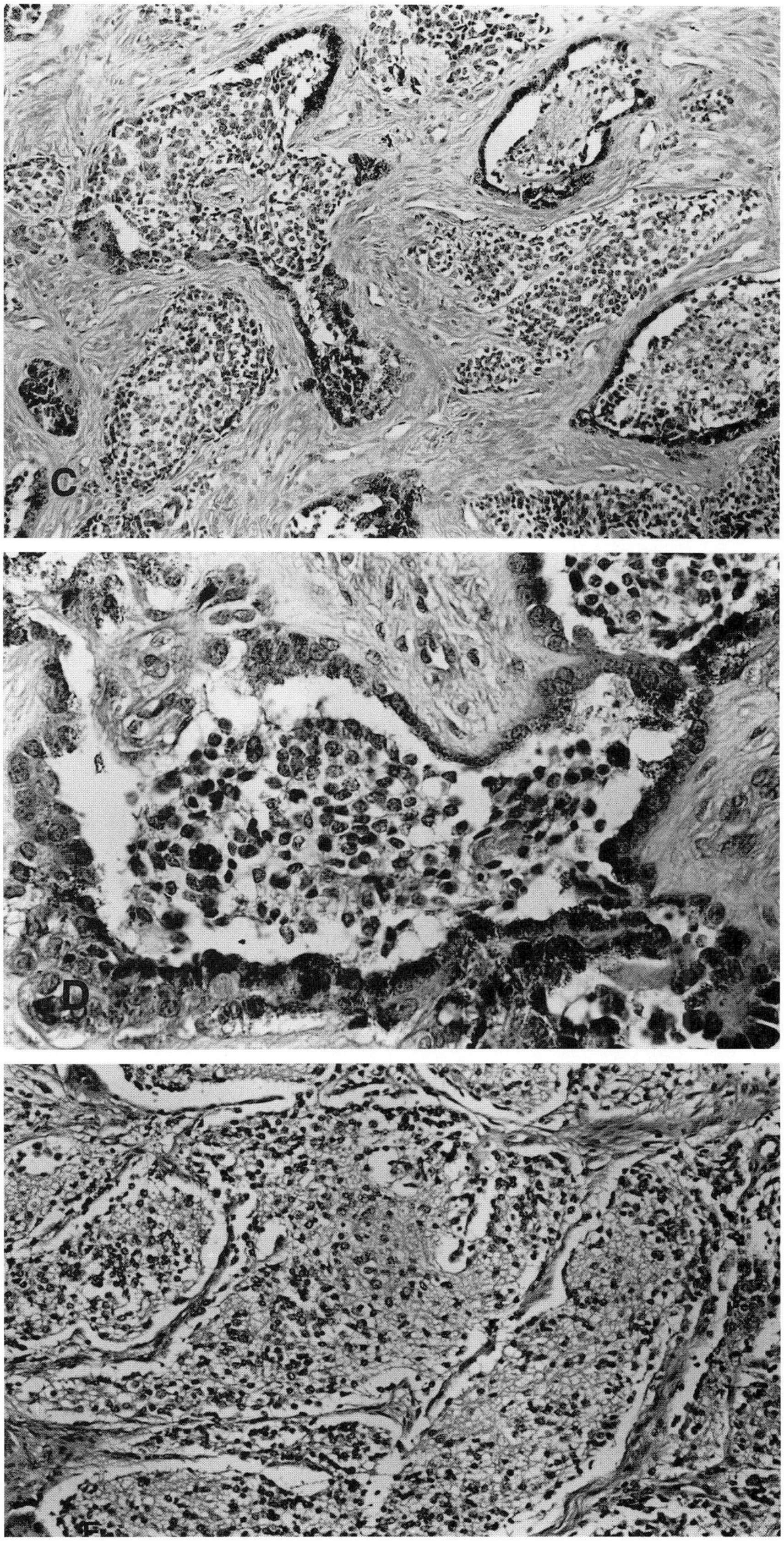

FIGURE 3.44. (*Continued*)

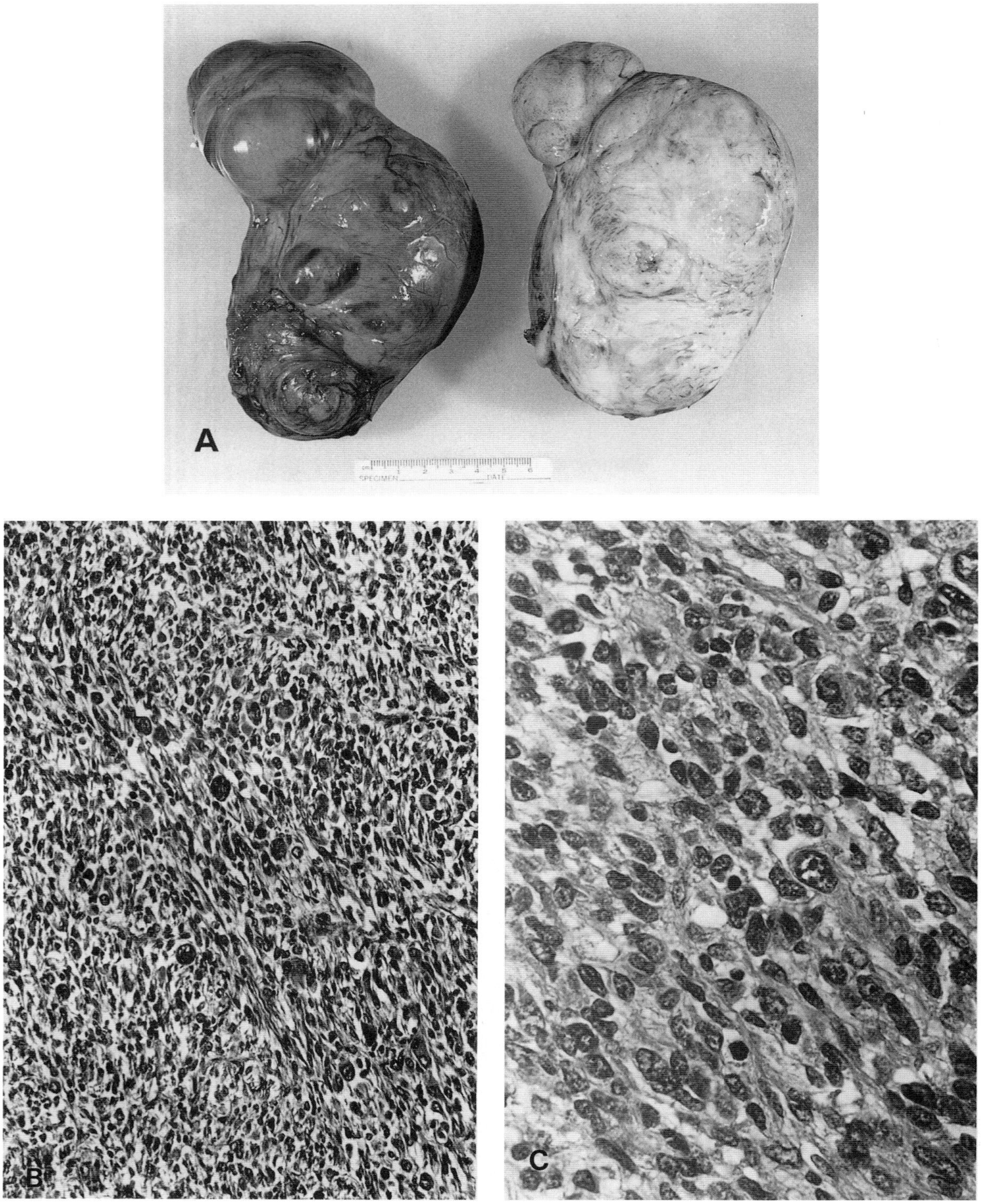

FIGURE 3.45. Ectomesenchymoma in a 6-month-old male, whose parents had noted an abdominal mass 2 months prior to hospitalization. (A) The tumor, 940 g, 18 × 11 cm^2, attached to the internal inguinal ring, has a white, bulging, whorled, lobular-appearing cut surface. (B) The tumor consists of spindle-shaped rhabdomyoblasts and larger, rounder ganglion cells. Trichrome; × 150. (C) Higher magnification (× 300) of the rhabdomyoblasts and ganglion cells. H&E.

Ectomesenchymoma (Gangliorhabdomyosarcoma)

Ectomesenchymoma is a neoplasm thought to be derived from remnants of migratory pluripotential neural crest cells, the so-called ectomesenchyme, which consists of both neuroectodermal and mesenchymal elements [106,108,123,124]. The former is composed of ganglioneuroma and/or neuroblastoma and Schwann cells, the latter of rhabdomyosarcoma [106]. Because of this biphasic pattern it has also been termed gangliorhabdomyosarcoma [108]. Ectomesenchymomas occur mainly in infants, comprising to 77% of the cases reported [123]. They occur in the head and neck but most have been intra-abdominal, involving the genitourinary system, perineum, retroperitoneum, and abdominal wall [13,106,110,123,124] (Figure 3.45). There are no specific cytogenetic findings [123].

The tumor is multilobate, often encapsulated with a pale tan to gray, bulging cut surface (Figure 3.45). Foci of necrosis and hemorrhage are present. The tumors vary in size from 3 to 18 cm in greatest dimension [123]. Microscopically the rhabdomyosarcoma component may be embryonal and/or alveolar type [123]. Over half the tumors contain mature ganglion cells with or without neuroblasts.

Immunohistochemical studies show that the rhabdomyoblasts are reactive with desmin- and muscle-specific actin antibodies and variably positive for myoglobin; the Schwann cells are reactive with S-100 protein; the ganglion cells and neuroblasts are reactive with neuron-specific enolase and synaptophysin antibodies [78,106,123]. Ultrastructural studies confirm the biphasic immunohistochemical findings [78,123].

Patients whose tumors are localized and resectable have the best prognosis, whereas those who have unresectable tumors or metastastases generally have a fatal outcome [78,123].

Polyphenotypic Small-Cell Tumor

Complex phenotypic tumors with neural, rhabdoid, skeletal muscle, and epithelial components occur. They include intra-abdominal small-cell desmoplastic tumors found in older children [124,125], and congenital facial tumors with PNET, glial, and and epithelial elements [115]. The orbit and soft tissues of the forearm are examples of other primary sites [13,126] (Figure 3.46). These small blue-cell tumors are immunoreactive with vimentin and cytokeratin, in addition to neuron-specific enolase and muscle-specific actin. With few exceptions most patients die from their malignancy within two years of diagnosis.

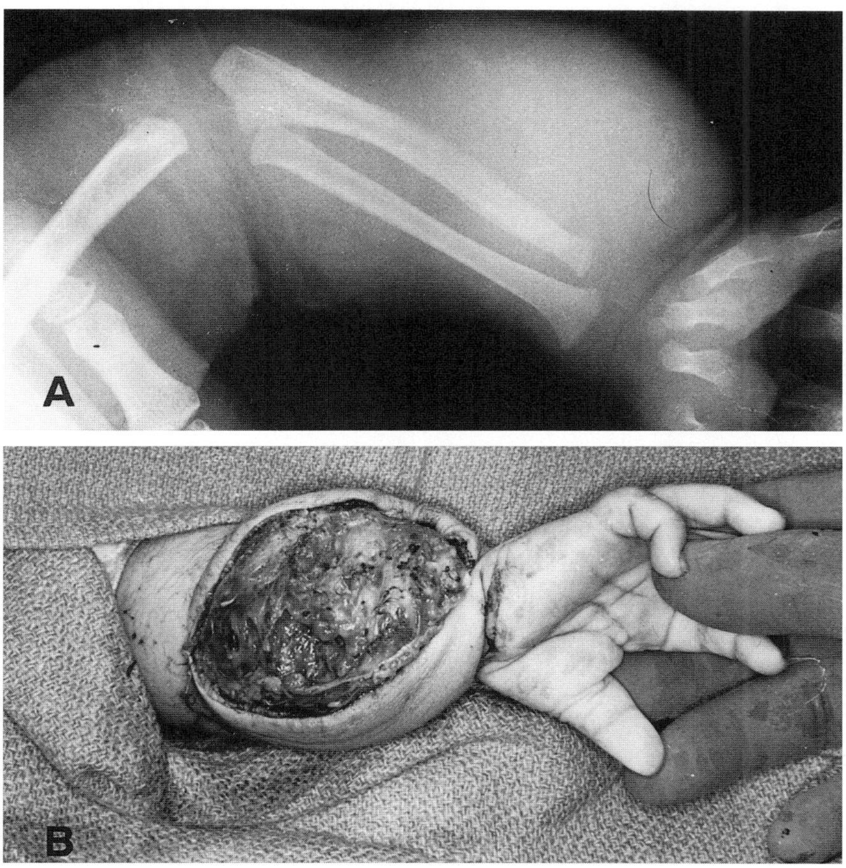

FIGURE 3.46. Polyphenotypic small-cell tumor in a 2-month-old male with a rapidly growing neoplasm of the forearm present since birth. (A) Radiogram depicting a large soft tissue mass. (B) Tumor is exposed in situ at surgery. (C) Grossly the tumor has a soft gelatinous appearance. (D) Tumor forms nests and cords of primitive-appearing, epithelial-like cells. Immunoperoxidase staining was weakly positive for neuron-specific enolase, vimentin, cytokeratin and actin. H&E; × 300. *(Continued)*

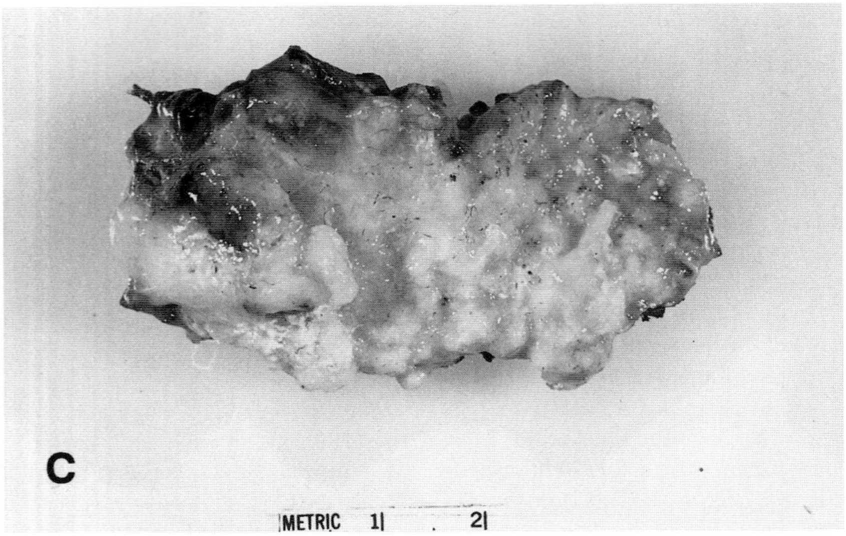

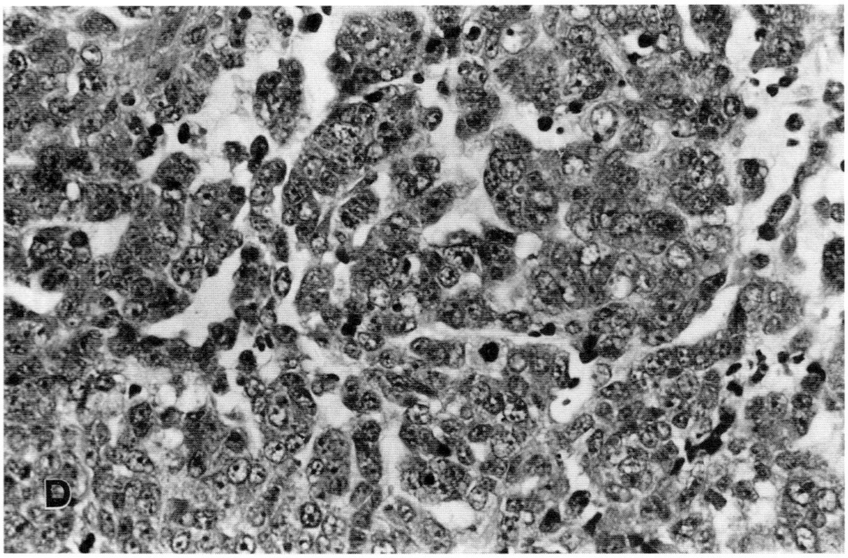

FIGURE 3.46. (*Continued*)

Triton Tumor

The triton tumor is another neoplasm with more than one line of differentiation. It is characterized by the coexistence of both nerve sheath tumor and skeletal muscle components [15,109, 127]. Both benign and malignant forms are described. Neurofibromatosis I is common in patients with this tumor [15,109]. Two fatal examples of congenital malignant triton tumor (malignant peripheral nerve sheath tumor with rhabdomyoblastic differentiation) occurring in the retroperitoneal area are described [109].

The diagnosis of malignant triton tumor depends on identifying both malignant peripheral nerve sheath tumor (malignant schwannoma) and rhabdomyosarcoma elements [15,109, 127]. The former consist of spindle- or comma-shaped cells with a scanty cytoplasm and elongated nuclei, which form alternating bundles, Antoni type A, and microcystic type B tissue, which are reactive with S-100 protein antibody. The skeletal muscle component has embryonal and/or alveolar rhabdomyoblasts immunoreactive with myoglobin, desmin, and actin. Electron microscopy reveals the presence of rhabdomyoblasts with basal lamina and thick and thin filaments with Z-band material [109,127]. Neither the site nor the age of the child apparently affects the outcome, for the prognosis for patients with malignant triton tumor is dismal [109,127]. Recurrence and metastases are the rule.

Proliferative Myositis and Fasciitis

Proliferative myositis and fasciitis are regarded as nonneoplastic, reparative lesions infrequently encountered in the newborn and infant; however, they should be considered in the differential diagnosis of a soft tissue mass [14]. Histologically they are characterized by marked cellularity, a brisk mitotic rate, nuclear atypia, and a myxoid appearance. The lesions are derived from myofibroblasts and histiocytes and are more common in adults and older children [14,15,46,128,129]. Mistakenly they have been called sarcomas because of their ominous microscopic features [15,46,128,129]. They are found in the neck, chest wall, back, and extremities in the young. The lesions tend to be small, usually less than 5 cm in diameter, and not well circumscribed [128,129]. Microscopically they are composed of fibroblasts and giant cells with vesicular nuclei, prominent nucleoli, and basophilic cytoplasm resembling ganglion cells, which are pathognomonic [14,15,129] (Figure 3.47). The ganglion-like cells stain positively with vimentin and actin and are focally positive with KP1 (CD68 antigen), consistent with myofibroblastic and histiocytic features, respectively [129]. Keratin, desmin, and S-100 immunoperoxidase are unreactive. Electron microscopy shows cells with fibroblastic, myofibroblastic, and histiocytic differentiation. The prognosis is excellent following local excision. Unnecessary aggressive therapy should be avoided [129].

Cranial fasciitis occurs in infants ranging in age from birth to 2 years and sometimes in older children [15,46,130,131] (Figure 3.48). The lesion usually begins as a rapidly growing firm soft tissue mass, which may or may not be painful [130,131]. Imaging studies reveal an irregular, lytic lesion with a sclerotic rim, varying in size from 1.5 to 9 cm, situated most commonly in the temporal and parietal areas of the cranium accompanied by an overlying soft tissue mass [130] (Figure 3.48A). Usually it involves only the outer table of the skull, with preservation of the inner table forming a saucer-like defect seen on tangential views [130,131]. Sometimes the lesion extends through the inner table into the underlying dura [131]. Cranial fasciitis is seldom bilateral [130]. The lesion clinically and on imaging studies may be confused with Langerhans cell histiocytosis [130].

Grossly the fasciitis consists of firm, gray-white, dense fibrous connective tissue; if small spicules of bone are present it may have a gritty sensation on section. Histological examination reveals small islands of cranial bone surrounded by spindle-shaped cells embedded in loose myxoid-appearing connective tissue stroma. Areas of dense fibrosis not unlike that found in proliferative fasciitis are present (Figure 3.48). Low mitotic activity, a storiform growth pattern, and mild chronic inflammation are other findings [131]. Excisional biopsy and curettage are curative but the lesion will recur if not adequately excised. The etiology is unknown.

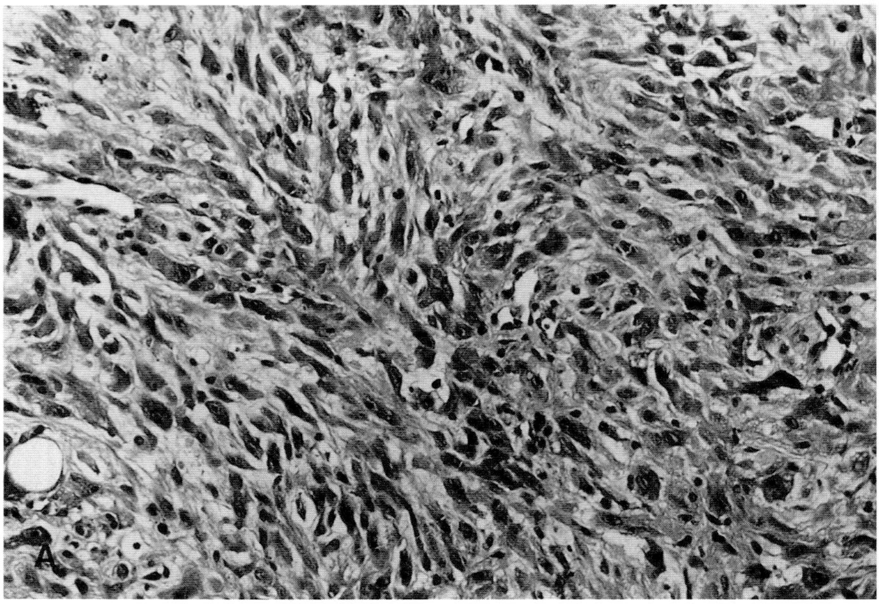

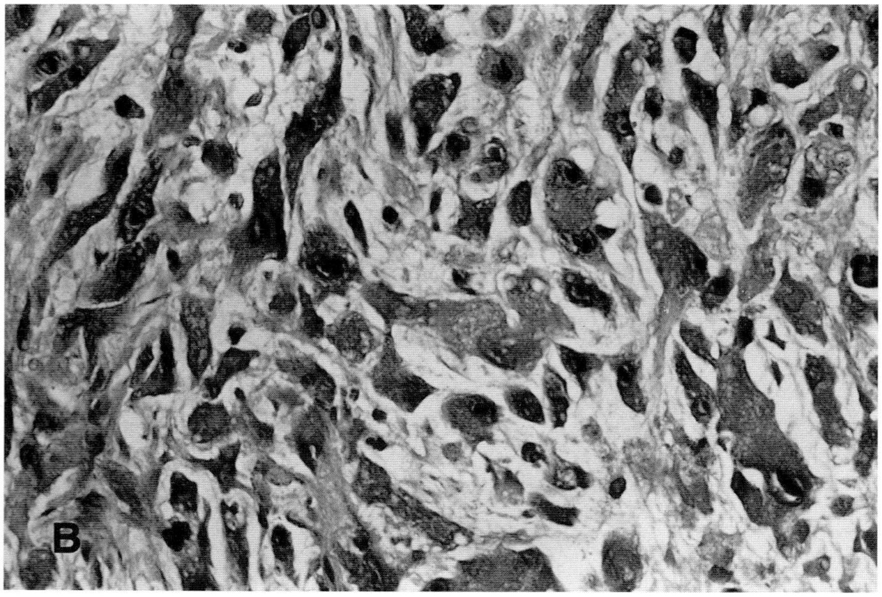

FIGURE 3.47. Proliferative fasciitis. (A) A 3-month-old female with a subcutaneous mass of the lower leg. Biopsy reveals fibroblasts and ganglion-like giant cells with vesicular nuclei, prominent nucleoli, and basophilic cytoplasms. In addition, there are a few scattered lymphocytes. H&E; ×100. (B) Higher magnification (× 600) showing the ganglion-like cells in more detail. H&E. (Courtesy of D. R. Dickson, M.D., Santa Barbara Cottage Hospital, Santa Barbara, CA.)

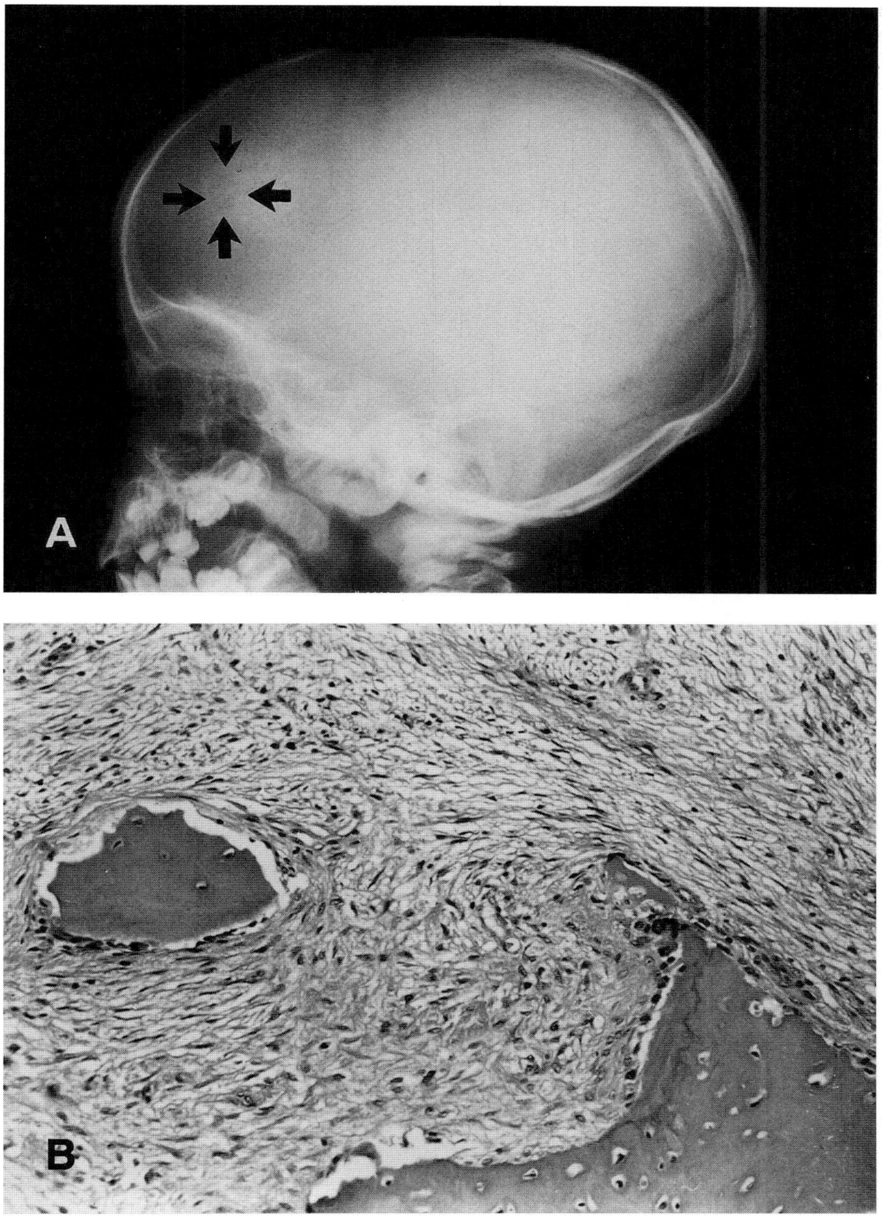

FIGURE 3.48. Cranial fasciitis. (A) Skull x-ray of a 1.5-year-old male with bilateral scalp masses depicts an irregular punched-out lesion (arrows) in the left temporal bone. Langerhans cell histiocytosis, which is the main consideration in the radiological differential diagnosis here, has a similar roentgenographic appearance. Osteomyelitis and metastatic neuroblastoma should be considered also in the differential. (B) Biopsy material showing fibroblastic reaction surrounding bone spicules. H&E; × 300. (From Adler R, Wong CA [130], with permission.)

References

1. Bader JL, Miller RW. US cancer incidence and mortality in the first year of life. *Am J Dis Child* 1979;133:157.
2. Barson AJ. Congenital neoplasia: the Society's experience. Abstracted. *Arch Dis Child* 1978;53:436.
3. Parkes SE, Muir KR, Southern L, et al. Neonatal tumours: a thirty-year population-based study. *Med Pediatr Oncol* 1994;22:309.
4. Berry PJ. Congenital tumours. In: Keeling JW, ed. *Fetal and Neonatal Pathology.* 2nd ed. Berlin: Springer-Verlag; 1993:273.
5. Campbell AN, Chan HSL, O'Brien A, et al. Malignant tumours in the neonate. *Arch Dis Child* 1987;62:19.
6. Dehner LP. Neoplasms of the fetus and neonate. In: Naeye RL, Kissane JM, Kaufman N, ed. *Perinatal Diseases.* Baltimore: Williams & Wilkins; 1981:286. International Academy of Pathology, Monograph No. 22.
7. Dehner LP. Some general considerations about the clinicopathologic aspects of soft tissue tumors. In: Coffin CM, Dehner LP, O'Shea PA, ed. *Pediatric Soft Tissue Tumors: A Clinical, Pathological, and Therapeutic Approach.* Baltimore: Williams & Wilkins; 1997:1.
8. Coffin CM, Dehner LP. Congenital tumors. In: Stocker JT, Dehner LP, eds. *Pediatric Pathology.* Vol I. Philadelphia: JB Lippincott; 1992:325.
9. Coffin CM, Dehner LP. Soft tissue tumors in the first year of life. A report of 190 cases. *Pediatr Pathol* 1990;10:509.
10. Isaacs H Jr. Neoplasms in infants: a report of 265 cases. *Pathol Annu* 1983;18(2):165.
11. Isaacs H Jr. Perinatal (congenital and neonatal) neoplasms: a report of 110 cases. *Pediatr Pathol* 1985;3:165.
12. Isaacs H Jr. *Tumors of the Newborn and Infant. Tumors of Soft Tissues.* St. Louis, MO: CV Mosby, 1991.
13. Isaacs H Jr. Soft tissue tumors. In: *Tumors of the Fetus and Newborn.* Philadelphia: WB Saunders; 1997. Major Problems in Pathology. Series; Vol 35.
14. Isaacs H Jr. Tumors. In: Gilbert-Barness E, ed. *Potter's Pathology of the Fetus and Infant.* Vol 2. St. Louis, MO: CV Mosby; 1997:1242–1339.
15. Enzinger FM, Weiss SW. *Soft Tissue Tumors.* 3rd ed. St. Louis, MO: CV Mosby; 1995.
16. Young JL, Miller RW. Incidence of malignant tumors in US children. *J Pediatr* 1975;86:254.
17. Isaacs H Jr: Congenital and neonatal malignant tumors: a 28-year experience at Children's Hospital of Los Angeles. *Am J Pediatr Hematol/Oncol* 1987;9:121.
18. Batcup G. Cancer in the very young child—pitfalls and problems for the pathologist. *Br J Cancer* 1992;66(suppl XVIII):S5.
19. Mueller BU, Mulliken JB. The infant with a vascular tumor. *Semin Perinatol* 1999;23:332.
20. Byard RW, Burrows PE, Izakawa T, et al. Diffuse infantile haemangiomatosis: clinicopathologic features and management problems in five fatal cases. *Eur J Pediatr* 1991;150:224.
21. Boon LM, Enjolras O, Mulliken JB. Congenital hemangioma: evidence of accelerated involution. *J Pediatr* 1996;128:329.
22. Coffin CM, Dehner LP. Vascular tumors in children and adolescents: a clinicopathologic study of 228 tumors in 222 patients. *Pathol Annu* 1993;28(1):97.
23. Coffin CM. Vascular tumors. In: Coffin CM, Dehner LP, O'Shea PA, eds. *Pediatric Soft Tissue Tumors.* Baltimore: Williams & Wilkins; 1997:40.
24. Kasabach HH, Merritt KK. Capillary hemangioma with extensive purpura: report of a case. *Am J Dis Child* 1940;59:1063.
25. Currie BG, Schell D, Bowring AC. Giant hemangioma of the arm associated with cardiac failure and the Kasabach-Merritt syndrome in a neonate. *J Pediatr Surg* 1991;26:734.
26. Galliani CA, Beatty JF, Grosfeld JL. Cavernous hemangioma of the lung in an infant. *Pediatr Pathol* 1992;12:105.
27. Sarkar M, Mulliken JB, Kozakewich HP, et al: Thrombocytopenic coagulopathy (Kasabach-Merritt phenomenon) is associated with kaposiform hemangioendothelioma and not with common infantile hemangioma. *Plast Reconstructr Surg* 1997;100:1377.
28. Enjolras O, Wassef M, Mazoyer E, et al. Infants with Kasabach-Merritt syndrome do not have true hemangiomas. *J Pediatr* 1997;130:631.
29. Gallagher PG, Mahoney MJ, Gosche JR. Cystic hygroma in the fetus and newborn. *Semin Perinatol* 1999;23:341.
30. Chervenak FA, Isaacson G, Tortora M. A sonographic study of fetal cystic hygromas. *J Clin Ultrasound* 1985;13:317.
31. Langer JC, Fitzgerald PG, DeSa D, et al.Cervical cystic hygroma in the fetus: clinical spectrum and outcome. *J Pediatr Surg* 1990;25:58.
32. Carson HJ, Taxy JB. Role of autopsy in congenital cystic hygroma. *Pediatr Pathol* 1994;14:183.
33. Brown M, Pysher T, Coffin CM. Lymphangioma and congenital pulmonary lymphangiectasis: a histologic, immunohistochemical, and clinico-pathologic comparison. *Mod Pathol* 1999;12:569.
34. Ordonez NG, Mackay B, el-Naggar AK, et al. Congenital hemangiopericytoma. An ultrastructural, immunocytochemical, and flow cytometric study. *Arch Pathol Lab Med* 1993;117:934.
35. Rodriguez-Galindo C, Ramsey K, Jenkins JJ, et al. Hemangiopericytoma in children and infants. *Cancer* 2000;88:198.
36. Jenkins JJ. Congenital malignant hemangiopericytoma. *Pediatr Pathol* 1987;7:119.
37. Bailey PV, Weber TR, Tracy TF Jr, et al. Congenital hemangiopericytoma: an unusual vascular neoplasm of infancy. *Surgery* 1993;114:936.
38. Enzinger FM, Smith MD. Hemangiopericytoma: an analysis of 106 cases. *Hum Pathol* 1976;7:61.
39. van Baarlan J, Drogtrop AP, Bax NMA. Congenital hemangiopericytoma. Report of a case. *Pediatr Pathol* 1988;8:109.
40. Gotte K, Hormann K, Schmoll J, et al. Congenital nasal hemangiopericytoma: intrauterine, intraoperative, and histologic findings. *Ann Otol Rhinol Laryngol* 1999;108:589.
41. Dardick I, Hammar S, Scheithauer BW. Ultrastructural spectrum of hemangiopericytoma: a comparative study of fetal, adult and neoplastic pericytes. *Ultrastruct Pathol* 1989;13:111.
42. Coffin CM, Dehner LP. Fibroblastic-myofibroblastic tumors in children: a clinicopathologic study of 108 examples in 103 patients. *Pediatr Pathol* 1991;11:559.
43. Baerg J, Murphy JJ, Magee F. Fibromatoses: clinical and pathologic features suggestive of recurrence. *J Pediatr Surg* 1999;34:1112.
44. Chung EB, Enzinger F M. Infantile fibrosarcoma. *Cancer* 1976;38:729.
45. Soule EH, Pritchard DJ. Fibrosarcoma in infants and children. A review of 110 cases. *Cancer* 1977;40:1711.

46. Coffin CM. Fibroblastic-myofibroblastic tumors. In Coffin CM, Dehner LP, O'Shea PA, eds. *Pediatric Soft Tissue Tumors. A Clinical, Pathological and Therapeutic Approach.* Baltimore: Williams & Wilkins; 1997:133.
47. Ayala AG, Ro JY, Goepfert H, et al. Desmoid fibromatosis: a clinicopathologic study of 25 children. *Semin Diagn Pathol* 3:138,1986.
48. Rosenberg HS, Stenback WA, Spjut HJ. The fibromatoses of infancy and childhood. *Perspect Pediatr Pathol* 1978;4:269.
49. Mukai M, Torikata C, Hisami I, et al. Immunohistochemical identification of aggregated actin filaments in formalin-fixed, paraffin embedded sections, I: A study of infantile digital fibromatosis by a new pretreatment. *Am J Surg Pathol* 1992;16:110.
50. Davids JR, Wenger DR, Mubarak SJ. Congenital muscular torticollis: sequela of intrauterine or perinatal compartment syndrome. *J Pediatr Orthoped* 1993;13:141.
51. Cheng JCY, Tang SP, Chen TMK. Sternocleidomastoid pseudotumor and congenital muscular torticollis in infants: a prospective study of 510 cases. *J Pediatr* 1999;134:712.
52. Chung EB, Enzinger FM. Infantile myofibromatosis. *Cancer* 1981;48:1807.
53. Wiswell TE, Davis J, Cunningham BE, et al. Infantile myofibromatosis: the most common fibrous tumor of infancy. *J Pediatr Surg* 1988;23:314.
54. Nagase H, Yamaguchi Y, Morinaga H, et al. Multicentric infantile myofibromatosis with spontaneous regression. *Pediatr Surg Int* 1993;8:84.
55. Gonzalez-Crussi F, Noronha R. Solitary fibromatosis in the newborn. Rare cause of neonatal intestinal obstruction. *Arch Pathol Lab Med* 1985;109:97.
56. Sonoda T, Itami S, Seguchi S, et al. Infantile myofibromatosis: report of two cases. *J Dermatol* 1994;21:508.
57. Mentzel T, Calonje E, Nascimento AG, et al. Infantile hemangiopericytoma versus infantile myofibromatosis: study of a series suggesting a continuous spectrum of infantile myofibroblastic lesions. *Am J Surg Pathol* 1994;18:922.
58. Fukasawa Y, Ishikura H, Takada A, et al. Massive apoptosis in infantile myofibromatosis: a putative mechanism of tumor regression. *Am J Pathol* 1994;144:480.
59. Enzinger FM. Fibrous hamartoma of infancy. *Cancer* 1965; 18:241.
60. Dickey GE, Sotelo-Avila C. Fibrous hamartoma of infancy: current review. *Pediatr Dev Pathol* 1999;2:236.
61. Popek EJ, Montgomery EA, Fourcroy JL. Fibrous hamartoma of infancy in the genital region: findings in 15 cases. *J Urol* 1994;152:990.
62. Dehner LP, Enzinger FM, Font RI. Fetal rhabdomyoma. An analysis of nine cases. *Cancer* 1972;30:160.
63. Chung EB, Enzinger FM. Benign lipoblastomatosis. An analysis of 35 cases. *Cancer* 1973;32:482.
64. Lack EE, Worsham GF, Callihan MD, et al. Gingival granular cell tumors of the newborn (congenital "epulis"): a clinical and pathologic study of 21 patients. *Am J Surg Pathol* 1981;5:37.
65. Ordonez NG, Mackay B: Granular cell tumor: a review of the pathology and histogenesis. *Ultrastruct Pathol* 1999;23:207.
66. Gonzalez-Crussi F, Wiederhold MD, Sotelo-Avila C. Congenital fibrosarcoma: presence of a histiocytic component. *Cancer* 1980;46:77.
67. Hays DM, Mirabal VQ, Karlan MS, et al. Fibrosarcomas in infants and children. *J Pediatr Surg* 1970;5:176.
68. Soule EH, Pritchard DJ. Fibrosarcoma in infants and children: a review of 110 cases. *Cancer* 1977;40:1711.
69. Kodet R, Stejskal J, Pilat D, et al. Congenital-infantile fibrosarcoma: a clinicopathological study of five patients entered on the Prague Children's Tumor Registry. *Pathol Res Pract* 1996; 192:845.
70. Knezevich SR, Garnett MJ, Pysher TJ, et al. *ETV6-NTRK3* gene fusions and trisomy 11 establish a histogenetic link between mesoblastic nephroma and congenital fibrosarcoma. *Cancer Res* 1998;58:5046.
71. Hamm CM, Pyesmany A, Resch L. Case report: congenital retroperitoneal fibrosarcoma. *Med Pediatr Oncol* 1997;28:65.
72. de Graaf JH, Timens W, Tamminga RYJ, et al. Deep juvenile xantogranuloma: a lesion related to dermal intermediate cells. *Hum Pathol* 1992;23:905.
73. Zuppan CW, Mierau GW, Campbell HL. Malignant fibrous histiocytoma in childhood: a report of two cases and a review of the literature. *Pediatr Pathol* 1987;7:303.
74. Enzinger FM. Angiomatoid malignant fibrous histiocytoma: a distinct fibrohistiocytic tumor of children and young adults simulating a vascular neoplasm. *Cancer* 1979;44:2147.
75. Cole CH, Magee JF, Gianoulis M, et al. Malignant fibrous histiocytoma in childhood. *Cancer* 1993;71:4077.
76. Costa MJ, Weiss SW. Angiomatoid malignant fibrous histiocytoma: a follow-up study of 108 cases with evaluation of possible histologic predictors of outcome. *Am J Surg Pathol* 1990; 14:1126.
77. Fanburg-Smith JC, Miettinen M. Angiomatoid "malignant" fibrous histiocytoma: a clinicopathologic study of 158 cases and further exploration of the phenotype. *Hum Pathol* 1999;30: 1336.
78. Mouton SCE, Rosenberg HS, Cohen MC, et al. Malignant ectomesenchymoma in childhood. *Pediatr Pathol Lab Med* 1996; 16:607.
79. Janney CG, Hurt MA Santa Cruz DJ. Deep juvenile xanthogranuloma: subcutaneous and intramuscular forms. *Am J Surg Pathol* 1991;15:150.
80. Freyer DR, Kennedy R, Bostrom BC, et al. Juvenile xanthogranuloma: forms of systemic disease and their clinical implications. *J Pediatr* 1996;129:227.
81. Kapadia SB, Meis JM, Frisman DM, et al. Fetal rhabdomyoma of the head and neck: a clinicopathologic and immunophenotypic study of 24 cases. *Hum Pathol* 1993;24:754.
82. Crotty PL, Nakhleh RF, Dehner LP. Juvenile rhabdomyoma: an intermediate form of skeletal muscle tumor in children. *Arch Pathol Lab Med* 1993;117:43.
83. Ruymann FB, Maddux HR, Ragab A, et al. Congenital anomalies associated with rhabdomyosarcoma: an autopsy study of 115 cases. A report from the Intergroup Rhabdomyosarcoma Study Committee (representing the Children's Cancer Study Group, the Pediatric Oncology Group, the United Kingdom Children's Cancer Study Group, and the Pediatric Intergroup Statistical Center). *Med Pediatr Oncol* 1988; 16:33.
84. Li FP, Fraumeni JF Jr. Rhabdomyosarcoma in children: epidemiologic study and identification of a familial cancer syndrome. *J Natl Cancer Inst* 1969;43:1365.

85. Parham DM. The molecular biology of childhood rhabdomyosarcoma. *Semin Diagn Pathol* 1994;11:39.
86. Ragab AH, Heyn R, Teft M, et al. Infants younger than 1 year of age with rhabdomyosarcoma. *Cancer* 1986;58:2606.
87. Qualman SJ, Coffin CM, Newton WA, et al. Intergroup Rhabdomyosarcoma Study: update for pathologists. *Pediatr Dev Pathol* 1998;1:550.
88. Leuschner I, Newton WA Jr, Schmidt D, et al. Spindle cell variants of embryonal rhabdomyosarcoma in the paratesticular region: a report of the Intergroup Rhabdomyosarcoma Study. *Am J Surg Pathol* 1993;17:221.
89. Gulbahce HE, Manivel JC. Congenital spindle cell rhabdomyosarcoma of the urinary bladder. *Pediatr Pathol Mol Med* 1999;18:207.
90. Tsokos M. The diagnosis and classification of childhood rhabdomyosarcoma. *Semin Diagn Pathol* 1994;11:26.
91. Parham DM, Shapiro DN, Downing JR, et al. Solid alveolar rhabdomyosarcomas with the t(2;13). Report of two cases with diagnostic implications. *Am J Surg Pathol* 1994;18:474.
92. Newton WA Jr, Gehan EA, Webber BL, et al. Classification of rhabdomyosarcomas and related sarcomas: pathologic aspects and proposal for a new classification—an Intergroup Rhabdomyosarcoma Study. *Cancer* 1995;76:1073.
93. Lobe TE, Wiener ES, Hays DM, et al. Neonatal rhabdomyosarcoma: the IRS experience. *J Pediatr Surg* 1994;29:1167.
94. Koscieiniak H, Harms D, Schmidt, et al. Soft tissue sarcomas in infants younger than 1 year of age: a report of the German Soft Tissue Sarcoma Study Group (CSW-81). *Med Pediatr Oncol* 1989;17:105.
95. Ragab AH, Heyn R, Teft M, et al. Infants younger than 1 year of age with rhabdomyosarcoma. *Cancer* 1986;58:2606.
96. Weeks DA, Beckwith JB, Mierau GW, et al. Rhabdoid tumor of the kidney: a report of 11 cases from the National Wilms' Tumor Study Pathology Center. *Am J Surg Pathol* 1989;13:439.
97. Parham DM, Weeks DA, Beckwith JB. The clinicopathologic spectrum of putative extrarenal rhabdoid tumors: an analysis of 42 cases studied with immuno-histochemistry and/or electron microscopy. *Am J Surg Pathol* 1994;18:1010.
98. Schmidt D, Leuschner I, Harms D, et al. Malignant rhabdoid tumor: a morphological and flow cytometric study. *Pathol Res Pract* 1989;184:202.
99. White FV, Dehner LP, Biegel JA, et al. Chromosome 22q11 abnormalities in congenital disseminated malignant rhabdoid tumor (MRT) [abstract]. *Mod Pathol* 1998;11:170A.
100. Hsueh C, Kuo T. Congenital malignant rhabdoid tumor presenting as a cutaneous nodule: report of 2 cases with review of the literature. *Arch Pathol Lab Med* 1998;122:1099.
101. Kodet R, Newton WA Jr, Sachs N, et al. Rhabdoid tumors of soft tissues: a clinicopathologic study of 26 cases enrolled in the Intergroup Rhabdomyosarcoma Study. *Hum Pathol* 1991;22:674.
102. Coffin CM. Lipoblastoma: an embryonal tumor of soft tissue related to organogenesis. *Semin Diagn Pathol* 1994;11:98.
103. Collins MH, Chatten J. Lipoblastoma/lipoblastomatosis: a clinicopathologic study of 25 cases. *Am J Surg Pathol* 1997;21:1131.
104. Dehner LP. Peripheral and central primitive neuroectodermal tumors: a nosologic concept seeking a consensus. *Arch Pathol Lab Med* 1986;110:997.
105. Hendrickson MR, Ross JC: Neoplasms arising in congenital giant nevi. Morphologic study of seven cases and review of the literature. *Am J Surg Pathol* 1981;5:109.
106. Karcioglu Z, Someren A, Mathes SJ. Ectomesenchymoma: a malignant tumor of migratory neural crest (ectomesenchyme) remnants showing ganglionic, Schwannian, melanocytic and rhabdomyoblastic differentiation. *Cancer* 1977;39:2486.
107. Qualman SJ, Green WR, Broval C, et al. Neurofibromatosis and associated neuroectodermal tumors: a congenital neurocristopathy. *Pediatr Pathol* 1986;5:65.
108. Kodet R, Kasthuri N, Marsden HB, et al. Gangliorhabdomyosarcoma: a histopathological and immunohistochemical study of three cases. *Histopathology* 1986;10:181.
109. Brooks JSJ, Freeman M, Enterline HT. Malignant "triton" tumors. Natural history and immunochemistry of nine new cases with literature review. *Cancer* 1985;55:2543.
110. Meis JM, Enzinger FM, Martz KL, et al. Malignant peripheral nerve sheath tumors (malignant schwannomas) in children. *Am J Surg Pathol* 1992;16:694.
111. Young S, Crocker D. Congenital midline bilateral ganglioneuromas associated with hydrops fetalis and neurofibromatosis. *Pediatr Pathol* 1993;13:115.
112. Tsokos M. Peripheral primitive neuroectodermal tumors: diagnosis, classification and prognosis. *Perspect Pediatr Pathol* 1992;16:27.
113. Schmidt D, Harms D, Burdach S: Malignant peripheral neuroectodermal tumours of childhood and adolescence. *Virchows Arch [Pathol Anat]* 1985;406:351.
114. Askin FB, Rosai J, Sibley RK, et al. Malignant small cell tumor of the thoracopulmonary region in childhood: a distinctive clinicopathologic entity of uncertain histogenesis. *Cancer* 1979;43:2438.
115. Hachitanda Y, Tsuneyoshi M, Enjoji M, et al. Congenital primitive neuroectodermal tumor with epithelial and glial differentiation: an ultrastructural and immunohistochemical study. *Arch Pathol Lab Med* 1990;114:101.
116. Chan GCF, Nicholls JM, Lee ACW, et al. Malignant peripheral neuroectodermal tumor in an infant with neurofibromatosis type 1. *Med Pediatr Oncol* 1996;26:215.
117. Dehner LP, Sibley RK, Sauk JJ Jr, et al. Malignant melanotic neuroectodermal tumor of infancy: a clinical, pathologic, ultrastructural and tissue culture study. *Cancer* 1979;43:1389.
118. Lamping KA, Albert DM, Lack E, et al. Melanotic neuroectodermal tumor of infancy (retinal anlage tumor). *Ophthalmology* 1985;92:143.
119. Pettinato G, Manivel C, d'Amore SG, et al. Melanotic neuroectodermal tumor of infancy: a reexamination of a histogenetic problem based on immunohistochemical, flow cytometric and ultrastructural study. *Am J Surg Pathol* 1991;15:233.
120. Khoddami M, Squire J, Zielenska M, et al. Melanotic neuroectodermal tumor of infancy: a molecular genetic study. *Pediatr Dev Pathol* 1998;1:295.
121. Ricketts RR, Majmudarr B. Epididymal melanotic neuroectodermal tumor of infancy. *Hum Pathol* 1985;16:416.
122. Kapadia SB, Frisman DM, Hitchcock CL, et al. Melanotic neuroectodermal tumor of infancy: clinicopathological, immunohistochemical and flow cytometric study. *Am J Surg Pathol* 1993;17:566.
123. Boue DR, Parham DM, Webber B, et al. Clinicopathologic study of ectomesenchymomas from Intergroup Rhabdomyo-

sarcoma Study Groups III and IV. *Pediatr Dev Pathol* 2000;3: 290.

124. Gonzalez-Crussi F, Crawford SE, Sun C-CJ. Intra-abdominal desmoplastic small-cell tumors with divergent differentiation: observations on three cases of childhood. *Am J Surg Pathol* 1990;14:633.
125. Ordonez NG, El-Naggar AK, Ro JY, et al. Intra-abdominal desmoplastic small cell tumor: a light microscopic, immunocytochemical, and flow cytometric study. *Hum Pathol* 1993;24:850.
126. Lyon DB, Dortzbach RK, Gilbert-Barness E. Polyphenotypic small-cell orbitocranial tumor. *Arch Ophthalmol* 1993;111: 1402.
127. Daimaru Y, Hashimoto H, Enjoji M. Malignant "triton" tumors. A clinico-pathologic and immunohistochemical study of nine cases. *Hum Pathol* 1984;15:768.
128. Pasquel P, Salazar M, Marvan E. Proliferative myositis in an infant: report of a case with electron microscopic observations. *Pediatr Pathol* 1988;8:545.
129. Meis JM, Enzinger FM. Proliferative fasciitis and myositis of childhood. *Am J Surg Pathol* 1992;16:364.
130. Adler R, Wong CA. Cranial fasciitis simulating histiocytosis. *J Pediatr* 1986;109:85.
131. Lauer D, Enzinger FM. Cranial fasciitis of childhood. *Cancer* 1980;45:401.

4 Tumors and Tumorlike Conditions of the Skin

Tumors and tumorlike conditions of the skin are not uncommon in the fetus and infant and include a wide variety of entities (Table 4.1). The vascular lesions, hemangioma and lymphangioma, are by far the most prevalent, but other conditions such as congenital melanocytic nevi, cutaneous hamartomas, and neoplastic-like infiltrations occur as well [1–5] (Table 4–1).

Malignant neoplasms may appear in the fetus and infant as cutaneous metastases, which may be the initial manifestation of the tumor [4,6,7]. Neuroblastoma and leukemia are foremost examples. Both rhabdomyosarcoma and rhabdoid tumor metastasize to the skin, forming cutaneous nodules [4–7].

Vascular Conditions

Hemangioma and lymphangioma are the main vascular lesions involving the skin of fetuses and infants. Hemangiomas are noted much more frequently than lymphangiomas and are operated upon less often than the latter, presumably because most regress or disappear spontaneously by adolescence [3, 4,8–10].

Hemangioma

Hemangioma, the most common tumorlike condition and birthmark of the newborn and infant, is found on any part of the skin, most often on the neck and face, as one or sometimes multiple lesions [1,2,4,8–14] (Figures 4.1–4.3). Hemangiomas are observed in 10% of infants but in only 2 to 4% of newborns, since many are not apparent at birth [1,12,14]. Females are affected more often than males (3:1). Of the various vascular lesions, the capillary hemangioma and the port wine nevus (Figure 4.1) are the ones observed most frequently during the first year of life [4,11,15].

The common hemangioma is classified into two main histological types, capillary and cavernous, on the basis of the caliber of the blood vessels involved [4,9]. Hemangiomas were discussed also in Chapter 3. The term "strawberry nevus or hemangioma" is applied to raised, verrucous, cutaneous vascular lesions with a reddish-purple color (Figure 4.3). They are usually mixed hemangiomas consisting of both capillary and cavernous components [4,9]. By age 7 years, at least 70% of the strawberry hemangiomas have regressed or disappeared completely [13]. Therefore conservative management is recommended.

Macular Stains

Both port wine stain and nevus flammeus are usually present at birth and are situated most often on the face and neck [3,4,13,15] (Figure 4.1). They occur in approximately 3 births per 1000 [14]. Initially they exhibit a pink appearance, but with age the port wine stain develops a bluish-red to purple tone, while becoming darker, irregular, thickened, and nodular. In contrast to the port wine stain, the nevus flammeus resolves with time, but some may remain and become lighter. Microscopic findings consist essentially of dilated capillaries in the dermis or no recognizable abnormality at all [3,4,13]. Individuals who have port wine stains of the eyelids, bilateral lesions, or unilateral port wine stains involving all three branches of the trigeminal nerve, should be screened for glaucoma and should be tested for central nervous system involvement [15].

Syndromes Associated with Vascular Conditions

Cutaneous vascular lesions, particularly macular stains and vascular malformations, may be an indication of a more serious underlying condition. This depends on the location of the birthmark and on whether structures beneath are affected [3,4,11,15]. Most vascular conditions described in syndromic newborns are not hemangiomas but instead are macular stains and vascular malformations (cavernous, venous, arterial, lymphatic, or combinations thereof) [15].

In the Sturge-Weber syndrome there is involvement of the forehead and upper eyelid in the distribution of the first branch of the trigeminal nerve, involving subjacent meninges and brain. The associated leptomeningeal hemangiomatosis is

TABLE 4.1. Tumors and tumorlike conditions of the skin in the fetus and infant.

Hemangioma
Capillary hemangioma
Port wine stain and nevus flammeus
Cavernous hemangioma
Lobular capillary hemangioma (pyogenic granuloma)
Kaposiform hemangioendothelioma
Tufted angioma
Lymphangioma
Neurofibroma
Digital fibromatosis
Myofibromatosis
Fibrous hamartoma of infancy
Smooth muscle hamartoma
Nevus lipomatosis
Granular cell tumor
Melanocytic nevus
Malignant melanoma
Epidermal nevus
Sebaceous nevus
Papilloma
Langerhans cell histiocytosis (Letterer-Siwe disease)
Urticaria pigmentosa (mast cell disease)
Juvenile xanthogranuloma
Dermal extramedullary erythropoiesis
Tumors metastatic to the skin
Neuroblastoma
Leukemia
Rhabdomyosarcoma
Rhabdoid tumor
Adrenocortical carcinoma
Choriocarcinoma

Source: Isaacs H Jr [5].

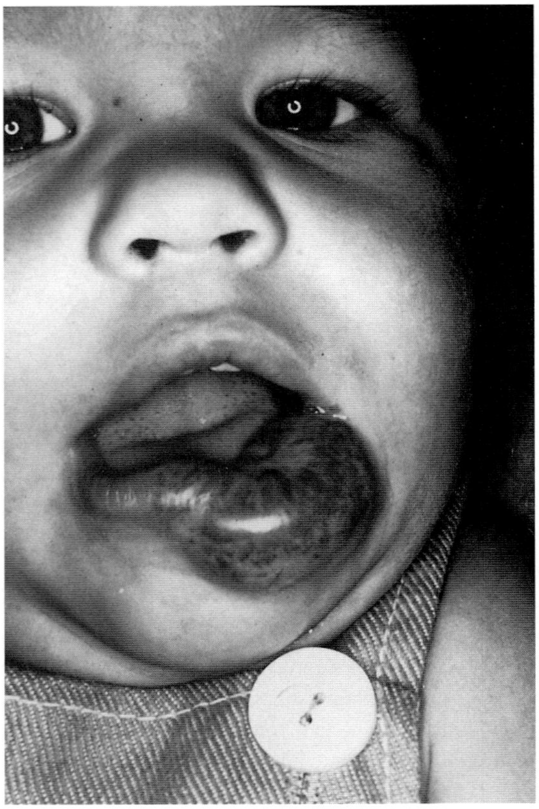

FIGURE 4.2. A 1-year-old male with a cutaneous capillary hemangioma presenting as a raised nodule on the lip at birth.

A

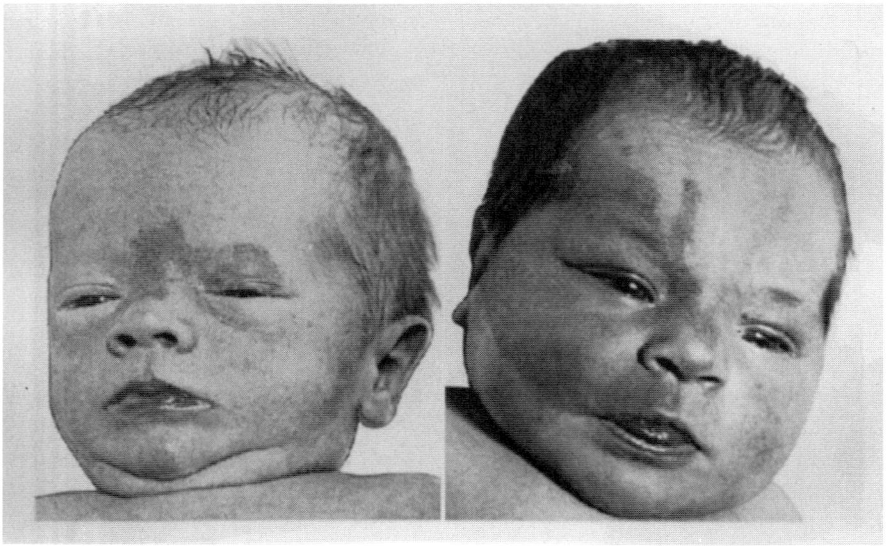

B

FIGURE 4.1. Macular stains. (A) Nevus flammeus involving the left eyelid and forehead in an infant aged 3 days. (B) Port wine stain (nevus venosus) involving the left cheek and forehead in an infant aged 2 days. Hemangiomas in this location are often associated with hemangiomas of the underlying meninges and brain. (From Isaacs H Jr [4], with permission.)

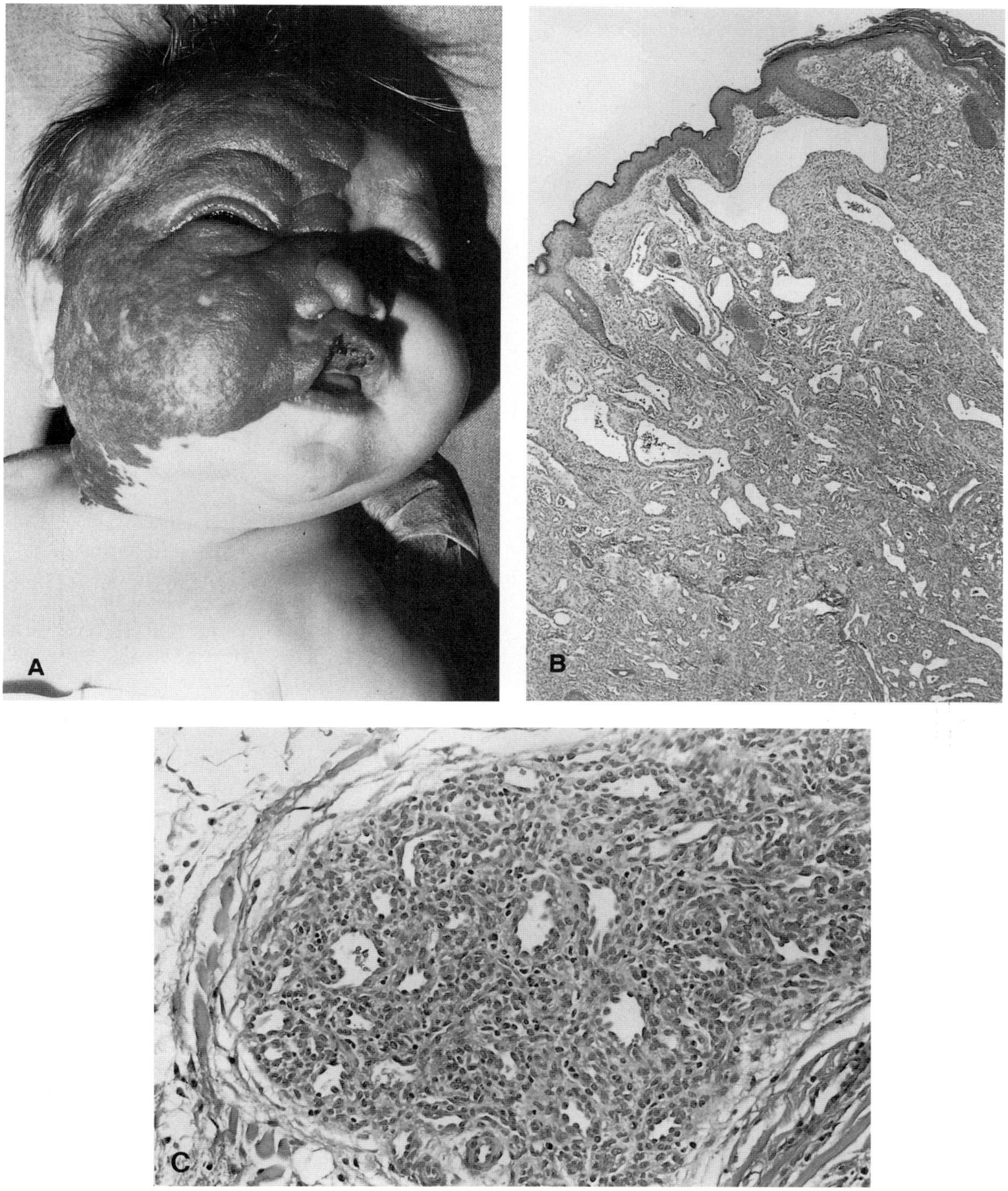

FIGURE 4.3. Strawberry hemangioma. (A) Newborn with an extensive capillary and cavernous hemangioma distributed over the scalp, face, and neck. In addition, the baby had a hemangioma of the larynx that was causing respiratory distress. (B) Low-power view (× 60) of the hemangioma showing the cavernous component within the dermis and the capillary component in the subcutaneous tissue. H&E. (C) The lobular pattern of the capillary hemangioma component within the deeper part of the tumor is shown. H&E; × 60.

manifested clinically in at least half the infants with this syndrome by a constellation of findings consisting of mental retardation, seizures, and hemiplegia with or without cerebral calcification [3,11]. Glaucoma is another important complication of the disease. When an individual has a hemangioma and/or a port wine stain present on an extremity accompanied by focal gigantism of the affected part, and often a bone deformity, the diagnosis of Klippel-Trenaunay-Weber syndrome should be entertained [9,15] (Figure 3.3B).

Infantile Kaposiform Hemangioendothelioma

About half of kaposiform hemangioendotheliomas are noted at birth or develop in early infancy, typically presenting as a red or purple cutaneous patch that rapidly enlarges, eventually becoming an indurated, deep soft tissue mass [16,17]. It is a distinctive form of hemangioma that may be associated with the Kasabach-Merritt syndrome [16,17].

Kaposiform hemangioendothelioma differs from the common childhood hemangioma in that the sexes are affected about equally and the former occurs more often on the trunk, in the retroperitoneum, and on the proximal portion of the extremities, and less frequently in the cervicofacial region. The most common locations are the retroperitoneum and upper extremities [16,17]. The large size and the infiltrative nature of these tumors result in complications of hemorrhage, obstruction, or respiratory compromise, making complete surgical excision difficult [16].

Two histological patterns, angiomatous and kaposiform, are described [16]. The former consists of rounded or dilated vascular spaces lined by flattened endothelial cellls and the latter of infiltrating fascicles, sheets, or nodules of spindled cells lining compressed (slitlike) or crescentic vascular channels. These channels are present about the periphery of the lobules. Mitoses are noted infrequently. Microthrombi, and hemosiderin deposits are variably present. It is the kaposiform pattern that differentiates this lesion from the common cellular hemangioma. Endothelial markers are present in the angiomatous areas but are variably expressed in the spindled cells [16]. When the tumor occurs in the retroperitoneum, which is the leading site in some studies, there is a high mortality rate, approaching 50% [16,17].

Tufted Angioma

Tufted angioma is a benign vascular lesion considered to be another form of hemangioma manifesting as a red or purple cutaneous patch [8,16–18]. Less than 20% are present at birth, and about half are diagnosed in the first year of life. The angioma is characterized by slow growth and the tendency to cover large areas of the trunk and neck. Other sites of involvement are the head and extremities. Patients with tufted angiomas may present with Kasabach-Merritt syndrome [17]. Histolologically the tumor consists of small lobules or vascular tufts scattered in a multiple "cannon ball–like" distribution in the dermis and subcutis (Figure 4.4). The endothelial cells are immunoreactive with *Ulex* lectin I and factor VIII, and the spindle cells between the vascular spaces are reactive with actin [18]. Endothelial cells and pericytes are demonstrated with electron microscopy. The histological features of kaposiform hemangioendothelioma and tufted angioma overlap. Rapid expansion may necessitate surgical excision [16,17].

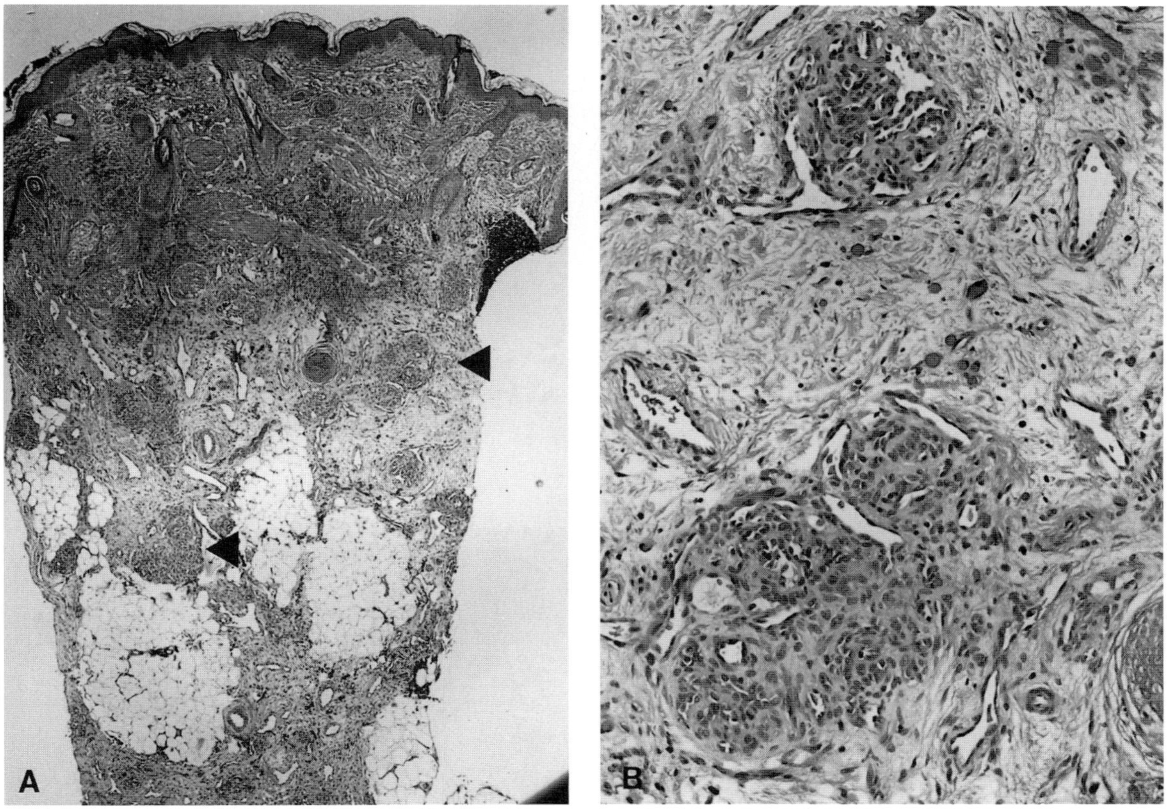

FIGURE 4.4. Tufted angioma in a 13-day-old female with a reddish-purple, slightly indurated area over the cheek. (A) Scattered small lobules of vascular tufts (arrows) are distributed throughout the dermis. H&E; × 48. (B) The lobules are composed of empty capillaries surrounded by dilated, crescent-shaped vascular channels. H&E; × 150.

Pyogenic Granuloma (Lobular Capillary Hemangioma)

Pyogenic granuloma (lobular capillary hemangioma) is a common acquired vascular lesion of the skin and mucous membranes in the pediatric age group [3,19,20]. It occurs during the first year of life and in the older individual typically as a single polypoid nodule that bleeds easily on palpation (Figure 4.5). The face and lip are common sites in the infant. Although it appears in areas of trauma or infection, it is considered to be a form of capillary hemangioma because it is practically indistinguishable from this entity microscopically [19,20].

The lesion consists of lobules of proliferating capillaries and fibroblasts often resembling granulation tissue. Characteristically there is an epithelial collarette at the base of the nodule, a helpful diagnostic feature that sets the lesion apart from some of the other vascular conditions just described (Figure 4.5). In contrast to the common capillary hemangioma, pyogenic granuloma, which is not an infectious process, rarely heals spontaneously and usually requires curettage or surgical excision [8,19,20].

Blue Rubber Bleb Nevus Syndrome

Known also as the Bean syndrome, Blue rubber bleb nevus syndrome is a rare condition characterized by distinctive vascular malformations of the skin, gastrointestinal tract, and less often in other organs, leading to intestinal bleeding and chronic anemia [3,21,22]. The vascular lesions are present at birth or early infancy, tending to increase in size and in number with age. Generally the condition is inherited as an autosomal dominant trait, but sporadic cases do occur. There is no sex predilection. Athough the tumors can occur throughout the body, they are found more often on the upper limbs, trunk, and perineum. The syndrome may be complicated by gastrointestinal bleeding and bony abnormalities as hemarthroses, pathological fractures, and skeletal bowing [22]. The lesions vary in diameter from a few millimeters to over 2 cm. They are soft dark blue, bulge from the skin surface, and are compressible. The hemangiomas may give the appearance of the so-called blueberry muffin baby (Table 4.2). Microscopically they show features of a cavernous hemangioma consisting of markedly dilated and congested vascular spaces lined by regular endothelial cells (Figure 3.4A).

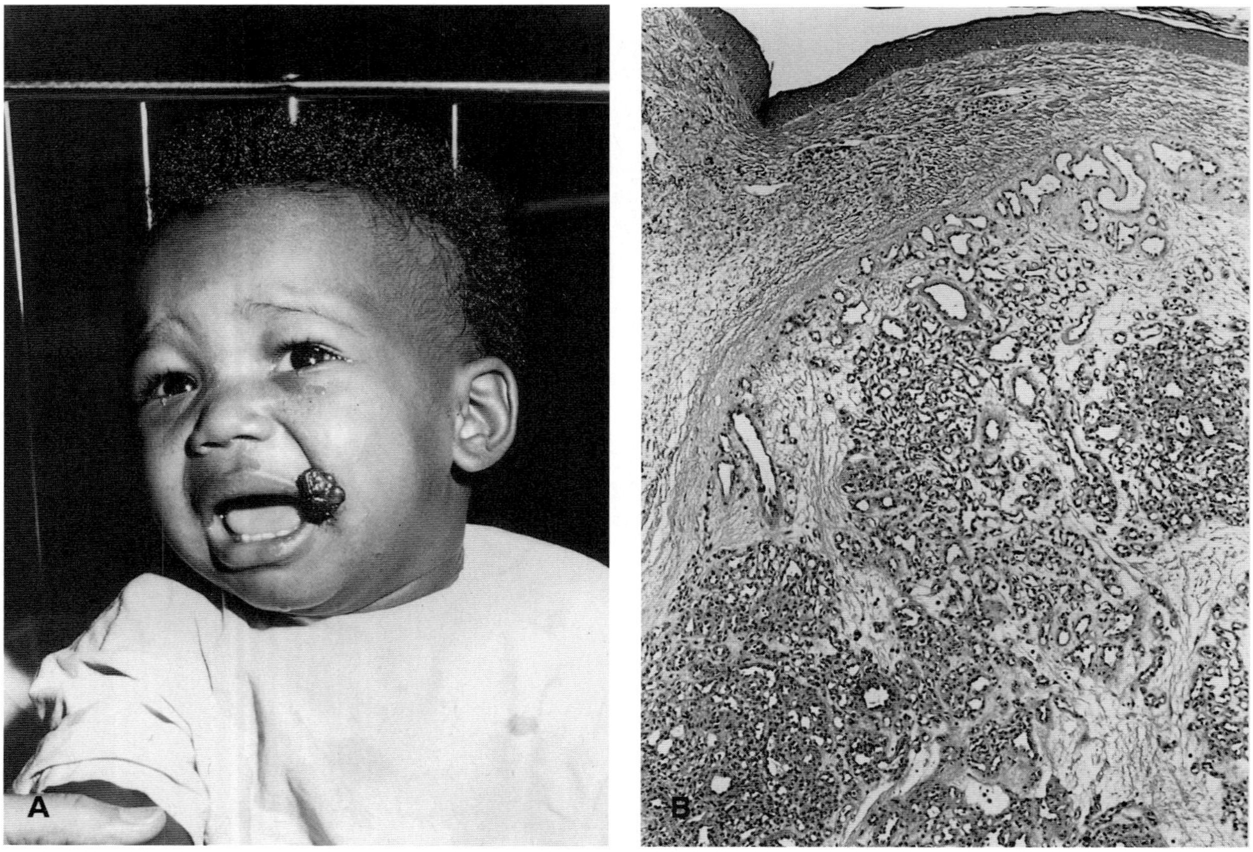

FIGURE 4.5. Lobular capillary hemangioma ("pyogenic granuloma"). (A) A 1-year-old male with a rapidly growing, polypoid mass on the left side of the cheek. The excised specimen measured $1.5 \times 1.5\ cm^2$ and had a red, fleshy-appearing cut surface. (B) Lobules of capillaries are separated by fibrous connective tissue. There is a notch in the skin suggestive of an epithelial collarette situated in the upper left corner of the photograph. Notice the similarity between the histological features of the pyogenic granuloma and those of the capillary hemangioma component depicted in Figure 4.3C. H&E; × 120.

TABLE 4.2. Differential diagnosis of the "blueberry muffin baby" (cutaneous blue nodules).

Dermal extramedullary erythropoiesis
Congenital infection
Rubella
Cytomegalovirus
Toxoplasmosis
Hemolytic disease of the newborn
Twin transfusion syndrome
Hereditary spherocytosis
Blue rubber bleb nevus syndrome
Neoplastic infiltrations of the skin
Neuroblastoma
Leukemia
Rhabdomyosarcoma
Choriocarcinoma

Source: Gottesfeld et al. [47]; Isaacs H Jr [5].

Lymphangioma

Lymphangiomas occur as superficial cutaneous and deep-seated soft tissue lesions, and regardless of their location consist histologically of dilated lymphatic channels lined by a single layer of regular endothelial cells [3,5,9]. The superficial ones, namely, cases of lymphangioma circumscriptum, composed of one or more small cutaneous vesicles, are seldom seen in the neonatal period and are noted more often later in infancy and childhood. The deeply situated cavernous lymphangiomas present clinically as rubbery, skin-colored nodules that may enlarge rapidly, producing a palpable mass in an extremity, or may form a large lymph-fluid-filled sac called a cystic hygroma, for example, in the neck, axilla, or mesentery [5] (Figures 3.6–3.10; see Chapter 3 for further discussion).

Epidermal Birthmarks

The two main epidermal birthmarks are the epidermal nevus (linear or verrucous epidermal nevus) and the sebaceous nevus, each having an incidence of about 0.2% [2]. The Becker nevus is a more uncommon member of this group. Linear epidermal nevus and nevus sebaceous syndromes are rare neurocutaneous syndromes characterized by epidermal nevi, epilepsy, and mental retardation [23]. Tuberose sclerosis, Sturge-Weber syndrome, and neurofibromatosis I are examples of other neurocutaneous syndromes.

Epidermal Nevus

The epidermal nevus occurs as local or systematized lesions noted at birth or developing later on in childhood. They present as flesh-colored to yellow brown, verrucous, oval or linear plaques that continue to enlarge and spread until puberty.

Microscopically these nevi have the appearance of a papilloma; namely, they display papillomatosis, hyperkeratosis, and acanthosis [3] (Figure 4.6). The epidermal nevus is associated not only with benign and malignant tumors that arise from within the nevus but also with an increased incidence of noncutaneous malformations within the central nervous system, eye, and skeleton. The epidermal nevus syndrome is characterized by the triad of large epidermal nevi, mental retardation, and epilepsy [2,23]. Skeletal anomalies, vascular malformations, and eye defects are described.

Becker's Nevus (Becker's Melanosis)

The pigmented variant of epidermal nevus named after Becker usually appears in childhood or adolescence, but it may be found at birth [3]. Clinically the lesion presents as a pigmented macule containing many hairs. Microscopically it shows acanthosis and diffuse, heavy pigmentation of epidermal cells. Moreover, Becker's nevus is described in association with smooth muscle hamartoma and neurofibroma [3].

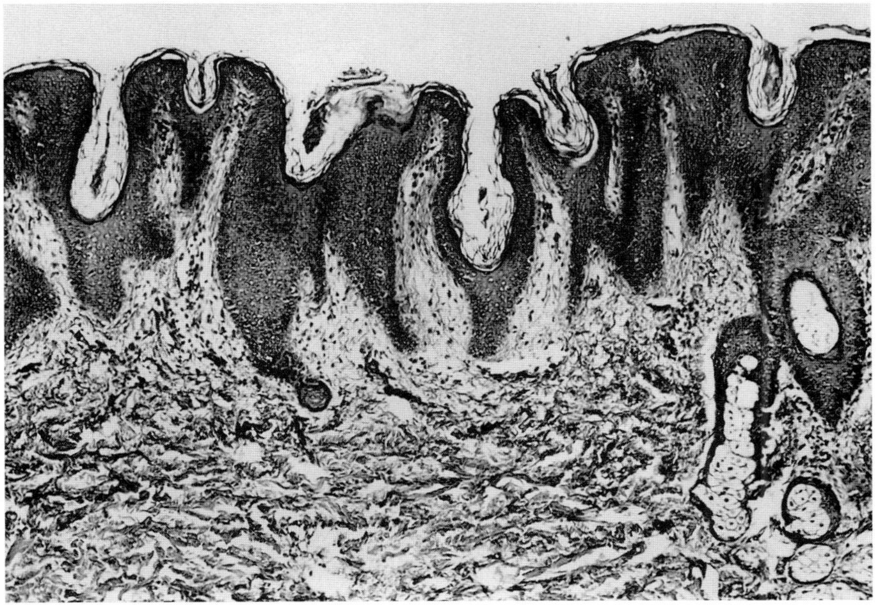

FIGURE 4.6. Giant linear epidermal nevus. This 3-year-old boy had multiple, large congenital epidermal nevi on the head, trunk, and extremities. The lesions involved the right side of the body, stopping abruptly at the midline, and did not increase in size after birth. No other congenital anomalies were present. The epidermal nevus has a verrucous appearance and appears darker than the surrounding skin. There is papillomatosis, hyperkeratosis, and acanthosis, features of a papillomatous process. H&E; × 100.

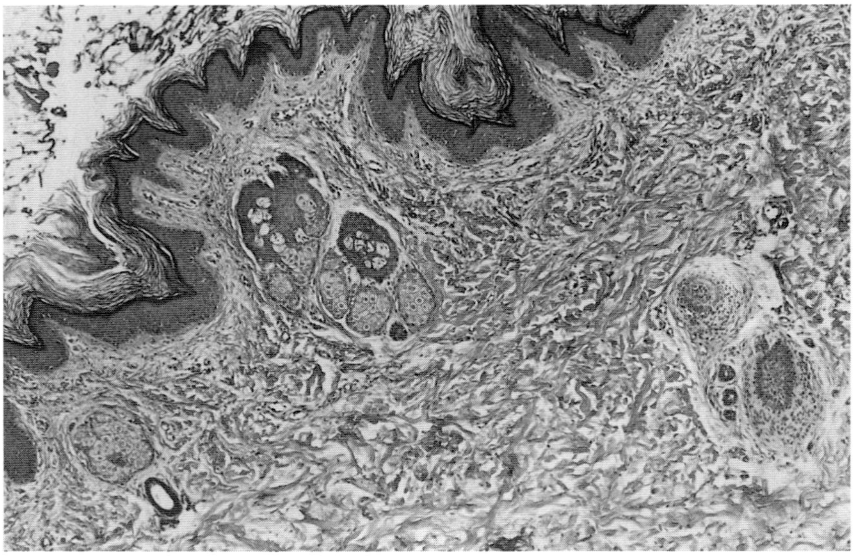

FIGURE 4.7. Nevus sebaceous. Typically the lesion presents at birth as a flat yellowish, hairless plaque on the scalp, forehead, or neck. Hyperkeratosis, irregular papillomatosis, and acanthosis are noted. Skin appendages are malformed, consisting of sebaceous glands without hair follicles. H&E; × 48.

Sebaceous Nevus

Sebaceous nevus occurs in infants as flat or slightly raised yellow-brown to pink, hairless plaques distributed over the scalp or forehead. The nevus tends to regress after birth, only to enlarge again at puberty. Biopsy of the lesion in the first year of life may show no diagnostic abnormality; later on, however, the skin appendages appear malformed and irregular and the sebaceous glands are increased in number [3,23] (Figure 4.7). Older individuals with nevus sebaceous are at risk for developing basal cell carcinoma and adnexal tumors arising from the lesion, and therefore removal is indicated. Moreover some patients with nevus sebaceous have central nervous system and other anomalies in common with both the epidermal nevus syndrome and tuberous sclerosis [3,11,23].

Cutaneous Hamartomas

Smooth Muscle Hamartoma

Smooth muscle hamartoma is usually noted at birth, manifesting as an irregular elevated patch with hypertrichosis and mild hyperpigmentation [3,24]. It is located primarily over the lumbar area. The incidence of the lesion is about 1:2600. Microscopically the main finding is a haphazard proliferation of smooth muscle bundles within the lower dermis (Figure 4.8). When hypertrichosis is present, prominent hair follicles are attached to the smooth muscle bundles. An association with Becker's melanosis has been suggested [3].

Examples of other "mesenchymal" hamartomas described in the newborn and infant skin include nevus lipomatosus,

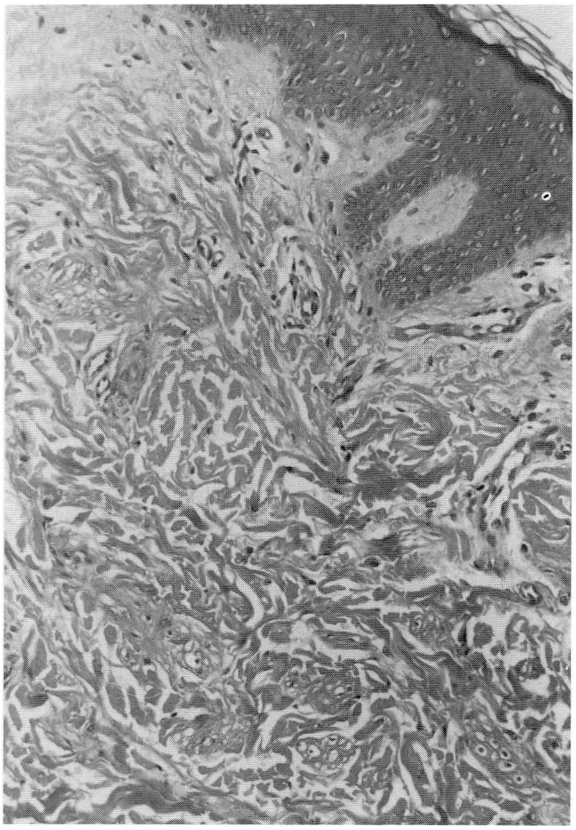

FIGURE 4.8. Smooth muscle hamartoma. The lesion probably represents overgrowths of arrector pili smooth muscle; typically it occurs in the lumbosacral region. Prominent bundles of smooth muscle are noted within the dermis. H&E; × 120.

connective tissue hamartomas, neural hamartomas composed of Schwann cells, and facial rhabdomyomatous mesenchymal hamartomas consisting of mature skeletal muscle and fat [3].

Melanocytic Nevi

Pigmented Nevus

The term "nevus" is used in a broad sense to include all abnormal growths in the skin; in a more restricted sense, however, it is limited to tumors containing nevus cells. Pigmented nevi also are called melanocytic or nevocellular nevi. Since most nevus cells contain melanin, most true nevi are pigmented, the color varying from light brown to black [4].

Although melanocytic nevi are common among adults, they are not often visible at birth. Approximately 1% of melanocytic nevi are congenital [4]. Nevi less than 2 cm in greatest dimension are far more common than the giant pigmented nevi, which are noted in less than 1 in 20,000 newborns [25,26].

Giant Congenital Melanocytic Nevi

Giant pigmented nevi are regarded as cutaneous developmental anomalies of the neural crest because they often contain both melanocytic (nevus cells) and neural elements, similar to those observed in association with neurofibromatosis [27–30]. Nevi may be located on the face, trunk, or extremities and may involve extensive areas of the skin, affecting an entire segment of the body [4] (Figures 4.9 and 4.10).

The histological features of congenital melanocytic nevi, giant or otherwise, are different in some respects from those of acquired nevi [4,25–27,31]. In the neonate, most congenital nevi are classified histologically as compound; that is, nevus cells are situated in the epidermis, epidermal-dermal junction, and dermis, but later on in childhood the nevus cells tend to be more intradermal in location [4]. The main finding observed in the congenital melanocytic nevus that distinguishes it from the acquired nevus is that nevus cells extend deep into the dermis and subcutaneous fat and connective tissue and surround skin appendages, as well as blood and lymphatic vessels, in an infiltrative fashion (Figure 4.10B). The cells, which are triangular or polyhedral, are arranged in small groups isolated by narrow connective tissue fasciculi. The nevus cells are reactive with S-100 protein and HMB-45. Some cells contain melanin. The amount of melanin increases after birth both in normal skin and in nevi and causes deepening of the color of the nevi or makes them visible for the first time. Maturation of nevus cells (i.e., they become smaller and rounder as they migrate deeper into the dermis) is an important histological feature of benign nevi [4].

A 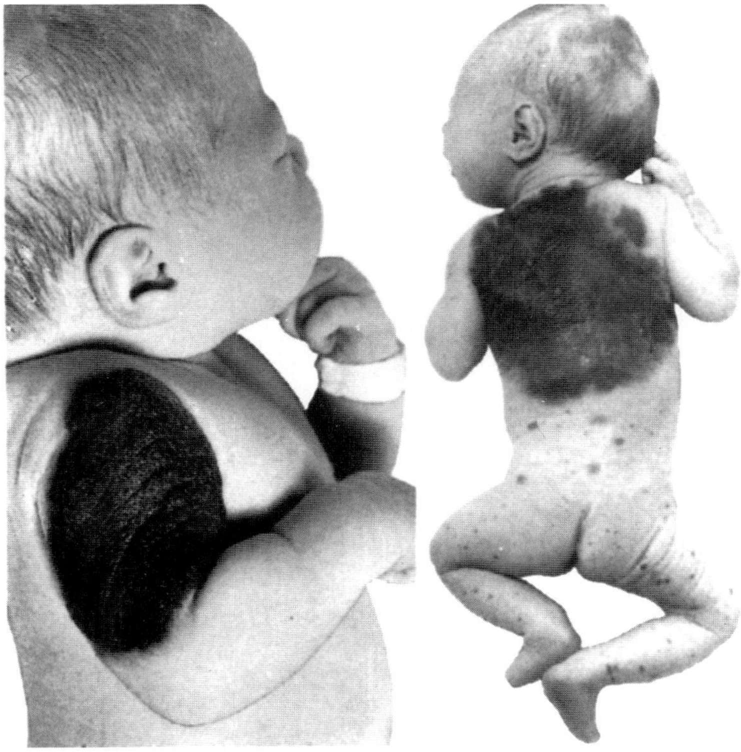 B

FIGURE 4.9. Congenital verrucous, hairy pigmented nevi. (A) Lesion localized to the upper arm. (B) Large lesion on the upper back with small, flat, pigmented nevi over the rest of the body. (From Isaacs H Jr [4], with permission.)

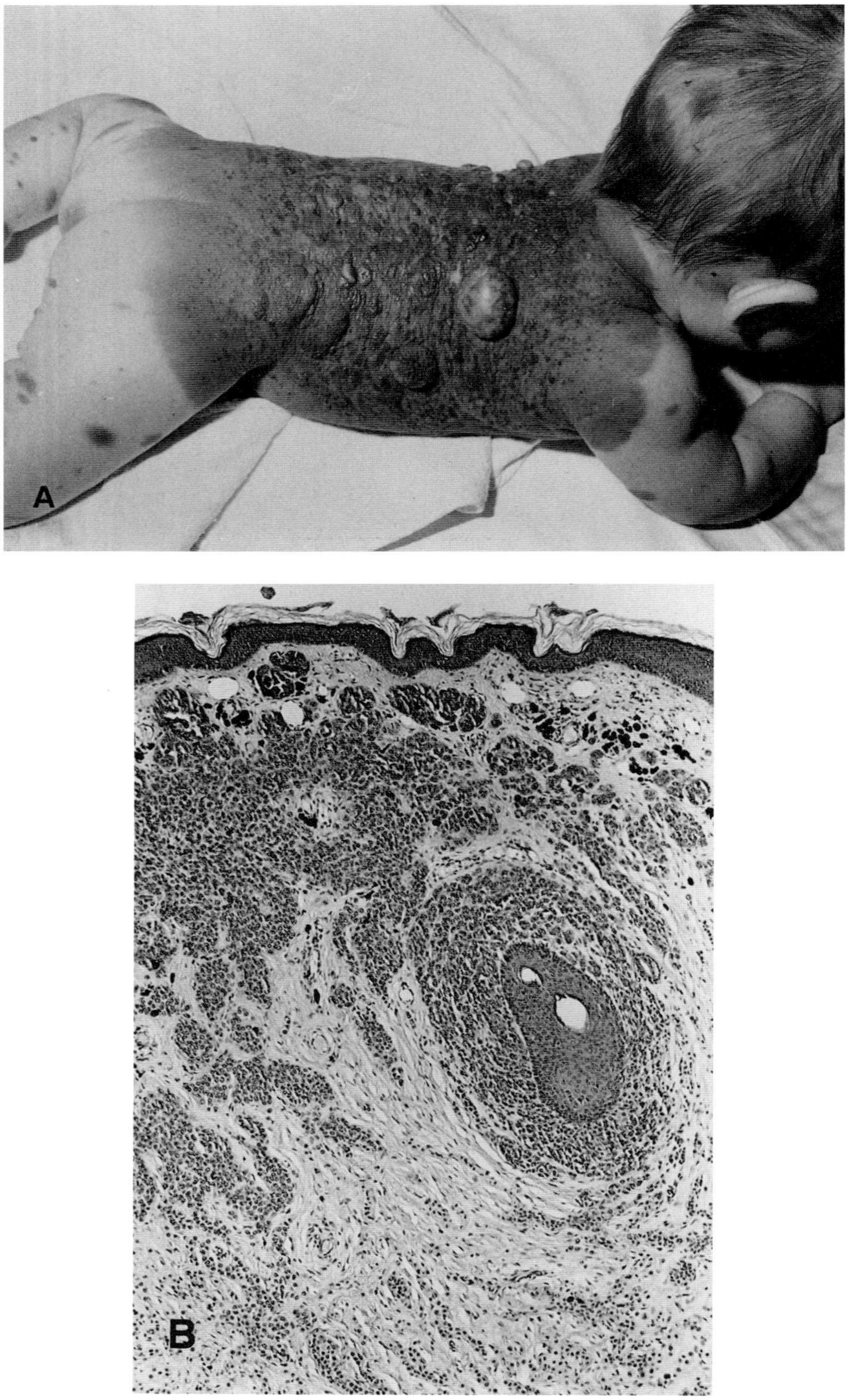

FIGURE 4.10. Giant congenital melanocytic nevus. (A) A 3-month-old male with a massive pigmented nevus involving most of the back, thorax, and abdomen. In addition there are multiple other pigmented nevi located on the extremities and scalp. (B) Biopsy reveals nests of pigmented and nonpigmented nevus cells situated in the dermis and deep in the subcutaneous tissue surrounding skin appendages, characteristic features of congenital melanocytic nevi. H&E; × 100. (C) The cut surface of a large nodule, 3 cm in diameter, on the right side of the back has a pale white gelatinous cystic appearance. (D) Sections of the nodule show pallisades of spindle shaped cells of a neuroid component. H&E; × 240.

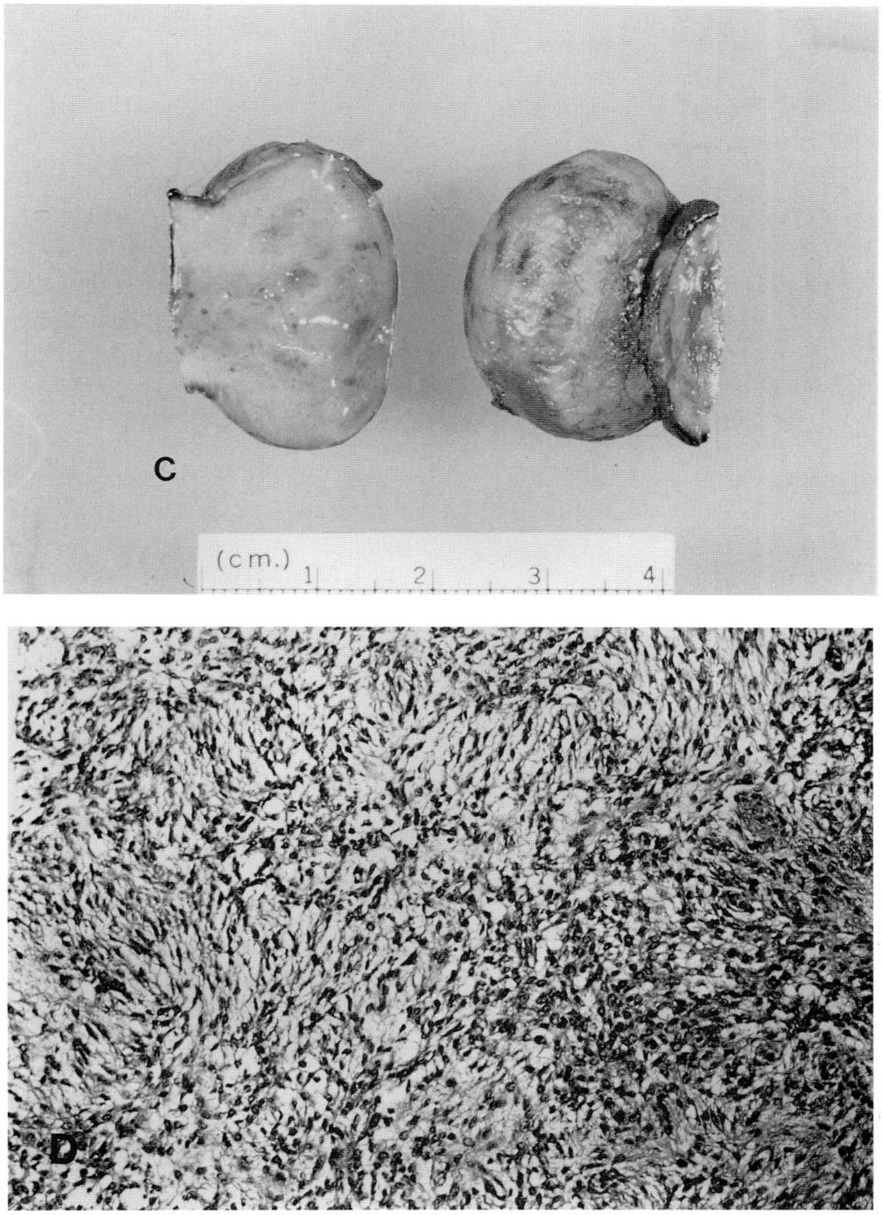

FIGURE 4.10. (*Continued*)

In addition to the melanocytic component already mentioned, the congenital nevus may contain blue nevus and neural elements, the latter in the form of Schwann cells, Meissner's corpuscles, Verocay bodies, neurofibroma, and ganglioneuroma [3,26,28,31] (Figures 4.10D and 4.11). An extra chromosome 7 (trisomy 7), described in malignant melanoma, is found in congenital pigmented nevi as well [31].

Malignant Melanoma

Nearly all nevi seen at birth are benign and grow only as the body grows, but a few have been reported that were considered to have been malignant even before birth [4,32]. The exact incidence of malignant melanoma arising from a giant pigmented nevus is difficult to determine from the literature, with published figures ranging from 3 to 12% [25,26,31–36].

Clinical Features

Congenital malignant melanoma can be divided into four main categories: melanoma transplacentally acquired from the mother, primary malignant melanoma developing antenatally *de novo*, melanoma originating from giant melanocytic nevus (Figure 4.12), and melanoma arising from neurocutaneous melanosis [26,32,35,37]. Transmission of melanoma to the mother from the affected fetus via the placenta has not been documented.

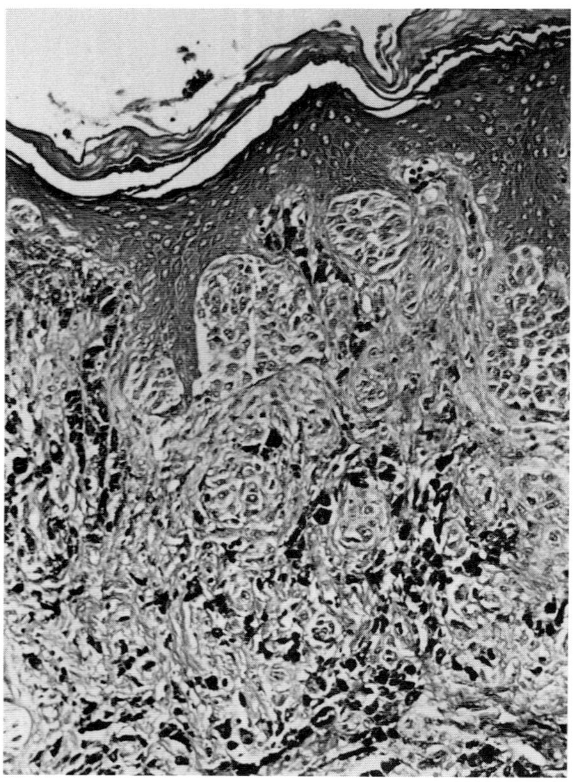

FIGURE 4.11. Blue nevus within a congenital melanocytic nevus. In addition to clusters of round to oval nevus cells within the epidermal-dermal junction, dermis, and subcutis, collections of darkly staining spindle cells are present. Congenital melanocytic nevi may contain tissues of three different types: pigmented nevus cells, neurocytic elements resembling neurofibroma, and blue nevus cells. H&E; × 120.

Neurocutaneous Melanosis

Neurocutaneous melanosis (melanocytosis, Touraine syndrome) is a rare congenital disorder in which giant or multiple congenital melanocytic nevi occur in association with benign or malignant melanocytic tumors of the leptomeninges [4,26,27,33,35,36,38]. This disorder comprises the fourth category in which congenital malignant melanoma occurs. Normally melanin-containing cells are present in the leptomeninges of individuals, particularly over the brain stem [27]. Cytological examination of the cerebrospinal fluid in patients with neurocutaneous melanosis is an important diagnostic procedure that reveals pigmented nevus cells with hyperchromatic nuclei and cytoplasmic projections [38,39].

Touraine syndrome is the ominous clinical manifestation of neurocutaneous melanosis characterized by increased intracranial pressure, seizures, developmental delay, and generally a poor prognosis [33,39]. Patients with neurocutaneous melanosis become symptomatic before 2 years of age. Eventually half the children develop leptomeningeal melanoma. Leptomeningeal melanoma infiltrates the leptomeninges and forms multiple nodules in the brain [39]. When this occurs frankly, malignant pigmented cells are observed in the cerebrospinal fluid.

Histopathology

When a melanoma originates from a congenital nevus, generally it is found at the epidermal-dermal junction and has the features of the ordinary form of malignant melanoma, that is, the superficial spreading type [3] (Figure 4.12B). When a melanoma in a giant pigmented nevus arises deep in the dermis, it is different from almost all other cutaneous melanomas [3,27]. In this setting the melanoma consists mostly of small undifferentiated "blastic cells" resembling lymphoblasts or neuroblasts and containing little or no melanin [3,27].

There are several different types of malignant neoplasm arising from congenital nevi other than melanoma [3,27–29,40, 41]. Liposarcoma, rhabdomyosarcoma, malignant blue nevus, neurosarcoma, and other undifferentiated spindle cell tumors presumably of neural origin are examples [40–42]. It is important to note that extraordinary differentiation is to be expected in neoplasms arising in congenital nevi, and surprisingly enough, alarming hypercellular tumors may not behave aggressively [3,28,31,34]. Because of this, histological findings do not always predict the outcome in the young infant [3].

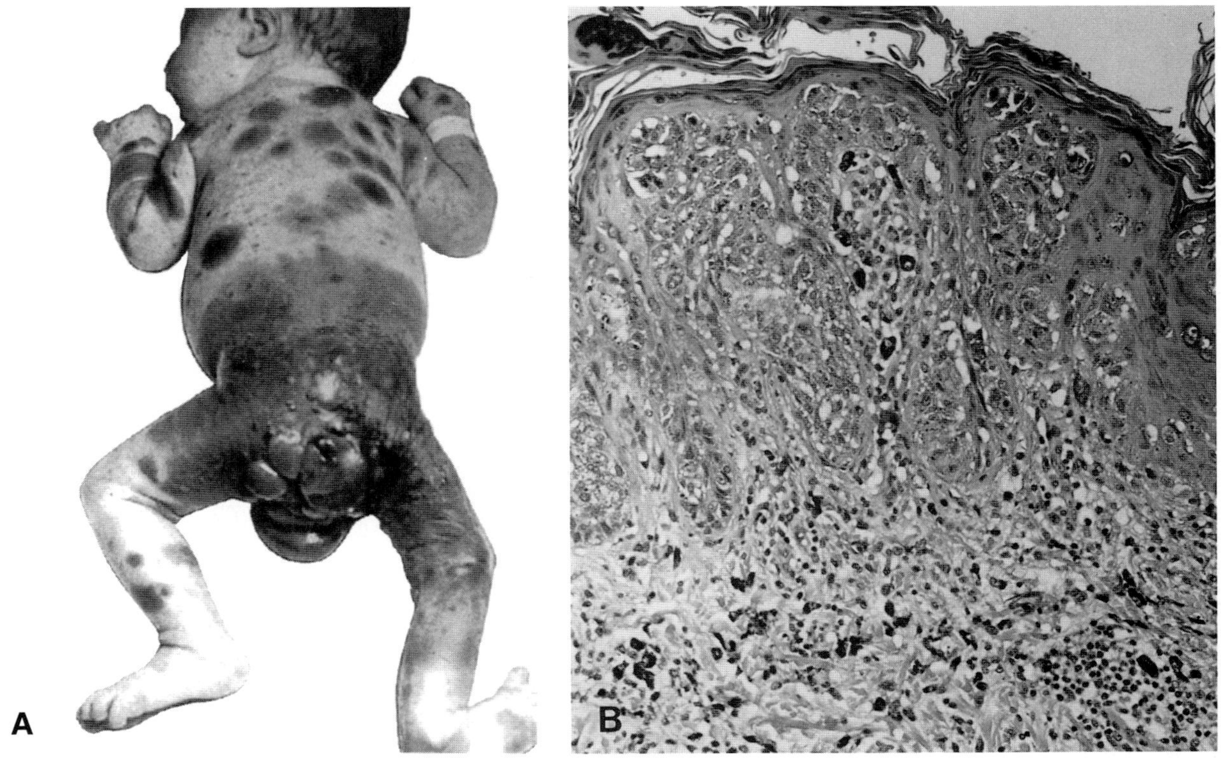

FIGURE 4.12. Malignant melanoma. (A) Widespread metastases were present over the body with tumor arising from the anal and genital region. Tumor cells were found also in the brain and liver. (From Isaacs H Jr [4], with permission.) (B) Skin with superficial spreading melanoma, which is one of the main forms of melanoma arising from congenital melanocytic nevi. Clusters of melanoma cells are present within the epidermal dermal junction with extension into the epidermis and dermis. The dermis beneath the tumor shows a chronic inflammatory infiltate intermingled with pigmented melanophages. Tumor from older child. H&E; × 120.

Generally, cellular proliferative nodules arising from dermal components of congenital pigmented nevi do not behave in a clinically aggressive fashion. Indications of a likely benign behavior in such nodules are the following: a low mitotic rate, absence of necrosis, lack of high grade nuclear atypia, and evidence of maturation in the form of "blending" or transitional forms between the cells in the nodule and the adjacent nevus cells [3]. If some of these worrisome features are present in slight degree, the descriptive diagnosis of "melanocytic tumor of uncertain potential" is suggested [3].

Neurofibromatosis

Neurofibromatosis (NF) includes several diseases, most of which are transmitted as an autosomal dominant trait [43–46]. The main form of neurofibromatosis recognized at birth is NF-1, which is the classic form, or Von Recklinghausen's disease, and is responsible for most of the cases. NF-2 is associated with bilateral acoustic neurofibromas. NF-5 is the localized or segmental form, and NF-6 is manifested only with café-au-lait spots. The gene locus for NF-1 has been localized to chromosome 17 and the locus for NF-2 to chromosome 22. The frequency of neurofibromatosis is about 1 in 3000 live births [44].

Skin findings in the newborn with types 1, 2, and 6 consist only of a few café-au-lait spots, which increase in number as the child grows older [43–45]. However one or more cutaneous neurofibromas may be present at birth, particularly with NF-1 (Table 4.3). These tumors are almost always plexiform neurofibromas [44]. The earlier the appearance of the plexiform neurofibromas, the more likely they will lead to serious neurological and other problems. Neurofibromas are discussed in more detail in Chapter 3.

TABLE 4.3. Skin lesions of the fetus and infant associated with neurofibromatosis.

Café-au-lait patches
Hyperpigmentation
Hypopigmentation
Axillary freckling
Neurofibroma
Hemangioma
Melanocytic nevi
Juvenile xanthogranuloma

Source: Elder D, et al. [3]; Listernick R, Charrow L [45]; Ricardi VM [44].

TABLE 4.4. Skin manifestations of tuberous sclerosis in the fetus and infant.

Lesion	Gross appearance	Histology
Angiofibroma	Red, smooth papules symmetrically distributed on nasolabial folds, cheeks, and chin	Dermal fibrosis, capillary dilatation
Fibroma	Raised, soft brown nodules on face, scalp, and beneath nails	Thick collagen bundles concentrically arranged around atrophic skin appendages; fibrosis
Shagreen patch	Raised and thickened areas of skin	Dense scarlike collagen in dermis resembling scleroderma; interwoven bundles of collagen and reduction of elastic fibers
Hypopigmentation	White, leaf-shaped hypopigmented areas of skin	Decreased pigmentation of melanocytes and keratinocytes

Source: Elder D, et al. [3]; Isaacs H Jr [5].

Tuberous Sclerosis

The four main cutaneous lesions associated with tuberous sclerosis are facial angiofibroma and fibroma, shagreen patch, and leaf-shaped areas of hypopigmentation [3] (Table 4.4). The earliest cutaneous manifestation of the disease is the maple leaf–shaped area of white hypopigmentation, which may be present at birth [3]. Later in childhood, when mental deficiency and seizures become apparent, the characteristic facial angiofibromas and other hamartomatous lesions begin to appear. Central nervous system, eye, and cardiac lesions of tuberous sclerosis are discussed in Chapters 8, 9, and 15, respectively.

Tumors and Tumorlike Infiltrations of the Skin

Malignant neoplasms may present in the fetus and infant as cutaneous metastases, which may be the first manifestation of the tumor [4]. Leukemia and neuroblastoma are the most common examples (Tables 4.1 and 4.2). Some nonneoplastic conditions may be mistaken for neoplastic disease. For example, subcutaneous fat necrosis of the newborn producing multiple nodules may occur several days after a difficult delivery [3,4] (Figure 4.13).

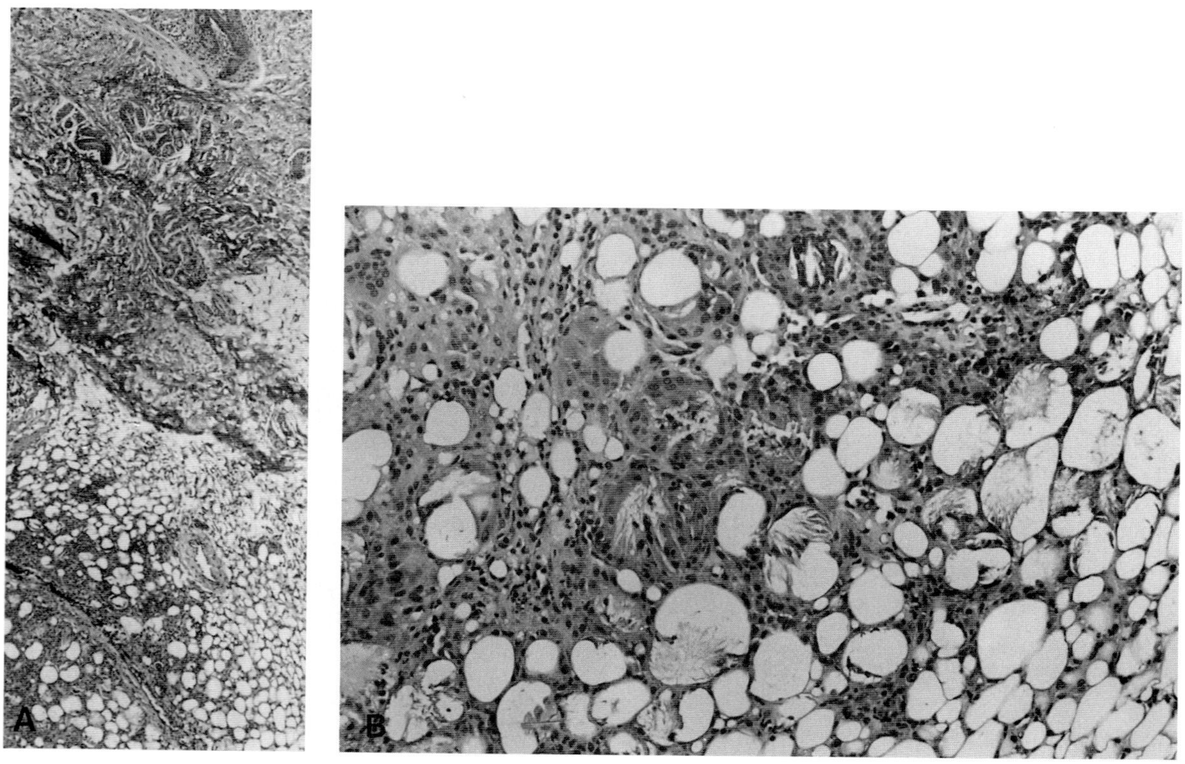

FIGURE 4.13. Subcutaneous fat necrosis of the newborn: 3-week-old female with progressive firm nodule formation on both upper arms and left thigh. There was a history of difficult delivery. (A) Low-power (× 30) view of the skin "punch biopsy" showing the distribution of the fat necrosis in the subcutis. H&E. (B) The fat is infiltrated by lymphocytes, macrophages, and foreign body type giant cells. Macrophages and multinucleated giant cells contain radial crystalline arrays of fat. H&E; × 300.

Leukemia

Leukemia cutis occurs in about a third of the infants with this disease and is often associated with a high leukocyte count [6,12,47,48] (see Chapter 6, Leukemia). When extensively involved, the skin has a nodular, indurated appearance. The natural course of congenital leukemia is not changed by the presence of leukemia cutis [6]. Skin biopsy reveals infiltrates of immature lymphoid, monocytic, or myeloid cells, depending on the type of leukemia, within the dermis and subcutaneous tissues (Figure 4.14). Occasionally ultrastructural and immunocytochemical studies are required to rule out other conditions that may present in a similar way (e.g., Langerhans histiocytosis, mastocytosis, metastatic neuroblastoma). Bone marrow examination is not always helpful in resolving the issue. In some instances the infant's leukemia and leukemia cutis may spontaneously resolve [12,47].

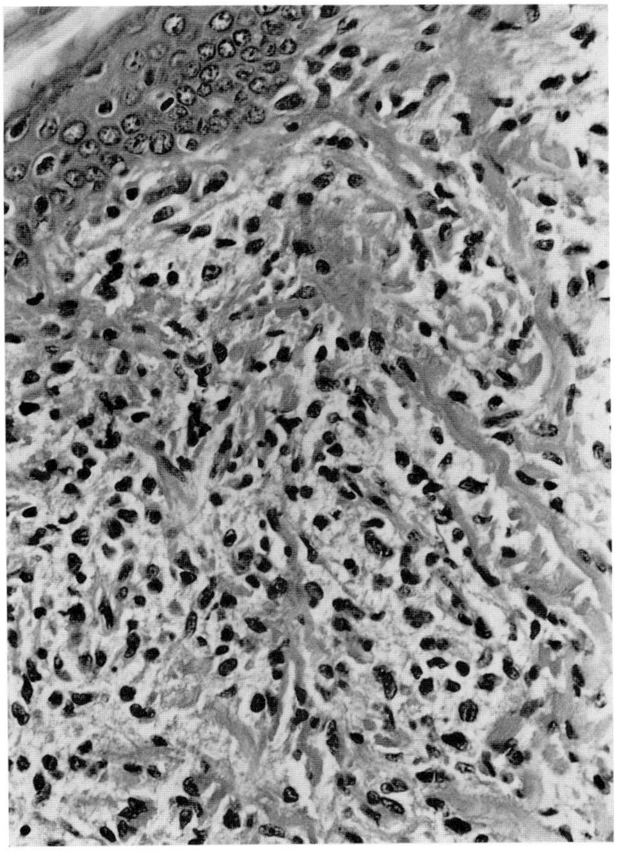

FIGURE 4.14. Leukemia cutis. A 1-month-old female with acute myeloid leukemia had a peripheral leukocyte count of 118,000 and several raised purplish nodules on the scalp. Biopsy of a scalp nodule reveals a dense cellular infiltrate composed of small round tumor cells in the dermis and subcutaneous tissue. The cells consist of round, oval, or indented nuclei some with prominent nucleoli. Myeloid markers were positive. H&E; $\times 250$.

Neuroblastoma

Multiple bluish cutaneous nodules, producing the appearance of the so-called blueberry muffin baby, may be the initial manifestation of metastatic neuroblastoma in the neonate, which like leukemia is seen in about a third of the cases [48,49] (Table 4.2) (Figure 4.15). Most affected infants are diagnosed as stage IV-S neuroblastoma, which is frequently associated with a favorable clinical outcome [50] (see Chapter 5, Neuroblastoma, for more detail). In addition to their skin lesions,

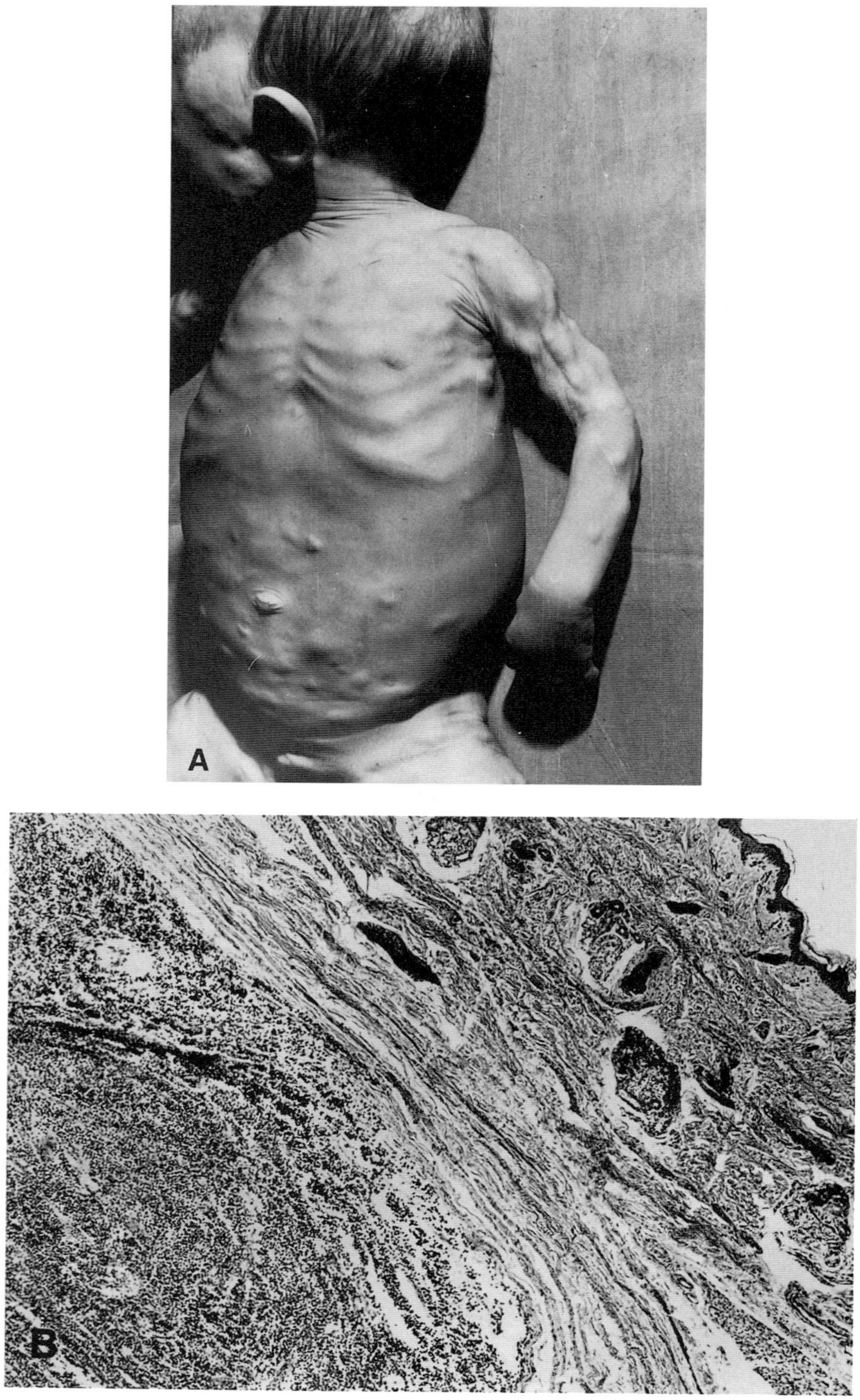

FIGURE 4.15. Skin nodules as the initial manifestation of congenital neuroblastoma ("blueberry muffin baby"). (A) A 3-month-old male with IV-S neuroblastoma and multiple cutaneous nodules distributed over the body. (B) Subcutaneous nodule showing a partially autolyzed small-cell malignant tumor. H&E; × 100.

typically the patients have a small primary tumor usually situated in the adrenal, with multiple metastases to the liver and bone marrow that tend to spontaneously resolve [50]. Histological examination of the cutaneous nodule reveals a uniform small-cell malignant tumor with or without Homer-Wright pseudorosettes (Figure 4.15). The tumor cells show neuron-specific enolase (NSE) reactivity and neurosecretory (dense core) granules and neurotubules with electron microscopy [5] (Table 3.2).

Rhabdomyosarcoma

Both primary and metastatic rhabdomyosarcoma of the skin are unusual [41]. Cutaneous alveolar rhabdomyosarcoma is composed of small to medium-sized, darkly staining cells with rounded nuclei and scant cytoplasm arranged in an alveolar pattern (Figure 4.16). The tumor is immunoreactive with vimentin, desmin, and muscle actin [41] (Table 3.2). Rhabdomyosarcoma metastasizes to the skin, forming bluish cutaneous nodules similar to those found in the "blueberry muffin baby" with neuroblastoma [41], with dermal erythropoiesis in Rh hemolytic disease disease of the newborn [51], or with congenital leukemia [6,47] (Table 4.2).

Rhabdoid Tumor

A rhabdoid tumor is a highly malignant tumor characterized clinically by rapid growth and early metastases and a high mortality rate in the newborn and infant. Rhabdoid tumor may present in the skin, particularly in the head and neck area, as a solitary primary tumor or as one or more metastatic skin nodules [7,52]. The histological appearance of the tumor is similar to the one occurring in the kidney (see Chapter 10, Renal Tumors; see Table 3.2 for differential diagnosis).

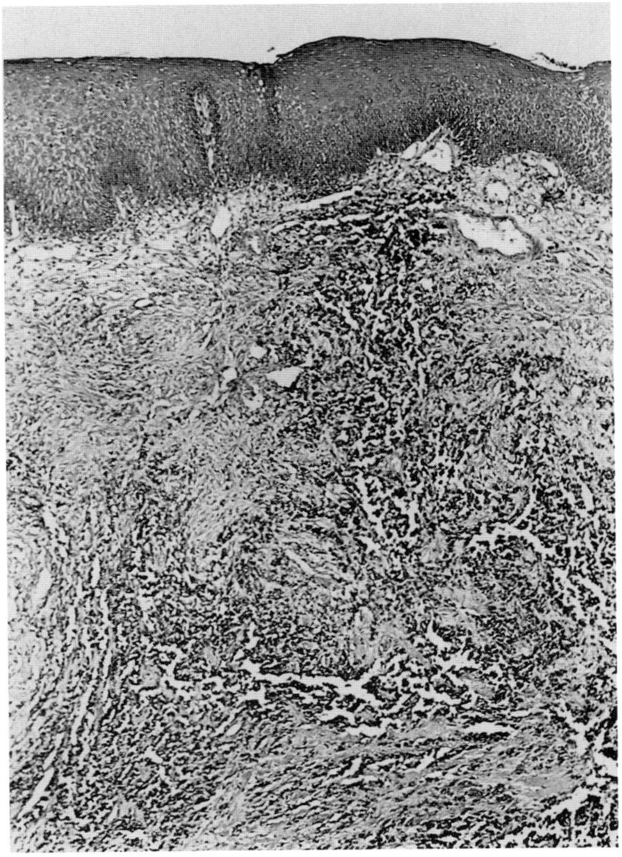

FIGURE 4.16. Rhabdomyosarcoma presenting as a skin metastasis in a 7-month-old male who had a biopsy of what was thought to be a pyogenic granuloma of the lip. The excised specimen measured 2.5 cm in diameter and had a pale tan-gray cut surface. The biopsy material shows dense infiltrates of small darkly staining neoplastic cells in the submucosa and underlying soft tissue. The tumor has a suggestive alveolar pattern. H&E; × 100. (See Figure 3.38 for other findings.)

Langerhans Cell Histiocytosis

Skin lesions are an important manifestation of Langerhans cell histiocytosis in the fetus and infant [4,5,53]. Typically the patients present with a salmon-colored macular papular rash, refractory to therapy, accompanied by anemia, thrombocytopenia, hepatosplenomegaly, lymphadenopathy, and general failure to thrive [4,5] (Figure 4.17). Sporadic cases of spontaneous regression of congenital cutaneous Langerhans cell histiocytosis are documented [54].

The diagnosis is established following a skin biopsy of a representative lesion, which microscopically shows variable numbers of Langerhans histiocytes infiltrating the subcutaneous tissue and dermis, and extending through the epidermis into the stratum corneum, a helpful diagnostic feature (Figure 4.17B). The Langerhans cell has a characteristic coffee bean–shaped nucleus, a small nucleolus, and an eosinophilic cytoplasm. Ultrastructural studies reveal the presence of diagnostic tennis racket–shaped Birbeck granules in the cytoplasm of the Langerhans cells, which are S-100 protein and T6 (CD1) thymocyte antigen immunoreactive [4,53,54] (see Chapter 7, Histiocytoses, for more detail).

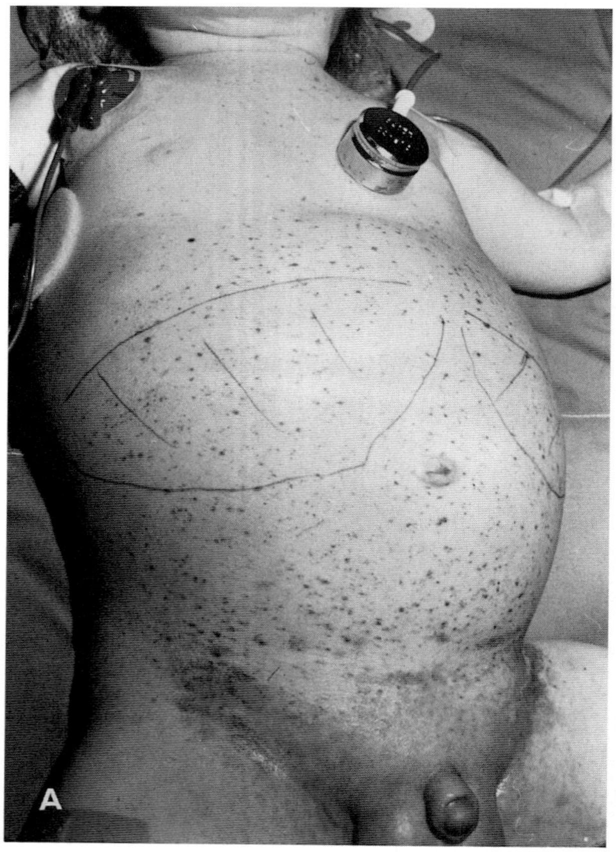

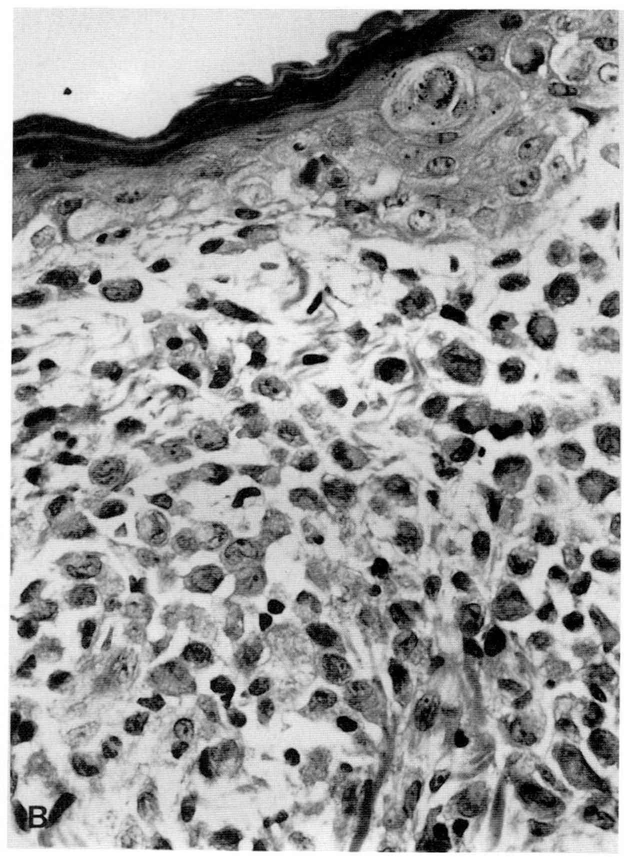

FIGURE 4.17. Disseminated Langerhans cell histiocytosis (Letterer-Siwe disease) in an 11-month-old male with a salmon-colored maculopapular rash, anemia, thrombocytopenia, and hepatosplenomegaly. (A) The maculopapular rash is particularly prominent over the diaper area. The enlarged liver and spleen are outlined on the abdomen. (B) Biopsy material from a skin lesion showing a dense cellular infiltrate composed of Langerhans histiocytes with characteristic "coffee bean," folded nuclei and prominent eosinophilic cytoplasms. The histiocytes typically extend into the epidermis in comparison to urticaria pigmentosa and juvenile xanthogranuloma, where the infiltrates stop abruptly at the epidermal junction. Langerhans histiocytes are S-100 and T6 thymocyte antigen positive and unlike the other entities just mentioned, contain Birbeck granules visible with electron microscopy. H&E; $\times 250$.

Juvenile Xanthogranuloma

Juvenile xanthogranuloma is a self-limited, benign disorder of histiocytes found more often in the newborn and infant than in the older child and adolescent [3,9,55–57]. One fifth of all infants with juvenile granuloma are affected at birth, and two thirds have onset of lesions prior to age 3 months [55]. Clinically xanthogranuloma presents as one or more yellow to reddish-brown cutaneous nodules usually less than 1 cm in diameter (Figure 4.18A). Most occur on the head and neck area and extremities; they spontaneously regress within 2 years or less after diagnosis [3,9,55]. Cutaneous xanthogranulomas may occur with similar lesions in other sites as the eye, lung, and deep soft tissues [3,9,55]. Ocular involvement is the most serious complication. Histiocytic infiltrates may involve the iris, ciliary body, sclera, or the entire orbit [58].

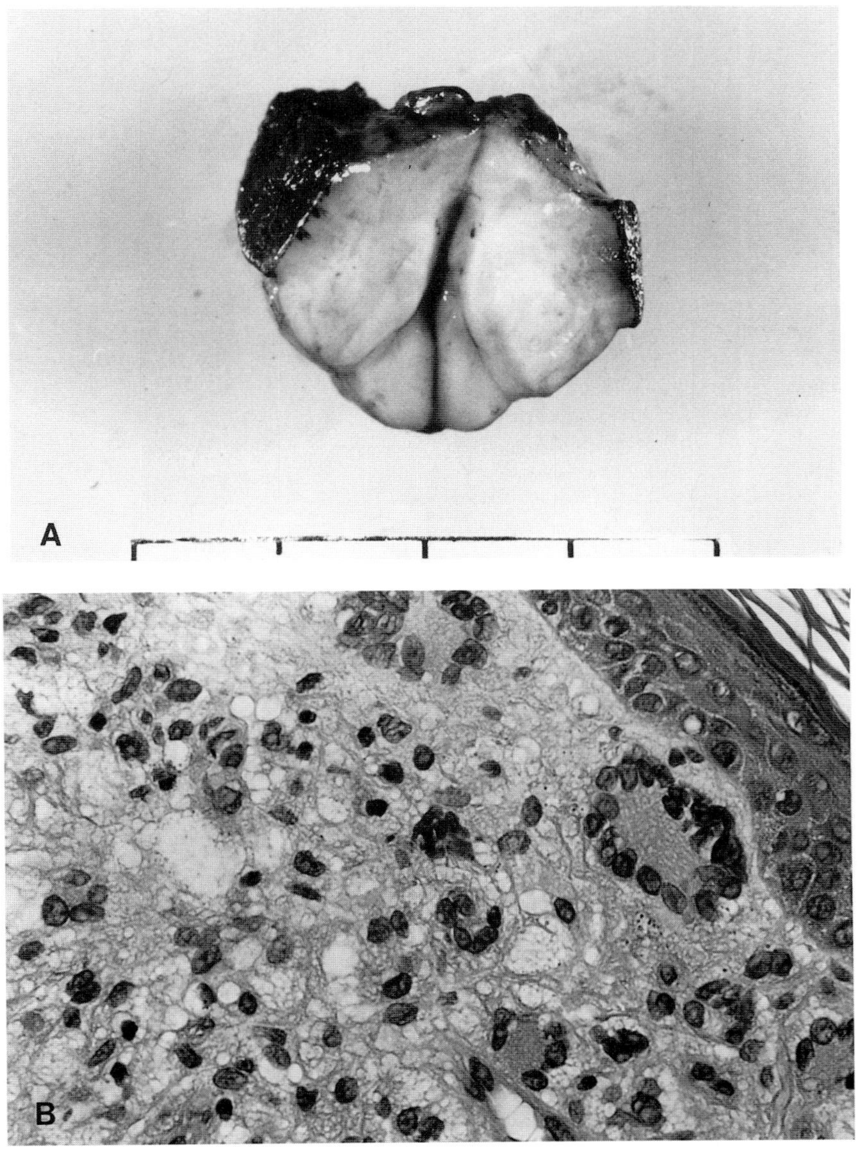

FIGURE 4.18. Juvenile xanthogranuloma, "classic" lesion. Biopsy of a scalp nodule removed from an 8-month-old male. (A) Gross specimen divided in half consists of a 1.2×1 cm^2 yellow nodule. (B) The section shows an intact epidermis without cellular infiltrates. The dermis is infiltrated by mature-appearing, vaculated histiocytes with regular nuclei. In addition, there are several multinucleated Touton giant cells with rows of nuclei situated at the periphery of the cell in a wreathlike fashion. The cells were unreactive with S-100. H&E; $\times$ 380.

5
Neuroblastoma

Neuroblastomas are derived from neuroblasts that arise in the neural crest and normally migrate downward to form ganglion cells and chromaffin tissue [1].

Incidence

Neuroblastoma is the most common malignant tumor of both fetuses and infants [1–7] (Table 1.1). The Third United States National Cancer Survey (1969–1971) recorded an incidence rate of 19.7 neuroblastoma cases per million live births per year in the neonate versus 4.7 for both leukemia and renal tumors [2]. Neuroblastoma comprised 21 (54%) of the 39 malignancies recorded in the survey for the neonatal period and 67 (34%) of 196 malignant tumors for infants [2].

Clinical Findings

Neuroblastoma is especially unique because of its very many different clinical manifestations in the young [3–8]. The tumor has been detected as an unsuspected finding in the fetus on routine prenatal sonography, appearing as a cystic and/or solid mass above upper pole of the kidney or as metastases without an identifiable primary tumor [9–12].

Hydrops fetalis with extensive placental metastases may be evident at birth [7,13–18] (Figure 5.1). Neonates with congenital neuroblastoma may have findings similar to those of erythroblastosis fetalis (hemolytic disease of the newborn), namely, hydrops, anemia, hepatosplenomegaly, jaundice, and increased numbers of nucleated erythrocytes on the peripheral blood smear [14]. Disseminated neuroblastoma is

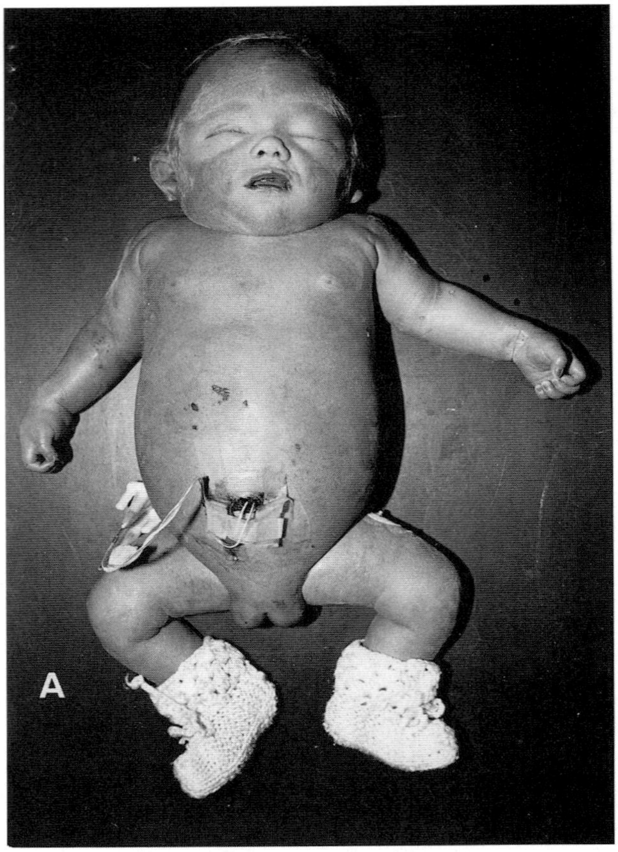

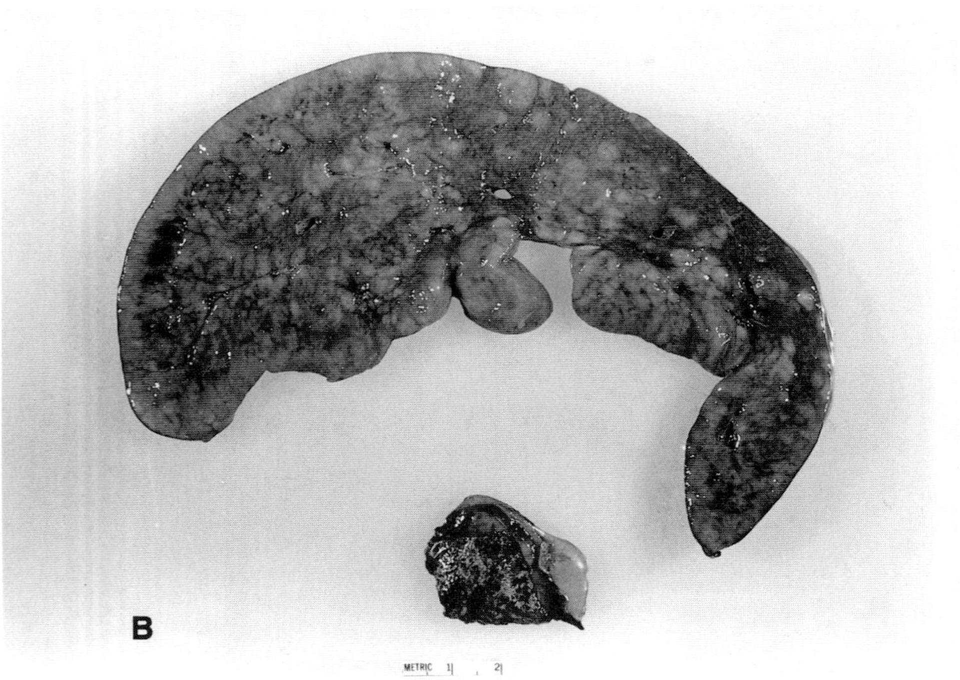

FIGURE 5.1. Disseminated neuroblastoma with hydrops fetalis. (A) A 1-week-old female with hydrops (generalized edema), ascites, and hepatomegaly. (B) Cut surfaces of the liver and adrenal with multiple gray nodular metastases. (C) Placental villous capillaries contain clusters of neuroblasts. H&E; × 600. (D) Liver showing extensive replacement by undifferentiated neuroblastoma. H&E; × 300. (E) Cerebral blood vessels are plugged with tumor emboli (arrows); there is necrosis of the adjacent cortical white matter. A cystic infarct is noted in the upper right-hand corner. H&E; × 300. (A–C from Isaacs H Jr [18], with permission.)

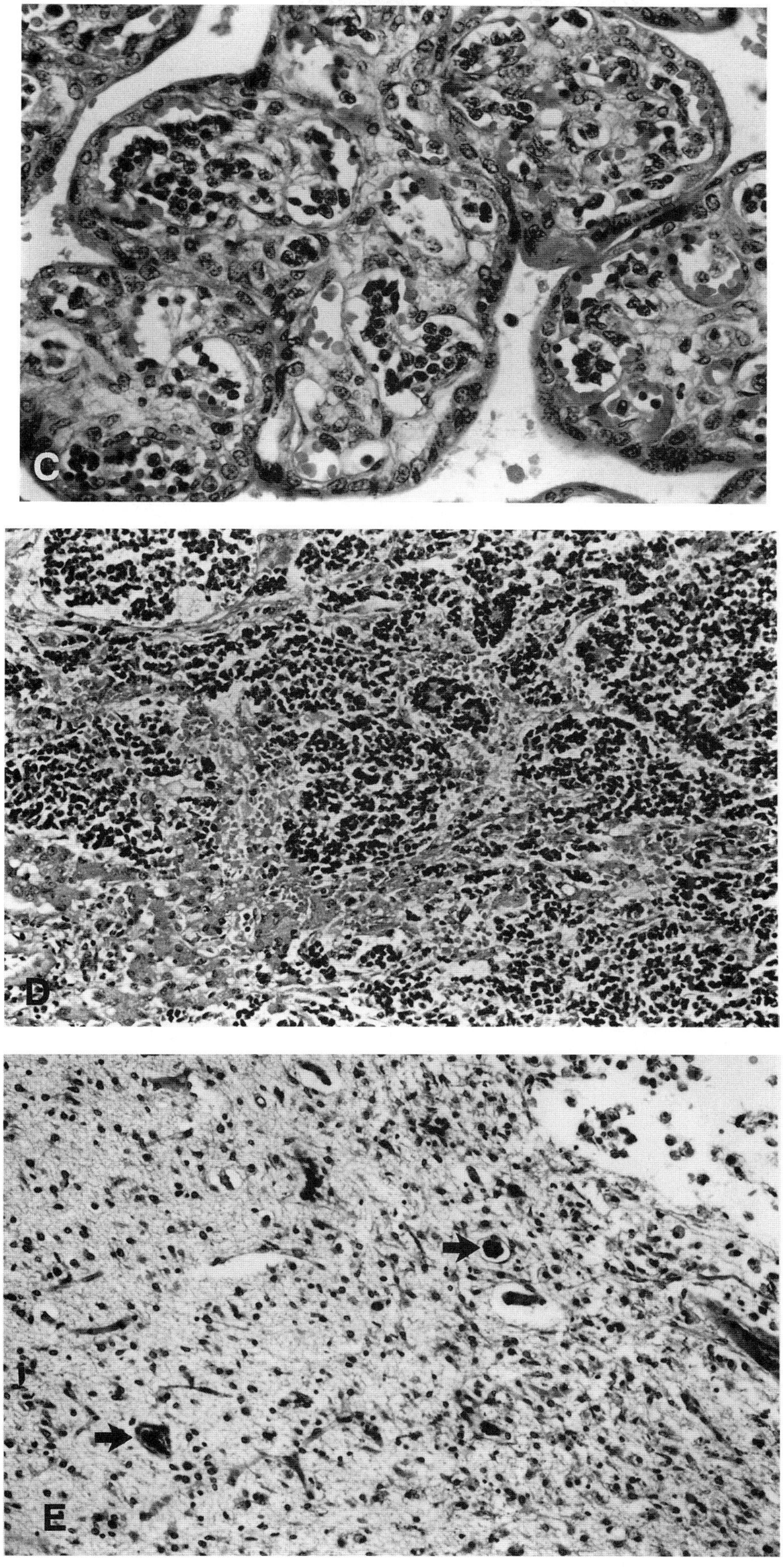

FIGURE 5.1. (*Continued*)

responsible for intrauterine death and stillbirth [13,14]. The placentas are histologically immature and greatly enlarged; their gross appearance suggests erythroblastosis fetalis because they weigh over a 1000 g and look bulky and edematous [13,14]. Neuroblasts are situated within blood vessels and in villous capillaries (Figure 5.1C).

Neuroblastoma is the malignant tumor most often found at birth, clinically presenting as an abdominal mass due to either an adrenal-retroperitoneal primary or to hepatomegaly secondary to extensive metastases [6,19–22] (Figures 5.2–5.5). Over half the tumors occur within the abdominal cavity and less frequently as a mass lesion in the mediastinum or neck, the former more commonly than the latter [4,19,22,23] (Figures 5.6–5.8).

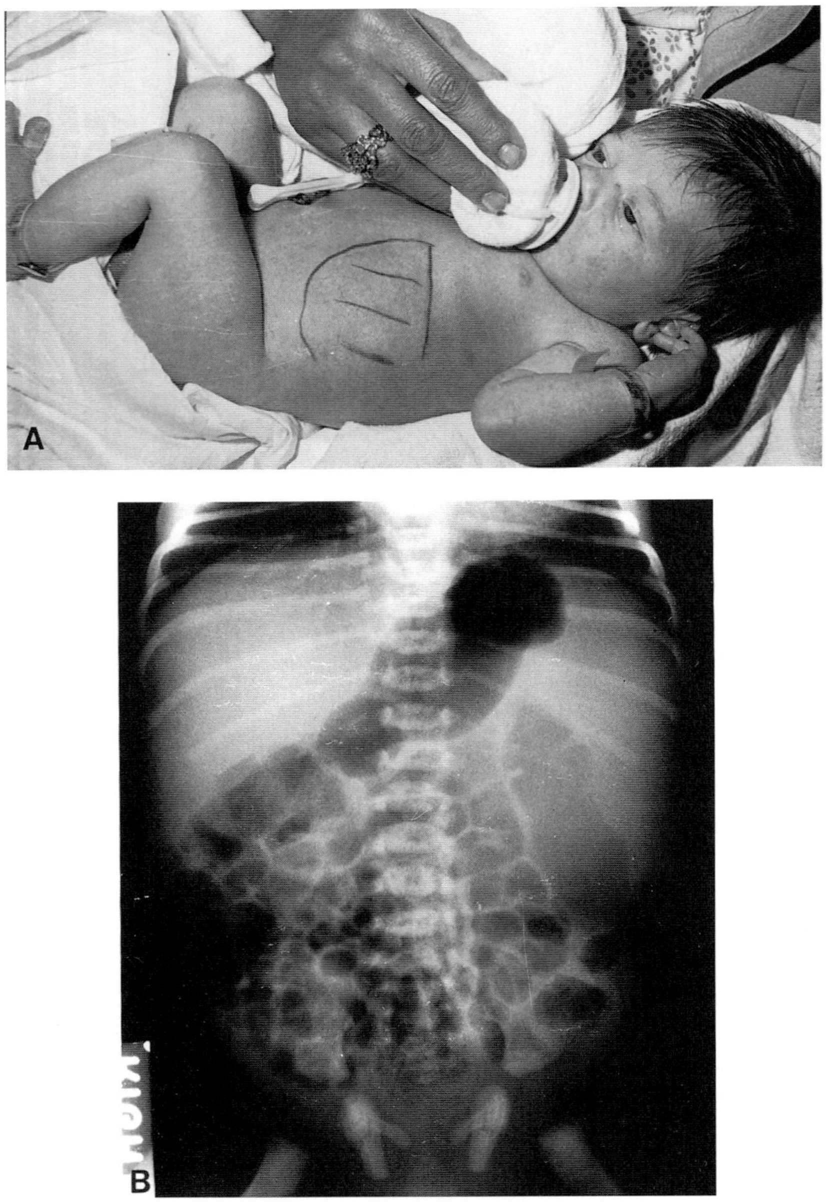

FIGURE 5.2. A 3-day-old male with stage IV-S neuroblastoma. (A) The left upper quadrant abdominal mass is outlined on the patient. (B) Abdominal radiograph reveals hepatomegaly due to multiple metastases. The adrenal tumor is not well visualized. (C) The 8 × 6 cm^2, 143 g tumor that replaced the left adrenal. The specimen has a diffuse, hemorrhagic cut surface, which is seen on the right. (D) The tumor consists of small, round, darkly staining cells forming Homer-Wright rosettes (arrowheads) containing fibrillary material (neuropil) in addition to the extensive hemorrhage (poorly differentiated, stroma poor, low MKI, "favorable histology"). H&E; × 750.

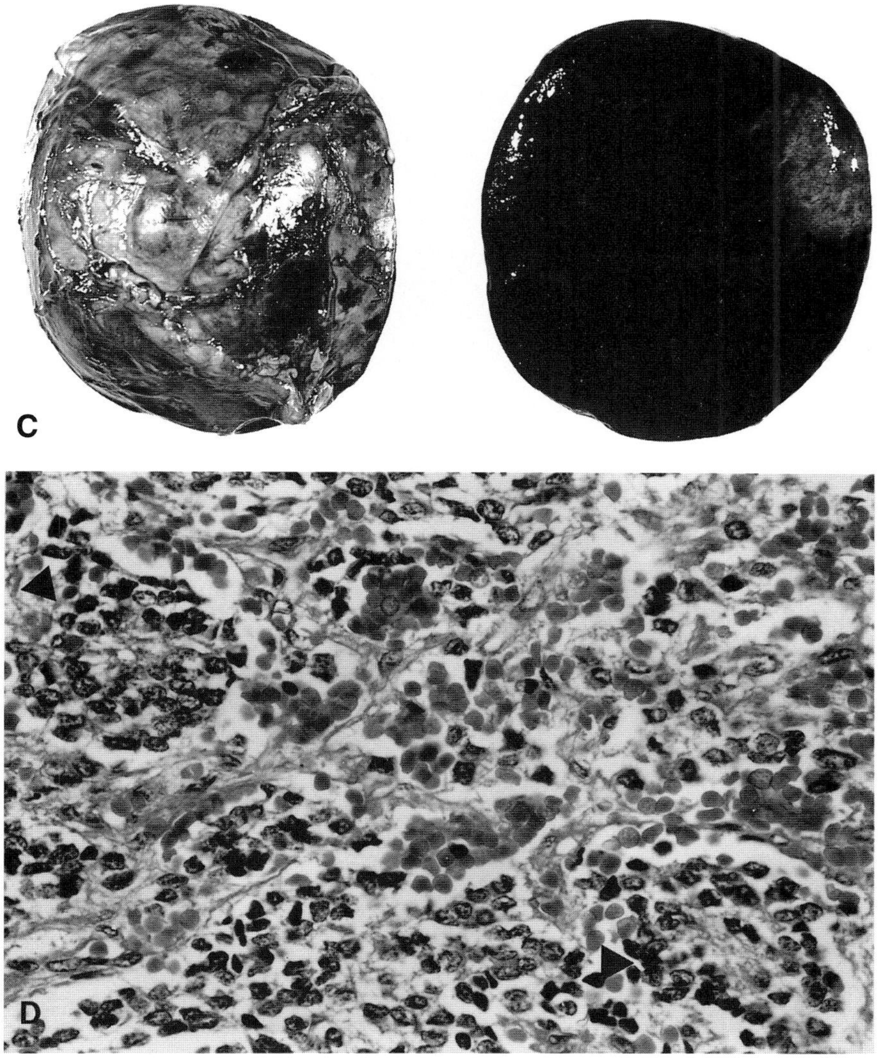

FIGURE 5.2. (*Continued*)

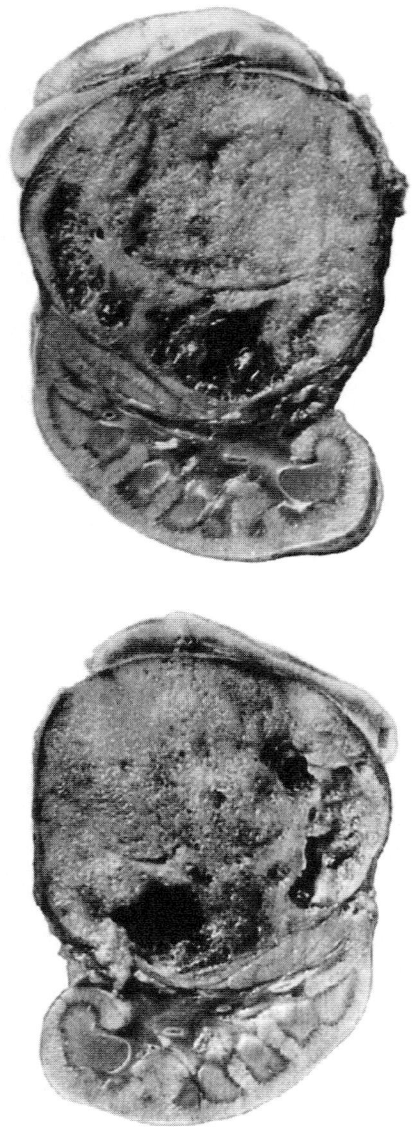

FIGURE 5.3. Adrenal neuroblastoma. Solid tumor, 4 cm in diameter, with areas of necrosis. Death from pulmonary hemorrhage at age 4 days. (From Isaacs H Jr [6], with permission.)

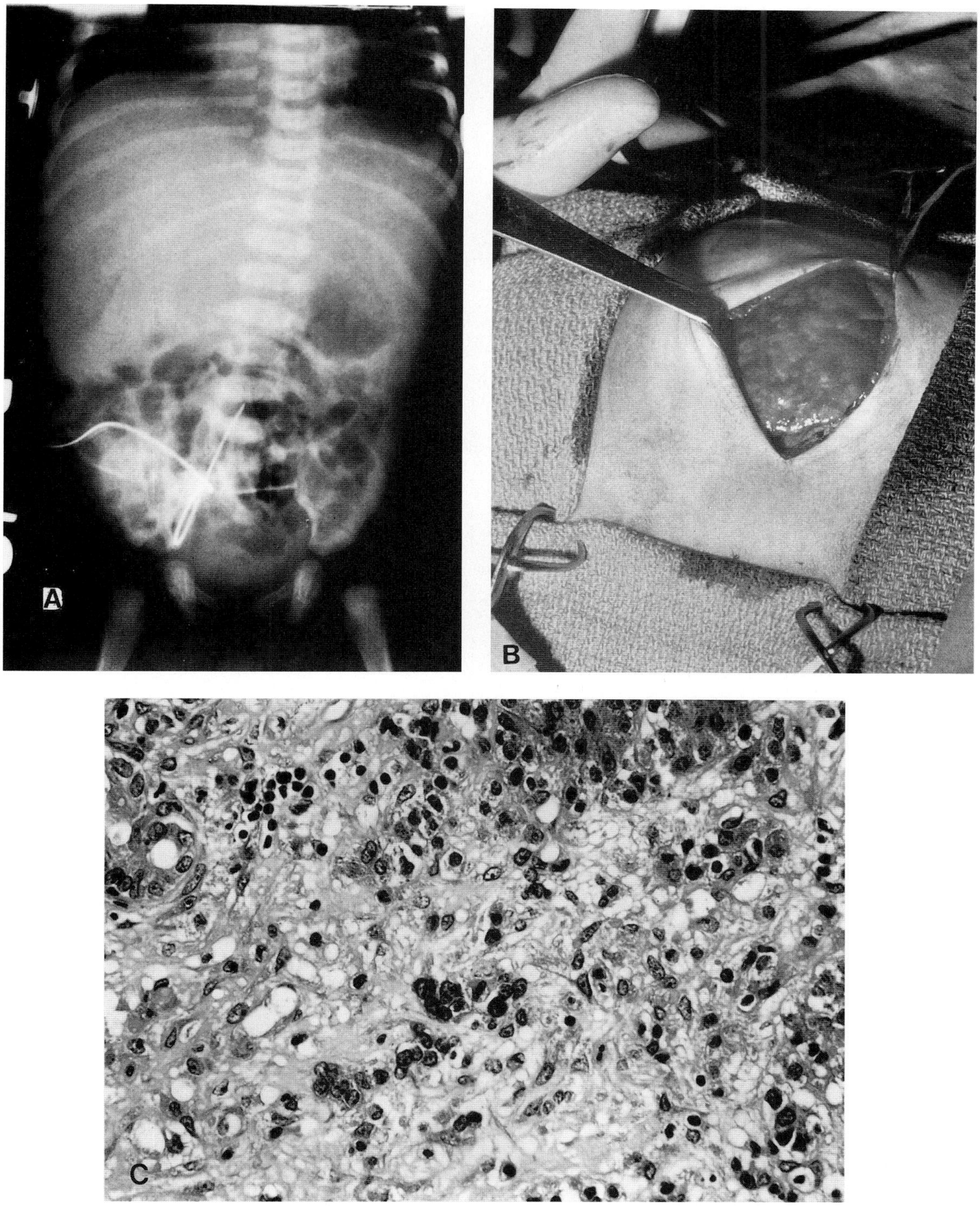

FIGURE 5.4. Stage IV-S neuroblastoma in a 6-day-old female, presenting as an abdominal mass due to neoplastic hepatomegaly. (A) Massive hepatomegaly depicted on an abdominal radiograph. (B) Exploratory laparotomy showing multiple tumor nodules replacing most of the liver. (C) Liver biopsy reveals extensive fibrosis, parenchymal loss, and tiny nests of residual neuroblasts, many of which are necrotic. Although the microscopic findings suggest spontaneous regression, infants with extensive metastatic liver disease may die either from hepatic failure or from pulmonary failure resulting from upward displacement of the diaphragm. H&E; × 600.

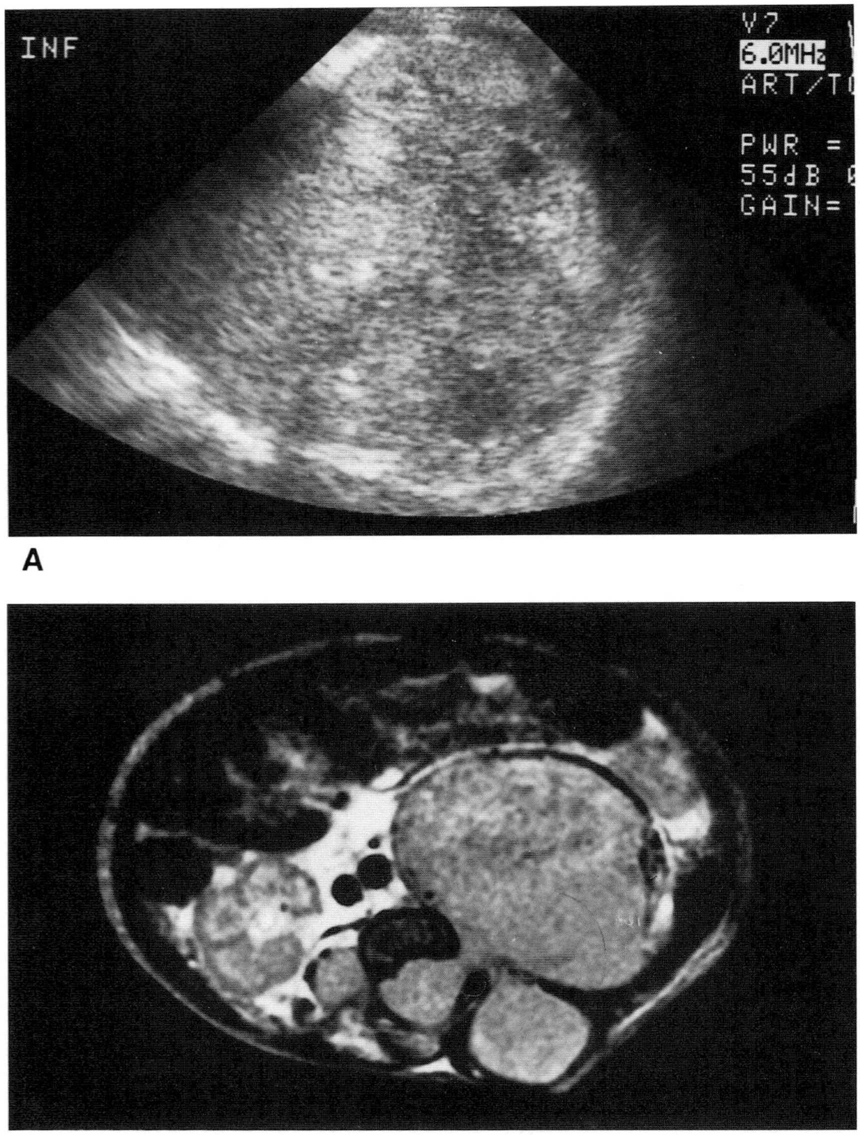

A

B

FIGURE 5.5. Retroperitoneal neuroblastoma. Newborn male with a retroperitoneal mass was paraplegic at birth. (A) Ultrasound scan reveals a large abdominal mass. (B) The tumor extends through the intervertebral foramina ("dumbbell" lesion). Other imaging studies showed a paraspinal mass going from T1 to L3 and liver metastases. (C) Low power (× 120) reveals a small-cell malignant tumor with a slightly lobular pattern and scant fibrillar material. Immunoperoxidase results were positive for neurofilament and negative for synaptophysin protein S-100, HBA71 antibodies, leukocyte common antigen (LCA), and muscle markers. There was no N-*myc* amplification, but aneuploidy was positive. The karyotype was 46,XY (undifferentiated, stroma poor, high MKI, unfavorable histology). H&E; × 120. (D) Electron photomicrograph (× 10,400) displays finely dispersed chromatin and a high nuclear-to-cytoplasmic ratio. Few organelles are present. (E) Higher magnification (× 17,800) reveals interdigitating cell processes with dense core (neurosecretory) granules and neurotubules. (Electron photomicrographs courtesy of Ann Peters, Department of Pathology, Children's Hospital San Diego.)

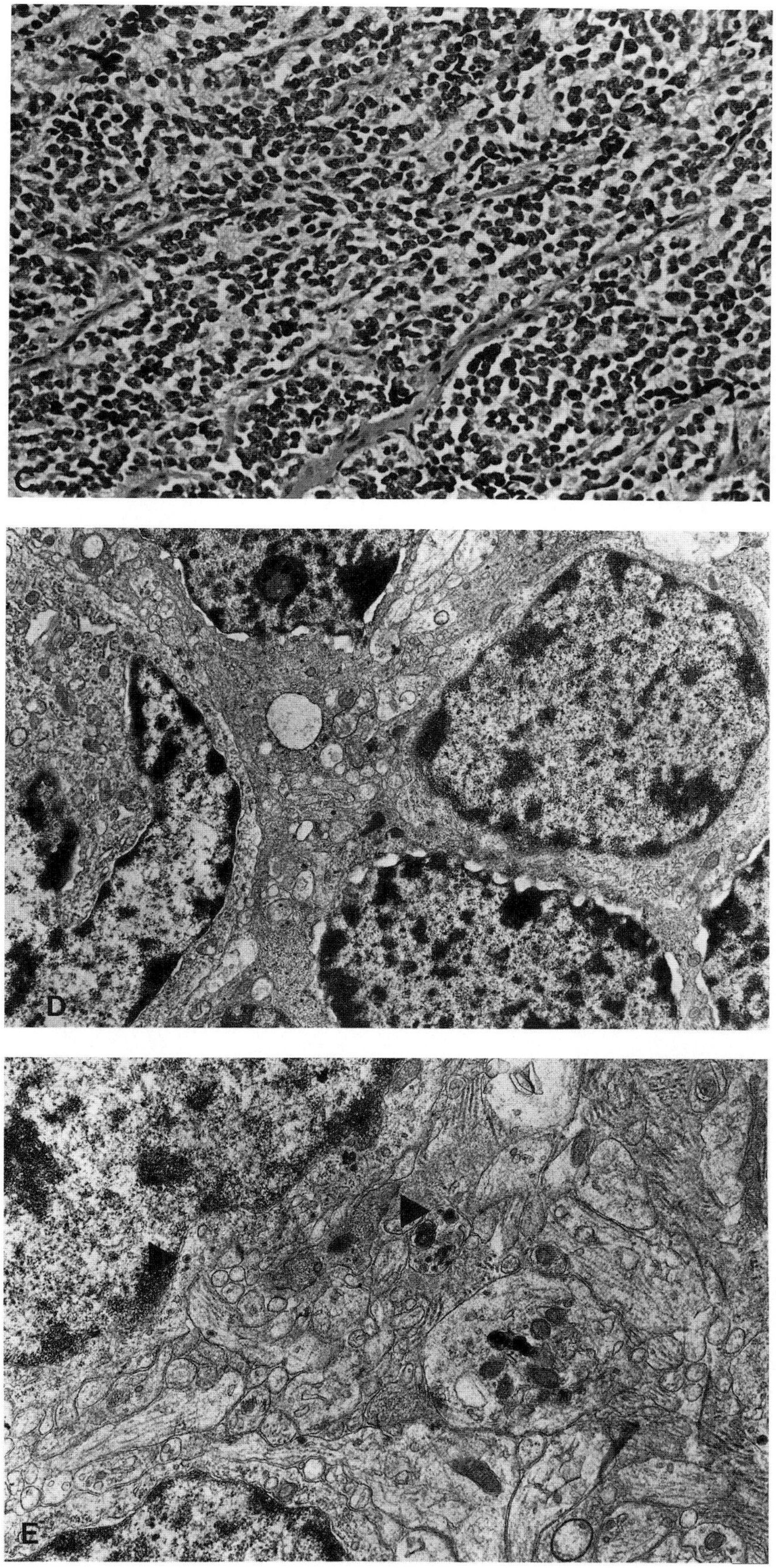

FIGURE 5.5. (*Continued*)

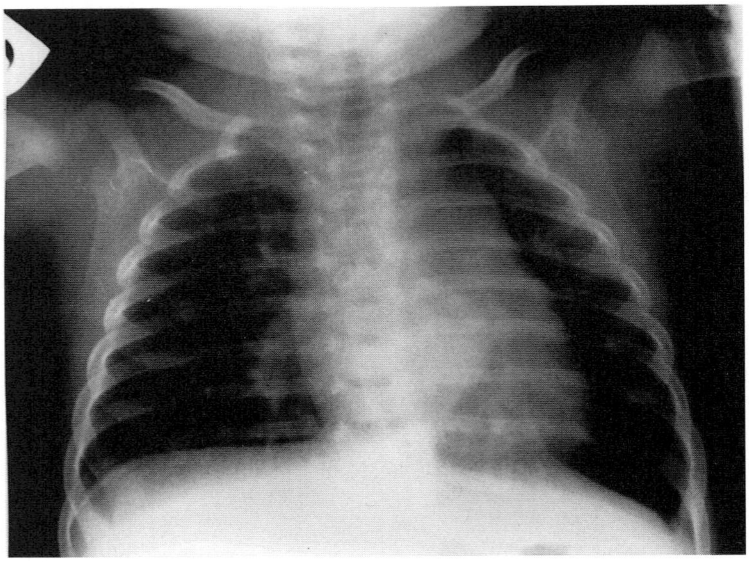

A

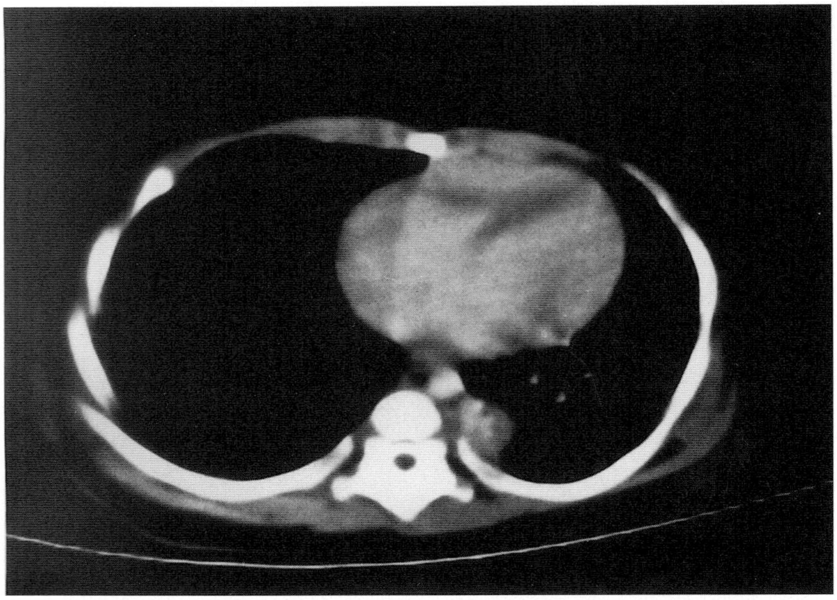

B

FIGURE 5.6. Mediastinal neuroblastoma in a 6-month-old female with a mediastinal mass noted as an incidental finding on a chest radiograph. (A) The radiograph reveals a posterior mediastinal mass with calcifications and suggestive hilar lymph node metastases. (B) Computerized tomographic scan outlining the prominent mediastinal mass. (C) The tumor, 5.1 g, 2.2×1.7 cm^2, has a dark red capsular suface (left side) and a soft white-pink cut surface (right) with focal hemorrhage and extensive calcifications. (D) The tumor has a lobular pattern with nests of neuroblasts and abundant neuropil surrounded by delicate fibrovascular septae. Foci of calcification, an additional favorable sign, are present (differentiating, stroma poor, low MKI, favorable histology). H&E; $\times$ 120.

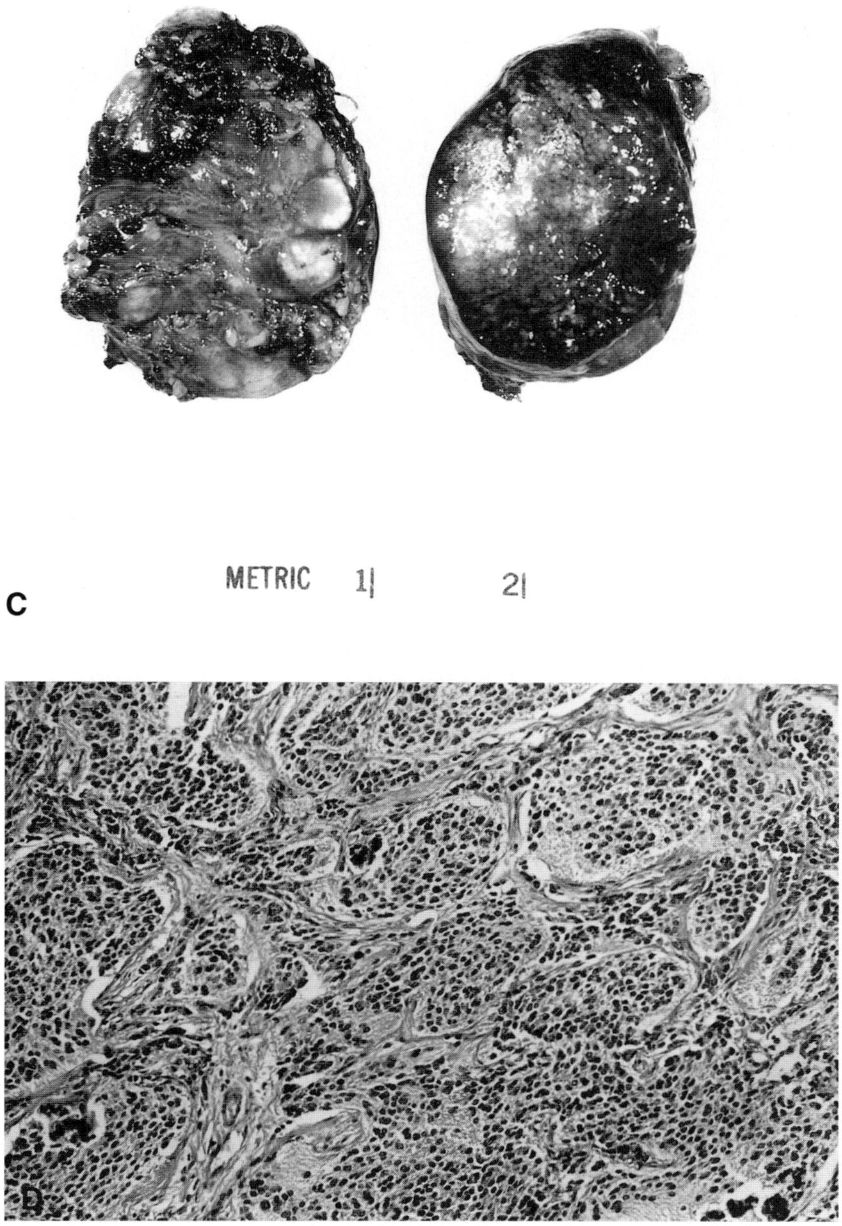

FIGURE 5.6. (*Continued*)

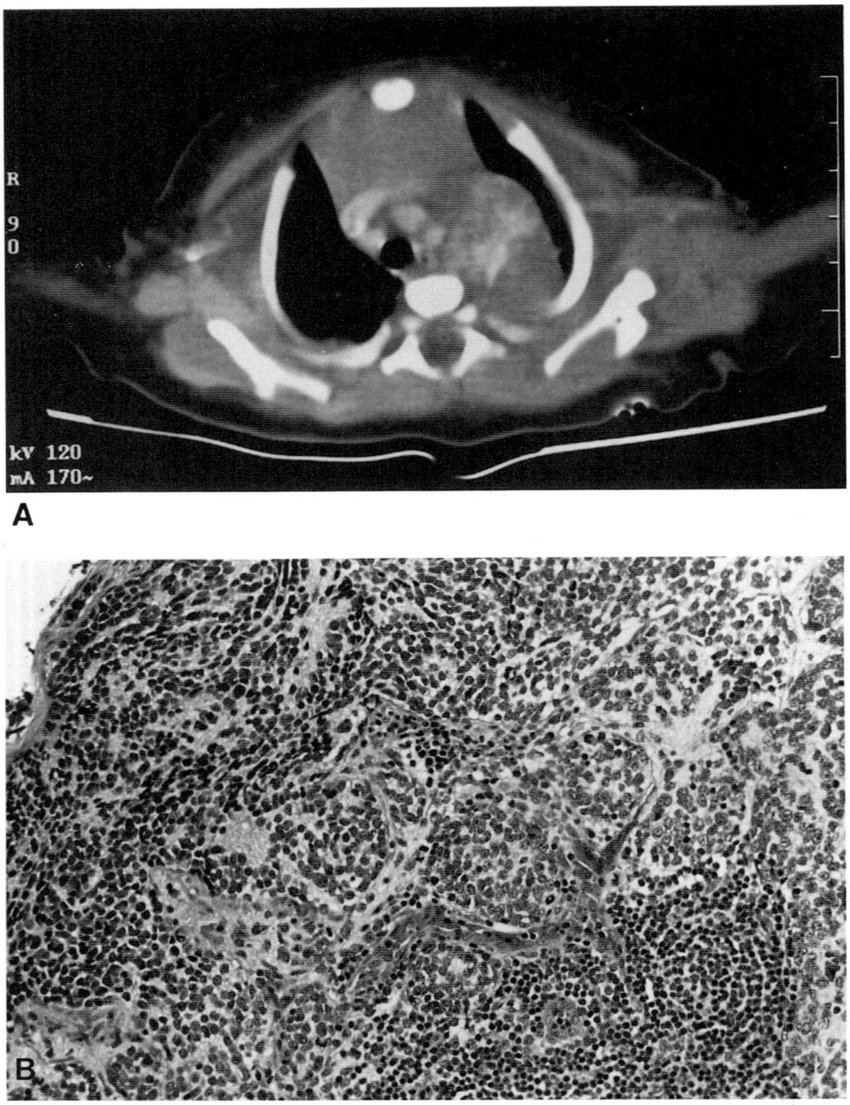

FIGURE 5.7. Posterior mediastinal neuroblastoma in a 6-day-old female with respiratory distress and a mediastinal mass. (A) Computerized tomographic scan shows a 3 cm, partially calcified posterior mediastinal mass situated in the thoracic paravertebral region and hilar lymph node metastases. (B) Posterior mediastinal lymph node with tumor metastases. Capsule of the node is situated in the upper left corner. Neuroblasts are slightly larger than lymphocytes within the lymphoid nodules. Tumor cells are surrounded by fine, delicate neuropil and form Homer-Wright rosettes, which aid in their identification. Low MKI, no N-*myc* amplification (differentiating, stroma poor, favorable histology). H&E; $\times$ 150.

FIGURE 5.8. Cervical neuroblastoma, stage I. (A) A 2-month-old male with a cervical mass and noisy respirations noted 1 month prior to hospitalization. The tumor, presumably arising from cervical sympathetic ganglia, is outlined on the patient. (B) The specimen, 4.5×3.5 cm^2, 20 g, has an irregular, light tan cut surface, which is on the left. (C) The tumor displays a lobular pattern, abundant neuropil, and focal ganglioneuroblasts with large vesicular nuclei, prominent nucleoli, and abundant cytoplasm (arrow) (differentiating, stroma poor, low MKI, favorable histology). H&E; $\times$ 600.

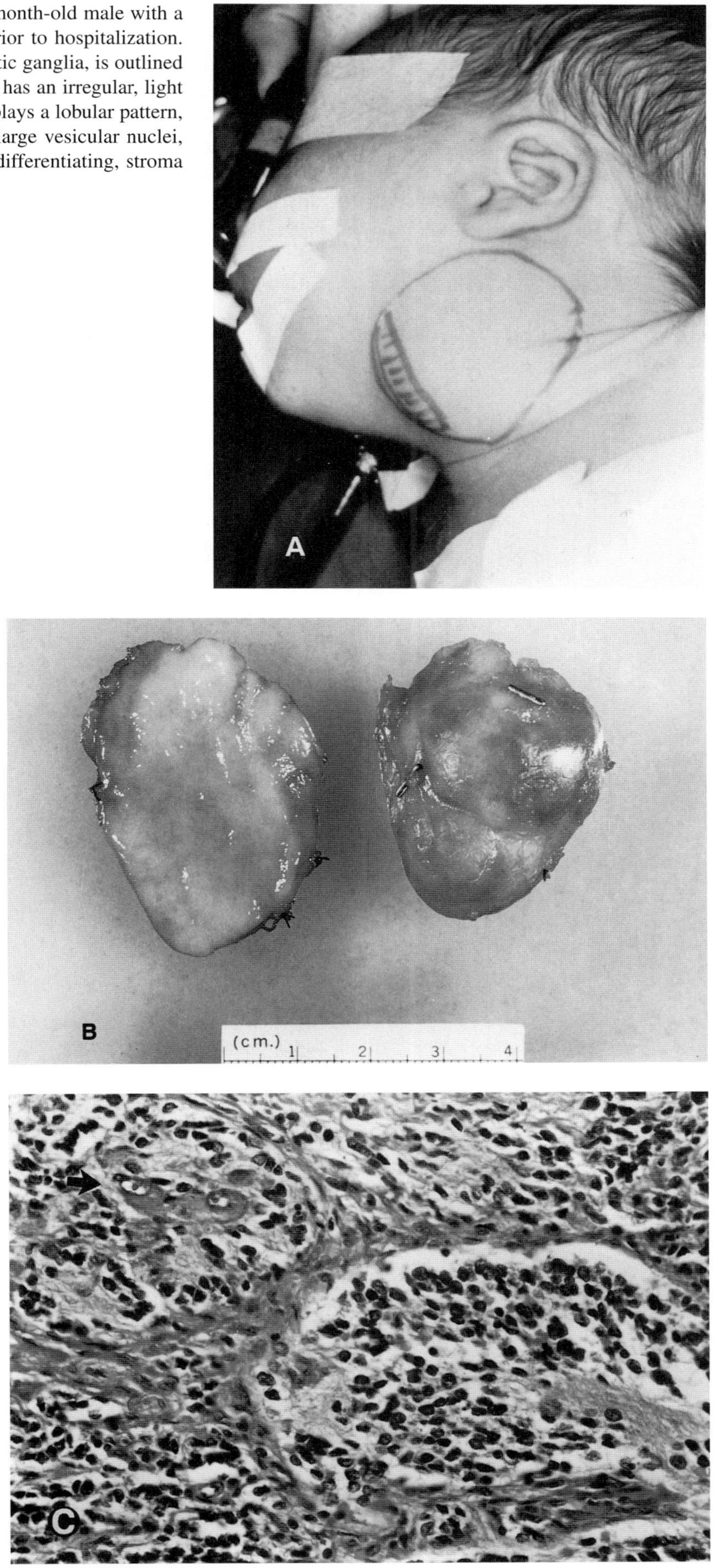

The initial clinical manifestations in greater than half the cases of neuroblastoma in newborns and infants are the result of metastatic disease [6,22]. In patients with congenital neuroblastoma, the adrenal is the leading primary site; over half have metastases at the time of diagnosis; and the main sites of metastases are the liver, bone marrow, and skin, in that order of frequency [22] (Figures 5.9 and 4.15). Clinical findings are summarized in Table 5.1.

Stage IV-S is a special form of metastatic neuroblastoma unique to the first year of life, and a favorable outcome would not be anticipated [24–29]. Patients designated as stage IV-S have small or undetectable primary neuroblastomas with metastases in one or more of the following sites: liver, skin, or bone marrow (not bone). The most common primary site is the adrenal, but other locations are the retroperitoneal area, medastinum, and pelvis, as well as multiple sites or no discernible primary site [23,27,29] (Figures 5.2 and 5.4). Multiple bluish cutaneous nodules producing a "blueberry muffin" appearance are noted in about a third of infants with stage IV-S neuroblastoma metastatic to the skin [30] (Figure 4.15). When the nodules are found, they are associated usually with liver metastases and a favorable prognosis [4,23,28]. Spontaneous regression and histological maturation are documented by serial biopsies or by excisions in patients with stage IV-S neuroblastoma and in those with stage I to II prenatally detected solitary adrenal tumors [6,11,25] (Figures 5.10 and 5.11).

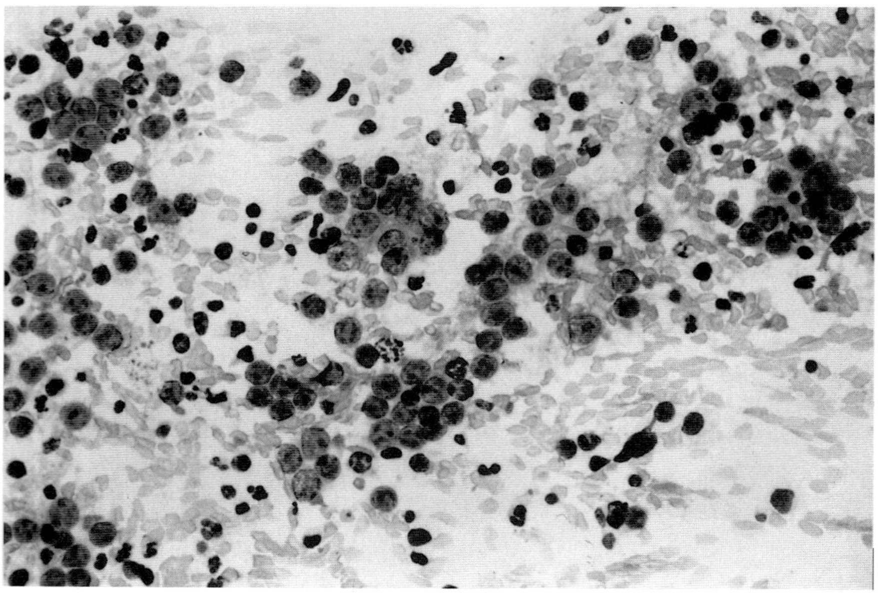

FIGURE 5.9. Bone marrow smear with neuroblastoma metastases. Giemsa-stained smear reveals small, round, darkly staining tumor cells forming syncytia. The bone marrow aspirate was from a neonate with stage IV-S neuroblastoma who had high urinary levels of vanillylmandelic acid and homovanillic acid. The bone marrow cytology and the urinary catecholamine findings were sufficient to establish the diagnosis. × 600.

TABLE 5.1. Clinical manifestations of neuroblastoma in the fetus and infant.

Abdominal mass
Neck mass
Hydrops fetalis
Cutaneous metastases ("blueberry muffin" appearance)
Progressive respiratory distress and pneumonia with bronchial obstruction from mediastinal tumor
Paralysis or weakness with spinal cord compression
Airway problems, stridor, and swallowing difficulties due to a large neck tumor
Metastasis to the umbilical cord leading to intrauterine death
Hemoperitoneum from spontaneous rupture of tumor during delivery
Neurogenic bladder
Hemiplegia with paraspinal "dumbbell" tumor
Congenital chylothorax
Cervical sympathetic ganglia involvement with Horner syndrome
Myoclonic encephalopathy
Visible dorsal paravertebral mass
Intestinal tumor associated with omphalocele and ileal atresia
Intraocular metastases
Scrotal mass

Source: Isaacs H Jr [6].

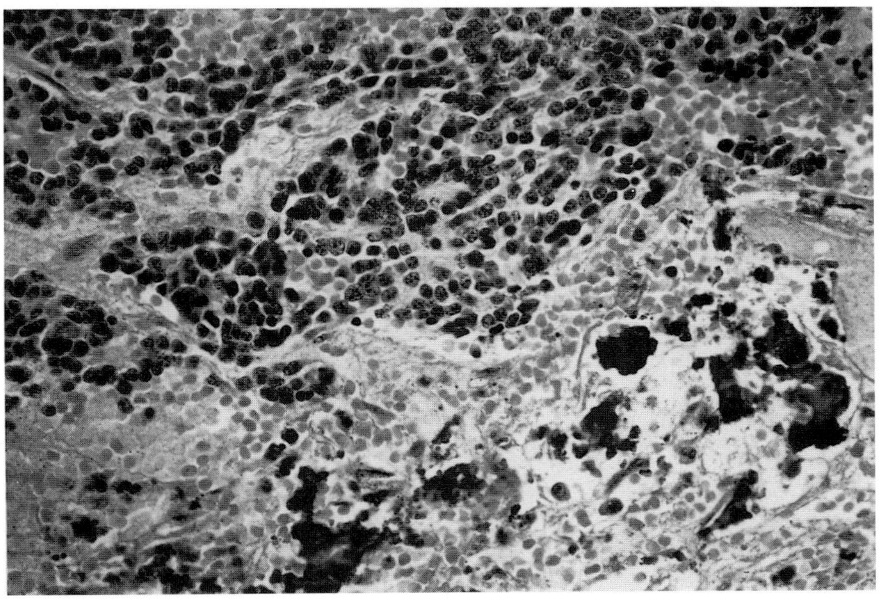

FIGURE 5.10. Spontaneous regression of congenital neuroblastoma, stage IV-S. Extensive necrosis, hemorrhage, and calcification are evidence of tumor regression. Apoptotic tumor cells are present. H&E; × 600.

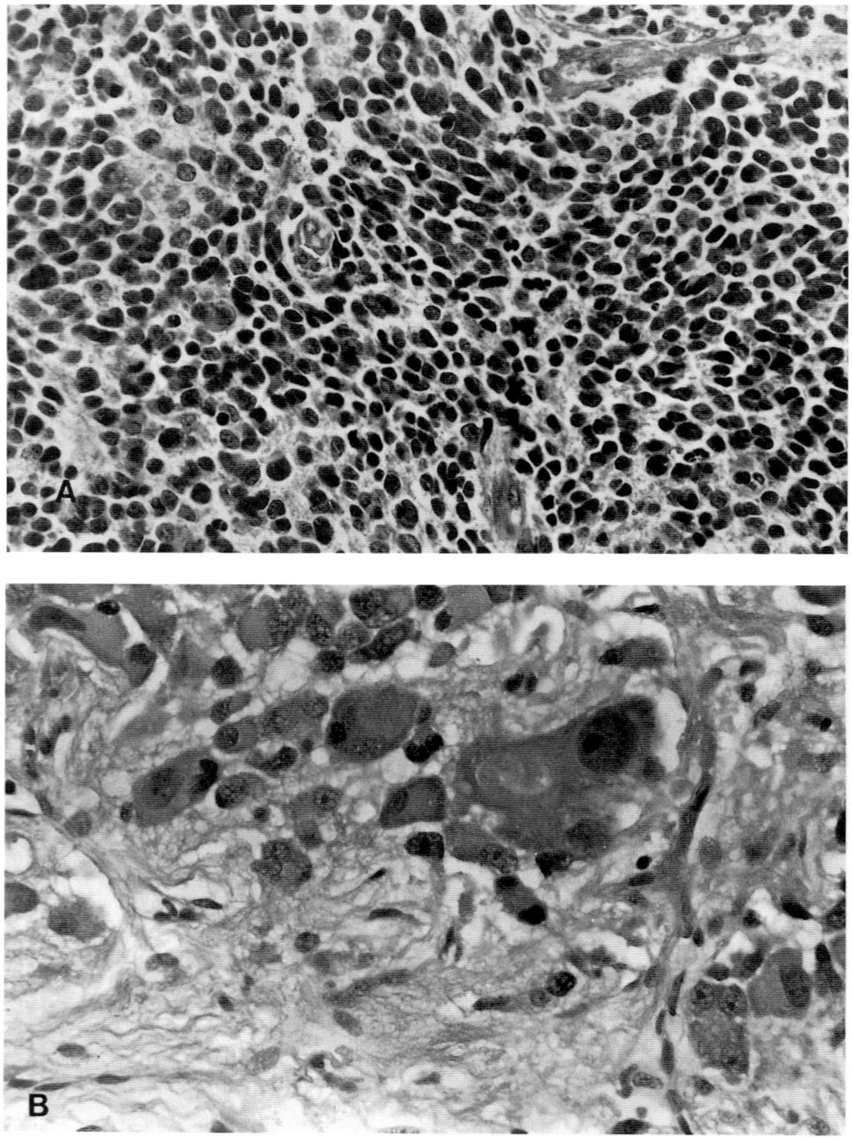

FIGURE 5.11. Maturation of neuroblastoma to ganglioneuroma in a 5-month-old male with a mediastinal mass and liver metastases. (A) Initial biopsy of the mediastinal mass reveals an undifferentiated neuroblastoma composed of small dark round to oval cells with inconspicuous cytoplasm and minimal intercellular fibrillar material (stroma poor, low MKI, favorable histology). The cells were immunoreactive with neuron-specific enolase (NSE) and showed weak focal reactivity with synaptophysin; the stained negatively with LCA and desmin. The patient was treated with chemotherapy (pediatric oncology group [POG] stage 4 protocol). H&E; × 300. (B) The patient's mediastinal mass was biopsied again 7 months later. The biopsy material shows maturation to ganglioneuroma. H&E; × 300.

TABLE 5.2. Favorable prognostic indicators in infants with neuroblastoma.

Feature[a]	Description
N-*myc*	1 copy (normal)
DNA ploidy	Hyperdiploid
1p LOH	Absent
14qLOH	Absent
TRK-A expression	High
Stage	Usually I, II, IV-S
3-year survival	95%

Source: Brodeur MG [40]; Lack EE [33].
[a] LOH, loss of heterozygosity; N-*myc*, N-*myc* oncogene; *TRK-A*, tyrosine kinase (receptor) growth factor.

TABLE 5.3. Neuroblastoma staging system proposed by the International Staging System Working Party.

Stage	Staging criteria
Stage I	Local tumor confined to the area of origin, complete gross excision, with or without microscopic residual disease; identifiable ipsilateral and contralateral lymph nodes negative microscopically
Stage IIa	Unilateral tumor with incomplete gross excision; identifiable ipsilateral and contralateral lymph nodes negative microscopically
Stage IIb	Unilateral tumor with incomplete gross excision with positive ipsilateral lymph nodes; contralateral lymph nodes negative microscopically
Stage III	Tumor infiltrating across the midline with or without regional lymph node involvement; or midline tumor with bilateral regional lymph node involvement
Stage IV	Dissemination of the tumor to distant lymph nodes, bone, bone marrow, liver, and/or other organs (except as defined under stage IV-S)
Stage IV-S	Localized primary tumor as defined for stage I or II with dissemination limited to liver, skin, and/or bone marrow

Source: Lack EE: Recommendations for the reporting of tumors of the adrenal cortex and medulla: Association of Directors of Anatomic and Surgical Pathology. *Virchows Arch* 1999;435:87. Used with permission.

Prognostic indicators in infants with neuroblastoma are outlined in Table 5.2 and the international staging system for this tumor in Table 5.3.

Pathology

There are three main forms of neuroblastic tumors: neuroblastoma, ganglioneuroblastoma, and ganglioneuroma [31–34]. Neuroblastomas can be classified as undifferentiated, poorly differentiated, or differentiating. Ganglioneuroblastomas are grouped into three subtypes: nodular, intermixed, and borderline [31–34].

Neuroblastoma is the foremost neuroblastic tumor of the fetus and infant [5,6] (Figures 5.1–5.7, 5.12). Ganglioneuroblastoma, consisting of ganglioneuroma comprising over half the tumor with a minor neuroblastoma component, is not as common (Figures 5.8 and 5.13). Ganglioneuroma,

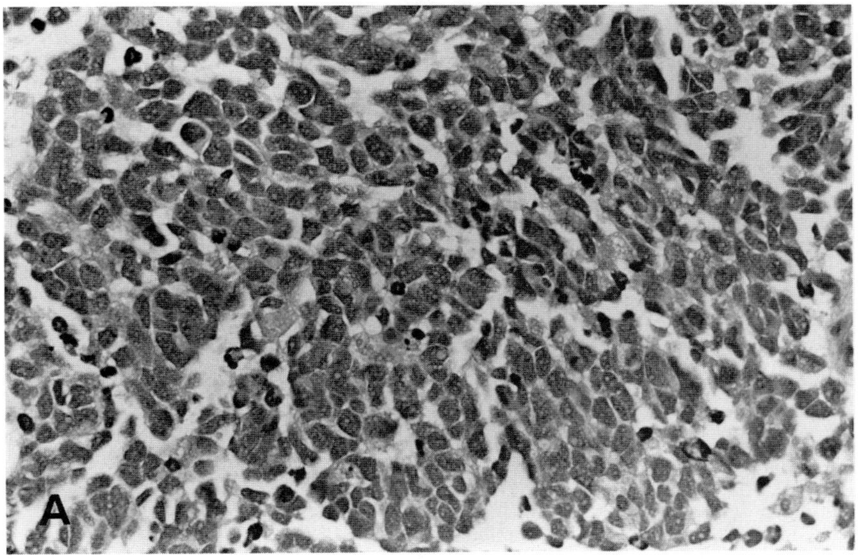

FIGURE 5.12. Undifferentiated neuroblastoma. (A) A small-cell malignant tumor without rosettes or neuropil (stroma poor, high MKI, N-*myc* amplification, unfavorable histology). H&E; ×150. (B) Imprint ("touch preparation") of the tumor. The cells are round to oval with a high nuclear-to-cytoplasmic ratio. Fine "peppery" chromatin pattern and one or more tiny nucleoli are present. (C) The cells are immunoreactive with NSE. In addition they are reactive with synaptophysin and focally reactive with S-100 protein, but stain negatively with desmin, actin, LCA, anti-β_2 microglobulin, and HBA71 antibodies. × 300. (D) Electron photomicrograph (× 42,000) reveals microtubules and dense core (neurosecretory) granules (arrows). (Electron photomicrograph courtesy of Ann Peters, Department of Pathology, Children's Hospital, San Diego.)

(Continued)

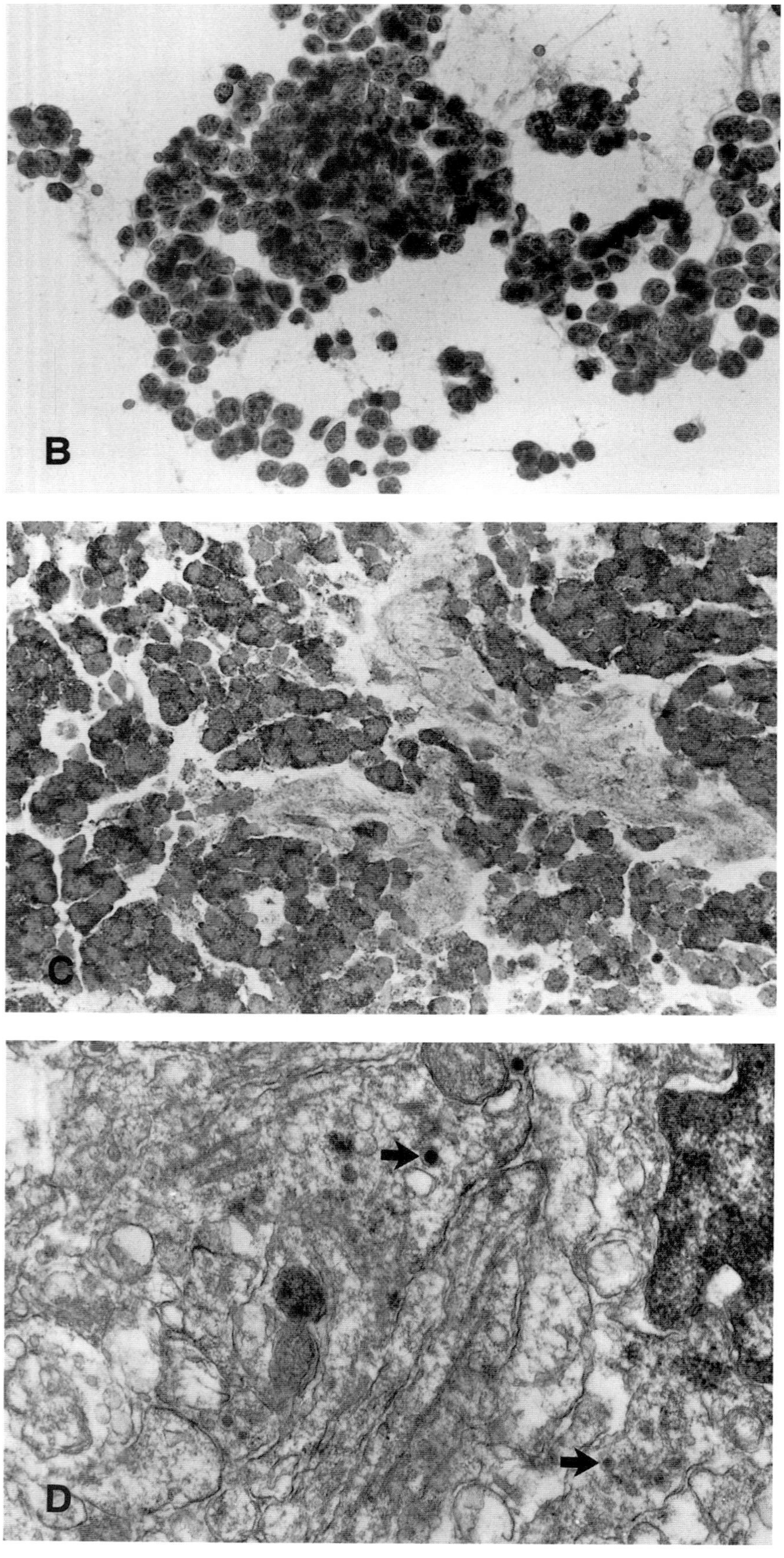

FIGURE 5.12. (*Continued*)

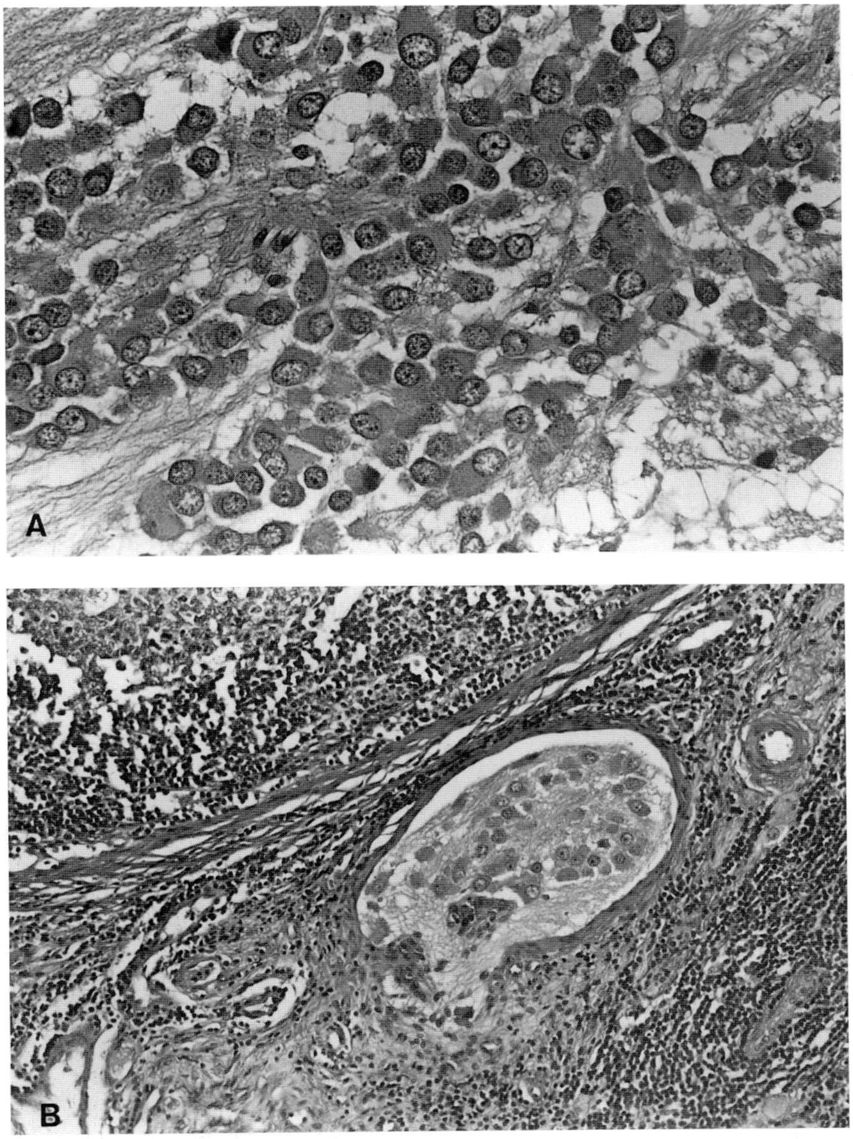

FIGURE 5.13. Cervical neuroblastoma. Neck mass in an 8-month old male with metastases to lymph nodes and bone marrow. (A) Biopsy of the neck tumor reveals maturing neuroblasts with round to slightly oval nuclei, a fine, peppery chromatin pattern, and prominent cytoplasms. No mitoses are noted. Delicate fibrillar neuropil is present. The cells were immunoreactive with NSE and protein S-100; also reported were tumor cytogenetics (46, XY) and N-*myc* amplification (normal) [differentiating neuroblastoma (ganglioneuroblastoma), stroma poor, low MKI, favorable histology]. H&E; × 300. (B) Section of a cervical lymph node biopsy reveals metastases within lymphatics. H&E; × 120.

composed of mature ganglion cells, Schwann cells, and fibrous connective tissue, seldom occurs as a primary tumor during the first year of life. However it is found in follow-up (serial) biopsy samples taken from infants with spontaneous regression, particulary in those with stages I, II, and IV-S disease, with or without treatment [5,25] (Figure 5.11).

Gross and microscopic features of neuroblastoma in the newborn and infant are similar to those seen in the older child [1,3,6,21,34]. Tumors in the adrenal gland vary in size from a few millimeters to several centimeters in diameter. They may be either solid or cystic (Figures 5.2 and 5.3). The adrenal cortex may persist and surround the tumor, or it may be visible only on one surface as a thin, orange-tan rind, while the rest is so attenuated that it cannot be distinguished on gross examination. The solid portions of the tumor are light gray to red-brown, depending on the amount of hemorrhage, soft, and somewhat granular in appearance. Flecks of cream-colored calcification are present in some tumors (Figure 5.6).

Microscopically the tumor consists of small neuroblasts with poorly demarcated boundaries whose round or oval hyperchromatic nuclei are proportionately large in relation to total cell volume (Figures 5.1, 5.2, 5.4–5.8, and 5.12). Fine fibrils (neuropil) either form a delicate interlacing network between the cells or occur in areas almost devoid of cells. Rosettes consisting of a small, pale, finely granular central area surrounded by a petal-like arrangement of cells (Homer-Wright rosettes) are variably present (Figure 5.2D). Hemorrhagic necrosis and calcification are noted particularly in larger tumors. Both bone marrow aspiration and bone marrow biopsy are recommended to obtain the highest yield of detection of metastatic neuroblastoma [35] (Figure 5.9).

Neuroblastoma is a classic example of one of the small, blue-cell tumors of infancy and childhood raising diagnostic problems associated with these tumors [7] (Figure 5.12). The light microscopic, immunohistochemical, and ultrastructural features of the small blue-cell tumors were listed in Table 3.2.

Shimada et al. propose a histopathological classification for neuroblastoma that correlates certain histological patterns and prognosis [36]. The histological criteria are based primarily on the evaluation of maturation of both neuroblastic and stromal elements and of nuclear morphology. Neuroblastomas are divided into two main categories, stroma rich and stroma poor; the terminology currently used in this system is "Schwannian stroma rich or poor" [32]. The mitosis-karyorrhexis index (MKI) of the neuroblasts is evaluated. The MKI appears to be the most important factor for prognosis particularly in children less than 1.5 years of age [36]. The data of Shimada and coworkers suggest that a low MKI is more significant than maturation for survival in this age group. Infants with stage I and IV-S disease, the most favorable prognostic groups, usually have in addition to a stroma-poor pattern, a low MKI and usually a normal N-*myc* copy. Usually the stroma-rich pattern is not observed in neuroblastomas removed from newborns and infants [7,32].

Another grading system devised by Joshi et al. seems easier to use [37]. The more familiar terms neuroblastoma and ganglioneuroblastoma are retained and used instead of stroma-poor neuroblastoma and stroma-rich neuroblastoma [36]. Undifferentiated neuroblastoma is classified as a subtype separate from poorly differentiated neuroblastoma, and "ganglioneuroblastoma" is reserved for a predominant ganglioneuroma component in addition to a minor neuroblastoma component [31,37].

Moreover, calcification and a low mitotic rate of less than 10 mitoses per high-power field are the two most important prognostic histological indicators [37] (Table 5.4). Histological grades 1, 2, and 3 are determined on the basis of the presence of both, any one, or none of these two histological indicators, respectively. Age (over or under 1 year), stage, DNA analysis, N-*myc* copy number, and the histological grade are taken into consideration to determine the final interpretation of high- and low-risk group categories. It was mentioned earlier that calcification within the tumor is a good prognostic sign. Infants with low grade tumors (grades 1 and 2) are categorized as a low-risk group and have the best prognosis [31,37].

TABLE 5.4. Pathologic features in the grading of neuroblastic tumors.

Differentiation of neuroblasts	Undifferentiated Poorly differentiated Differentiating
Mitosis-karyorrhexis index	Low (≤2% cells) Intermediate (2–4% of cells) High (≥4% of cells)
Stroma (neuroma-like or schwannian; not neurofibrillary)	Poor Rich
Calcifications	Present Absent

Source: Lack EE [33]. Used with permission.

The ultrastructural features of neuroblastoma are described by several investigators [7,34,38,39]. The findings depend on the degree of cytological differentiation of the tumor in question (Figures 5.5, 5.12, and 5.14). Neuroblastomas with the densely cellular, primitive, "small blue-cell tumor" appearance of undifferentiated neuroblastoma on light microscopy may be difficult to distinguish by electron microscopy from such other small-cell malignant tumors as the undifferentiated sarcomas, alveolar rhabdomyosarcoma, and leukemia-lymphoma (Figure 5.12) (Table 3.2). The presence of dense core or neurosecretory granules, which are found also in small numbers in the primitive neuroectodermal tumor, suggests the diagnosis by electron microscopy [38] (Figure 5.12). The more differentiated tumors show in addition a network or neuropil composed of cells with neural processes, microtubules, neurofilaments, and cell attachments some resembling synapses [38,39] (Figure 5.14).

Ultimately immunoproxidase studies are required to confirm the diagnosis. The tumor cells are reactive with neuron-specific enolase (NSE), synaptophysin, neurofilament, and focally positive for S-100 protein; they stain negatively for desmin, actin, leukocyte common antigen, cytokeratin, and the Ewing antigen [6,34] (Figure 5.12C) (Table 3.2).

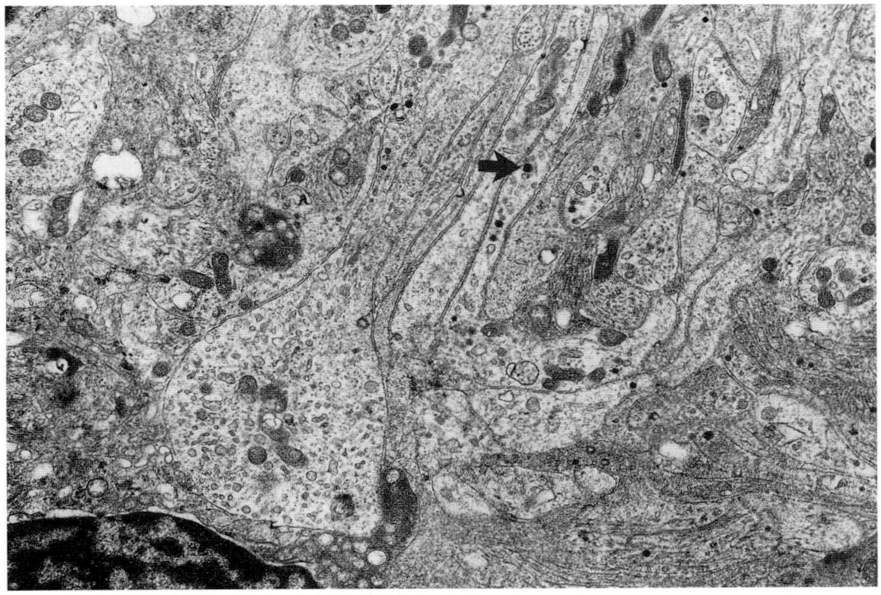

FIGURE 5.14. Ultrastructure of a differentiating neuroblastoma: neuropil composed of a network of neurites containing neurotubules and neurosecretory granules (arrow). One neurite directly adjacent to the nucleus has a bulbous swelling suggestive of synaptic bouton. × 20,000. (Courtesy of Darkin Chan, Department of Pathology, Children's Hospital, Los Angeles.)

Spontaneous Regression

Spontaneous regression of neuroblastomas is not uncommon, particularly in newborns (a mean age of 3 months) [3,25,26]. It is manifested either by maturation of the tumor or by foci of degeneration, necrosis, and eventual disappearance [20] (Figure 5.10). Spontaneous regression of neuroblastoma occurs more often than for any other pediatric neoplasm [6,40], but its exact incidence is unknown [25]. Massive apoptosis of tumor cells characterized by condensed nuclear fragments and eosinophilic cytoplasm is observed in neuroblastomas and may be related to tumor regression [41] (Figure 5.10). It was mentioned that serial biopsy samples taken from stage IV-S neuroblastoma patients may show evidence of spontaneous regression [6,25] (Figure 5.11). Similar trends are observed among congenital retinoblastomas (see Chapter 9, Tumors of the Eye).

Neuroblastoma in Situ

Neuroblastoma in situ occurs exclusively in the fetus and young infant and is defined as microscopic, incidentally found nests of neuroblastic cells situated in the adrenal medulla and not associated with either grossly visible tumor or metastases [42] (Figure 5.15). The frequency in which small collections

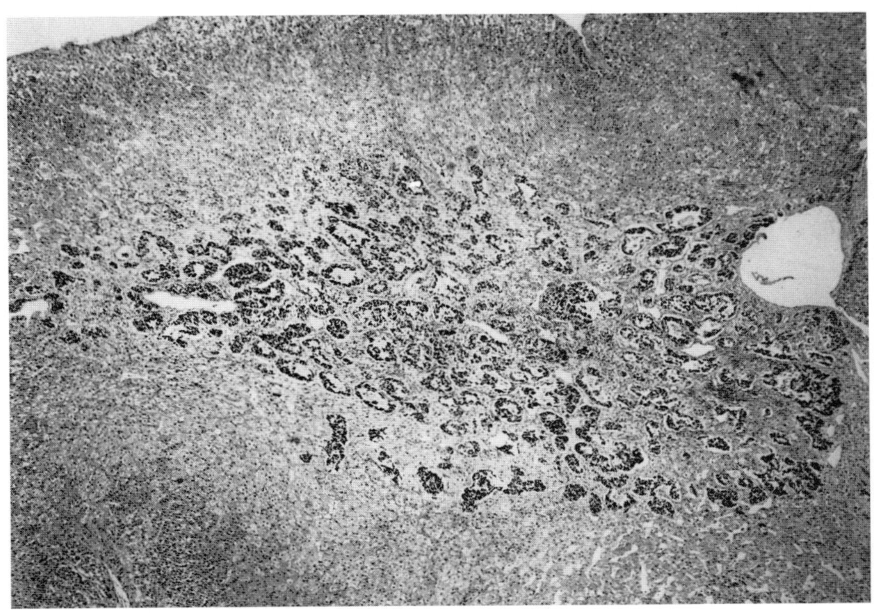

FIGURE 5.15. Neuroblastoma in situ. The adrenal contains a central area of neuroblasts not invading the capsule. This was an incidental postmortem microscopic finding in a 2-day-old, 3 kg male who died from the hypoplastic left heart syndrome. H&E; × 60.

of neuroblasts are found in newborns suggests that maturation and cessation of growth of potentially lethal tumors is a relatively common occurrence [42–44]. The incidence of neuroblastoma in situ reported differs widely varying from 1 in 10 to 1 in 500 necropsies [5,42–44].

Congenital heart disease, tracheoesophageal fistula, genitourinary tract anomalies, and trisomy 18 are found in fetuses and newborns at postmortem examination in as many as 73% of patients with neuroblastoma in situ [5,42,44]. Such collections of cells have been observed also in pre- and postnatally detected cystic adrenals, and these cystic neuroblastomas are associated with a good prognosis [6,9,12,45] (Figure 5.16).

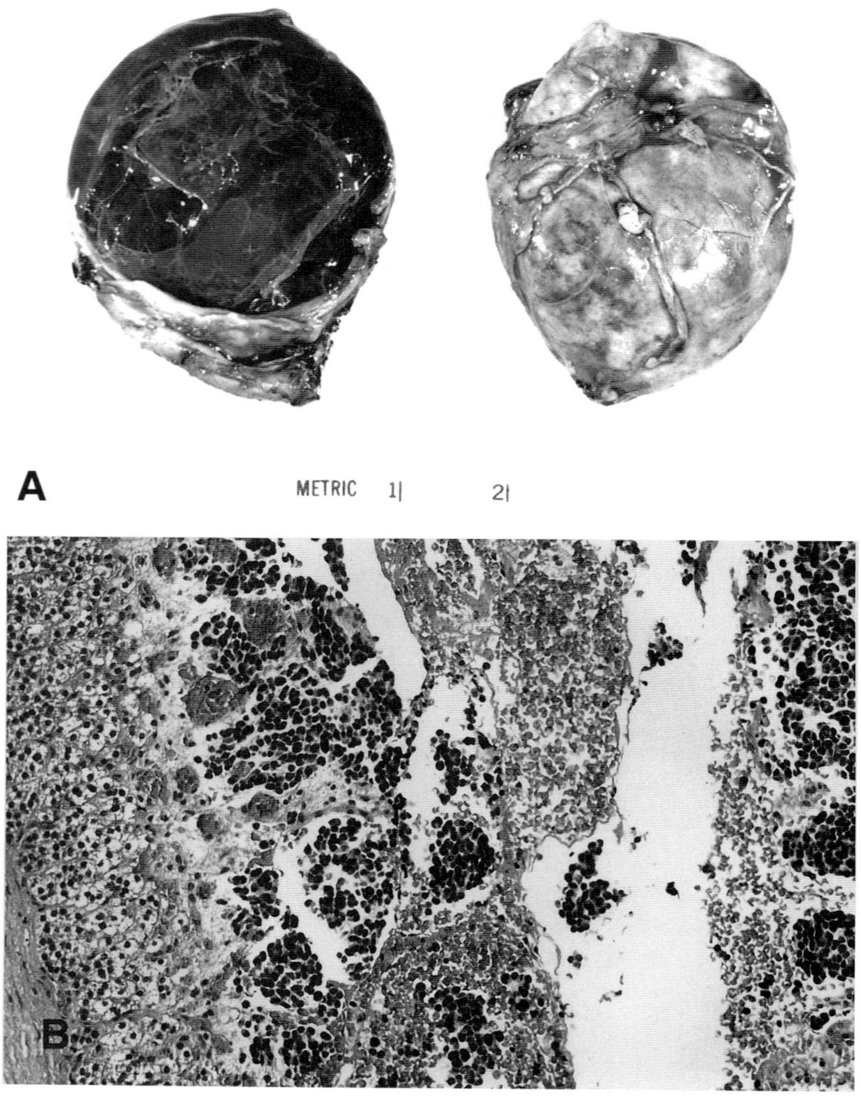

FIGURE 5.16. Cystic adrenal with hemorrhage and neuroblastoma in situ. A cystic lesion thought to be adrenal hemorrhage was detected on routine prenatal sonography. (A) The cut surface (left) of the adrenal reveals a large area of hemorrhage. (B) Small, dark round neuroblasts are situated within the adrenal cortex beneath the capsule. No capsular invasion is present. Usually prenatal cystic neuroblastoma in situ has a good prognosis. H&E; × 150.

Ganglioneuroma

Ganglioneuromas of the adrenal area seldom occur in the fetus and infant [5,46]. Typically they are tumors of older children and adults. Ganglioneuromas are more highly differentiated than neuroblastoma and are considered to be the benign counterpart of the latter [31,34].

Histologically ganglioneuroma consists of mature or slightly immature ganglion cells having an abundant pink cytoplasm and a round nucleus with prominent nucleolus; they are surrounded by interlacing bundles of Schwann cells, the stroma resembling a neurofibroma [31,34]. Mitotic figures should not be observed in ganglioneuromas. In the young they may occur spontaneously or after therapy for neuroblastoma. Spontaneous and therapy-induced maturation of neuroblastoma to ganglioneuroma is documented [7,25–28,46] (Figure 5.11B). This maturation process occurs also in metastatic sites. Nevertheless, growth of ganglioneuroma may cause functional problems requiring surgical intervention [46]. Although some are unresectable, ganglioneuromas are associated with a long-term disease-free survival [46].

Prognosis

The age of the patient and the stage of the disease are the two most important prognostic factors. The prognosis for infants and newborns is far better than that for older children. Several reasons are given for this observation: (1) infants present more often with a treatable localized lesion (a lower stage) compared with the older child [8,23] (see Table 5.3 for staging of neuroblastoma); (2) more infants and newborns are diagnosed with stage IV-S disease, which exhibits a tendency to spontaneously regress with little or no therapy [8,25,29,47]; (3) those with antenatally detected neuroblastoma represent a subset of patients who generally have a good outcome [48]; and (4) neuroblastomas appear to be biologically and genetically different from those occurring in the older child [40].

References

1. Bolande RP. The neurocristopathies. A unifying concept of diseases arising in neural crest maldevelopment. *Hum Pathol* 1974; 15:409.
2. Bader JL, Miller RW. US cancer incidence and mortality in the first year of life. *Am J Dis Child* 1979;133:157.
3. Dehner LP. Neoplasms of the fetus and neonate. In: Naeye RL, Kissane JM, Kaufman N, eds. *Perinatal Diseases*. Baltimore: Williams & Wilkins; 1981:286. International Academy of Pathology, Monograph No. 22.
4. Broadbent VA. Malignant disease in the neonate. In: Roberton NRC, ed. *Textbook of Neonatology*. 2nd ed. Edinburgh: Churchill Livingston; 1992:879.
5. Isaacs H Jr. Perinatal (congenital and neonatal) neoplasms: a report of 110 cases. *Pediatr Pathol* 1985;3:165.
6. Isaacs H Jr. Tumors. In: Gilbert-Barness E, ed. *Potter's Pathology of the Fetus and Infant.* Vol 2. St Louis, MO: CV Mosby; 1997:1248.
7. Isaacs H Jr. Neuroblastoma. In *Tumors of the Newborn and Infant*. St Louis, MO: Mosby-Year Book; 1991.
8. Isaacs H Jr. Neuroblastoma. In: *Tumors of the Fetus and Newborn*. Philadelphia: WB Saunders; 1997:130. Major Problems in Pathology. Series; Vol 35.
9. Atkinson GO, Zaatari GS, Gay BB, et al. Cystic neuroblastoma in infants: radiographic and pathologic features. *Am J Radiol* 1986;146:113.
10. Jaffa AJ, Many A, Hartoov J, et al. Prenatal sonographic diagnosis of metastatic neuroblastoma: report of a case and review of the literature. *Prenat Diagn* 1993;13:73.
11. Acharya S, Jayabose S, Kogan SJ, et al. Prenatally diagnosed neuroblastoma. *Cancer* 1997;80:304.
12. Kozakewich HPW, Perez-Atayde AR, Donovan MJ, et al. Cystic neuroblastoma: emphasis on gene expression, morphology, and pathogenesis. *Pediatr Dev Pathol* 1998;1:17.
13. Benirschke K, Kaufman P. *Pathology of the Human Placenta*. 4th ed. New York: Springer-Verlag; 2000:471.
14. Anders D, Kindermann G, Pfeifer U. Metastasizing fetal neuroblastoma with involvement of the placenta simulating fetal erythroblastosis. Report of two cases. *J Pediatr* 1973;82:50.
15. Moss TJ, Kaplan L. Association of hydrops fetalis with congenital neuroblastoma. *Am J Obstet Gynecol* 1987;132:905.
16. Ohyama M, Kobayashi S, Aida N, et al. Congenital neuroblastoma diagnosed by placental examination. *Med Pediatr Oncol* 1999;33:430.
17. Lynn AAA, Parry SI, Morgan MA, et al. Disseminated congenital neuroblastoma involving the placenta. *Arch Pathol Lab Med* 1997;121:741.
18. Isaacs H Jr. Congenital malignant tumors. In: Reed GB, Claireaux AE, Bain AD, eds. *Diseases of the Fetus and Newborn: Pathology, Radiology and Genetics*. London: Chapman Hall; 1989:132.
19. Campbell AN, Chan HSL, O'Brien A, et al. Malignant tumours in the neonate. *Arch Dis Child* 1987;62:19.
20. Bolande RP. Developmental pathology. *Am J Pathol* 1979;94: 627.
21. Isaacs H Jr. Neoplasms in infants: a report of 265 cases. *Pathol Annu* 1983;18(2):165.
22. Schneider KM, Becker JM, Krasna IH: Neonatal neuroblastoma. *Pediatrics* 1965;36:359.
23. Grosfeld JL, Rescoria FJ, West KW, et al. Neuroblastoma in the first year of life: clinical and biologic factors influencing outcome. *Semin Pediatr Surg* 1993;2:37.
24. Evans AE, D'Angio GJ, Randolph J. A proposed staging for children with neuroblastoma. Children's Cancer Study Group A. *Cancer* 1971;27:374.
25. Haas D, Ablin AR, Miller C, et al. Complete pathologic maturation and regression of stage IV-S neuroblastoma without treatment. *Cancer* 1988; 62:818.
26. Stephenson SR, Cook BA, Mease AD, Ruymann FB. The prognostic significance of age and pattern of metastases in stage IV-S neuroblastoma. *Cancer* 1986;58:372.
27. Hachitanda Y, Hata J-I. Stage IV-S neuroblastoma: a clinical, histological and biological analysis of 45 cases. *Hum Pathol* 1996; 27:1135.

28. Evans AE, Chatten J, D'Angio JG, et al. A review of 17 IV-S neuroblastoma patients at the Children's Hospital of Philadelphia. *Cancer* 1980;45:833.
29. Evans AE, Baum E, Chard R. Do infants with stage IV-S neuroblastoma need treatment? *Arch Dis Child* 1981; 56:271.
30. Gottesfeld E, Silverman RA, Coccia PF, et al. Transient blueberry muffin appearance of a newborn with congenital monoblastic leukemia. *J Am Acad Dermatol* 1989;21:347.
31. Joshi VJ, Silverman JF. Pathology of neuroblastic tumors. *Semin Diagn Pathol* 1994;11:107.
32. Shimada H, Ambros IM, Dehner LP, et al. Terminology and morphologic criteria of neuroblastic tumors: recommendations by the International Neuroblastoma Pathology Committee. *Cancer* 1999;86:349.
33. Lack EE. Recommendations for the reporting of tumors of the adrenal cortex and medulla: Association of Directors of Anatomic and Surgical Pathology. *Virchows Arch* 1999;435:87.
34. Lack EE. Tumors of the adrenal gland and extra-adrenal paraganglia. In: *Atlas of Tumor Pathology*. 3rd series. Fascicle 19. Washington, DC: Armed Forces Institute of Pathology; 1997.
35. Aronica PA, Pirrotta VT, Yunis EJ, et al. Detection of neuroblastoma in the bone marrow: biopsy versus aspiration. *J Pediatr Hematol Oncol* 1998;20:330.
36. Shimada H, Chatten J, Newton WA Jr, et al. Histopathologic prognostic factors in neuroblastic tumors: definition of subtypes of ganglioneuroblastoma and an age-linked classification of neuroblastomas. *J Natl Cancer Inst* 1984;73:405.
37. Joshi VV, Cantor AB, Altshuler G, et al. Age-linked prognostic categorization based on a new histologic grading system of neuroblastomas. *Cancer* 1992;69:2197.
38. Romansky SG, Crocker DW, Shaw KNF. Ultrastructural studies on neuroblastoma. Evaluation of cytodifferentiation and correlation of morphology and biochemical survival data. *Cancer* 1978; 42:2392.
39. Taxy JB. Electron microscopy in the diagnosis of neuroblastoma. *Arch Pathol Lab Med* 1980;104:355.
40. Brodeur GM. Molecular pathology of human neuroblastomas. *Semin Diagn Pathol* 1994;11:118.
41. Ikeda H, Hirato J, Akami M, et al. Massive apoptosis detected by in situ DNA end labeling in neuroblastoma. *Am J Surg Pathol* 1996;20:649.
42. Beckwith JB, Perrin EV. In situ neuroblastomas: a contribution to the natural history of neural crest tumors. *Am J Pathol* 1963; 43:1089.
43. Bolande RP. Neoplasia of early life and its relationships to teratogenesis. In: Rosenberg HS, Bolande RP, eds. *Perspectives in Pediatric Pathology.* Vol 3. Chicago: Year Book Medical Publishers, 1976:145.
44. Guin HG, Gilbert FE, Jones B. Incidental neuroblastoma in infants. *Am J Clin Pathol* 1969; 51:126.
45. Ho PTC, Estroff JA, Kozakewich H, et al. Prenatal detection of neuroblastoma: a ten-year experience from the Dana-Farber Cancer Institute and Children's Hospital. *Pediatrics* 1993;92: 358
46. Hayes FA, Green AA, Bhaskar NR. Clinical manifestations of ganglioneuroma. *Cancer* 1989;63:1211.
47. Moppett J, Haddadin I, Foot ABM (United Kingdom Children's Cancer Study Group). Neonatal neuroblastoma. *Arch Dis Child Fetal Neonatal Ed* 1999;81:F134.
48. Granata C, Fagnani AM, Gambini C, et al. Features and outcome of neuroblastoma detected before birth. *J Pediatr Surg* 2000; 35:88.

6
Leukemia

The leukemias are a heterogeneous group of hematologic neoplasms characterized by the infiltration of abnormal hematopoietic cells in the bone marrow, peripheral blood, various organs, and tissues. Following this clonal overgrowth of leukemic cells, the patient eventually develops the hematological sequelae of bone marrow failure manifested by anemia, neutropenia, and thrombocytopenia.

The clinical findings and biological features of leukemia in the fetus and infant are distinctively different from those of older children and adults. Moreover, the disease in this age group has an unexplained natural tendency to undergo spontaneous remissions lasting months or sometimes years [1–5]. Despite the therapeutic advances in the treatment of acute leukemia in childhood, the prognosis for newborns with this disease regardless of therapy is discouraging [6–15].

With few exceptions, fetal and infant leukemias are classified as acute because they have a predominance of immature hematopoietic precursors. Most are acute nonlymphoblastic (ANLL): as high as 90% in some series in comparison to acute lymphoblastic leukemia (ALL) [14–18]. Congenital chronic myelogenous leukemia is the subject of only a few case reports [19,20].

Hematopoiesis

Development of blood and blood vessels begins within the blood islands of the yolk sac and to a lesser extent in the chorion, body stalk, and allantois at approximately 19 days [21]. Cells within the center of the blood islands become the blood-forming cells, the hematopoietic stem cells, while those at the periphery form the endothelial cells, which line vessels. Hematopoietic stem cells are carried in the blood from the yolk sac and populate the liver, bone marrow, spleen, lymph nodes, and other sites (e.g., thymus and kidney). Blood formation begins in the liver during the fifth week, and this organ becomes the main source of erythrocytes until near the end of fetal life (Figure 6.1). By the end of the third month hematopoiesis is found in the bone marrow, which becomes the major site of erythropoiesis during the latter part of the third trimester [21].

Blood cells formed within the yolk sac and in the liver during the second trimester are predominantly erythropoietic elements, many of which are large and nucleated having a megaloblastic appearance (Figure 6.1). Leukocytes (granulocytes, histiocytes, lymphocytes, and megakaryocytes) appear in the liver and yolk sac at about 5 weeks and a few weeks later in the spleen, lymph nodes and bone marrow [21].

Clinical Features

The prenatal diagnosis of leukemia should be a serious consideration when hepatosplenomegaly, hydrops, and polyhydramnios are detected by ultrasonography [22]. The diagnosis is established by fetoscopy and fetal blood sampling [22–25]. Leukemia has been diagnosed in fetuses with Down syndrome [22,24,25]. Antenatal leukemia is an important cause of stillbirth [10,26–30]. The affected stillborn fetus is often hydropic and macerated, and the placenta is significantly enlarged and edematous, having the appearance of erythroblastosis fetalis.

The clinical signs and symptoms of neonatal leukemia are variable. Occasionally the disease is characterized by a rapid downhill course, but it may be unpredictable. Some neonates show signs of leukemia at birth and die shortly thereafter, while others appear normal following delivery but develop clinical and hematological problems later. In another group leukemia is not discovered until the third to sixth week of life, with a history suggestive of hematological abnormalities dating back a few weeks earlier [9].

Bleeding tendencies, hepatosplenomegaly, and leukemic cutaneous infiltrates (leukemia cutis) are characteristic clinical manifestations of perinatal leukemia [14,15,26] (Figure 6.2). Lymphadenopathy is seldom noted. Cutaneous involvement is observed in stillborns with the disease and is the initial or only presenting sign in a third of newborns, particularly those with acute myeloid leukemia, M4 or M5. Leukemia

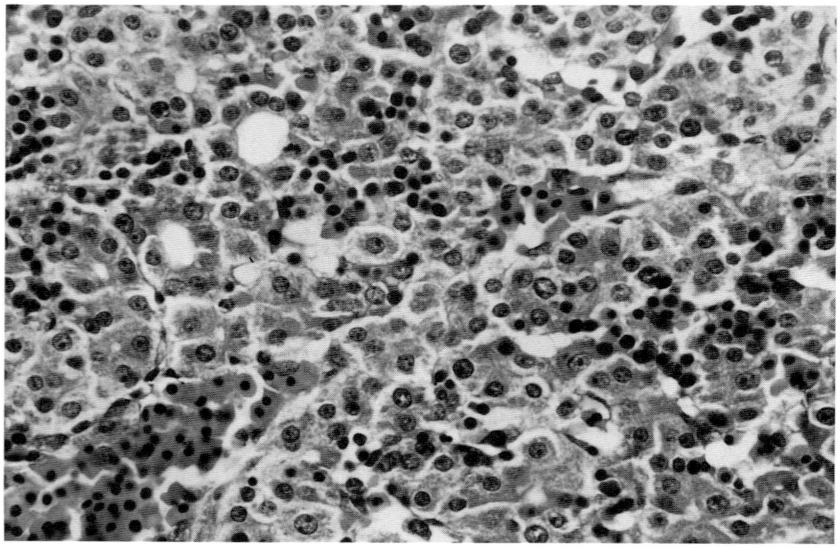

FIGURE 6.1. Fetal liver with hematopoiesis. Liver from 16 mm fetus (approximately 6 weeks gestation) showing extensive hematopoiesis. Sinusoids are filled with enlarged nucleated erythrocytes and nests of smaller nucleated erythroblasts. H&E; × 300.

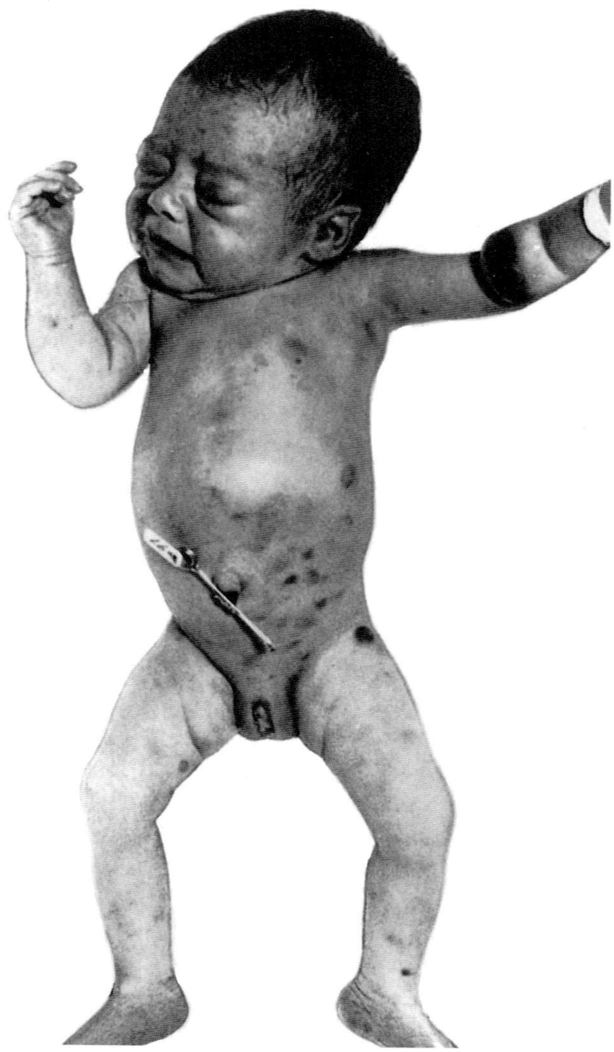

FIGURE 6.2. Perinatal leukemia. Multiple flat and elevated areas of involvement of the skin; the largest area is in the left antecubital fossa. Death occurred at 2 days. (From Penchansky L, Browansky IL, Gilbert Barness E [17], with permission.)

TABLE 6.1. Conditions causing petechiae and purpura in the newborn.

Congenital rubella
Toxoplasmosis
Cytomegalovirus
Platelet antibodies
Amegakaryocytic thrombocytopenia
Congenital leukemia
Hemolytic anemia (e.g., hereditary spherocytosis)
Twin-to-twin transfusion
Fetal-maternal transfusion
Isoimmunization

Source: Isaacs H Jr [73]. Used with permission.

cutis may precede other signs of leukemia by as much as 4 months [31,32]. Other skin manifestations include petechiae and ecchymoses related to thrombocytopenia (Table 6.1) (Figure 6.2).

When respiratory distress is present at birth it is attributed to pulmonary hemorrhage secondary to thrombocytopenia and/or to extensive leukemic infiltration complicated by atelectasis [10]. A bulging fontanelle is a sign of meningeal involvement or an intracranial bleed secondary to thrombocytopenia [15]. In the group of patients whose leukemia is discovered toward the end of the neonatal period and later in infancy, the signs and symptoms tend to be more vague; for example, the baby may have a low grade fever, pallor, lethargy, hepatosplenomegaly, a bleeding tendency, diarrhea, or general "failure to thrive" [15]. Sepsis and hemorrhage are the most common causes of death in infants, which is the same as in older children [10].

Generally infants with leukemia, either ANLL or ALL, present with findings associated with an unfavorable outcome. Hyperleukocytosis, hepatosplenomegaly, central nervous system involvement, hypogammaglobulinemia, and thrombocytopenia are seen more often than in older children with leukemia [14,33]. Infants with ALL experience a very high bone marrow and central nervous system relapse rate and an increased incidence of skin involvement in comparison to the rest of the pediatric age group [18,33].

Cytogenetics

Certain chromosomal anomalies, syndromes, and malformations are found in assocation with congenital leukemia [6,11,34,35]. By far the most common chromosomal anomaly, trisomy 21, occurs in patients with Down syndrome. Other chromosomal defects described are trisomy 8, 19 and 13, Turner syndrome (45,X0), Klinefelter syndrome (47,XXY); 1;6, 9;11, 8;16 translocations, monosomy 7, mosaic trisomy 9 and chromosome 11 (11q23) anomalies are further examples [6,8,11,34,36]. Congenital defects or syndromes described in association with leukemia include Bloom syndrome, Diamond-Blackfan anemia, Klippel-Feil syndrome, Fanconi anemia, Ellis-van Creveld syndrome, neurofibromatosis I, Schwachman syndrome, TAR syndrome (thrombocytopenia and absent radii), Marfan syndrome, and various cardiac malformations [35,37,38].

Fetal and infant leukemias characteristically are associated with abnormal karyotypes, most often translocations (Table 6.2). Rearrangements of chromosomal band 11q23 are frequent in both ALL and ANLL, comprising more than 70% of the observed chromosomal abnormalities in this age group [3,7,16,39] (Figure 6.3). The *MLL* gene, a *Drosophila trithorax* gene homologue, is located at the 11q23 breakpoint where the translocations occur in infant leukemias. Molecular rearrangements of *MLL* are present in most cases of ANLL and are strongly correlated with monocytic or myelomonocytic phenotypes [16].

TABLE 6.2. Common immunophenotypes and karyotypes in congenital leukemias.

	Leukemias[a]		
	ALL (L1 and L2)	AML (M4 and M5)	AMKL (M7)
Immunophenotype			
Common	CD 10 (CALLA) − surface immunoglobulin (SIg) + CD19 (very early B cell) + HLA-DR + TdT +	CD33 + CD13 + CD14 +	IIb/IIIa + GPIIIa + Platelet GpIb + Factor VIII antigen +
Variably present	CD33 (myeloid) + CD9, CD24, CD9, CD24, CD45 +	HLA-DR +	CD33 + CD13, HLA-DR + CD34 +
Karyotype			
Common	t(4;11)(q21;q23)	t(11;19)(q23;p13) t(9;11) (p21;q23)	t(1;22) (p13;q13)
Less common	t(9;11)(p2;q23) t(11;19) (q23;p13)	Many others involving 11q23 and various partners	
Immunglobulin genes	Heavy chain rearranged Light chain and TCR germline	Heavy chain germ line Light chain and TCR germ line	

[a] TCR, T-cell receptors; HLA-DR, human leukocyte antigen-DR; TdT, terminal DNA polymerase.
Source: Sande JE, et al. [14]. Used with permission.

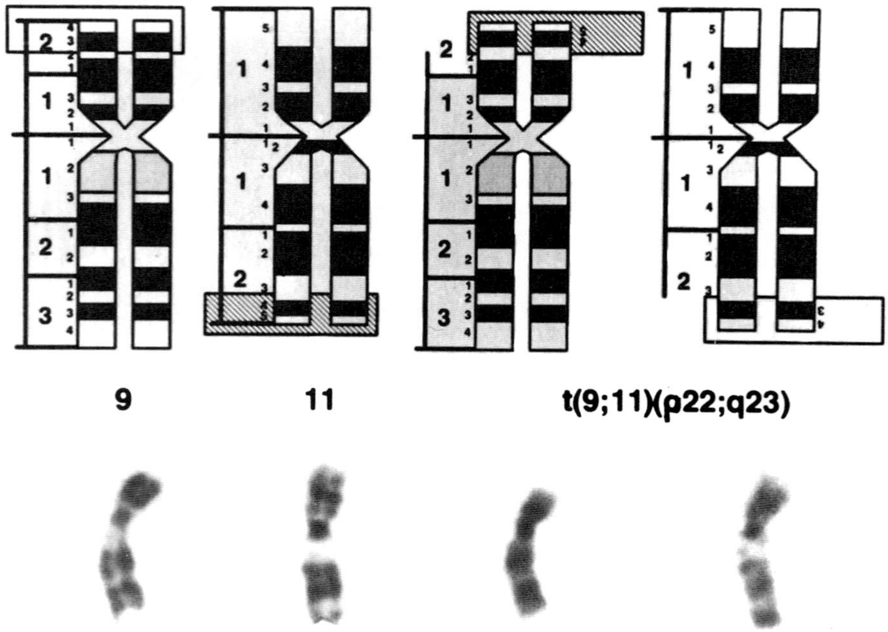

FIGURE 6.3. An 11q23 translocation. t(9;11)(p22;q23) is a frequent type of 11q translocation observed in ANNL monocytic leukemia (FAB M5). (From Heim S, Mittelman F. Acute nonlymphocytic leukemia. *Cancer Cytogenetics.* New York: AR Liss; 1987. With permission.)

Hyperleukocytosis, organomegaly, and central nervous system involvement are the initial clinical findings in infants with the 11q23 translocation [13,40]. Leukemic cell immunophenotyping shows a high incidence of coexpression of lymphoid- and myeloid-associated antigens with early B-cell markers (T and CALLA negative, HLA/DR, TdT, and CD19 positive) and immunoreactive myeloid markers CD33, C14, and CD15 consistent with a mixed pre-B/myeloid phenotype [7,40–42]. Some so-called null-cell leukemias fall into this category and represent mixed lineage or biphenotypic leukemias [40,41]. Apparently the leukemic cells are capable of differentiating along either myeloid or lymphoid lines. Moreover infants with null-cell leukemias who have relapsed after successful treatment have shown lineage switches of their leukemic cells from lymphoid to myeloid [13,42,43]. It is likely that these events contribute to the difficulties in treatment and to the very poor prognosis of neonatal ALL [3,13,14].

Acute megakaryoblastic leukemia (AMKL) is one of the main types of leukemia occurring in infants with Down syndrome [6,14]. In addition to the trisomy 21, their leukemic cells may show another chromosomal anomaly, a t(1;22) (p13;q13) translocation [8,44–47] (Figure 6.4). Since this translocation is not found in any other morphological subtype of AML or ALL, the t(1;22) anomaly could be used as a marker for identification of AMKL [6]. Similar to other leukemias, hepatosplenomegaly, anemia, and thrombocytopenia are the initial findings. However, extensive hepatic and bone marrow fibrosis are the unique histopathological features of infant AMKL and t(1;22) [8]. The striking hepatosplenomegaly and infiltration with cohesive tumor cells in lymph nodes and liver can be mistaken easily for other small-cell malignant neoplasms, particularly metastatic neuroblastoma [8,17] (Table 3.2).

Erythroleukemia occurs also in patients with Down syndrome and in patients with trisomies 8 and 19 [48]. Erythroleukemia may occur concomitantly with acute megakaryoblastic leukemia [49].

In contrast to Down syndrome, acute monoblastic (M5) or myelomonoblastic (M4) leukemia are the most common forms of congenital leukemia [5,18]. The affected infants often present with with leukemia cutis. Translocations involving 11q23 [e.g., t(9;11) (p21;q23), and t(8;16) (p11;p13)] are associated with M4 and M5 AML [16,50,51] (Table 6.2).

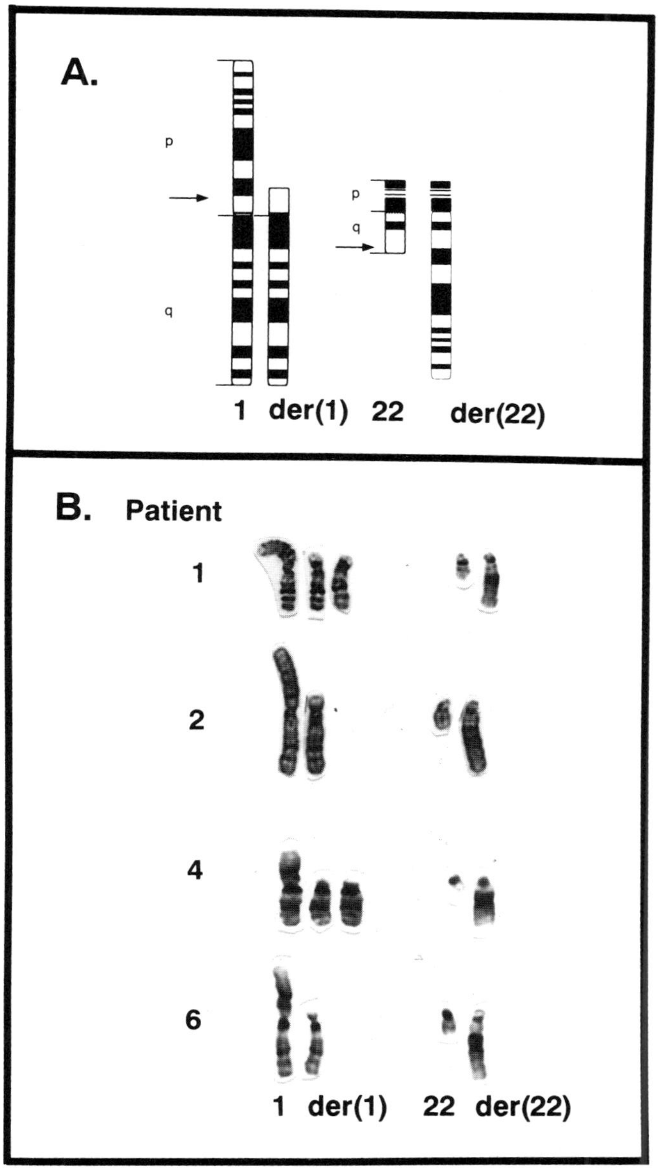

FIGURE 6.4. The t(2;22)(p13;q13) rearrangement associated with acute megakaryoblastic leukemia. (A) Diagram showing the G-banded pattern of the derivative (der) chromosomes and normal homologues. The arrows indicate the approximate positions of the breakpoints. (B) Representative partial karyotypes from four patients. (Courtesy of Dr. Andrew Carroll from Carroll A, et al. The t(1;22) (p13;q13) is nonrandom and restricted to infants with acute megakaryoblastic leukemia: a Pediatric Oncology Group study. *Blood* 1991;78:748. With permission.)

Differential Diagnosis

The predominance of ANLL during the perinatal period contributes to the difficulty in distinguishing the disease from the more common conditions causing leukemoid reactions [11,14] (Table 6.3). Infections and hemolytic disease of the newborn stimulate both erythroid and myeloid hematopoiesis in the sites where they normally occur in fetal life, and therefore exclusion of leukemoid (leukoerythroblastic) reactions and blood group incompatibility is important but sometimes difficult [11].

Leukocytosis, thrombocytopenia, and hepatosplenomegaly often accompany congenital infections as rubella, cytomegalovirus, toxoplasmosis, syphilis, and bacterial sepsis. Severe hemolytic disease of the newborn (immune erythroblastosis fetalis) may masquerade as leukemia (Figure 6.5). Neonates with this disorder, if severe enough, have hydrops fetalis, hepatosplenomegaly, skin nodules with dermal hematopoiesis, numerous erythroblasts in the peripheral blood smear, and sometimes thrombocytopenia [11].

Hepatosplenomegaly and cutaneous nodules are seen in newborns with disseminated neuroblastoma, stage IV-S (Figure 4.15). Peripheral blood counts are usually normal, but when the bone marrow is involved, one sees characteristic small syncytia of neuroblastoma cells, reactive with neuron-specific enolase, which are distinguishable from the bone marrow replaced by leukemic cells (see Chapter 5). Moreover, urinary catecholamines are usually elevated. Dermal erythropoiesis with a "blueberry muffin" appearance is found more often in congenital infections (e.g., rubella, cytomegalovirus) and in hemolytic disease of the newborn than with either leukemia or neuroblastoma [31,32].

Three main criteria are required to establish the diagnosis of congenital leukemia: the presence of immature hematopoietic cells in the peripheral blood and bone marrow, infiltration of nonhematopoietic tissues by these cells, and exclusion of conditions that should be considered in the differential diagnosis in this age group, namely, hemolytic disease of the newborn, congenital infections, and neuroblastoma [5,9,26] (Table 6.3).

TABLE 6.3. Differential diagnosis of congenital leukemia.

Infectious disease
Congenital cytomegalovirus
Congenital rubella
Toxoplasmosis
Congenital herpes virus
Congenital syphilis
Hepatitis
Human immunodeficiency virus
Intrauterine parvovirus infection
Bacterial
Listeria monocytogenes
Hematologic
Immune processes
Rh factor in blood incompatibility
Other major or minor blood group incompatiblities
Nonimmune processes
α-Thalassemia
Congenital dyserythropoietic anemia
Diamond-Blackfan anemia
Intrauterine or birth-related hypoxia
Twin-to-twin transfusion
Malignant tumors
Neuroblastoma
Histiocytoses
Langerhans cell histiocytosis
Hemophagic lymphohistiocytosis

Source: Sande JE, et al. [14]. Used with permission.

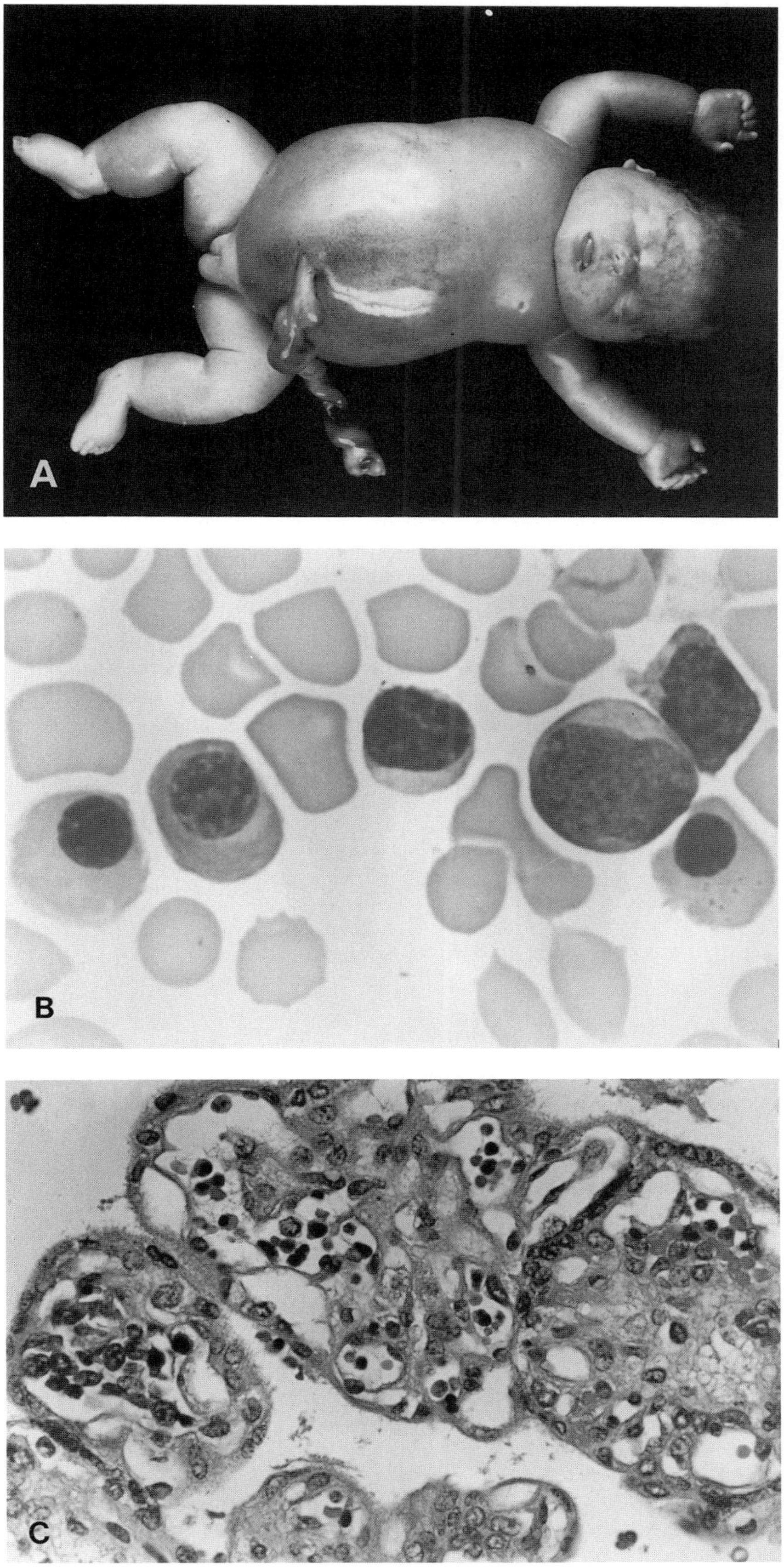

FIGURE 6.5. Erythroblastosis due to Rh incompatibility. (A) Stillborn infant with hydrops fetalis. (B) Peripheral smear from a newborn with erythroblastosis fetalis showing immature erythrocytes consisting of erythroblasts and normoblasts. Anisocytosis, poikilocytosis, and polychromasia are evident. Hematological values: hemoglobin 10 g/Dl, leukocyte count 39×10^9/L, platelet count 158×10^9/L. Wright-Giemsa; $\times$ 1200. (C) Placenta from hydropic newborn with erythroblastosis fetalis due to Rh incompatibility. Placenta was enlarged (780 g), pale, and edematous. Microscopic section reveals edematous villi with large numbers of nucleated erythrocytes within villous capillaries. H&E; $\times$ 380.

Congenital Leukemia and Down Syndrome

Down syndrome is associated with a significantly increased incidence of leukemia in both infants and children [46]. The frequency of leukemia is at least 14 times greater in individuals with Down syndrome than among the general population, and it is even greater in newborns [52]. The association is noted in 25% of neonates with ANNL and Down syndrome [15].

Approximately 10% of newborns with Down syndrome develop an enigmatic myeloproliferative disorder that is clinically and morphologically indistinguishable from leukemia, often megakaryoblastic in type; this condition is called transient leukemia, transient myeloproliferative disorder, transient congenital leukemia, or ineffective regulation of myelopoiesis, among other terms [47,53] (Figure 6.6). The disorder has been detected also prenatally [55]. The peripheral leukocyte counts range from 25,000 to several hundred thousand, and bone marrow examination reveals 30 to 70% blasts. Hepatosplenomegaly and thrombocytopenia are present. The hematological picture may return to normal after 1 to 4 months with only supportive care. Some patients with this so-called transient leukemia who die of other conditions related to the Down syndrome (e.g., cardiopulmonary disease, bowel obstruction, infection) lack microscopic evidence of leukemia at postmortem examination, while others relapse and die months or years later. The question of whether this transient disease represents leukemia or a disorder in regulation of myelopoiesis has not been settled [54]. However recent evidence suggests that this "transient leukemia" is actually true leukemia based on the following observations: it is a clonal proliferation, it can be fatal and tissue infiltration by leukemic cells occurs; and the morphological features of the neonate's blasts are the same as at the time of recurrence [18,45]. Moreover in the first 4 years of life, approximately 25% of patients who recover will develop acute megakaryoblastic leukemia which if not treated is fatal [45,46].

Often cases of ANLL in the newborn with Down syndrome are classified as an unusual form of leukemia, acute megakaryoblastic leukemia (M7) [6,45–47]. Acute lymphoid leukemia occurs mostly in older children with Down syndrome. Megakaryoblasts are diagnosed on the basis of their morphology on light and electron microscopy and by immunohistochemical markers [47,56] (Table 6.4). The AMKL cell is considered to be a hematopoietic progenitor cell capable of differentiating into three cell lines, namely, megakaryocytic, erythrocytic, and basophilic [46]. In fact, some Down syndrome neonates develop such erythroid abnormalities as dyserythropoiesis and erythroleukemia and basophilic proliferations [46,57]. Moreover there is an increased frequency of neonatal polycythemia in Down syndrome.

Much controversy exists over the natural history and treatment of transient leukemia of the newborn with Down syndrome. In most instances the megakaryoblasts disappear from

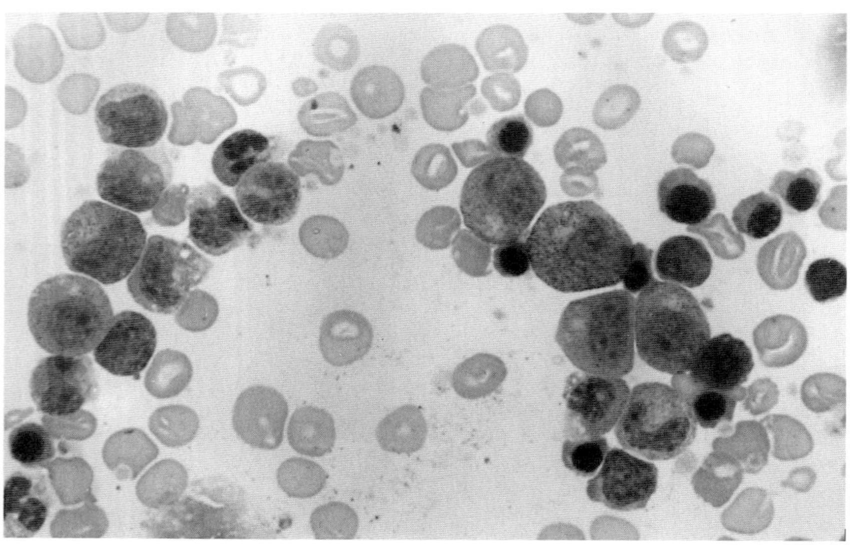

FIGURE 6.6. Transient myeloproliferative disorder with Down syndrome in a 2-day-old male. Workup for imperforate anus revealed a leukocyte count of 31×10^9/L and hemoglobin of 12.1. The peripheral smear revealed 30% immature granulocytes, 17% bands, 31% lymphocytes, 8% monocytes, 1% eosinophils, and 10% blasts; 20 nucleated red blood cells per 100 white blood cells; reduced platelets. Bone marrow histochemical studies: nonspecific esterase and acid phosphatase reactive; peroxidase, sudan black B, PAS, and chloroacetate esterase negative. After a few weeks the patient's blood picture returned to normal without therapy. He is alive and well at age 15 years without recurrence of leukemia. Bone marrow aspirate showing predominantly myeloid elements and early erythroid forms. Wright-Giemsa; $\times$ 600.

TABLE 6.4. Morphological and cytochemical features of ALL and ANLL.

Type of leukemia	FAB[a]	Wright-Giemsa stain morphology[b]	Cytochemistry[c]									
			PAS	Sudan	MP	CAE	ANB	ANA	Gly	VIII	PPO	IIb
Acute lymphoid	L1	Predominantly small cells High N:C ratio Indistinct 1 to 2 nucleoli Course nuclear chromatin Blue cytoplasm	+	–	–	–	–	–	–	–	–	–
	L2	Large cells, vary in size Lower N:C ratio than L1 Prominent nucleoli Stippled chromatin Blue cytoplasm	+	–	–	–	–	–	–	–	–	–
Acute myeloid	M1-M3	Low N:C ratio Auer rods ± 2 to 5 nucleoli Granules present Fine nuclear chromatin Blue-gray cytoplasm	–	+	+	+	–	+	–	–	–	–
Acute myelo-monocyte	M4	Myeloblasts and monoblasts[d]	–	+ +	+	+	±	+	–	–	–	–
Acute monocytic	M5	Low N:C ratio; Nuclei lobular or reniform; cells resemble monocytes; Fine nuclear chromatin	–	± ±	±	–	+	+	–	–	–	–
Acute erythro-leukemia	M6	Bizarre erythroblasts Multiple lobation of nucleus; multiple nuclei Blast cells with ragged cytoplasmic borders Vacuolated blue cytoplasms	±	±	+	–	–	–	+	–	–	–
Acute mega-karyoblastic leukemia	M7	Blasts with high N:C ratio; cytoplasmic blebs	+	–	–	–	–	±	–	+	+	+

[a] French-American-British morphologic classification.
[b] N:C, nuclear to cytoplasmic.
[c] PAS, periodic acid–Schiff; Sudan, sudan black B; MP, myeloperoxidase; CAE, chloroacetate esterase; ANB, α-naphthylbutyrate esterase; ANA, α-naphthyl acetate esterase; Gly, glycophorin A; VIII, factor VIII; PPO, platelet peroxidase (by electron microscopy); IIb, glycoprotein IIb/IIIa.
[d] See the description for both acute myeloid and acute monocytic leukemias in this table.
Source: Allan RR et al. [48]; Altman AJ [65,65]; Behm FG [68]; Bennett JM, et al. [65]; Isaacs H Jr [11]; Penchansky L; Brownsky IL, Gilbert-Barness E [17].

the blood within the first 2 months of life, and essentially the blood picture returns to normal [46,47]. However about 25% of these infants eventually go on to develop AMKL 5 to 30 months later. Therefore one could conclude that the proliferation of blast cells in transient leukemia also represents a preleukemic phase. The main problem for the clinician is in predicting the outcome in cases of "transient" leukemia and how to treat these patients. In other words, is the leukemia transient or not? From what has been said, it is practically impossible to determine with any accuracy what the final outcome will be in a newborn with Down syndrome (and even the outcome in newborns without Down syndrome and normal chromosomes) and leukemia. Therefore it appears that conservative, watchful management is definitely indicated [2,15,18].

Phenotypically normal infants with mosaic trisomy 21 karyotypes may develop transient leukemia [58,59]. Spontaneous remission is uncommon in neonates who have leukemia and normal karyotypes, but this unusual event does occur [2,4,5,31,60] (Figure 6.7).

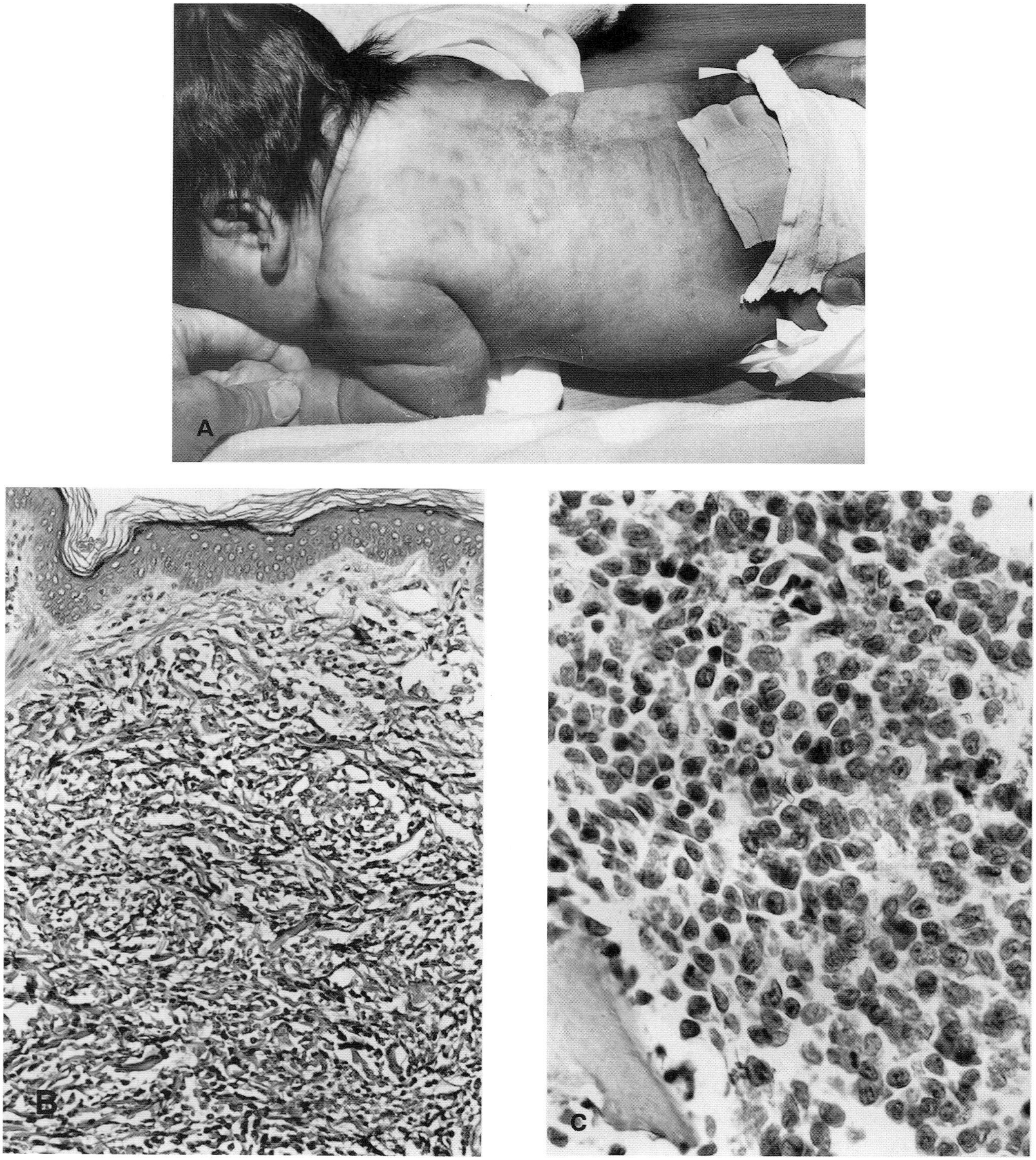

FIGURE 6.7. Transient congenital leukemia. (A) Newborn male with multiple raised cutaneous nodules distributed over most of his body. His karyotype, 46,XY, was normal. The initial leukocyte count was 117×10^9/L and hemoglobin was 12.7 g/Dl. (B) Skin biopsy taken of one of the nodules showing diffuse dermal and subcutaneous leukemic cell infiltration. H&E; × 300. (C) The bone marrow is replaced by myeloblasts. H&E; × 970.

Pathology

Placentas from fetuses with leukemia are enlarged (>700 g), pale, and edematous, having gross and microscopic appearance similar to erythroblastosis fetalis (hemolytic disease of the newborn) [11,27–29,48,51,62]. Microscopically the uniformly enlarged villi are composed of fetal blood vessels and villous stroma filled with leukemic blast cells, which usually do not invade the intervillous space of the maternal circulation. Since leukemia, neuroblastoma, and hepatoblastoma, the fetal tumors known to involve the placenta, are small blue-cell malignancies, it may be difficult to distinguish them by H&E-stained sections alone, particularly if the affected fetus is hydropic, macerated, or stillborn. If the tissue is well preserved, ultrastructural and immunohistochemical studies [e.g., leukocyte common antigen (leukemias in general), platelet peroxidase, and platelet glycoprotein GPIIIa (acute megakaryoblastic leukemia), epithelial membrane antigen (hepatoblastoma), and neuron-specific enolase (neuroblastoma)]. Wright-Giemsa-stained touch preparations should be examined for cellular morphology; surface markers and chromosomal analysis are definitely indicated if the material is fresh [11].

The fetus or infant with leukemia at necropsy shows pallor and other external signs of anemia and such manifestations of bleeding as petechiae and ecchymoses and oozing from needle puncture sites, various orifices, and the umbilicus (Figure 6.2).

Significant organomegaly is present. The liver may weigh more than double its normal expected weight, and the spleen greater than four times normal; these organs appear to occupy much of the abdominal cavity [11] (Figure 6.8B). Renomegaly secondary to leukemic infiltration is common feature of the disease. The kidneys weigh as much as twice their normal expected weight and display tense, tan-gray smooth capsules, characteristic peripelvic hemorrhage, and bulging cut surfaces with partial or total obliteration of the normal cortical-medullary architecture by gray to tan infiltrates (Figure 6.8E). Grossly the enlarged spleen shows a swollen, purple-gray smooth capsule and tan to red-brown bulging cut surfaces without either well-defined trabecular markings or white pulp lymphoid nodules, which are normally absent in the newborn (Figure 6.8D). Irregular dark red splenic infarcts may be present. Typically the liver has a pale, tan smooth swollen capsule and a similar cut surface with gray to red markings representing leukemic infiltrates and hemorrhage, respectively (Figure 6.8C). Lymphadenomegaly is uncommon in the untreated or partially treated infant. When involved, the lymph nodes are light tan to gray, swollen, with small foci of hemorrhage. Lymphoid nodules are not observed [11].

Microscopic examination reveals varying degrees of infiltration of the involved organs and tissues by immature lymphoid or myeloid elements, depending on the type of leukemia and whether treatment was given. The lymph nodes, spleen, and bone marrow in particular show diffuse infiltration by leukemic cells accompanied by variable numbers of young erythroid elements, namely, normoblasts and erythroblasts. Extramedullary organs as the lungs, kidneys, skin, and gastrointestinal tract are similarly involved. Lymphoid atrophy of normal lymphoid tissue is present particularly after chemotherapy [11].

The histological diagnosis of perinatal leukemia must be made with caution based only on the finding of hematopoietic infiltrates in the liver, spleen, and lymph nodes because these organs are normally the sites of extramedullary hematopoiesis in the fetus and neonate [9] (Figure 6.1). Moreover infiltrates of immature hematopoietic cells occur also in leukemoid reactions and in hemolytic disease of the newborn. Demonstration of blast cells in tissues not usually the site of fetal extramedullary hematopoiesis is requisite for the diagnosis [9].

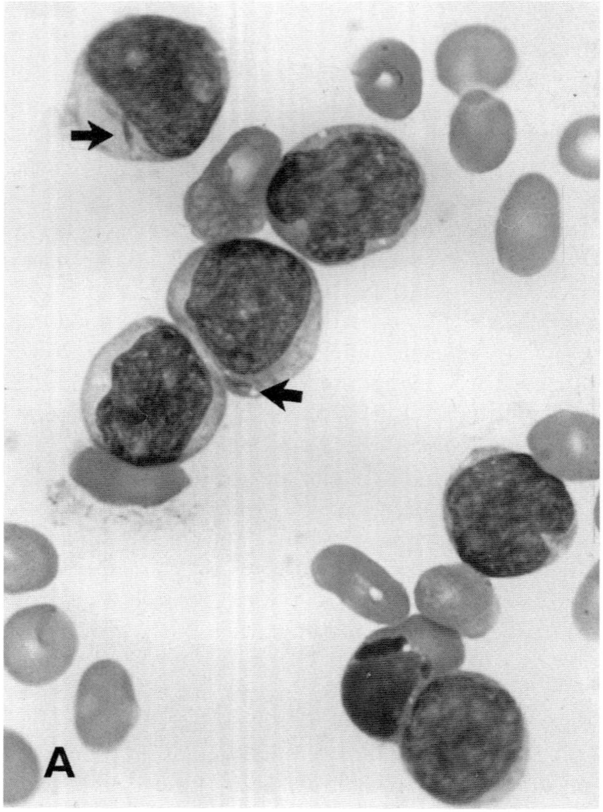

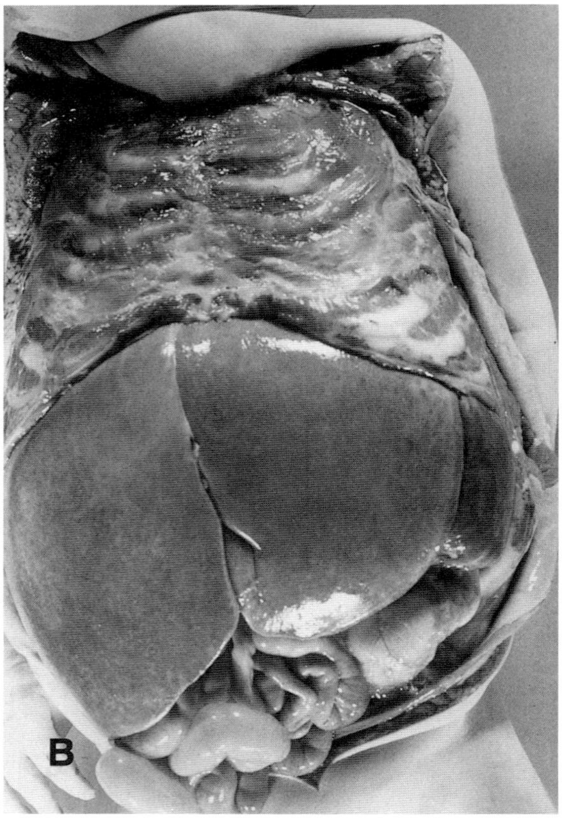

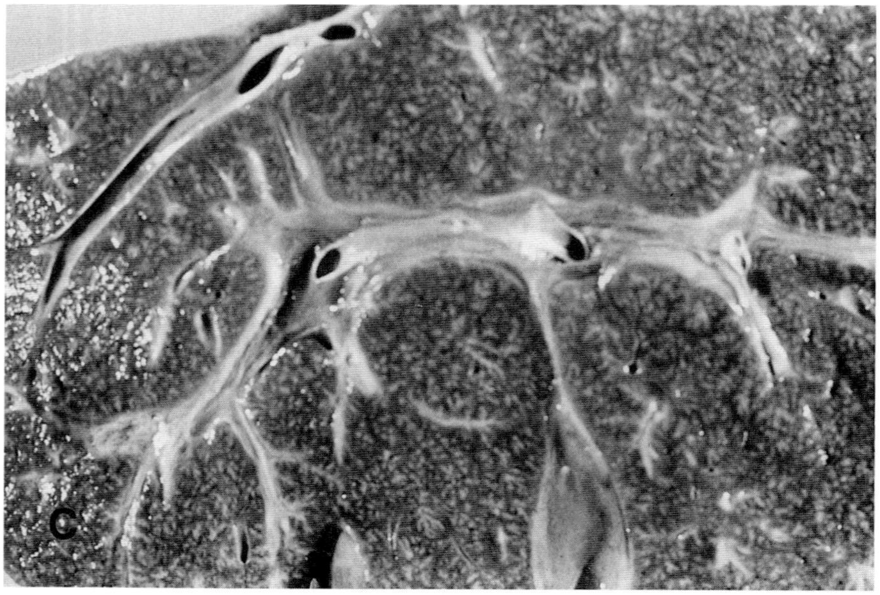

FIGURE 6.8. Acute myeloid leukemia in a 6-month-old male with a 1-week history of fever and lethargy. Laboratory findings revealed an anemia (hemoglobin 5.6 g/Dl), leukopenia (WBC 28 × 10^9/L: 94% blasts and 6% neutrophils) and thrombocytopenia (platelets 32 × 10^9/L). (A) Bone marrow aspirate with myeloblasts showing features consistent with M2 FAB morphology; the leukemic cells have a relatively low nuclear-to-cytoplasmic ratio, nuclei with a fine chromatin pattern, and two to three prominent nucleoli. Auer rods are present in two cells (arrows). Wright-Giemsa; × 800. Cytochemical findings: sudan black, myeloperoxidase, and nonspecific esterase reactive; PAS negative. (B) Postmortem examination revealed multiple petechiae and ecchymoses over the skin and viscera, hepatosplenomegaly, renomegaly, and lymphadenomegaly. The photograph depicts in situ hepatosplenomegaly. (C) Cross section of the liver showing extensive whitish leukemic infiltrates. (D) Cut surface of the spleen with diffuse leukemic infiltration and an infarct on the left (dark area). (E) Capsular and cut surfaces of the kidney with central peripelvic hemorrhage and blurring of the cortical-medullary architecture by leukemic infiltrates. (F) Microscopic section of the liver reveals expansion of the portal tract by leukemic cells, which are noted also within the adjacent sinusoids. H&E; × 600.

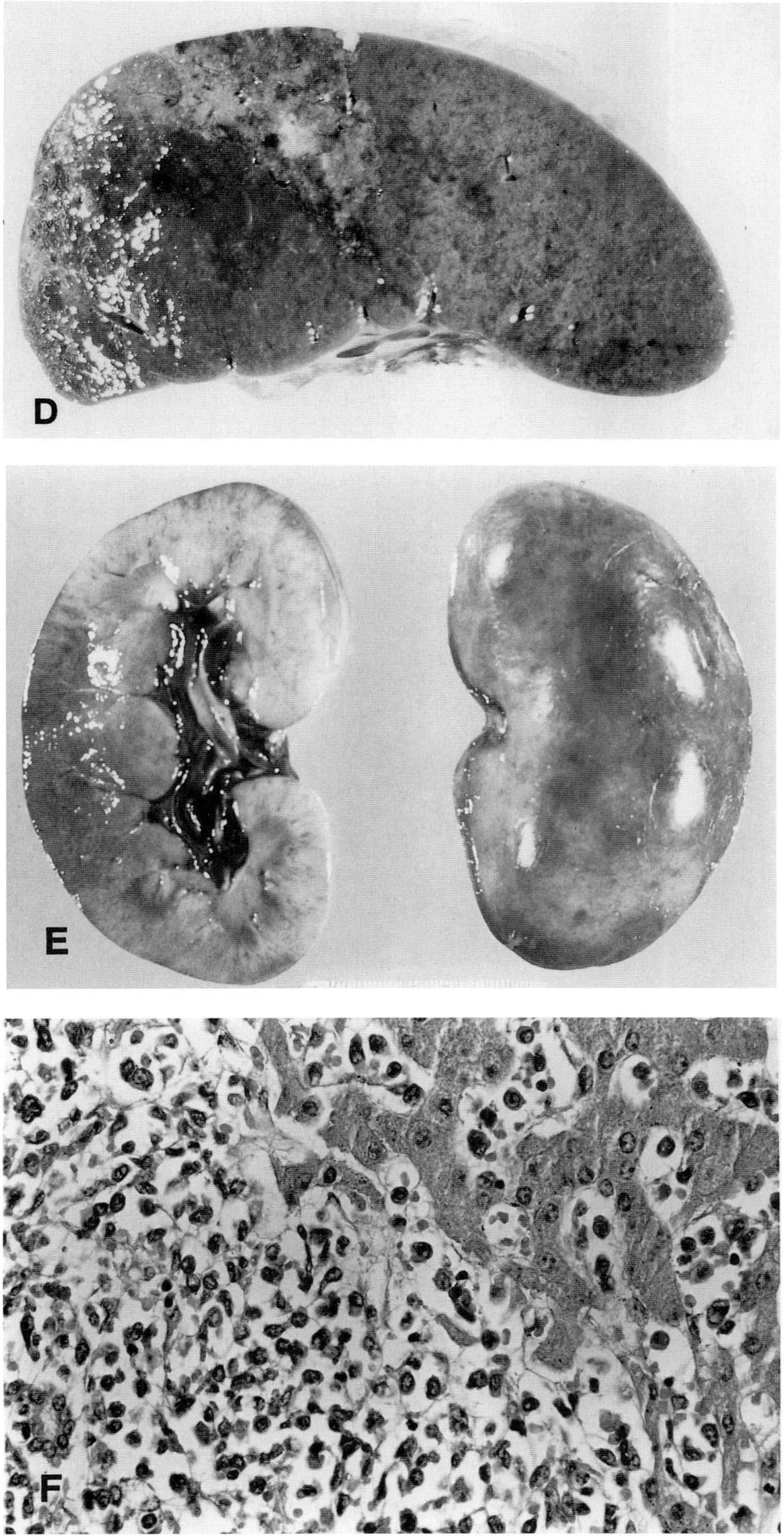

FIGURE 6.8. (*Continued*)

Multiple petechiae and ecchymoses and nodular leukemic infiltrates are the cutaneous manifestations of congenital leukemia [26,31,32,62,63] (Figures 6.2 and 6.9A). Leukemia cutis is noted in about a third of newborns and may be the first sign of the disease [63]. Skin involvement is particularly common with acute monocytic leukemia, which is the most frequent form of ANNL in the newborn. The firm red-brown to gray-blue nodular infiltrates may involve over half the body and range in size from a few millimeters to several centimeters in diameter, giving the baby a "blueberry muffin" appearance (Table 4.2). The histological findings consist of infiltrates of myeloid, lymphoid, or monocytic cells, depending on the type of leukemia, within the dermis and subcutaneous tissue (Figures 6.7B and 6.9A). In contrast to Langerhan's cell histiocytosis usually the epidermis is intact and not invaded by leukemic cells; the infiltrates are confined to the dermis and subcutis [11,62].

Cytological Diagnosis

When the diagnosis of acute leukemia is established, it is important particularly for therapeutic and prognostic reasons to determine whether the leukemia is lymphocytic or nonlymphocytic [48,64,65]. The next step is to decide the subclassification of ALL or ANNL to which the leukemia belongs. ALL and ANNL are diagnosed and classified on the basis of morphological findings such as the presence of granules or Auer rods and nuclear, nucleolar, and cytoplasmic appearance, and by cytochemical stains. Traditionally the international French-American-British classification (FAB) based on these findings is used to classify the leukemias [66] (Table 6.4). The FAB classification subdivides ALL into three groups (L1, L2, and L3) and ANNL into seven groups (M1–M7) (Table 6.4) [66]. Most infant ALL cases fall into the L1 or L2 FAB morphology [14].

Surface marker analysis accurately characterizes the cell types in leukemias [65,67,68]. Monoclonal antibodies for myeloid and lymphoid surface antigens are used to distinguish ANNL from ALL and to further subtype the leukemic cells [65,67,68] (Table 6.4). Most newborn and infant ALL cells express an early pre-B-cell phenotype manifested by HLA/DR reactivity and nonreactivity for CALLA, T-cell, and B-cell antigens [67,69–71]. An L1 and/or L2 FAB morphology is noted most often in this age group (Figure 6.9C). Acute monocytic leukemia (M5) (52%) is the most common form of ANNL in infants, followed by acute myelomonocytic leukemia (M4) (30%) and acute myeloblastic leukemia (M1) (17%) [11] (Figure 6.10). Mixed myeloid and lymphoid lineages are not uncommon in fetus and infant leukemias. The presence of glycophorin A in leukemic cells is considered diagnostic of erythroleukemia [46,48,68].

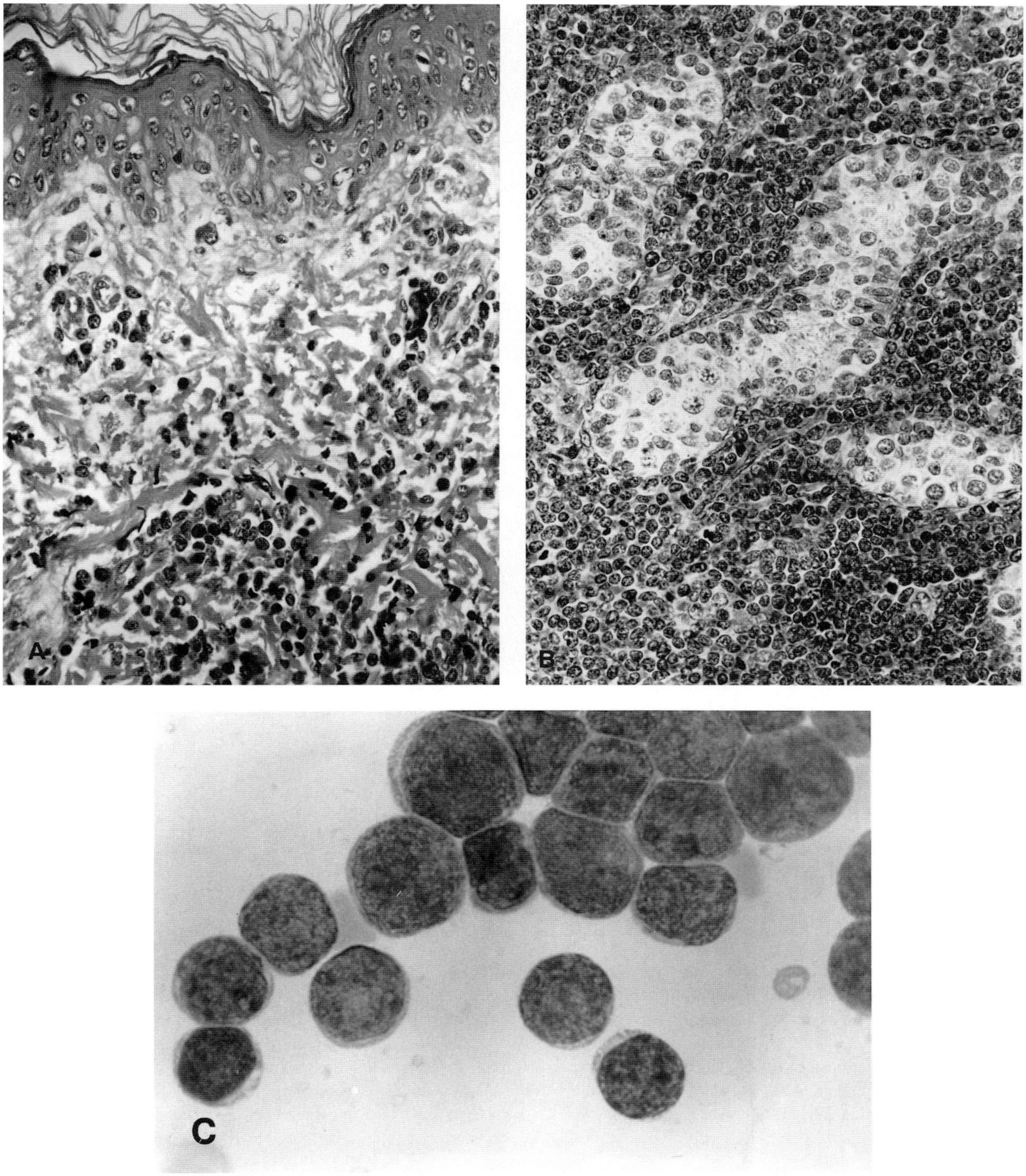

FIGURE 6.9. Acute lymphoid leukemia (ALL) in a 3-month-old male with anemia (hemoglobin 5.8 g/Dl) and a leukocytosis of 25 $\times$ 10^9/L with a differential of 96% lymphoblasts and 4% neutrophils. Platelet count was 104 $\times$ 10^9/L. Scattered bluish-purple nodules less than 1 cm in diameter were distributed over the chest, back, groin, and scalp, and bilateral testicular enlargement were noted on physical examination. His karyotype was normal. (A) Skin biopsy material showing leukemic cells in the dermis and subcutaneous tissue. H&E; $\times$ 600. (B) Testis biopsy material with extensive interstitial leukemic cell infiltrates. H&E; $\times$ 600. (C) An imprint ("touch preparation") of the testis biopsy showing small lymphoblasts with a high nuclear-to-cytoplasmic ratio, clumped chromatin pattern, and one or two small nucleoli consistent with FAB L1 morphology. Similar cells found in the bone marrow contained PAS-positive granules and were CALLA, peroxidase, sudan black, and a-naphthyl esterase negative. The anti-DR antibody was positive, but no T-cell antibodies were detected. Interpretation of the findings was that the lymphoblasts were of early pre-B-cell lineage falling into the so called pre-B-cell ALL category. Wright-Giemsa; $\times$ 800.

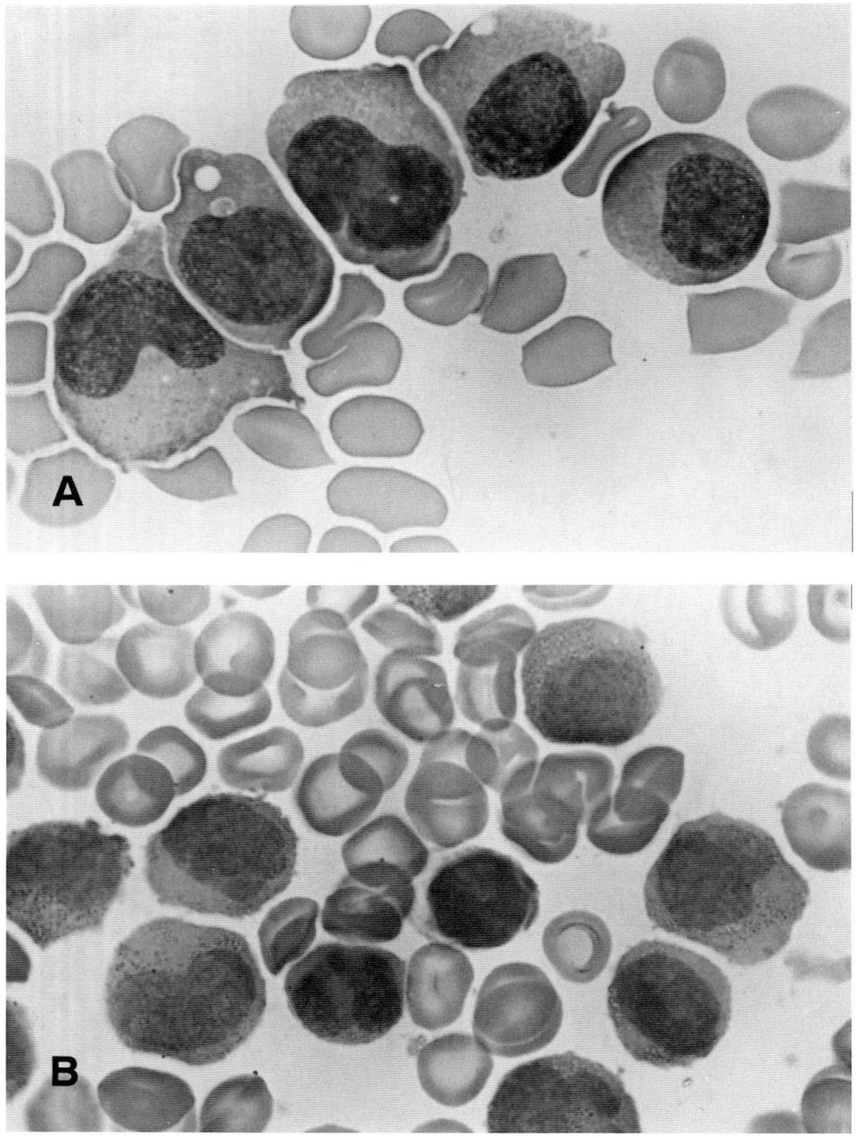

FIGURE 6.10. Acute myelomonocytic leukemia in a 1-month-old female with a leukocytosis of 50 X 10^9/L, 84% blasts, and a hemoglobin of 10.6 g/Dl. (A) Bone marrow smear reveals myeloid and large monocytoid forms, the latter, with abundant (pale gray-blue) cytoplasm, nuclear lobulations, and folds, chromatin varying from delicate to coarse, occasional nucleoli, and cytoplasmic vacuoles and pseudopodia (FAB M4-M5). Wright-Giemsa; × 1200. (B) Nonspecific esterase (× 1500) is positive. (C) Chloroacetate esterase (× 1200) is positive. (D) Acid phosphatase (× 1500) is positive. Negative cytochemical tests: peroxidase, sudan black B. Positive myeloid markers: CD14, CD15, and CD33. CD4 (an antigen found on monocytes and T-helper lymphocytes) was also reactive. The B-cell marker CD19 was positive (93% of blasts) while other B-cell markers and TdT were unreactive. Chromosome analysis revealed a t(11;9) balanced translocation. The data are consistent with acute myelomonocytic leukemia with B-cell coexpression.

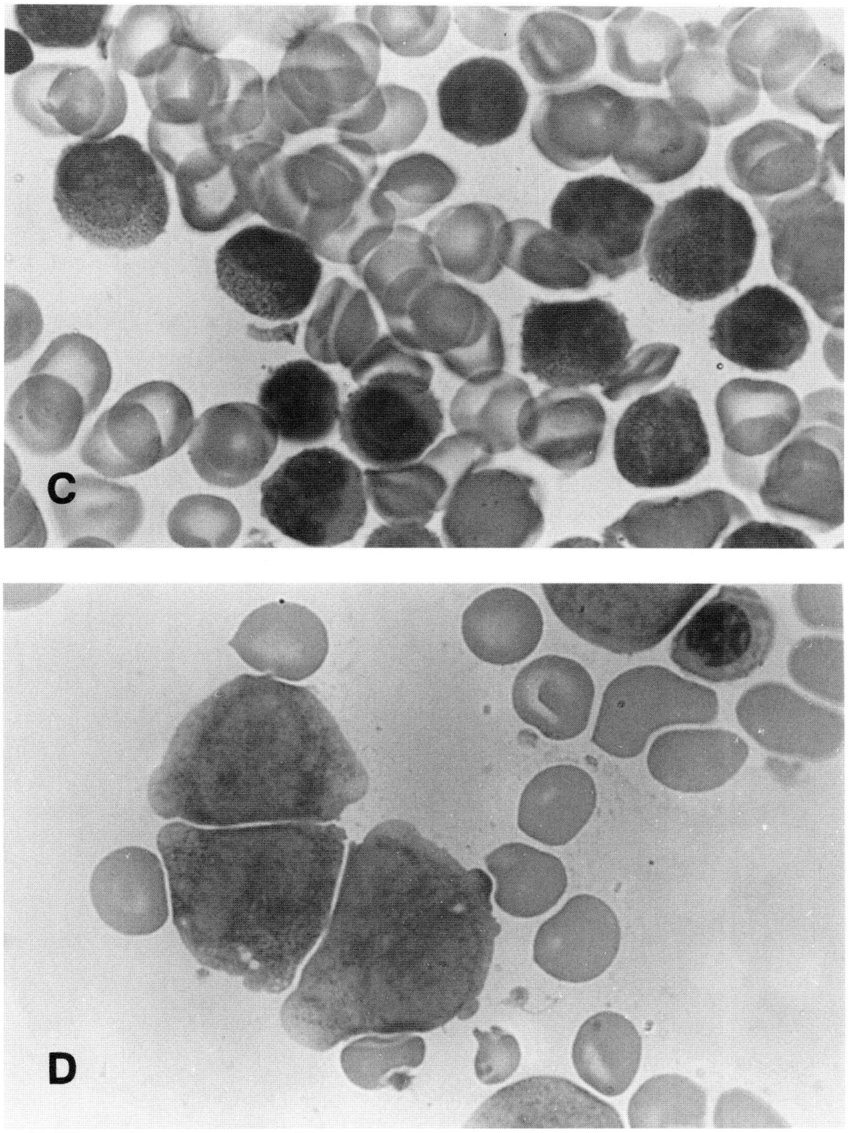

FIGURE 6.10. (*Continued*)

Since immature cells of the megakaryocytic series in smears and in tissues frequently lack distinctive morphological features that aid in their recognition, examples of AMKL have been classified as acute undifferentiated leukemia [72]. The diagnosis of acute megakaryoblastic leukemia is established on the basis of the appearance of the blasts with cytoplasmic blebs on Wright-Giemsa-stained smears, the platelet peroxidase reaction by electron microscopy (which is also positive in erythroleukemic blasts), and reactivity with monoclonal platelet-specific antibodies [4,8,56,68,72] (Figure 6.11). Megakaryoblasts in Wright-Giemsa-stained smears display a variable morphology conforming to the M7 category according to the FAB classification [4,8]. Small- and large-cell blasts with cytoplasmic projections or blebs suggest the diagnosis. The former have round nuclei, prominent nucleoli, stippled chromatin, and a moderate basophilic cytoplasm without granules and the latter a scant cytoplasm, a round nucleus with stippled chromatin, and an inconspicuous nucleolus (Figure 6.11).

Megakaryoblasts in blood and marrow smears react with platelet glycoprotein GPIIb/IIIa antibody [8,44,68]. Moreover factor VIII (von Willebrand factor), CD43, and *Ulex europaeus* lectin I are expressed also in paraffin section material [8]. However the blasts are nonreactive with CALLA (CD10), TdT, OKM1, myeloperoxidase and α-naphthyl butyrate esterase [72]. The AMKL cells are immunoreactive with CD34, HLA-DR, and the platelet factors CD61and CD 41 (glycoprotein IIa/IIb) [44]. The syncytial appearance of AMKL cells in the marrow and the infiltrative nature in other organs accompanied by fibrosis often suggest a nonhematopoietic metastatic malignancy, and markers as neuron-specific enolase, synaptophysin, desmin, and cytokeratin may be required to exclude neuroblastoma, rhabdomyosarcoma, primitive neuroectodermal tumor (PNET), and other small-cell malignant tumors [4,8,72] (see Table 3.2).

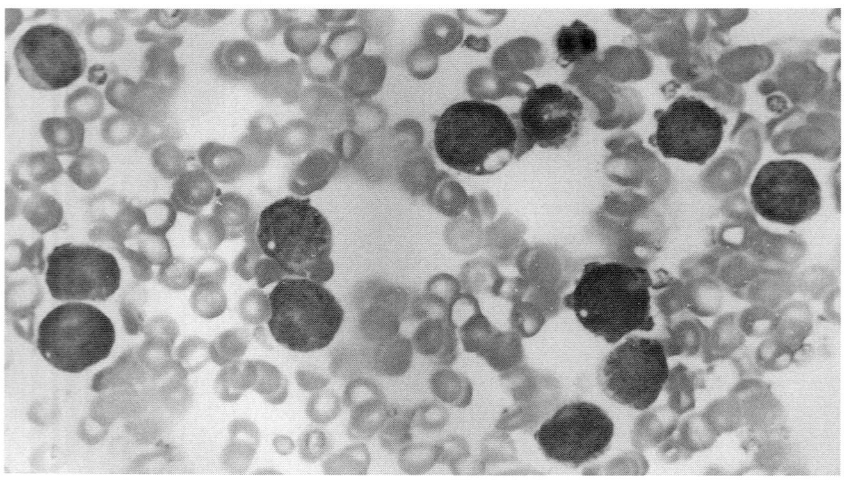

FIGURE 6.11. Acute megakaryoblastic leukemia in an 11-month-old male who presented with fever and leukocytosis of 19.6×10^9/L. He had a normal karyotype. Bone marrow aspirate revealed 87% blasts consisting of small- and large-cell blasts with suggestive cytoplasmic projections (blebs). The former have round nuclei, prominent nucleoli, stippled chromatin, and a moderate (basophilic) cytoplasm without granules; the latter have a scant cytoplasm, a round or double nucleus with stippled chromatin, and inconspicuous nucleoli. Cytochemistry: PAS and acid phosphatase positive; CAE, sudan black B, and peroxidase negative; CD13, CD33, and CD34 positive; platelet antibodies CD 41, CD42a, and CD61 positive. Wright-Giemsa; $\times$ 600.

References

1. Chu J-Y, O'Connor DM, Gale GB, et al. Congenital leukemia: two transient regressions without treatment in one patient. *Pediatrics* 1983;71:277.
2. Lampkin, BC, Peipon, JJ, Price, JK, et al. Spontaneous remission of presumed congenital acute nonlymphoblastic leukemia (ANLL) in a karyotypically normal neonate. *Am J Pediatr Hematol Oncol* 1985;7:346.
3. Pui C-H, Raimondi SC, Murphy SB, et al. An analysis of leukemic cell chromosomal features in infants. *Blood* 1987;69:1289.
4. Penchansky L, Wollman MR, Gartner JC, et al. Spontaneous remission of infantile acute nonlymphocytic leukemia for 11 years in a child with a normal karyotype. *Cancer* 1993;71:1928.
5. Dinulos JG, Hawkins DS, Clark BS, et al. Spontaneous remission of congenital leukemia. *J Pediatr* 1997;131:300.
6. Carroll A, Civin C, Schneider N, et al. The t(1;22) (p13;q13) is nonrandom and restricted to infants with acute megakaryoblastic leukemia: a Pediatric Oncology Group study. *Blood* 1991;78: 748.
7. Kaneko Y, Shikano T, Maseki N, et al. Clinical characteristics of infant acute leukemia with or without 11q23 translocations. *Leukemia* 1988;2:672.
8. Chan WC, Carroll A, Alvarado CS, et al. Acute megakaryoblastic leukemia in infants with t(1;22)(p13;q13) abnormality. *Am J Clin Pathol* 1992;98:214.
9. Isaacs H Jr. Perinatal (congenital and neonatal) neoplasms: a report of 110 cases. *Pediatr Pathol* 1985;3:165.
10. Isaacs H Jr. Congenital and neonatal malignant tumors: a 28-year experience at Children's Hospital of Los Angeles. *Am J Pediatr Hematol Oncol* 1987; 9:121.
11. Isaacs H Jr. Leukemia. In: *Tumors of the Fetus and Newborn.* Philadelphia: WB Saunders; 1997:150. Major Problems in Pathology Series; Vol 35.
12. Parkes SE, Muir KR, Southern L, et al. Neonatal tumours: a thirty-year population based study. *Med Pediatr Oncol* 1994;22: 309.
13. Pui C-H, Behm FG, Downing JR, et al. 11q23/MLL Rearrangement confers a poor prognosis in infants with acute lymphoblastic leukemia. *J Clin Oncol* 1994;12:909.
14. Sande JE, Arceci RJ, Lampkin BC. Congenital and neonatal leukemia. *Semin Perinatol* 1999;23:274.
15. Broadbent VA. Malignant disease in the neonate. In: Roberton NRC, ed. *Textbook of Neonatology.* 2nd ed. Edinburgh: Churchill Livingston; 1992:879.
16. Sorensen PHB, Chen C-S, Smith FO, et al. Molecular rearrangements of *MLL* gene are present in most cases of infant acute myeloid leukemia and are strongly correlated with monocytic or myelomonocytic phenotypes. *J Clin Invest* 1994;93:429.
17. Penchansky L, Browarsky IL, Gilbert-Barness E. Hematopoietic system. In: Gilbert-Barness E, ed. *Potter's Pathology of the Fetus and Infant.* Vol 2. St. Louis, MO: CV Mosby; 1997:936.
18. Lampkin BC. The newborn infant with leukemia. *J Pediatr* 1997; 131:176.
19. Clark RH, Taylor LL, Wells RJ. Congenital juvenile chronic myelogenous leukemia: case report and review. *Pediatrics* 1984; 73:324.
20. Hazani A, Barak Y, Berant M, et al. Congenital juvenile myelogenous leukemia: therapeutic trial with interferon alpha-2. *Med Pediatr Oncol* 1993;21:73.
21. Oyer CE, Singer DB. The hematopoietic system. In: Wigglesworth JS, Singer DB, eds. *Textbook of Fetal and Perinatal Pathology.* 2nd ed. Boston: Blackwell Science; 1998:1143.
22. Nicolaides KH, Rodeck CH, Lange I, et al. Fetoscopy in the assessment of unexplained fetal hydrops. *Br J Obstet Gynaecol* 1985;92:671.
23. Donnenfeld AE, Scott SC, Henselder-Kimmel M, et al. Prenatally diagnosed non-immune hydrops caused by congenital transient leukemia. *Prenat Diagn* 1994;14:721.
24. Fourcar K, Friedman K, Llewellyn A, et al. Prenatal diagnosis of transient myeloproliferative disorder via percutaneous umbilical blood sampling: report of two cases in fetuses affected by Down's syndrome. *Am J Clin Pathol* 1992;97:584.
25. Zerres K, Schwanitz G, Niesen M, et al. Prenatal diagnosis of acute non-lymphoblastic leukemia in Down syndrome. *Lancet* 1990;335:117.
26. Pierce MI. Leukemia in the newborn infant. *J Pediatr* 1979;54: 691.
27. Gray ES, Balch NJ, Kohler H, et al. Congenital leukemia: an unusual cause of stillbirth. *Arch Dis Child* 1986; 61:1001.
28. Benirschke K, Kaufman P. Maternal diseases complicating pregnancy. In: *Pathology of the Human Placenta.* 4th ed. New York: Springer-Verlag; 2000:542.
29. Las Heras J, Leal G, Haust MD. Congenital leukemia: report of a case with ultrastructural study. *Cancer* 1986;58:2278.
30. Greenbaum BH, Steinfeld J, Obermeyer-Imaizumi S, et al. Congenital monoblastic leukemia as a cause of third-trimester loss: the role of monoclonal antibody testing. *J Perinatol* 1996;16:202.
31. Gottesfeld E, Silverman RA, Coccia PF, et al. Transient blueberry muffin appearance of a newborn with congenital monoblastic leukemia. *J Am Acad Dermatol* 1989;21:347.
32. Resnik KS, Brod BB. Leukemia cutis in congenital leukemia. *Arch Dermatol* 1993;129:1301.
33. Reaman G, Zeltzer P, Bleyer WA, et al. Acute lymphoblastic leukemia in infants less than one year of age: a cumulative experience of the Children's Cancer Study Group. *J Clin Oncol* 1985;3: 1513.
34. Gale GB, Toledano SR. Congenital acute lymphocytic leukemia in a newborn with Klinefelter syndrome. *Am J Pediatr Hematol Oncol* 1984;6:338.
35. Sharief N, Kingston JE, Wright VM, et al. Acute leukemia in an infant with Marfan's syndrome: a case report. *Pediatr Hematol Oncol* 1991;8:323.
36. Carney LA, Kinney JS, Higgins RR, et al. X;6 translocation in a child with congenital acute lymphocytic leukemia. *Cancer* 1992;69:799.
37. Bader JL, Miller RM. Neurofibromatosis and childhood leukemia. *J Pediatr* 1978; 92:925.
38. Dehner LP. Neoplasms of the fetus and neonate. In: Naeye RL, Kissane JM, Kaufman N, eds. *Perinatal Diseases.* Baltimore: Williams & Wilkins; 1981:286. International Academy of Pathology; Monograph No. 22.
39. Super HJG, Rothberg PG, Kobayashi H, et al. Clonal, nonconstitutional rearrangements of the *MLL* gene in infant twins with acute lymphoblastic leukemia: in utero chromosome rearrangement of 11q23. *Blood* 1994;83:641.
40. Katz F, Malcolm S, Gibbons B, et al. Cellular and molecular studies on infant null acute lymphoblastic leukemia. *Blood* 1988; 71:1438.
41. Nagasaka M, Maeda S, Maeda H, et al. Four cases of t(4;11)

acute leukemia and its myelomonocytic nature in infants. *Blood* 1983;61:1174.
42. Miura M, Yachie A, Hashimoto I, et al. Coexistence of lymphoblastic and monoblastic populations with identical mixed lineage leukemia gene rearrangements and shared immunoglobulin heavy chain rearrangements in leukemia developed in utero. *J Pediatr Hematol Oncol* 2000;22:81.
43. Ridge SA, Cabrera ME, Ford AM, et al. Rapid intraclonal switch of lineage dominance in congenital leukaemia with an *MLL* gene rearrangement. *Leukemia* 1995;9:2023.
44. Ng K-C, Tan A-M, Chong Y-Y, et al. Congenital acute megakaryoblastic leukemia (M7) with chromosomal t(1;22)(p13;q13) translocation in a set of identical twins. *J Pediatr Hematol Oncol* 1999;21:428.
45. Zipursky A, Poon A, Doyle J. Leukemia in Down syndrome: a review. *Pediatr Hematol Oncol* 1992;9:139.
46. Zipursky A, Rose T, Skidmore M, et al. Hydrops fetalis and neonatal leukemia in Down syndrome. *Pediatr Hematol Oncol* 1996;13:81.
47. Zipursky A, Brown E, Christensen H, et al. Leukemia and/or myeloproliferative syndrome in neonates with Down syndrome. *Semin Perinatol* 1997;21:97.
48. Allan RR, Wadsworth LD, Kalousek DK, et al. Congenital erythroleukemia: a case report with morphological, immunophenotypic and cytogenetic findings. *Am J Hematol* 1989;31:114.
49. Mori T, Kaneko H, Kumagai MA, et al. Congenital leukaemia with a mixed phenotype of megakaryoblasts and erythroblasts: a case report and characterization of the blasts. *Br J Haematol* 1997;96:740.
50. Bernstein R, Pinto MR, Spector I, et al. A unique 8;16 translocation in two infants with poorly differentiated monoblastic leukemia. *Cancer Genet Cytogenet* 1987;24:213.
51. Hanada T, Ono I, Minosaki Y, et al. Translocation t(8;16)(p11;p13) in neonatal acute monocytic leukemia. *Eur J Pediatr* 1991;150:323.
52. Levitt GA, Stiller CA, Chessells JM. Prognosis of Down's syndrome with acute leukaemia. *Arch Dis Child* 1990;65:212.
53. Barnett PLJ, Clark ACL, Garson OM. Acute nonlymphocytic leukemia after transient myeloproliferative disorder in a patient with Down syndrome. *Med Pediatr Oncol* 1990;18:347.
54. Becroft DMO, Zwi LJ. Perinatal visceral fibrosis accompanying the megakaryoblastic leukemoid reaction of Down syndrome. *Pediatr Pathol* 1990;10:397.
55. Baschat AA, Wagner T, Malisius R, et al. Prenatal diagnosis of a transient myeloproliferative disorder in trisomy 21. *Prenat Diagn* 1998;18:731.
56. Kojima S, Matsuyama T, Sato T, et al. Down's syndrome and acute leukemia in children: an analysis of phenotype by use of monoclonal antibodies and electron microscopic platelet peroxidase reaction. *Blood* 1990;76:2348.
57. Worth LL, Zipursky A, Christensen H, et al. Transient leukemia with extreme basophilia in a phenotypically normal infant with blast cells containing pseudodiploid clone, 46,XY 1(21)(q10). *J Pediatr Hematol Oncol* 1999;21:63.
58. Ridgway D, Beda GI, Magenis E, et al. Transient myeloproliferative disorder of the Down type in the normal newborn. *Am J Dis Child* 1990;144:1117.
59. Doyle JJ, Thorner P, Poon A, et al. Transient leukemia followed by megakaryoblastic leukemia in a child with mosaic Down syndrome. *Leukemia Lymphoma* 1995;17:345.
60. Grundy RG, Martinez A, Kempski H, et al. Spontaneous remission of congenital leukemia: a case for conservative treatment. *J Pediatr Hematol/Oncol* 2000;22:252.
61. Fox H. Non-trophoblastic tumours of the placenta. In: *Pathology of the Placenta*. 2nd ed. London: WB Saunders; 1997:375. Major Problems in Pathology Series; Vol. 7.
62. Attal H, Kowal-Vern A, Husain AN, et al. Newborn infant with multiple purple skin nodules. *Arch Dermatol* 1996;132:341.
63. Francis JS, Sybert VP, Benjamin DR. Congenital monocytic leukemia: report of a case with cutaneous involvement, and review of the literature. *Pediatr Dermatol* 1989;6:306.
64. Altman AJ. Cytologic diagnosis of the acute nonlymphoid leukemias, I: Morphologic, cytochemical, and ultrastructural features. *Am J Pediatr Hematol Oncol* 1985;7:21.
65. Altman AJ. Cytologic diagnosis of the acute nonlymphoid leukemias, II: Flow cytometry, surface markers, cytogenetics, and use of cell culture techniques. *Am J Pediatr Hematol Oncol* 1985; 7:156.
66. Bennett JM, Catovsky D, Daniel M-T et al. Proposals for the classification of acute leukemias. French-American-British (FAB) co-operative group. *Br J Haematol* 1976;33:451.
67. Dinndorf PA, Reaman G. Acute lymphoblastic leukemia in infants: evidence for B cell origin of disease by use of monoclonal antibody phenotyping. *Blood* 1986;68:975.
68. Behm FG. Malignancies of the bone marrow. In: Parham DM, ed. *Pediatric Neoplasia: Morphology and Biology.* Philadelphia: Lippincott- Raven; 1996:449.
69. Ludwig W-D, Bartram CR, Harbott J, et al. Phenotypic and genotypic heterogeneity in infant acute leukemia, I: Acute lymphoblastic leukemia. *Leukemia* 1989;3:431.
70. Spier CM, Kjeldsberg CR, O'Brien R, et al. Pre-B-cell acute lymphoblastic leukemia in the newborn. *Blood* 1984;64:1064.
71. Felix CA, Reaman GH, Korsmeyer SJ, et al. Immunoglobulin and T cell receptor gene configuration in acute lymphoblastic leukemia of infancy. *Blood* 1987;70:536.
72. Windebank KP, Tefferi A, Smithson WA, et al. Acute megakaryocytic leukemia (M7) in children. *Mayo Clin Proc* 1989;64:1339.
73. Isaacs H Jr. Tumors and tumor-like conditions of the skin. In: *Tumors of the Fetus and Newborn.* Philadelphia: WB Saunders; 1997:109. Major Problems in Pathology Series; Vol 35.

7
Histiocytoses

The histiocytoses represent a spectrum of conditions characterized by the infiltration and accumulation of cells of the monophagocytic system within the blood and tissues. They are a diverse group of disorders ranging from cytologically benign proliferations of mature cells to such clearly malignant entities as acute monocytic leukemia (Table 7.1). Lymphocytes may be a part of the proliferative process to a variable extent. Many of the signs and symptoms of the histiocytoses may be the result of the functional activity of these cells, and abnormal or altered regulation of histiocyte activity may be important in the pathogenesis of these diseases. Each of the histiocytic disorders may present with overlapping clinical and pathological findings, making diagnosis difficult in some instances [1–7a].

TABLE 7.1. Histiocyte disorders of the fetus and infant.

Langerhans cell histiocytosis
Disseminated
Focal
Pure cutaneous histiocytosis (Hashimoto-Pritzker syndrome)
Juvenile xanthogranuloma
Sinus histiocytosis with massive lymphadenopathy (Rosai-Dorfman disease)
Hemophagocytic syndromes
Familial erythrophagocytic lymphohistiocytosis
Infection (reactive) associated hemophagocytic syndrome
Viral
Bacterial
Fungal
Parasitic
Malignant disorders
Acute monocytic leukemia
Histiocytic lymphoma
Chronic granulomatous disease of childhood
Graft-versus-host disease
Omenn disease
X-linked lymphoproliferative syndrome (Duncan disease)
Chediak- Higashi syndrome (accelerated phase)
Lysosomal storage diseases (e.g., Gaucher disease, Niemann-Pick disease, mucopolysaccharidoses)

Source: Isaacs H Jr [7a], with permission.

Not all diseases in which histiocytic infiltrates occur are classified as histiocytoses. The infiltrates may represent a secondary process or a reaction to a condition with a known etiology. Mycobacterial infections, histoplasmosis, graft-versus-host disease, chronic granulomatous disease of childhood, the X-linked lymphoproliferative syndrome, and the lysosomal storage diseases are a few examples [1,3,6,7a,b] (Figure 7.1).

Ontogeny of the Monophagocytic System

The monocyte is the precursor of the macrophage and of related phagocytic cells that comprise the mononuclear phagocyte system. Cells of the monocytic and granulocytic series originate from a common bone marrow progenitor cell. Within the marrow the monocyte matures through several stages from the monoblast, the promonocyte, and finally the monocyte. The normal bone marrow contains from 1 to 3% monocytes and their precursors [1,3,6].

When a monocyte leaves the bloodstream and enters the tissues, it becomes a wandering macrophage (histiocyte) and subsequently undergoes further differentiation to form a variety of mononuclear phagocytic cells known by various names depending on their tissue location. For example, they become Kupffer cells in the liver, alveolar macrophages in the lung, sinusoidal lining cells in the spleen, and Langerhans cells in the skin. Macrophages that have accumulated at a site of chronic inflammation (e.g., in a tuberculous lymph node), can differentiate further into epithelioid histiocytes, or they may fuse to form multinucleated giant cells. Moreover macrophages can divide in tissues [1,2,7a,8].

The properties and function of macrophages vary according to their location. Primarily they are responsible for ingestion and killing of microorganisms. Macrophages have cytoplasmic lysosomal granules that contain hydrolytic enzymes, the antibacterial enzyme lysozyme, and Fc receptors for antigen-antibody complexes. Following exposure to antigens (e.g., from an invading microorganism), T lymphocytes

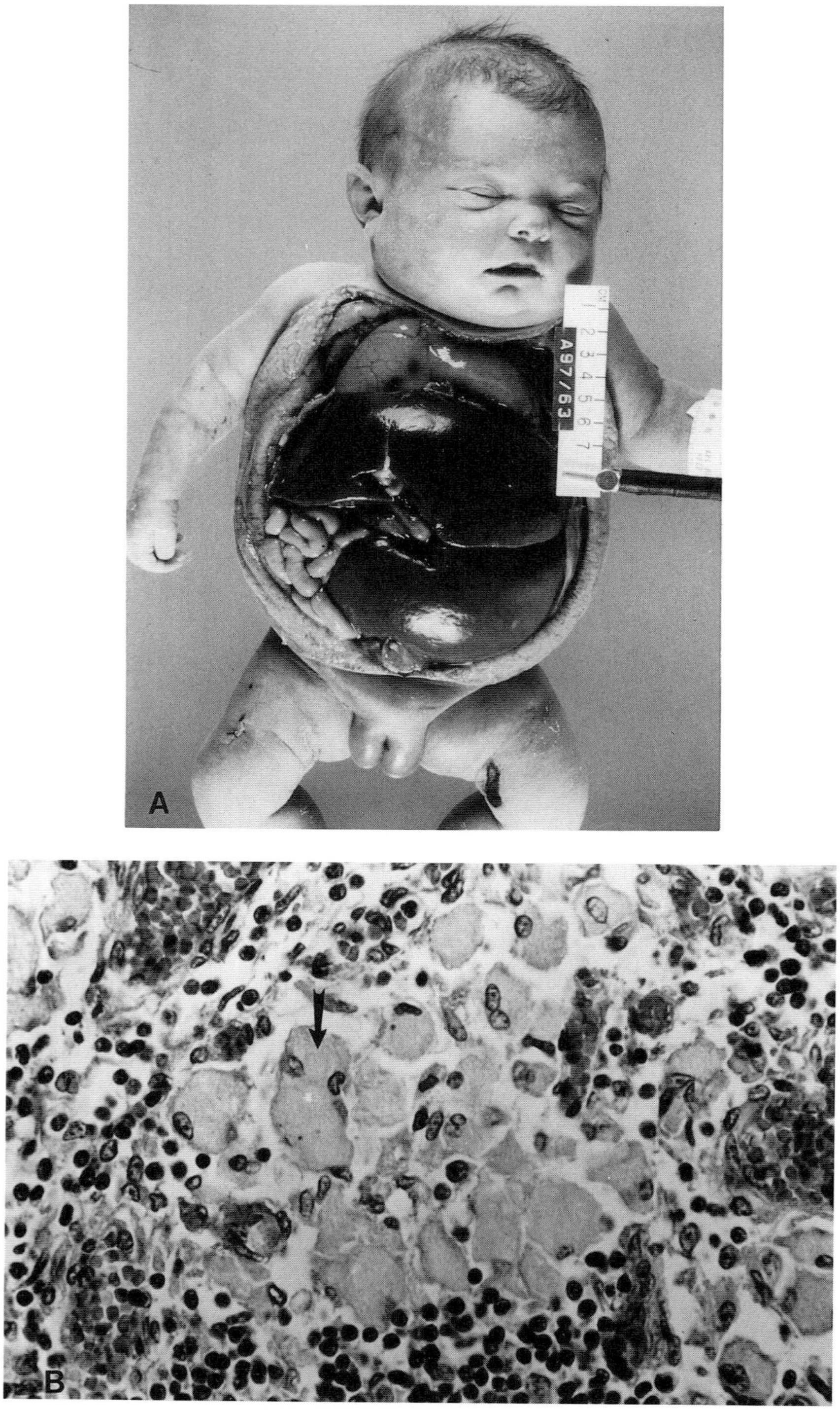

FIGURE 7.1. Histiocytosis due to lysosomal storage disease. (A) A 15-day-old hydropic female with the lethal infantile form of Gaucher disease (type 2). The liver and spleen are massively enlarged. (B) The liver shows a histiocytosis consisting of large cells with eccentric small, oval nuclei and abundant cytoplasm, having a wrinkled appearance consistent with Gaucher cells. The spleen, lymph nodes, and bone marrow were similarly involved. H&E; × 300. (From Sharma R, Monteiro C, et al. *Am J Perinatol* 2000;17:201. Used with permission.)

liberate lymphokines (soluble factors) such as interferon, which activate or assist the macrophage in disposal of the invader. Macrophages process antigen for presentation to lymphocytes in immune responses, thus initiating antibody formation. Moreover activated macrophages secrete cytokines having important biological functions, such as interleukin 1, which plays a role in cell activation and in the body's febrile reaction, and the tumor necrosis factor, which participates in host defense against tumor growth. Macrophages clear from the blood effete erythrocytes, noxious substances, and cellular debris as they pass through organs of the monophagocytic system (e.g., in the liver and spleen) [1,3].

Cells of the mononuclear phagocytic system are divided on the basis of function into two main groups, phagocytic cells (antigen-processing cells) and dendritic cells (antigen-presenting cells) [3,8].

Classification of Histiocytosis

The Histiocyte Society, an international organization, classified the childhood histiocytoses into three main classes or categories according to their histopathological characteristics [3]. Class I included the Langerhans cell histiocytosis group, which had been called histiocytosis X (Letterer-Siwe disease, Hand-Schuller-Christian syndrome, eosinophilic granuloma), self-healing histiocytosis, pure cutaneous histiocytosis, Langerhans cell granulomatosis, and "nonlipid reticuloendotheliosis." Disorders of the non-Langerhans histiocytes, namely, familial erythrophagocytic lymphohistiocytosis and the infection associated hemophagocytic syndrome, fell into the class II histiocytoses. Class III histiocytoses included malignancies of mononuclear phagocytes: acute monocytic leukemia, malignant histiocytosis, and histiocytic lymphoma [2,3,5].

In 1997 the Histiocyte Society revised the earlier classification in an attempt to correlate more closely the histiocytes of various types with their biological behavior [3] (Table 7.2).

TABLE 7.2. Classification of histiocytic disorders.

Disorders of varied biological behavior
Dendritic cell related
Langerhans cell histiocytosis
Juvenile xanthogranuloma
Macrophage related
Hemophagocytic syndromes
Familial erythrophagocytic lymphohistiocytosis
Infection (reactive) associated hemophagocytic syndrome
Sinus histiocytosis with massive lymphadenopathy (Rosai-Dorfman disease)
Malignant disorders
Monocyte related
Acute monocytic leukemia
Dendritic cell– or macrophage-related histiocytic sarcoma (lymphoma)[a]

[a] Histogenesis and classification remains disputed (? occur during the first year of life).
Source: Favara BE, et al. [2]. Abstracted from Histiocyte Society classification.

This current classification consists of two main categories: disorders of varied biological behavior and malignant disorders [2] (Table 7.2). According to this scheme, the types of histiocytoses encountered in the infant are (1) the dendritic cell disorders: Langerhans' cell histiocytosis and juvenile xantogranuloma and; (2) the macrophage disorders: the hemophagocytic syndromes (familial and infection-associated lymphohistiocytosis) and Rosai-Dorfman disease, all of which belong to the the varied biological behavior disorder category. Monocytic leukemia falls into the second main category, malignant disorders [2,6,9–11]. The histogenesis of malignant histiocytosis and histiocytic lymphoma-sarcoma remains unclear and unsettled; these rare, nebulous tumors are found primarily in adults and are the subject of an isolated case report in the first year of life [12].

Langerhans' Cell Histiocytosis

"Langerhans cell histiocytosis" is the term now substituted for the older generic one, "histiocytosis X," which included the clinical triad of Letterer-Siwe disease, Hand-Schuller-Christian syndrome, and eosinophilic granuloma [3,7a,13, 14]. Histologically all three conditions are characterized by granuloma-like lesions composed of Langerhans histiocytes [1,14–16]. Although it behaves like a malignancy particularly in the young, Langerhans cell histiocytosis is regarded by some investigators not as a true neoplasm but instead as a disorder of immune regulation [16,17]. On the other hand, others feel that this histiocytosis is a neoplastic rather than a reactive process [18]. Molecular clonality studies suggest that Langerhans cell histiocytosis is a monoclonal disease [1,19].

The Langerhans cell is a member of the monophagocytic system originating from the bone marrow and migrating primarily to the skin and to other organs. It is one of three types of dendritic cell occurring in the epidermis; the others are the melanocyte and the indeterminate dendritic cell [20].The Langerhans cell is found also in extracutaneous sites such as lymph nodes, thymus, oral mucosa, and vagina. Normally the cell functions in the processing of antigens, for example, those of viruses, bacteria, and tumor cells, coming into contact with the skin, and in the presentation of antigens to T lymphocytes [19–21]. Studies show that unlike samples from normal controls, Langerhans cell histiocytes from lesions of patients with disseminated Langerhans cell histiocytosis are unable to present alloantigens to T cells in vitro [21].

Disseminated Langerhans cell histiocytosis (Letterer-Siwe disease) may present in the perinatal period and during the first year of life [14,15]. The disease occurs in the first month of life in 2 to 15% of children with this condition [14,15, 22–25]. Skin lesions are the first manifestation from birth, and the time interval before systemic symptoms occur varies from days to months [22,24,25].

Clinical Findings

Disseminated Langerhans cell histiocytosis, formerly known as Lettere-Siwe disease or disseminated histiocytosis X, is the generalized form affecting primarily infants and children less than 2 years of age. Involvement of two or more organ systems is required for the diagnosis. The peak incidence of the disseminated histiocytosis is during the first 6 months of life [23].

The initial clinical manifestation, which may be apparent at birth, is a seborrheic, eczematoid, occasionally hemorrhagic rash distributed over the scalp, face, abdomen, and diaper areas (Figures 4.17 and 7.2). Subsequently hepatosplenomegaly, lymphadenopathy, and generally "failure to thrive" become evident. Hematological studies reveal an anemia, a variable leukocytosis, abnormal clotting factors, and thrombocytopenia, all of which portend disseminated disease and a poor prognosis [17]. Imaging studies reveal pulmonary infiltrates becoming more progressive as the disease progresses. Histiocytes are readily identified in bronchoalveolar lavage fluid with immunostaining and by electron microscopy [5]. Skeletal surveys show lytic ("punched out") lesions in the skull and long bones (see Chapter 17, Tumors and Tumorlike Conditions of Bone). The diagnosis is definitely established by a skin or lymph node biopsy [14,17]. Usually the bone marrow is involved at the time of diagnosis.

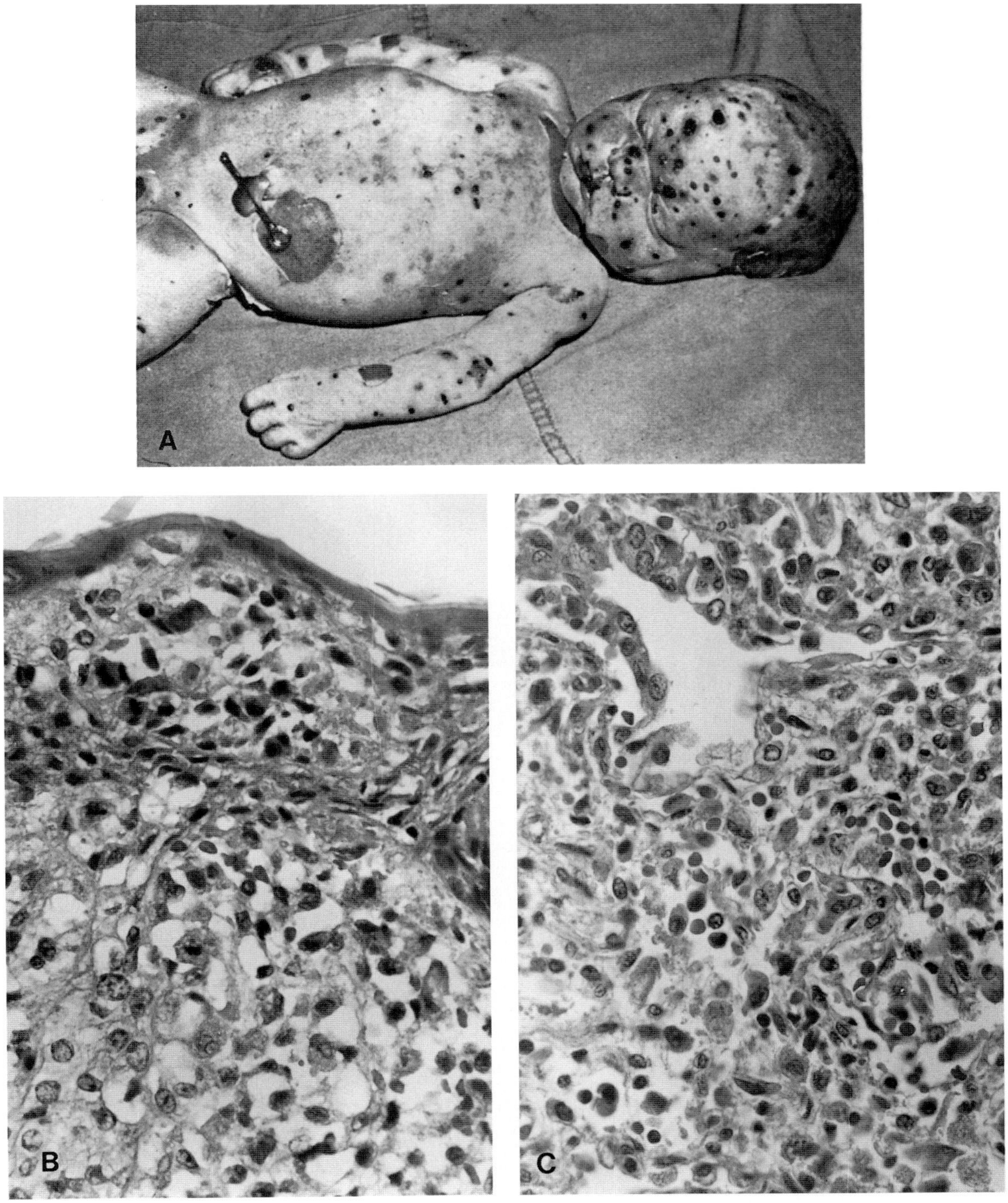

FIGURE 7.2. Congenital disseminated Langerhans cell histiocytosis (Letterer-Siwe disease). (A) Term stillborn with widely distributed hemorrhagic, crusted, or scaling nodules. In addition, a focal, purpuric macular rash is present. Diagnosis was established by skin and lymph node biopsies. (From Ahnquist G, Holyoke JB. *J Pediatr* 1960;57:897. Used with permission.) (B) Disseminated Langerhans cell histiocytosis (Letterer-Siwe disease) from an identical 5-day-old female with a maculopapular rash and severe respiratory distress. There was a maternal history of polyhydramnios. The skin shows erosion of the stratum corneum, as well as invasion of the edematous epidermis and dermis by Langerhans cell histiocytes. Multiple other organs were involved. H&E; × 380. (C) The pulmonary interstitium, alveolar ducts, and alveoli are diffusely infiltrated by Langerhans cell histiocytes with characteristic infolded, coffee bean–like nuclei. Pulmonary involvement is an important cause of death in newborns and infants with this disease. H&E; × 380.

Pathology

The gross appearance varies with the site and the age of the lesion. Specimens curetted from bones are soft, red, hemorrhagic, and necrotic. When visible, soft grayish yellow to red granuloma-like nodules, usually multiple, are found in the liver, spleen, and lymph nodes. Older lesions are firmer because of fibrosis and are tan-yellow instead of grayish-red owing to the presence of numerous foamy histiocytes [7a].

Disseminated Langerhans cell histiocytosis rarely causes stillbirth [22]. Widely distributed hemorrhagic, crusted, or scaling cutaneous nodules composed of proliferating histiocytes may be present at birth (Figure 7.2); similar infiltrates are found in the lungs, brochi, liver, spleen, lymph nodes, thymus, and bones at necropsy.

The Langerhans cell has a characteristic infolded or coffee bean–shaped nucleus, a small nucleolus, and eosinophilic cytoplasm (Figure 7.3A–C). Some lesions show nuclear atypia

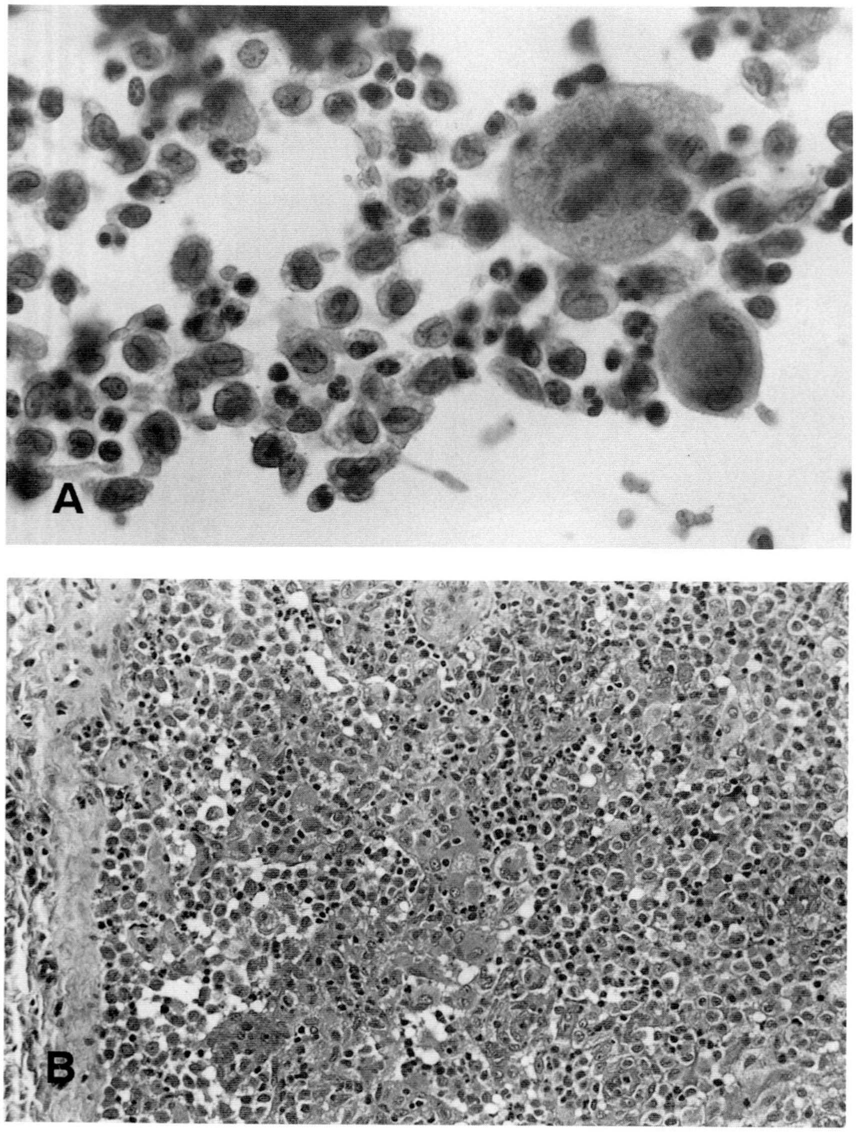

FIGURE 7.3. Langerhans cell histiocytosis, lymph node biopsy. (A) Imprint ("touch preparation") revealing histiocytes with scant eosinophilic cytoplasms and characteristic, folded, coffee bean–like nuclei containing one or more tiny nucleoli. The multinucleated giant cells are composed of similar-appearing nuclei. H&E; × 480. (B) Histiocytes flooding the subcapsular and cortical sinusoids. H&E; × 150. (C) Higher power view (× 480) showing the histiocytes in more detail. H&E. (D) The cytoplasms stain uniformly positive with S-100 protein. H&E; × 300. (E) Racket-shaped Birbeck granule demonstrated by electron microscopy, × 28,500. (Electron photomicrograph courtesy of Ann Peters, Department of Pathology, Children's Hospital, San Diego.)

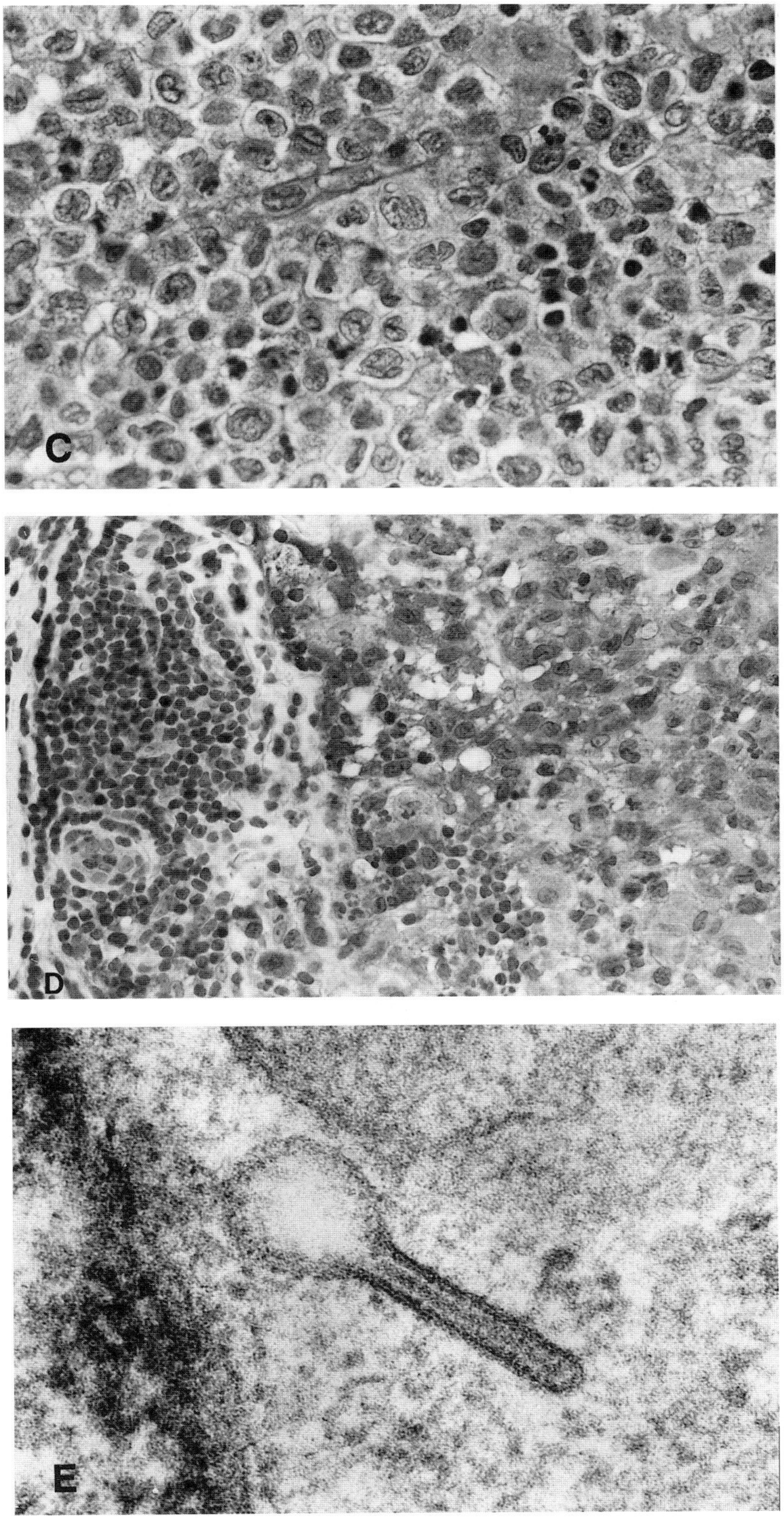

FIGURE 7.3. (*Continued*)

and mitotic figures, but these findings do not correlate with prognosis [16]. Extensive infiltration of histiocytes and occasionally eosinophils are the typical histological findings. Histiocytic erythrophagocytosis is seldom noted. The Langerhans cell histiocyte has immunohistochemical and ultrastructural properties distinctly different from those of the normal monocyte and macrophage [1,3,18]. The histiocytes are reactive with S-100 and CD1a (OKT6) and CD45 surface antigens (Table 7.3). Langerhans cell histiocytes react with placental alkaline phosphatase, a proliferation marker Ki-S1, and a nuclear antigen Ki-67 [18]. Both Ki-S1 and Ki-67 are indicators of neoplastic proliferation.

Ultrastructurally the cytoplasm of the Langerhans histiocyte contains characteristic racket-shaped Birbeck granules, which are requisite for the diagnosis [1] (Figure 7.3E). The significance and function of these structures are unknown. Sometimes they are difficult to find, and multiple thin sections may be required. The granules are observed also in Langerhans cells normally present in the epidermis but not in the intermediate dendritic cells or melanocytes [4]. Hashimoto, Pritzker, and their colleagues described myelinated laminated and nonmyelinated dense bodies in addition to Birbeck granules in a patient with congenital self-healing Langerhans cell histiocytosis [26,27].

Skin and bone are the tissues most often involved at the time of presentation [13,23]. In the skin the histiocytes are situated primarily in the dermis directly beneath the epidermal-dermal junction. Characteristically they invade the epidermis and extend into the stratum corneum (Figure 7.2B). The differential diagnosis of cutaneous infiltrates in the newborn includes leukemia cutis, extramedullary hematopoiesis, urticaria pigmentosa (mast cell disease), and metastatic malignancies such as rhabdomyosarcoma and neuroblastoma [7a] (Tables 4.1 and 4.2).

Skeletal surveys reveal lytic lesions involving the skull, ribs, pelvis, and occasionally the long bones and vertebra [28]. In most instances a solitary lesion occurring in the skull or other bone is a self-limited process even in the young and is cured by excision and curettage (see Chapter 17). The curettings show granuloma-like collections of Langerhans cell histiocytes and multinucleated giant cells, extensive necrosis with destruction of bone, eosinophilic "abscesses," and variable numbers of lymphocytes and plasma cells. On the other hand older, "end-stage" or healing lesions contain histiocytic foam cells, which are nonreactive for S-100 and fibrosis, and therefore may resemble a benign fibrous histiocytoma.

Solitary osseous lesions rarely herald the onset of soft tissue and visceral involvement and a fatal outcome. Moreover infants with multiple bone lesions may survive [13,15]. Hand-Schuller-Christian syndrome is a clinical manifestation of multifocal Langerhans cell histiocytosis involving primarily the cranial bones and adjacent structures including the pituitary. The syndrome is very uncommon, for only a few patients, usually older infants and children, actually fulfill the diagnostic criteria of exophthalmos, diabetes insipidus, and histiocytic infiltrates in membrane bone [6].

TABLE 7.3. Identifying features of histiocytes.

Macrophage	Lysosome, CD45, CD14, and CD68 positive S-100, CD1a and factor XIIIa negative Birbeck granules absent
Indeterminate cell	CD45, S-100, and CD1a positive Factor XIIIa negative Birbeck granules absent
Langerhans cell	CD45, S-100, and CD1a positive CD14 and factor XIIIa negative Birbeck granules present
Interdigitating dendritc cell	CD45 and S-100 positive CD14, CD1a, and factor XIIIa negative Birbeck granules absent
Dermal dendrocyte	Factor XIIIa, CD45, and CD68 positive S-100 and CD1a negative Birbeck granules absent
Follicular dendritic cell	KiM4, CD21, and CD35 positive S-100 variable CD45 negative Birbeck granules absent

Source: Favara BE, et al. [2]. Used with permission.

Exophthalmos results from a space-occupying histiocytic lesion in the orbit, and otitis media with drainage follows destructive infiltrates in the mastoid and petrous portion of the temporal bone. The lesions tend to be less aggressive with less cellularity and less monotony of cell type than in Letterer-Siwe disease, but they exhibit more fibrosis and evidence of healing [7a]. The clinical course is usually protracted, and the mortality is approximately half that observed for the disseminated disease.The least aggressive of the three forms, eosinophilic granuloma, typically occurs in a single bone in an older child but may be found in adolescents and adults in the skin and lungs.

Infiltration of the pulmonary interlobular septa and alveolar spaces produces airway obstruction and eventually the formation of cystically dilated, emphysematous areas. Moreover, Langerhans cell histiocyte infiltration of the lungs is a significant cause of death in young infants [14,24,29] (Figure 7.2C). In the lymph nodes the histiocytic infiltrate replaces and obscures the normal architecture (Figure 7.3). The splenic red pulp is heavily infiltrated, but the white pulp and trabeculae usually are spared. Bone marrow elements are replaced by Langerhans cell histiocytes. Granuloma-like infiltrates are found the pancreas, gonads, liver, and salivary glands and sometimes in the brain and meninges [14,15,30]. The liver findings consist of variable degrees of periportal and sinusoidal histiocytic infiltration and fibrosis progressing to focal biliary cirrhosis [30,31].

A pure cutaneous form of Langerhans cell histiocytosis, also known as self-healing histiocytosis or the Hashimoto-Pritzker syndrome, is usually congenital and is associated with a good prognosis [17,26,27,32–34]. Surprisingly enough, the skin lesions, which may be extensive, undergo rapid spontaneous regression.

Ultrastructural and immunohistochemical properties of the histiocytes from patients with congenital self-healing histiocytosis are similar to both normal epidermal Langerhans cells and Langerhans histiocytosis cells [33]. Two types of histiocyte have been described, one with and one without Birbeck granules [34]. Nevertheless, patients with self-healing histiocytosis should be followed carefully for progression of disease [33].

Prognosis

The clinical outcome cannot be predicted from the histological appearance alone [14–16]. The clinical findings, particularly the age of the patient and extent of organ involvement at the time of diagnosis, are the most reliable indicators for predicting prognosis [6,16,35]. Organ dysfunction may be more important than the number of organs involved. Bone marrow involvement at the time of diagnosis is associated with an unfavorable outcome [13].

Disseminated Langerhans cell histiocytosis patients with multiple sites of involvement and organ dysfunction have a poor prognosis, with less than 40% survival [13]. Those who fall into this category are usually less than one year of age and comprise less than 15% of the total. The overall mortality rate of children under 2 years of age at diagnosis is 50 to 60%, while that of children over 2 years decreases to 15% [6].

Hemophagocytic Syndromes

The hemophagocytic syndromes are the most common of the macrophage-related disorders of varied biological behavior and are responsible for significant morbidity and mortality [2,3]. Hemophagocytic lymphohistiocytosis is the major disease of this group, which encompasses most of the patients with macrophage-related disorders previously called class II histiocytoses. Hemophagocytic lymphohistiocytosis includes primary (familial) and secondary (infection- or reactive-associated) types. The term "hemophagocytic syndrome" is used to include both these entities [2]. There is a striking similarity between primary and secondary hemophagocytic lymphohistiocytosis, for the clinical, biochemical, and pathological findings are almost identical (Table 7.1).

Primary Hemophagocytic Lymphohistiocytosis

Primary hemophagocytic lymphohistiocytosis (familial hemophagocytic lymphohistiocytosis) is a rapidly fatal disease characterized by fever, hepatosplenomegaly, and pancytopenia frequently occurring in siblings [5,8,36–43]. The disease occurs primarily in infants and young children. Most patients are less than 6 months of age at the time of diagnosis, and 10 to 15% are neonates [38,41]. The German/Austrian Registry for Histiocytic Disorders reports an annual incidence of 1 per million children under 15 years of age [41].

Familial hemophagocytic lymphohistiocytosis is a disorder of unknown etiology that affects the monophagocytic system [2,6]. It has an autosomal recessive pattern of inheritance, but at least 25% of the cases are nonfamilial [37,38,41]. Nonspecific chromosomal abnormalities are described [44].

Clinical Findings

The classical presenting signs are fever, hepatosplenomegaly, and pancytopenia [9,39,40]. Other manifestations of this rare, usually fatal disease are irritability, anorexia, wasting, and "failure to thrive." Hepatic dysfunction and bleeding diathesis are frequently present. Typically there is a short period of normal development and absence of symptoms after birth, although fever or hepatomegaly already may be evident [9,40].

Stupor, or coma, and seizures with hemiplegia and cranial nerve palsies are the ominous neurological signs [36,45]. The onset of nervous system involvement heralds progression of the disease, which is reflected by an elevated level of cerebrospinal fluid protein and pleocytosis [36,37]. In addition, laboratory studies reveal hypofibrinogenemia and hypertriglyceridemia [3,9,39]. Thrombocytopenia is accompanied by severe bleeding, and the granulocytic series may show a maturation arrest. The triad of elevated triglycerides, increased liver enzymes, and low fibrinogen levels suggests the diagnosis [40].

Frequently the diagnosis of familial hemophagocytic lymphohistiocytosis is difficult to establish during life; it may be apparent only after postmortem findings have prompted the physician to review a family history, which discloses that a sibling had a similar disease [9,40].

Pathology

The histopathological hallmarks of familial homophagocytic lymphohistiocytosis are extensive lymphohistiocytic infiltration and erythrophagocytosis within the liver, spleen, lymph nodes, bone marrow, and leptomeninges [9,30,38,40,42,46,47] (Figure 7.4). The spleen and liver are massively enlarged, up to 10 times and 2 times normal, respectively. Thymic involution and depletion of lymphoid tissue in lymph nodes and spleen are characteristic findings. Pulmonary infiltration is predominantly centrilobular, peribronchial, and perivascular [30]. The submucosa of the tracheal and major bronchi show extensive involvement. Many different organs are affected [30,38,47].

The main neuropathological finding is focal or diffuse leptomeningeal lymphohistiocytic infiltration with erythrophagocytosis [36,48]. The arachnoid, choroid plexus, and Virchow-Robin perivascular spaces are affected in about a third to half of the cases [9,38,45]. Perivascular cuffing with extension of

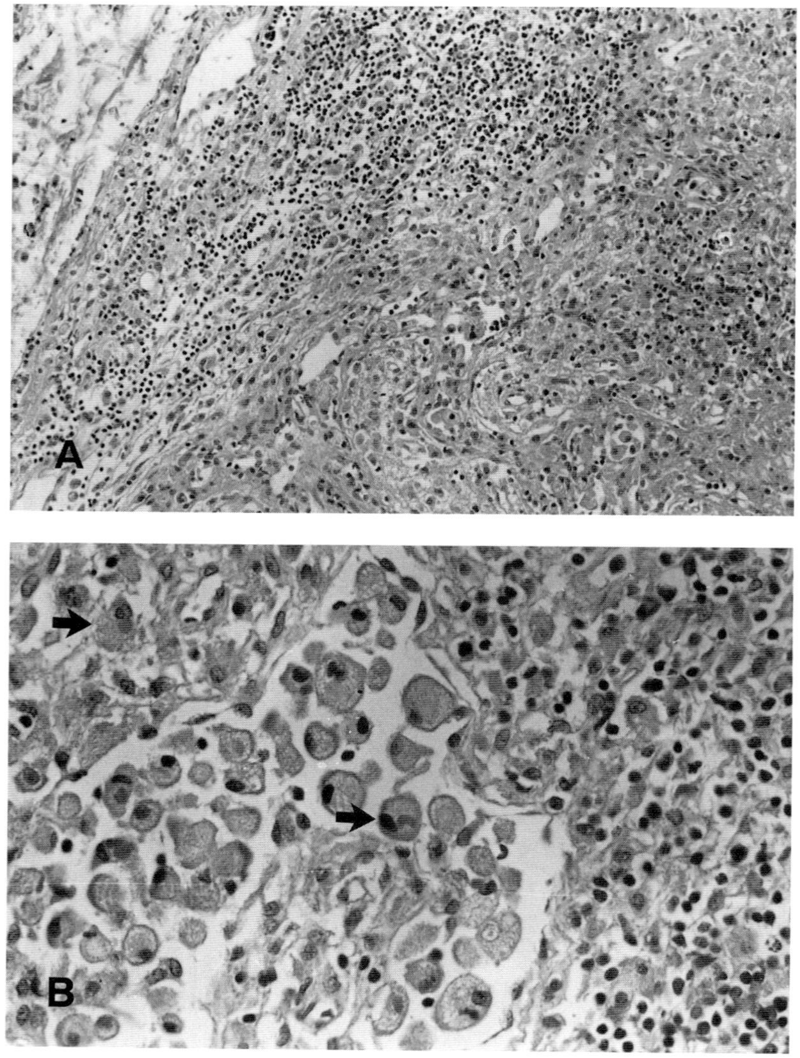

FIGURE 7.4. Familial erythrophagocytic lymphohistiocytosis. Lymph node from a 1-year-old male with fever, hepatosplenomegaly, hemolytic anemia, hypofibrinogenemia, and elevated liver enzymes. Female sibling died from a similar condition. (A) The lymph node shows lymphoid depletion and infiltrates of lymphocytes and histiocytes. H&E; × 380. (B) The lymphocytes and macrophages appear to be reactive without atypical features. Histiocytic hemophagocytosis is present (arrow). H&E; × 300. (C) The spleen is heavily infiltrated by lymphocytes and histiocytes. Lymphoid follicles are not recognizable. H&E; × 150. (D) Normal marrow elements are replaced by lymphocytes and histiocytes, the latter displaying hemophagocytosis. (E) The cerebellum exhibits perivascular cuffing by lymphocytes and histiocytes. Focal parenchymal necrosis is present. H&E; × 150.

lymphocytes and histiocytes into adjacent cerebral or cerebellar cortex and brain stem occurs (Figure 7.4E). Diffuse cerebral infiltration by histiocytes and multifocal necrosis are found in the most severely involved brains [36,48]. The severity of the central nervous system lesions seems to be related to the duration of the illness [36]. The brain is affected much more than the rest of the body in some patients [36].

The histiocytes and lymphocytes within the infiltrates appear to be reactive but show no cellular atypia, abnormal mitotic activity, or other evidence of malignancy. Histiocytic phagocytosis of erythrocytes and sometimes lymphocytes and platelets is a characteristic finding. The proliferative histiocytes in familial erythrophagocytic lymphohistiocytosis have immunohistochemical and ultrastructural properties of normal reactive histiocytes; that is, they express CD14, CD45, CD68, and HLA-DR. They lack the S-100 protein and CD1a surface antigen reactivity and Birbeck granules characteristic of Langerhans cell histiocytes [43,46] (Table 7.3). Ultrastructural studies confirm the reactive nature of the histiocytes and lymphocytes and the absence of Birbeck granules [38,43].

Diagnostic biopsy is recommended for an enlarged lymph node. The findings of lymphohistiocytic infiltrates, hemo-

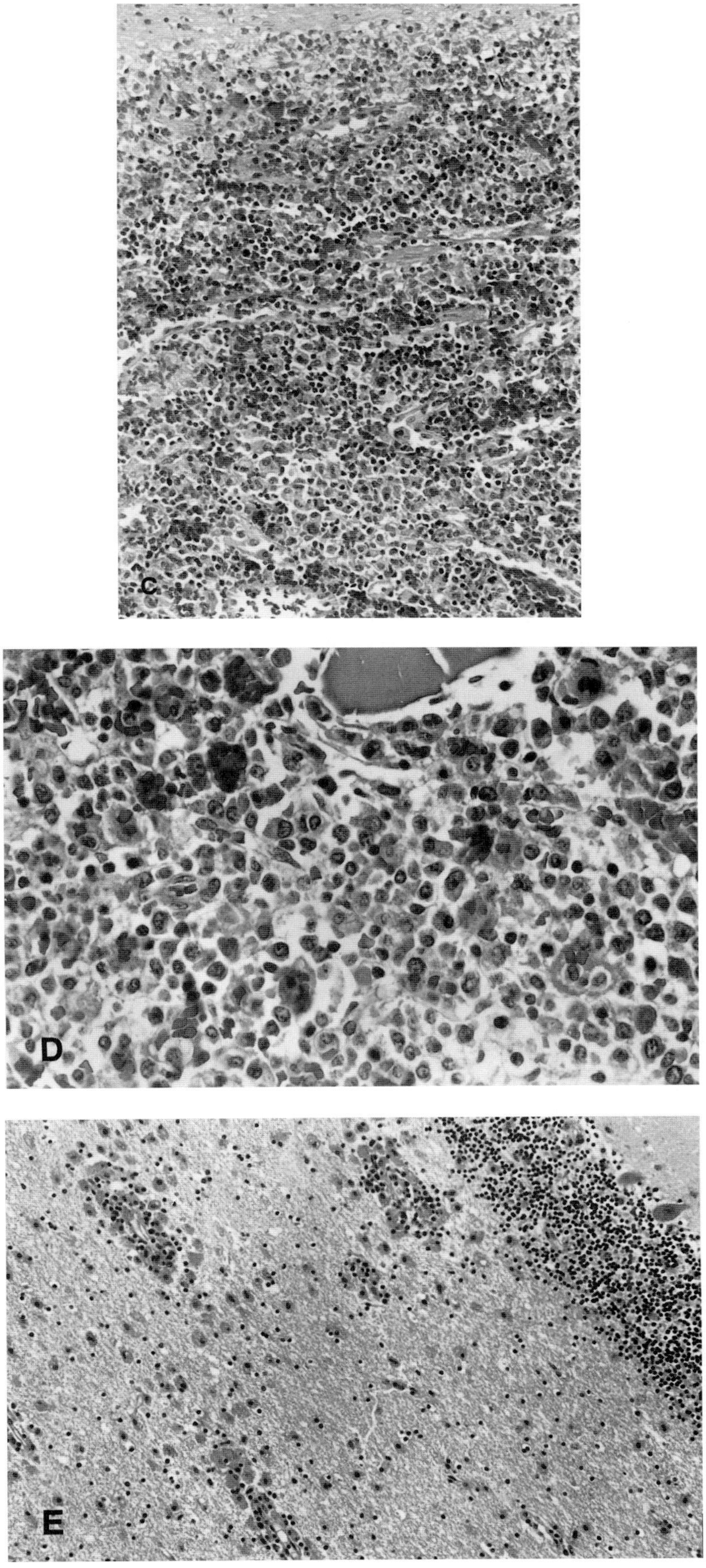

FIGURE 7.4. (*Continued*)

phagocytosis, and lymphoid depletion suggest the diagnosis [9]. However it should be remembered that these findings are rather nonspecific and that hemophagocytosis is present also in a variety of other conditions, including infections, hemolytic anemia, and graft-versus-host disease. Lymphohistiocytic infiltrates with lymphoid depletion are found also in infection-associated hemophagocytosis, the X-linked lymphoproliferative syndrome, and the accelerated phase of Chediak-Higashi syndrome [11,40]. In addition, Langerhans cell histiocytosis, leukemia, nonspecific inflammation, and extramedullary hematopoiesis should be considered in the histological differential diagnosis [40]. Splenic fine-needle aspiration biopsy can be used for demonstrating histiocytes with hemophagocytosis [39].

Secondary Hemophagocytic Lymphohistiocytosis (Infection-Associated Hemophagocytic Syndrome)

Infection-associated hemophagocytic syndrome was first reported in 1979 by Risdall et al., who described a disease characterized by a generalized histiocytic proliferation and hemophagocytosis associated with a systemic viral infection [11]. Cytomegalovirus, herpes simplex virus, adenovirus, enterovirus, and Epstein-Barr virus are examples of the viral agents implicated [11,49–52]. The disease occurs also with other infections (e.g., bacterial, fungal, parasitic) [11,53,54]. The syndrome is described in immunocompromised and in normal individuals but more commonly in the former [11,55]. Infection-associated hemophagocytic syndrome is probably the most common histiocytic disorder in the infant and occurs throughout all ages, including adulthood [11]. Other terms applied to this lymphohistiocytosis are "reactive" and malignancy-associated hemophagocytic syndrome [2,6].

The clinical manifestations of the infection-associated hemophagocytic syndrome are practically the same as familial hemophagocytic lymphohistiocytosis: fever, failure to thrive, anemia, hepatosplenomegaly, and variable lymphadenomegaly [37,53]. Leukopenia, abnormal liver function tests, and a coagulopathy are the laboratory findings [11]. Bilateral pulmonary infiltrates may be present on chest radiographs.

Familial hemophagocytic lymphohistiocytosis is the entity causing the most difficulty in the differential diagnosis [40]. There is considerable overlap between the two diseases, making distinction between them difficult. In contrast to the familial form, often there is an underlying immune deficiency (iatrogenic or otherwise), a proven infection, recovery in most instances, and usually no family history in the infection-associated hemophagocytic syndrome. Patients are reported who were thought initially to have familial hemophagocytic lymphohistiocytosis but in whom eventually a viral agent was recovered by tissue culture, electron microscopy, or sensitive nucleic acid probes [40,50,51]. Some investigators believe that the two diseases are essentially the same and that the distinction between them is superficial [51]. The term "hemophagocytic syndrome" has been applied to include both entities [2].

The diagnosis of infection-associated hemophagocytic syndrome is suggested by bone marrow examination. Bone marrow smears reveal prominent histiocytic hemophagocytosis [11,56] (Figure 7.5C, D). The reactive-appearing histiocytes contain phagocytized erythrocytes and platelets. They show

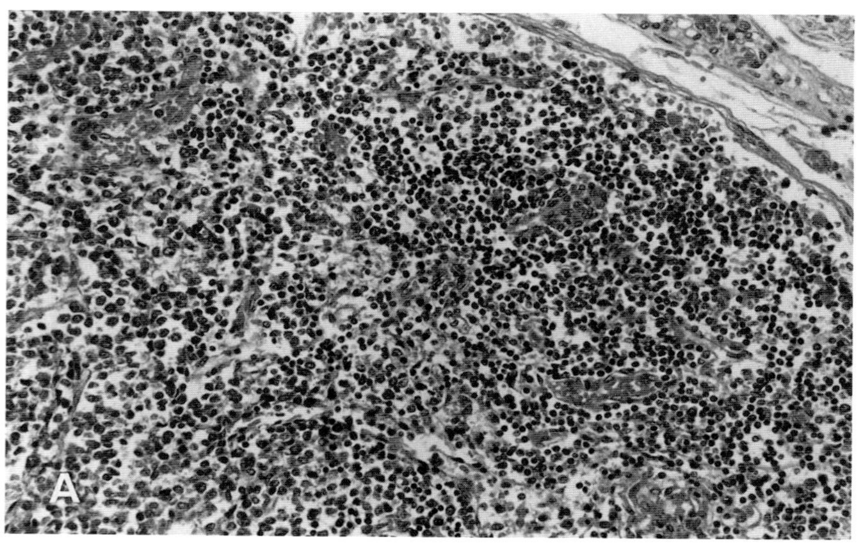

FIGURE 7.5. Infection-associated hemophagocytic syndrome in a 6-week-old female with fever, lethargy, vomiting, and diarrhea. Other findings included hepatosplenomegaly, anemia, thrombocytopenia, abnormal results from coagulation studies, and elevated liver enzymes. (A) Lymph node is replaced by lymphocytes and histiocytes. H&E; × 150. (B) The infiltrate is composed of reactive lymphocytes and histiocytes. H&E; × 380. (C) Postmortem bone marrow aspirate shows only lymphocytes and histiocytes. Wright-Giemsa; × 300. (D) Section of rib lacks normal marrow elements. Lymphocytes and histiocytes, the latter with phagocytosed erythrocytes, comprise the cell population. H&E; × 380.

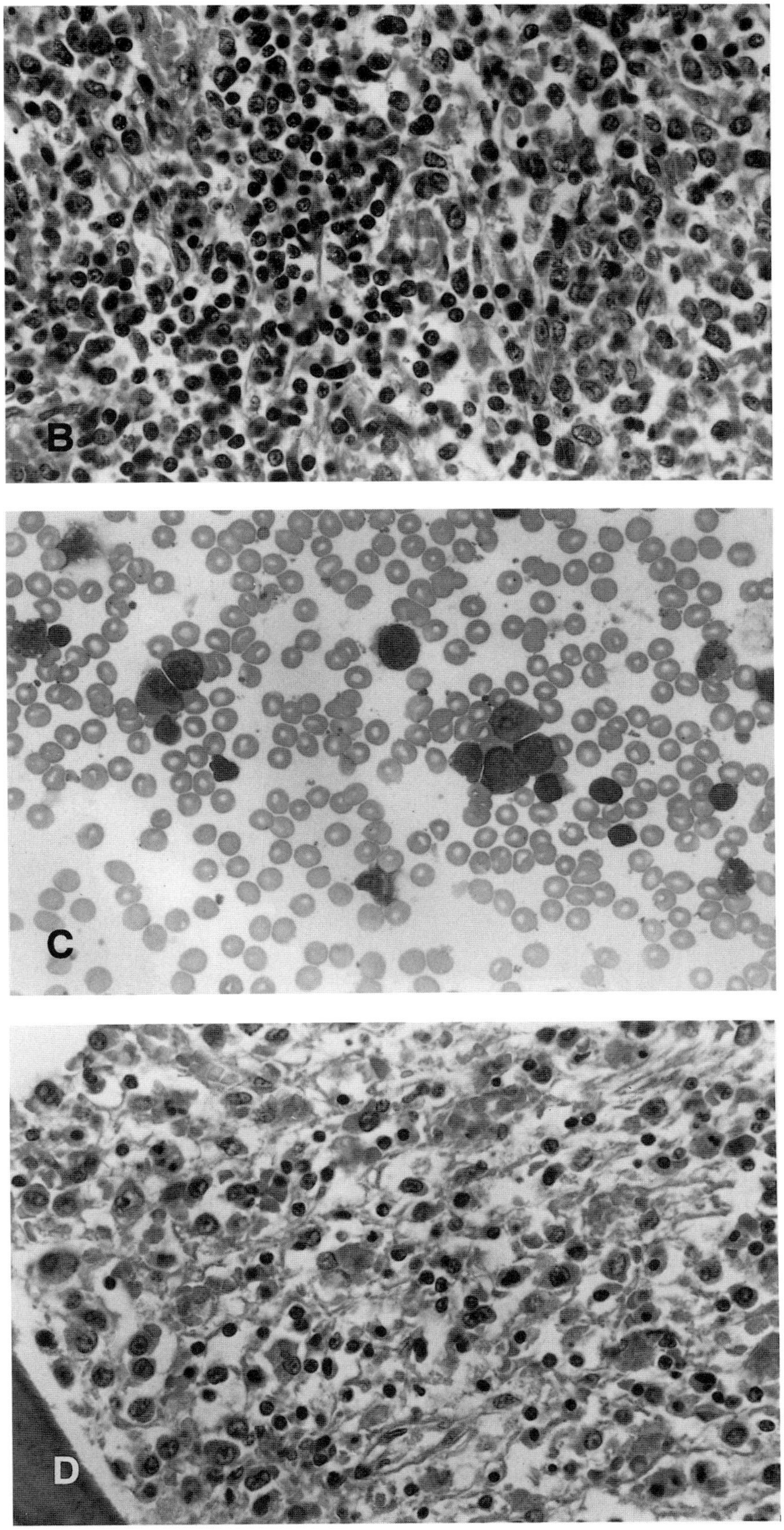

FIGURE 7.5. (*Continued*)

the cytological, ultrastructural, and histochemical features of normal histiocytes, in contrast to those of the Langerhans cell and acute monocytic leukemia. Atypical lymphocytes ("virocytes") and immunoblasts may be present. Atypical lymphocytes have large often reniform nuclei, prominent nucleoli, and a characteristic blue cytoplasm with white fluffy inclusions giving the cytoplasm a "snow storm-like" appearance. The peripheral blood smears exhibit varying degrees of cytopenia with or without atypical lymphocytes. Lymph nodes and spleen are depleted of lymphoid elements, have typically pale burned-out germinal centers, and are infiltrated by hemophagocytic histiocytes [11,38] (Figure 7.5A). Early in the disease lymph nodes may show a florid immunoblastic proliferative response similar to that observed in the early stage of human immunodeficiency virus (HIV) infections. Liver biopsy reveals prominent portal infiltrates composed of lymphocytes, erythrophagocytic histiocytes, and sometimes immunoblasts. If a viral or other infectious agent is found, the diagnosis is established [53,57,58].

Sinus Histiocytosis with Massive Lymphadenopathy

The syndrome of sinus histiocytosis with massive lymphadenopathy was established as a clinicopathological entity in 1972 by Rosai and Dorfman who described a benign, chronic, massive enlargement of cervical lymph nodes accompanied by fever, leukocytosis, and hyperglobulinemia [59]. Since 1972 several other articles on the subject have appeared [60–63]. By far the largest and most comprehensive review is by Fourcar, Rosai, and Dorfman, who analyzed 423 cases [63]. Fewer than 10 examples are reported in the newborn and infant [61].

The most common presentation is cervical adenopathy, which is usually bilateral. Laboratory findings are a normochromic anemia, leukocytosis, lymphopenia, polyclonal hyperglobulinemia, and an elevated erythrocyte sedimentation rate. Approximately 10% of the patients with sinus histiocytosis with massive lymphadenopathy have an underlying immune defect, and almost half with this association die from their immune problem [61].

Marked expansion of lymph node sinusoids by collections of histiocytes and lymphocytes with focal effacement of the nodal architecture are the histological findings. The histiocytes are benign appearing, having abundant cytoplasm and regular nuclei. Leukocytophagocytosis (emperipolesis), where lymphocytes and plasma cells are phagocytized by histiocytes, is a distinguishing histological feature essential for the diagnosis. Reactive, vacuolated histiocytes, numerous plasma cells, and extensive fibrosis are present. The lesions are focally destructive and affect also extranodal sites such as nasal mucosa, salivary gland, orbit, skin, and bones [60–63].

The histiocytes in this disease show cytological and immunohistochemical findings similar to those of normal histiocytes and macrophages [60]. In addition, placental alkaline phosphatase and S-100 are variably expressed by the histiocytes [60]. Molecular clonality studies suggest that Rosai-Dorfman disease is a polyclonal disorder [2].

Malignant Histiocytosis

The malignant histiocytoses are considered to be true neoplastic disorders of histiocytes. Malignant histiocytosis, acute monocytic leukemia, and histiocytic lymphoma are members of this group, which was classified as class III by the Histiocyte Society [3]. Of the three, acute monocytic leukemia is the major one occurring in the fetus and infant. Malignant histiocytosis and histocytic lymphoma are extremely rare diseases found almost exclusively in adults, although sporadically an infant case is reported [11,64].

Acute Monocytic Leukemia

According to the French-American-British (FAB) classification of leukemia, the criteria required for the diagnosis of acute monocytic leukemia, in addition to other findings of leukemia, are the presence of over 30% leukemic monoblasts or more mature leukemic monocytes (FAB M5) and fewer than 20% leukemic granulocytic precursors in the bone marrow [65]. Leukemic monocytes are characteristically α-naphthyl esterase and myeloid surface markers CD13, CD33, CD14 reactive; 11q23 translocation such as t(9;11) are common [66].

Typically infants with acute monocytic leukemia present with a significant leukocytosis and a high frequency of involvement of nonhematopoietic tissues, particularly the skin. Hepatosplenomegaly and lymphadenomegaly are found initially in about half the cases [66,67]. This malignancy was discussed in more detail in Chapter 6.

Other Histiocytosis

Omenn Disease

Omenn disease is an uncommon familial immunodeficiency disorder of the lymphoid and monophagocytic systems. Eosinophilia, a widespread eczematous rash often present at birth, hepatosplenomegaly, recurrent infections, and generalized lymphadenopathy with histiocytic infiltrates and abnormal T cells characterize the disease [12,68–71]. It occurs primarily in newborns and in infants less than three 3 months of age and is transmitted as an autosomal recessive trait.

Severe lymphoid depletion in the presence of numerous histiocytes and eosinophils is the principal finding in the lymph nodes, spleen, and bone marrow [12,69–71]. The lymph node architecture is effaced and lacks a distinct cortex without follicle formation [71]. Diffuse hyperplasia of S-100 positive interdigitating reticulum cells is described. Lymphocytes are predominantly C8 suppressor type, while depletion of B lymphocytes is noted [71]. The thymus shows practically no lymphoid elements, and Hassel's corpuscles are absent. The histiocytes appear to be reactive; they stain positively for esterase, acid phosphatase, and OKM1 monoclonal antibody [69,70]. In addition, CD1a-positive histiocytes and granulocytes are found in the infiltrates. Erythrophagocytosis is absent.

The etiology of this unusual condition is unknown. An immunodeficiency with a defect primarily involving T cells is one proposal, and graft-versus-host disease associated with a primary cellular immunodeficiency is another [12,69]. With few exceptions, Omenn disease is fatal by one year of age [69]. However recovery following bone marrow transplantation has been described [5,70].

Chediak-Higashi Syndrome

The cardinal features of Chediak-Higashi syndrome, a rare autosomal recessive disease, are decreased pigmentation of hair and eyes (partial oculocutaneous albinism), photophobia, nystagmus, frequent bacterial infections, neurological dysfunction, and characteristic giant lysosmal granules in leukocytes and other granule-containing cells. Death occurs from infection or from a so-called accelerated phase of the disease, usually before 7 years of age. The accelerated phase is characterized by hepatosplenomegaly, lymphadenopathy, anemia, pancytopenia, and abnormal liver function tests similar to those found in patients with the infection-associated hemophagocytic syndrome [72,73]. Moreover Epstein-Barr virus infection has been implicated in patients with the accelerated phase of Chediak-Higashi disease, a situation perhaps analogous to the X-linked lymphoproliferative syndrome and this virus [72,74].

Although neonatal presentation of Chediak-Higashi syndrome is unusual, patients may experience recurrent infections beginning in the first few months of life. Nevertheless the diagnosis is established prenatally or at birth when the characteristic large lysosomal granules within leukocytes are identified on the peripheral blood smear or bone marrow [72,73] (Figure 7.6).

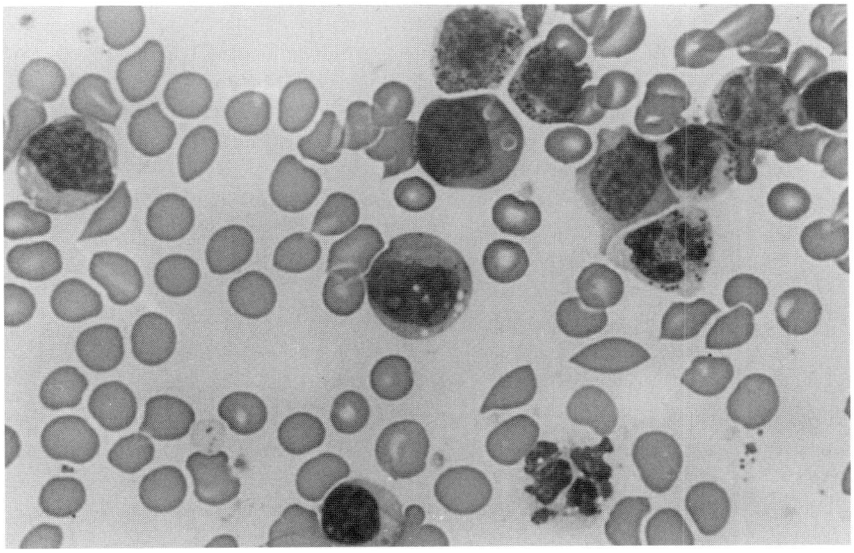

FIGURE 7.6. Chediak-Higashi syndrome in a 5-year-old female with oculocutaneous albinism, gray hair, developmental delay, susceptibility to infection, and neutropenia. Lymphocytes contain diagnostic large eosinophilic lysosomal inclusions. Prominent irregular cytoplasmic granules are present in neutrophils and other myeloid elements. Wright-Giemsa; × 750. (Courtesy of Glenn Billman, M.D., Children's Hospital, San Diego.)

Juvenile Xanthogranuloma

Juvenile xanthogranuloma may occur at birth, presenting as one or more yellow nodules ranging in size from one to several centimeters, particularly when the subcutaneous tissue is involved [7,75,76]. The head and neck and extremities are common sites. Occasionally significant morbidity and mortality result from extensive extracutaneous systemic involvement [75]. Cutaneous lesions are present in less than half with disseminated disease. The most common extracutaneous sites in decreasing order are the subcutaneous soft tissues, liver/spleen, lung, eye/orbit, and oropharynx [75].

The dermal dendrocyte is considered to be the cell of origin of juvenile xanthogranuloma [2,77]. Dermal dendrocytes are found normally in the upper dermis and are phenotypically different from the indeterminate cell and Langerhans cell (Table 7.3).

Microscopically the tumor consists of histiocytes with eosinophilic or foamy cytoplasms, Touton multinucleated giant cells, and variable numbers of lymphocytes and eosinophils. The histiocytes appear to be reactive, and they lack Birbeck granules, S-100 reactivity, and the CD1a (T6) surface markers characteristic of Langerhans cells [76,78]. Sudan black B and oil red O stains are variably lipid positive. Early lesions in the young infant may lack the typical Touton multinucleated giant cells and may appear to be hypercellular, complicating the diagnostic problem. Solitary giant xanthogranuloma and benign cephalic histiocytosis are considered to be variants of juvenile xanthogranuloma [77]. For further discussion, the reader is referred to Chapters 3 and 4 (Soft Tissue Tumors and Tumors and Tumorlike Conditions in Bone, respectively).

References

1. Favara BE. Histiocytosis syndromes and related disorders of early infancy: a pathologist's perspective. *Pediatr Hematol Oncol* 1989;6:213.
2. Favara BE, Feller AC, Pauli M, et al. Contemporary classification of histiocytic disorders. *Med Pediatr Oncol* 1997;29:157.
3. Writing Group of the Histiocyte Society. Histiocytosis syndromes in children. *Lancet* 1987;1:208.
4. Chu T, D'Angio GJ, et al. Histiocytosis syndromes in children. *Lancet* 1987;1:208.
5. Stephan JL. Histiocytoses. *Eur J Pediatr* 1995;154:600.
6. Huang F, Arceci R. The histiocytoses of infancy. *Semin Perinatol* 1999;23:319.
7. (a) Isaacs H Jr. Histiocytoses. In: *Tumors of the Fetus and Newborn*. Philadelphia: WB Saunders; 1997:169. Major Problems in Pathology. Series; Vol 35. (b) Sharma R, Hudak ML, Perszyk AA, et al. Perinatal lethal form of Gaucher's disease presenting with hemosiderosis. *Am J Perinatol* 2000;17:2019.
8. Lasser A. The mononuclear phagocytic system: a review. *Hum Pathol* 1983;14:108.
9. Janka GE. Familial hemophagocytic lymphohistiocytosis. *Eur J Pediatr* 1983;140:221.
10. Nesbit ME Jr, O'Leary M, Dehner LP, et al. The immune system and the histiocytosis syndromes. *Am J Pediatr Hematol Oncol* 1981;3:141.
11. Risdall RJ, McKenna RW, Nesbit ME, et al. Virus-associated hemophagocytic syndrome; a benign histiocytic proliferation distinct from malignant histiocytosis. *Cancer* 1979;44:993.
12. Hong R, Gilbert EF, Opitz JM. Omenn disease: termination in lymphoma. *Pediatr Pathol* 1985;3:143.
13. Raney RB Jr, D'Angio GJ. Langerhans cell histiocytosis (histiocytosis X): experience at the Children's Hospital of Philadelphia, 1970–1984. *Med Pediatr Oncol* 1989;17:20.
14. Isaacs H Jr. Perinatal (congenital and neonatal) neoplasms: a report of 110 cases. *Pediatr Pathol* 1985;3:165.
15. Isaacs H Jr. Neoplasms in infants: a report of 265 cases. *Pathol Annu* 1983;18(2):165.
16. Risdall RJ, Dehner LP, Duray P, et al. Histiocytosis X (Langerhans histiocytosis). Prognostic role of histopathology. *Arch Pathol Lab Med* 1983;107:59.
17. Broadbent VA. Malignant disease in the neonate. In: Roberton NRC, ed. *Textbook of Neonatology*. 2nd ed. Edinburgh: Churchill Livingston, 1992:879.
18. Hage C, Willman CL, Favara BE, et al. Langerhans cell histiocytosis (histiocytosis X): immunophenotype and growth fraction. *Hum Pathol* 1993;24:840.
19. Willman CL. Detection of clonal histiocytes in Langerhans cell histiocytosis: biology and clinical significance. *Br J Cancer* 1994;70(suppl XIII):S29.
20. Yu RCH, Morris JF, Pritchard J, et al. Defective alloantigen-presenting capacity of 'Langerhans cell histiocytosis cells.' *Arch Dis Child* 1992;67:1370.
21. Bjercke S, Elgo J, Braathen L, et al. Enriched epidermal Langerhans cells are potent antigen presenting cells for T cells. *J Invest Dermatol* 1984;83:286.
22. Ahnquist G, Holyoke JB. Congenital Letterer-Siwe disease (reticuloendotheliosis) in a term stillborn infant. *J Pediatr* 1960;57:897.
23. Nezelof C, Frileux-Herbert F, Cronier-Sachot J. Disseminated histiocytosis X: analysis of prognostic factors based on a retrospective study of 50 cases. *Cancer* 1979;44:1824.
24. Vade A, Hayani A, Pierce KL. Congenital histiocytosis X. *Pediatr Radiol* 1993;23:181.
25. Dehner LP. Neoplasms of the fetus and neonate. In: Naeye RL, Kissane JM, Kaufman N, eds. *Perinatal Diseases*. Baltimore: Williams & Wilkins; 1981:286. International Academy of Pathology Monograph 22.
26. Hashimoto K, Pritzker MS. Electron microscopic study of reticulohistiocytoma: an unusual case of congenital self-healing reticulohistiocytosis. *Arch Dermatol* 1973;107:263.
27. Hashimoto K, Griffin D, Kohsbaki M. Self-healing reticulohistiocytosis: a clinical histologic and ultrastructural study of the fourth case in the literature. *Cancer* 1982;49:331.
28. Senac MO Jr, Isaacs H, Gwinn JL. Primary lesions of bone in the 1st decade of life: retrospective survey of biopsy results. *Radiol* 1986;160:491.
29. Scully RE, Mark EJ, McNeely WF, et al. Case records of the Massachusetts General Hospital, Case 18-1996. *N Engl J Med* 1996;334:1591.
30. Landing BH. Lymphohistiocytosis in childhood: pathologic comparison with fatal Letterer-Siwe disease (disseminated visceral histiocytosis X). *Perspect Pediatr Pathol* 1987;9:48.
31. Heyn RM, Hamoudi A, Newton WA Jr. Pretreatment liver biopsy

in 20 children with histiocytosis X: a clinicopathologic correlation. *Med Pediatr Oncol* 1990;18:110.
32. Dehner LP, Bamford JT, McDonald EC. Spontaneous regression of congenital cutaneous histiocytosis X: report of a case with discussion of nosology and pathogenesis. *Pediatr Pathol* 1983;1:99.
33. Pujol RM, Moreno A, et al. Childhood self-healing histiocytosis X. *Pediatr Dermatol* 1988; 5:97.
34. Kanitakis J, Zambruno G, Schmitt D, et al. Congenital self-healing histiocytosis (Hashimoto-Pritzker): an ultrastructural and immunohistochemical study. *Cancer* 1988;61:508.
35. Rivera-Luna R, Martinez-Guerra G, Altamirano-Alvaraez E, et al. Langerhans cell histiocytosis: clinical experience with 124 patients. *Pediatr Dermatol* 1988;5:145.
36. Akima M, Sumi SM. Neuropathology of familial erythrophagocytic lymphohistiocytosis. *Hum Pathol* 1984;15:161.
37. Arico M, Janka G, Fischer A, et al. Hemophagocytic lymphohistiocytosis. Report of 122 children from the International Registry. *Leukemia* 1996;10:197.
38. Gilbert EF, ZuRhein GM, Wester SM, et al. Familial hemophagocytic lymphohistiocytosis: report of four cases in two families and review of the literature. *Pediatr Pathol* 1985;3:59.
39. Henter J-I, Elinder G: Familial erythrophagocytic lymphohistiocytosis: clinical review based on the findings in seven children. *Acta Paediatr Scand* 1991;80:269.
40. Janka GE. Familial hemophagocytic lymphohistiocytosis: Diagnostic problems and differential diagnosis. *Pediatr Hematol Oncol* 1989;6:219.
41. Janka GE. Familial hemophagocytic lymphohistiocytosis: therapy in the German experience. *Pediatr Hematol Oncol* 1989;6: 227.
42. Popper HH, Zenz W, Mache C, et al. Familial haemophagocytic lymphohistiocytosis: a report of three cases with unusual lung involvement. *Histopathology* 1994;25:439.
43. Wieczorek R, Greco MA, McCarthy K, et al. Familial erythrophagocytic lymphohistiocytosis: immunophenotypic, immunohistochemical and ultrastructural demonstration of the relation to sinus histiocytes. *Hum Pathol* 1986;17:55.
44. Kletzel M, Gollin SM, Gloster ES, et al. Chromosome abnormalities in familial hemophagocytic lymphohistiocytosis. *Cancer* 1986;57:2153.
45. Rettwitz W, Sauer O, Burow HM, et al. Neurological and neuropathological findings in familial erythrophagocytic lymphohistiocytosis. *Brain Dev* 1983;5:322.
46. Goldberg JC, Nezelof C. Familial lymphohistiocytosis: the pathologist's view. *Pediatr Hematol Oncol* 1989;6(3):199.
47. Ost A, Nilsson-Ardnor S, Henter J-I. Autopsy findings in 27 children with haemophagocytic lymphohistiocytosis. *Histopathology* 1998;32:310.
48. Henter J-I, Nennesmo I. Neuropathologic findings and neurologic symptoms in twenty-three children with hemophagocytic lymphohistiocytosis. *J Pediatr* 1997;130:358.
49. Chen R-L, Su I-J, Lin K-H, et al. Fulminant childhood hemophagocytic syndrome mimicking histiocytic medullary reticulosis. *Am J Clin Pathol* 1991;96:171.
50. Danish EH, Dahms BB, Kumar ML. Cytomegalovirus associated hemophagocytic syndrome. *Pediatrics* 1985;75:280.
51. McClain K, Gehrz R, Grierson H, et al. Virus-associated histiocytic proliferations in children: frequent association with Epstein-Barr virus and congenital and acquired immunodeficiencies. *Am J Pediatr Hematol Oncol* 1988;10:196.
52. Barre V, Marret S, Mendel I, et al. Enterovirus-associated haemophagocytic syndrome in a neonate. *Acta Paediatr* 1998;87: 469.
53. Arya S, Hong R, Gilbert EF. Reactive hemophagocytic syndrome. *Pediatr Pathol* 1985;3:129.
54. Campo E, Condom E, Miro MJ, et al. Tuberculosis-associated hemophagocytic syndrome. A systemic process. *Cancer* 1986; 58:2640.
55. Junker AK, Chan KW, Massing BG. Clinical and immune recovery from Omenn syndrome after bone marrow transplantation. *J Pediatr* 1989;114:596.
56. Chan JKC, Ng CS, Law CK, et al. Reactive hemophagocytic syndrome: a study of 7 fatal cases. *Pathology* 1987;19:43.
57. Hoang MP, Dawson DB, Rogers ZR, et al. Polymerase chain reaction amplification of archival material for Epstein-Barr virus, cytomegalovirus, human herpesvirus 6, and parvovirus B19 in children with bone marrow hemophagocytosis. *Hum Pathol* 1998;29:1074.
58. Chen TK, Nesbit ME, McKenna R, et al. Histiocytic medullary reticulosis in acute lymphocytic leukemia of T cell origin. *Am J Dis Child* 1976;130:1262.
59. Rosai J, Dorfman RF. Sinus histiocytosis with massive lymphadenopathy: a pseudolymphomatous benign disorder: analysis of 34 cases. *Cancer* 1972;30:11742.
60. Eisen RN, Buckley PJ, Rosai J. Immunophenotypic characterization of sinus histiocytosis with massive lymphadenopathy (Rosai-Dorfman disease). *Semin Diagn Pathol* 1990;7:74.
61. Fourcar E, Rosai J, Dorfman RF, et al. Immunologic abnormalities and their significance in sinus histiocytosis with massive lymphadenopathy. *Am J Clin Pathol* 1984;82:515.
62. Fourcar E, Rosai J, Dorfman RF. Sinus histiocytosis with massive lymphadenopathy: an analysis of 14 deaths occurring in a patient registry. *Cancer* 1984;54:1834.
63. Fourcar E, Rosai J, Dorfman R. Sinus histiocytosis with massive lymphadenopathy (Rosai-Dorfman disease): review of the entity. *Semin Diagn Pathol* 1990;7:19.
64. Schouten TJ, Hustinx TWJ, Scheres JMJC, et al. Malignant histiocytosis: clinical and cytogenetic studies in a newborn and a child. *Cancer* 1983;52:1229.
65. Bennett JM, Catovsky D, Daniel M-T, et al. Proposals for the classification of acute leukemias. French-American-British (FAB) Co-operative Group. *Br J Haematol* 1976;33:451.
66. Sande JE, Arceci RJ, Lampkin BC. Congenital and neonatal leukemia. *Semin Perinatol* 1999;23:274.
67. Lampkin BC. The newborn infant with leukemia. *J Pediatr* 1997;131:176.
68. Omenn GS. Familial reticuloendotheliosis with eosinophilia. *N Engl J Med* 1965;273:427.
69. Jouan H, Le Diest F, Nezelof C. Omenn's syndrome—pathologic arguments in favor of a graft-versus-host pathogenesis: a report of nine cases. *Hum Pathol* 1987;18:1101.
70. Junker AK, Chan KW, Massing BG. Clinical and immune recovery from Omenn syndrome after bone marrow transplantation. *J Pediatr* 1989;114:596.
71. Martin JV, Willoughby PB, Giusti V, et al. The lymph node patholgy of Omenn's syndrome. *Am J Surg Pathol* 1995;19: 1082.
72. Rubin CM, Burke BA, McKenna RW, et al. The accelerated phase of Chediak-Higashi syndrome: an expression of the virus-associated hemophagocytic syndrome? *Cancer* 1985;56:524.

73. Barak Y, Nir E. Chediak-Higashi syndrome. *Am J Pediatr Hematol Oncol* 1987;9:42.
74. Merino F, Henle W, Ramirez-Duque R. Chronic active Epstein-Barr virus infection in patients with Chediak-Higashi syndrome. *J Clin Immunol* 1986;6:299.
75. Freyer DR, Kennedy R, Bostrom BC, et al. Juvenile xanthogranuloma: forms of systemic disease and their clinical implications. *J Pediatr* 1996;129:227.
76. Sonoda T, Hashimoto H, Enjoji M. Juvenile xanthogranuloma: clinicopathologic analysis and immunohistochemical study of 57 patients. *Cancer* 1985;56:2280.
77. Zelger BG, Zelger B, Steiner H, et al. Solitary giant xanthogranuloma and benign cephalic histiocytosis—variants of juvenile xanthogranuloma. *Br J Dermatol* 1995;133:598.
78. Seo S, Min KW, Mirkin LD. Juvenile xanthogranuloma: ultrastructural and immunocytochemical studies. *Arch Pathol Lab Med* 1986;110:911.

8
Brain Tumors

Brain tumors occur less often in the first year of life than in the older child and adolescent [1–3]. The peak age incidence in children is between 5 and 8 years [4]. Central nervous system neoplasms are the leading solid tumor in children and adolescents and are surpassed only by leukemia and lymphoma in frequency [4]. During the perinatal period brain tumors are diagnosed less often than teratomas, neuroblastoma, and leukemia [5–8] (Table 1.1). However they cause over 10% of the deaths due to neoplasms in this age period [5,9,10].

Central nervous system tumors of the fetus and infant are not entirely the same as those occurring in the older child and adolescent in several respects. First the location is different, for roughly two thirds of brain tumors in the former occur above the tentorium, the reverse is true for older individuals [9,11,13,14–19]. Second the various histologic types occur with dissimilar frequency; for example, intracranial teratomas comprise at least one third of some neonatal series reported [8,13,16–20] (Table 8.1).

Definite associations exist between brain tumors and hereditable diseases and other conditions. Neurofibromatosis and tuberous sclerosis are prime examples [9,12,21–26] (Tables 8.2 and 8.3). Brain tumors occur in association with certain renal tumors in infants and children, for example, cerebral primitive neuroectodermal tumor (PNET) and Wilms tumor and rhabdoid tumor [12,25,26]. One report cites cerebral PNET and rhabdoid tumor of the liver in a newborn [27].

Clinical Findings

The clinical findings in the fetus or infant with a brain tumor differ from those of an older child or adult [1,2,8,11,12,17,28] (Table 8.4). They may be deceptively minimal and readily missed or misininterpreted owing to the extraordinary resiliency of the immature brain and the infant skull's ability to expand. Moreover, enlargement of the cranial cavity defers the onset of increased intracranial pressure and its neurological symptoms [1]. Intracranial hemorrhage, chronic subdural hematoma, neurological deficits, unexplained hydrocephalus, and distortion of the cranium with bony defects are examples of initial findings. Signs of increased intracranial pressure such as vomiting, lethargy, craniomegaly, separation of cranial sutures, and bulging fontanelles are observed in the infant with a rapidly growing neoplasm. Less than 20% of tumors presenting in the first year of life are diagnosed before or during delivery. Increased intracranial pressure, dystocia, preterm delivery, and stillbirth are the most common manifestations of congenital brain tumors [13,16,29]. Some brain tumors (e.g., teratomas) grow enormously in utero, resulting in dystocia and/or stillbirth. Large tumors cause fetal hydrops or rupture of the cranium, or require decompression of the skull to permit vaginal delivery [9,28,30].

The presenting signs and symptoms may be nonlocalizing. Irritability, nuchal rigidity, paresis, cranial nerve palsies, ab-

TABLE 8.1. Histological type of 225 congenital brain tumors[a].

Type	Total number	Percentage of total
Teratoma	101	44.9
Neuroepithelial	96	42.5
Astrocytoma	36	16.0
Medulloblastoma	17	7.6
Choroid plexus papilloma	16	7.1
Ependymoma	7	3.1
Unclassified gliomas	5	2.2
Spongioblastoma	4	1.8
Medulloepithelioma	3	1.3
Ganglioglioma	2	0.9
Pineoblastoma	2	0.9
Subependymoma	2	0.9
Choroid plexus carcinoma	1	0.4
Mixed malignant glioma	1	0.4
Mesenchymal	12	5.4
Meningioma	7	3.1
Sarcoma	5	2.2
Craniopharyngioma	15	6.8
Hemangioblastoma	1	0.4

[a] Tumors were defined as congenital if they produced signs or symptoms at or before birth in this study.
Source: Abstracted from Raisanen JM, Davis RL. Congenital brain tumors. *Pathology* 1993;2:103.

TABLE 8.3. Ages at which lesions of tuberous sclerosis may first appear.

Age	Lesion/Sign
Between 20th week of gestation and birth	Cardiac rhabdomyoma; subependymal nodule/tumor[a]
Perinatal period	Multiple renal cysts (ultrasound)
	Hydrops fetalis
	Seizures in utero
	Wolf-Parkinson-White (WPW) syndrome
Newborn period	Forehead fibrous plaques
	Partial seizures with or without generalization
	Abdominal distension/uremia
Infancy (< 1 year of age)	Infantile spasms/West syndrome
	Cutaneous hypomelanotic nodules
	Retinal hamartomas
	Regression of social-adaptive behavior
	Delayed motor development

[a] By CT or MRI.
Source: Abstracted from Gomez MR [24].

TABLE 8.2. Neurofibromatosis type I central nervous system findings.

- Plexiform neurofibroma
- Malignant peripheral nerve sheath tumors
- Schwannoma
- Gliomas
  - Astrocytoma of optic pathway
  - Astrocytomas of hypothalmus, thalmus, cerebrum, cerebellum, and spinal cord[a]
  - Ependymoma
  - Medulloblastoma
- Meningioma
- Aqueductal stenosis
- Hydrocephalus
- Cerebrovascular abnormalities
- Mental retardation
- Seizures

[a] Astrocytomas are usually pilocytic, while some are diffuse fibrillary or of varying grades [31].
Source: Burger PC, Scheithauer BW [21]; Riccardi VM [23].

TABLE 8.4. Neonatal brain tumors: presenting signs and symptoms (n = 200).

	Presentation at birth		
Signs and symptoms	Teratoma (%)	Others (%)	Presentation in first 3 months (%)
Dystocia	43	23	
Stillbirth	43	19	
Prematurity	30	19	
Macrocephaly and/or bulging fontanelle	58	71	70
Local skull swelling	10	8	3
Seizures		2	14
Vomiting		1	33
Proptosis	8	1	

Source: Data abstracted from Wakai S, Arai T, Nagai M [16].

normal eye movements, seizures, and "failure to thrive" are examples [1,12,14,16,17,30–32]. Developmental delay, the loss of a previously acquired skill, such as rolling over, sitting up, or crawling, especially in the presence of an abnormally large head, are signs that definitely suggest the need for imaging studies of the brain [1,28,30].

Pathology

Intracranial teratomas comprise approximately one third of the total reported cases in newborns, and astrocytoma, medulloblastoma, ependymoma, and choroid plexus papilloma are the major neuroepithelial tumors [9,11,12,14,18–20,33,34] (Table 8.1).

Both immunohistochemistry and electron microscopy are indispensable for the identification and classification of brain tumors [21,22,26,35] (Table 8.5).The immunohistochemical markers used specifically for the evaluation of central nervous system neoplasms are the glial fibrillary acidic protein (GFAP) and the neurofilament protein (NFP) antibodies [22,26,37]. The GFAP is helpful in distinguishing between glial and nonglial neoplasms and for the identification of astrocytic elements and ependymomas [22,26]. The neurofilament proteins are intermediate filament proteins found in neurons and their processes and are usually resticted to these cells [22,26]. Monoclonal antibodies directed against neuroectodermal-associated antigens include neuron-specific enolase (NSE) and the S-100 protein (Table 8.5).

TABLE 8.5. Immunohistochemical methods used in the diagnosis of central nervous system tumors.

Antibody	Diagnostic use
Glial fibrillary acidic protein (GFAP)	Astrocytoma, ependymoma
S-100 protein	Tumors of neural crest origin: gliomas, nerve sheath and choroid plexus tumors, melanoma, granular cell tumor
Synaptophysin	Tumors with neurosecretory differentiation: neuroblastoma, medulloblastoma
Chromogranin	Neuroendocrine tumors
Neuron-specific enolase (NSE)	Meningiomas, ependymomas, choroid plexus papilloma, medulloblastoma, schwannoma
Neurofilament antibodies	Tumors of neural origin: ganglioneuroma, neuroblastoma, medulloblastoma, ganglioglioma
Epithelial membrane antigen	Epithelial tumors, chordoma, meningioma, choroid plexus tumors
Cytokeratins	Epithelial tumors, craniopharyngioma, choroid plexus tumors, chordoma, some germ cell tumors
Placental alkaline phosphatase, human chorionic gonadotropin, α-fetoprotein	Germ cell tumors
Vimentin	Meningioma, nerve sheath tumors, medulloblastoma, astrocytoma
S-antigen	Retinoblastoma
Nestin	Medulloblastoma, ependymoma

Source: Bigner DD, McLendon RE, Bruner JM [22]; Timperly WR [36].

Astrocytoma

Astrocytoma, the leading neuroglial tumor found throughout infancy and childhood, is a neoplasm derived from and composed of astrocytes showing various degrees of differentiation. As a group they differ in their gross and histological features as well as in their site of origin and clinical manifestations [9,21,22] (Figures 8.1–8.5).

Astrocytomas in fetuses and infants usually occur outside the cerebellum and above the tentorium. The cerebral hemisphere, optic nerve, thalamus-hypothalamus, mesencephalon, and pons are the main primary sites in the young. Those arising from the cerebral hemisphere are often large, involve more than one lobe, and displace the lateral ventricle (Figure 8.4). Hydrocephalus is the major presenting finding. Some congenital astrocytomas present with spontaneous intracerebral hemorrhage, and therefore an underlying neoplasm should be considered when a full-term infant develops a cerebral bleed [16,18,38,39]. Polyhydramnios secondary to hypothalamic dysfunction, congenital hydrocephalus, dystocia, and stillbirth have been attributed to this brain tumor [40,41].

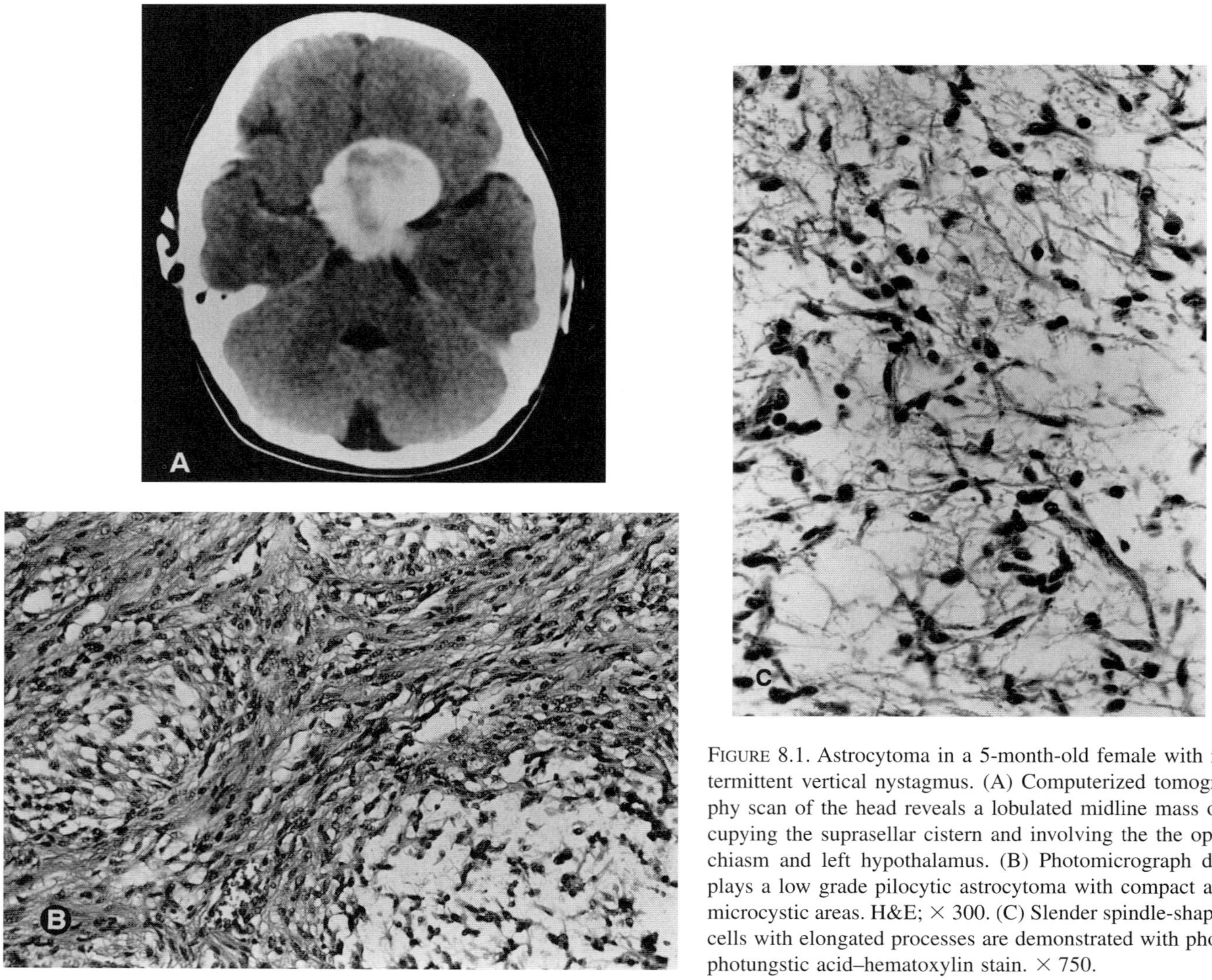

FIGURE 8.1. Astrocytoma in a 5-month-old female with intermittent vertical nystagmus. (A) Computerized tomography scan of the head reveals a lobulated midline mass occupying the suprasellar cistern and involving the the optic chiasm and left hypothalamus. (B) Photomicrograph displays a low grade pilocytic astrocytoma with compact and microcystic areas. H&E; × 300. (C) Slender spindle-shaped cells with elongated processes are demonstrated with phosphotungstic acid–hematoxylin stain. × 750.

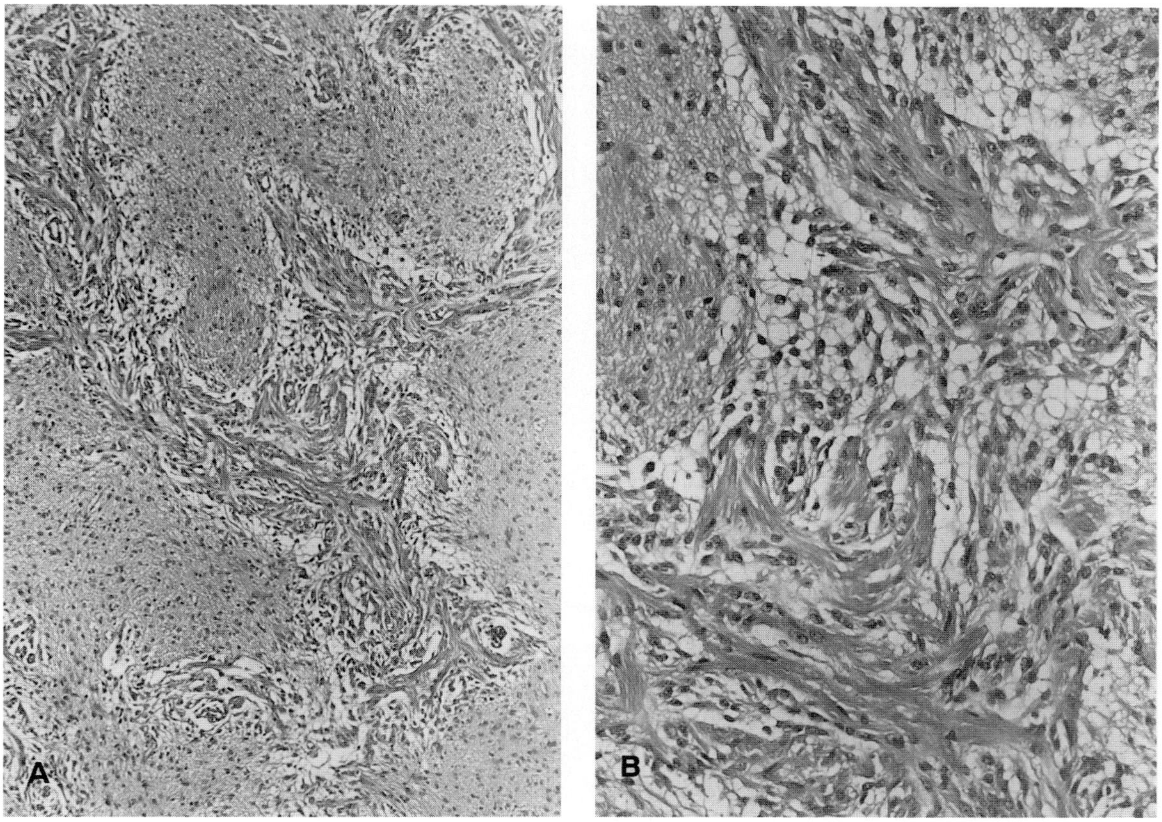

FIGURE 8.2. Optic glioma in a 7-month-old male who presented with seizures, facial twitching, and right eye deviation. Computerized tomography scan revealed obstructive hydrocephalus and a 3×4 cm^2 suprasellar, solid, and cystic mass arising from the optic chiasm, compressing the third ventricle and producing enlargement of right optic foramen. (A) The cerebral cortex is infiltrated by a pilocytic astrocytoma having histological features similar to those described in Figure 8.1B. H&E; $\times$ 48. (B) Higher magnification ($\times$ 150) showing the biphasic pattern consisting of compact fascicles of fibrillated, elongated bipolar cells alternating with more loosely arranged microcystic areas. H&E.

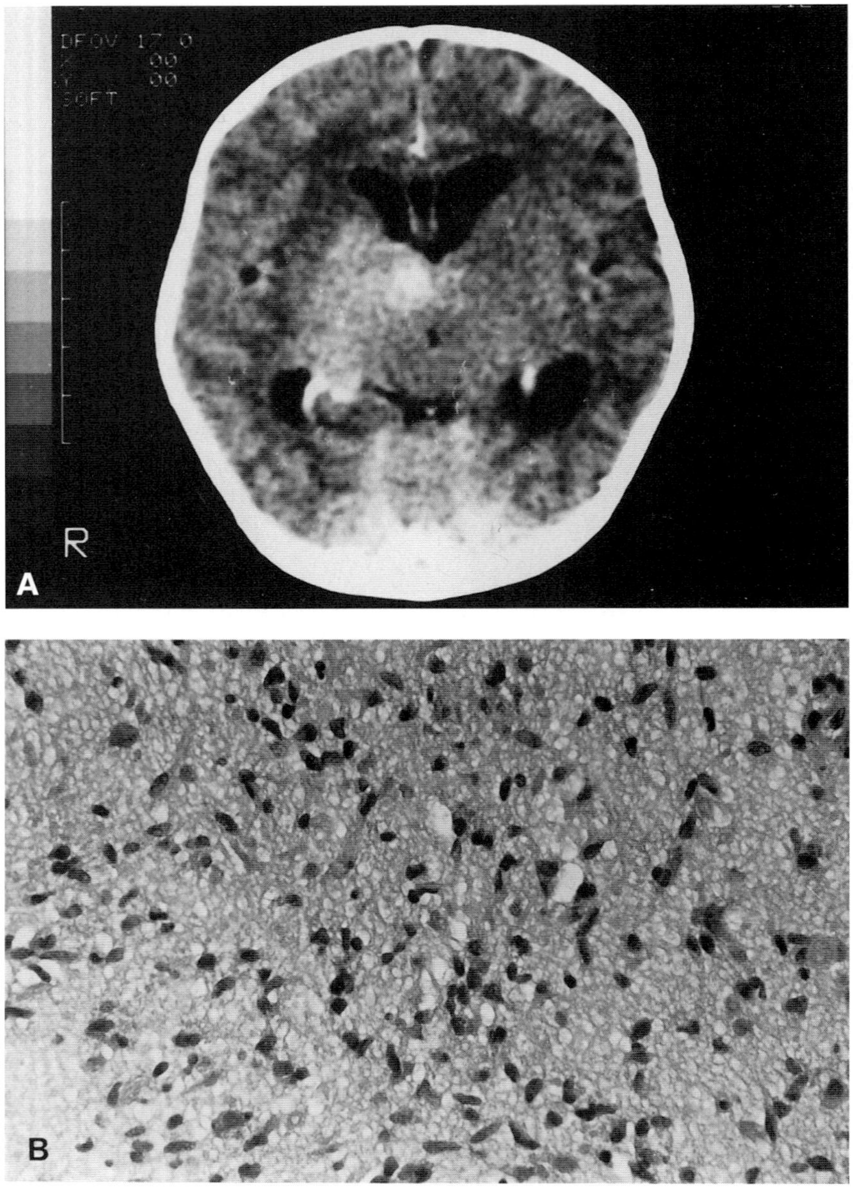

FIGURE 8.3. Anaplastic astrocytoma in a 2-month-old male with hydrocephalus and a mass in the left and right thalamus. (A) Computerized tomography scan shows the infiltrative thalamic mass. (B) The tumor is hypercellular with moderate nuclear and cytoplasmic pleomorphism. The nuclei are darkly staining; some have an angulated appearance. H&E; × 150. (Courtesy of Ignacio Gonzalez, M.D., Department of Pathology, Children's Hospital, Los Angeles.)

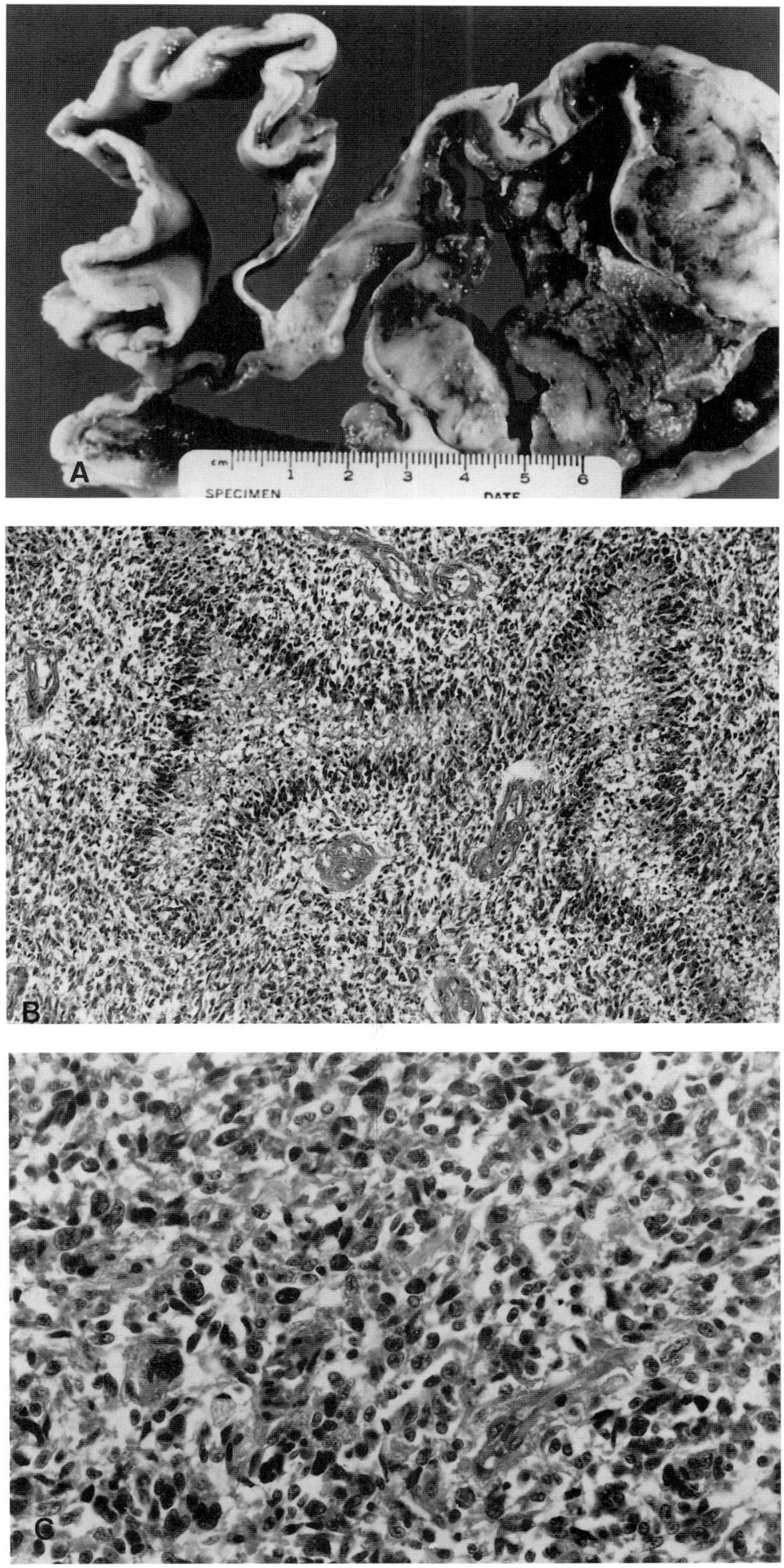

FIGURE 8.4. Glioblastoma multiforme in an 8-month-old female with congenital hydrocephalus. (A) The tumor originated from the right parietal lobe, crossed the midline, and invaded the left cerebral hemisphere and ventricles. Large necrotic, hemorrhagic tumor masses are noted in the cerebral cortex, basal ganglia, and ventricular wall. Severe hydrocephalus is present. (B) The densely cellular tumor is composed of small, pleomorphic cells forming palisades around necrotic central areas. Microvascular proliferation is present. H&E; × 120. (C) Higher magnification (× 300) reveals small cells displaying hyperchromatism, pleomorphism, and nuclear angulation. There are one or two tumor giant cells. H&E. (B and C courtesy of Ignacio Gonzalez, M.D., Department of Pathology, Children's Hospital, Los Angeles.)

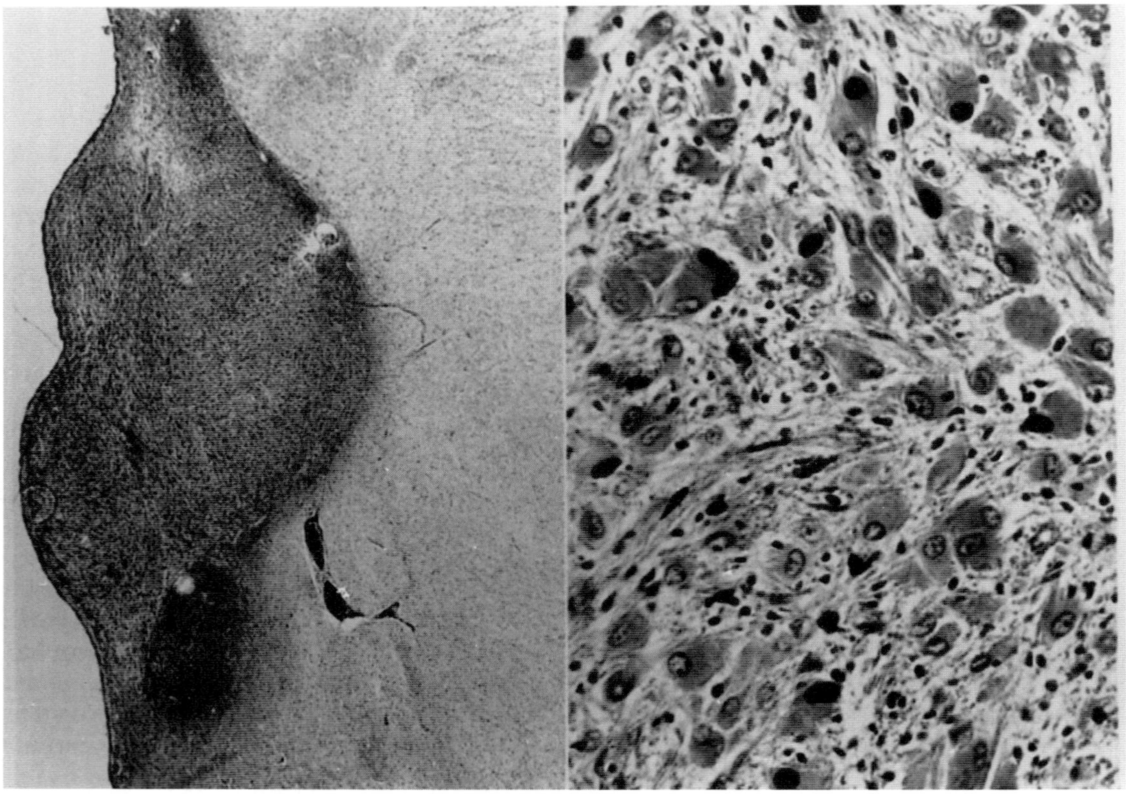

FIGURE 8.5. Subependymal giant cell astrocytoma. Tuberous sclerosis in an infant aged 22 days. (A) Cellular subependymal nodule protruding into the lateral ventricle. (B) The tumor consists of large astrocytic cells with abundant cytoplasm, eccentric, slightly vesicular nuclei and large nucleoli, surrounded by numerous fibrillary processes. (From Isaacs H Jr [9], with permission.)

Astrocytomas of the fetus and infant vary in their appearance and in their histological classification from low grade to malignant (Table 8.6). The *grade 1 pilocytic astrocytoma* consists of small bipolar and stellate cells with scanty processes forming loose and compact areas and microcysts (Figures 8.1 and 8.2). Rosenthal fibers and eosinophilic granular bodies are important findings. No anaplastic changes are present, and mitoses are uncommon. Tumor cells show strong GFAP staining [21,22,40].

Anaplastic astrocytomas are astrocytomas of intermediate grade of malignancy roughly corresponding to grades 2 and 3. Microscopically they show cytoplasmic and nuclear pleomorphism, hypercellularity, mitotic activity, and to a degree, the anaplasia of glioblastoma; however, they lack the palisading necrosis, hemorrhage, or vascular proliferation that is observed in the latter [22,41] (Figure 8.3).

Glioblastoma multiforme occurs in the cerebral hemispheres and basal ganglia of fetuses and newborns [19,22,40,42,43]. The salient histological features are a densely cellular tumor, marked pleomorphism, multinucleated giant cells with hyperchromatic nuclei, microvascular proliferation, and neoplastic cells forming palisades around central necrotic foci [22] (Figure 8.4). GFAP immunoreactivity is characteristically present.

Subependymal giant cell astrocytoma is defined by large fusiform or pyramidal cells with prominent eosinophilic cytoplasms typically found in the brains of infants and children with tuberous sclerosis (Figure 8.5). Immunohistochemical studies reveal variable staining with GFAP and synaptophysin, but NSE generally is reactive [21,22].

Desmoplastic cerebral astrocytoma, a variant associated with a good prognosis, typically occurs in the first year of life [8,44,45]. The tumor has been detected prenatally by sonography. The desmoplastic astrocytoma tends to be a large, cystic hemispheric tumor involving the cerebral cortex and adjacent leptomeninges.

TABLE 8.6. The World Health Organization (WHO) classification of astrocytoma.

Astrocytoma (Grade 1)	tumor composed primarily of pilocytic astrocytes
Astrocytoma (Grade 2)	applies to low grade, diffuse astrocytomas
Astrocytoma (Grade 3)	refers to anaplastic (malignant) astrocytoma
Astrocytoma (Grade 4)	includes glioblastoma multiforme and variants

Source: Burger PC, Scheithauer BW [21].

Microscopic examination reveals a cellular spindle cell tumor with a prominent collagenous stroma. Numerous bipolar and spindle-shaped cells with eosinophilic cytoplasms and cells with long processes are present within the stroma. The astrocytic cells form bundles and nests separarated by thick bands of collagen [44,45]. Reticulin stain shows thick reticulin fibers surrounding nests of individual cells. Tumor cells and their processes are strongly reactive with GFAP and the collagen fibers with Masson's trichrome stain [45]. Infantile desmoplastic ganglioglioma is the main consideration in the differential diagnosis because the two tumors are histologically similar. Complete resection of the tumor correlates with a good outcome.

Primitive Neuroectodermal Tumors

The term "primitive neuroectodermal tumor" is applied to a group of small-cell malignant tumors of the central and peripheral nervous systems and soft tissues [29,35,46–50]. The neural crest is proposed as the site of origin of these highly malignant neoplasms. PNETs occur primarily in the pediatric age group and are characterized by similar biological behavior in the central nervous system regardless of their primary site. The tumor has been diagnosed antenatally by ultrasound and during delivery as the result of dystocia and obstructed labor [29,50]. Characteristically they are highly aggressive and metastasize widely throughout the cerebrospinal fluid pathways, seeding the meninges of the brain and spinal cord. They occur in several locations including most commonly the cerebellum (where they are called medulloblastoma), the cerebral hemispheres (cerebral neuroblastoma), the pineal gland (pineoblastoma), the brain stem, the spinal cord, the olfactory nerve, and the retina (retinoblastoma) [29,35,48]. Some metastasize locally to extracranial tissues and to distant organs [51,52a].

Microscopically the PNETs consist of small, darkly staining cells that appear to be poorly differentiated and display a high mitotic rate. Other components such as astrocytic, ependymal, neuronal, retinal, or melanocytic cells may be present depending on the location of the tumor [35,49] (Figure 8.6). Mesenchymal elements (e.g., skeletal muscle, fibrous tissue, cartilage, epithelial elements) are noted in some. PNETs are variably immunoreactive with vimentin, NSE, and GFAP [50].

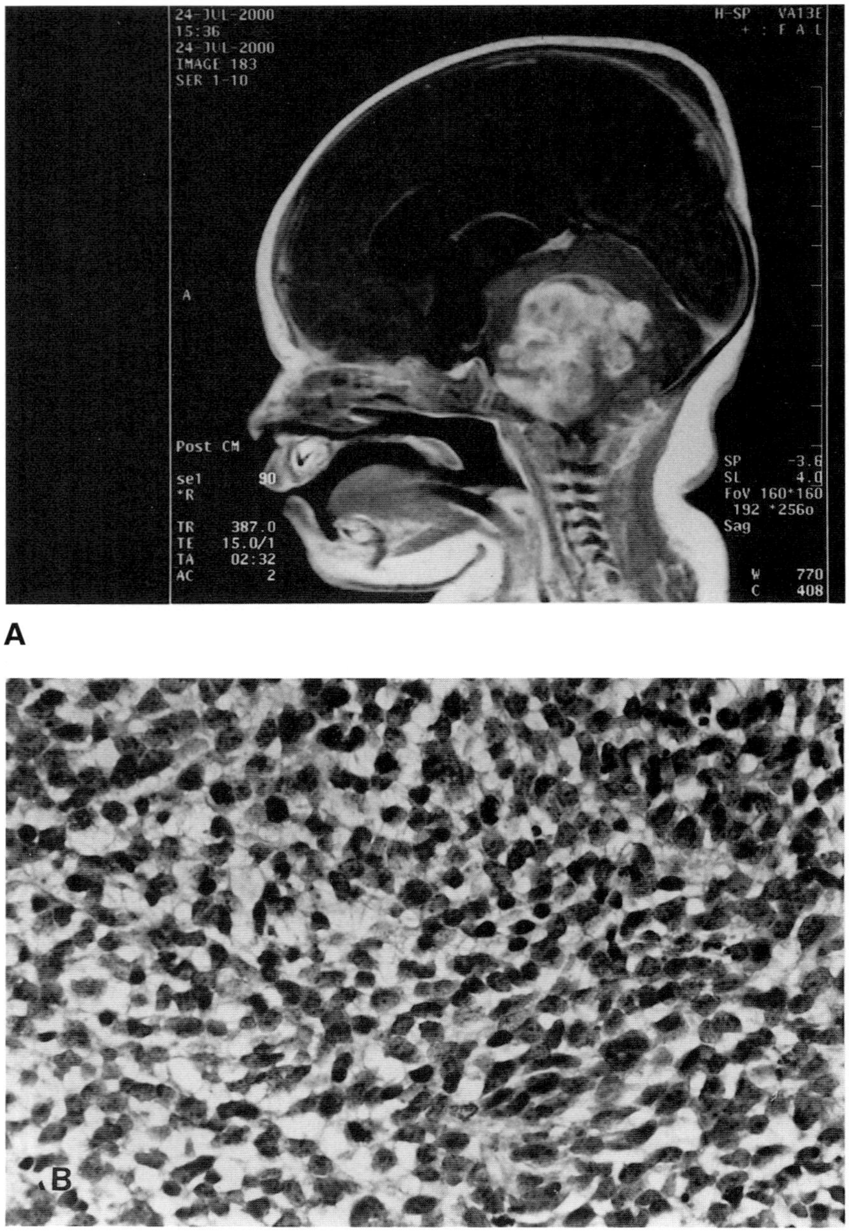

FIGURE 8.6. Primitive neuroectodermal tumor with divergent differentiation in a 1-day-old female with hydrocephalus and a large posterior fossa mass. (A) Magnetic resonance image depicting the posterior fossa tumor displacing the cerebellum and adjacent structures. The ventricles are markedly dilated. (B) The hypercellular tumor consists primarily of small darkly staining cells with round or oval nuclei and scant cytoplasm. H&E; × 380. (C) There are areas of neuroblastic differentiation consisting of tumor cells surrounded by neutropil. H&E; × 380. (D) The tumor also displays foci of ependoblastic differentiation characterized by rosette formation. H&E; × 380. (Courtesy of Hal Meltzer, M.D., Department of Neurosurgery, Children's Hospital, San Diego.)

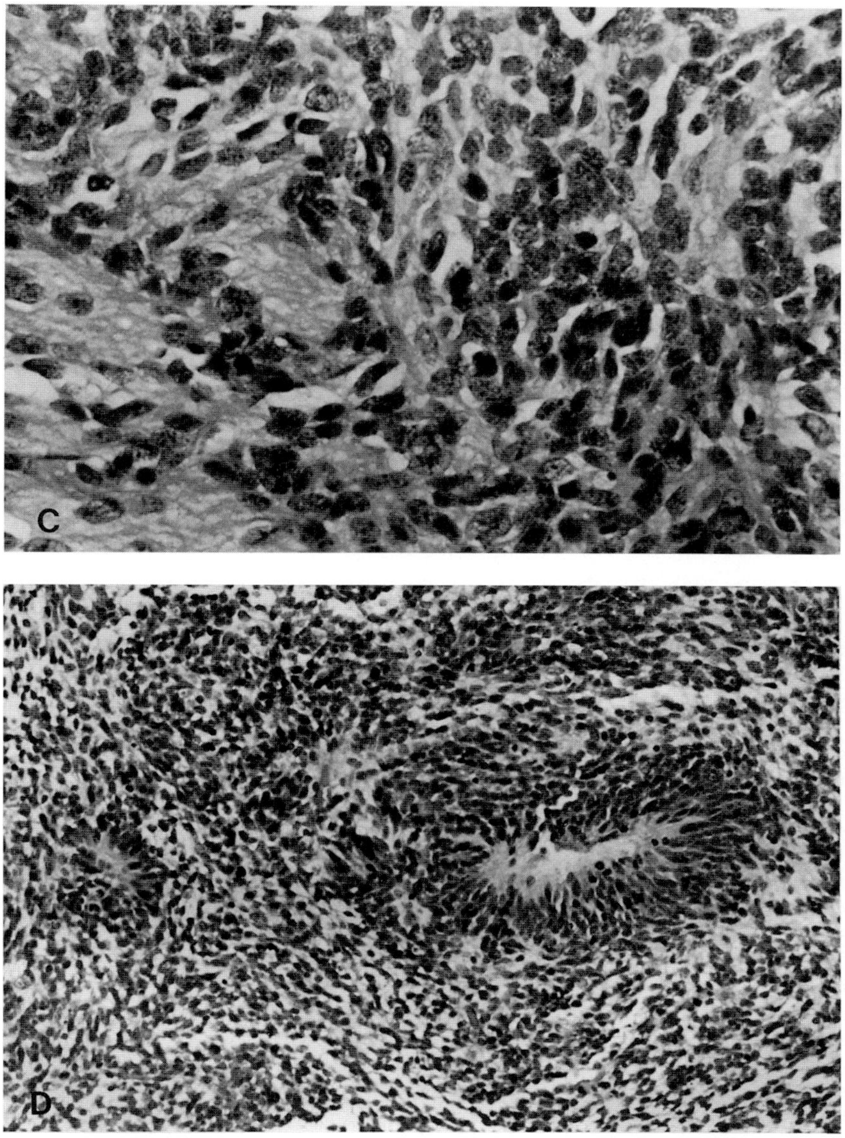

FIGURE 8.6. (*Continued*)

Medulloblastoma

Medulloblastoma ("cerebellar PNET") is second in frequency to astrocytoma in several studies [1,8,9,13,32,49] (Tables 8.1 and 8.7). Most infratentorial tumors found in newborns and infants are histologically malignant, since medulloblastoma is the main posterior fossa neoplasm diagnosed in this age group [2,12]. Patients with this malignancy have a relatively high incidence of stillbirth, hydrocephalus, and congenital anomalies [33,52,53].

Medulloblastomas originate from the vermis of the cerebellum and grow into the fourth ventricle and adjacent cerebellar hemispheres. Subsequently obstructive hydrocephalus and leptomeningeal seeding along the cerebrospinal axis ensue [25,35, 51]. At this time the cerebrospinal fluid (CSF) cytological examination may reveal the presence of tumor cells. If there is vascular invasion, the tumor enters the bloodstream and metastasizes, in 5 to 18% of cases, to organs outside the central nervous system, primarily to the liver, lungs, and bone marrow and sometimes to the lymph nodes [51,52b]. Therefore bone marrow biopsy and CSF cytology should be an integral part of the initial metastatic workup before therapy is begun [9].

Histologically medulloblastomas consist of small, darkly staining cells with varying amounts of intercellular pink-staining fibrillar material (Figure 8.7). Well-defined Homer-Wright pseudorosettes are present in less than half the specimens. Extensive necrosis and moderate mitotic activity are other findings [9,21,54].

Electron microscopic studies show that medulloblastoma shares some ultrastructural features in common with neuroblastoma [8,35]. The tumor is composed of small, primitive cells surrounded by cytoplasmic processes containing neurofilaments and neurotubules. In a few specimens dense core granules measuring 100 nm in diameter are found, but they are seen in much smaller numbers in medulloblastoma than in peripheral neuroblastoma (Figure 8.7).

TABLE 8.7. Survival rates for 204 fetal and newborn brain tumors.

Tumor	Number	Living/Dead	Lost to follow-up	Survival (%)
Teratoma	49	3/46		6
Astrocytoma	46	12/34		26
Choroid plexus tumors[a]	31	16/13	2	52
Primitive neuroectodermal tumors[b]	30	3/27		10
Craniopharyngioma	14	3/11		21
Meningeal tumors[c]	12	3/9		25
Ependymoma[d]	9	1/8		11
Rhabdoid tumor	3	0/3		0
Medulloepithelioma	2	0/2		0
Pineoblastoma	2	0/2		0
Oligodendroglioma	2	0/2		0
Hypothalamic hamartoma	2	1/1		50
Hemangioblastoma	2	1/1		50
Total	204	43/158	2	21[e]

[a] Includes 4 papillary carcinomas (2 of 4 survivors).
[b] Includes 20 medulloblastomas (1 of 20 survivors).
[c] Includes 3 ependymoblastomas (no survivors).
[d] Includes 9 sarcomas (2 survivors) and 3 meningiomas (2 survivors).
[e] Overall survival rate: 43/202 × 100 = 21%.
Source: Isaacs H Jr [8].

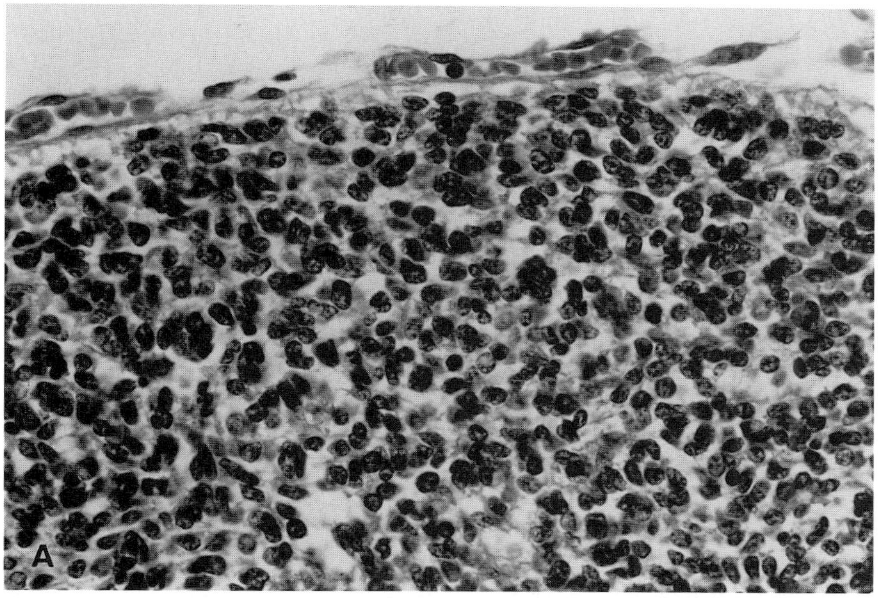

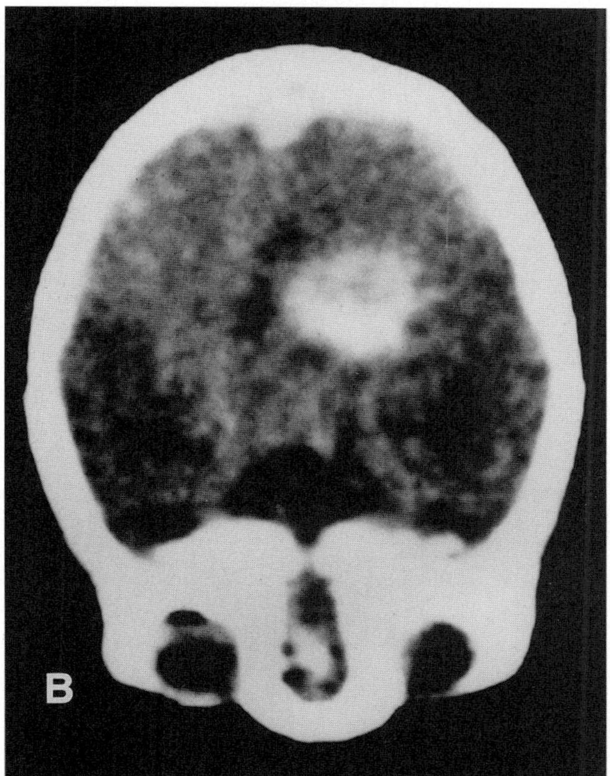

FIGURE 8.7. Medulloblastoma ("cerebellar PNET"). (A) Specimen from the cerebellum of a 650 g, 10-hour-old male infant of 26 weeks gestation showing the developing external granular layer in an immature, but normally otherwise brain. Compare the histology of this photomicrograph with Figure 8.7C. Superficial Biopsy samples excised from the cerebellum can be misinterpreted as medulloblastomas on frozen section. H&E; × 750. (B) Computerized tomography scan of a newborn with a several week history of poor feeding and vomiting and terminally bulbar paresis. There is a large enhancing mass within the medial aspect of the left cerebellar hemisphere. (C) Small-cell malignant tumor with a fibrillary background and suggestive pseudorosette formation. Several mitotic figures are present. H&E; × 750. (D) Ultrastructural findings consist of primitive-appearing cells with irregular nuclei. The cytoplasm contains a few organelles, mostly mitochondria and scattered ribosomes. Few dense core granules are seen (arrow). Microtubules are noted in some cytoplasmic processes. Electron micrograph; × 9000.

(*Continued*)

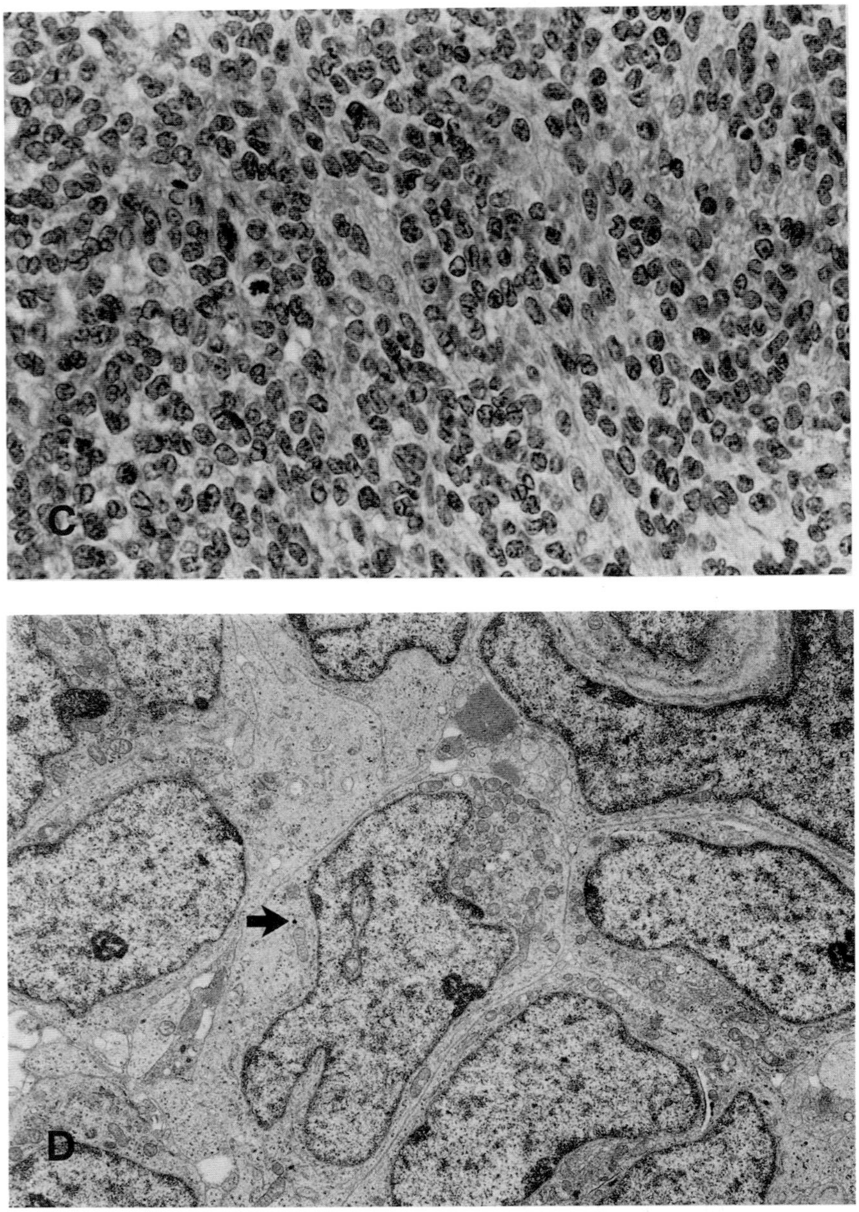

FIGURE 8.7. (*Continued*)

Cerebral Neuroblastoma (Cerebral PNET)

Cerebral PNET is a very uncommon, highly malignant small-cell tumor occurring primarily in young children and infants and characterized by early recurrence, intra- and extracranial metastasis, and high mortality [35,47,48,55,56]. The exact origin and histogenesis of this tumor are uncertain. The subependymal plate of the cerebrum is the proposed site (Figure 8.8A). A fourth of all cerebral PNETs occur before age 2; one of the youngest patients whose case was recorded was 28 days of age [21,26,35].

Cerebral PNETs tend to be large, with extensive cystic necrosis and hemorrhage. They occupy much of a cerebral hemisphere. The tumor is soft and friable with focal firmer gliotic areas. Microscopic examination reveals a highly cellular, small-cell tumor as the major component (Figures 8.8 and 8.9). Cerebral PNETs show as the peripheral neuroblastoma areas with a similar histological appearance, including Homer-Wright rosettes and pink fibrillar neuropil. Focal ganglion cell differentiation, a peripheral astrocytic component, and a prominent connective tissue stroma are additional findings [48,55] (Figure 8.8). Tumor cells are reactive with synaptophysin and neurofilament antibodies. Focal collections of hypertrophic astrocytes are immunoreactive with GFAP [55].

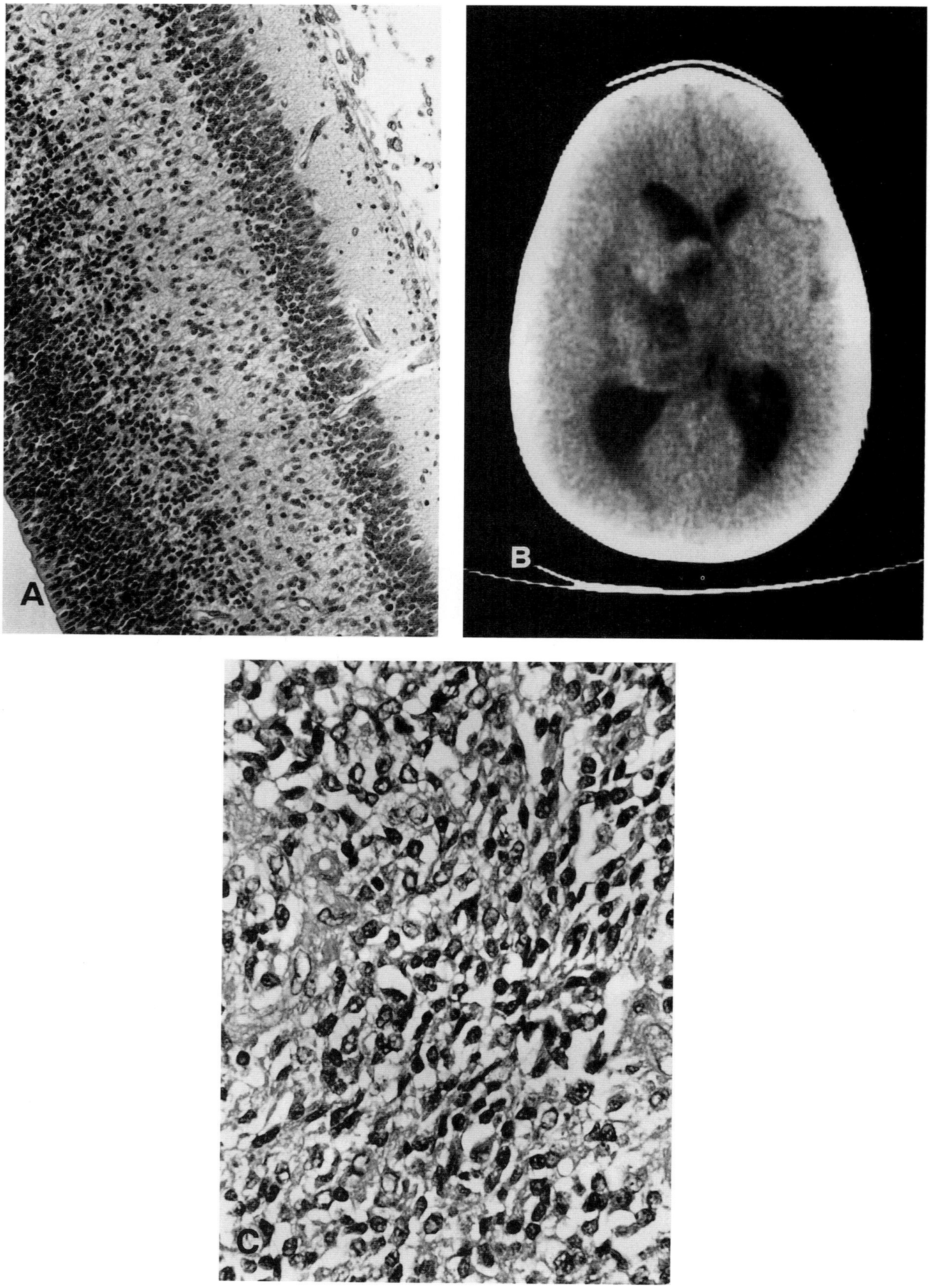

FIGURE 8.8

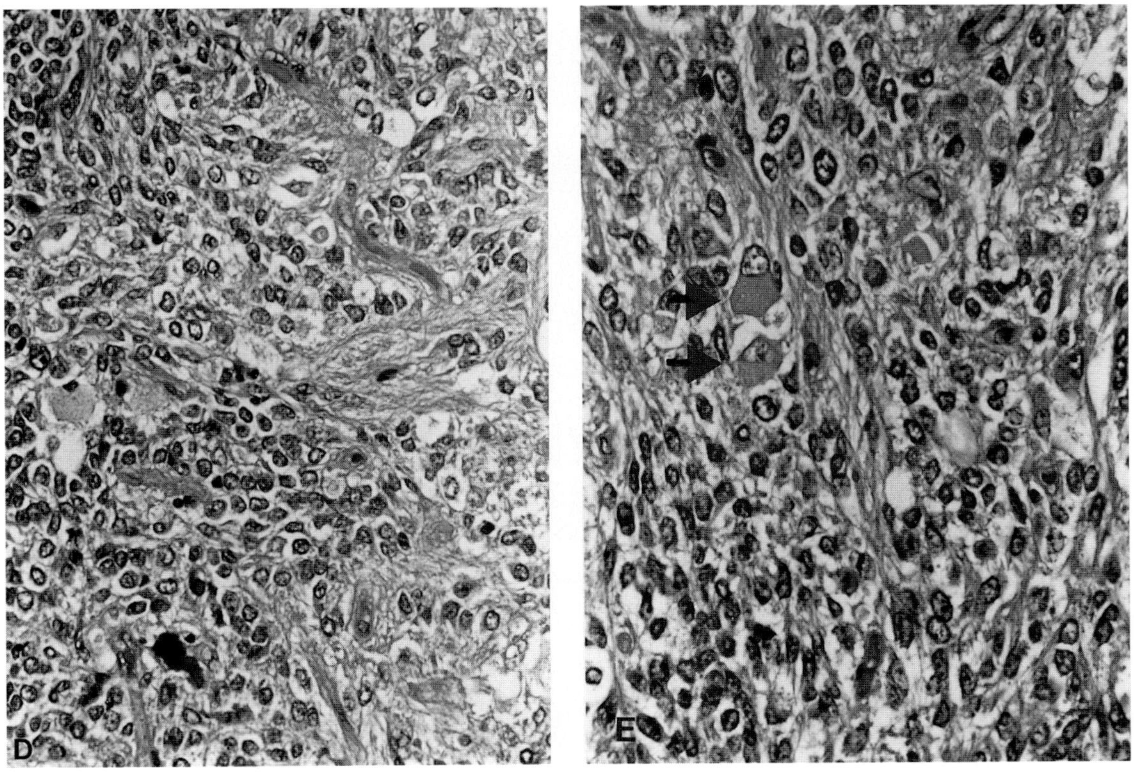

FIGURE 8.8. Cerebral neuroblastoma (primitive neuroectodermal tumor). (A) Development of the cerebral hemisphere (forebrain), estimated 12 to 14 weeks gestation. The wall of the cerebrum has differentiated into four layers beginning right to left: molecular (beneath the pia-arachnoid), primordial cortical gray, intermediate, and ependymal layers (adjacent to the lateral ventricle). Inner part of the intermediate layer contains numerous migrating neuroblasts. H&E; × 150. (B) Computerized tomography scan of an 11-month-old male with vomiting and paresis of the left arm. A large heterogeneous enhancing mass is present in the right frontoparietal area with irregular low density areas in the right basal ganglia. The lateral ventricles are dilated. (C) The tumor consists of sheets of small, poorly differentiated, darkly staining cells. H&E; × 750. (D) Areas of fibrosis (desmoplasia) and calcification (black-staining material) are present. Masson trichrome; × 600. (E) Foci of astrocytic differentiation (arrows). H&E; × 750.

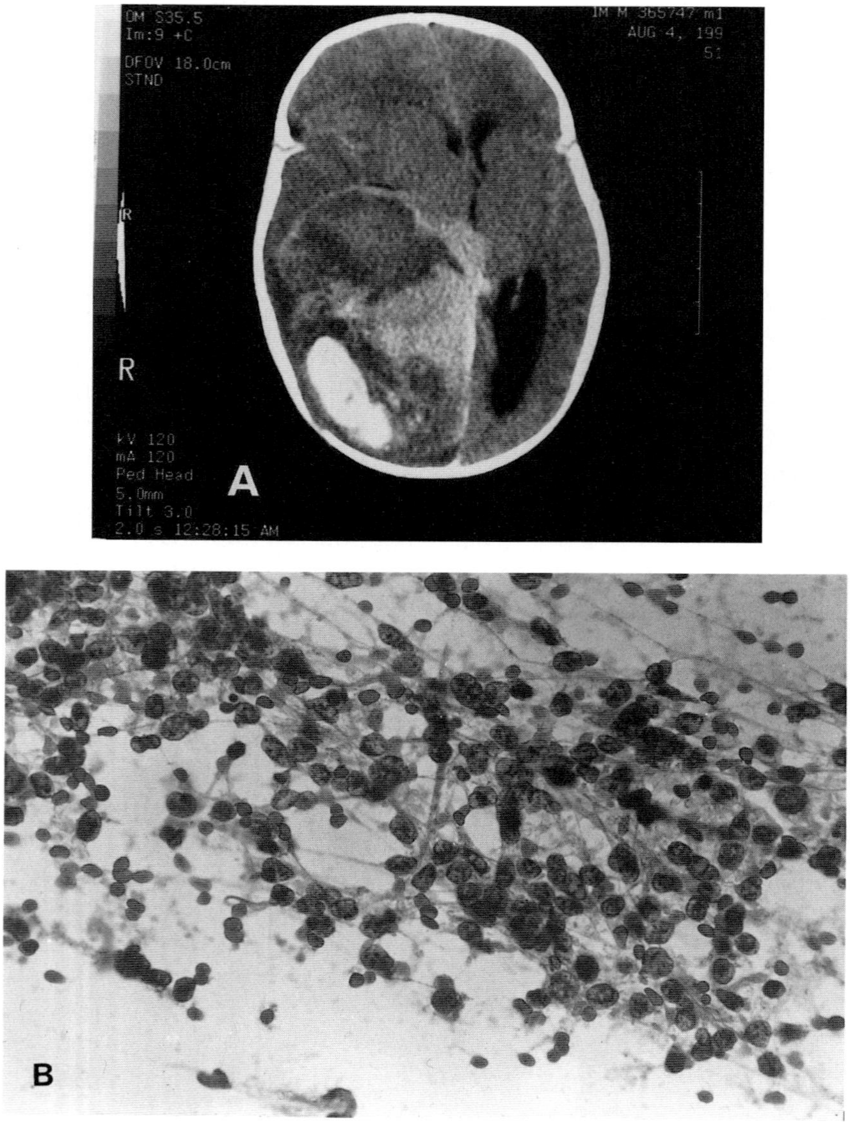

FIGURE 8.9. Cerebral primitive neuroectodermal tumor in a 6-week-old male with a 2-day history of vomiting and increasing lethargy. (A) Computerized tomography scan reveals a large heterogeneous, enhancing mass with areas of hemorrhage and necrosis occupying much of the parieto-occipital lobes. Herniation across the midline and dilatation of the left lateral ventricle are noted. (B) Small round to oval cells with darkly staining nuclei and delicate processes are seen on imprint cytology. H&E; × 750. (C) Section reveals a hypercellular, small-cell malignant tumor with variable nuclei and indistinct cell borders surrounded by fibrillar material. Positive immunostaining for neuron-specific enolase, S-100, and GFAP suggest a neuroectodermal origin with glial differentiation. Cells cultured from this tumor yielded a normal male karyotype 46,XY. H&E; × 480. (D) An area of palisading necrosis (arrow). PNET is one of the few tumors other than glioblastoma to show necrosis with palisading. H&E; × 150. (E) Electron photomicrograph (× 48,400) depicting primitive-appearing cells with microfilaments. Some suggestive neurosecretory granules were found in other fields. (Electron photomicrographs courtesy of Ann Peters, Department of Pathology, Children's Hospital, San Diego.)

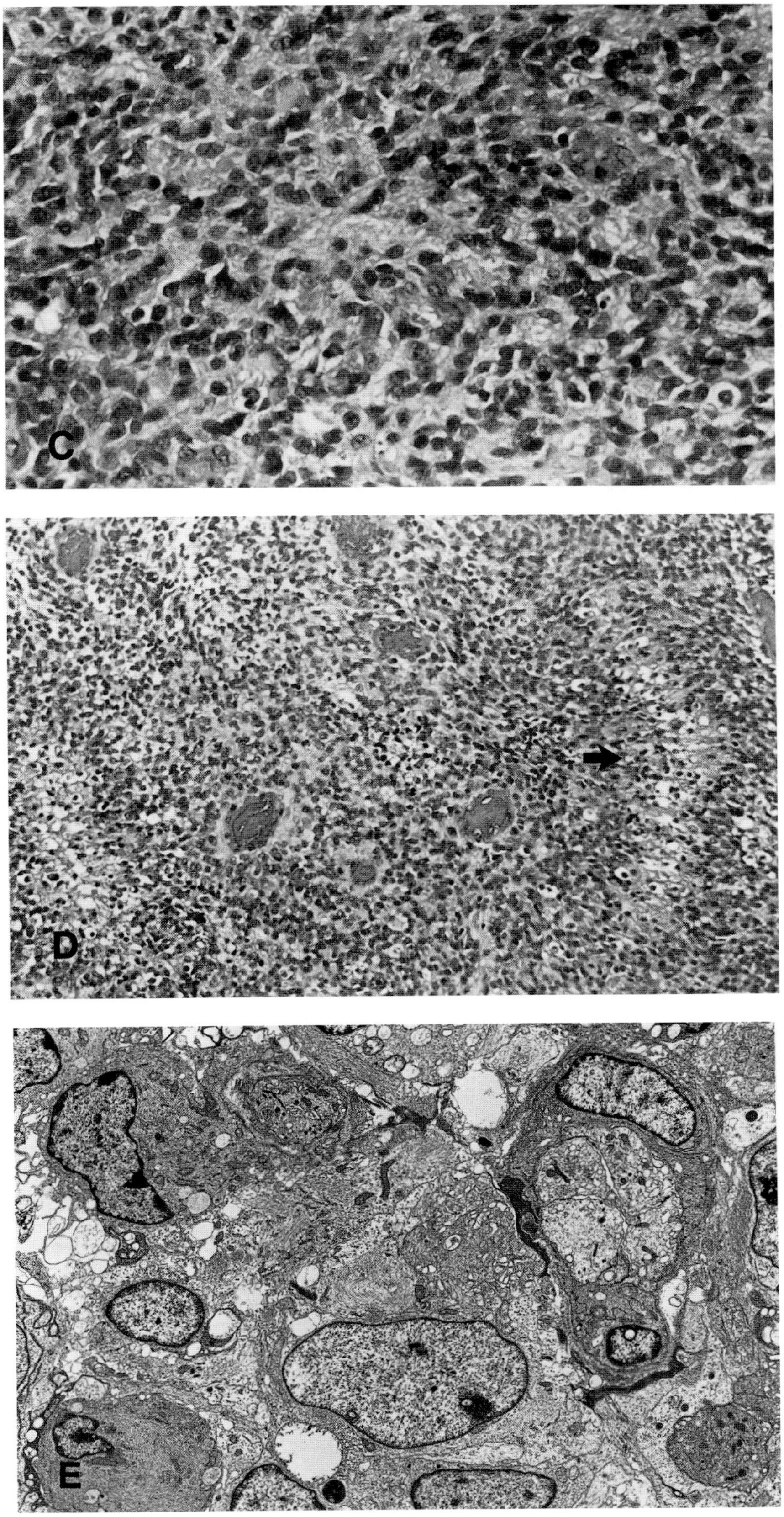

FIGURE 8.9. (*Continued*)

Ganglioglioma

Ganglioglioma is an uncommon central nervous system neoplasm found in patients of all ages including the newborn; it comprises less than 5% of all brain tumors diagnosed in infants, who frequently present with seizures [57–60]. The tumors vary in size and are well circumscribed and finely granular on cross section. Histologically they consist of collections of mature, sometimes atypical-appearing ganglion cells and astrocytes surrounded by a network of glial fibers and fibrosis (Figure 8.10). Some tumors contain microcysts and foci of calcification.

Desmoplastic infantile ganglioglioma is a distinctive supratentorial neoplasm arising most often in the frontal and parietal areas. They appear on CT scans as cystic masses with a solid cortical component within the cerebrum situated next to the meninges [57,60] (Figure 8.11A). The tumors are defined by their very large size, cystic appearance on imaging studies, marked desmoplasia, and the presence of astrocytic and ganglionic elements.

Microscopically desmoplastic infantile ganglioglioma consists of a prominent fibrous tissue component composed of fascicles of spindle-shaped fibroblastic cells displaying a storiform growth pattern and desmoplasia. Variable numbers of glial, neuronal, and sometimes primitive neuroectodermal components are present [60] (Figure 8.11). Schwann cells are noted in some tumors. Astrocytic cells stain positively for GFAP. The neural component is reactive with NSE and S-100 protein but unreactive with GFAP. The fibroblastic areas do not react with these antibodies, but focally positive macrophage antigens (Ki-M1P) are noted in some [60]. Electron microscopy depicts astrocytic cells containing intermediate filaments and pericellular basal lamina and cells showing neuronal differentiation with dense core granules. The fibroblastic cells have marginated chromatin, ribosomes, intermediate filaments, and Golgi apparatus and are surrounded by collagen—features similar to those described for fibrous histiocytoma [57,60]. Patients with desmoplastic ganglioglioma have a favorable clinical outcome following complete surgical resection [57].

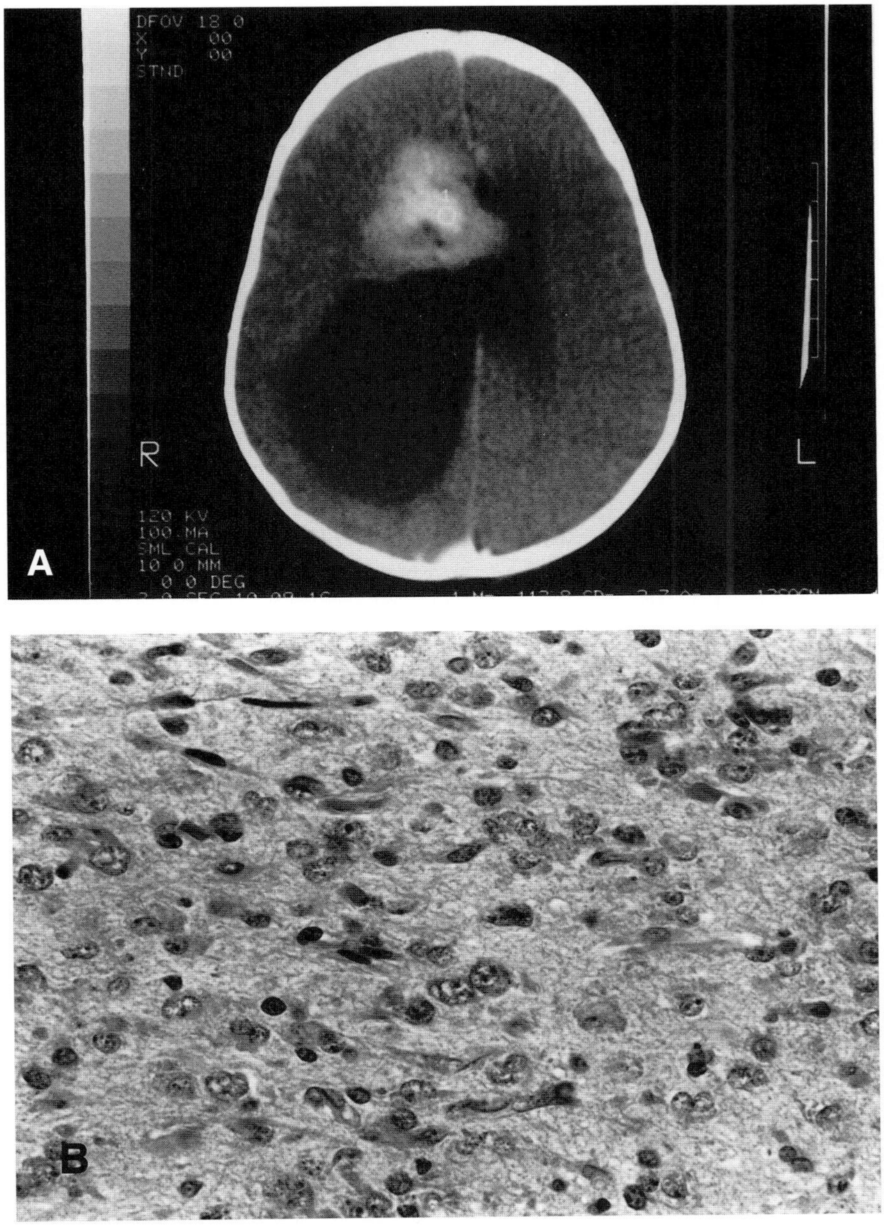

FIGURE 8.10. Cerebral ganglioglioma in a 7-week-old male with macrocephaly, bulging tense fontanelles, conjugate eye deviation to the right, and left homonymous hemianopsia. (A) Computerized tomography scan depicts a large enhancing mass occupying the right frontal lobe and extending into the adjacent ventricle. Obstruction of the foramen of Monroe and dilatation of the right lateral ventricle are present. (B) The tumor consists of atypical-appearing ganglion cells with slightly vesicular nuclei, one or two large nucleoli, and prominent eosinophilic cytoplasms accompanied by hypertrophic astrocytes. H&E; × 750.

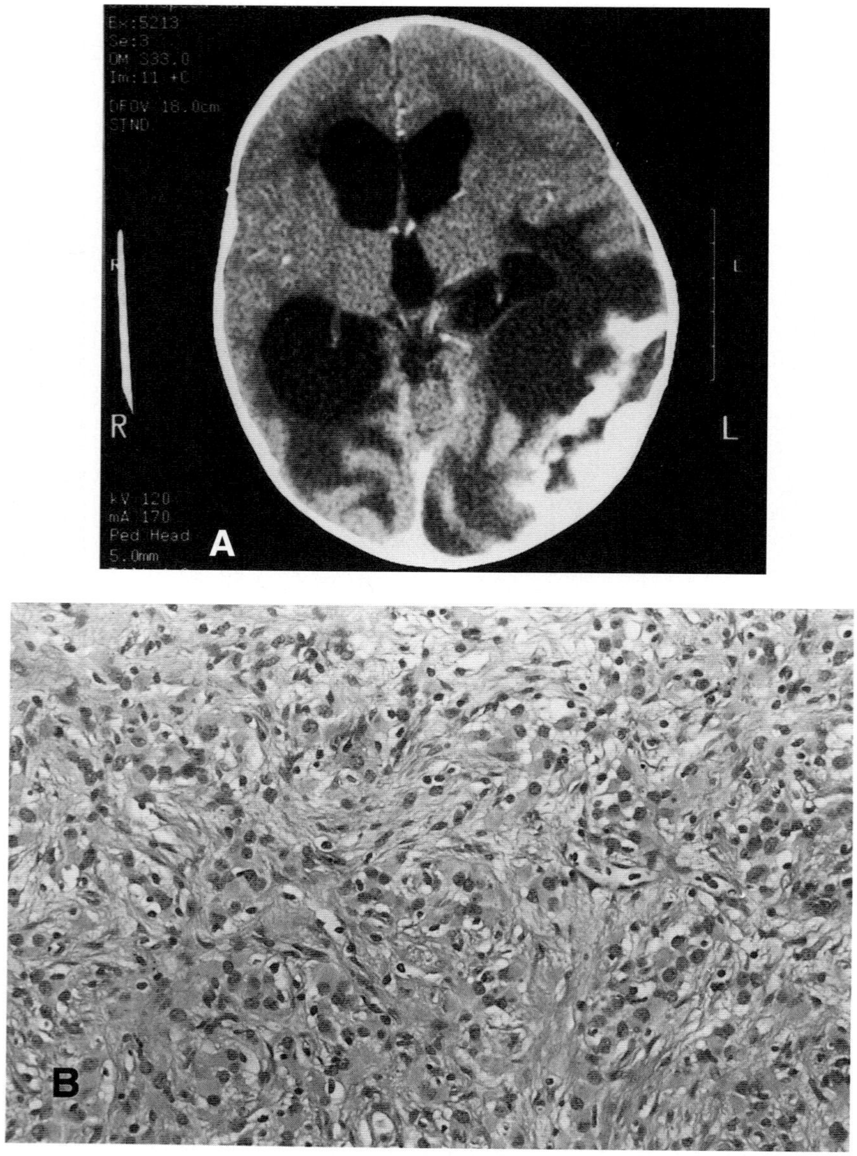

FIGURE 8.11. Cerebral infantile desmoplastic ganglioglioma in a 5-month-old female with macrocephaly and vomiting. (A) Computerized tomography scan reveals a large, superficial, contrast-enhancing cystic mass within the left parieto-occipital lobes and marked ventriculomegaly. (B) Clusters of ganglion cells are surrounded by a fibrous tissue spindle cell component. H&E; × 120. (C) Dense desmoplastic areas consisting of spindle-shaped cells comprise most of this tumor. H&E; × 120. (D) Ganglion cells are round with slightly vesicular nuclei and prominent nucleoli. H&E; × 300. (E) Reticulin stain (× 150) demonstrating collagen tissue surrounding ganglion cells. Immunohistochemical findings: vimentin, S-100, and GFAP positive; ganglion cells focally reactive with synaptophysin and NSE.

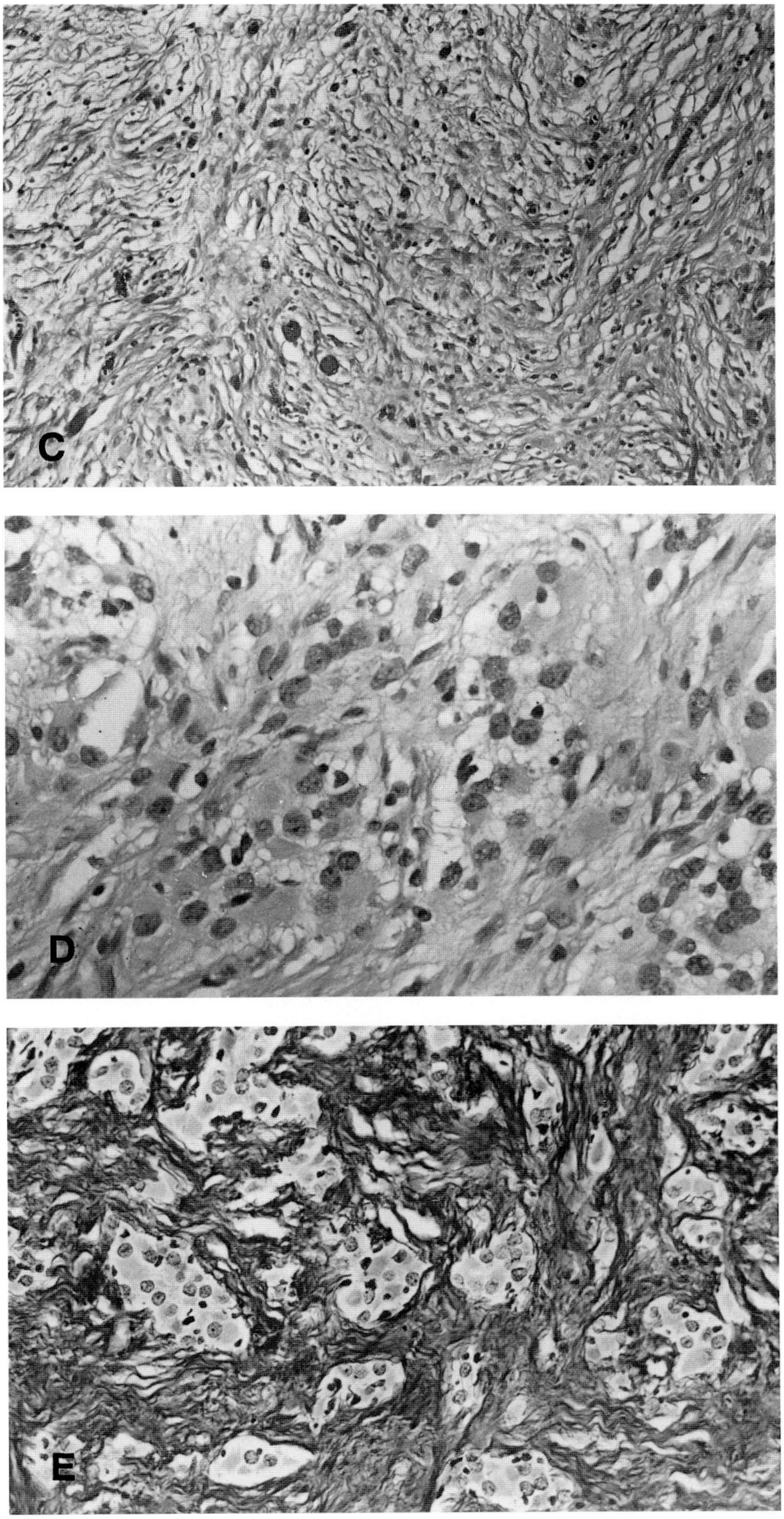

FIGURE 8.11. (*Continued*)

Ependymoma

Ependymoma is the third most frequently diagnosed brain tumor in infants (16% of the total) and fourth in frequency in the newborn (10% of the total) [8]. It comprises 5 to 10% of all central nervous system tumors in children [61]. Most ependymomas of the fetus and infant arise from the wall of the fourth ventricle [61,62]. Occasionally they are found near the lateral ventricle, the commonest site in adults.

Patients with ependymal tumors present with hydrocephalus and signs of increased intracranial pressure. The tumors cause dystocia, stillbirth, and spontaneous intracerebral hemorrhage [61–63]. Central nervous system metatastases may be present at the time of diagnosis [62].

Ependymomas are tumors derived from and consisting of ependymal cells that originate from within or near the ependymal lining of the ventricles or central canal of the spinal cord [22]. The presence of cilia and blepharoplasts, demonstrated by electron microscopy and with the phosphotungstic acid–hematoxylin (PTAH)-stain by light microscopy, distinguish them from other central nervous system tumors. Ependymomas in the young are most often classified as the cellular type and are composed of rather uniform, darkly staining cells forming ependymal perivascular rosettes (Figures 8.12 and 8.13). Halos or clear zones distributed about blood vessels surrounded by tumor cells are helpful diagnostic findings, especially on frozen section. Mitoses are not uncommon. Ependymoblastomas are extremely rare neoplasms and are discussed elsewhere [8,21,22].

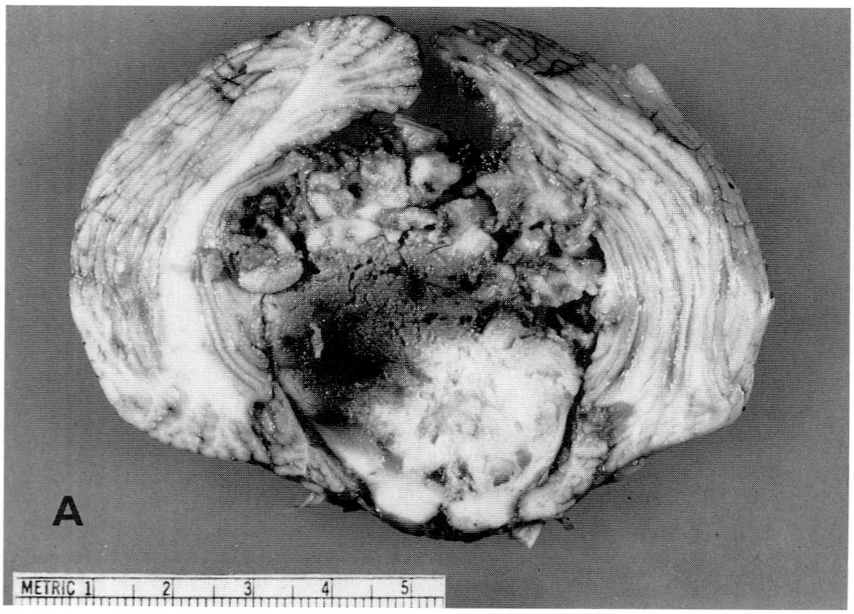

FIGURE 8.12. Ependymoma in a 9-month-old male with macrocephaly, vomiting, and marasmus occurring over a 5-month period. Imaging studies revealed severe hydrocephalus (the mantle was only 1.5 cm thick), aquaduct stenosis, and a large midline cerebellar mass. (A) Gross photograph shows a $7 \times 5 \times 5$ cm^2 midline cerebellar tumor occupying the fourth ventricle, invading both hemispheres, the mesencephalon, the pons, and the medulla. The immediate cause of death was a 3×3 cm^2 recent hemorrhage within the fourth ventricle, filling the aquaduct of sylvius. (B) Severe hydrocephalus is seen in a more proximal coronal section. (C) Low magnification ($\times$ 64) reveals a cellular, rather uniform small-cell neoplasm forming perivascular rosettes. H&E. (D) Higher magnification ($\times$ 750) of the ependymal rosettes. H&E.

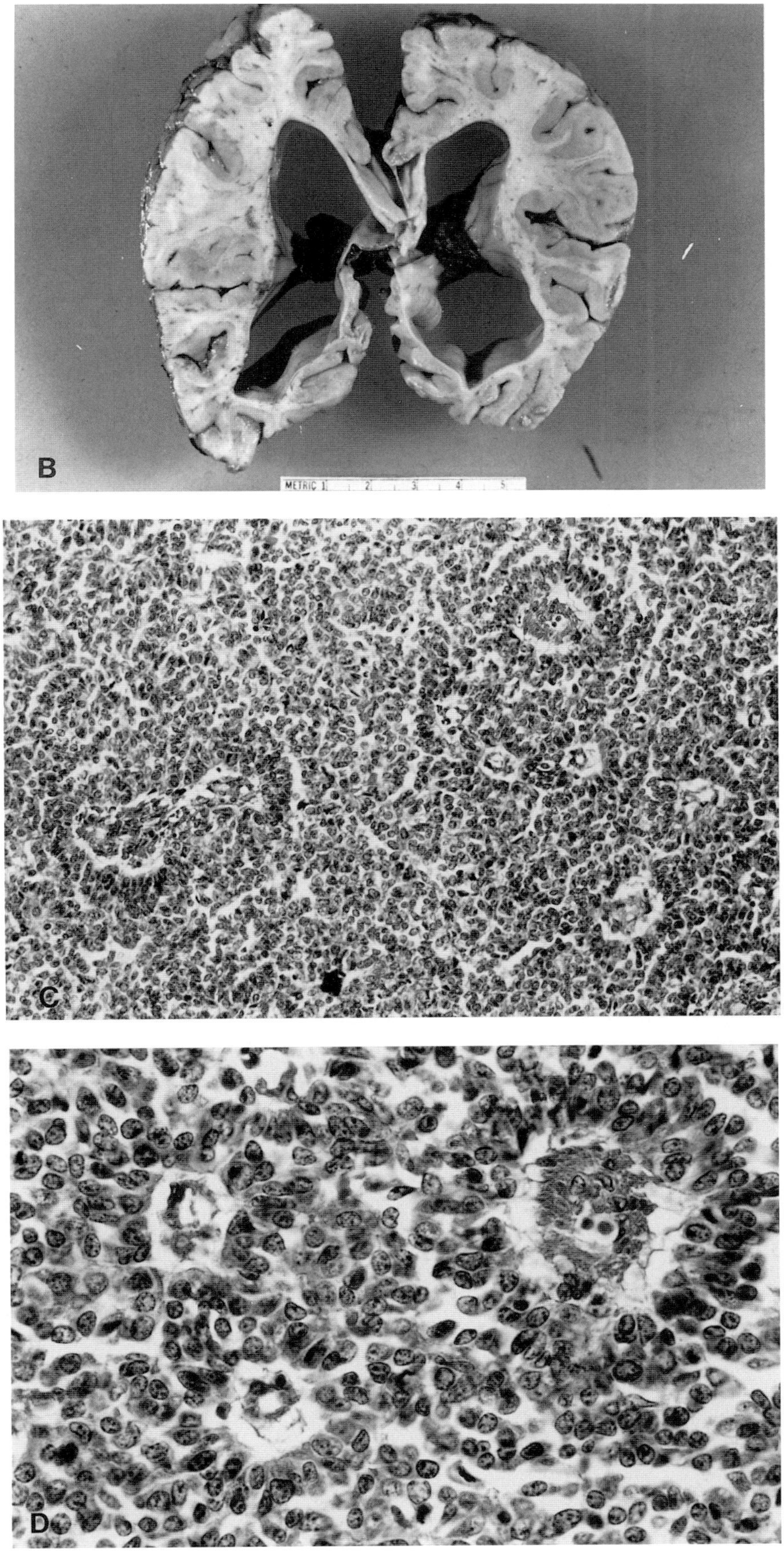

FIGURE 8.12. (*Continued*)

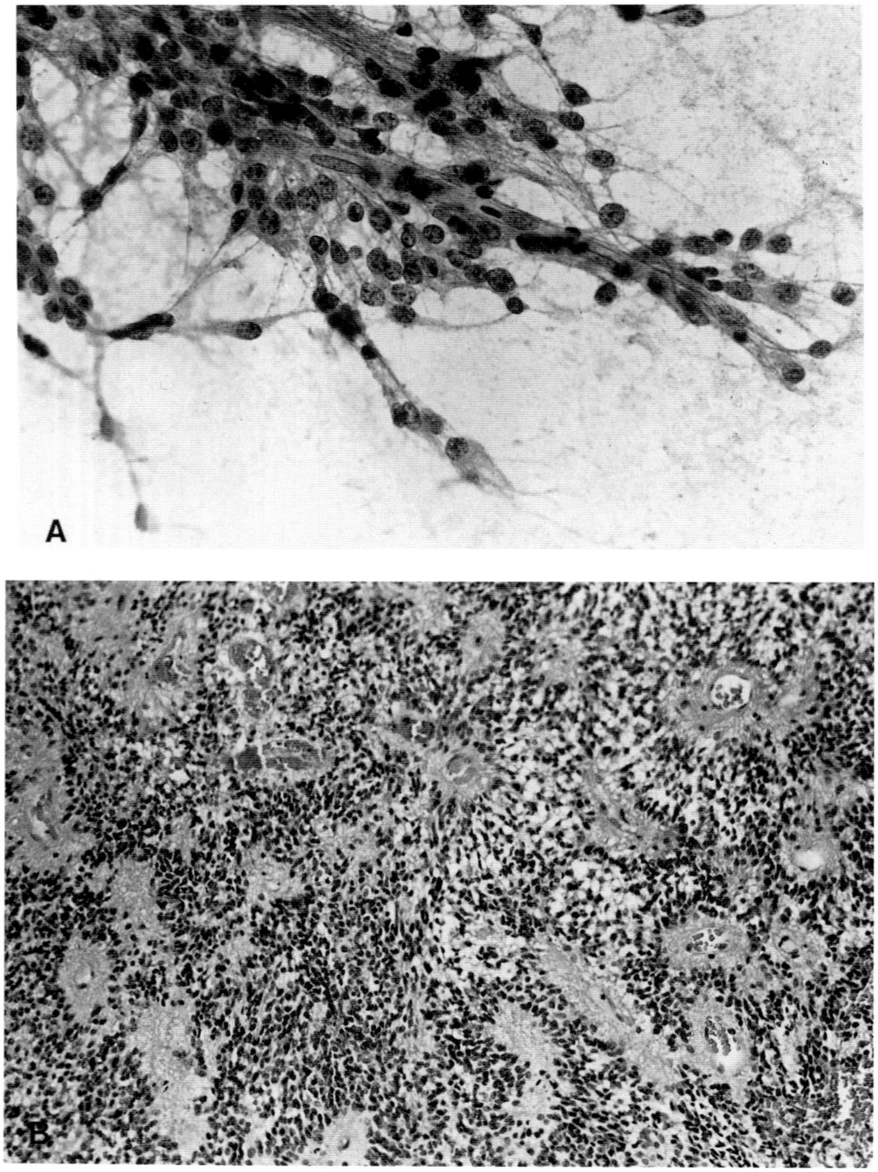

FIGURE 8.13. Ependymoma of the fourth ventricle. (A) Cytological preparation shows fine processes extending from tumor cells and elongated nuclei. H&E; × 300. (B) Perivascular pale zones are surrounded by tumor cells (ependymal rosettes), a helpful diagnostic aid particularly on frozen section. H&E; × 300.

Choroid Plexus Papilloma

Choroid plexus papillomas are tumors consisting of mature epithelial cells that line the ventricular choroid plexuses [22]. About half in the pediatric age group are diagnosed during the first year of life [2,64,65] (Tables 8.1 and 8.7). Choroid plexus paplillomas comprise 10% of all brain tumors in infants and 5% of all perinatal brain tumors [8]. They are the cause of rapid onset of hydrocephalus in the fetus and neonate [2,6,66,67] and stillbirth [34].

Typically the papilloma grows into the lateral ventricle, producing a large space-occupying nodular lesion readily demonstrated by imaging studies [65] (Figures 8.14 and 8.15). Occasionally these tumors arise from the third or fourth ventricle [67,68]. Large quantities of cerebrospinal fluid are formed by the tumor, resulting in marked dilatation of the entire ventricular system which may be present at birth.

The papilloma has a finely nodular, pink, cauliflower-like gross appearance, and microscopically the tumor displays numerous papillary formations covered by uniform, regular cuboidal to columnar epithelial cells resembling choroid plexus (Figure 8.14). Immunohistochemical findings are an expression of both epithelial (cytokeratin) and glial (GFAP) markers; like normal choroid plexus, microvilli, cilia, and zonula adherens junctions are noted ultrastructurally [64,69, 70]. Complete surgical resection results in cure. However the procedure may be difficult and complicated by fatal hemorrhage because the tumors are so highly vascular [31,65,70].

Fewer than one third of choroid plexus carcinomas are diagnosed in the first year of life [70–72]. Most occur in the lateral ventricles. Widespread dissemination throughout the cerebrospinal subarachnoid space is a fatal complication [70–72]. The histological distinction between papilloma and carcinoma may prove difficult (Figure 8.15).The diagnosis of malignancy depends on several histological criteria: poorly formed papillary structures, infiltration into adjacent tissues, hypercellularity, pleomorphic nuclei, increased mitotic activity, vascular proliferation, and necrosis [22,69,71]. Immunohistochemical studies may be of some help in establishing the diagnosis of malignancy; for example, when the carcinoembryonic antigen (CEA) stain is positive and the S-100 protein negative, the choroid plexus tumor is probably malignant [69,71]. Total surgical resection offers the best chance for cure [70]. Currently the use of chemotherapy in patients with choroid plexus carcinoma is of unproven benefit [70,72].

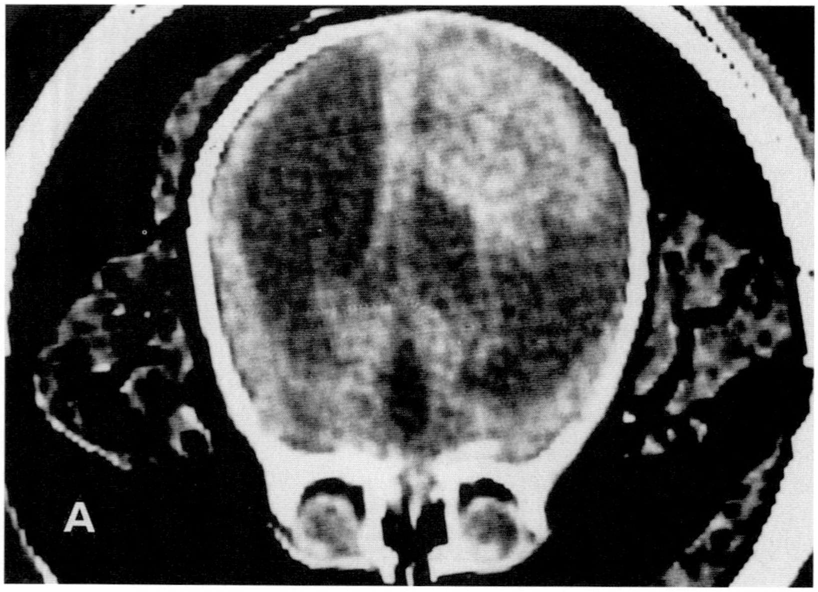

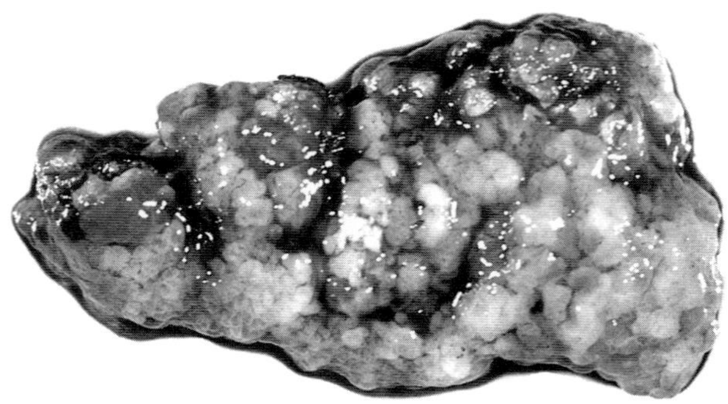

B

FIGURE 8.14. Congenital choroid plexus papilloma. Rapidly progressive hydrocephalus, a bulging fontanelle, and vomiting were the presenting signs in this 1-month-old female. (A) Computerized tomography scan reveals a large, enhancing mass partially filling the left lateral ventricle and severe hydrocephalus. (B) Two cauliflower-like masses, 5.5 × 2.7 and 4.7 × 3.5 cm^2, were excised from the left lateral ventricle. One is depicted in the gross photograph. (C) Low power (× 48) view showing the overall architecture and resemblance to choroid plexus. H&E. (D) Edematous papillary structures are lined by small cells with regular, round to oval darkly staining nuclei. H&E; × 600.

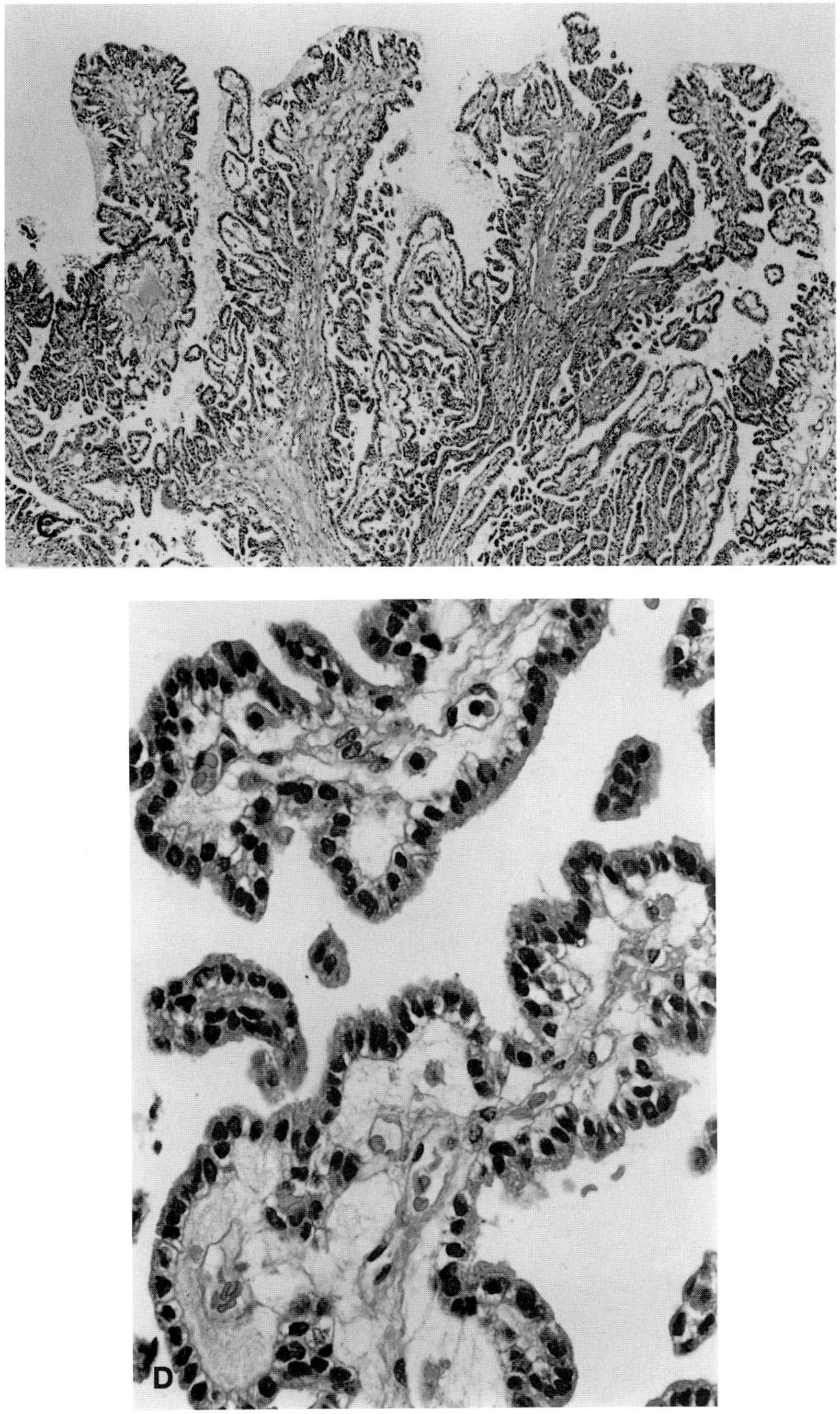

FIGURE 8.14. (*Continued*)

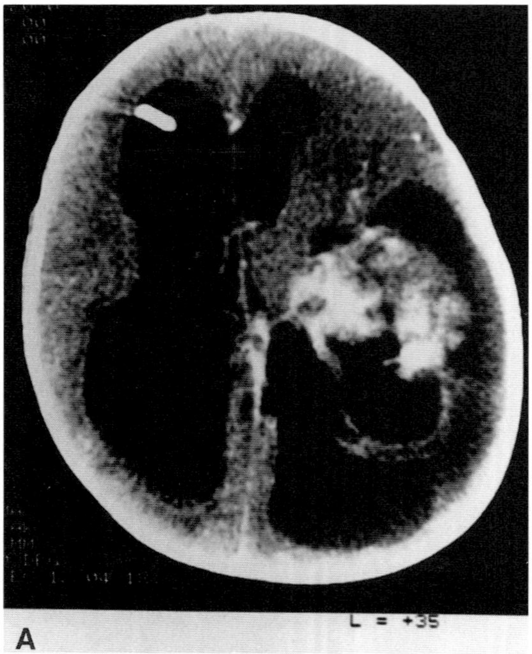

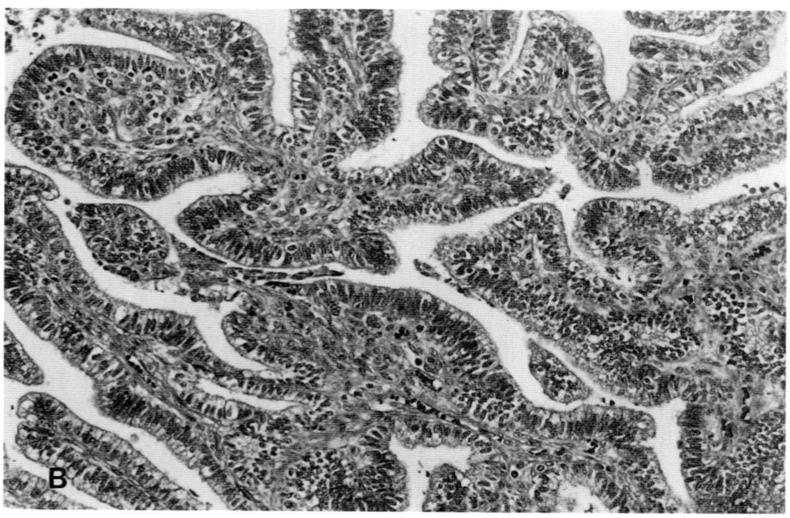

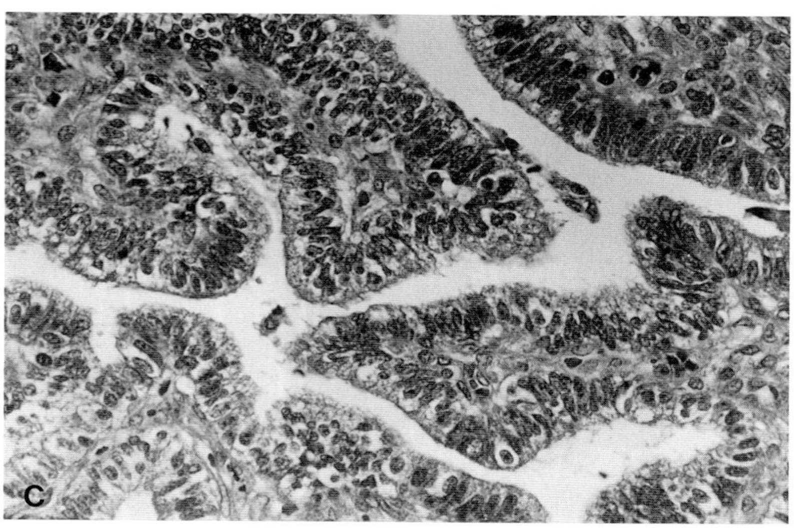

FIGURE 8.15. Choroid plexus papilloma with atypical changes in a 2-month-old male with hydrocephalus. (A) Computerized tomography scan depicts a large nodular mass partially filling the left lateral ventricle. Severe hydrocephalus is present. (B) Low power (× 150) view reveals only slight resemblance to normal choroid plexus. The papillary formations have a more blunted appearance. H&E; × 150. (C) Moderate cytological atypia, increased nuclear/cytoplasmic ratio and cellularity are present (cf. Figure 8.14D). Invasion into the underlying stroma is not apparent. H&E; × 380. (B and C courtesy of Ignacio Gonzalez, M.D., Department of Pathology, Children's Hospital, Los Angeles.)

Intracranial Germ Cell Neoplasms

Teratoma

Overall, teratoma is the major intracranial germ cell tumor in the first year of life and is the main neonatal brain tumor in several studies [9,11,16,33,35] (Tables 8.1 and 8.7). Teratomas arise from several locations within the central nervous system: the pineal region, the hypothalamic area, the suprasellar region, and the cerebral hemispheres [34,73–78]. The tumor may erode through the skull into the orbit or extend into the neck [76,77]. In the past most intracranial teratomas were diagnosed at postmortem examination, but more recently the diagnosis is readily established in utero or in the perinatal period by sonography or other imaging techniques [76,77].

Intracranial teratomas manifest clinically in a variety of ways. Many are detected incidentally on "routine" prenatal ultrasonography (Figure 8.16A). The large head filled with tumor prevents passage through the maternal birth canal, causing dystocia. Mothers of fetuses with intracranial teratomas may present with a sudden onset of a large uterus for dates in otherwise normal pregnancies [77]. The rapid growth in maternal girth is attributed to the accompanying polyhydramnios or tumor size.

Sonography reveals an intracranial mass containing solid and cystic areas and sometimes foci of calcification, with distortion or destruction of the normal brain architecture [74,77]. The tumor causes stillbirth and perinatal death [16,33,74,77] (Figure 8.16).

Intracranial teratomas typically are large and cystic with solid areas replacing the most of the brain [74,77] (Figure 8.16C). Histologically they consist of both mature components from all three germ layers and immature neuroglial elements, the usual tissues found in teratomas observed elsewhere in this age group [74,76,78]. Sometimes immature neuroglial elements in infant teratomas comprise most of the tumor and should not be misinterpreted as malignant. Practically all infants with antenatally diagnosed intracranial teratomas die before or shortly after birth (mean survival 21 days) [77]. The prognosis worsens with increasing tumor size and decreasing gestational age at diagnosis [77].

Choriocarcinoma

Choriocarcinoma is a rare and unusual malignant germ cell neoplasm occurring either alone or in association with an intracranial teratoma, or as the result of a placental primary tumor that has metastasized to the brain [18,28] (Figures 2.33 and 2.34). Choriocarcinoma may manifest in the young infant as a distinctive syndrome characterized by widespread metastases, bleeding tendencies, and elevated chorionic gonadotropin levels. The increased hormone levels result from a placental primary tumor that has invaded the chorionic villous vessels, entered the umbilical vein, and spread throughout the fetal circulation [9,79] (see Chapter 2, Germ Cell Tumors, for further discussion).

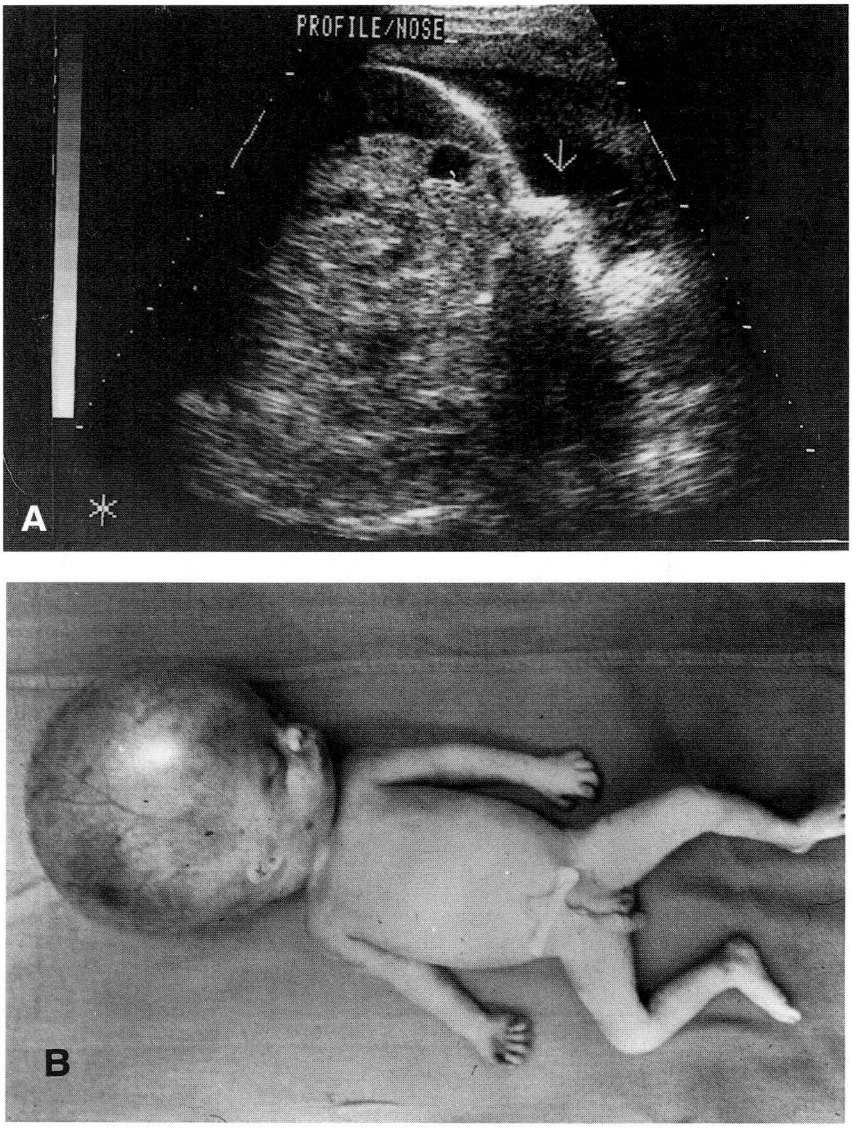

FIGURE 8.16. Intracranial teratoma. (A) Sonogram at 31 weeks gestation. Coronal view displays a large cystic and solid mass filling the cranial vault. The arrow is pointed at the nose. The forehead is the curved structure above the arrow and the mouth and chin are below it. (B) The macrocephalic infant at birth. (C) The intracranial contents consist of a variety of cystic and solid tissues; there was no recognizable brain. (D) The tumor consists of cartilage, squamous and respiratory epithelium, and immature neuroglial elements. H&E; × 48. (E) Immature neuroglial elements, epithelium, and cartilage viewed at higher magnification (× 150). H&E. (A, Courtesy of Val Catanzarite, M.D., Department of Neonatology, Mary Birch-Sharp Hospitals, San Diego. B, C, D, and E courtesy of Arturo Mendoza, M.D., Department of Pathology, Sharp Hospital, San Diego.) (From Weyerts LK, et al. *J Med Genet* 1993;30:880. Used with permission.)

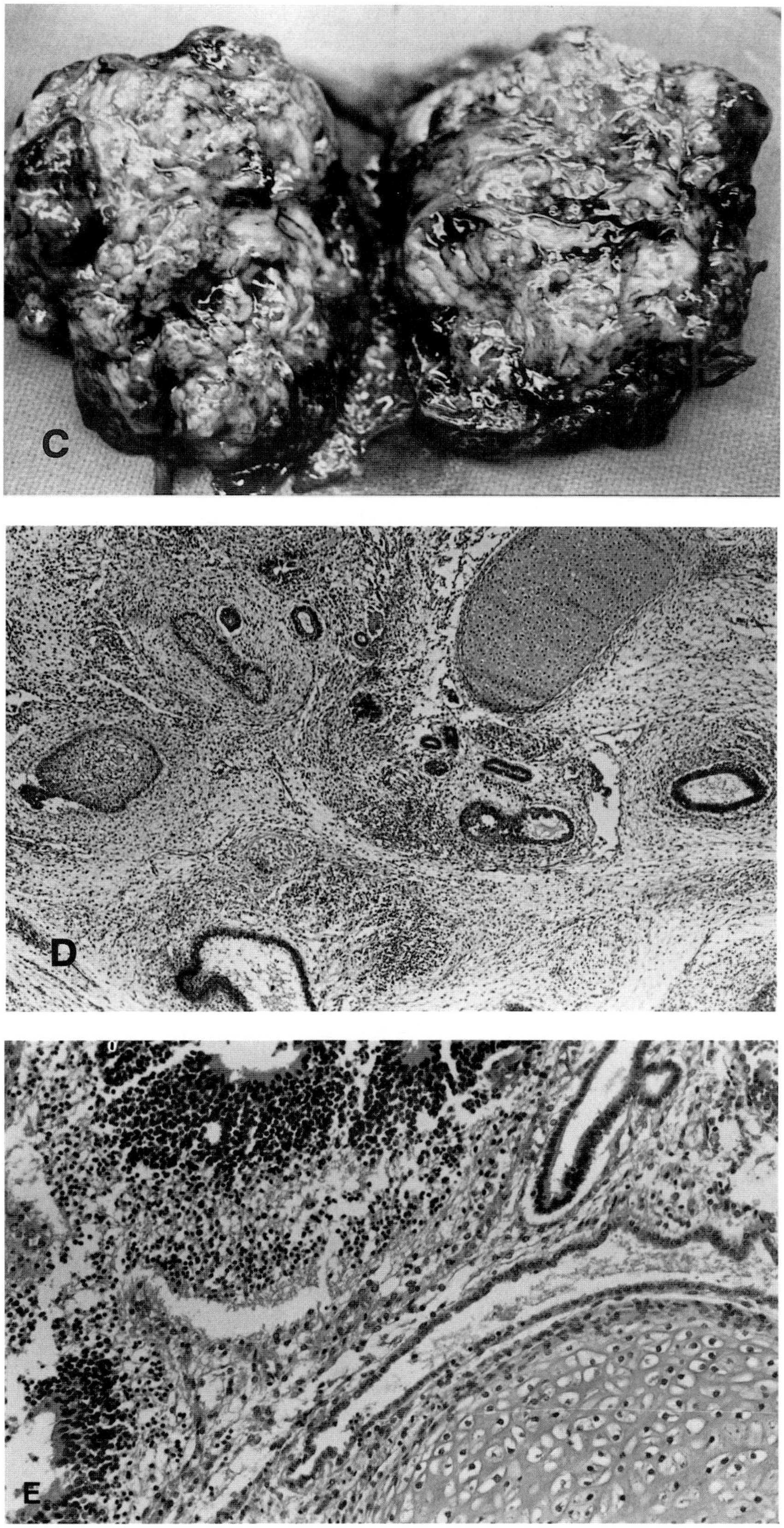

FIGURE 8.16. (*Continued*)

Craniopharyngioma

Although craniopharyngioma is a relatively common tumor of childhood, comprising 7% of intracranial neoplasms, it seldom occurs in the fetus and infant [22,80] (Tables 8.1 and 8.7). Because of its epithelial morphology and location intimately related to the pituitary gland, the tumor is thought to arise from Rathke's pouch, an ectodermal diverticulum developing from the roof of the mouth. Macrocephaly due to hydrocephalus and/or the tumor size has been the most frequent clinical presentation in the fetus and newborn [80]. Polyhydramnios is variably present. Imaging studies reveal a large, heterogeneous, calcified, cystic mass in the suprasellar region. Some tumors completely replace all normal brain tissue, not unlike a teratoma [9,80].

Microscopically craniopharyngiomas display both squamous and adamantinomatoid patterns with keratin pearls and foci of calcification [22,80] (Figure 8.17). Stacked arrays of plump, necrobiotic squames are an important feature of the adamantinomatous pattern [22].

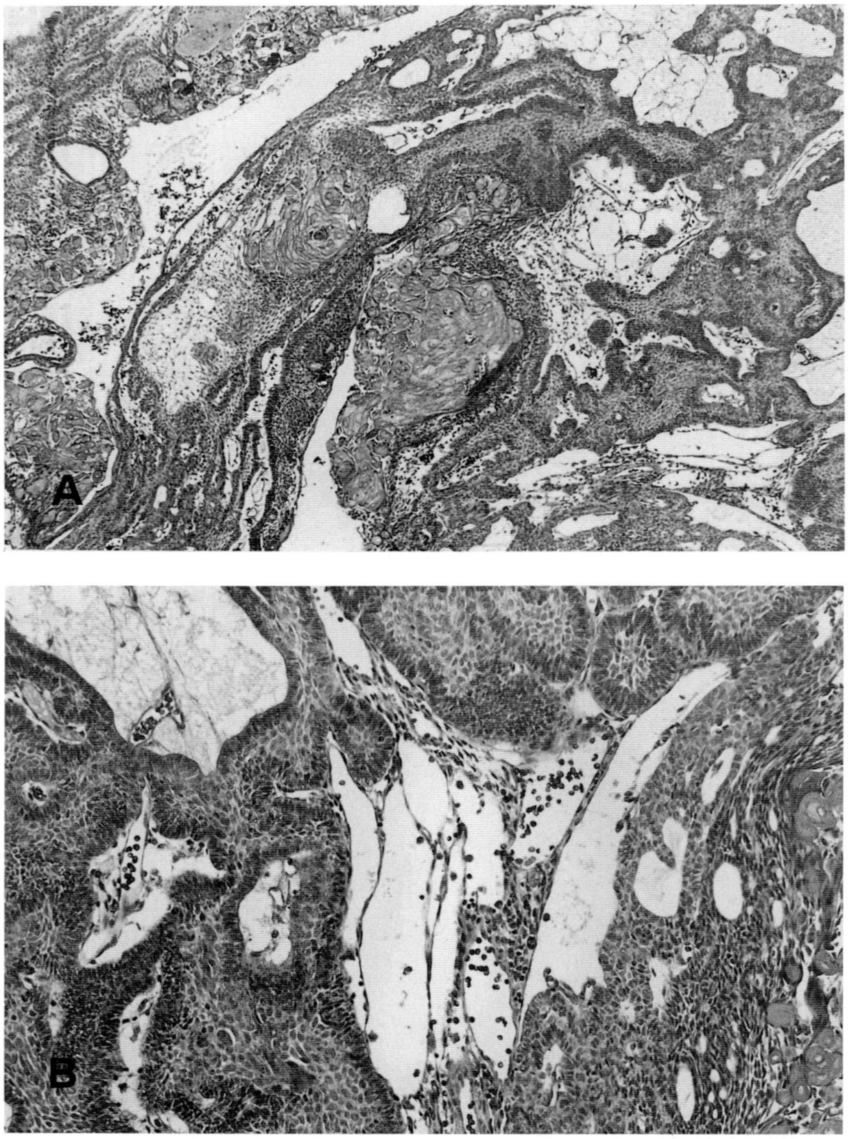

FIGURE 8.17. Craniopharyngioma in a 3-month-old female with upward gaze nystagmus and a suprasellar mass (with enhancement) detected on CT scan. (A) The tumor consists of epithelial cells forming compact masses or lining cysts. H&E; × 48. (B) Nests of epithelial cells are surrounded by a layer of columnar "basal" cells separated by a myxoid stroma consisting of stellate cells forming the "adamantinomatous" pattern. Clusters of keratin pearls are noted in the right side of the photomicrograph. H&E; × 120.

Atypical Teratoid/Rhabdoid Tumor

Atypical teratoid/rhabdoid tumors occur primarily in infancy and early childhood and are invariably fatal [81–85]. In the central nervous system these highly malignant neoplasms are histologically similar to those found in the kidney, soft tissues, and other sites [86,87]. Moreover congenital examples with hydrocephalus and hydranencephaly are reported [83, 85]. Since expression of multiple lineages in these central nervous system tumors suggests a teratomatous lesion, the term "atypical teratoid/rhabdoid tumor" was coined by Rorke et al. [81]. The posterior fossa, particularly the cerebellum, is the major site of origin. In the latter location atypical teratoid/rhabdoid tumors show clinical and imaging findings similar to those of medulloblastoma (Figure 8.18). Cerebral hemispheres and brain stem are additional primary sites [81,83,84]. Monosomy 22 is associated with the atypical teratoid/rhabdoid tumor [83].

Grossly atypical teratoid/rhabdoid tumors consist of gray-tan friable tissue with areas of necrosis [82]. They are distinguished histologically from medulloblastoma and other PNETs by the presence of rhabdoid cells, which have a pink cytoplasm, a round vesicular nucleus with a large nucleous, and a cytoplasmic eosinophilic paranuclear body composed of intermediate filaments (Figure 8.18). Atypical teratoid/rhabdoid tumors have both rhabdoid and PNET components, the latter having the appearance of a "small blue-cell" tumor [81–85,88]. Occasionally epithelial glands and Flexner-Wintersteiner rosettes are present. The tumors display a complex polyphenotypic immunophenotype. Vimentin, epithelial membrane antigen, cytokeratin, glial fibrillary acidic protein, and cytokeratin are immunoreactive [83]. Smooth muscle actin, chromogranin, and synaptophysin are variably reactive. By electron microscopy the presence of circumscribed bundles of intermediate filaments forming tight whorls, corresponding to the eosinophilic cytoplasmic inclusions observed by light microscopy, are requisite for the diagnosis of the rhabdoid cells [22].

Recognition of the rhabdoid tumor component is very important because this finding is predictive of a prognosis much worse than that for medulloblastoma [81,82]. Extensive brain invasion, recurrence, and early CSF spread contribute to an overall survival of only 6 to 11 months [81–83].

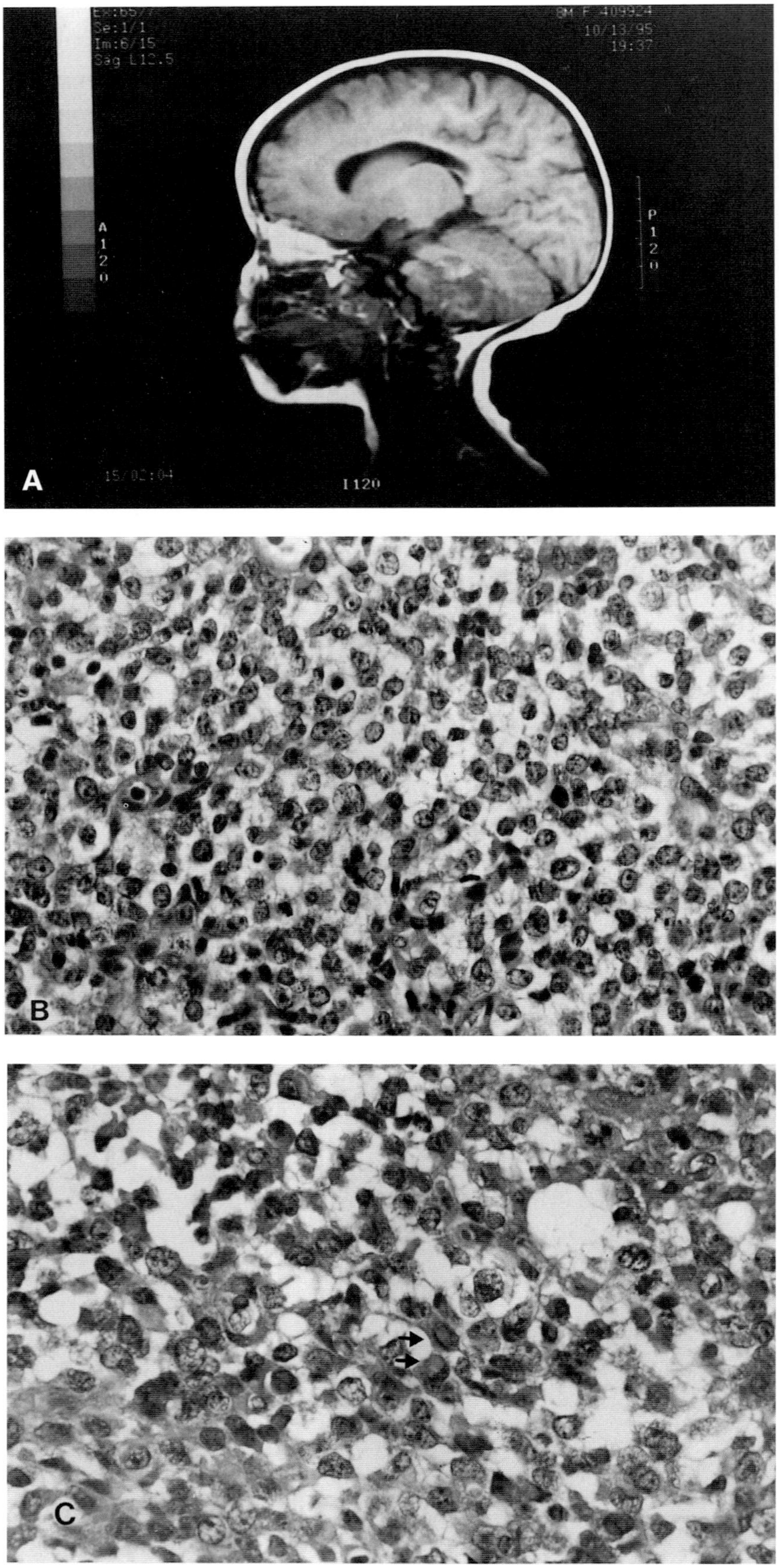

FIGURE 8.18. Atypical teratoid/rhabdoid tumor in an 8-month-old female with a posterior fossa mass. (A) Magnetic resonance imaging displays a 3 cm tumor in the left cerebellar peduncle that enhances slightly. (B) The tumor shows a varied histological appearance. Frequently vacuolated cells contain round to oval nuclei with a fine chromatin pattern and a prominent nucleolus. H&E; × 300. (C) Larger oval to rectangular cells with eccentic nuclei and prominent nucleoli. Some contain cytoplasmic eosinophilic inclusions representing packets of intermediate filaments (arrows). Immunoperoxidase studies: neuron-specific enolase and epithelial membrane antigen stain strongly positive; keratin and actin are focally weakly reactive; GFAP and synaptophysin are negative. Cultured tumor cells yielded a 46,XX karyotype. H&E; × 380.

Medulloepithelioma

The cerebral medulloepithelioma is a neoplasm of early childhood (average age of 2 years); congenital cases have been reported [88–90]. The paraventricular region of the cerebrum is a common site, but the tumors may arise from almost any location in the brain and spinal cord. Increasing head size, seizures, strabismus, hemiparesis, lethargy, and coma are the presenting signs in infants [89]. These highly malignant neoplasms extensively invade structures locally and spread along the cerebrospinal fluid pathways not unlike the primitive neuroectodermal tumors described earlier.

Medulloepithelioma is a unique "embryonal tumor" that histologically recapitulates the primitive medullary epithelium of the medullary plate and neural tube. The tumor forms papillary and tubular structures composed of crowded pseudostratified columnar cells delineated by well-defined basement membranes and in addition shows divergent differentiation ranging from embryonal to mature neuroepithelial and neuronal cell types [35,88–90]. They consist of primitive medullary epithelium as described earlier with foci of astrocytes, oligodendroglia, ependymal cells, neuroblasts, and mature ganglion cells in varying proportions. The medulloepitheliomatous component is not immunoreactive with glial fibrillary acidic protein, neuron-specific enolase, or S-100 protein [90]. However this component shows vimentin and microtubule-associated protein type 5 reactivity similar to what is found in primitive neural tube cells.

Meningeal Tumors

The meninges are the primary site of a variety of benign and malignant mesenchymal tumors of childhood ranging from meningiomas to sarcomas and melanomas [91]. In one survey 1.8% of all intracranial tumors of childhood were meningeal in origin (vs 14% for adults) [91]; meningeal sarcoma was the diagnosis in two infant cases presenting as rapidly growing occipital masses.

To date fewer than 80 cases of meningioma have been reported in infants and fewer than 10 in newborns [9,91–95]. Meningiomas of the young are distinctly different from those diagnosed in older age groups [94]. They occur more often in males; they are more frequently convex rather than infratentorial in location. The major signs and symptoms in this age group are macrocephaly, tense fontanelle, cranial asymmetry, vomiting, seizures, and diffuse spasticity [95]. Some tumors attain a large size; for example, one mengioma removed from a 5-month-old male weighed 600 g [92]. The histological type is usually fibroblastic, consisting of uniform spindle cells with a prominent vascular stroma [94]. Moreover, several other histological types have been described, including angioblastic, meningothelial, and a form unique to children, the sclerosing meningioma, composed of whorling collagen bundles and cells with perinuclear halos [93]. Surgery is the treatment of choice. The prognosis is good provided the tumor is completely excised [92,94].

Vascular Malformations of the Brain

Although histologically benign, intracranial vascular conditions cause significant morbidity and mortality in the first year of life. Arteriovenous malformations, Sturge-Weber syndrome, cavernous hemangioma, and neonatal hemangiomatosis are examples [6,8,96–99].

Cyanosis and congestive heart failure are the signs of large intracranial arteriovenous malformations. Because of their clinical findings they may be misinterpreted as congenital heart disease [6].

The Sturge-Weber syndrome (flat facial hemangiomata, meningeal hemangiomata with seizures) is characterized by intractable seizures and a port wine stain hemangioma distributed over the side of the face, forehead, and parietal area (96,97). Computerized tomography scans reveal cerebral atrophy and focal calcifications (Figure 8.19). Diffuse hemangiomatosis of the leptomeninges with focal involvement of gray and white matter of the temporal and occipital lobes are present [8].

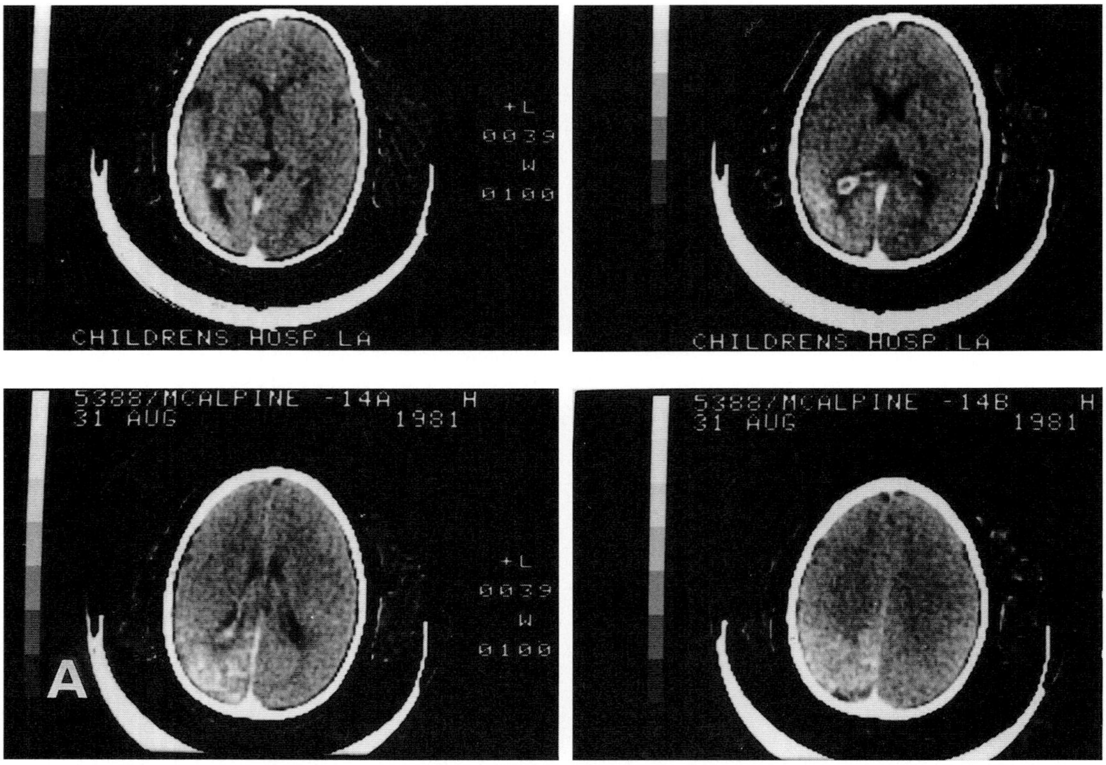

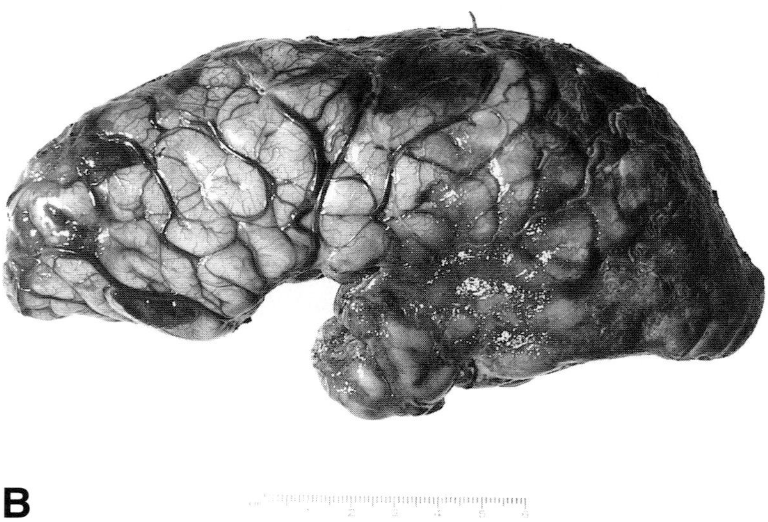

B

FIGURE 8.19. Sturge-Weber syndrome in a 1-year-old male with a "port wine stain" hemangioma distributed over the left side of the face, forehead, and parietal area, and intractable seizures present since birth. (A) Computerized tomography scan reveals left cerebral atrophy and focal calcifications. (B) The cortical surface of the left cerebral hemisphere displays thickening and discoloration of the leptomeninges due to a diffuse hemangioma situated over the temporal, parietal, and occipital areas. (C) Cut surface of the cerebral hemisphere showing diffuse thickening of the leptomeninges and focal lesions in the cortical gray and white matter. (D) Photomicrograph of the temporal lobe depicting a leptomeningeal hemangioma composed of numerous thin-walled vessels lined by a single layer of flattened endothelium and separated by variable amounts of loose connective tissue. Gliosis and focal neuronal necrosis appear in the adjacent gray and white matter. H&E; × 120.

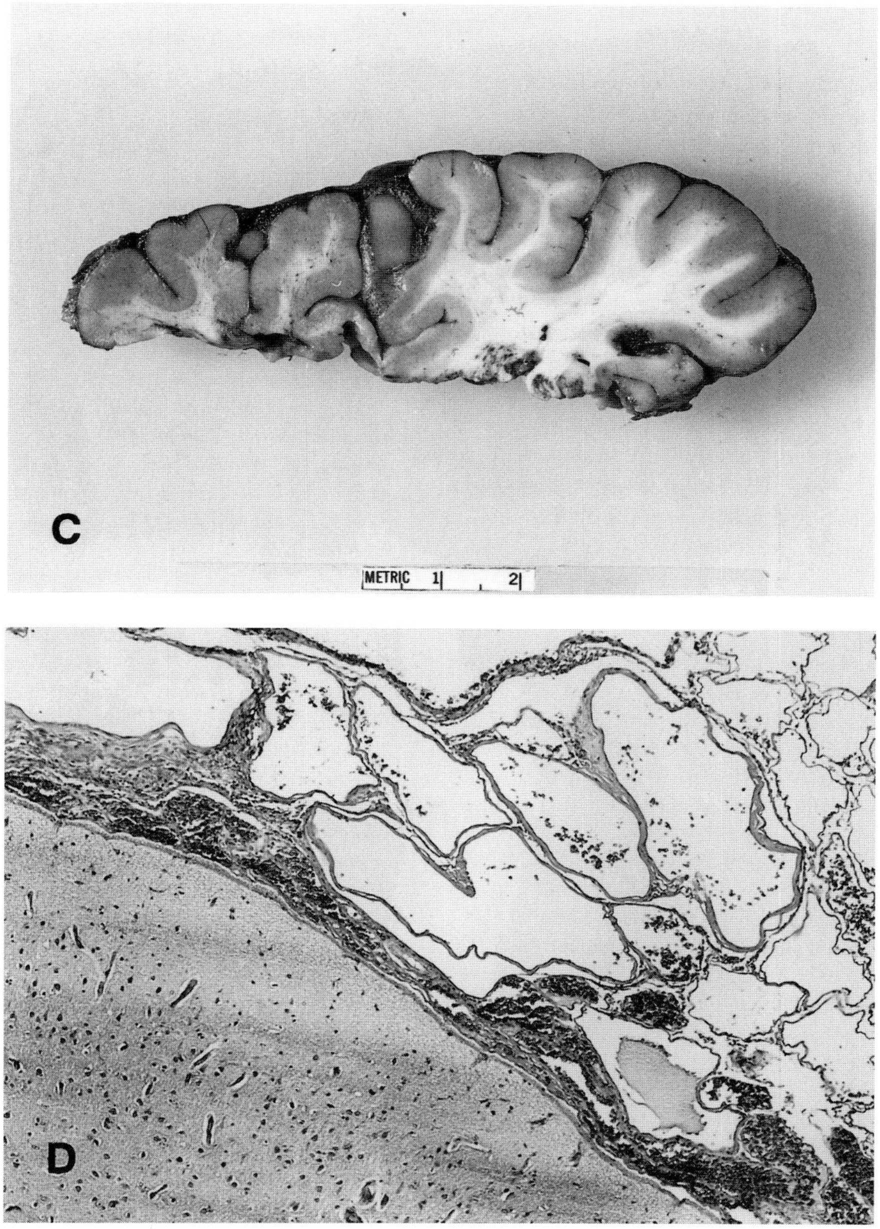

FIGURE 8.19. (*Continued*)

Prognosis for Brain Tumors

Young patients with brain tumors, especially neonates, generally have a poor prognosis. The survival rate has somewhat improved with newer imaging and neurosurgical techniques. Outcome is related to the size and location of the tumor, the histology, surgical resectability, and the condition of the infant at the time of diagnosis [17]. For example, patients with a choroid plexus papilloma or a low grade astrocytoma have a relatively good prognosis in comparison to those with primitive neuroectodermal tumors (e.g., medulloblastoma, cerebral PNET) [16,17] (Table 8.7).

Surgery remains the treatment of choice. Radiotherapy in the newborn and infant is not recommended because of the effect of this form of therapy on the immature brain and other developing tissues. Currently the use of chemotherapy for brain tumors in this young age group is under investigation [13,17].

References

1. Farwell JR, Dohrmann GJ, Flannery JT. Intracranial neoplasms in infants. *Arch Neurol* 1978;35:353.
2. Jooma R, Hayward RD, Grant DN. Intracranial neoplasms during the first year of life: analysis of one hundred consecutive cases. *Neurosurgery* 1984; 14:31.
3. Tomita T, McLone DG. Brain tumors during the first twenty-four months of life. *Neurosurgery* 1985;17:913.
4. Young JL, Miller RW. Incidence of malignant tumors in US children. *J Pediatr* 1975; 86:254.
5. Isaacs, H Jr. Congenital and neonatal malignant tumors: a 28-year experience at Children's Hospital of Los Angeles. *Am J Pediatr Hematol Oncol* 1987;9(2):121.
6. Isaacs H Jr. Perinatal (congenital and neonatal) neoplasms: a report of 110 cases. *Pediatr Pathol* 1985;3:165.
7. Isaacs H Jr. Brain tumors. In: *Tumors of the Newborn and Infant.* St. Louis, MO: Mosby-Year Book; 1991:82.
8. Isaacs H Jr. Brain tumors. In: *Tumors of the Fetus and Newborn.* Philadelphia: WB Saunders; 1997:187. Major Problems in Pathology. Series; Vol 35.
9. Isaacs H Jr. Tumors. In: Gilbert-Barness E, ed. *Potter's Pathology of the Fetus and Infant.* Vol 2. St. Louis, MO: CV Mosby; 1997:1248.
10. Bader JL, Miller RW. US cancer incidence and mortality in the first year of life. *Am J Dis Child* 1979;133:157.
11. Oi S, Matsumoto S, Choi JU, et al. Brain tumors diagnosed in the first year of life in five Far-Eastern countries. Statistical analysis of 307 cases. *Child's Nerv Syst* 1990;6:79.
12. Janisch W. Zur Epidemiologie der primaren Geschwulste des Zentralnervensystems im ersten Lebensjahr. *Arch Geschwulstforsch* 1985;55:489.
13. Di Rocco C, Iannelli, Ceddia A. Intracranial tumors of the first year of life: a cooperative survey of the 1986–1987 Education Committee of the ISPN. *Child's Nerv Syst* 1991;7:150.
14. Campbell AN, Chan HSL, O'Brien A, et al. Malignant tumours in the neonate. *Arch Dis Child* 1987;62:19.
15. Janisch W, Schreiber D, Martin H, Gerlach H. Primäre intrakranielle Tumoren als Todesursache bei Feten und Sauglingen. *Zentralbl Allg Pathol* 1984;192(2):75.
16. Wakai S, Arai T, Nagai M. Congenital brain tumors. *Surg Neurol* 1984;21:597.
17. Mazewski CM, Hudgins RJ, Reisner A, et al. Neonatal brain tumors: a review. *Semin Perinatol* 1999;23:286.
18. Dehner LP. Neoplasms of the fetus and neonate. In: Naeye RL, Kissane JM, Kaufman N, eds. *Perinatal Diseases.* Baltimore: Williams & Wilkins; 1981:286. International Academy of Pathology, Monograph No. 22.
19. Sabet LM. Congenital glioblastoma multiforme associated with congestive heart failure. *Arch Pathol Lab Med* 1982;106:31.
20. Rickert CH. Neuropathology and prognosis of foetal brain tumors. *Acta Neuropathol* 1999;98:567.
21. Burger PC, Scheithauer BW. Tumors of the central nervous system. In: *Atlas of Tumor Pathology.* 3rd ser; Fascicle 10. Washington, DC: Armed Forces Institute of Pathology; 1994.
22. Bigner DD, McLendon RE, Bruner JM, eds. *Russell and Rubinstein's Pathology of Tumors of the Nervous System.* 6th ed. London: Arnold; 1998.
23. Riccardi VM: Type 1 neurofibromatosis and the pediatric patient. *Curr Probl Pediatr* 1992;22:66.
24. Gomez MR. Definition and criteria for diagnosis. In: Gomez MR, Sampson JR, Whittemore VH, eds. *Tuberous Sclerosis Complex.* 3rd ed. Oxford: Oxford University Press; 1999:18.
25. Ehret M, Jacobi G, Hey A, et al. Embryonal brain neoplasms in the neonatal period and early infancy. *Clin Neuropathol* 1987; 6:218.
26. Bonnin JM, Rubenstein LJ, Palmer NF, et al. The association of embryonal tumors originating in the kidney and in the brain: a report of seven cases. *Cancer* 1984;54:2137.
27. Chang C, Ramirez N, Sakr WA. Primitive neuroectodermal tumor of the brain associated with malignant rhabdoid tumor of the liver: a histologic, immunohistochemical and electron microscopic study. *Pediatr Pathol* 1989;9:307.
28. Kelly DL Jr, Kushner J, McLean WT. Neonatal intracranial choriocarcinoma. *J Neurosurg* 1971;35:465.
29. Inwald D, Kempley S, Hird M. Congenital primitive neuroectodermal tumour presenting as obstructed labour. *Arch Dis Child Fetal Neonatal Ed* 1998;78:F222.
30. Jooma R, Kendall B, Hayward R. Intracranial tumors in neonates: a report of seventeen cases. *Surg Neurol* 1984;21:165.
31. Asai A, Hoffman HJ, Hendrick EB, et al. Primary intracranial neoplasms in the first year of life. *Child's Nerv Syst* 1989;5:230.
32. Ellams ID, Neuhauser G, Agnoli AL. Congenital intracranial neoplasms. *Child's Nerv Syst* 1986;2:165.
33. Werb P, Scurry J, Ostor A, et al. Survey of congenital tumors in perinatal necropsies. *Pathology* 1992;24:247. 8.23
34. Jellinger K, Sunder-Plassmann M. Connatal intracranial tumors. *Neuropaediatrie* 1973;4:46.
35. Becker LE, Hinton D. Primitive neuroectodermal tumors of the central nervous system. *Hum Pathol* 1983;14:538.
36. Timperly WR. ACP Best Practice No. 158. Neuropathology. *J Clin Pathol* 2000;53:255.
37. Kleihues P, Kiessling M, Janzer RC. Morphological markers in neuro-oncology. *Curr Top Pathol* 1987;77:307.
38. Boop FA, Chadduck WM, Sawyer J, et al. Congenital aneurysmal hemorrhage and astrocytoma in an infant. *Pediatr Neurosurg* 1991–92;17:44.

39. Tekkok IH, Ventureyra ECG. Spontaneous intracranial hemorrhage of structural origin during the first year of life. *Child's Nerv Syst* 1997;13:154.
40. Geraghty AV, Knott PD, Hanna HM. Prenatal diagnosis of fetal glioblastoma multiforme. *Prenat Diagn* 1989;9:613.
41. Roosen N, Deckert MD, Nikolaus N, et al. Congenital anaplastic astrocytoma with favorable prognosis: case report. *J Neurosurg* 1988;69:604.
42. Mazzone D, Magro G, Lucenti A, et al. Report of a case of congenital glioblastoma multiforme: an immunohistochemical study. *Child's Nerv Syst* 1995;11:311.
43. Lee DY, Kim YM, Yoo SJ, et al. Congenital glioblastoma diagnosed by fetal sonography. *Child's Nerv Syst* 1999;15197.
44. de Chadarevian J-P, Parrisapu JV, Faerber EN. Desmoplastic cerebral astrocytoma of infancy: light microscopy, immunocytochemistry and ultrastructure. *Cancer* 1990;66:173.
45. Louis DN, Von Deimling A, Dickersin GR, et al. Desmoplastic cerebral astrocytomas of infancy: a histopathologic, immunohistochemical, ultrastructural and molecular genetic study. *Hum Pathol* 1992;23:1402.
46. Dehner LP. Peripheral and central primitive neuroectodermal tumors. *Arch Pathol Lab Med* 1986;110:997.
47. Horten BC, Rubinstein LJ. Primary cerebral neuroblastoma: A clinicopathological study of 35 cases. *Brain* 1976;99:735.
48. Kosnik EJ, Boesel CP, Bay J, et al. Primitive neuroectodermal tumors of the central nervous system in children. *J Neurosurg* 1978;48:741.
49. Rorke LB. The cerebellar medulloblastoma and its relationship to primitive neuroectodermal tumors. *J Neuropathol Exp Neurol* 1983;42:1.
50. Friese M, Kern M, Schroder. A prenatally diagnosed PNET with an extensive tumor mass in the brain. *Clin Neuropathol* 1998; 17;271 Abstract P9.
51. McComb JG, Davis RL, Isaacs H Jr. Extraneural metastatic medulloblastoma in childhood. *Neurosurgery* 1981;9:548.
52. (a) Mitchell CS, Wood BP, Shimada H. Neonatal disseminated primitive neuroectodermal tumor. *Am J Roentgenol* 1994;162: 1160. (b) Asai M, Kitamura H, Yanagibashi T, et al. Case of acrania associated with congenital medulloblastoma. *Eur J Obstet Gynecol Reprod Biol* 1998;81:115.
53. Kadin ME, Rubinstein LJ, Nelson JS. Neonatal cerebellar medulloblastoma originating from the fetal external granular layer. *J Neuropathol Exp Neurol* 1970;29:593.
54. Katsetos CD, Burger PC. Medulloblastoma. *Semin Diagn Pathol* 1994;11:85.
55. Eberhart CG, Brat DJ, Cohen KJ, et al. Pediatric neuroblastic brain tumors containing abundant neuropil and true rosettes. *Pediatr Dev Pathol* 2000;3:346.
56. Yu I-T, Ho DM, Wong T-T, et al. Congenital cerebral primitive neuroectodermal tumor with astrocytic differentiation and extracranial metastases. *Child's Nerv Syst* 1990;6:179.
57. VandenBerg SR. Desmoplastic infantile ganglioglioma and desmoplastic cerebral astrocytoma of infancy. *Brain Pathol* 1993; 3:275.
58. Smith NM, Carli MM, Hanieh A, et al. Gangliogliomas in childhood. *Child's Nerv Syst* 1992;8:258.
59. Paulus W, Schlote W, Perentes E, et al. Desmoplastic supratentorial neuroepithelial tumours of infancy. *Histopathology* 1992 21:43.
60. Parisi JE, Scheithauer BW, Priest JR, et al. Desmoplastic infantile ganglioglioma (DIG), a form of gangliogliomatosis. *J Neuropathol Exp Neurol* 1992;51:365. Abstract.
61. Kudo H, Oi S, Tamaki N, et al. Ependymoma diagnosed in the first year of life in Japan in collaboration with the International Society for Pediatric Neurosurgery. *Child's Nerv Syst* 1990;6: 375.
62. Comi AM, Backstrom JW, Burger PC, et al. Clinical and neuroradiologic findings in infants with intracranial ependymomas. *Pediatr Neurol* 1998;18:23.
63. Ernestus R-I, Schroder R, Klug N. Spontaneous intracerebral hemorrhage from an unsuspected ependymoma in early infancy. *Child's Nerv Syst* 1992;8:357.
64. Ho DM, Wong T-T, Liu H-C. Choroid plexus tumors in childhood: histopathologic study and clinico-pathological correlation. *Child's Nerv Syst* 1991;7:437.
65. Tomita T, McLone DG, Flannery AM. Choroid plexus papillomas of neonates, infants and children. *Pediatr Neurosci* 1988; 14:23.
66. Isaacs H Jr. Neoplasms in infants: a report of 265 cases. *Pathol Annu* 1983;18(2):165.
67. Lippa C, Abroms IF, Davidson R, et al. Congenital choroid plexus papilloma of the fourth ventricle. *J Child Neurol* 1989; 4:127.
68. Costa JM, Ley L, Claramunt E, et al. Choroid plexus papillomas of the III ventricle in infants. Report of three cases. *Child's Nerv Syst* 1997;13:244.
69. Coffin CM, Wick MR, Braun JT, et al. Choroid plexus neoplasms: clinicopathologic and immunohistochemical studies. *Am J Surg Pathol* 1986;10:394.
70. Gianella-Borradori A, Zeltzer PM, Bodey B, et al. Choroid plexus tumors in childhood: response to chemotherapy, and immunophenotypic profile using a panel of monoclonal antibodies. *Cancer* 1992;69:809.
71. Paulus W, Janisch W. Clinicopathologic correlations in epithelial choroid plexus neoplasms: a study of 52 cases. Acta Neuropathol 1990;80:635.
72. Packer RJ, Perilongo G, Johnson D, et al. Choroid plexus carcinoma of childhood. *Cancer* 1992;69:580.
73. Rueda-Pedraza ME, Heifetz SA, Sesterhenn IA, et al. Primary intracranial germ cell tumors in the first two decades of life. *Perspect Pediatr Pathol* 10:160,1987.
74. Odell JM, Allen JK, Badura RJ, et al. Massive congenital intracranial teratoma: a report of two cases. *Pediatr Pathol* 1987; 7:333.
75. Washburne JF, Magann EF, Chauhan SP, et al. Massive congenital intracranial teratoma with skull rupture at delivery. *Am J Obstet Gynecol* 1995;173:226.
76. Rostad S, Kleinschmidt-DeMasters BK, Manchester DK. Two massive congenital intracranial immature teratomas with neck extension. *Teratology* 1985;32:163.
77. Alagappan A, Shattuck KE, Rowe T, et al. Massive intracranial immature teratoma with extracranial extension into oral cavity, nose, and neck. *Fetal Diagn Ther* 1998;13:321.
78. Felix I, Becker LE. Intracranial germ cell tumors in children: an immunohistochemical and electron microscopic study. *Pediatr Neurosurg* 1990–91;16:156.
79. Witzleben CL, Bruninga G. Infantile choriocarcinoma: a characteristic syndrome. *J Pediatr* 1968;73:374.

80. Kultursay N, Gelal F, Mutluer S, et al. Antenatally diagnosed neonatal craniopharyngioma. *J Perinatol* 1995;15:426.
81. Rorke LB, Packer RJ, Biegel JA. Central nervous system atypical teratoid/rhabdoid tumors of infancy and childhood: definition of an entity. *J Neurosurg* 1996;85:56.
82. Yachnis AT, Neubauer D, Muir D. Characterization of a primary central nervous system atypical teratoid/rhabdoid tumor and derivative cell line: immunophenotype and neoplastic properties. *J Neuropathol Exp Neurol* 1998;57:961.
83. Burger PC, Yu I-T, Tihan T, et al. Atypical teratoid/rhabdoid tumor of the central nervous system: a highly malignant tumor of infancy and childhood frequently mistaken for medulloblastoma. *Am J Surg Pathol* 1998;22:1083.
84. Behring B, Bruck W, Goebel HH, et al. Immunohistochemistry of primary central nervous system malignant rhabdoid tumors: report of five cases and review of the literature. *Acta Neuropathol* 1996;91:578.
85. Velasco ME, Brown JA, Kini J, et al. Primary congenital rhabdoid tumor of the brain with neoplastic hydranencephaly. *Child's Nerv Syst* 1993;9:185.
86. Biggs PJ, Garen PD, Powers JM. et al. Malignant rhabdoid tumor of the central nervous system. *Hum Pathol* 1987;18:332.
87. Weeks DA, Beckwith JB, Mierau GW, et al. Rhabdoid tumor of the kidney: a report of 11 cases from the National Wilms' Tumor Study Pathology Center. *Am J Surg Pathol* 1989;13:439.
88. Scheithauer BW, Rubinstein LJ. Cerebral medulloepithelioma: report of a case with multiple divergent neuroepithelial differentiation. *Child's Brain* 1979;5:62.
89. Molloy PT, Yachnis AT, Rorke LB, et al. Central nervous system medulloepithelioma: a series of eight cases including two arising in the pons. *J Neurosurg* 1996;84:430.
90. Khoddami M, Becker LE. Immunohistochemistry of medulloepithelioma and neural tube. *Pediatr Pathol Lab Med* 1997;17:913.
91. Nakamura Y, Becker LE. Meningeal tumors of infancy and children. *Pediatr Pathol* 1985;3:341.
92. Lopez MJ, Olivares JL, Ramos F, et al. Giant meningioma in a 5-month-old infant. *Child's Nerv Syst* 1988;4:112.
93. Davidson GS, Hope JK. Meningeal tumors of childhood. *Cancer* 1989;63:1205.
94. Benli K, Cataltepe O, Oge HJ, et al. Giant congenital meningioma in a newborn. *Child's Nerv Syst* 1990;6:462.
95. Chadduck WM, Boop FA, Blankenship JB, et al. Meningioma and sagital cranisosynostosis in an infant: case report. *Neurosurgery* 1992;30:441.
96. Oakes WJ. The natural history of patients with the Sturge-Weber syndrome. *Pediatr Neurosurg* 1992;18:287.
97. Tallman B, Tan OT, Morelli JG, et al. Location of port-wine stains and the likelihood of ophthalmic and/or central nervous system complications. *Pediatrics* 1991;87:323.
98. Sakai N. Yamada H, Nishimura Y, et al. Intracranial cavernous angioma in the 1st year of life and a review of the literature. *Child's Nerv Syst* 1992; 8:49.
99. Capin DM, Gottleib SM, Rosman NP. Central nervous system hemangiomatosis in early childhood. *Pediatr Neurol* 1997;17:365.

9 Tumors of the Eye

Retinoblastoma

Retinoblastoma, the principal intraocular neoplasm of infancy and childhood, originates from the retina of one or both eyes as a single mass or multifocal tumors. Although it is relatively infrequent in the pediatric age group overall in comparison to other neoplasms, it is responsible for significant morbidity and mortality during the first year of life [1–7]. Retinoblastoma occurs in infants and newborns, and far advanced disease already may be present at birth [1,2,8–12]. The tumor has been detected in utero by ultrasonography as early as 21 weeks of gestation [11].

Cytogenetics

The retinoblastoma gene, a suppressor gene, is situated at the 13q14 chromosome locus, and deletions or mutations of both gene copies at this site are believed to be responsible for the formation of tumors [13,14]. The two alleles normally found at the 13q14 locus provide double protection against carcinogenesis [13]. This observation seems to apply not only to retinoblastoma but also to certain other childhood tumors such as neuroblastoma, Wilms tumor, and osteosarcoma [13].

Genetic factors are considered in the diagnosis and treatment of individuals with retinoblastoma and their families [2,13]. Patients with this tumor are divided into two categories: hereditary and nonhereditary ("sporadic"). The nonhereditary form, resulting from two somatic mutations, comprises 60% of the cases; the hereditary form, resulting from one germinal mutation and one somatic mutation, accounts for the remainder [14,15]. Only 5 to 6% of hereditary cases have a family history of either unilateral or bilateral tumors. Unilateral nonhereditary disease is present in 60%; 15% are hereditary and unilateral; 25% are hereditary and bilateral. The bilateral cases are always hereditary. Bilateral disease is far more prevalent in neonates and infants, almost 70% [2]. Bilateral retinoblastoma is a dominant condition, and therefore patients are heterozygous for chromosome 13 deletion and the genotype (*Rb+/rb−*) is transmitted to approximately half their offspring [13]. Because so many retinal cells are at risk, one of three cells in each eye will undergo a second sporadic tumorigenic inactivation mutation, resulting in loss of all tumor suppression and the appearance of tumor [13].

Clinical Findings

Infants diagnosed with retinoblastoma present with differences in laterality, stage, signs, and symptoms, and they respond to treatment differently from older children [2]. The clinical findings depend upon when the tumor is discovered. The diagnosis of bilateral retinoblastoma is made more often than unilateral disease during the first year of life [1,2]. The inherited predisposition to retinoblastoma in bilateral cases shortens the time to the appearance of the tumor. The highest percentage of patients (58%) to develop new tumors are diagnosed before the age of 3 months, and 95% are diagnosed before age 2 years [2].

Leukocoria or white pupillary reflex is the dominant early finding, noted in almost 60% of infants [1] (Table 9.1) (Figure 9.1). Funduscopic examination at this time reveals one or more yellow-white, creamy, rather smooth shaped lesions on the retina. Strabismus is the presenting sign in 20% of cases in infants older than 6 months [2]. Less common findings are anisocoria, heterochromia, cataract, swelling, nystagmus, orbital cellulitis, and retinal detachment [1,13,16]. Although children with retinoblastoma usually present with leukocoria and strabismus, those diagnosed in the first 3 months are generally seen because of a family history of the tumor rather than leukocoria [2].

When the optic nerve is involved by tumor, rubeosis iridis (the formation of blood vessels and fibrosis on the surface of the iris) is observed [13]. Later findings with advanced disease are proptosis, perforation of the globe, and blindness. When metastases occur outside the globe, the patient in time develops a large, disfiguring orbital mass with or without adjacent bone and paranasal sinus invasion. Irritability, vomiting, seizures, and coma are ominous signs of central nervous system metastases.

TABLE 9.1. Clinical presentation of 71 infants less than 3 months of age with retinoblastoma.

Finding	Number of patients
Family history	21
Leukocoria	41
Strabismus	3
Miscellaneous[a]	6

[a] Miscellaneous findings were anisocoria, heterochromia, tearing, swelling, proptosis, cataract, nystagmus, and perforation of the globe.

Source: Abramson DH, et al. [1]. Copyrighted 1983, American Medical Association. Used with permission.

Less than 5% of children with retinoblastoma have the 13q (del) syndrome, which is defined by, in addition to the chromosomal defect, a variety of abnormalities, including microcephaly, mental retardation, hypotonia, imperforate anus, malformed digits, and growth failure [17].

Intraocular retinoblastoma is classified according to the Reese-Ellsworth staging system [18] (Table 9.2). According to the New York Hospital–Cornell Medical Center infant retinoblastoma study of 401 patients, the most common stage of diagnosis was group V [2]. Of infants diagnosed with bilateral tumors, 77% of eyes were group V; 84% of eyes with unilateral tumors were group V also. In the first 3 months of life, 71% of bilateral affected eyes and 83% of unilateral affected eyes had group V tumors. Therefore this study indicates that earlier age of detection does not necessarily ensure an earlier stage of ocular involvement, contrary to what one would expect. Unfortunately despite this earlier detection the presenting eye is usually full of tumor and thus cannot be salvaged [2].

TABLE 9.2. Criteria for staging of retinoblastoma.

Group I (very favorable)
A. Solitary tumor, less than 4 disk diameters in size, at or behind the equator
B. Multiple tumors, none over 4 disk diameters in size, all at or behind the equator
Group II (favorable)
A. Solitary lesion, 4 to 10 disk diameters in size, at or behind the equator
B. Multiple tumors, 4 to 10 disk diameters in size, behind the equator
Group III (doubtful)
A. Any lesion anterior to the equator
B. Solitary tumors larger than 10 disk diameters behind the equator
Group IV (unfavorable)
A. Multiple tumors some larger than 10 disk diameters
B. Any lesion extending anterior to the ora serrata
Group V (very unfavorable)
A. Massive tumors involving over half the retina
B. Vitreous seeding

Source: Ellsworth, RM [18]. Used with permission.

Differential Diagnosis

Retinoblastoma can mimic other conditions involving the eye, particularly hemorrhage and inflammation [1,2,13,19]. Hypopyon (pus in the anterior chamber), orbital cellulitis (painful red eye), and hyphema (hemorrhage into the anterior chamber) are examples. Coats disease (vascular malformation or hemangioma of the retina with retinal detachment), retinopathy of prematurity (Figure 9.2), and persistent hyperplastic primary vitreous (fibrovascular membrane or plaque partially covering the retina) are conditions most easily mistaken for retinoblastoma [1,13,14]. Toxoplasmosis, *Toxocara canis* in-

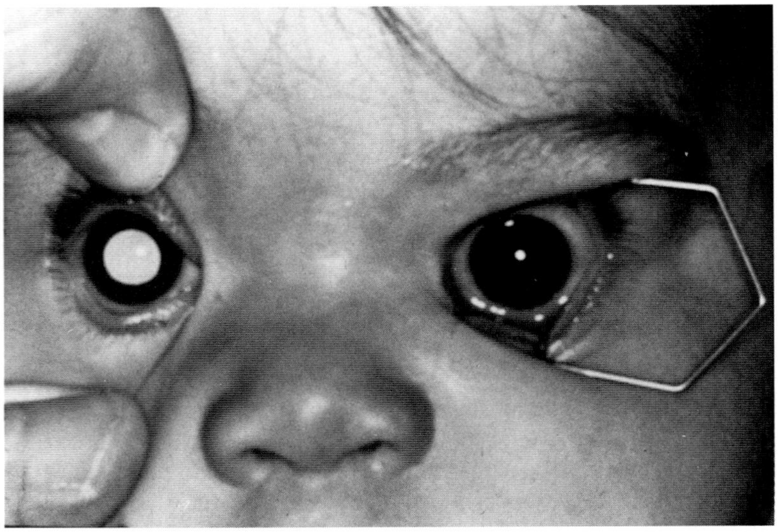

FIGURE 9.1. Leukocoria. White pupillary reflex (leukocoria) is the most common finding in the infant with retinoblastoma. Leukocoria of the right eye of this 9-month-old female was noted by her mother 2 weeks prior to diagnosis. (Courtesy of Linn Murphree, M.D., and Mr. Ben Szirth, Department of Ophthalmology, Children's Hospital, Los Angeles.)

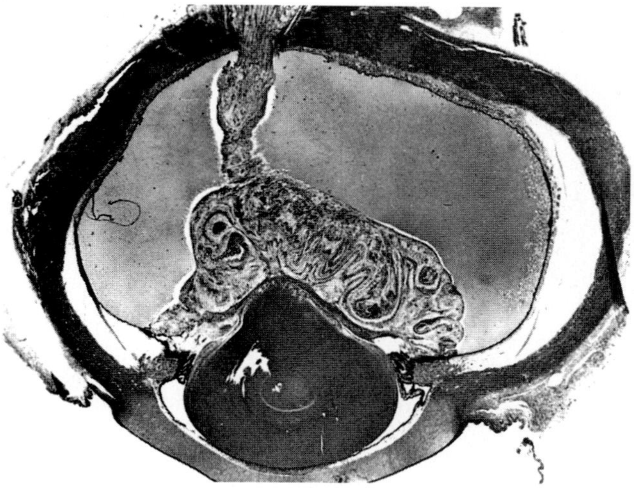

FIGURE 9.2. Retinopathy of prematurity. The retina is detached and bound into a firm fibrous mass adherent to the posterior surface of the lens together with a mass of abnormally prolferated blood vessels. Infant weighed 1100 g at birth. Eye removed because of erroneous diagnosis of retinoblastoma at age 6 months. (From Potter EL, Craig JM [60], with permission.)

Large and mostly necrotic tumors tend to be less well demarcated and sometimes have the appearance of thick white paste within the vitreous cavity. Retinal detachment is found in both the endophytic and exophytic growth patterns, particularly in advanced disease, but it may be seen earlier in the former, where the tumor grows toward the choroid from the retina. Most retinoblastomas are exophytic; that is, they grow anteriorly into the vitreous cavity from the retina (Figures 9.4C and 9.5A). The rare diffuse infiltrating form is characterized by diffuse infiltration of both the retina and choroid by tumor without a well-defined mass [19]. Choroidal invasion per se is not always an unfavorable prognostic sign, but it is frequently seen when the optic nerve is involved [13,27].

Careful examination and taking proper microscopic sections of the optic nerve are mandatory in the evaluation of a retinoblastoma specimen, since the presence or absence of tumor at the line of resection determines the outcome. When tumor invades the nerve for a distance of 10 mm or more, it will eventually enter the subarachnoid space and lead to fatal dissemination of tumor cells into the cerebrospinal fluid, with seeding of the leptomeninges at the base of the brain and spinal cord (Figure 9.5B,E,F). In approximately 40% of cases

fection, and other forms of severe uveitis have been confused with this tumor [19]. Retinal astrocytoma can mimic this malignancy, and the former would be more likely in a baby with tuberous sclerosis [20]. Medulloepithelioma can present with similar findings [19].

Spontaneous Regression

Occasionally spontaneous regression occurs in infants and children with retinoblastoma, as with neuroblastoma. The process is heralded clinically by inflammation of the involved eye and microscopically by hemorrhage, necrosis, lymphocytic infiltrates, and vascular changes, resulting in retinal gliosis of varying degree, phthisis bulbi (shrinkage, fibrosis, and disorganization of the eye), or even a normal eye [13,21]. Although neither the exact incidence nor the cause of spontaneous regression is known, it is found more often in patients with bilateral retinoblastoma [21–23].

Pathology

Light and electron microscopic findings suggest that retinoblastoma develops from the retinal neural precursor cells of photoreceptor cells [13,19,23–26] (Figure 9.3). Retinoblastoma arises from the retina of one or both eyes, most often as a single mass, but occasionally as multifocal lesions.

The involved eye on cross section shows a well-defined soft gray-white mass in the retina containing tiny white flecks of calcification [13,19,22,25] (Figure 9.4C). When there are multifocal primary lesions or foci of vitreous seeding, they appear as white nodules or plaques on the retina or choroid.

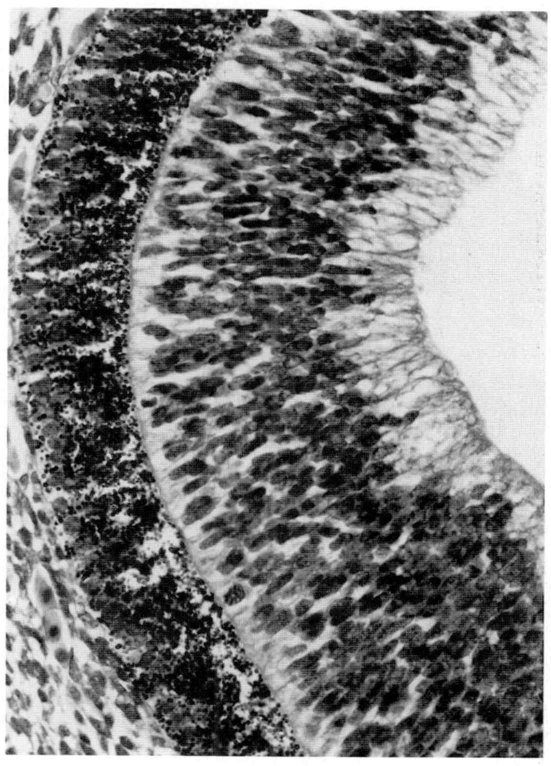

FIGURE 9.3. Developing retina. The external layer of the optic cup contains retinal pigment and the inner layer the sensory (neural) retina. At this stage of development the neural retina consists of four zones: proliferative (next to the pigment layer), external neuroblastic, transient fiber, and internal neuroblastic zone. From a fetus of 8 to 10 weeks estimated gestation. H&E; × 750.

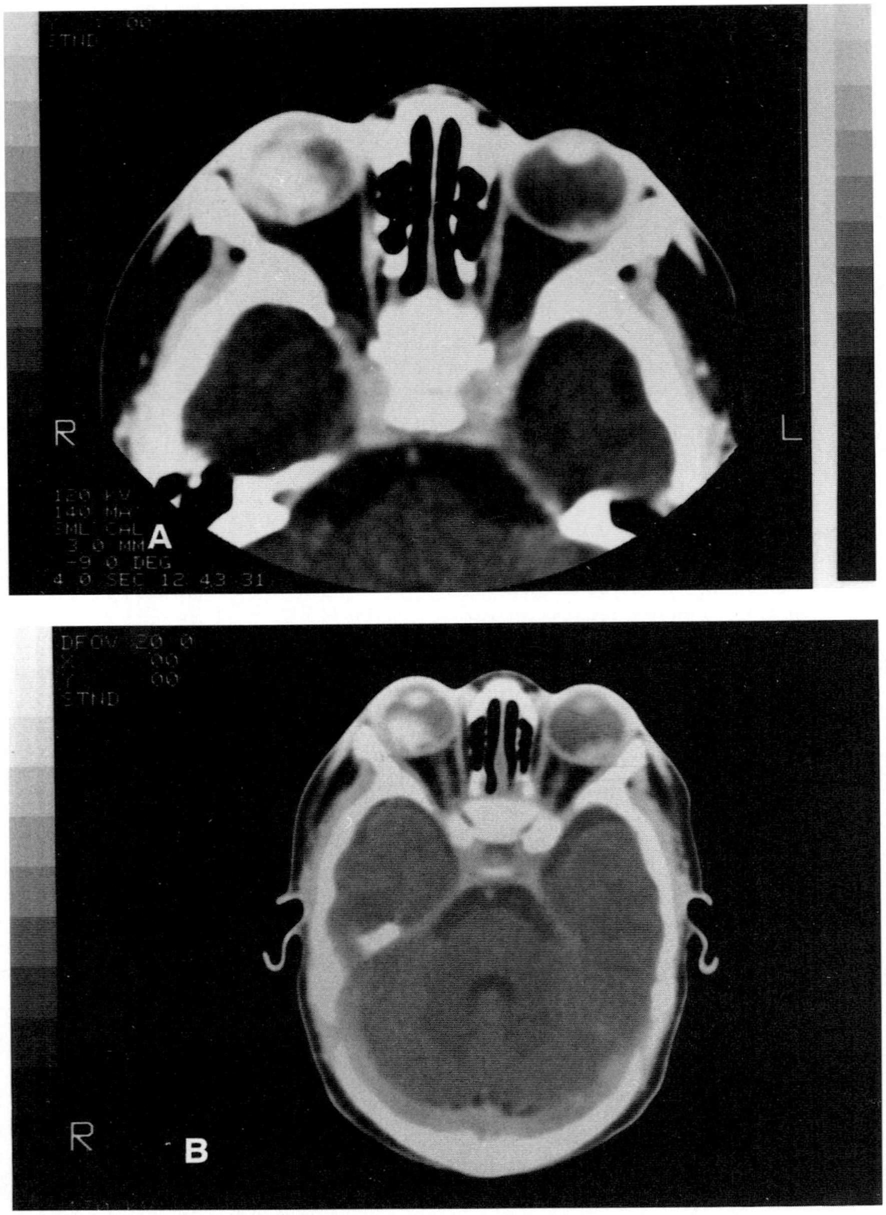

FIGURE 9.4. Congenital bilateral retinoblastoma in a 1.5-month-old female who presented with leukocoria of the right eye. (A) Computerized tomography and (B) magnetic resonance imaging scans demonstrate tumors in both eyes. (C) The globe of the right eye is hemisected, revealing a gray-white tumor mass, 11 × 13 mm^2, filling most of the vitreous cavity and containing white flecks of calcification. The stub of the optic nerve is situated on the posterior external surface of the globe on the right. Microscopically it was not involved. At the time of the enucleation the left eye was noted to have a 12 × 12 mm^2 disk diameter lesion in the posterior pole and four additional smaller ones. (D) The tumor is a differentiated retinoblastoma composed of small round cells with scant cytoplasms and round to oval nuclei with a fine peppery chromatin pattern containing one or two small nucleoli. Several Flexner-Wintersteiner rosettes are present. H&E; × 750. (E) Electron photomicrograph (× 18,900) of a Flexner-Wintersteiner rosette with a clear space in the center surrounded by four immature-appearing tumor cells. Clusters of mitochondria and scattered microtubules are present in the cytoplasm. In the left lower corner of the photograph there is a primitive cell junction. (Electron photomicrograph courtesy of Darkin Chan, Deparment of Pathology, Children's Hospital, Los Angeles.)

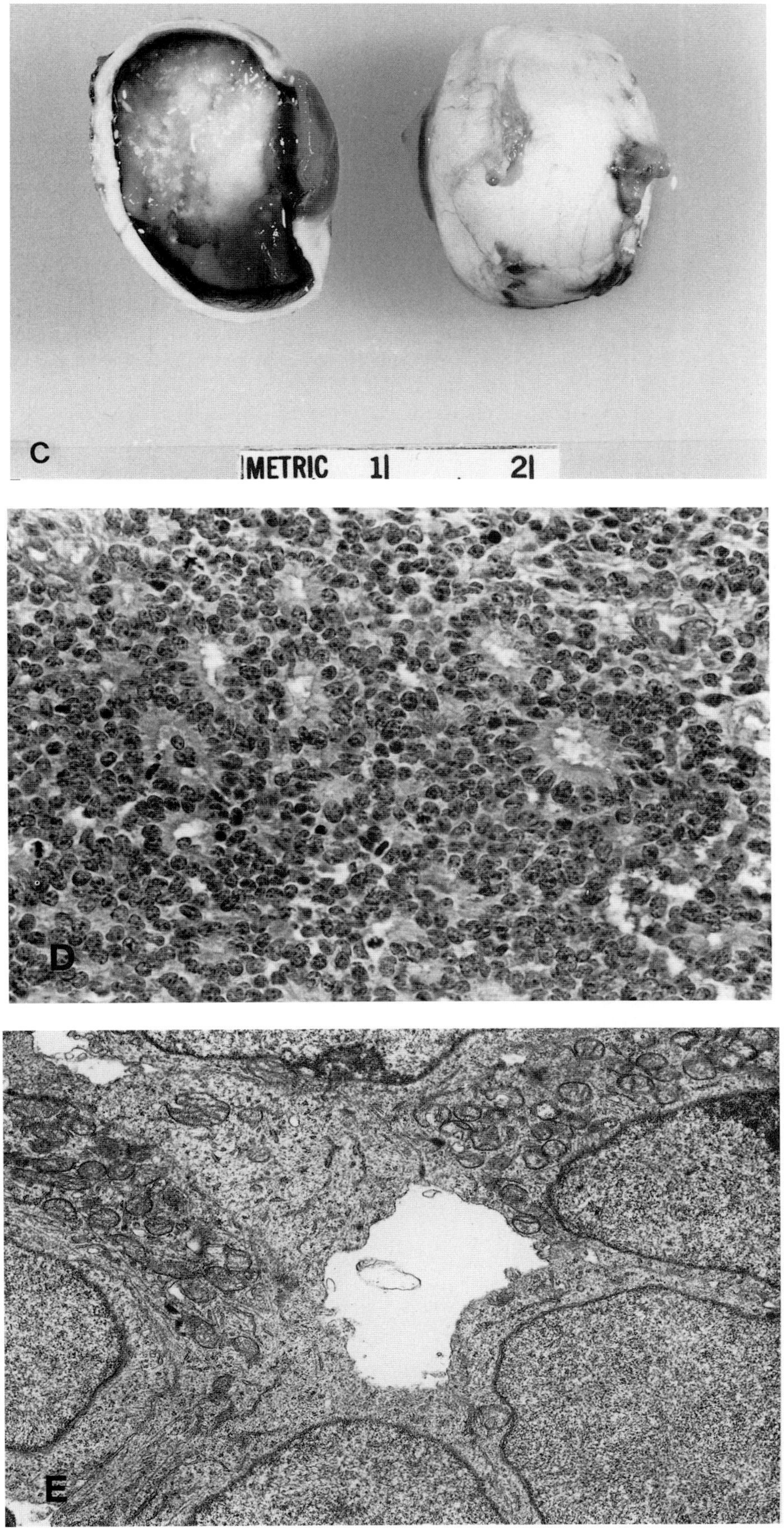

FIGURE 9.4. (*Continued*)

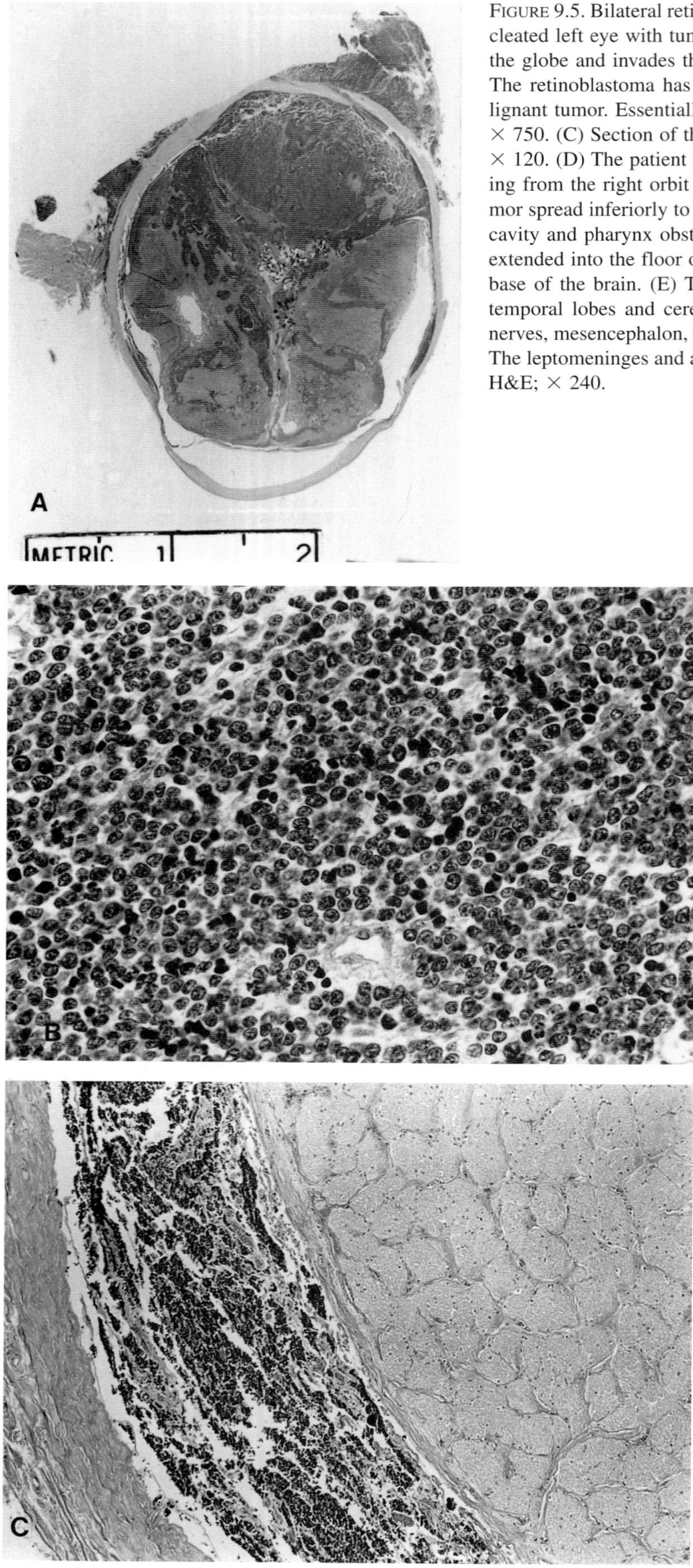

FIGURE 9.5. Bilateral retinoblastoma with metastases in a 1-year-old male. (A) Enucleated left eye with tumor filling the entire vitreous cavity. The tumor perforates the globe and invades the surrounding periorbital tissue (Ellsworth group V). (B) The retinoblastoma has the appearance of a poorly differentiated small-cell malignant tumor. Essentially no photoreceptor cell differentiation is apparent. H&E; × 750. (C) Section of the optic nerve showing subarachnoid tumor spread. H&E; × 120. (D) The patient at postmortem examination has a fungating mass protruding from the right orbit and involving the right maxilla and frontal bone. The tumor spread inferiorly to the right maxilla and medially and posteriorly to the nasal cavity and pharynx obstructing these structures. It invaded the right orbital plate, extended into the floor of the anterior and middle cranial fossae, and involved the base of the brain. (E) The leptomeninges of the inferior and medial frontal and temporal lobes and cerebellar hemispheres are infiltrated by tumor. The cranial nerves, mesencephalon, pons, and medulla are encased by metastatic deposits. (F) The leptomeninges and adjacent insular cerebral cortex are invaded by tumor cells. H&E; × 240.

FIGURE 9.5. (*Continued*)

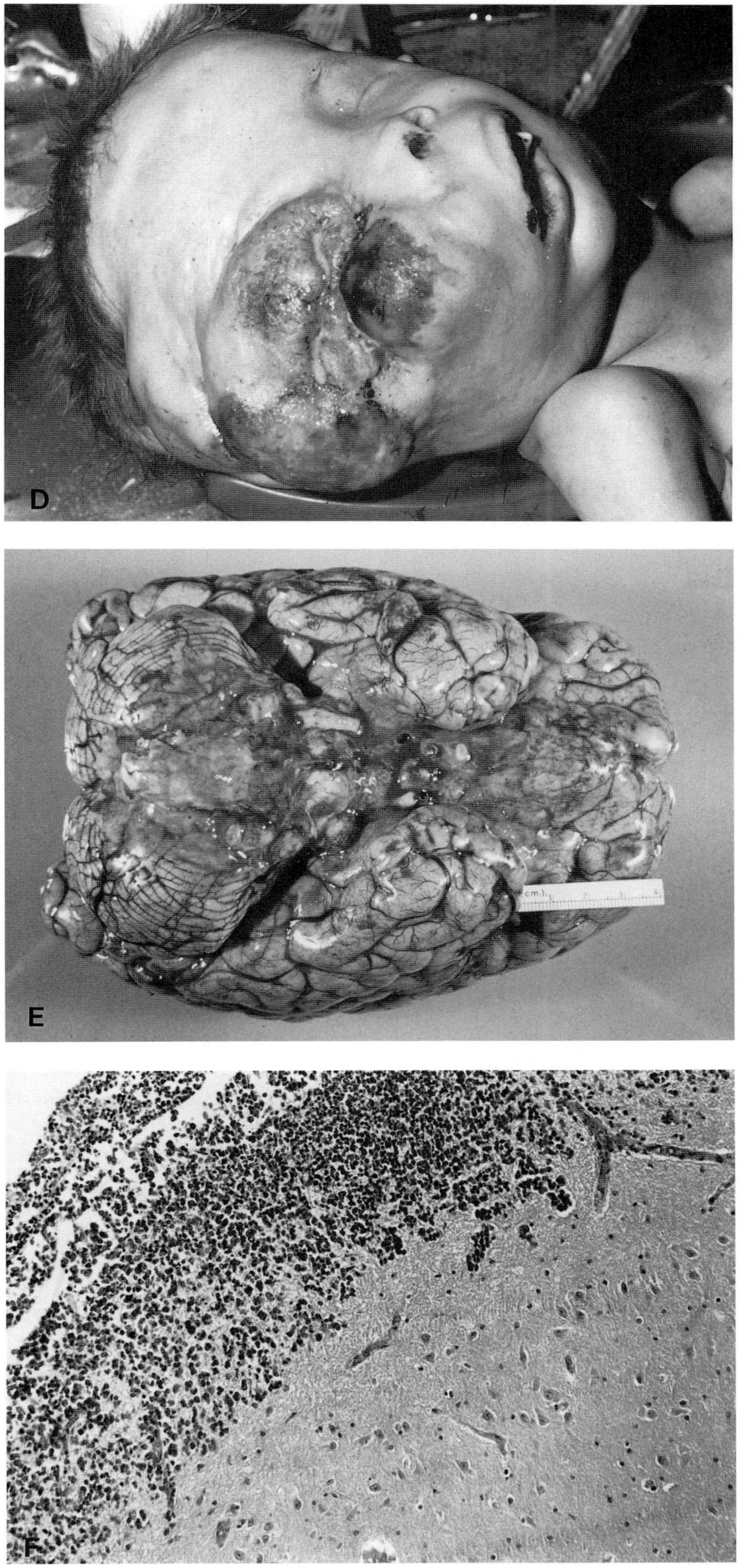

choroidal vessels are invaded by tumor cells, which then enter the circulation and metastasize primarily to the bone marrow, lymph nodes, and liver [25]. In advanced disease with extensive involvement of the eye, the tumor perforates the globe and invades the periorbital soft tissues and bone, forming a large disfiguring mass [7] (Figure 9.5D).

Retinoblastoma is one of the "small blue-cell tumors" of infancy and childhood and thus would fall into the category of "primitive neuroectodermal tumor" because it shares certain histological and clinical findings (e.g, microfilaments, neurotubules, neurosecretory granules), with such other PNETs as neuroblastoma and medulloblastoma [15,22]. The main difference from the other PNETs is that the more differentiated retinoblastomas show photoreceptor cell differentiation.

Typically the tumor is very cellular, consisting of sheets of small, round to oval, darkly staining cells with hyperchromatic basophilic nuclei and scant cytoplasm. Extensive necrosis and calcification and numerous mitoses usually are present. Although little or no recognizable stroma is noted, there is a prominent vascular network coursing through the tumor, with more viable cells forms sheaths about small vessels. Necrosis and calcification may be so extensive that a diagnosis cannot be established from histological findings alone (Figure 9.6C). Cytodifferentiation, the amount varying with each specimen, is manifested by the formation of Homer-Wright rosettes consisting of a ring of tumor cells situated about a central area containing pink fibrillary material (Figure 9.7) and the Flexner-Wintersteiner rosettes composed of a single layer of cells present about a well-defined inner circular membrane with a clear zone in the center (Figures 9.4D and 9.8A). Although several different kinds of tumor contain Homer-Wright rosettes, only the Flexner-Wintersteiner rosette is considered to be pathognomonic of retinoblastoma, since it is related to photoreceptor cells and is present in about half the specimens [15,19,23,24]. Fleurettelike structures composed of clusters of cells having a "fleur-de-lis"–like arrangement, representing still further cytodifferentiation, are observed rarely [22,23,25,26,28]. Fleurettes are present in retinocytomas, the most differentiated form of retinoblastoma [28]. By electron microscopy, the cells forming the Flexner-Wintersteiner rosettes show at their luminal borders clusters of mitochondria, zonula adherens–like cell junctions, and characteristic cilia.

Immunohistochemical studies performed on both retinoblastoma and human retina show certain features in common, further supporting the concept of the histogenesis of this tumor [14,15,29,30]. For example, neuron-specific enolase (NSE) is immunoreactive in both tumor cells and in human retina neurons. NSE determination is especially helpful in establishing the diagnosis in the bone marrow when there is a question of metastatic disease. However it should be remembered that neuroblastoma (like other PNETs) is NSE immunoreactive also and that this procedure will not differentiate between these tumors. In addition, the retinal S antigen, a component of photoreceptor cell outer segment, is present in the tumor and normal retina [13].

Three findings associated with an unfavorable outcome are optic nerve invasion beyond the lamina cribosa, choroidal involvement, and orbital extension [1,22,27,31]. Retinoblastomas metastasize in four ways: direct extension, dispersion (into the vitreous and cerebrospinal fluids), hematogenous dissemination, and lymphatic spread [13]. Generally retinoblastoma metastases appear much less differentiated than the intraocular primary tumor (Figure 9.8). Rosettes are hard to identify and poorly formed, and fleurettes are never found [19]. Postmortem studies reveal that the routes of metastases of retinoblastoma in addition to optic nerve and choroid invasion are, in decreasing order, the cerebrospinal fluid, bone marrow, orbit-to-orbit extension through the meninges, and liver [32]. Once retinoblastoma has spread beyond the eye, essentially there is no survival (Figure 9.5).

FIGURE 9.6. Retinoblastoma with extensive necrosis and calcification in a 2-month-old female who presented with a swollen left eye several days prior to hospital admission. (A) Computerized tomography scan reveals bilateral retinal tumors, the left larger than the right. (B) The hemisected left eye shows necrotic and calcified tumor filling most of the vitreous cavity. (C) The tumor is totally necrotic and calcified. No recognizable "viable" tumor cells are evident. H&E; × 150.

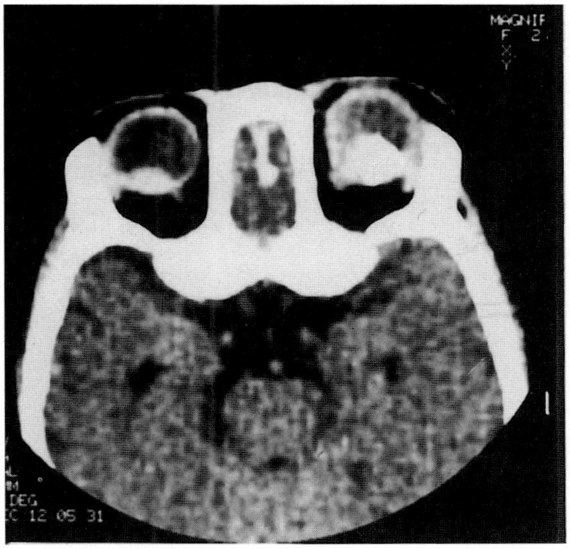

A

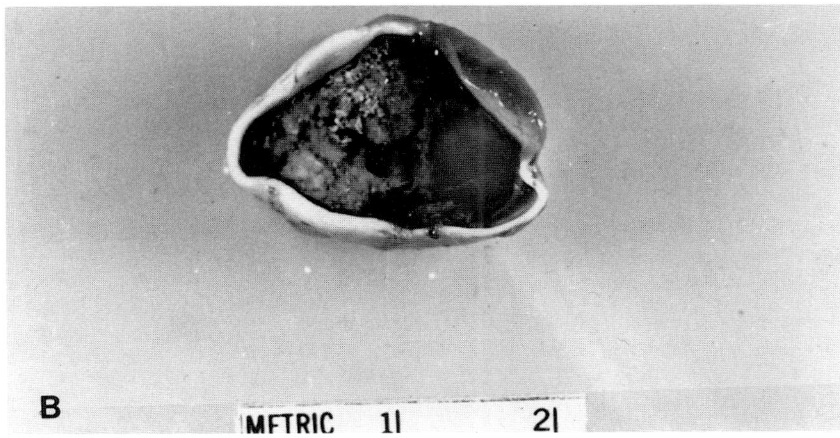

B

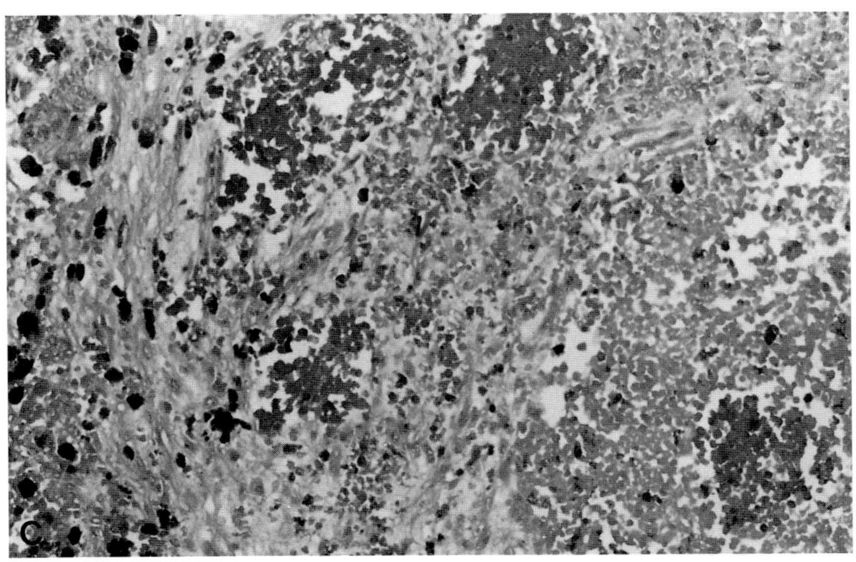

C

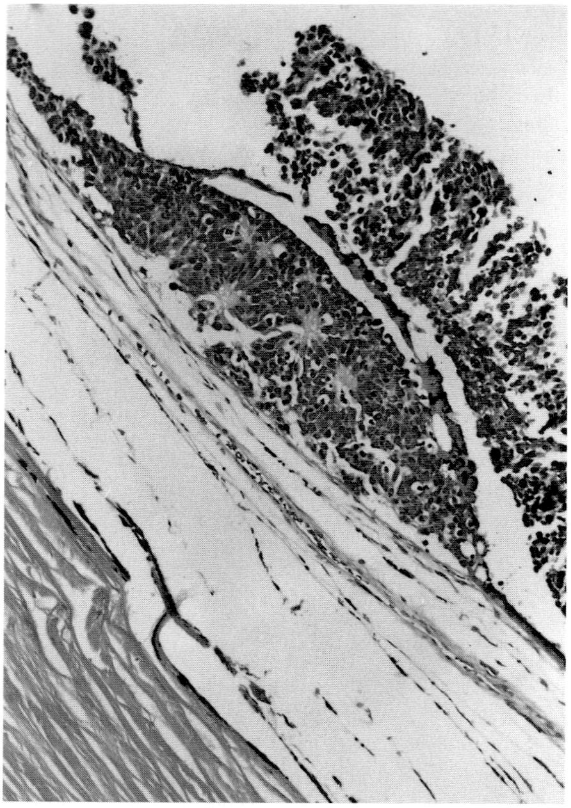

FIGURE 9.7. Retinoblastoma, choroidal seeding. Nodules of tumor cells seeding onto the surface of the choroid. Several Homer-Wright rosettes are present. H&E; × 120.

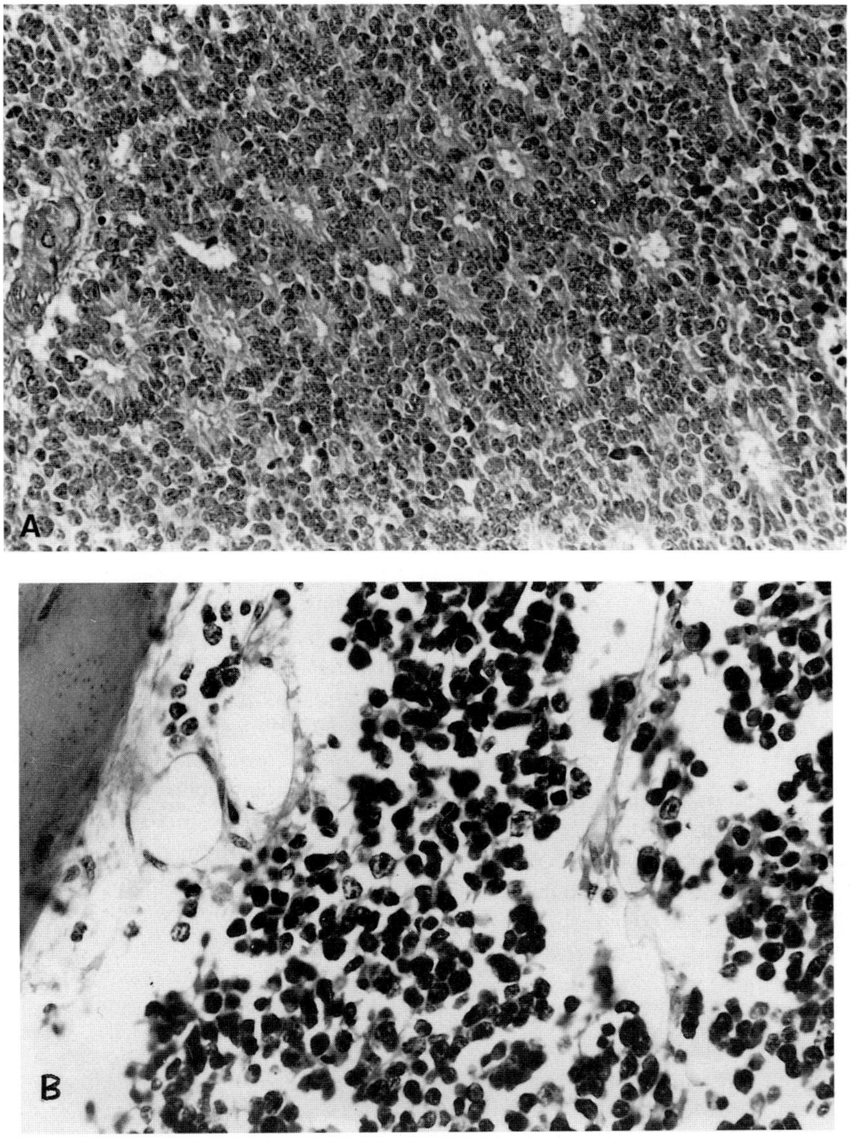

FIGURE 9.8. Retinoblastoma, metastases are generally less differentiated than the primary tumor. (A) Photomicrograph of a well-differentiated retinoblastoma from an infant with bilateral tumors. Well-defined Flexner-Wintersteiner rosettes are present. H&E; × 300. (B) Postmortem bone marrow biopsy sample taken several months later, after surgery and chemotherapy. The tumor consists of small, darkly staining cells without evidence of photoreceptor differentiation. H&E; × 300.

Prognosis

The extent of the tumor, or stage of the disease at the time of diagnosis, is the most important factor with regard to the salvage of the eye(s) and prognosis [1,13,22,27,32]. Usually the prognosis is poor for patients with massive tumors involving over half the globe and in those with vitreous seeding (Ellsworth group 5) [18] (Table 9.2). A family history of retinoblastoma leads to earler recognition of the disease but does not increase the probability of salvaging the eye(s); however prognosis and advanced stage are the same for infants with a family history and those without [2]. It was mentioned that the two most important findings determining survival are optic nerve involvement and extraorbital spread [1,13,27, 31,32]. A definite relationship between cytodifferentiation of the tumor and prognosis has not been confirmed [13,19].

When distant metastases or local extension occurs outside the eye, the mortality rate approaches 100% regardless of the form of therapy [1,2,13]. The length of survival after diagnosis of metastatic retinoblastoma ranges from 6 months to 1 year [13,33]. Second malignancies, particularly with treated bilateral retinoblastomas, are not uncommon; patients are at risk for developing osteosarcoma and other sarcomas [13,19,34].

Trilateral Retinoblastoma

Bilateral retinoblastoma associated with retinoblastoma in the pineal has been termed "trilateral retinoblastoma" [35,36]. Pineoblastomas unassociated with retinoblastomas occasionally show photoreceptor differentiation [37]. Extraocular extension or local invasion into the central nervous system from the ocular retinoblastoma or from distant metastases has been proven in the trilateral syndrome [35,37,38]. Bilateral and pineal retinoblastomas may be identical in histological appearance with or without photoreceptor differentiation [37]. Some trilateral lesions have the histological appearance of an undifferentiated small blue-cell tumor.

This ultimately lethal condition is uncommon. The incidence of trilateral retinoblastoma in patients with retinoblastoma is reported at 3% [36,38]. Most patients with trilateral retinoblastoma had bilateral tumors and a positive family history or both [39]. In trilateral cases a positive family history is obtained in over two thirds, and the age of presentation of retinoblastoma, 7 months, is earlier than that for classic bilateral retinoblastoma, 15 months [39]. Trilateral retinoblastoma has been reported in infants as young as age 4 months who present with a calcified suprasellar mass and increased intracranial pressure due to obstructive hydrocephalus [39]. Mean survival from diagnosis of trilateral disease to death is about 7 months, and all patients die with cerebrospinal metastases [38,39].

Medulloepithelioma

Medulloepithelioma, an extremely rare neoplasm thought to arise from the medullary epithelium lining the optic cup, that begins in early embryonic life as an outpouching of the neural tube. It is a tumor of neuroepithelium occurring most often in the ciliary body and/or iris and infrequently in the optic nerve or retina [19,40–42]. The main initial clinical findings are a cyst or mass in the iris, anterior chamber, or ciliary body, and glaucoma. Medulloepithelioma has been reported in infants [42,43].

Histologically the tumor is composed of sheets and cords of small, blue, neuroblastic-like cells ("medullary epithelium") forming tubular structures and cyst formations [40]. The netlike appearance formed by anastomosing cords of medullary epithelium is termed the diktyomatous pattern (from the Greek word dikty or "net"). When heterotopic tissues (e.g., skeletal muscle, cartilage, neuroglial tissue) are present, the tumor is classified as a teratoid medulloepithelioma [19,40].

The neuroepithelium may differentiate toward retinal pigmented epithelium, nonpigmented and pigmented ciliary epithelium, neurons, and neuroglia [19]. The tumor is defined as malignant when it is composed mostly of undifferentiated cells resembling retinoblastoma or if the heteroplastic tissue is anaplastic with a high mitotic rate [19]. About 20% of eyes with medulloepithelioma also show histological evidence of hyperplastic primary vitreous [40].

Medulloepithelioma is the main primary ocular tumor that histologically resembles retinoblastoma and thus may create a diagnostic problem. Undifferentiated foci in some malignant medulloepitheliomas contain cells that are indistinguishable from those of retinoblastoma [19].

Medulloepitheliomas are usually slow growing and associated with a favorable outcome, provided they are localized to the eye and completely excised [41,42]. However the tumors are potentially fatal when they perforate the globe and extend into the orbit. Death results from intracranial metastases. Therefore the most important prognostic factor is the presence or absence of orbital invasion [19,40].

Astrocytic Tumors of the Retina

Tuberous Sclerosis

Astrocytic tumors or hamartomas of the retina or optic nerve occur in association with tuberous sclerosis and are found in almost half the patients with this disease [20,44]. Moreover, they are observed in children with neurofibromatosis and in those who do not have either one of these phakomatoses. Early the lesions present as small, flat transparent retinal thickenings and later on in childhood and adulthood as raised nodular and calcified whitish masses within the retina or optic disk [19,45]. The astrocytic lesions arise from the nerve fiber and ganglion cell layers, grow slowly in childhood, are confined to the eye, and generally do not invade adjacent structures.

Astrocytic tumors may be congenital and bilateral, and affected infants may present with seizures and retinal hamartoma(s) [44]. Obviously funduscopic examination should be performed on any baby with epileptic seizures or other findings suggestive of tuberous sclerosis [44,45].

Large retinal astrocytic hamartomas in patients with tuberous sclerosis may resemble retinoblastoma through the funduscope, especially if the mass is large, calcified, and accompanied by vitreous seeding. Although the ophthalmic presentation is suggestive of retinoblastoma, radiological evidence of subependymal hamartomas and/or cardiac rhabdomyomas pathognomonic of tuberous sclerosis help to establish the correct diagnosis [46]. Characteristic skin lesions also may be present at birth.

Histological findings consist of elongated, spindle-shaped (pilocytic) astrocytes with small, oval nuclei surrounded by a fine meshwork of glial fibers [19]. Foci of calcification may be present. Giant astrocytes are found in the tumors (hamartomas) of patients with tuberous sclerosis; they occur both the retina and brain [20].

Neurofibromatosis

Retinal tumors or hamartomas of astrocytic origin occur in association with neurofibromatosis type 1 in 14% of the patients afflicted with the disease [20]. Less than half the infants with neurofibromatosis have physical signs at birth, consisting of multiple café-au-lait spots, plexiform neurofibromas, and hamartomas of the ocular structures [47]. Neurofibromatosis may involve the optic nerve and the entire uveal tract (the iris, clilary body, and choroid) [19] (Figure 9.9). When involved, the choroid shows diffuse thickening due to the presence of spindle-shaped Schwann cells, ganglion cells, nerve fibers, and ovoid bodies composed of Schwann cells and melanocytes [48] (Figure 9.9D). Although Lisch nodules, which are focal collections of melanocytic cells situated on the anterior border of the iris, are the most common intraocular abnormality seen in adults, they are seldom found in children less than 5 years of age [47]. Plexiform neurofibroma of the eyelid correlates with a high incidence of congenital glaucoma (patient depicted in Figure 9.9).

Astrocytoma is the foremost intracranial tumor associated with neurofibromatosis type I, and optic glioma is the most common histological form of astrocytoma [19,47]. Optic gliomas arise from the optic nerve(s) and chiasm, consist of pilocytic astrocytes, and vary histologically from benign to malignant. Magnetic resonance imaging scans through the orbits can detect the presence of optic nerve gliomas, which appear as thickenings of the entire nerve, usually a solid fusiform enlargement of the nerve. Tumors of the chiasm appear as a mass in the suprasellar region [47] (Figure 8.2) (see Chapter 8, Brain Tumors, for further discussion of astrocytoma).

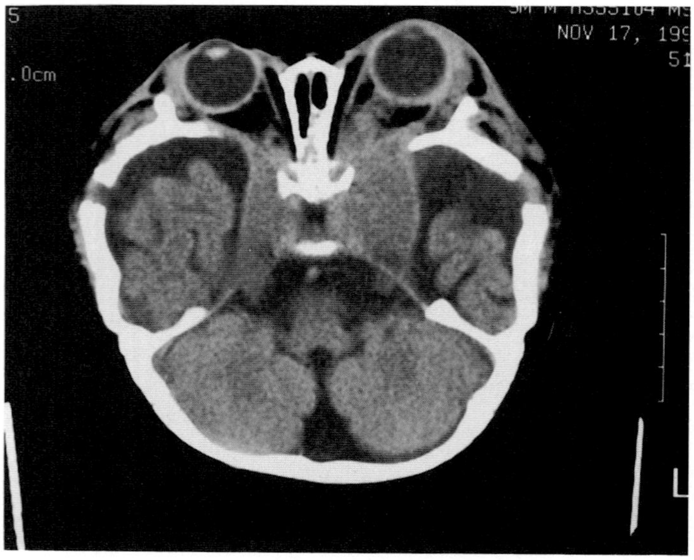

A

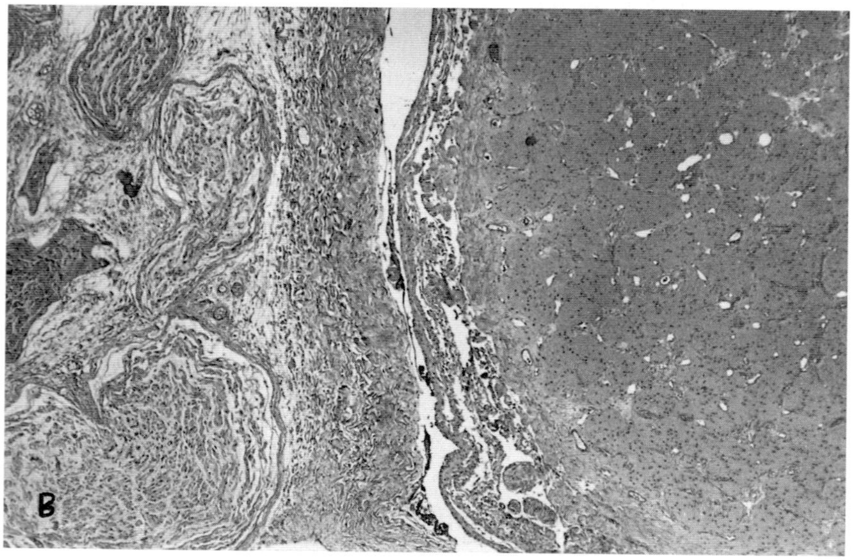

FIGURE 9.9. Neurofibromatosis of the eye, optic nerve, and orbit. (A) Magnetic resonance image at age 5 months reveals tumor within the right orbit, cavernous, and paranasal sinuses. The left eye is exophthalmic. The eye was enucleated at age 21 months because of the severe, disfiguring proptosis and glaucoma. (B) The meninges are thickened. There is a plexiform neurofibroma adjacent to the optic nerve. H&E; × 36. (C) Microscopic section of the ciliary ganglion reveals diffuse infiltration by neurofibroma composed of wavy elongated cells with regular oval or fusiform nuclei within a fibrillar matrix. H&E; × 300. (D) Section of sclera and retina. The sclera is diffusely thickened by collections of melanocytes, spindle-shaped Schwann cells, and fibrocytes. The sclera contains a small nerve with a plexiform neurofibroma (upper right-hand corner). H&E; × 120.

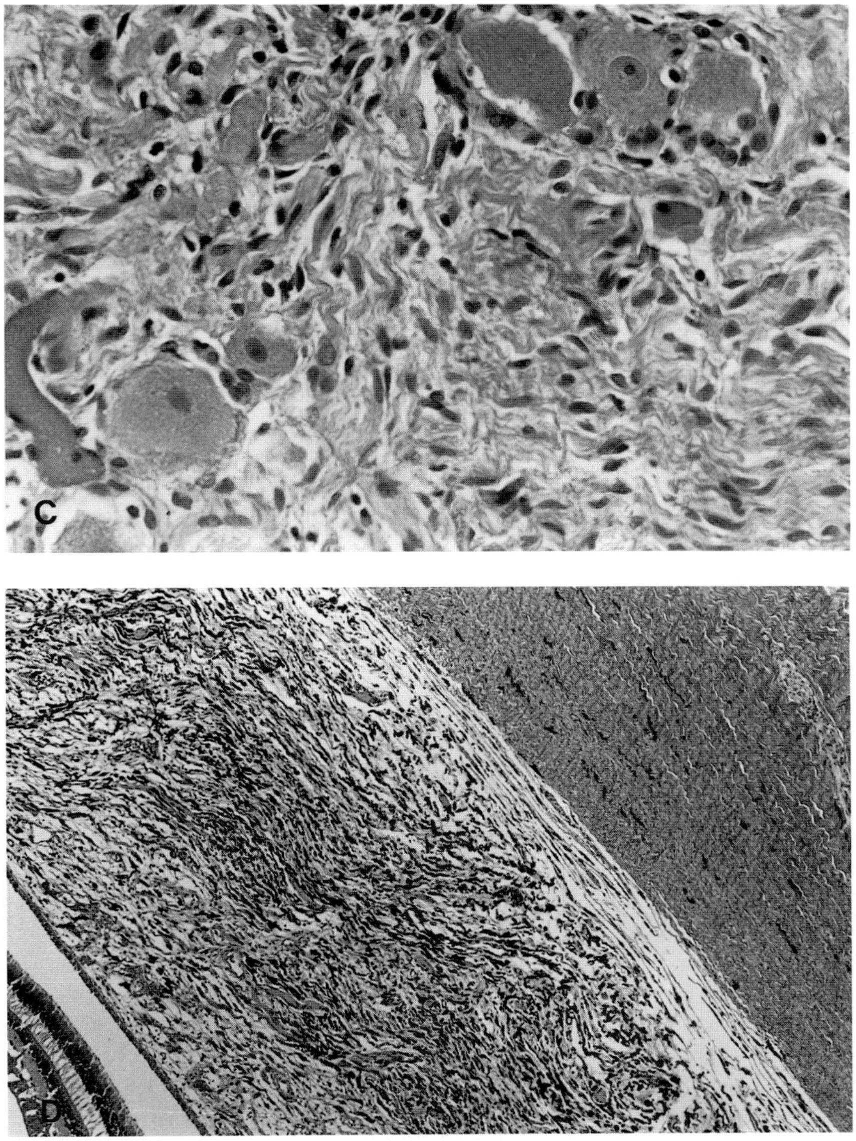

FIGURE 9.9. (*Continued*)

Orbital Tumors

Typically tumors of the orbit present in the fetus and infant as abnormal protrusions of the eyeball (exophthalmos). Other causes are retrobulbar hemorrhage and congenital anomalies of the eye (e.g., cystic microphthalmia, Figure 9.10, and congenital cystic eye). Encephalocele causes proptosis at birth in addition to trauma with retrobulbar hemorrhage [49].

Although orbital tumors of the fetus and infant are uncommon, a variety of neoplasms arise from this site. The major ones are teratoma, hemangioma, neurofibroma, rhabdomyosarcoma, and metastatic neuroblastoma. Orbital teratoma is discussed in Chapter 2, Germ Cell Tumors. Much rarer examples of orbital tumors are rhabdoid tumor [50,51], polyphenotypic small-cell tumor [52], myofibromatosis [49,53], fibrosarcoma [54], malignant peripheral nerve sheath tumor [55], lipoblastoma [56], and melanotic neuroectodermal tumor [57].

Rhabdomyosarcoma of the eyelid and orbit is the subject of a few case reports [19,58,59] (Figure 9.11). This highly malignant tumor accounts for most soft tissue sarcomas in infants, and it is second in frequency to fibrosarcoma in the newborn [7,19]. Most rhabdomyosarcomas diagnosed during the first year of life are of the embryonal histological type, which has a much better prognosis than the alveolar form [59] (see Chapter 3, Soft Tissue Tumors, for further discussion).

FIGURE 9.10. Cystic microphthalmia in a 2-week-old female with an orbital mass discovered on routine prenatal ultrasound examination. (A) Patient with a large left orbital mass, which clinically was thought to be a teratoma. (B) Magnetic resonance imaging reveals a cystic and solid mass occupying the left orbit. There were no recognizable ocular structures either by imaging studies or by gross inspection of the specimen. However, an empty cystic cavity, 2 cm in diameter, was present. (C) The wall of the cyst consists of neuroglial tissue and dysplastic retinal elements with pseudorosette formations. Portions of sclera and choroid and ciliary body were identified microscopically within the mass. Along the left border of the photograph there is a portion of immature retina bordering the cyst. H&E; × 300. (Courtesy of Lynne Bird, M.D., Department of Dysmorphology and Genetics, Children's Hospital, San Diego.)

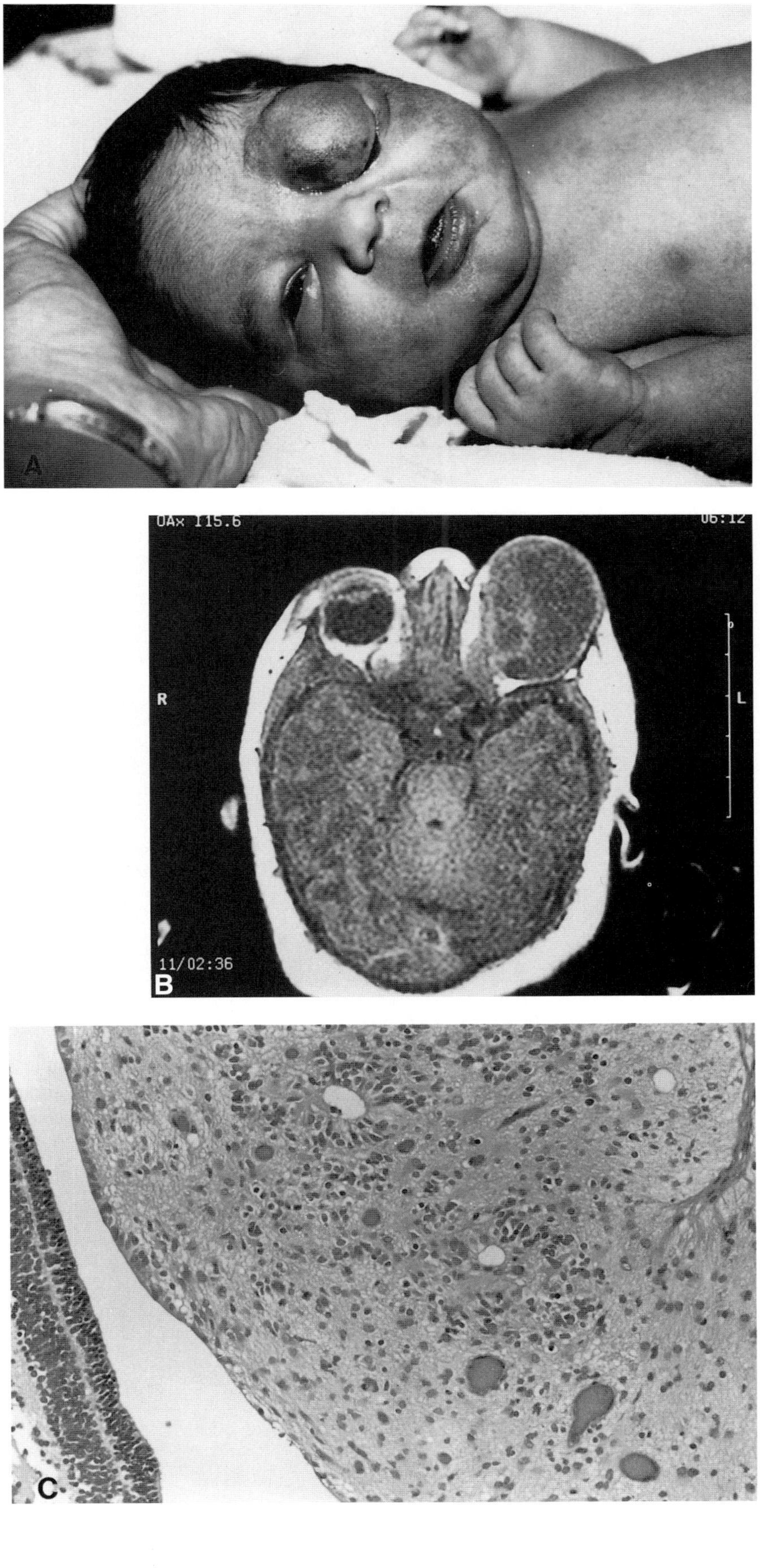

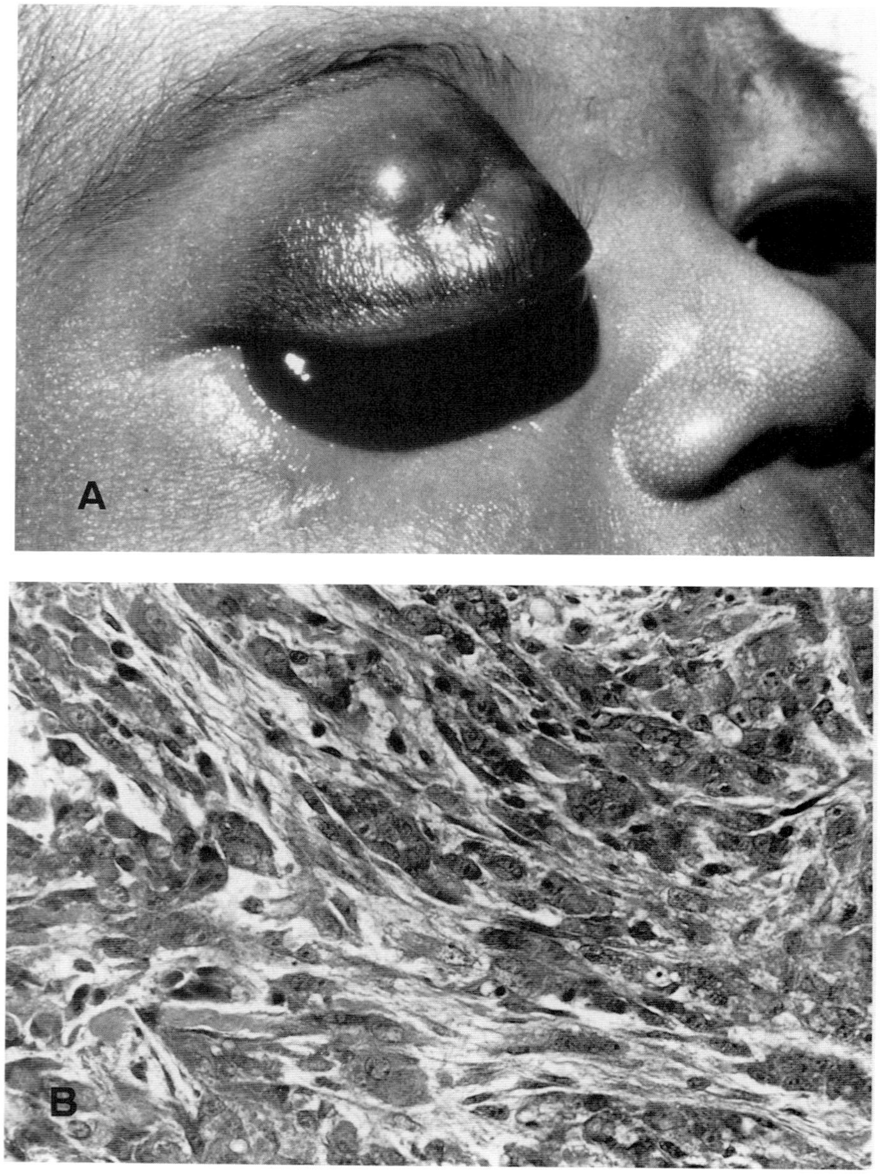

FIGURE 9.11. Rhabdomyosarcoma of the eyelid. This 6-day-old male presented with a nodular mass involving the right upper eyelid which was excised. (A) The patient shortly after birth with the eyelid tumor and underlying hemorrhagic, chemotic conjunctiva. (B) Photomicrograph of a right preauricular lymph node metastasis removed 9 days after the initial surgery. Large spindle-shaped cells consistent with rhabdomyoblasts are present. H&E; × 600.

References

1. Abramson DH, Notterman RB, Ellsworth RM, et al. Retinoblastoma treated in infants in the first six months of life. *Arch Ophthalmol* 1983;101:1362.
2. Abramson DH, Servodidio CA. Retinoblastoma in the first year of life. *Ophthalmic Paediatr Genet* 1992;13:191.
3. Bader JL, Miller RW. US cancer incidence and mortality in the first year of life. *Am J Dis Child* 1979;133:157.
4. Isaacs H Jr. Neoplasms in infants: a report of 265 cases. *Pathol Annu* 1983;18(2):165.
5. Isaacs H Jr. Congenital and neonatal malignant tumors: a 28-year experience at Children's Hospital of Los Angeles. *Am J Pediatr Hematol Oncol* 1987;9:121.
6. Isaacs H Jr. Retinoblastoma. In: *Tumors of the Newborn and Infant*: St. Louis, MO: Mosby-Year Book; 1991:109.
7. Isaacs H Jr. Tumors of the eye. In: *Tumors of the Fetus and Newborn.* Philadelphia: WB Saunders; 1997: 229. Major Problems in Pathology Series. Vol 35.
8. Plotsky D, Quinn G, Eagle R Jr, et al. Congenital retinoblastoma: a case report. *J Pediatr Ophthalmol Strabismus* 1987;24:120.
9. Campbell AN, Chan HSL, O'Brien A, et al. Malignant tumours in the neonate. *Arch Dis Child* 1987;62:19.
10. Parkes SE, Muir KR, Southern L, et al. Neonatal tumours: a thirty-year population-based study. *Med Pediatr Oncol* 1994;22:309.
11. Maat-Kievit JA, Oepkes D, Hartwig NG, et. al. A large retinoblastoma detected in a fetus at 21 weeks gestation. *Prenat Diagn* 1993;13:377.
12. Isaacs H Jr. Perinatal (congenital and neonatal) neoplasms: a report of 110 cases. *Pediatr Pathol* 1985;3:165.
13. Murphree AL, Munier FL. Retinoblastoma. In: Ryan S, ed. *Retina*. Vol I. 2nd ed. St Louis, MO: CV Mosby; 1994:571.
14. Friend SH, Dryja TP, Weinberg RA. Oncogenes and tumor suppressing genes. *N Engl J Med* 1988;318:618.
15. Gallie BL, Hinton D. Retinoblastoma: a prototype childhood malignancy. In: Finegold M, ed. *Pathology of Neoplasia in Children and Adolescents.* Philadelphia: WB Saunders, 1986:419. Major Problems in Pathology Series; Vol 18.
16. Shields JA, Shields CL, Suvarnamani C, et al. Retinoblastoma manifesting as orbital cellulitis. *Am J Ophthalmol* 1991;112:442.
17. Gilbert F, Potluri VR, et al. Retinoblastoma, chromosome abnormalities and oncogene expression. *Ophthalmol Paediatr Genet* 1987;8:3.
18. Ellsworth RM. The practical management of retinoblastoma. *Trans Am Ophthalmol Soc* 1969;67:462.
19. McLean IW, Burnier MN, Zimmerman LE, Jakobiec FA. Tumors of the eye and ocular adenexa. In: *Atlas of Tumor Pathology*. 3rd ser; Fascicle 12. Washington, DC: Armed Forces Institute of Pathology; 1994:97.
20. Ulbright TM, Fulling KH, Helveston EM. Astrocytic tumors of the retina: differentiation of sporadic tumors from phakomatosis-associated tumors. *Arch Pathol Lab Med* 1984;108:160.
21. Gangwar DN, Jain MSIS, Gupta A, et al. Bilateral spontaneous regression of retinoblastoma with dominant transmission. *Ann Opthalmol* 1982;14:479.
22. Sang DN, Albert DM. Retinoblastoma: clinical and histopathologic features. *Hum Pathol* 1982;13:133.
23. Shuangshoti S, Chaiwun B, Kasantikul V. A study of 39 retinoblastomas with particular reference to morphology, cellular differentiation and tumor origin. *Histopathology* 1989;15:113.
24. Popoff NA, Ellsworth RM. The fine structure of retinoblastoma. In vivo and in vitro observations. *Lab Invest* 1971;25:389.
25. Rootman J, Carruthers JDA, Miller RR. Retinoblastoma. *Perspect Pediatr Pathol* 1987;10:208.
26. Ts'o M, Zimmerman LE, Fine BS. The nature of retinoblastoma, I: Photoreceptor differentiation: a clinical and histopathologic study; II: Photoreceptor differentiation: an electronmicroscopic study. *Am J Ophthalmol* 1970;69:339.
27. Redler LD, Ellsworth RM. Prognostic importance of choroidal invasion in retinoblastoma. *Arch Ophthalmol* 1973; 90:294.
28. Margo C, Hidayat A, Kopelman J, et al. Retinocytoma: a benign variant of retinoblastoma. *Arch Ophthalmol* 1983;101:1519.
29. Donoso LA, Shields CL, Lee EY. Immunohistochemistry of retinoblastoma. a review. *Ophthalmic Pediatr Genet* 1989;10:3.
30. Terenghi G, Polak JM, Ballesta J, et al. Immunocytochemistry of neuronal and glial markers in retinoblastoma. *Virchows Arch [Pathol Anat]* 1984;404:61.
31. Magramm I, Abramson DH, Ellsworth RM. Optic nerve involvement in retinoblastoma. *Ophthalmology* 1989;96:217.
32. Shields JA, Shields CL. Current management of retinoblastoma. *Mayo Clin Proc* 1994;69:50.
33. MacKay CJ, Abramson DH, Ellsworth RM. Metastatic patterns of retinoblastoma. *Arch Ophthalmol* 1984;102:391.
34. Dickman PS, Barmada M, Gollin SM, et al. Malignancy after retinoblastoma: secondary cancer or recurrence. *Hum Pathol* 1997;28:200.
35. Bader JL, Meadows AT, Zimmerman LE, et al. Bilateral retinoblastoma with ectopic intracranial retinoblastoma: trilateral retinoblastoma. *Cancer Genet Cytogenet* 1982;5:203.
36. Amoaku WMK, Willshaw HE, Parkes SE, et al. Trilateral retinoblastoma: a report of five patients. *Cancer* 1996;78:858.
37. Johnson DL, Chandra R, Fisher WS, et al. Trilateral retinoblastoma: ocular and pineal retinoblastoma. *J Neurosurg* 1985;63:367.
38. Pesin SR, Shields JA. Seven cases of trilateral retinoblastoma. *Am J Ophthalmol* 1989;107:121.
39. Lueder GT, Judisch F, Wen B-C. Heritable retinoblastoma and pinealoma. *Arch Ophthalmol* 1991;109:1707.
40. Broughton WL, Zimmerman LE. A clinicopathologic study of 56 cases of intraocular medulloepitheliomas. *Am J Ophthalmol* 1978;85:407.
41. Canning CR, McCartney AC, Hungerford J. Medulloepithelioma (diktyoma). *Br J Ophthalmol* 1988;72:764.
42. Shields JA, Eagle RC Jr, Shields CL, et al. Congenital neoplasms of the nonpigmented ciliary epithelium (medulloepithelioma). *Ophthalmology* 1996;103:1998.
43. Steinkuller PG, Font RL. Congenital malignant teratoid neoplasm of the eye and orbit: a case report and review of the literature. *Ophthalmology* 1997;104:38.
44. Shami MJ, Benedict WL, Myers M. Early manifestations of retinal hamartomas in tuberous sclerosis. *Am J Ophthalmol* 1993;115:539.
45. Williams R, Taylor D. Tuberous sclerosis. *Surv Ophthalmol* 1985;30:143.
46. Mullaney PB, Jacquemin C, Abboud E, et al. Tuberous sclerosis in infancy. *J Pediatr Ophthalmol Strabismus* 1997;34:372.
47. Riccardi VM. Type 1 neurofibromatosis and the pediatric patient. *Curr Probl Pediatr* 1992;22:66.

48. Kurosawa A, Kurosawa H. Ovoid bodies in choroidal neurofibromatosis. *Arch Ophthalmol* 1982;100:1939.
49. Waeltermann JM, Huntrakoon M, Beatty EC Jr, et al. Congenital fibromatosis (myofibromatosis) of the orbit: a rare cause of proptosis at birth. *Ann Ophthalmol* 1988;20:394.
50. Rootman J, Damji KF, Dimmick JE. Malignant rhabdoid tumor of the orbit. *Ophthalmology* 1989;96:1650.
51. Stidham DB, Burgett RA, Davis MM, et al. Congenital malignant rhabdoid tumor of the orbit. *J Am Assoc Pediatr Ophthalmol Strabismus* 1999;3:318.
52. Lyon DB, Dortzbach RK, Gilbert-Barness E. Polyphenotypic small-cell orbitocranial tumor. *Arch Ophthalmol* 1993;111:1402.
53. Shields CL, Husson M, Shields JA, et al. Solitary intraosseous infantile myofibroma of the orbital roof. *Arch Ophthalmol* 1998; 116:1528.
54. Weiner JM, Hidayat AA. Juvenile fibrosarcoma of the orbit and eyelid. A study of five cases. *Arch Ophthalmol* 1983;101: 253.
55. Fezza JP, Wolfley DE, Flynn SD. Malignant peripheral nerve sheath tumor of the orbit in a newborn: a case report and review. *Pediatr Ophthalmol Strabismus* 1997;34:128.
56. Adams RJ, Drwiega PJ, Rivera CA. Congenital orbital lipoblastoma: a pathologic and radiologic study. *Pediatr Ophthalmol Strabismus* 1997;34:194.
57. Lamping KA, Albert DM, Lack E, et al. Melanotic neuroectodermal tumor of infancy (retinal anlage tumor). *Ophthalmology* 1985;92:143.
58. Gormley PD, Thompson J, Aylward GW, et al. Congenital undifferentiated sarcoma of the orbit. *J Pediatr Ophthalmol Strabismus* 1994;31:59.
59. Kodet R, Newton WA Jr, Hamoudi AB, et al. Orbital rhabdomyosarcomas and related tumors in childhood: relationship of morphology to prognosis—an Intergroup Rhabdomyosarcoma Study. *Med Pediatr Oncol* 1997;29:51.
60. Potter EL, Craig JM. *Pathology of the Fetus & Infant.* 3rd edition. Chicago: Year-Book Medical Publishers; 1976:590.

10
Renal Tumors

Pediatric renal tumors consist of separate distinct entities with different biological behaviors and responses to treatment [1–4]. The National Wilms Tumor Study (NWTS) has demonstrated an important relationship between the histology of the various renal tumors and their prognosis [1,2].

Wilms tumor, congenital mesoblastic nephroma, and rhabdoid tumor, in that order, are the principal neoplasms of the kidney occurring during the first year of life [5–11]. In the fetus and newborn mesoblastic nephroma is the most common one, followed by Wilms tumor and rhabdoid tumor [5,6,11–15]. The typical clinical presentation of a renal tumor is an abdominal mass [8,9]. However, nonneoplastic diseases of the kidney are far more prevalent, with hydronephrosis and renal cystic disease accounting for roughly 40% of all abdominal masses found in newborns [13,16] (Figures 10.1–10.4) (Table 1.2).

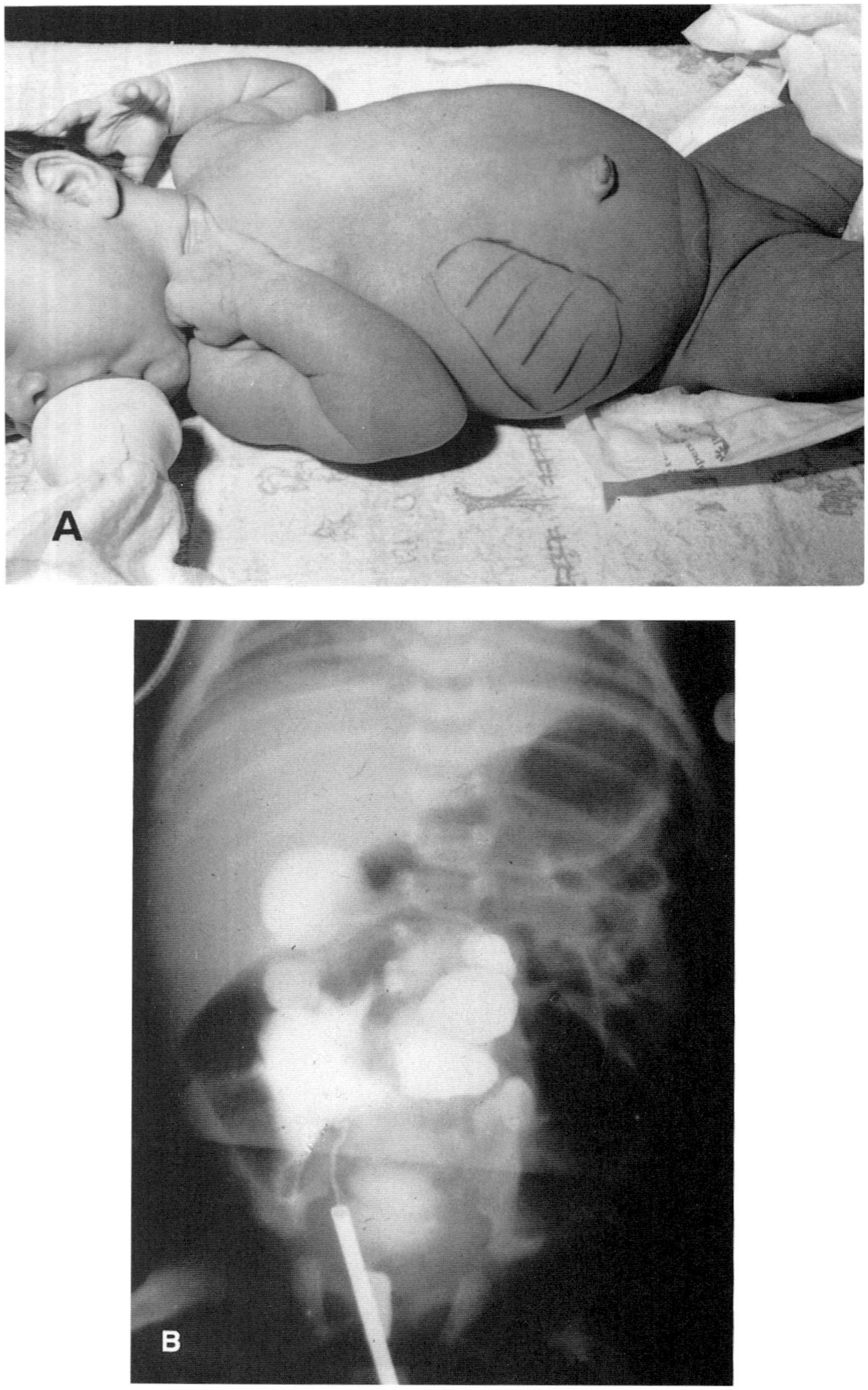

FIGURE 10.1. Hydronephrosis presenting as an abdominal mass. This malformation is one of the most frequent ones detected on routine antenatal abdominal sonography. (A) A 1-week-old female with an abdominal mass. (B) Retrograde pyelogram revealing the dilated renal pelvis outlined by injected contrast medium. (C) The dilated pelvis and stub of ureter are situated on the left side of the kidney. (D) The opened specimen reveals the dilated pelvis and caliyces. A ureteropelvic junction obstruction (stenosis) was found and was believed to be the cause of the hydronephrosis.

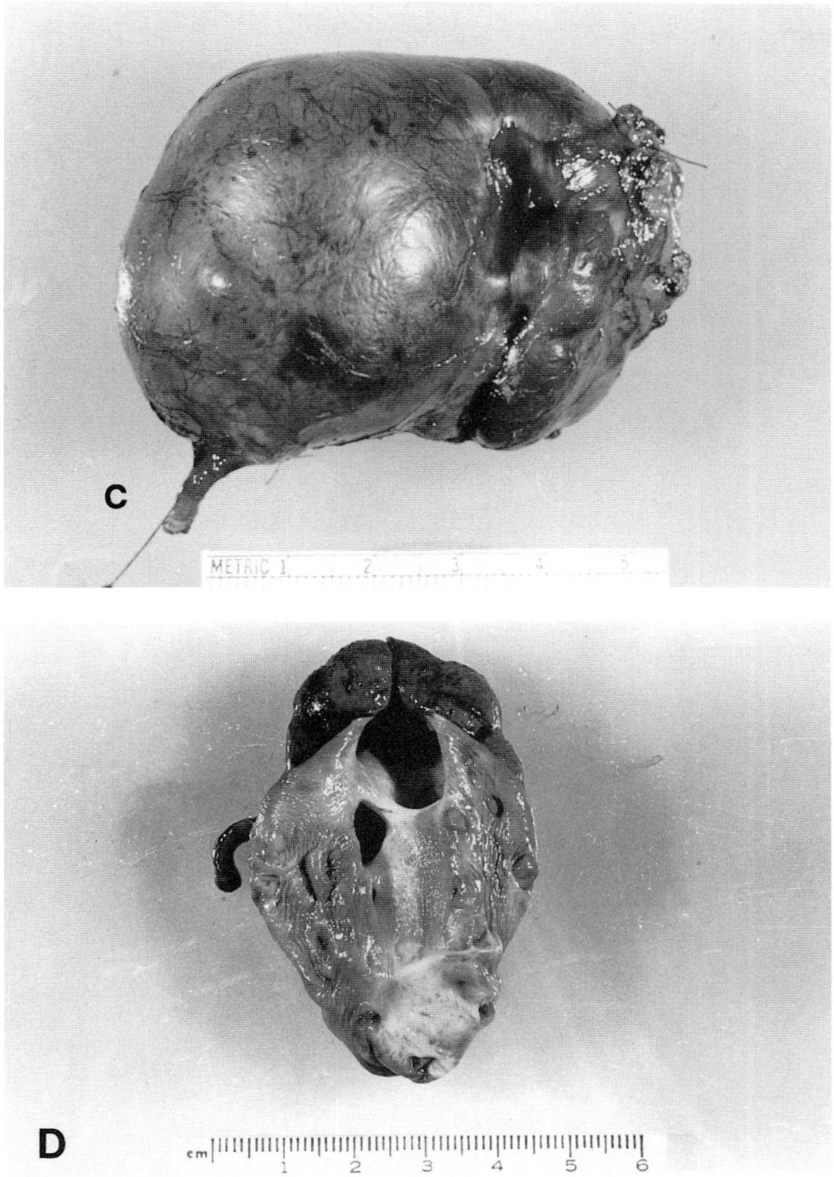

FIGURE 10.1. (*Continued*)

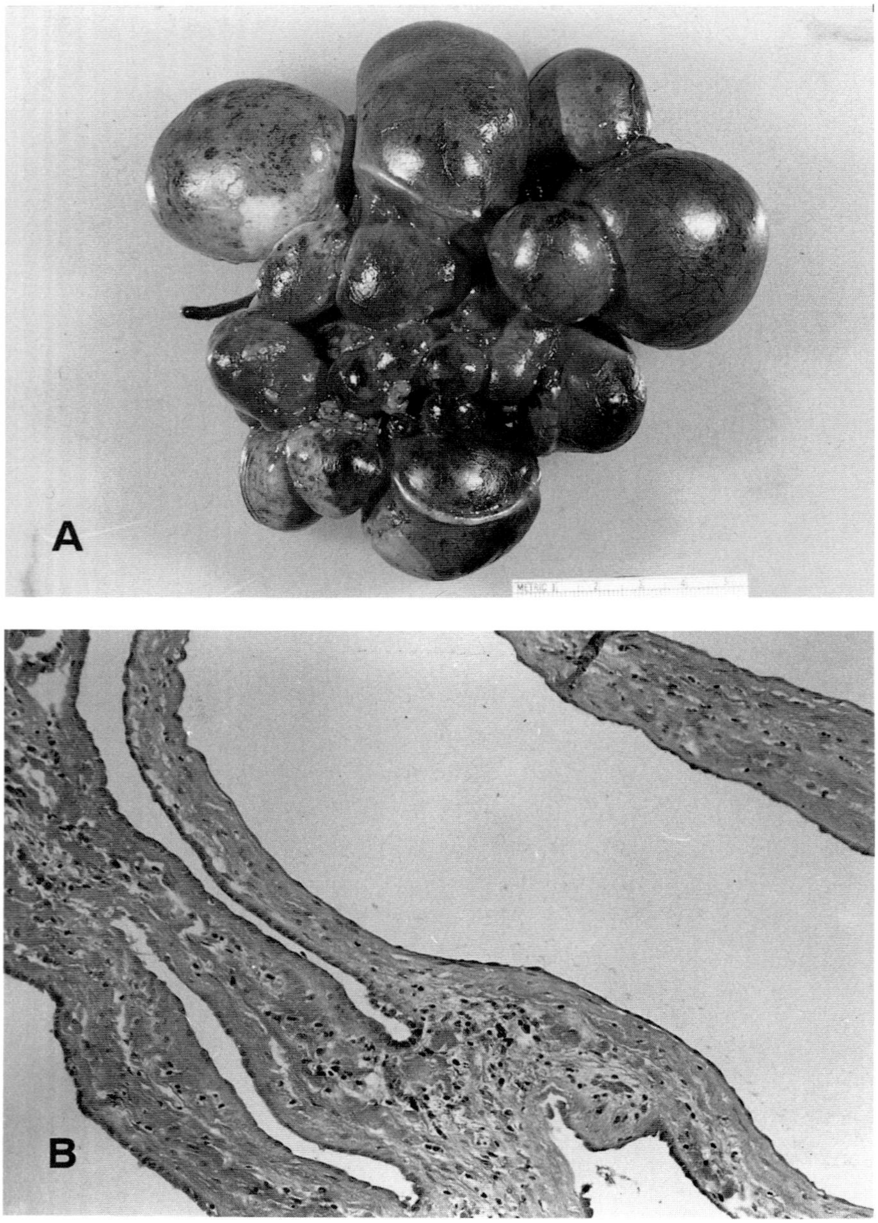

FIGURE 10.2. Multicystic dysplastic kidney in a 4-month-old female who presented with an abdominal mass, which was confirmed on imaging studies. (A) Almost the entire left kidney is replaced by multiple grapelike cysts of various sizes filled with clear straw-colored fluid. The specimen weighed 368 g and measured $13 \times 9 \times 5$ cm^3. An atretic ureter is attached to the cystic kidney on the left. (B) The cysts are lined by cuboidal and flattened epithelial cells. H&E; $\times$ 50. Compare this micrograph with Figure 10.25B of a multilocular cyst and note the close resemblance between the two. Foci of residual renal tissue, cartilage, and medullary dysplasia (tubules lined by primitive urothelium and surrounded by collars of mesenchyme), not shown here, are found in cystic dysplastic kidneys and not within multilocular cysts.

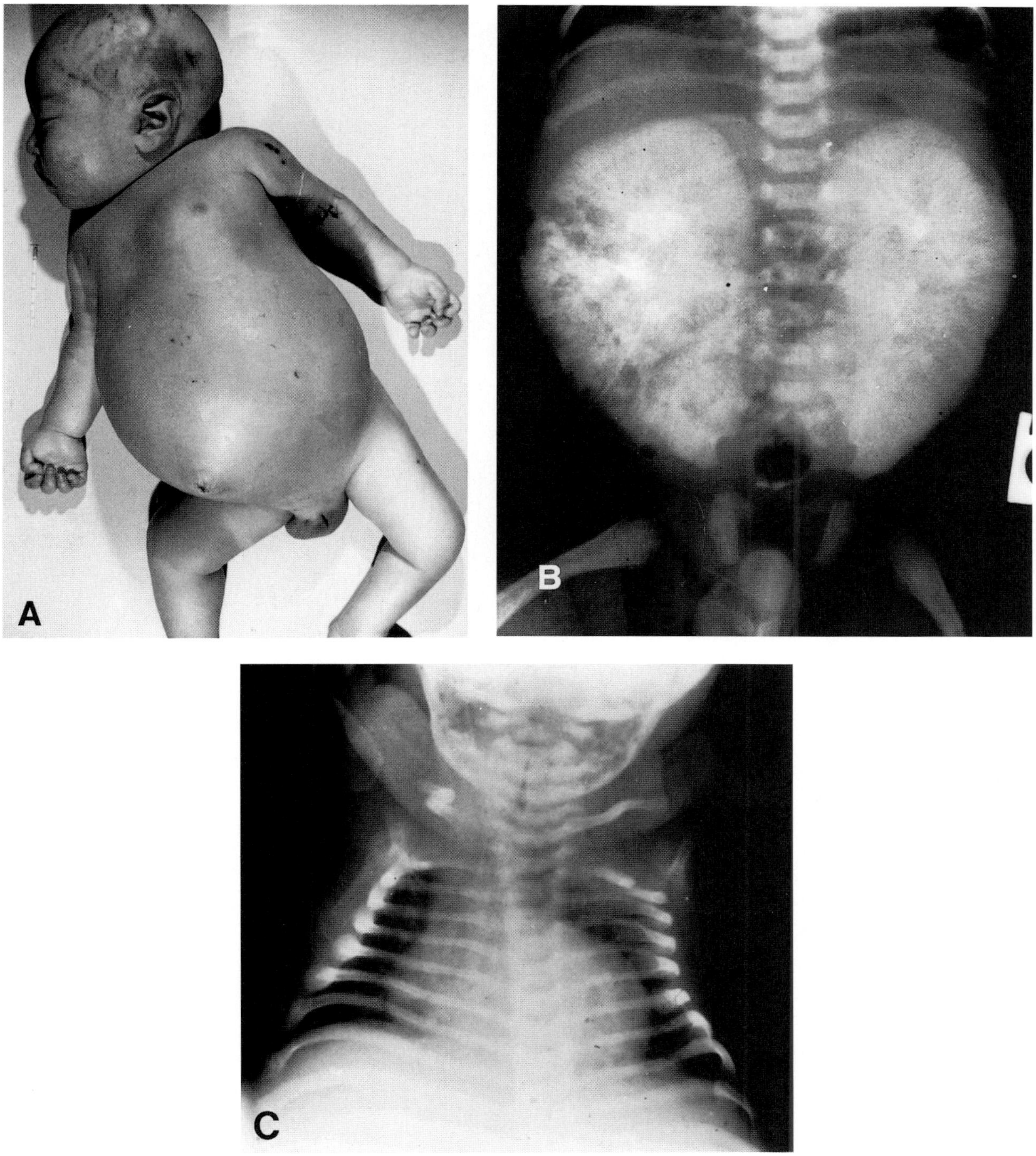

FIGURE 10.3. Infantile polycystic disease of kidney and liver (autosomal recessive). (A) A 3-month-old male with bilateral flank masses noted at birth who died of renal and pulmonary failure complicated by *E. coli* sepsis. The abdomen is distended by markedly enlarged kidneys and liver. (B) Intravenous pyelogram displaying bilateral renomegaly and cystically tubules filled with dye. (C) Chest radiograph shows an abnormally small thoracic cavity and hypoplastic lungs. (D) Radially arranged, fusiform cysts are situated in the cortex and medulla. The blackened cortical areas represent foci of acute pyelonephritis due to *E. coli*. (E) Cystically dilated collecting ducts are present. H&E; × 48.

(*Continued*)

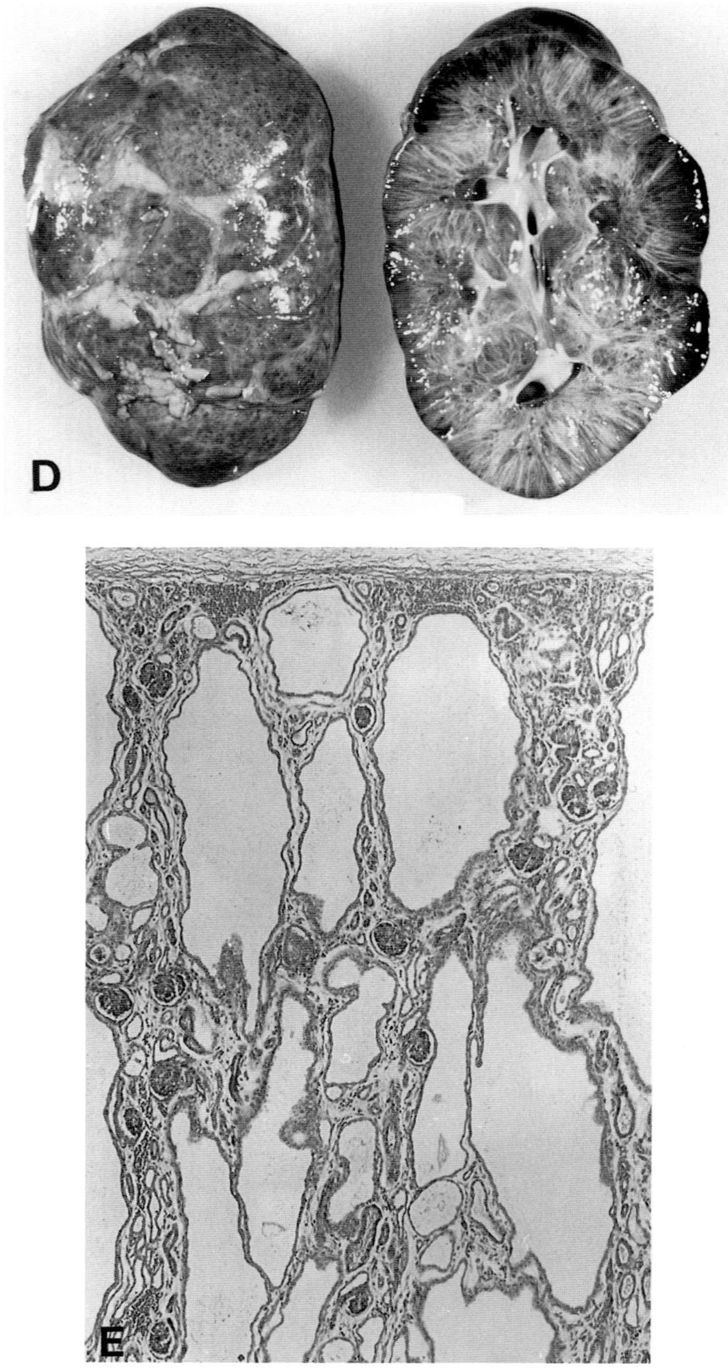

FIGURE 10.3. (*Continued*)

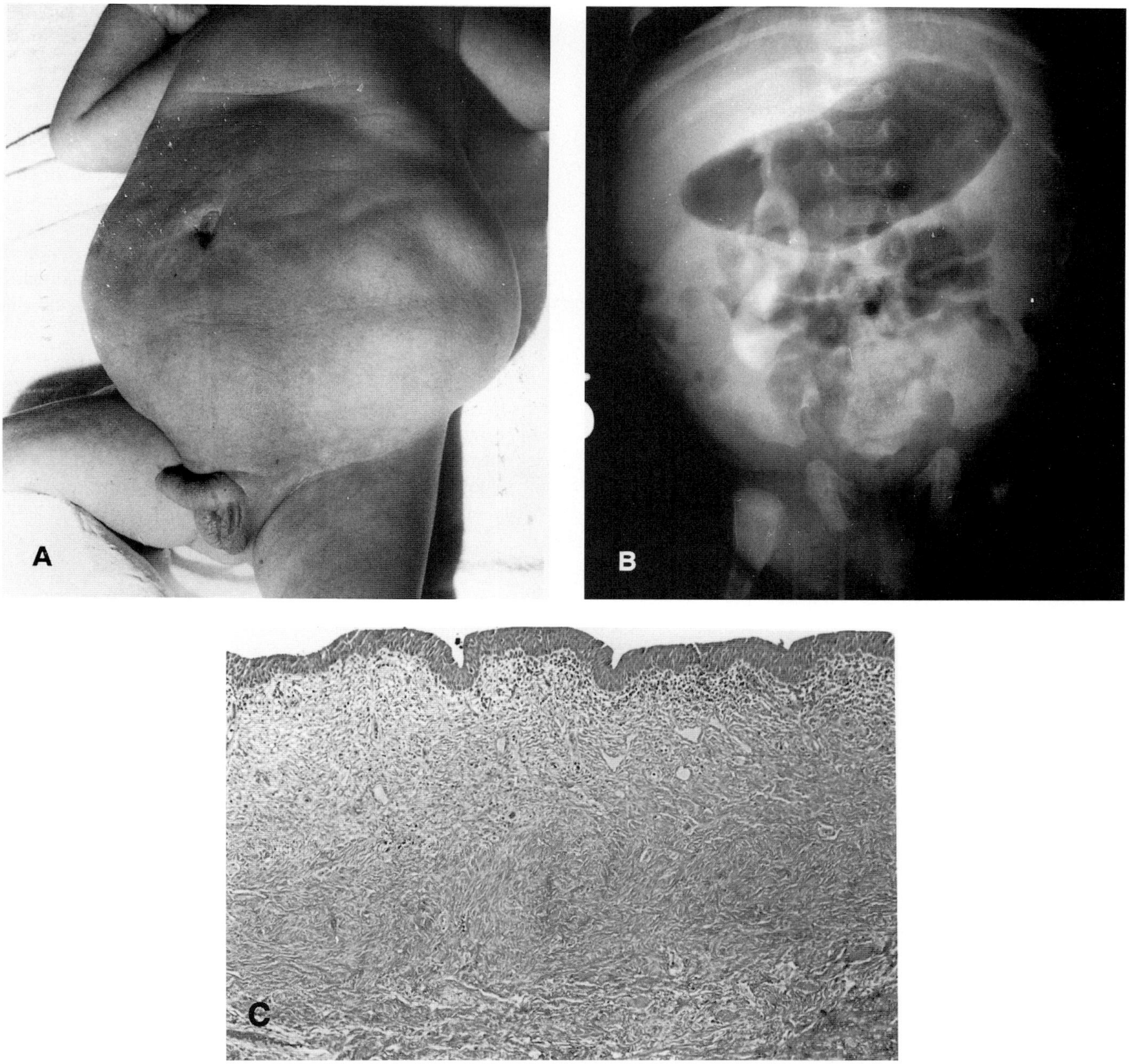

FIGURE 10.4. Prune belly syndrome (megacystis, megaureter, and absence of abdominal musculature). (A) Infant with a distended, scaphoid, "doughy" abdomen and bilateral undescended testes. (B) Intravenous pyelogram depicting a markedly dilated ureter and enlarged urinary bladder. (Radiograph courtesy of Melvin O. Senac Jr., M.D., Department of Radiology, Children's Hospital, San Diego.) (C) Portion of megaureter from a 1-year-old male with this syndrome shows intramural fibrosis and markedly decreased smooth muscle. Skeletal muscle was not found in a full-thickness biopsy sample taken from the abdominal wall. H&E; $\times$ 48.

TABLE 10.1. Classification of fetal and infant renal tumors.

Congenital mesoblastic nephroma
Classic mesoblastic nephroma
Cellular mesoblastic nephroma
Wilms tumor
Classic triphasic
Blastemal
Epithelial (monomorphous)
Fetal rhabdomyomatous
Botryoid tumor of the renal pelvis
Wilms tumor with anaplasia[a]
Nephroblastomatosis complex
Cystic renal tumors
Cystic nephroma (multilocular cyst)
Cystic partially differentiated nephroblastoma
Multilocular cystic nephroma (cystic Wilms tumor)
Metanephric stromal tumor
Rhabdoid tumor of kidney
Clear cell sarcoma of kidney
Ossifying renal tumor of infancy
Renal cell carcinoma[a]

[a] Usually does not occur in infants.
Source: Isaacs H Jr [12].

The author's proposed classification for renal tumors is given in Table 10.1, and Table 10.2 lists distinguishing gross and microscopic features.

Wilms Tumors

Only 10 to 20% of all Wilms tumors are diagnosed during the first year of life [10,14,17]. Wilms tumor seldom occurs in the neonate, being usually found in the older infant [7,14, 16,18,19]. Prenatal detection by sonography is reported [20,21]. Typically it presents clinically as an abdominal mass. The tumor is associated with certain congenital defects and malformation syndromes (Table 10.3).

Karyotypic analyses of Wilms tumors show a deletion in the short arm of chromosome 11, del 11p13 [22,23]. Chromosome 11 deletions occur in the leukocytes as well as in the tumor and can be used to identify individuals and families at risk for developing the tumor and to the confirm an equivocal histological diagnosis (e.g., some poorly differentiated Wilms tumors present diagnostic problems).

TABLE 10.2. Renal tumors and tumorlike conditions of the fetus and infant: distinguishing features.

Tumor	Gross	Histopathology[a]
Mesoblastic nephroma		
Classic	Nonencapsulated, noncircumscribed, light tan; fingerlike projections; resembles a leiomyoma	Monomorphous uniform spindle-shaped cells surrounding normal parenchymal elements; rare mitoses; vim +, act +; EM: spindle cells with features of fibroblasts and myofibroblasts
Cellular	Encapsulated, pale gray, circumscribed, cystic, gelatinous	Monomorphous spindle-shaped cells; hypercellular, mitoses, slight nuclear atypia; vim +, act ±; EM: primitive spindle-shaped cells with features of fibroblasts and myofibroblasts; trisomy 11
Wilms tumor	Encapsulated, well circumscribed Light tan-gray, soft; cysts, necrosis, Hemorrhage	*Epithelial:* primitive tubules and glomeruli scant stroma; del 11p13 *Blastemal:* small round to oval cells; many mitoses *Triphasic:* epithelial, blastemal, mesenchymal elements (e.g., skeletal muscle, fibroblasts) *Fetal rhabdomyomatous:* triphasic histology with with over 40% skeletal muscle *Anaplasia[b]:* bizarre mulberry-like giant cells; huge mitoses
Nephroblastomatosis complex nephrogenic rests (see Figure 10.13)		
Perilobar (PLNR)[c]	Small tan-gray nodules present beneath the capsule, visible depending on their size	Nodules of small, dark embryonic-appearing tubules and glomeruli resembling the nephrogenic zone of the developing kidney
Intralobar (ILNR)	Small tan-gray nodules present in the cortex or medulla, visible depending on their size	Nodules of small, dark embryonic-appearing tubules and glomeruli resembling the developing kidney
Nephroblastomatosis Perilobar (PLNR)	Tan-gray nodules present beneath the capsule; kidney may be diffusely enlarged with irregular nodular capsular surfaces	Nodules of small, dark embryonic-appearing tubules and glomeruli resembling the nephrogenic zone of the developing kidney
Intralobar (ILNR)	Tan-gray nodules present in the cortex or medulla; kidney may be diffusely enlarged	Nodules of small, dark embryonic-appearing tubules and glomeruli resembling the developing kidney

(continued)

TABLE 10.2. (*Continued*)

Tumor	Gross	Histopathology[a]
Nephroblastomatosis complex *(continued)*		
Combined (PLNR + ILNR)	Tan-gray confluent nodules present beneath the capsule, in the cortex and medulla; kidneys diffusely enlarged	Nodules of small, dark embryonic-appearing tubules, glomeruli and blastema resembling the developing kidney
Universal (panlobar)	Hyperlobulated, cerebriform appearance; no recognizable cortex or medulla; kidney diffusely enlarged	Diffuse hyperplastic-appearing tubules, glomeruli and blastema occupying cortex and medulla; tubulointerstitial dysplasia
Metanephric stromal tumor	Tan lobulated, cystic and fibrous mass centered in the renal medulla	Variegated cellular stromal tumor with spindled or stellate cells forming a nodular pattern, onion skin collarette formation around entrapped tubules, angiodysplasia of of intratumoral vessels, juxtaglomerular cell hyperplasia of entrapped glomeruli
Multilocular cyst of kidney (cystic nephroma)	Multilocular cysts with thin fibrous septa, contain clear fluid; do not communicate with renal pelvis	Regular low cuboidal epithelium lining cysts. No metanephric elements in cyst walls; no atypia or mitoses
Cystic partially differentiated nephroblastoma	Multilocular cysts with thin fibrous septa, contain clear fluid; does not communicate with renal pelvis	Regular low cuboidal epithelium lining cysts; immature or mature metanephric elements and skeletal muscle in cyst walls
Multilocular cystic nephroma (cystic Wilms tumor)	Multilocular cyst with a nodule of Wilms tumor; Wilms tumor containing a multilocular cyst	Regular low cuboidal or atypical epithelium lining cysts; Wilms tumor histopathology as described above
Rhabdoid tumor of kidney	Encapsulated, tan-gray well circumscribed, large areas of cystic necrosis and hemorrhage	Monomorphous population of polygonal cells with abundant eosinophilic cytoplasm, vesicular nucleus with single big nucleolus; red, globular cytoplasmic body consists of intermediate filaments by EM
Clear-cell sarcoma of kidney	Encapsulated, well circumscribed; hemorrhagic and cystic	Monomorphous population of round to oval cells with water-clear cytoplasm, open (vesicular) nuclei, fine chromatin, and tiny nucleoli; few mitoses; rows of cells separated by delicate fibrous septa; palisading and vascular branching capillary patterns; vim +, NSE –, cytok – des –, act –; EM: primitive small cells with sparse organelles and fluffy intra- and extracellular deposits
Ossifying renal tumor of infancy	Calcified mass projecting into dilated calyx	Osteoid trabeculae surrounded by osteoblasts and spindle-shaped cells

[a] vim, vimentin; act, actin; cytok, cytokeratin; des, desmin; NSE, neuron-specific enolase; EM, electron microscopy.
[b] "Unfavorable histology" associated with a poor prognosis; anaplasia is rare in infancy.
[c] Found in association with bilateral Wilms tumor and congenital anomalies.
Source: Isaacs H Jr [12].

TABLE 10.3. Malformations and syndromes associated with Wilms tumor.

Aniridia	Nephroblastomatosis	VATER association[e]
Hemihypertrophy	Beckwith–Wiedemann syndrome[a]	Bloom syndrome[f]
Genitourinary anomalies	WAGR syndrome[b]	Prader-Willi syndrome[g]
Hypospadias	Drash syndrome[c]	Fraser syndrome[h]
Undescended testes	Perlman syndrome[d]	
Ambiguous genitalia		

[a] Omphalocele, macroglossia, visceromegaly, and gigantism.
[b] **W**ilms tumor, **a**nirdia, **g**enitourinary abnormalities, and mental **r**etardation: the 11p deletion syndrome.
[c] Ambiguous genitalia, nephritis, and Wilms tumor.
[d] Fetal gigantism, renal hamartomas, and nephroblastomatosis with Wilms tumor.
[e] **V**ertebral malformations, **a**nal atresia, **t**racheo**e**sophageal fistula, limb dysplasias, and **r**enal anomalies.
[f] Growth failure, sun-sensitive facial telangiectasia, and defective immunity.
[g] Obesity, short stature, hypogonadism, and mental retardation.
[h] Cryptophthalmos, malformed ears, aberrant umbilicus, and incomplete development of external genitalia.
Source: Bolande RP [5], Bove [6], Isaacs H Jr [12], Murphy WM, Beckwith JB, Farrow GM [26].

Embryonic nephrogenic tissue normally differentiates into the metanephric blastema, the precursor of nephrons and mesenchymal stroma (Figure 10.5). Abnormal proliferation of the metanephric blastema may cause epithelial and/or stromal renal tumors, found almost exclusively in young patients [13].

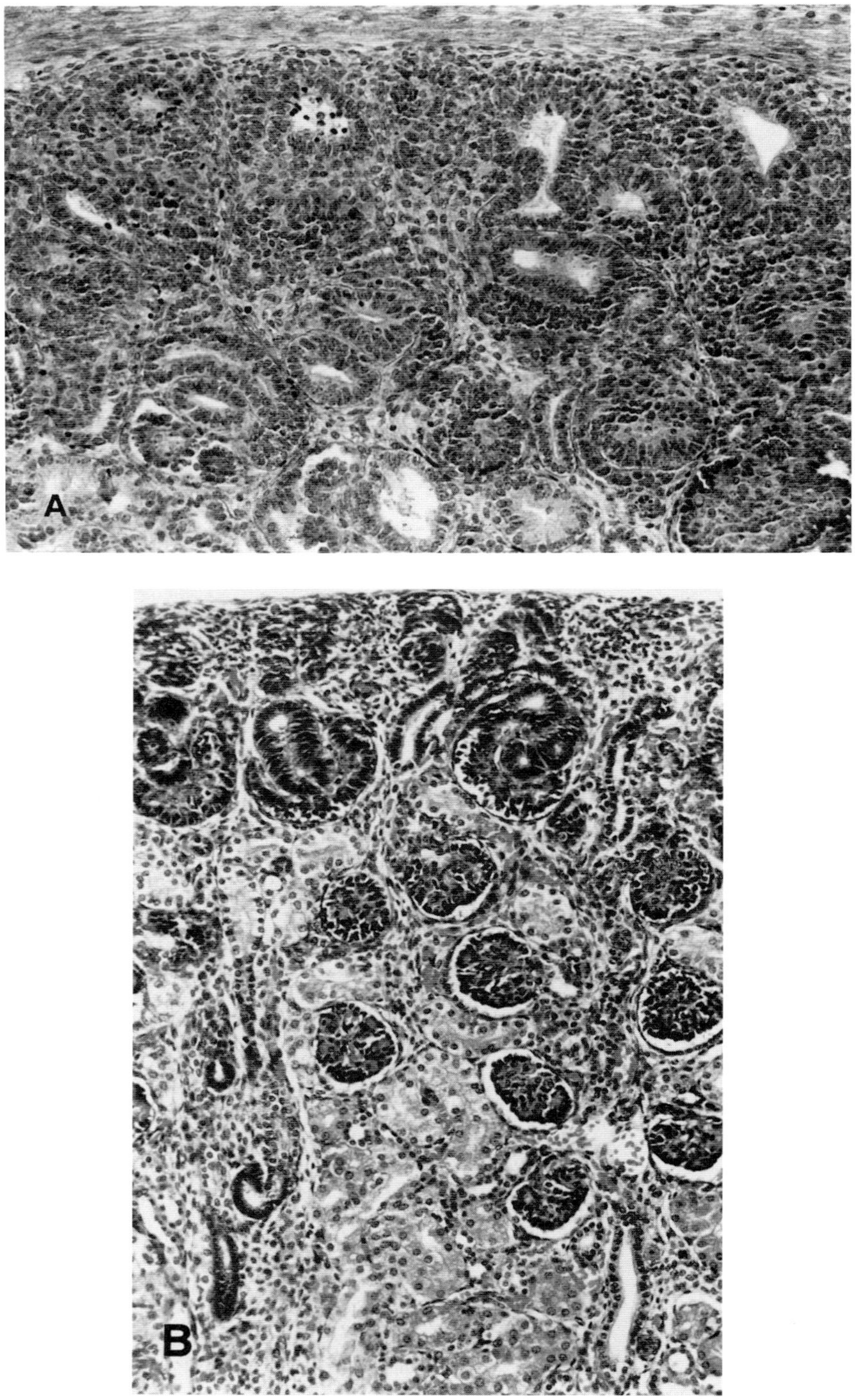

FIGURE 10.5. Development of the kidney. (A) Glomeruli and tubules are beginning to form in the cellular metanephric blastema of the nephrogenic zone situated beneath the renal capsule. From an embryo of estimated 10 to 12 weeks gestation. H&E; × 60. (B) At the top of the photograph beneath the capsule there is a row of immature glomeruli and tubules arising from the metanephric blastema. Three or four layers of definitive glomeruli are noted beneath. Section taken from a 650 g fetus. H&E; × 200.

Pathology

Wilms tumor on section has a pale tan to light gray cut surface with or without areas of necrosis or cyst formation. The tumor is encapsulated and well circumscribed (Figures 10.6, 10.8, 10.9, 10.10). Rupture of the capsule with evidence of spillage and/or extension of tumor through the capsule is present in some specimens, particularly in the larger ones. Renal vein invasion is noted in about one fourth of the time [13].

Microscopically most Wilms tumors found in infants show the classic "triphasic" histological pattern consisting of epithelial elements with varying degrees of differentiation, spindle-shaped blastema cells, and interstitial fibroblastic or myoblastic stromal cells (Figures 10.6 and 10.7). Some

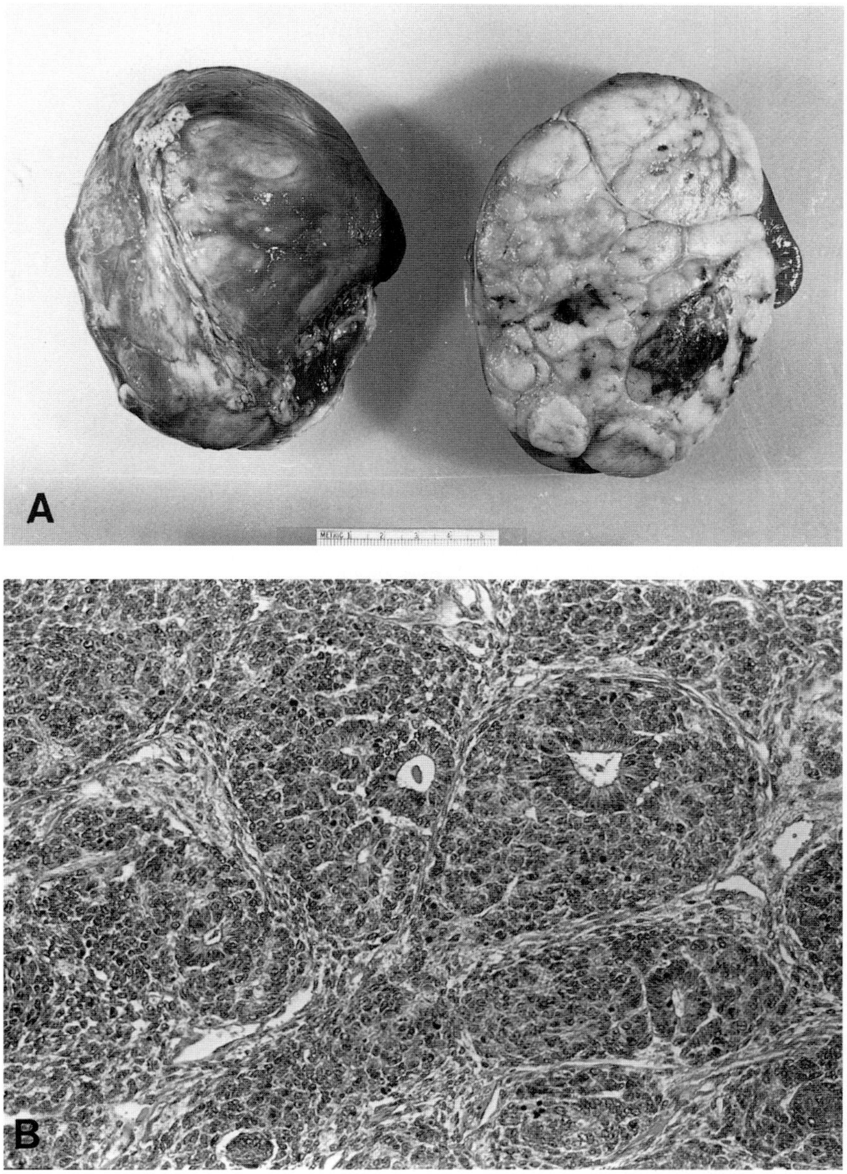

FIGURE 10.6. Wilms tumor in a 1-year-old male. A pediatrician, while evaluating the patient for an upper respiratory infection, found an abdominal mass. This is the most common physical finding in a child with a Wilms tumor. (A) The kidney, 658 g, $13 \times 11 \times 9$ cm^3, is replaced by a soft, tan-gray lobular neoplasm that is encapsulated and well circumscribed. Areas of hemorrhagic necrosis are present. (B) The tumor consists of blastemal, epithelial tubular, and fibrous connective tissue components. The "triphasic" histological pattern (epithelial, blastemal, and mesenchymal stromal components) is the most common one seen in Wilms' tumors in children overall, whereas the epithelial histology is more frequently noted in infants. Notice how the spindle-shaped stromal cells blend imperceptibly with the small, round, more darkly staining blastemal cells. This microscopic feature is helpful when one is attempting to make the diagnosis, for example, on frozen section. H&E; $\times$ 300.

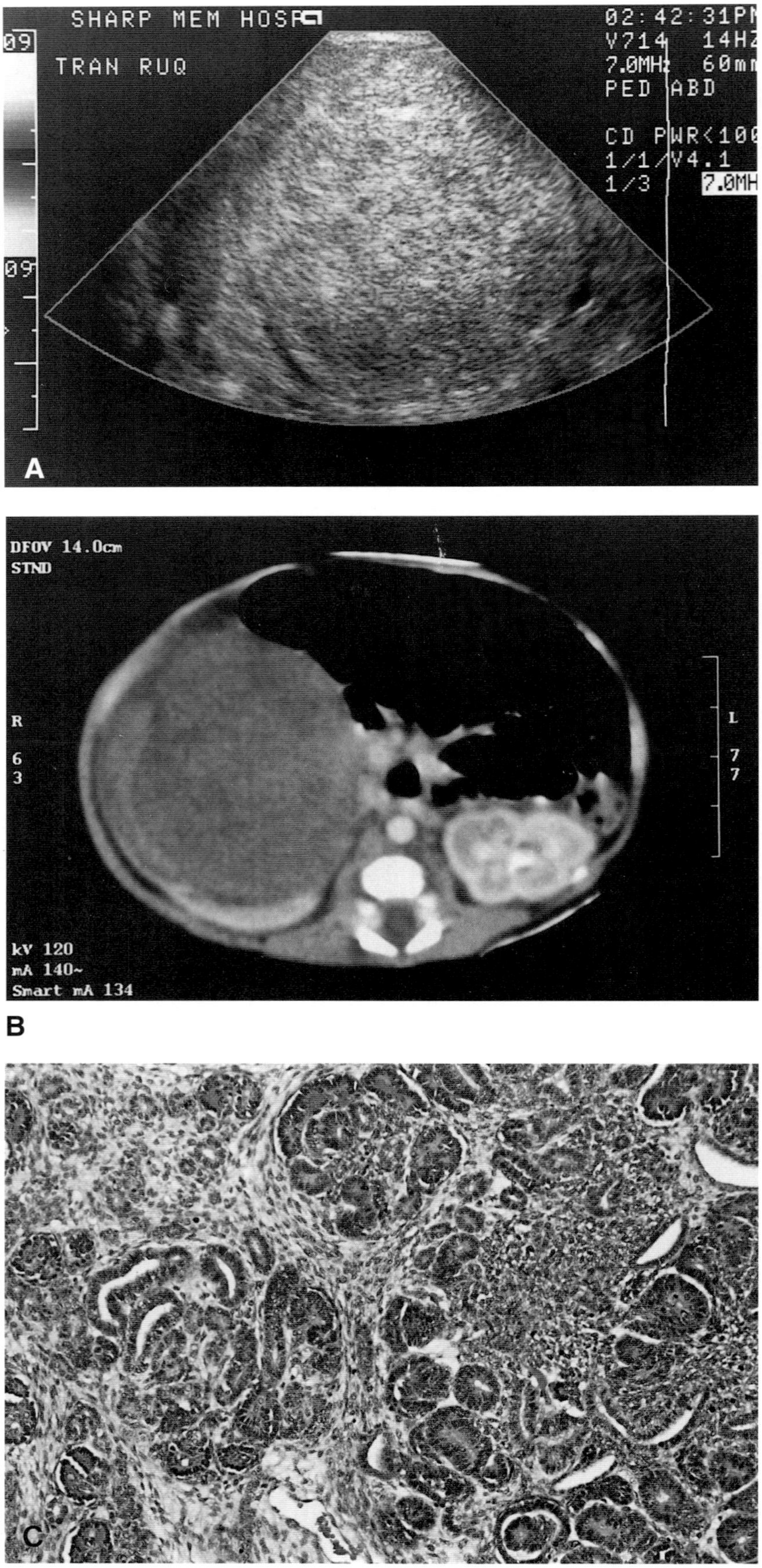

FIGURE 10.7. Congenital Wilms tumor ($6 \times 5 \times 4\ cm^3$, stage 1) in a 6-day-old female, one of twins, with an abdominal mass detected on prenatal sonography at 34 weeks gestation. (A) Prenatal ultrasound showing the abdominal mass. (B) Abdominal CT scan reveals the tumor replacing most of the right kidney with the classic peripheral "claw sign" characteristic of Wilms tumor. (C) The tumor is well differentiated, with a triphasic histological pattern that resembles developing kidney (cf. Figure 10.5B). H&E; $\times$ 60.

tumors are composed predominantly of either epithelial tubular or blastemal components (Figures 10.8 and 10.9). The prognosis for a blastemal cell pattern tend to be less favorable than that for differentiated stromal or epithelial cells [24,25] (Figure 10.9). Anaplastic features—that is, the presence of large, mulberry-like cells with hyperchromatic nuclei and multipolar mitotic figures—are associated with an adverse prognosis and occur more often in older children [1,2,26] (Figure 10.11). However both focal and diffuse anaplasia are described occasionally in infants [25]. The mortality associated with diffuse anaplasia increases rapidly with increasing stage and the type of cells involved by anaplasia [25,27]. Focal anaplasia in a tumor is of prognostic value as well.

Some unique histological patterns of Wilms tumor are noted primarily in the first year of life. The fetal rhabdomyomatous nephroblastoma (tumors with a prominent skeletal muscle component) and botryoid tumors of the renal pelvis showing a triphasic pattern, with or without skeletal muscle differentiation, are examples [26,28,29] (Figures 10.10 and 10.12).The prognosis associated with these microscopic variants is essentially the same as for the typical Wilms tumor with favorable histology.

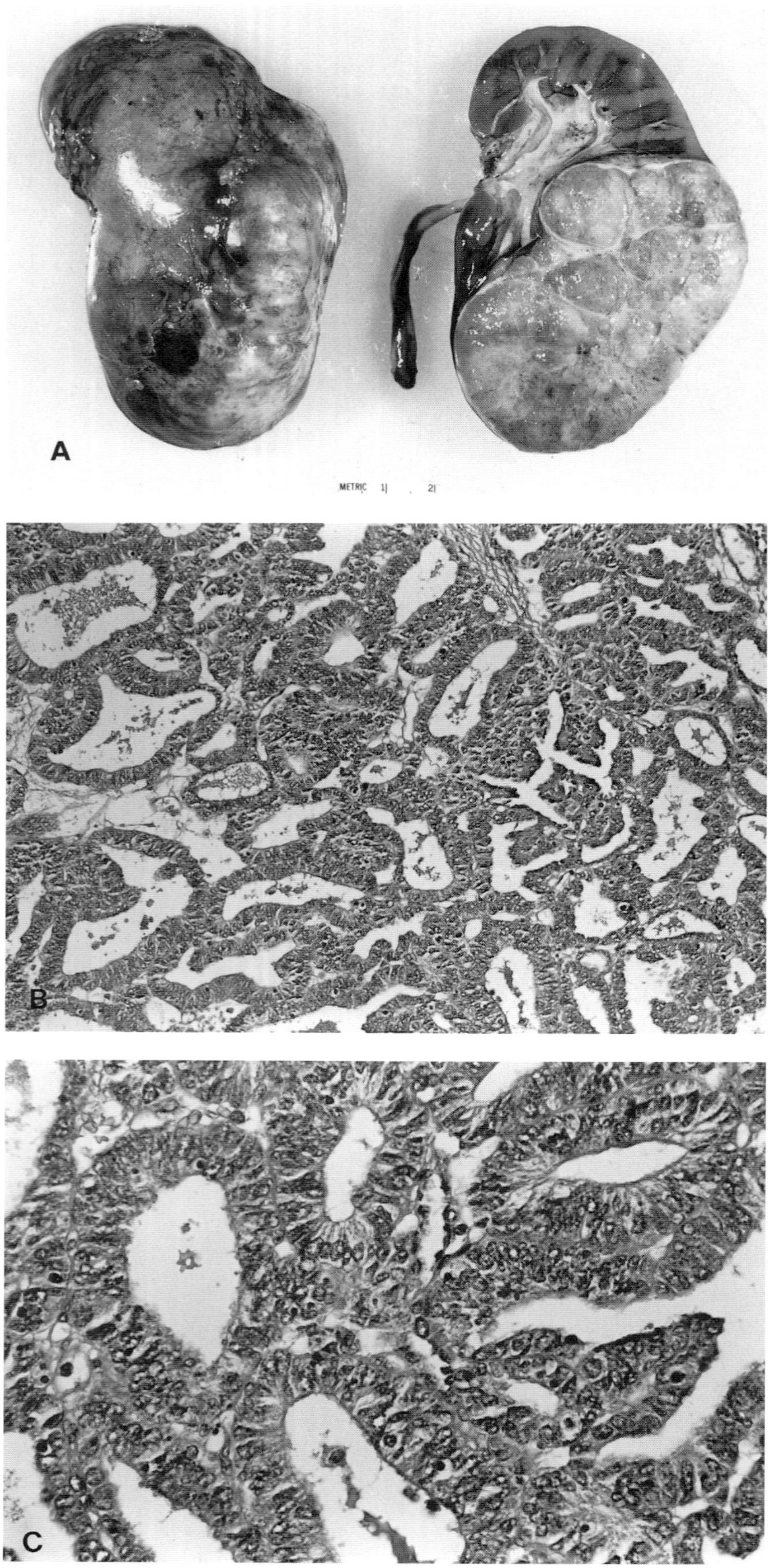

FIGURE 10.8. Wilms tumor. (A) Renal tumor, 250 g and 9×5.5 cm^2, excised from an 11-month-old male with an abdominal mass. The cut surface of the kidney on the right shows a bulging, lobulated, encapsulated, and well-circumscribed neoplasm with a pale, slimy translucent appearance. (B) Low power view ($\times$ 240) revealing predominantly epithelial tubular elements with very little stroma in between. H&E. (C) Higher magnification ($\times$ 600) showing "back-to-back" tubular formations composed of vacuolated columnar cells with pleomorphic nuclei and moderate mitotic activity. H&E.

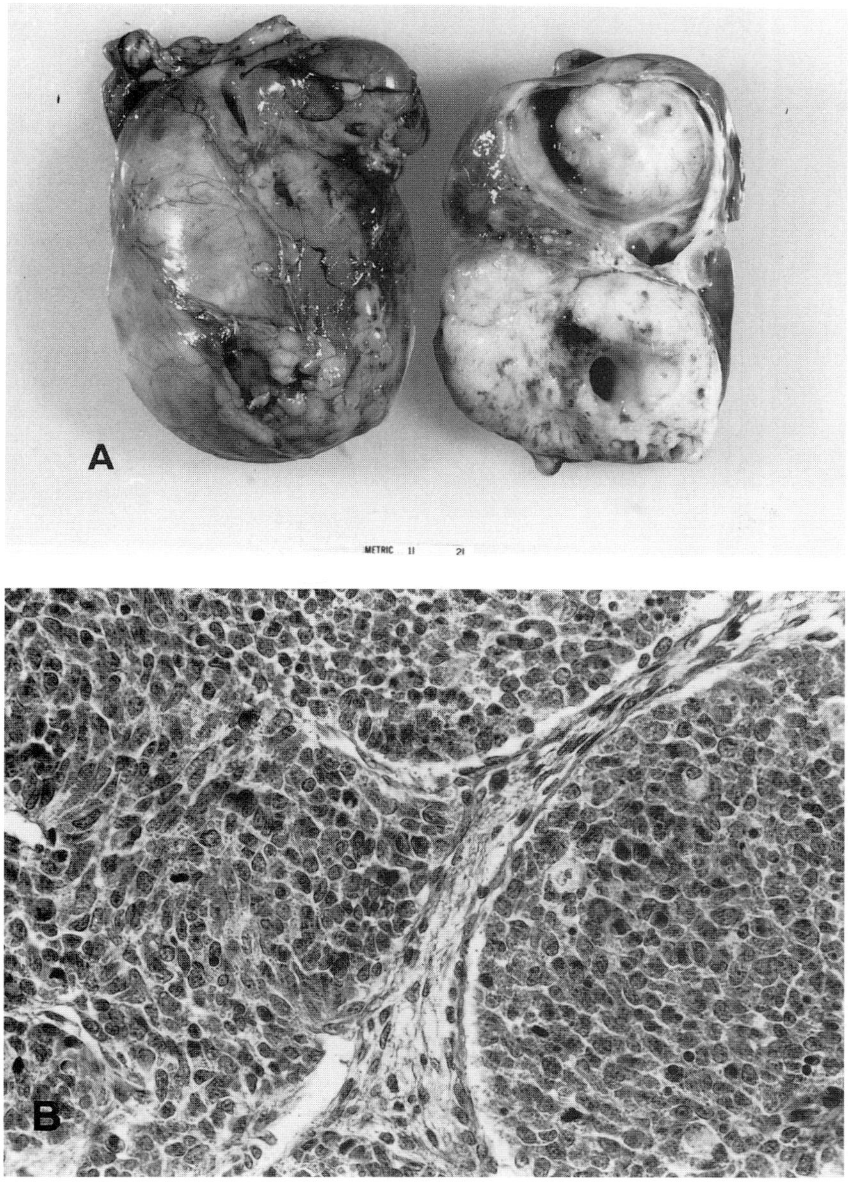

FIGURE 10.9. Wilms tumor. (A) A 192 g, 9.5 × 7 cm^2, solid and cystic neoplasm removed from a 4-month-old female, who presented with an abdominal mass. (B) Lobules of small round to oval blastemal cells with a high nuclear to cytoplasmic ratio are separated by delicate fibrovascular septa. Sometimes the blastemal pattern has the appearance of a "small-cell malignant tumor." H&E; × 300.

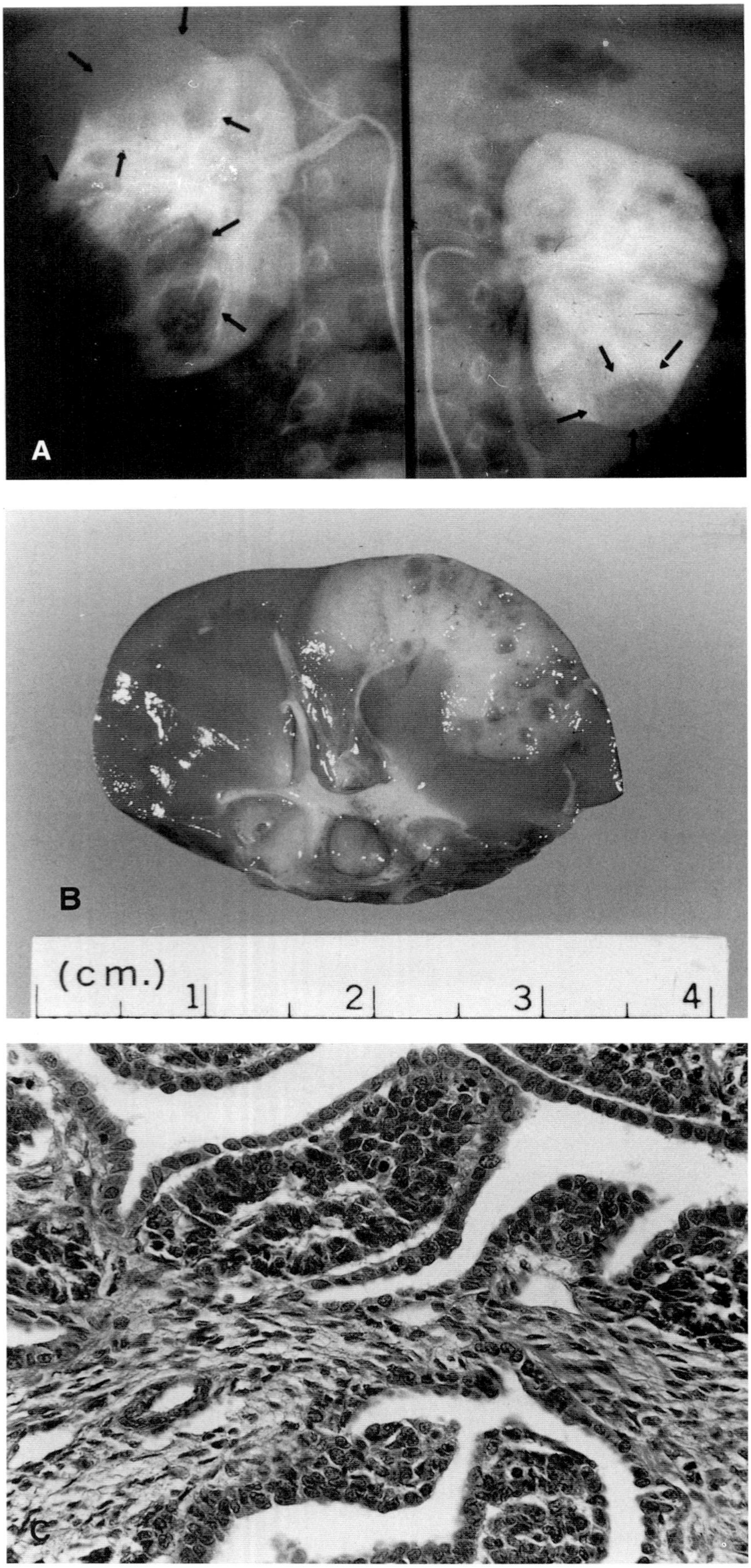

FIGURE 10.10. Bilateral Wilms tumor in a 1-year-old male with congenital anomalies often associated with this malignancy. The clinical presentation of hypospadias, bilateral undescended testes, and aniridia prompted an investigation of the patient for the possibility of Wilms tumor. (A) Arteriogram demonstrating two large masses in the right kidney and a smaller one in the left. The lesions are outlined by arrows. The baby had a total right nephrectomy, since this kidney was more involved, and a partial left nephrectomy. (B) A 2×1 cm^2 tumor nodule situated beneath the capsule of the lower pole of the left kidney has a light tannish-gray, cystic gelatinous appearance, resembling the two larger tumors in the right kidney. (C) All three tumors had the same histological appearance, with triphasic and epithelial patterns. The "botryoid" polypoid epithelial component consists of papillary formations lined by single rows of cuboidal cells and nests of more atypical-appearing epithelial cells beneath. H&E; $\times$ 600.

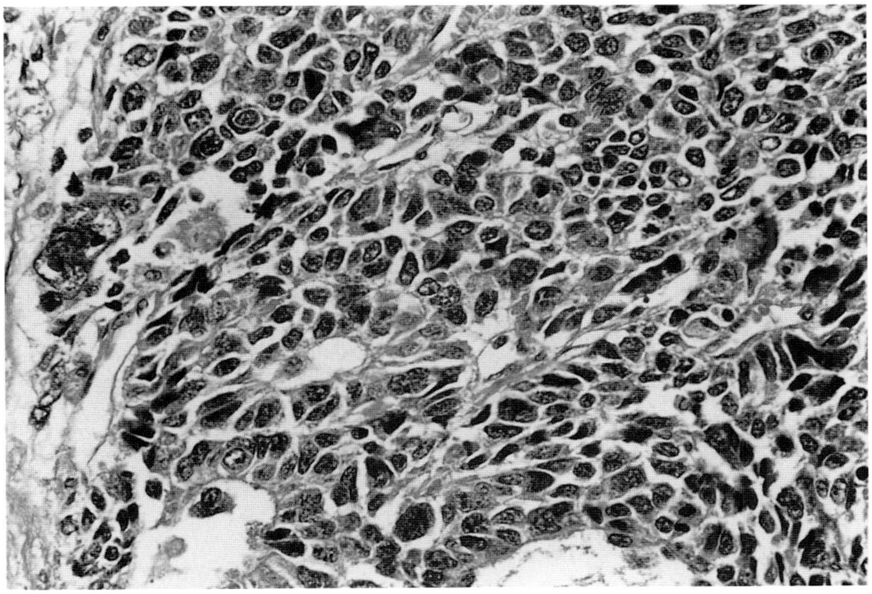

FIGURE 10.11. Wilms tumor with "anaplasia." Wilms tumor with "unfavorable histology" from a 5-year-old girl. There are ill-defined tubule formations composed of anaplastic, darkly staining cells showing marked pleomorphism in cell size and shape. Bizarre appearing giant cells and large mitotic figures are present. H&E; × 200.

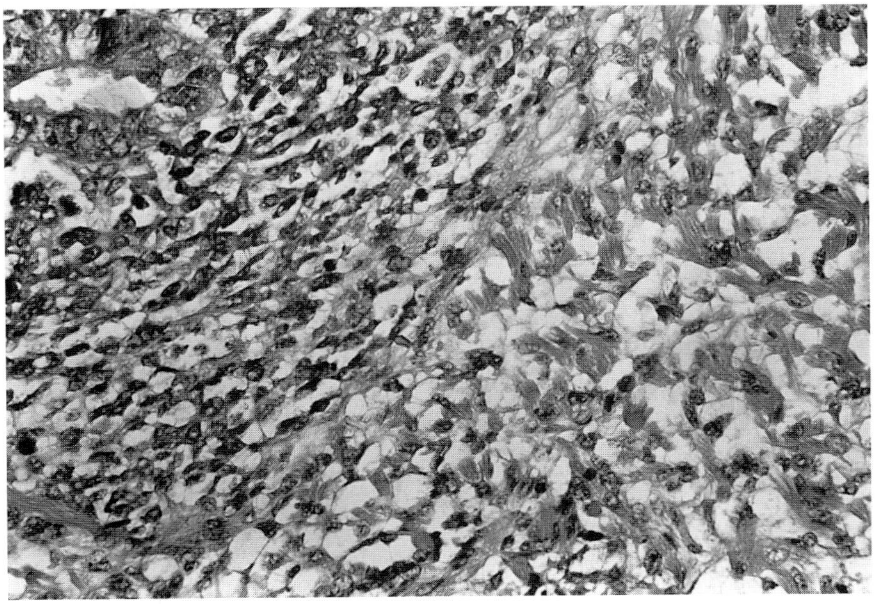

FIGURE 10.12. Wilms tumor with a predominantly skeletal muscle component ("fetal rhabdomyomatous" Wilms tumor). Rhabdomyoblasts with cytoplasmic cross-striations (right side of the field) are bordered by blastemal cells and an embryonic tubule formation in the left upper corner. The renal tumor, 845 g, 15 × 11 cm^2, was removed from a 10-month-old male. The renal pelvis was extensively involved by tumor. H&E; × 600.

TABLE 10.4. Wilms tumor staging definitions (NWTS-5).

Stage	Definition
I	Tumor confined to the kidney and is completely excised
II	Tumor extends beyond the kidney (i.e., capsular invasion), but is completely excised; presurgical biopsy and local tumor spillage are also included in this stage
III	Gross residual tumor, and/or involved surgical margin, and/or tumor in regional nodes
IV	Hematogenous metastases
V	Bilateral Wilms tumors

Source: Beckwith JB [27].

Most Wilms tumors occurring in infants are classified as stage I or II and have a "favorable histology" (i.e., triphasic or epithelial), which probably accounts for their good prognosis in this age group [14,17] (Table 10.4). The incidence of bilateral Wilms tumors (stage V) is reported as 5 to 10%, with figures ranging from 2 to 50%, depending on the study [5,8,18,19,30]. Patients with bilateral tumors tend to be younger at diagnosis and to have an increased frequency of congenital malformations, nephrogenic rests, and favorable histology (epithelial or less often a triphasic histological pattern). Although bilateral Wilms tumors generally have a good prognosis with proper treatment, they present a therapeutic dilemma to balance curative surgical resection with preservation of renal tissue [30].

Nephrogenic Rests and Nephroblastomatosis

"Nephrogenic rests" and "nephroblastomatosis" are terms used to designate a group of renal lesions resembling the metanephric blastema (nephrogenic zone) of the developing kidney [5,26,31–34] (Figure 10.5). Nephrogenic rests are regarded as abnormally persistent nests of embryonal cells, representing microscopic malformations (dysplasias) of the developing kidney [32]. Nephrogenic rests are described in association with a variety of syndromes (Table 10.5). The lesions may be single, multiple, or bilateral in distribution and may be seen either grossly or microscopically depending on their size. Nephrogenic rests are divided into two types: perilobar (PLNR) and intralobar (ILNR); nephroblastomatoses are of four types: perilobar (PLNR, multifocal and diffuse) and intralobar (ILNR, deep cortical and panrenal [32] (Figure 10.13). Occurring at the lobar (cortical) surface, PLNR is the most common form of nephrogenic rest (Figure 10.14), whereas ILNR is less common and is found anywhere in the cortex or medulla. Combined nephroblastomatosis designates both PLNR and ILNR. Panrenal (panlobar, diffuse) nephroblastomatosis is the rarest form and is usually fatal; microscopically there is no recognizable renal cortex or medulla [35]. Diffuse nephroblastomatosis has been detected antenatally by sonography [36].

The fourth National Wilms Tumor Study showed a high incidence of nephrogenic rests in kidneys with Wilms tumors, both unilateral and bilateral (29% for the former and 99% for the latter), and a high risk of developing a Wilms tumor in the contralateral kidney when nephrogenic rests were found. Beckwith lists the risk figures as 26% for PLNR alone, 59% for ILNR alone, and 94% for combined ILNR and PLNR [34].

Most nephrogenic rests regress and do not develop into a Wilms tumor [34], a situation analogous to neuroblastoma in situ [5,34]. The importance of the rests, in addition to being potentially malignant (perhaps they represent the first mutational event in carcinogenesis), is that they are associated with certain syndromes, trisomies, and congenital defects [5,13,31] (Table 10.5). Nephrogenic rests may be detected before evolution to Wilms tumor by serial imaging studies [6]. The type of rest is associated with certain histological types of Wilms tumor; for example, the perilobar rests are seen in conjunction with the blastemal or epithelial Wilms tumors with little stroma, and the intralobar rests are seen with Wilms tumors having a prominent stroma, embryonic nephronlike elements, and heterologous cell types. Actual progression of nephrogenic rests to nephroblastomatosis to Wilms tumor has been documented by serial biopsies [37].

Occasionally it is difficult or impossible to distinguish between a true Wilms tumor and nephroblastomatosis. Neither the size of the lesion nor the microscopic appearance is always helpful in this distinction [34,38]. This distinction is particularly difficult in attempting to decide whether primitive nephrogenic structures (e.g., in a biopsy of a bilateral Wilms tumor treated with chemotherapy) represent nephrogenic rests, residual Wilms tumor, recurrent Wilms tumor, or de novo tumor growth.

TABLE 10.5. Conditions associated with nephroblastomatosis.

Wilms tumor
Hemihypertrophy
Drash syndrome (see note *c*, Table 10.3)
Thanatophoric dwarfism
Klippel-Trenaunay syndrome[a]
Trisomies 13 and 18
Perlman syndrome (see note *d*, Table 10.3)
Beckwith-Wiedemann syndrome (see note *a*, Table 10.3)
Congenital heart disease (high incidence)

[a] Cutaneous hemangiomas, varicose veins, and bony and soft tissue hypertrophy.

Source: Bove KE [6], Isaacs H Jr [12], Beckwith JB [32, 34].

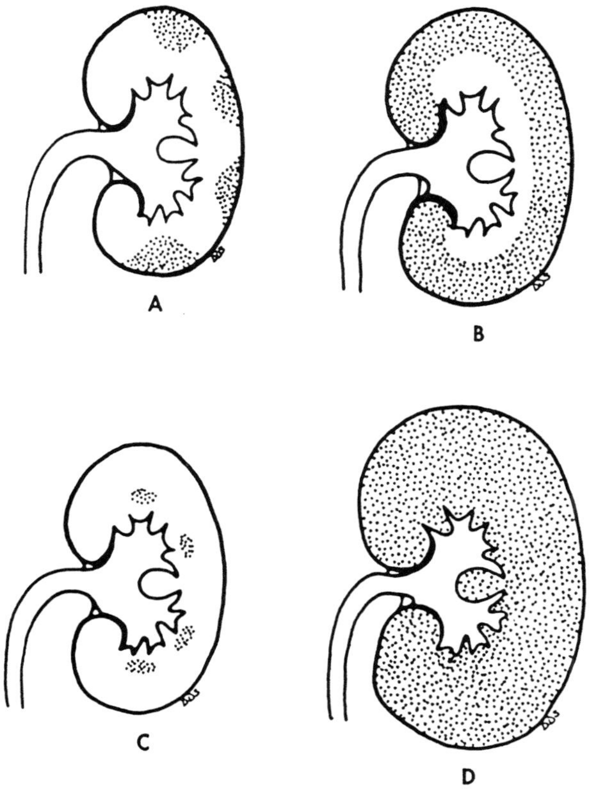

FIGURE 10.13. Location of different types of nephroblastomatosis: (A) perilobar multifocal, (B) perilobar diffuse, (C) intralobar (deep cortical), and (D) panrenal. (From Isaacs H Jr. [13], with permission.)

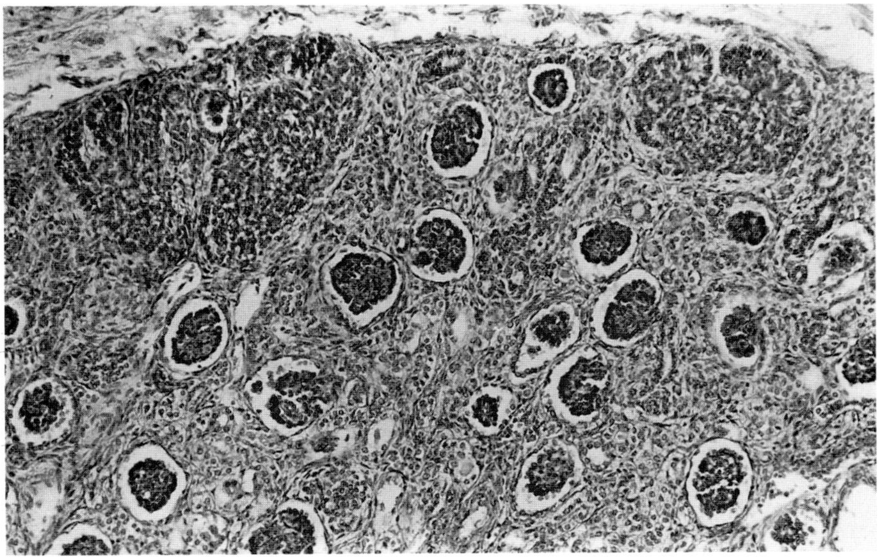

FIGURE 10.14. Nephrogenic rests. Well-circumscribed nodules of darkly staining nephrogenic rests are situated beneath the renal capsule at the top of the photomicrograph (perilobar nephrogenic rests). Early tubule and glomerulus formation is evident. This was an incidental postmortem microscopic finding in both kidneys of a 3-week-old, 1984 g, female of 35 weeks gestation with the Pierre-Robin syndrome and terminal bronchopneumonia. In addition, malformations of the genitourinary tract including a hypoplastic right kidney with a double collecting system and hydronephrosis were found. H&E; × 240.

Congenital Mesoblastic Nephroma

Congenital mesoblastic nephroma, clearly distinct from Wilms tumor, is the leading tumor of the kidney occurring in the newborn and follows Wilms tumor in frequency overall during the first year of life [4,7,8,10]. Occasionally the tumor is observed in patients with hemihypertrophy and the Beckwith-Wiedemann syndrome [13].

Typically patients with congenital mesoblastic nephroma present with an abdominal mass that is discovered on physical examination or by routine antenatal ultrasound scan [4,10,12,39,40] (Figures 10.15–10.18). The tumor's characteristic sonographic findings are different from those of Wilms tumor. The abdominal mass in the fetus may be large enough to produce dystocia [4].

Pathology

Often the diagnosis of congenital mesoblastic nephroma can be made from its distinctive gross appearance, which is clearly different from that of Wilms tumor, clear-cell sarcoma, rhabdoid tumor, and cystic conditions. Congenital mesoblastic nephroma occupies half or more of the involved kidney and on cut section reveals an unencapsulated mass having a whorled pattern similar to a uterine leiomyoma [4,10,41] (Figures 10.15 and 10.16). Fingerlike projections of tumor extend into the adjacent renal parenchyma. Although unusual, extension into the renal vein, vena cava, extrarenal soft tissues, and cystic variants occur [26,41]. Only a few bilateral examples have been documented [4].

Microscopically congenital mesoblastic nephroma shows a proliferation of uniform spindle-shaped cells, demonstrated by electron microscopy to be myofibroblasts and fibroblasts. The tumor cells infiltrate the renal parenchyma and appear to push aside the glomeruli and tubules rather than invading them [41,42] (Figures 10.15 and 10.16). Mitotic figures and small islands of cartilage may be present. Both vimentin and actin are immunoreactive, and desmin is focally expressed [41]. Ultrastructural and immunohistochemical findings suggest that congenital mesoblastic nephroma shares certain histological properties with infantile fibromatosis and other myofibroblastic tumors of childhood [41,42] (see Chapter 3, Soft Tissue Tumors). For example, *ETV6-NTRK3* gene fusions and trisomy 11 are found in both mesoblastic nephroma and congenital fibrosarcoma, suggesting a histogenetic link between these two tumors [43,44].

The *cellular congenital mesoblastic nephroma* is a variant characterized histologically by hypercellularity, nuclear atypia, and numerous mitotic figures [45–47] (Figures 10.17 and 10.18). It has a softer, more cystic and sometimes hemorrhagic cut surface than classic mesoblastic nephroma and appears to be more circumscribed without the interdigitating periphery (Figures 10.17 and 10.18). Large cystic cellular mesoblastic nephromas weighing as much as 1300 g occur in older infants [45]. Among infants less than 3 months of age who have mesoblastic nephroma, the cellular variant is found in almost a third and apparently has no prognostic significance in this younger age group, provided *the lesion is completely removed*. However in older infants and children the risk of recurrence or metastases is significantly increased [26,47]. Surgical resection is the treatment of choice [48].

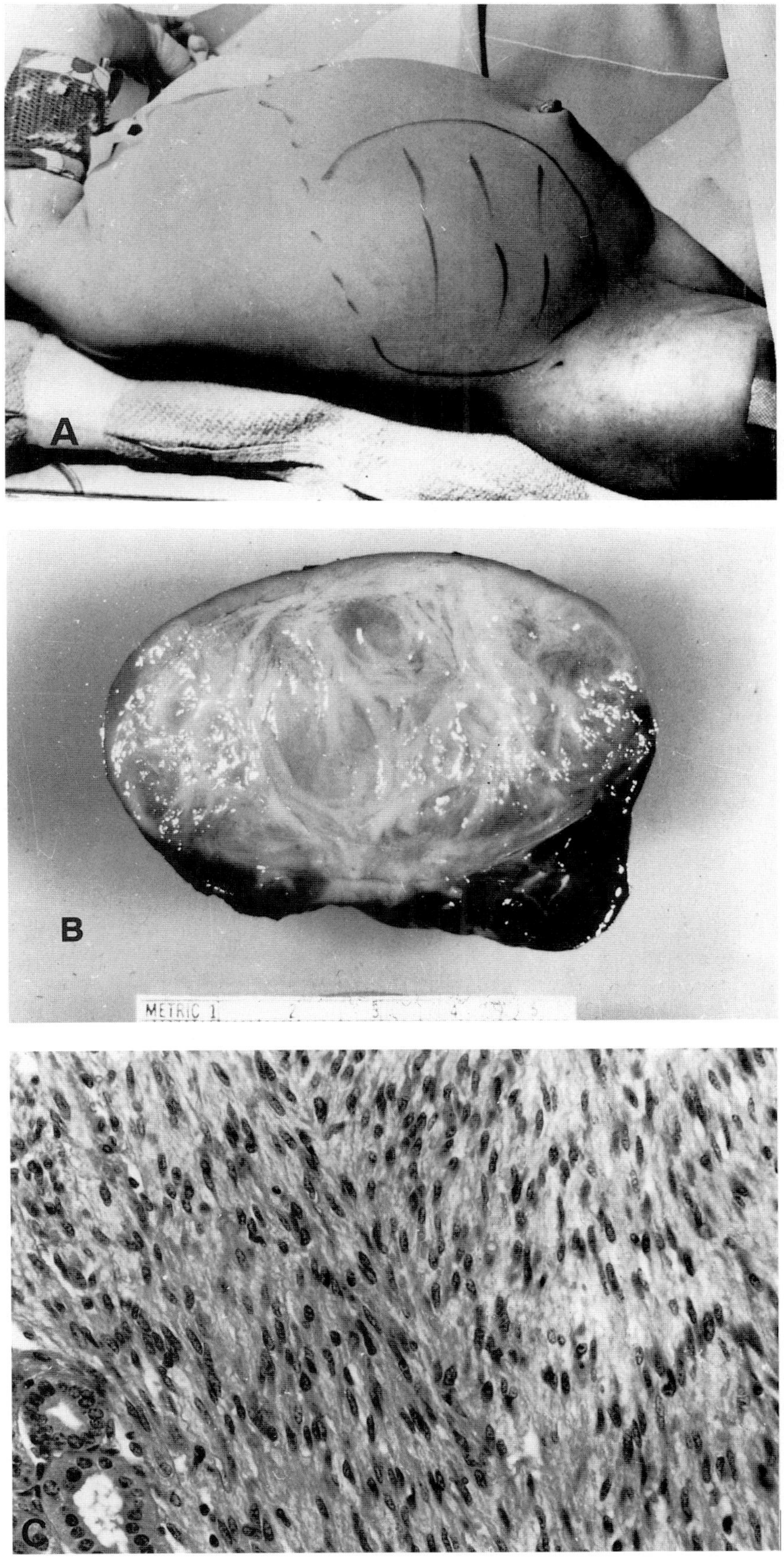

FIGURE 10.15. Classic congenital mesoblastic nephroma. (A) A 1-week-old female with an abdominal mass, the most common physical finding in this condition, and a maternal history of polyhydramnios. (B) The tumor, 94 g, 8×5 cm^2, replaces the lower two thirds of the kidney and has a light tan-yellow whorled appearance. It is nonencapsulated, with small, fingerlike projections extending into adjacent hemorrhagic kidney. (C) The tumor is composed of uniform, small spindle-shaped cells surrounding "entrapped" tubules. H&E; $\times$ 300. (From Isaacs H Jr. [8], with permission.)

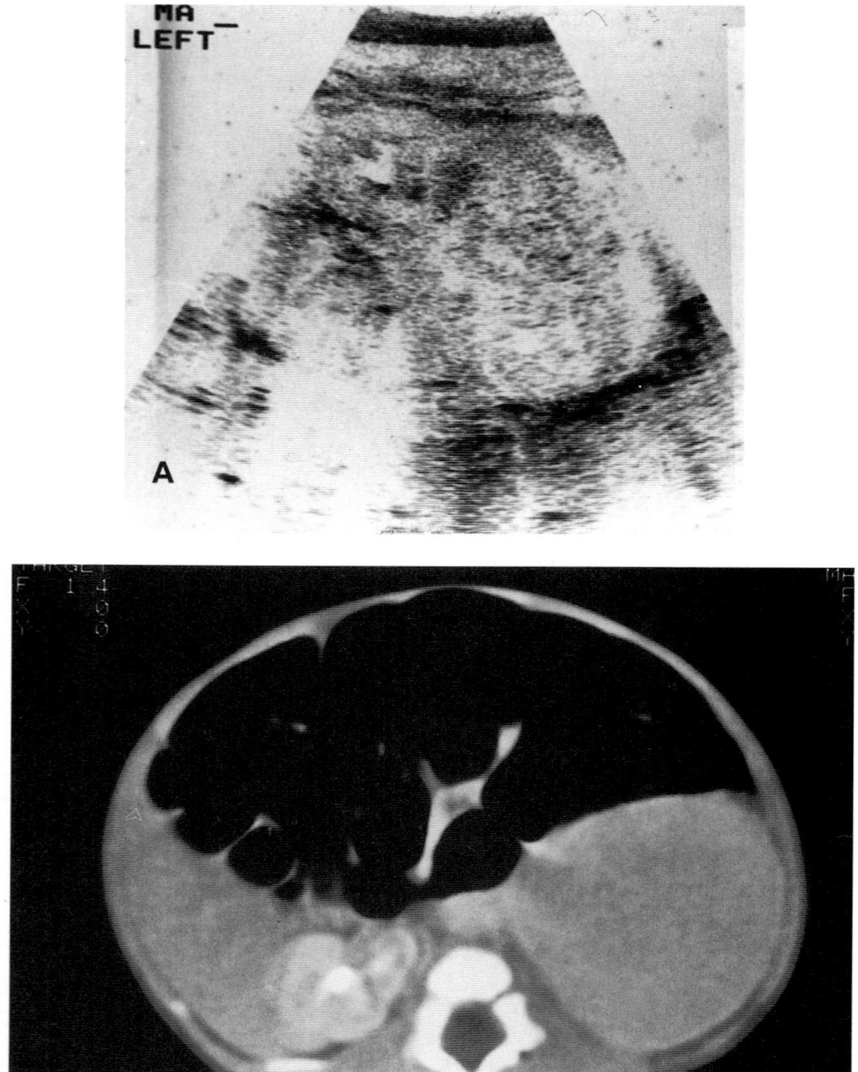

FIGURE 10.16.

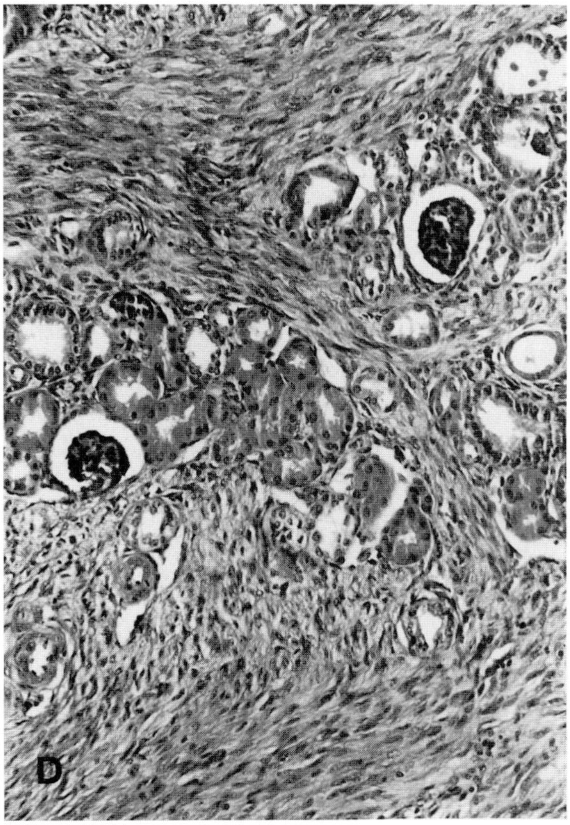

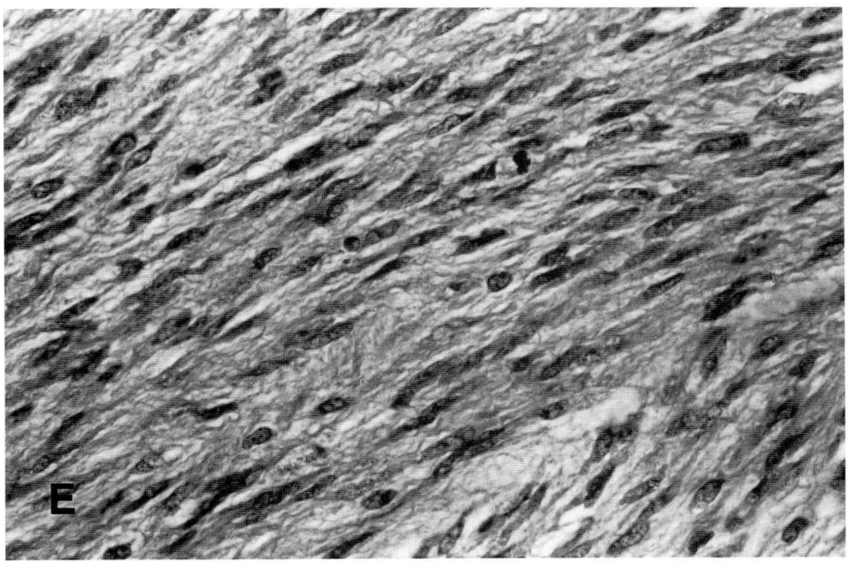

FIGURE 10.16. (*Continued*) Classic congenital mesoblastic nephroma. (A) Sonogram of a 5-day-old male with a left flank mass noted at birth demonstrating the lesion. (B) Computerized tomography scan confirms the sonographic findings. (C) The tumor, 50 g, 5×5 cm^2, has a whorled appearance similar to a uterine leiomyoma and involves the pelvis and calyces. (D) Regular spindle-shaped cells encircle tubules and glomeruli. H&E; $\times$ 300. (E) Tumor cells are elongated, spindle-shaped, with regular oval to fusiform nuclei. Mitoses are seldom noted. H&E; $\times$ 380.

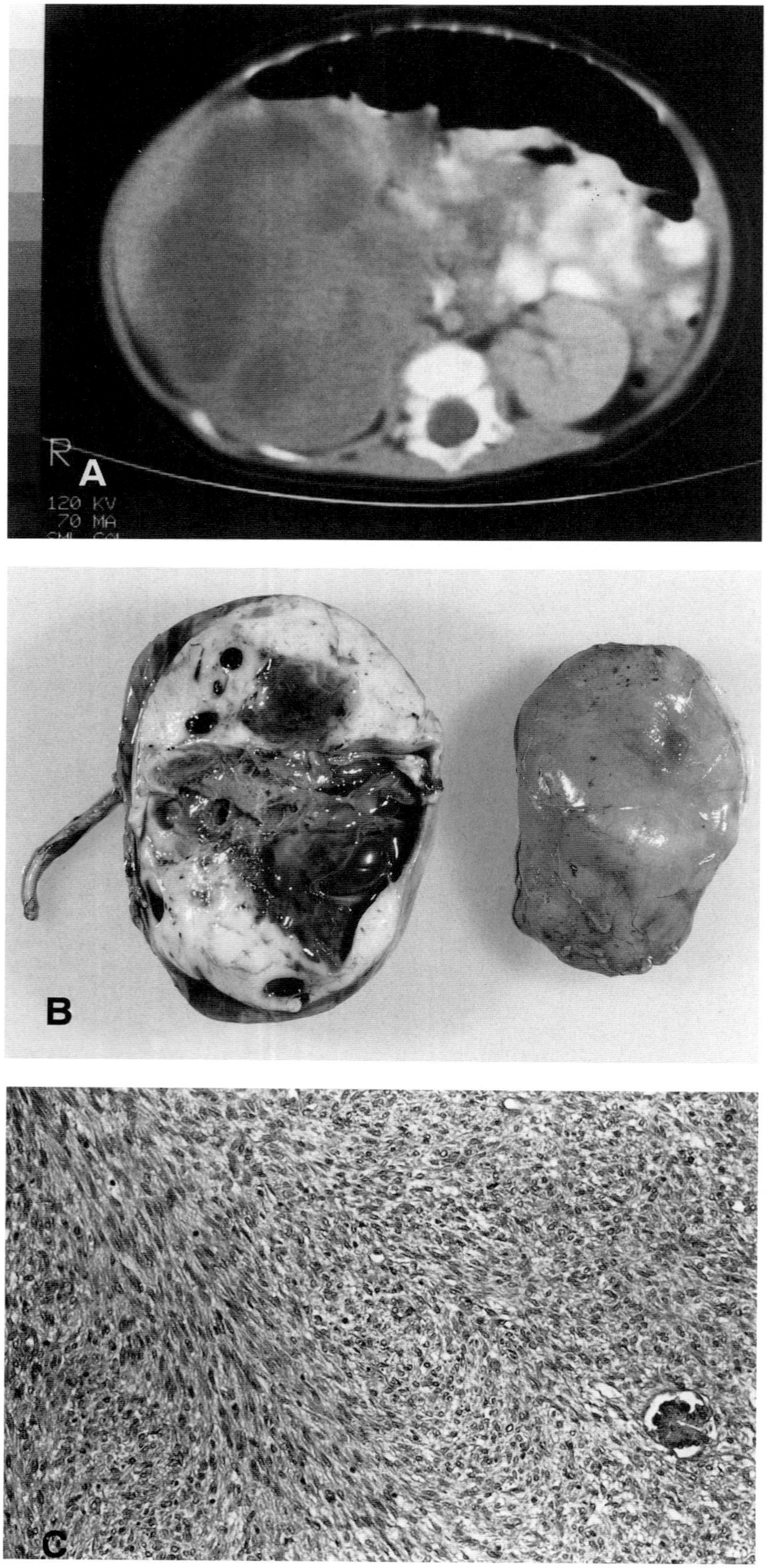

FIGURE 10.17. Cellular congenital mesoblastic nephroma. (A) Computerized tomography scan of a 4-month-old female with an abdominal mass revealing a large right renal lesion composed of both solid and cystic areas. (B) The right kidney, 254 g, 10×8 cm^2, is replaced by a soft, light tan-gray tumor with a prominent cystic component. In contrast to the "classic" congenital mesoblastic nephroma described in Figures 10.15 and 10.16, this tumor is encapsulated, circumscribed, cystic, and hemorrhagic. (C) Spindle-shaped cells with plump oval or fusiform nuclei are separated by a slightly vacuolated fibrillary stroma. It displays an interdigitating ("herringbone") growth pattern. H&E; ×150. (D) The tumor is more cellular than the "classic" mesoblastic nephroma described in Figures 10.15 and 10.16, and the nuclear atypia and mitotic activity are greater. H&E; × 300. (E) Electron photomicrograph (× 3800) reveals spindle cells with features consistent with fibroblasts. Cytoplasmic fibrillar material is present. (Electron photomicrograph courtesy of Ann Peters, Department of Pathology, Children's Hospital, San Diego.) (F) Karyotype reveals a trisomy 11, which is described for this tumor [43,44]. (Karyotype courtesy of James Mascarello, Ph.D., Department of Genetics, Children's Hospital, San Diego.)

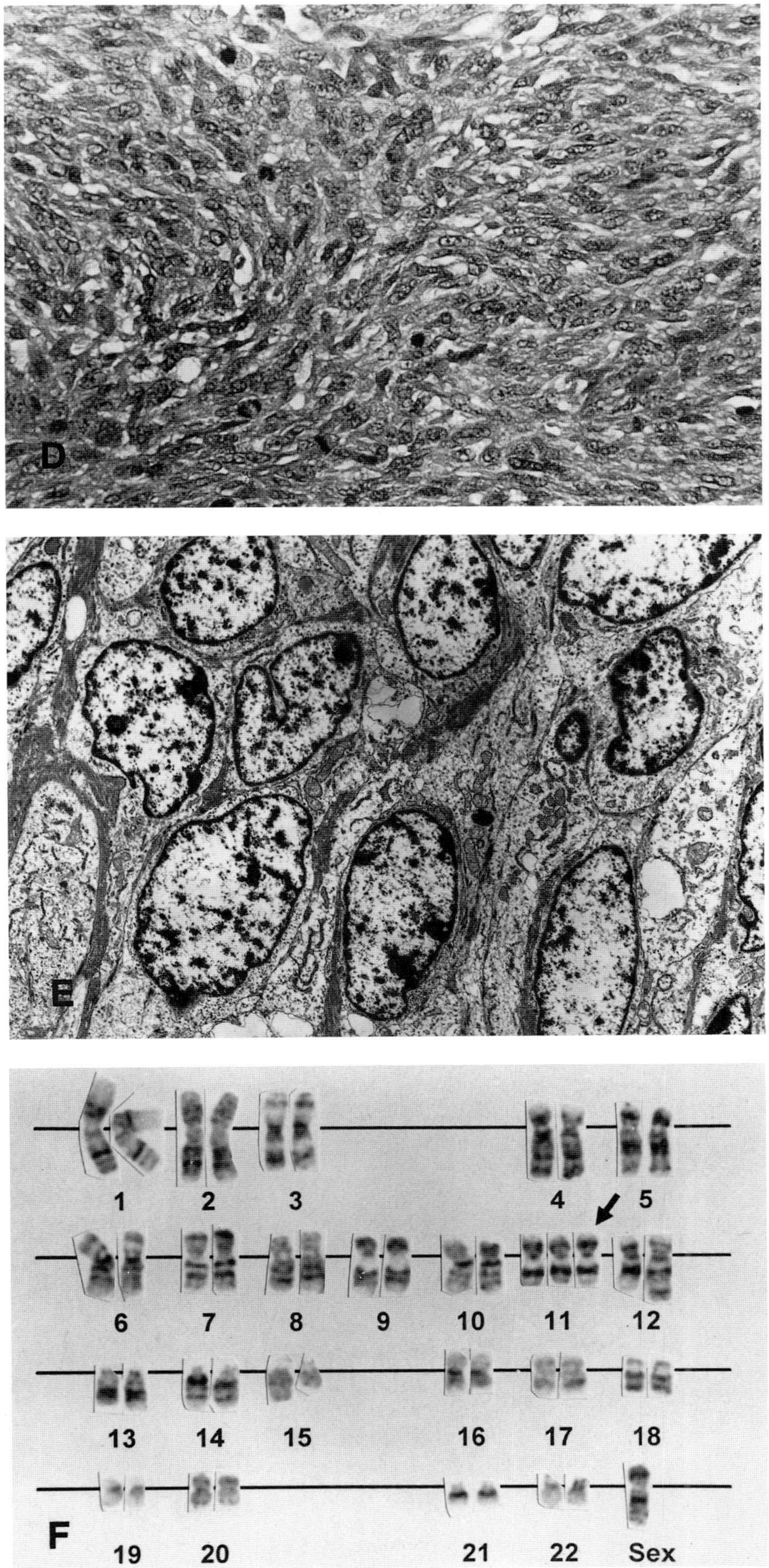

FIGURE 10.17. (*Continued*)

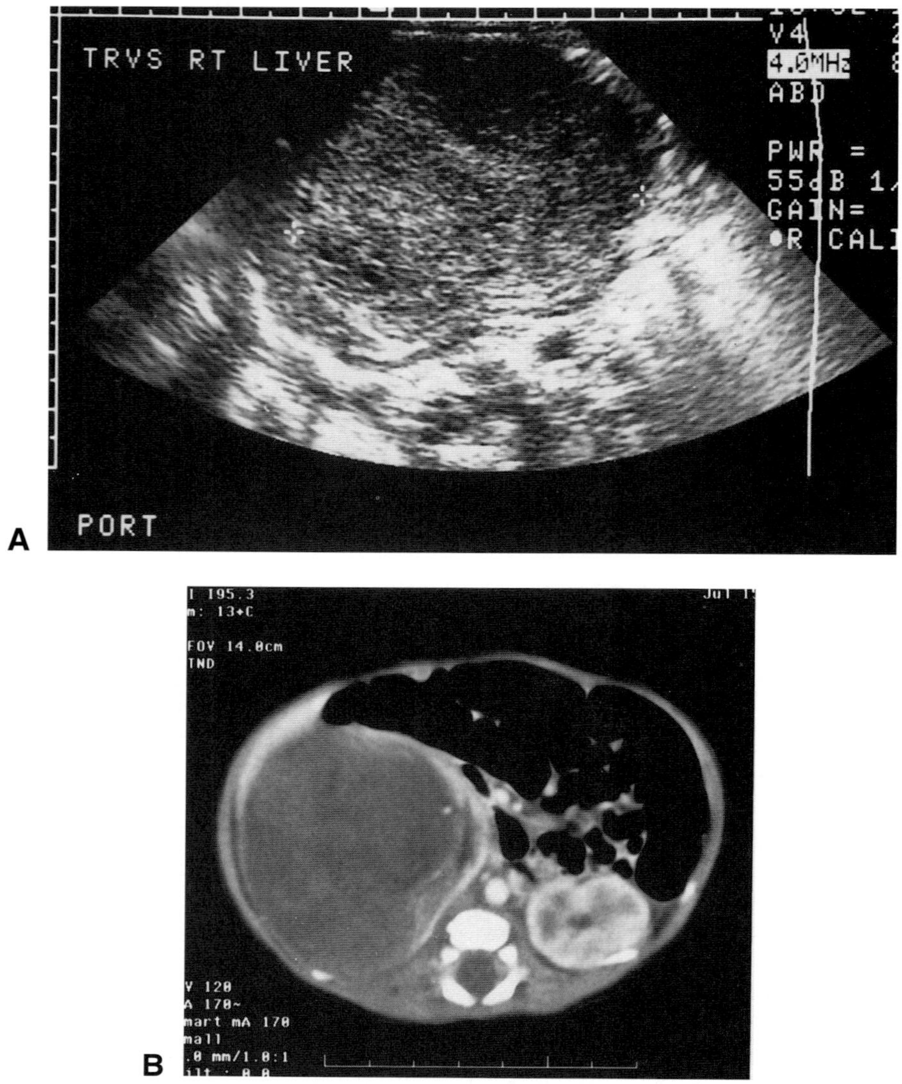

FIGURE 10.18. Congenital cellular mesoblastic nephroma. (A) Sonogram depicts a right abdominal mass. (B) Computerized tomography scan reveals a large mass replacing most of the kidney. (C) The tumor, 122 g, 6.5 × 5.5 cm^2, has a white-yellow, gelatinous cut surface with foci of hemorrhagic necrosis; it occupies almost the entire kidney except for a small portion of renal tissue in the lower pole and hilar region. (D) The hypercellular tumor shows an interdigitating growth pattern. The cells are small round to spindle-shaped cells with hyperchromatic, pleomorphic nuclei and a few mitoses. H&E; × 380.

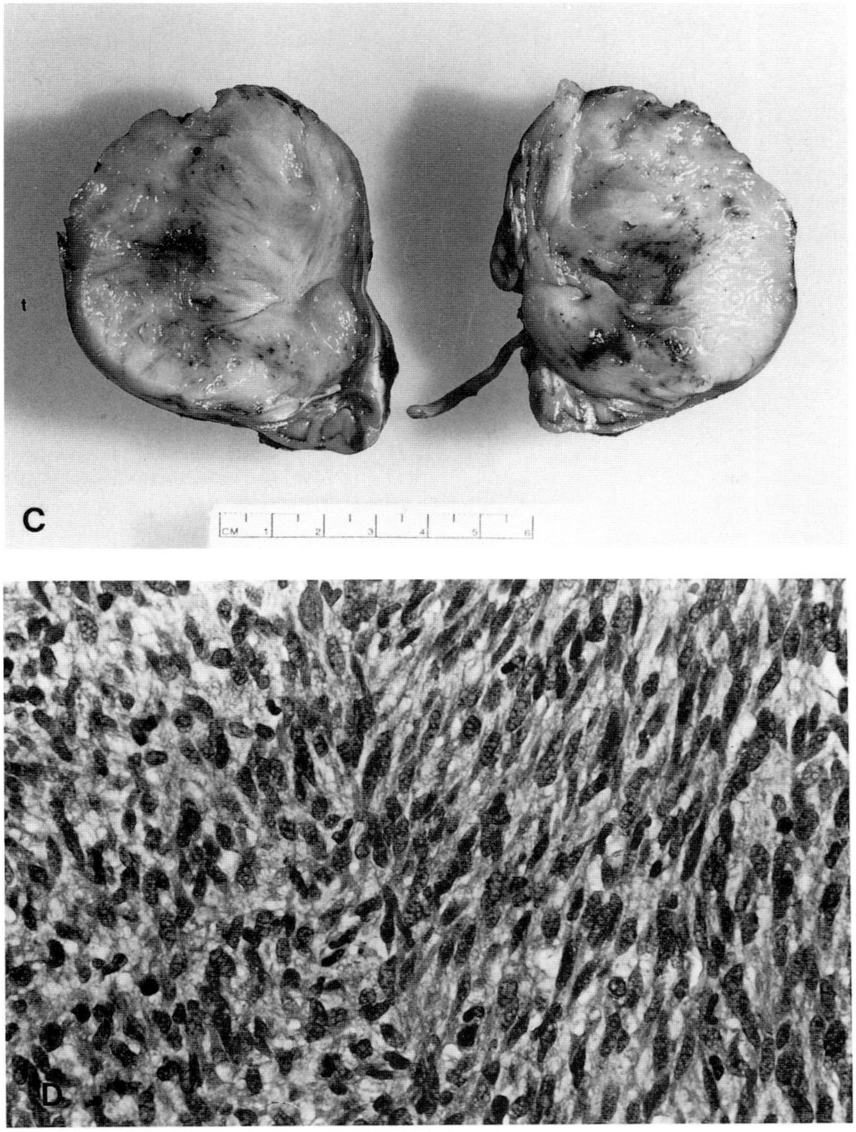

FIGURE 10.18. (*Continued*)

Rhabdoid Tumor of the Kidney

Rhabdoid tumor of the kidney was given its name because microscopically it resembles rhabdomyosarcoma, although it does not show skeletal muscle markers either by immunohistochemical studies or by electron microscopy [1–3] (Figures 10.19–10.22). This highly aggressive, malignant neoplasm, characterized clinically by early metastases and a high mortality rate, occurs primarily in the infant and newborn (60–70% of all rhabdoid tumors of the kidney) [3,13,14,18,49,50]. Rhabdoid tumor is the second most common malignant renal tumor in the neonate following Wilms tumor, surpassing clear-cell sarcoma [17].

Pathology

The cut surface of the tumor is soft, friable, gray-pink to tan with focal cystic necrosis and hemorrhage [50]. The rhabdoid tumor shows more necrosis and hemorrhage than the typical Wilms tumor [1,13,26] (Figure 10.20). In a young child the former may be as large as 15 cm across, replacing most of the kidney [10].

The rhabdoid tumor "classic pattern" consists of a monomorphous population of cells composed of large round to oval cells with prominent eosinophilc cytoplasm, large vesicular nucleus, characteristically a big single nucleolus, and round hyaline cytoplasmic inclusions [1,26,50] (Figure 10.19). Ultrastructurally the inclusions are composed of a meshwork of intermediate filaments. Immunoperoxidase studies reveal that the tumor cells are vimentin and cytokeratin immunoreactive but are nonreactive for the skeletal muscle markers myoglobin and desmin; neuron-specific enolase (NSE) is variably reactive [42,50]. Cell cultures of rhabdoid tumor show mesenchymal features, and unlike neuroblastoma and Wilms tumors, the cells express the c-*myc* rather than the n-*myc* oncogene [51].

Additional morphological appearances of rhabdoid tumor are described: organoid, sclerosing, spindled, lymphomatoid, and vascular patterns [3] (Figures 10.20–10.22). These histological variations may or may not be present in addition to the classical pattern.

The vascular variant can be confused with clear-cell sarcoma of the kidney, since it consists of nests of tumor cells with clear, vacuolated cytoplasm separated by very prominent vascular septa. Although cytoplasmic inclusions may not be observed, the large nucleoli and denser chromatin differ from typical nuclei of clear-cell sarcoma [3,50]. The spindled pattern consists of interlacing bundles of densely packed spindle cells resembling fibroblastic tumors (e.g., cellular mesoblastic nephroma) (Figure 10.22). The sclerosing pattern is composed of a dense hyaline stroma resembling chondoid or osteoid matrix containing small groups of cells [3].

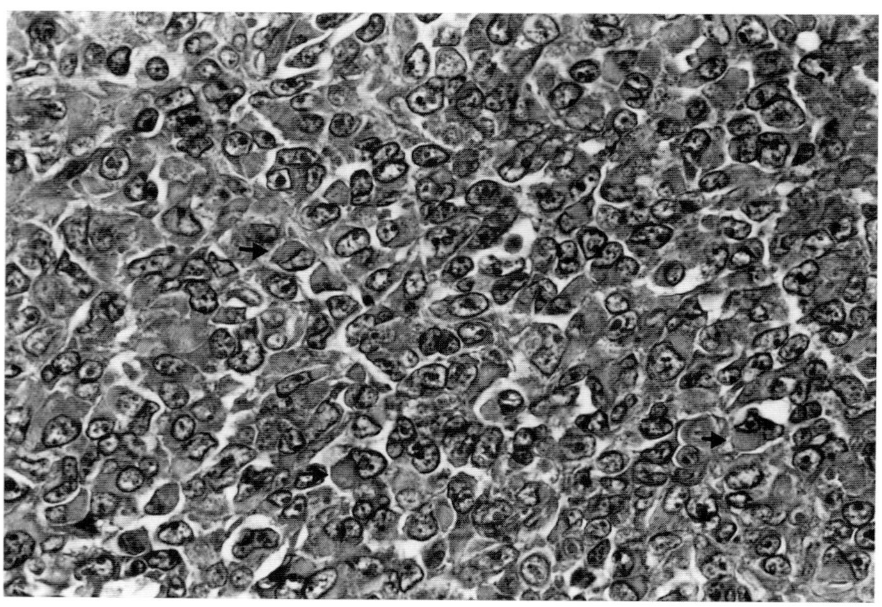

FIGURE 10.19. Rhabdoid tumor of the kidney, classic pattern, in a 2-year-old girl with gross hematuria and an abdominal mass. Teardrop-shaped cells have vesicular nuclei and contain a single, large nucleolus. Within the cytoplasm there are ill-defined round to oval hyaline inclusions (arrows) composed of intermediate filaments best demonstrated by electron microscopy. H&E; × 250.

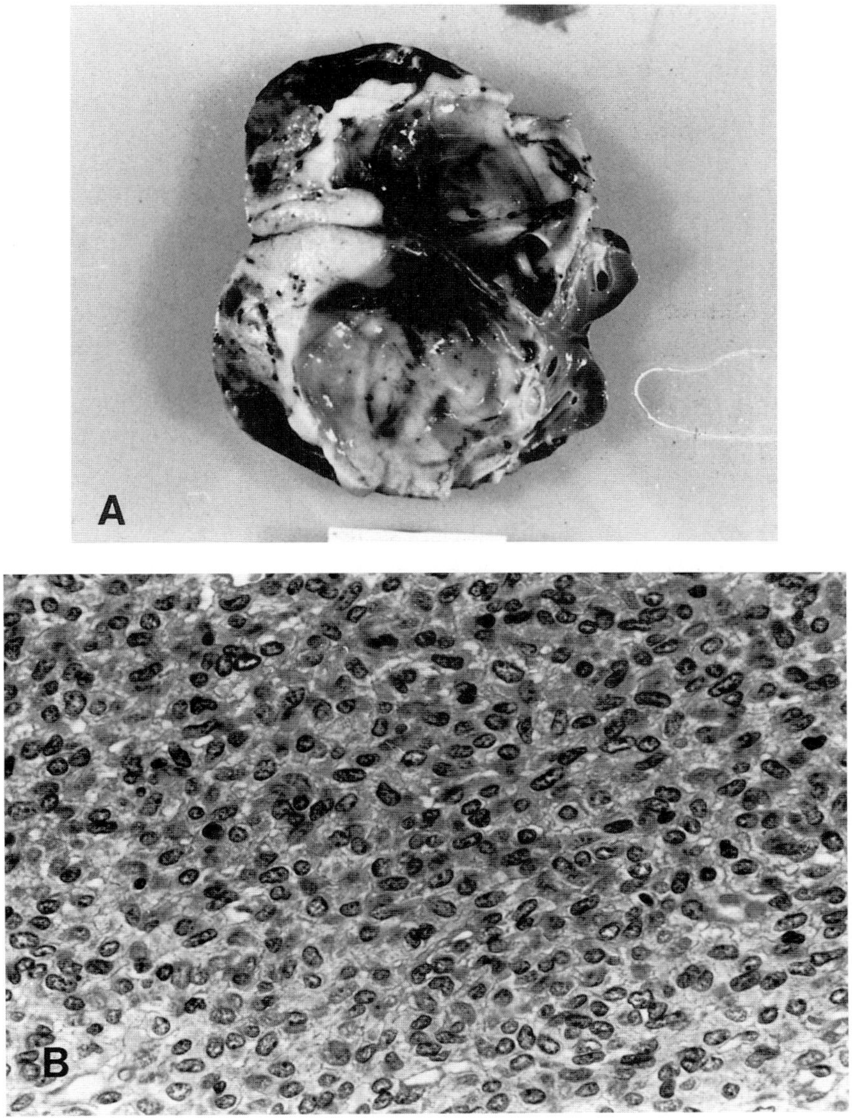

FIGURE 10.20. Rhabdoid tumor of the kidney, lymphomatoid pattern, in a 1.5-month-old male with an abdominal mass. (A) The tumor, 517 g, $13 \times 8 \times 1.8$ cm^3, replaces most of the kidney. The specimen is divided in half and spread open revealing a bulging, light tan-gray soft mass with extensive hemorrhagic necrosis. (B) Solid sheets of noncohesive tumor cells with small to moderately large nucleoli and scant cytoplasm devoid of inclusions mimic a large cell lymphoma [3]. H&E; $\times$ 380.

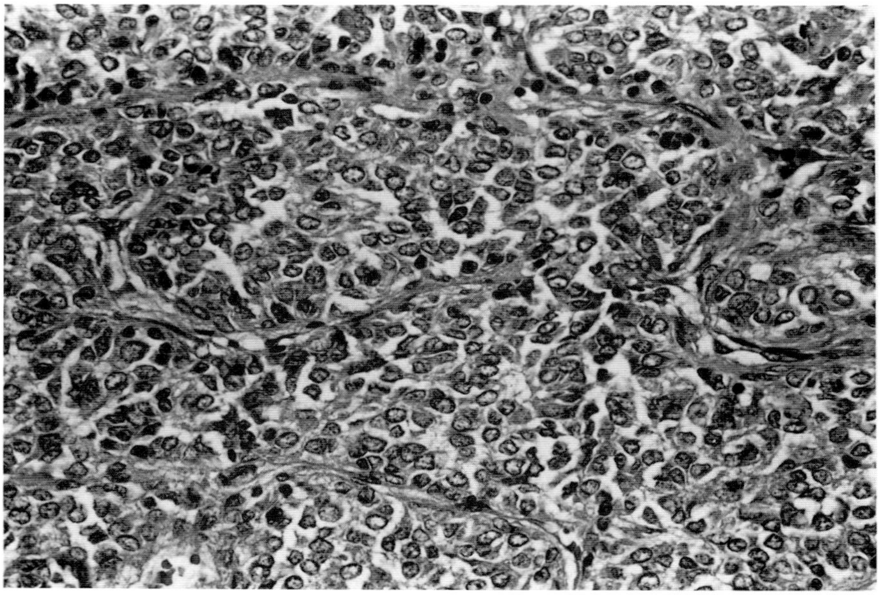

FIGURE 10.21. Rhabdoid tumor of the kidney, organoid pattern, in a 1-year-old male with a renal mass and liver metastases. Delicate fibrovascular septa separate lobules of round to oval cells with vesicular nuclei and small nucleoli [3]. This histological pattern of rhabdoid tumor may be confused with clear-cell sarcoma. H&E; × 200.

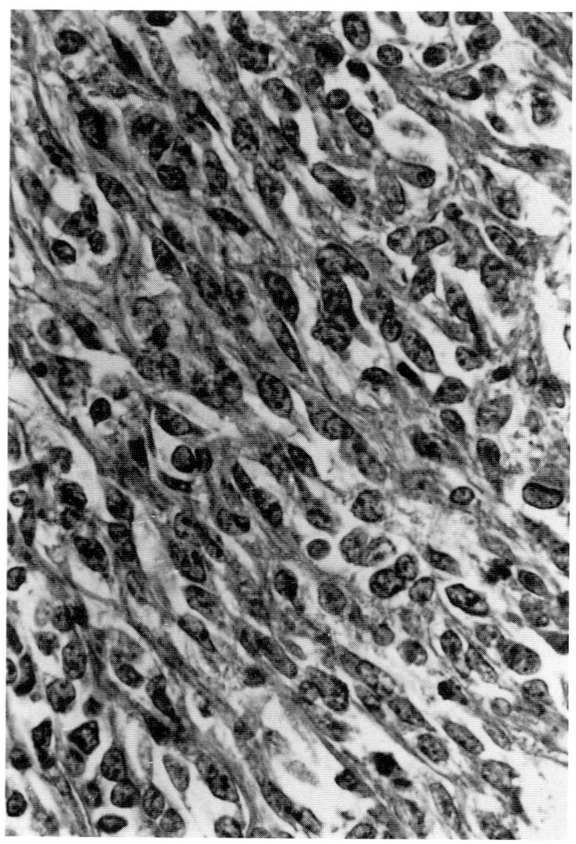

FIGURE 10.22. Rhabdoid tumor of the kidney, spindled pattern, in a 1-month-old female with a posterior neck mass present at birth and a rapidly enlarging, nonresectable renal tumor, which was biopsied. Atypical spindle-shaped cells resembling rhabdomyoblasts but without cross-striations are present. There are a few cells with large vesicular nuclei and nucleoli. H&E; × 400. (From Isaacs H Jr. [10], with permission.)

Clear-Cell Sarcoma of the Kidney

Clear-cell sarcoma of the kidney (CCSK) was found in 2.5% of the patients enrolled on the National Wilms Tumor Study [2]. The tumor has a tendency to metastasize to bones and carries a poor prognosis [2,14,26,52,53]. Several cases are described in newborns and infants [12,26,53,54]. Clear-cell sarcoma of the kidney occurs mostly in males but with rare exceptions is not associated with any specific congenital anomaly, clinical condition, or malformation syndrome [26,52,54].

Pathology

The gross features of CCSK are similar to a Wilms tumor, but bilateral renal clear-cell sarcomas are not described [13,26,53] (Figures 10.23 and 10.24). Microscopically the tumor displays a characteristic monomorphous appearance consisting of rows of round to oval clear-staining cells separated by delicate septa (Figures 10.23 and 10.24). The nuclei are small and irregular with fine chromatin or a vesicular or clear appearance ("open nuclei"). Mitoses are rare. There is a distinctive vascular component composed of branching capillaries within thin fibrous septa. Cyst formation is not uncommon. The tumor appears to surround or push aside preexisting glomeruli and tubules, which have an embryonic metaplastic appearance. Although the neoplasm is called a sarcoma, neither the ultrastructural nor the immunoperoxidase findings support this concept [42,52,53]. At present its histogenesis is unknown.

Several other histological patterns are described for clear-cell sarcoma of the kidney. The palisaded "Verocay body" pattern consists of parallel rows of cells resembling a schwannoma (Figure 10.24C); the sclerosing pattern has areas of fibrosis and stromal hyalinization mimicking osteoid; the epithelioid trabecular pattern shows an epithelial trabecular arrangement resembling tubules; and the spindle cell pattern with spindle cells produces storiform formations simulating fibrohistiocytic neoplasms (Figure 10.24C) [42,52,53].

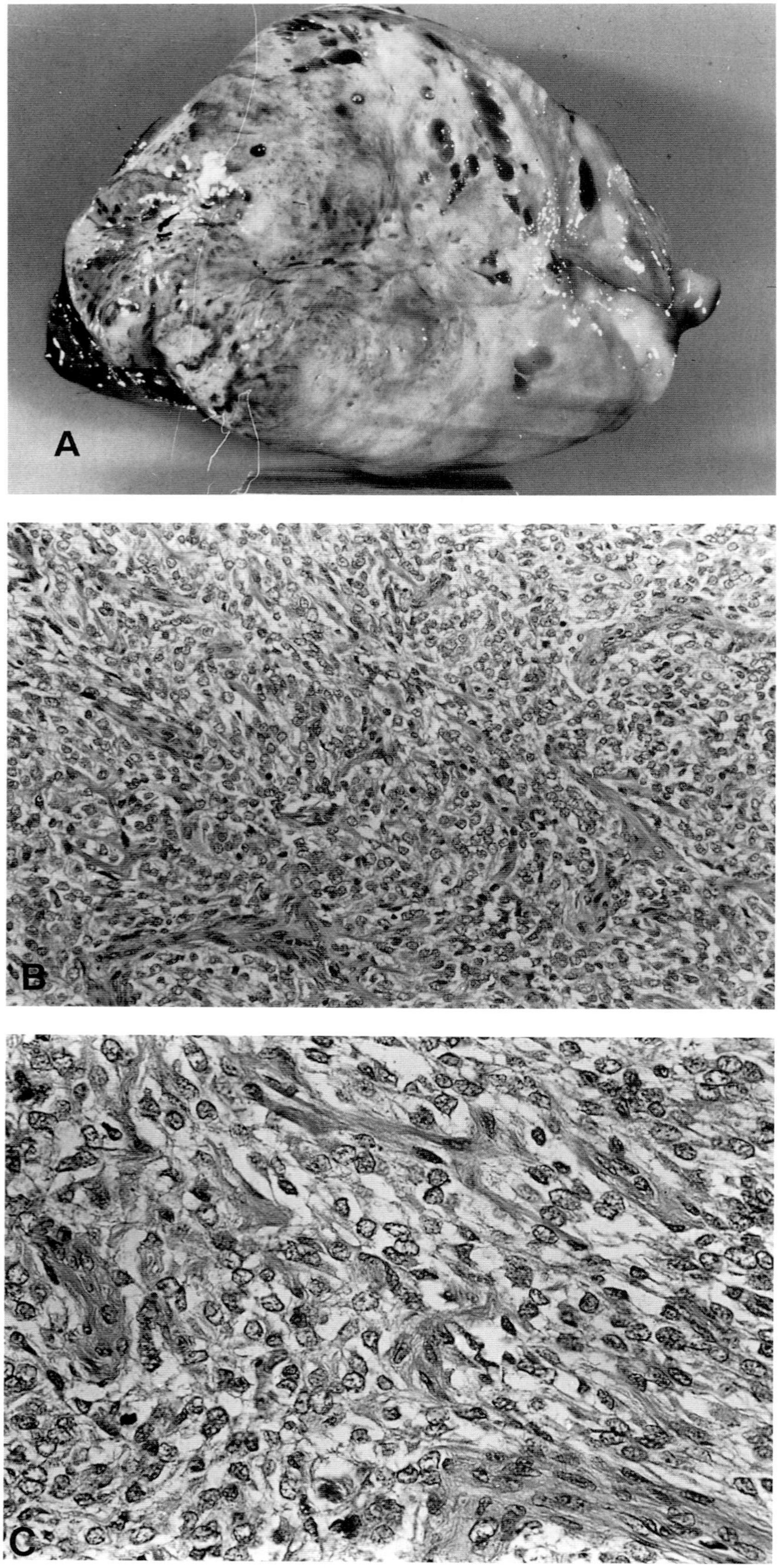

FIGURE 10.23. Clear-cell sarcoma of the kidney in a 15-month-old male with an abdominal mass. (A) The 735 g, 16.5 × 11.5 cm^2, right renal tumor has a soft gray-white to gray-pink cut surface. It invaded the renal hilum and adjacent liver. (B) The tumor consists of fields of cells with "open" (empty) nuclei within a pale myxoid stroma. Note the prominent branching vascular pattern, which is a diagnostic feature of this neoplasm. H&E; × 100. (C) Higher magnification (× 200) revealing polygonal cells consisting of water-clear cytoplasms, and round to oval "open" nuclei with finely dispersed chromatin and inconspicuous nucleoli. H&E.

FIGURE 10.24. Clear-cell sarcoma of the kidney in a 4-month-old female with a right abdominal mass discovered by her pediatrician on routine physical examination. Solid and cystic areas were noted not only on imaging studies but also on gross inspection of the tumor. (A) Most of the kidney is replaced by the tumor, 254 g, $10 \times 8 \times 7$ cm^2, which has light tan-gray, gelatinous, hemorrhagic cut surface. The renal sinus was invaded (stage II). (B) Cytological touch preparation displays tumor cells having a spermlike configuration embedded in mucoid material. The nuclei show a fine chromatin pattern and one or more tiny nucleoli. (C) The interdigitating spindle cells form a storiform (cartwheel) growth pattern. The tumor shows nuclear palisading suggestive of Antoni A areas of a schwannoma (arrows). H&E; $\times$ 150.

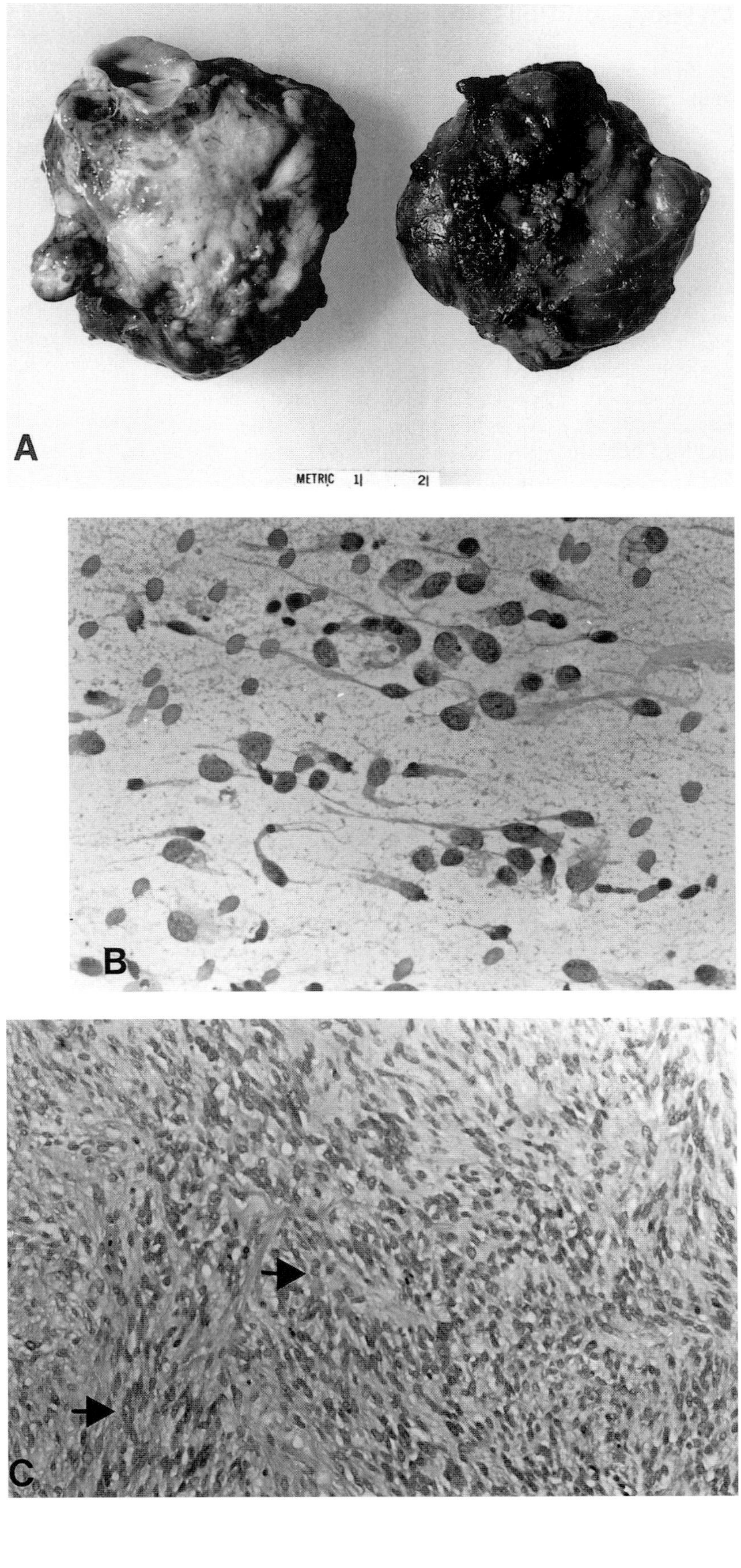

Cystic Conditions of the Kidney

Cystic renal lesions occur in fetuses and infants and include both true tumors and neoplastic-like conditions of probable developmental origin [13,26,55–61]. Hydronephrosis and multicystic kidney are responsible for most abdominal masses in this age group (Table 1.2). Members of this group include the multilocular cyst (MLC), cystic partially differentiated nephroblastoma (CPDN), and cystic Wilms tumor (multilocular cystic nephroma) (MLCN). Cystic renal lesions may antedate the classical features of tuberous sclerosis in the infant and may be mistaken clinically and radiologically for a neoplasm [55]. Multilocular cyst, cystic partially differentiated nephroblastoma, and cystic Wilms tumor appear to be related histogenetically and perhaps could be considered to be a continuum with MLC at the benign end, CPDN in the middle, and MLCN at the malignant end [26,56–61].

Multiocular Cyst of the Kidney

Multilocular cyst of the kidney presents clinically as an abdominal mass and is often diagnosed preoperatively by imaging studies. Most multilocular cysts are unilateral, but bilateral examples are described [60]. The lesion consists of a multicystic mass situated in the upper or lower pole of the kidney, well demarcated from the renal pelvis by a discrete light tan to white capsule (Figures 10.25 and 10.26).

The cysts vary in size, contain clear fluid, and have thin fibrous septa, which by definition lack nephrogenic elements on microscopic examination [5,8,26,56,57]. Multilocular cyst has been described in association with both mesoblastic nephroma and Wilms tumor [61].

Cystic partially differentiated nephroblastoma or cystic nephroma is another multicystic condition grossly identical to MLC but microscopically different because it contains immature or mature nephrogenic (metanephric) elements within the fibrous septa that separate the cysts [26,56,57,59]. Cystic partially differentiated nephroblastoma occurs both in neonates and in older children.

Cystic Wilms tumor is the third member of this spectrum of cystic "neoplastic" conditions of the kidney. A well-differentiated Wilms tumor with focal or multifocal cystic change sometimes creates a difficult problem in the differential diagnosis of a multicystic renal mass. Although unusual, a multilocular cyst may occur in the same kidney with a Wilms tumor (13,61).

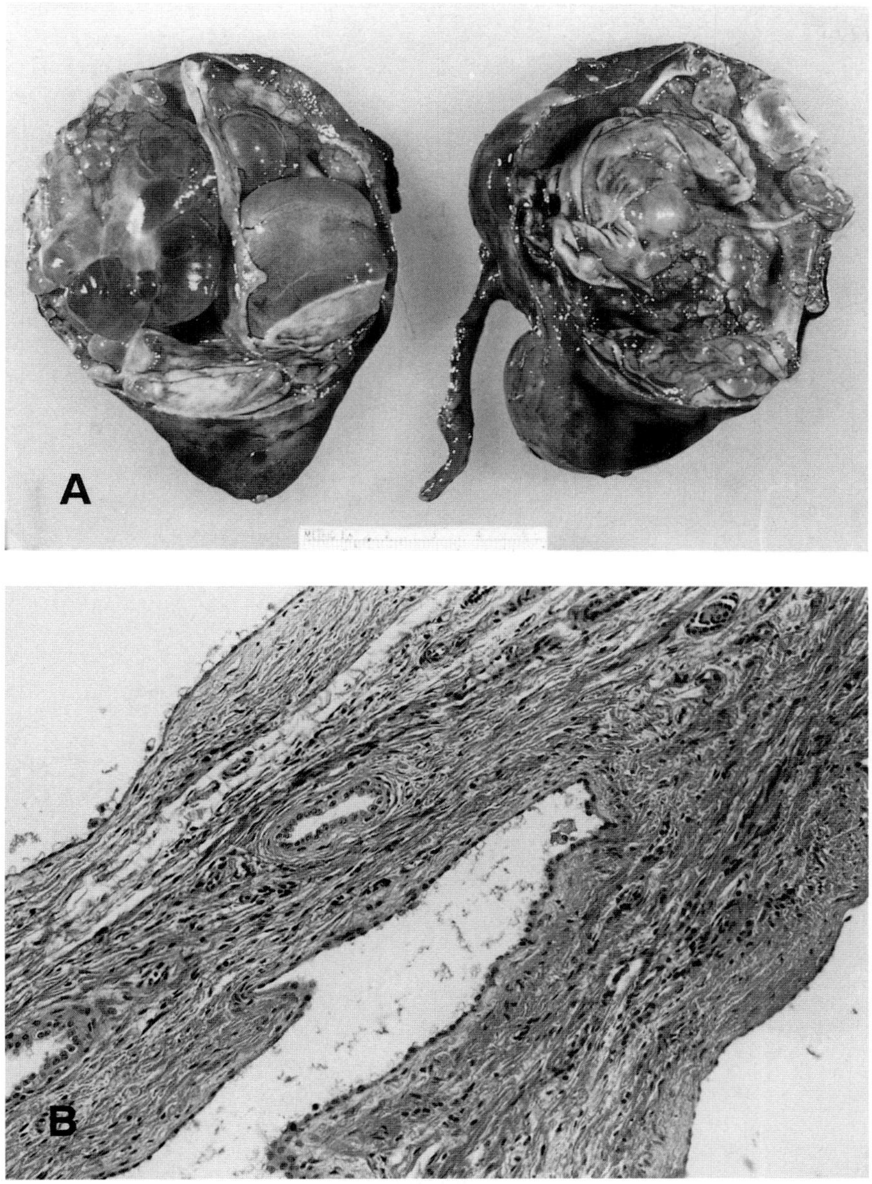

FIGURE 10.25. Multilocular cyst of the kidney in a 2-year-old girl with a left abdominal mass. (A) The specimen, 405 g, 13 × 10 cm^2, consists of a muliloculated cyst occupying most of the kidney. The lesion does not communicate with the renal pelvis. The cysts range from 0.2 cm up to 4 cm in diameter and contain clear straw-colored fluid. The lesion is well encapsulated and surrounded by a capsule 2 to 3 mm thick. (B) Cysts of various sizes are separated by fibrous connective tissue septa and are lined by cuboidal and flattened epithelium. No nephrogenic elements are present within the cyst walls. H&E; × 120. (From Fobi M, Mahour GH, Isaacs H Jr. Multilocular cyst of the kidney. *J Pediatr Surg* 1979;14:282. Used with permission.)

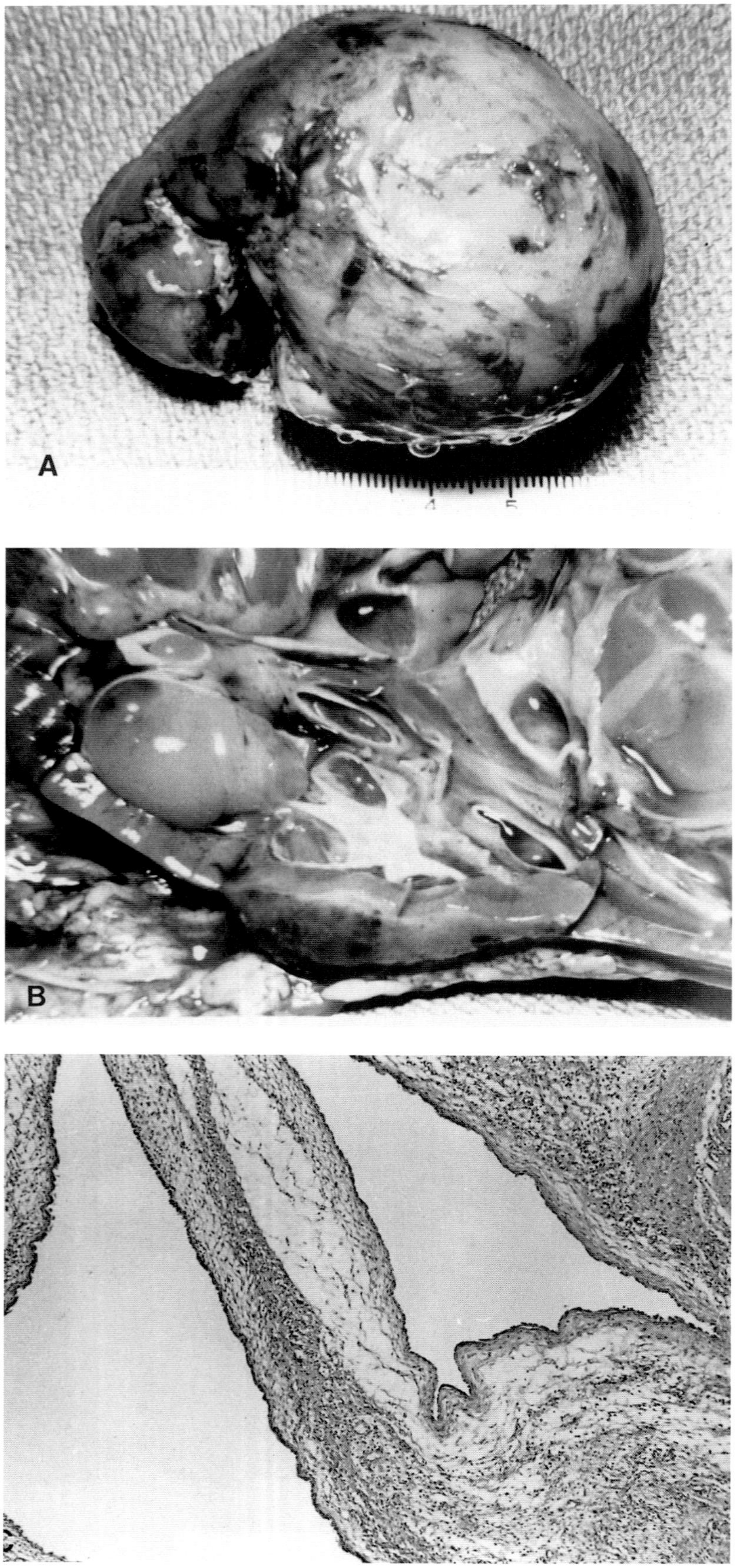

FIGURE 10.26. Multilocular cyst of the kidney in a 1-year-old with an abdominal mass. (A) The 6 × 5 cm^2 cystic lesion, which was situated within the upper and middle third of the right kidney. (B) The cut surface reveals multiple cysts of various sizes containing clear fluid. (C) The cysts are separated by slightly vascular fibrous connective tissue and lined by flattened epithelial cells. H&E; × 50. (From Isaacs H Jr. [8]. Used with permission.)

Ossifying Renal Tumor of Infancy

Ossifying tumor of the kidney consists mostly of bone and typically presents with hematuria [62–64]. These benign lesions occur as a calcified, polypoid mass arising from the upper or middle pole collecting system. Fewer than 20 cases have been reported [62–64]. Infants, usually male, present with gross hematuria and a calcified renal lesion demonstrated by imaging studies [63,64]. Ossifying renal tumor is considered a benign neoplastic condition rather than a reactive process [62].

Grossly the specimens consist of a rock-hard mass, 2 to 3 cm in diameter, attached to a renal papilla by a broad pedicle [62–64] (Figure 10.27). Histologically ossifying renal tumors are composed of three main components: osteoid, osteoblast, and spindle cells. The tumor contains a central nidus of mineralized osteoid admixed with polygonal cells having a plump nucleus, large nucleolus, and eosinophilic to amphophilic cytoplasm, features consistent with osteocytes or osteoblasts (Figure 10.27C,D). Islands of osteoid are surrounded by mesenchyme composed of spindle-shaped cells that blend imperceptibly with the surrounding renal medulla. The mesenchymal cells may have the appearance of a mesoblastic nephroma (Figure 10.27E). The polygonal cells appear to originate from the spindle cells. Variable perivascular fibrosis is present. Mitotic figures and cytologic atypia are absent [62–64]. Immunohistochemical studies reveal strong reactivity of the tumor cells with vimentin and no staining with either epithelial or vascular markers, which react only with adjacent renal tubules and blood vessels [64].

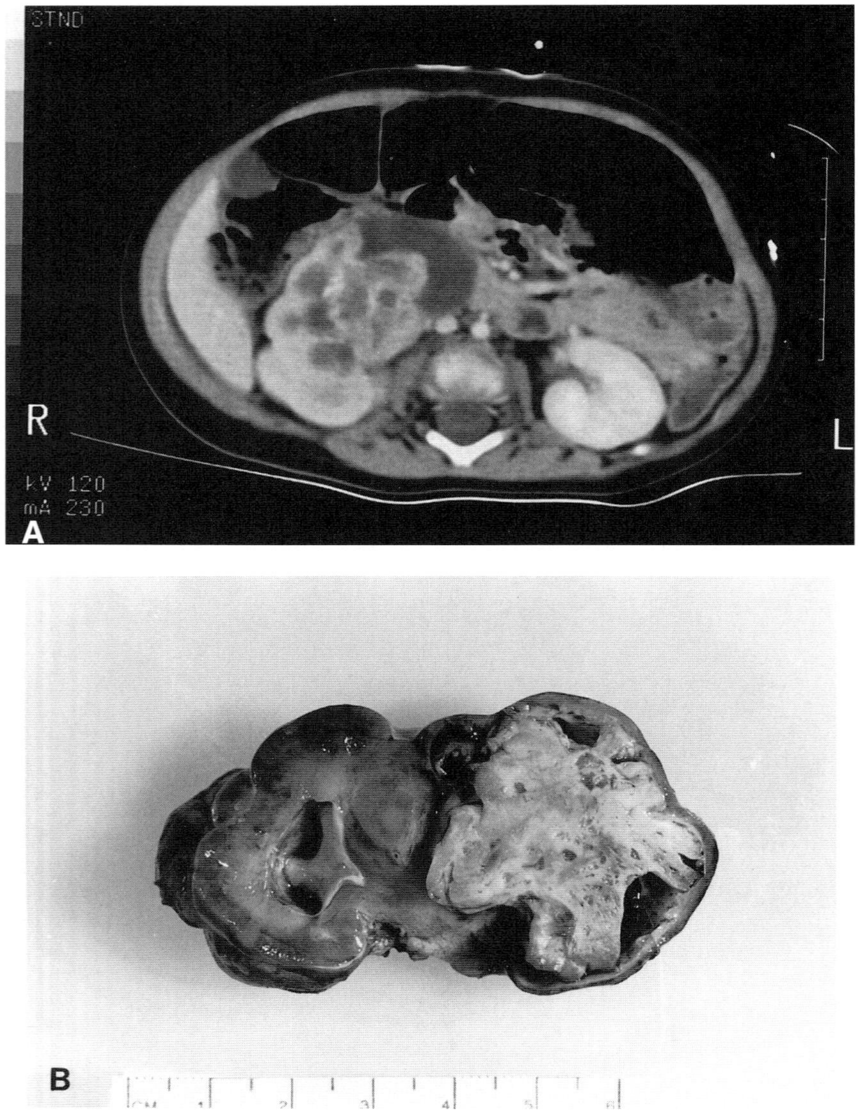

FIGURE 10.27. Ossifying renal tumor of infancy in a 4-month-old male with an abdominal mass and hematuria. (A) Magnetic resonance image showing a partially calcified mass arising from the right kidney. (B) The kidney, 45 g, contains an irregular, firm, gray-white mass, 3.5 × 3.0 × 2.5 cm^3, situated within the middle third and lower pole. The pelvis and calyces are dilated and the renal papillae blunted. Chronic pyelonephritis was present also. (C) Junction between an island of bone and nests of regular spindle-shaped cells. H&E; × 150. (D) Osteoid trabeculae are separated by clusters of osteoblasts and osteocytes with small, regular round nuclei. H&E; × 300. (E) An area adjacent to bone consisting of interdigitating small, regular spindle cells histologically resembling congenital mesoblastic nephroma. Note the lack of mitoses and nuclear atypia in all the microscopic sections illustrated. H&E; × 150.

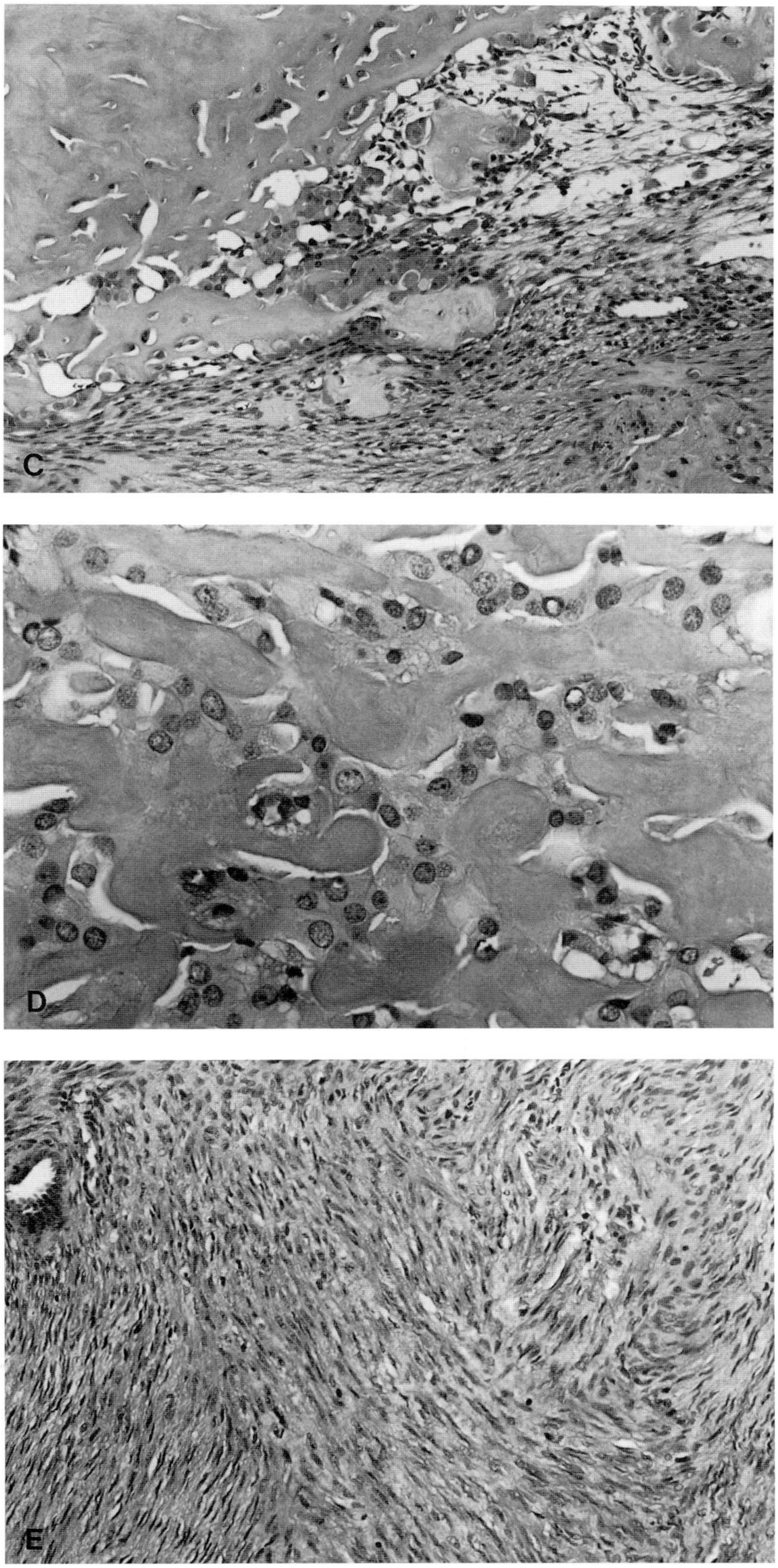

FIGURE 10.27. (*Continued*)

References

1. Beckwith JB, Palmer N. Histopathology and prognosis of Wilms' tumor. Results of the first National Wilms' Tumor Study. *Cancer* 1978;41:1937.
2. Beckwith JB. Wilms' tumor and other renal tumors of childhood: a selective review from the National Wilms' Tumor Study Pathology Center. *Hum Pathol* 1983;14:481.
3. Weeks DA, Beckwith JB, Mierau GW, et al. Rhabdoid tumor of the kidney. A report of 111 cases from the National Wilms' Tumor Study Pathology Center. *Am J Surg Pathol* 1989;13:439.
4. Bolande RP, Brough AJ, Izant RJ. Congenital mesoblastic nephroma of infancy. A report of eight cases and the relationship to Wilms' tumor. *Pediatrics* 1967;40:272.
5. Bolande RP. Developmental pathology. *Am J Pathol* 1979;94:627.
6. Bove KE. Wilms' tumor and related abnormalities in the fetus and newborn. *Semin Perinatol* 1999;23:310.
7. Dehner LP. Neoplasms of the fetus and neonate. In: Naeye RL, Kissane JM, Kaufman N, eds. *Perinatal Diseases*. Baltimore: Williams & Wilkins; 1981:286. International Academy of Pathology, Monograph No. 22.
8. Isaacs H Jr. Neoplasms in infants: a report of 265 cases. *Pathol Annu* 1983;18(2):165.
9. Isaacs H Jr. Congenital and neonatal malignant tumors: a 28-year experience at Children's Hospital of Los Angeles. *Am J Pediatr Hematol Oncol* 1987;9:121.
10. Isaacs H Jr. Perinatal (congenital and neonatal) neoplasms: a report of 110 cases. *Pediatr Pathol* 1985;3:165.
11. Isaacs H Jr. Renal tumors. In: *Tumors of the Newborn and Infant.* St. Louis, MO: Mosby Year Book; 1991:121.
12. Isaacs H Jr. Renal tumors. In: *Tumors of the Fetus and Newborn.* Philadelphia: WB Saunders; 1997:244. Major Problems in Pathology Series. Vol 35.
13. Isaacs H Jr. Tumors. In: Gilbert-Barness E, ed. *Potter's Pathology of the Fetus and Infant.* Vol 2. St. Louis, MO: CV Mosby; 1997:1302.
14. Marsden HB, Lawler W. Primary renal tumors in the first year of life. *Virchows Arch* [*A*] 1983;399:1.
15. Koop, CE. Abdominal mass in the newborn infant. *N Engl J Med* 1973;289:570.
16. Gordon B, Manivel JC, Gonzalez R, et al. Synchronous bilateral Wilms' tumor in a neonate. *Urology* 1996;47:409.
17. D'Angio GJ, Evans A, Breslow N, et al. The treatment of Wilms' tumor: results of the second National Wilms' Tumor Study. *Cancer* 1981;47:2302.
18. Hrabovsky EE, Othersen HB, deLorimer A Jr, et al. Wilms' tumor in the neonate: a report from the National Wilms' Tumor Study. *J Pediatr Surg* 21:385,1986.
19. Campbell AN, Chan HSL, O'Brien A, et al. Malignant tumours in the neonate. *Arch Dis Child* 1987;62:19.
20. Applegare KE, Ghei M, Perez-Atayde AR. Prenatal detection of a Wilms' tumor. *Pediatr Radiol* 1999;29:65.
21. Beckwith JB. Prenatal detection of a Wilms' tumor [editorial; comment]. *Pediatr Radiol* 1999;2964.
22. Koufos A, Hansen MF, Copeland NG, et al. Loss of heterozygosity in three embryonal tumors suggests a common pathogenetic mechanism. *Nature* 1985;316:330.
23. Coppes MJ, Haber DA, Grundy PE. Genetic events in the development of Wilms' tumor. *N Engl J Med* 1994;331:586.
24. Chatten J. Epithelial differentiation in Wilms' tumor: a clinicopathologic appraisal. *Perspect Pediatr Pathol* 1976;3:225.
25. Faria P, Beckwith JB, Mishra K, et al. Focal versus diffuse anaplasia in Wilms tumor—new definitions with prognostic significance. A report from the National Wilms Tumor Study Group. *Am J Surg Pathol* 1996;20:909.
26. Murphy WM, Beckwith JB, Farrow GM. Tumors of the kidney, bladder, and related urinary structures. In: *Atlas of Tumor Pathology.* 3rd ser; Fascicle 11. Washington, DC: Armed Forces Institute of Pathology; 1994.
27. Beckwith JB. National Wilms Tumor Study: an update for pathologists. *Pediatr Dev Pathol* 1998;1:79.
28. Gonzalez-Crussi F, Hsueh W, Ugarte N. Rhabdomyogenesis in renal neoplasia of childhood. *Am J Surg Pathol* 1981;5:525.
29. Weinberg AG, Currarino G, Hurt GE Jr. Botryoid Wilms' tumor of the renal pelvis. *Arch Pathol Lab Med* 1984;108:147.
30. Tomlinson GS, Cole CH, Smith NM. Bilateral Wilms' tumor: a clinicopathologic review. *Pathology* 1999;31:12.
31. Bove KE, McAdams JA. The nephroblastomatosis complex and its relationship to Wilms' tumor: a clinicopathologic treatise. In: Rosenberg HS, Bolande RP, eds. *Perspectives in Pediatric Pathology.* Vol 3. Chicago: Year Book Medical Publishers, 1976:185.
32. Beckwith JB. Nephrogenic rests and the pathogenesis of Wilms tumor: developmental and clinical considerations. *Am J Med Genet* 1998;79:268.
33. deChadarevian J, Fletcher BD, Chatten J, et al. Massive infantile nephroblastomatosis: a clinical, radiological, and pathological analysis of four cases. *Cancer* 1977;39:2294.
34. Beckwith JB. Precursor lesions of Wilms' tumor: clinical and biological implications, *Med Pediatr Oncol* 1993;21:158.
35. Bolande RP. Benignity of neonatal tumors and concept of cancer regression in early life. *Am J Dis Child* 1971;122:12.
36. Ambrosino MM, Hernanz-Schulman M, Horii SC, et al. Prenatal diagnosis of nephroblastomatosis in two siblings. *J Ultrasound Med* 1990;9:49.
37. Kulkarni R, Baile MD, Bernstein J. Progression of nephroblastomatosis to Wilms' tumor. *J Pediatr* 1980;96:178.
38. Beckwith JB. Personal communication.
39. Haddad B, Haziza J, Touboul C, et al. The congenital mesoblastic nephroma: a case report of prenatal diagnosis. *Fetal Diagn Ther* 1996;11:61.
40. Howey DD, Farrell EE, Sholl J, et al. Congenital mesoblastic nephroma: prenatal ultrasonic findings and surgical excision in a very-low-birth-weight infant. *J Clin Ultrasound* 1985;13:506.
41. Pettinato G, Manivel JC, Wick MR et al. Classical and cellular (atypical) congenital mesoblastic nephroma: a clinicopathologic, ultrastructural, immunohistochemical and flow cytometric study. *Hum Pathol* 1989;20:682.
42. Mierau GW, Beckwith JB, Weeks DA. Ultrastructure and histogenesis of the renal tumors of childhood: an overview. *Ultrastruct Pathol* 1987;11:313.
43. Knezevich SR, Garnett MJ, Pysher TJ, et al. *ETV6-NTRK3* gene fusions and trisomy 11 establish a histogenetic link between mesoblastic nephroma and congenital fibrosarcoma. *Cancer Res* 1998;589:5046.
44. Mascarello JT, Cajulis TR, Krous HF, et al. Presence or absence of trisomy 11 is correlated with histologic subtype in congenital mesoblastic nephroma. *Cancer Genet Cytogenet* 1994;77:50.
45. Kelly DR. Cystic cellular mesoblastic nephroma. *Pediatr Pathol* 1985;4:157.

46. Beckwith JB,Weeks DA. Congenital mesoblastic nephroma. When should we worry? *Arch Pathol Lab Med* 1986;110:98.
47. Gormley TS, Skoog SJ, Jones RV, et al. Cellular congenital mesoblastic nephroma: what are the options. *J Urol* 1989;142:479.
48. Howell CG, Otherson HB, Kiviat NE, et al. Therapy and outcome in 51 children with mesoblastic nephroma: a report of the National Wilms' Tumor Study. *J Pediatr Surg* 1982;17:826.
49. Hsueh C, Kuo T. Congenital malignant rhabdoid tumor presenting as a cutaneous nodule: report of 2 cases with review of the literature. *Arch Pathol Lab Med* 1998;122:1099.
50. Vujanic GM, Sandstedt B, Harms D, et al. Rhabdoid tumour of the kidney: a clinicopathological study of 22 patients from the International Society of Paediatric Oncology (SIOP) nephroblastoma file. *Histopathology* 1996;28:333.
51. Gansler T, Gerald W, Anderson G, et al. Characterization of a cell line derived from rhabdoid tumor of the kidney. *Hum Pathol* 1991;22:259.
52. Sotelo-Avila C, Gonzalez-Crussi, Sadowinski S, et al. Clear cell sarcoma of the kidney: a clinicopathologic study of 21 patients with long term follow-up evaluation. *Hum Pathol* 1986;16:1219.
53. Argani P, Perlman EJ, Breslow NE, et al. Clear cell sarcoma of the kidney: a review of 351 cases from the National Wilms' Tumor Study Group Pathology Center. *Am J Surg Pathol* 2000;24:4.
54. Uzoaru I, Podbielski FJ, Chou P. et al. Familial adenomatous polyposis coli and clear cell sarcoma of the kidney. *Pediatr Pathol* 1993;13:133.
55. Miller ID, Gray ES, Lloyd DL. Unilateral cystic disease of the neonatal kidney: a rare presentation of tuberous sclerosis. *Histopathology* 1989;14:529.
56. Andrews MJ Jr, Askin FB, Fried FA, et al. Cystic partially differentiated nephroblastoma and polycystic Wilms' tumor: a spectrum of related clinical and pathologic entities. *J Urol* 1983;129: 577.
57. Domizio P, Risdon RA. Cystic renal neoplasms of infancy and childhood: a light microscopical, lectin histochemical and immunohistochemical study. *Histopathology* 1991;19:199.
58. Beckwith JB, Kiviat NB. Multilocular renal cysts and cystic renal tumors. *Am J Roentgenol* 1980;136:435.
59. Joshi VV, Beckwith JB. Pathologic delineation of the papillonodular type of cystic partially differentiated nephroblastoma: a review of 11 cases. *Cancer* 1990;66:1568.
60. Chatten J, Bishop HC. Bilateral multilocular cysts of the kidney. *J Pediatr Surg* 1977;12:749.
61. Ganick DJ, Gilbert EF, Beckwith JB, et al. Congenital cystic mesoblastic nephroma. *Hum Pathol* 1981;112:1039.
62. Chatten J, Cromie WJ, Duckett JW. Ossifying tumor of infantile kidney: report of two cases. *Cancer* 1980;45:609.
63. Sotelo-Avila C, Beckwith JB, Johnson JE. Ossifying renal tumor of infancy: a clinopathologic study of nine cases. *Pediatr Pathol* 1995;15:745.
64. Wheeler RA, Moore IE, Atwell JD. An infantile ossifying renal tumour. *Pediatr Surg Int* 1994;9:135.

11 Liver Tumors

Tumors and tumorlike conditions of the liver are uncommon in the fetus and infant [1]. The incidence of malignant liver tumors during the first year of life as determined from the third National Cancer Survey (1960–1969) was 7.5 per million live births per year [2]. Of a total of 196 neoplasms of various types recorded, only 8 (4%) were hepatic malignancies. They ranked seventh as the cause of infant deaths from cancer in the study.

Hemangioma is the most common of primary hepatic tumors and tumorlike conditions in both infants and children, followed by mesenchymal hamartoma and hepatoblastoma [1,3–9]. The majority of hepatic neoplasms are metastatic rather than primary if the total number of surgical and autopsy specimens are considered [3]. Neuroblastoma is the tumor that most often metastasizes to the liver in newborns and infants, leukemia is second, and renal tumors are third [3,5,6]. Rhabdomyosarcoma, rhabdoid tumor, and yolk sac tumor are examples of other malignancies that metastasize to the liver in this age group [1,3,5,6].

The classification of liver tumors and tumorlike conditions is given in Table 11.1.

Hepatoblastoma

During the first year of life hepatoblastoma is the leading primary malignant hepatic tumor [1,4,7,10–14]. In the entire pediatric population, over half the tumors are diagnosed in infants and fewer than 10% are found in neonates [1,7,11,14].

Various congenital anomalies and malformation syndromes are associated with hepatoblastoma [4,14–16] (Table 11.2). Hemihypertrophy, the Beckwith-Wiedemann syndrome, and intestinal adenomatous polyposis are cited most often [7,10, 12,16,17]. Some of the same congenital malformations and syndromes occur also in patients with Wilms tumor [16] (Table 10.3).

Hepatoblastoma typically presents as an upper abdominal mass arising as a solitary tumor more often from the right lobe of the liver than the left [1,8]. The tumor has been detected on prenatal ultrasound examination [9,19]. Other clinical findings are respiratory distress and/or gastrointestinal symptoms secondary to a massively enlarged liver, spontaneous rupture with hemoperitoneum, anemia and jaundice, sexual precocity in males, polyhydramnios and nonimmune fetal hydrops [3,4,9,19] (Table 11.2.). In addition to neuroblastoma and leukemia, hepatoblastoma is one of the pediatric malignancies that metastasizes to the placenta and causes fetal death [20].

Practically all patients with hepatoblastoma have significantly elevated serum α-fetoprotein level, which is a valuable diagnostic tumor marker [7,9,12]. However, it should be pointed out that serum α-fetoprotein is normally increased during the first few months of life (see Table 2.3 for normal serum levels in newborns and infants). Moreover, levels are elevated also in patients with yolk sac tumor.

Grossly hepatoblastomas consist of a single round mass, replacing a part or almost the entire lobe of the liver and measuring up to 10 cm or more (see later: Figures 11.2, 11.6, and 11.8). The tumor is usually well circumscribed and well demarcated from the adjacent normal liver by a whitish-gray, fibrous pseudocapsule. The cut surface is light tan or cream colored, soft in consistency, with or without gray myxoid areas and yellow necrotic zones. Tiny white foci of calcification are noted in the mixed hepatoblastomas [1,3,4,8,12].

Hepatoblastoma is classified histologically into two main types, epithelial or mixed, depending on whether mesenchymal elements are present [3,4,7,8,13,14,21–24] (Table 11.3). The epithelial components occurring in both the epithelial and mixed hepatoblastomas are subdivided further into four main cell types designated as fetal, embryonal, macrotrabecular, and small-cell undifferentiated ("anaplastic"). The fetal cell type, as the name implies, resembles fetal liver cells (Figure 11.1) and consists of small regular cells with minimal nuclear atypia and a low mitotic rate forming narrow cords 1 or 2 cells wide, with bordering sinusoids and prominent foci of hematopoiesis (Figure 11.2). In addition the fetal type

TABLE 11.1. Classification of liver tumors and tumorlike conditions in the fetus and infant.

Hemangioma[a]
Capillary hemangioma ("hemangioendothelioma")
Cavernous hemangioma
Kaposiform hemangioendothelioma
Mesenchymal hamartoma
Solitary unilocular cyst
Adenoma[b]
Focal nodular hyperplasia[b]
Hepatoblastoma
Epithelial
Fetal
Embryonal
Mixed: epithelial and mesenchymal
Anaplastic
Hepatocellular carcinoma[b]
Germ cell tumors
Teratoma
Yolk sac tumor
Rhabdoid tumor
Sarcoma[b]
Rhabdomyosarcoma
Undifferentiated (embryonal) sarcoma
Metastatic neoplasms
Neuroblastoma
Leukemia
Malignant renal tumors (Wilms tumor, rhabdoid tumor, and clear-cell sarcoma)
Yolk sac tumor

[a] With or without arteriovenous malformation.
[b] Generally observed in older children.
Source: Isaacs H Jr. [19].

TABLE 11.2. Conditions associated with hepatoblastoma.

Aicardi syndrome
Beckwith-Wiedemann syndrome
Hemihypertrophy
Wilms tumor
Polyposis coli syndromes
Isosexual precocity
Fetal alcohol syndrome
Nonimmune hydrops fetalis
Cystathioninuria
α_1-Antitrypsin deficiency
Maternal contraceptive use
Trisomy 18
Umbilical hernia
Osteoporosis
Adenomatoid transformation of the renal epithelium
Placental metastases
Spontaneous rupture with hemoperitoneum
Anemia and jaundice
Polyhydramnios

Source: Isaacs H Jr [19]; Stocker JT, Ishak KG [30].

TABLE 11.3. Histological classification of hepatoblastoma.

Epithelial
Fetal pattern
Embryonal pattern
Macrotrabecular pattern
Small-cell undifferentiated ("anaplastic") pattern
Mixed epithelial and mesenchymal
Mixed with teratoid components[a]

[a] Tissues from all three germ layers.
Source: Conran RM, et al. [21].

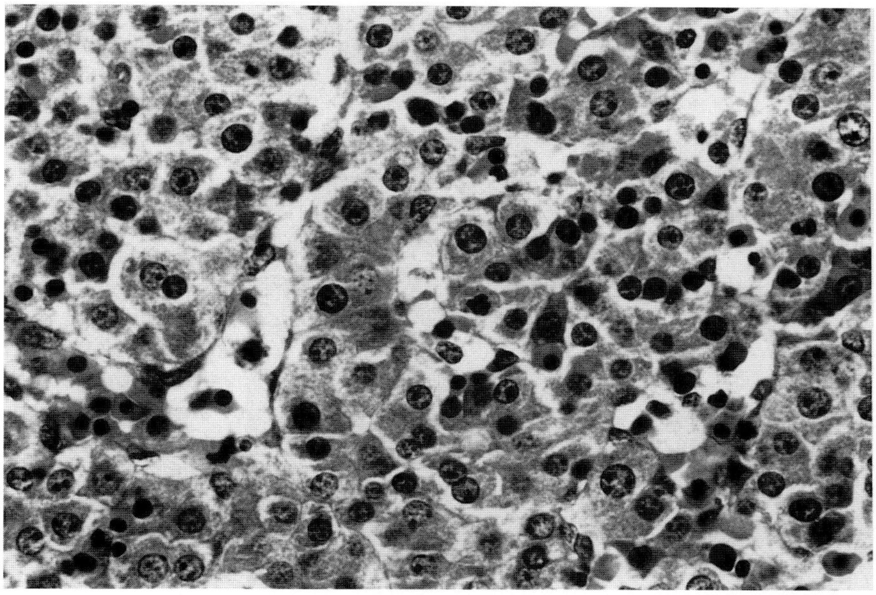

FIGURE 11.1. Developing liver, 12 mm human embryo. H&E; × 600. Hepatoblasts are cells of variable size with regular, round nuclei and one or two prominent nucleoli. Irregularly arranged early liver cell cord formations are bordered by sinusoids containing nucleated erythrocytes (erythroblasts).

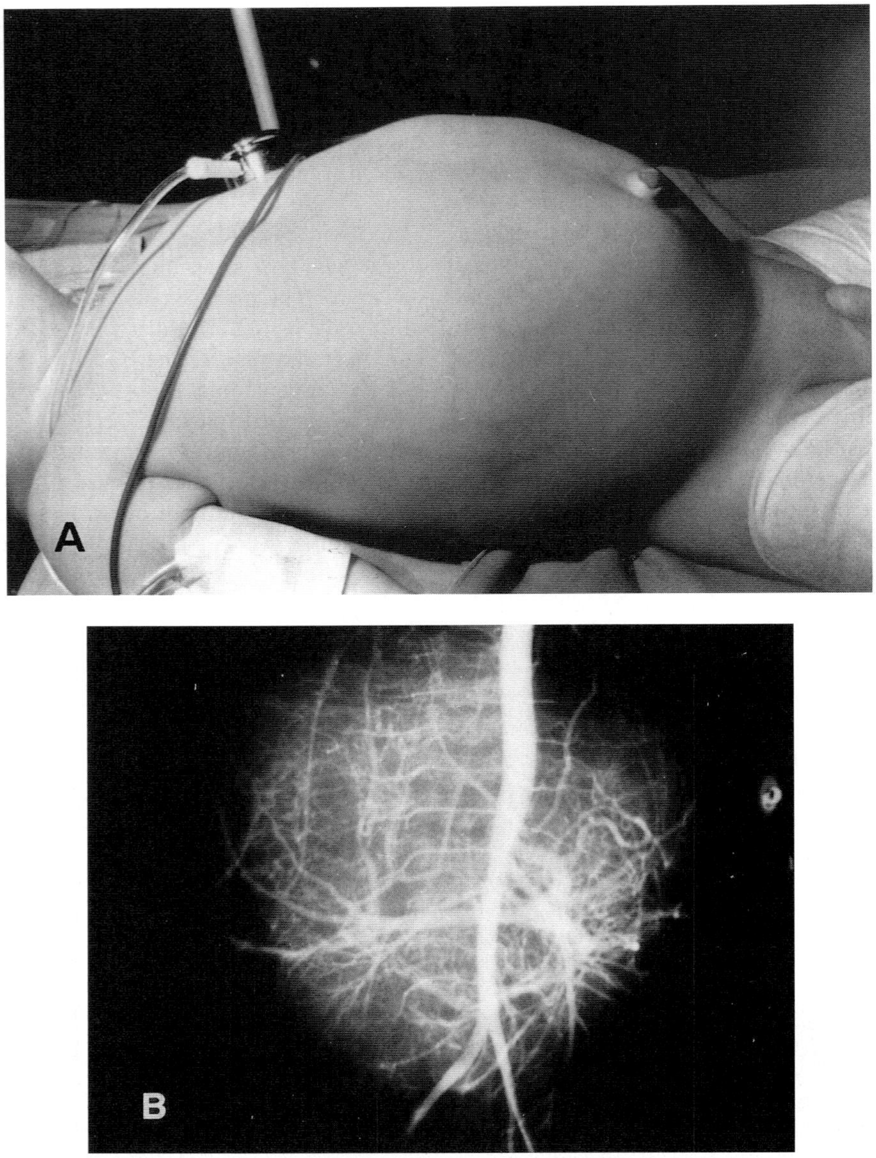

FIGURE 11.4. Hepatoblastoma, mixed with fetal and embryonal epithelial components, in a 5-month-old male with a 3-month history of progressive abdominal distension. (A) The huge abdominal mass arising from the liver. (B) Arteriogram reveals involvement of almost the entire liver. (C) Liver biopsy specimen shows a mixed hepatoblastoma consisting of both osteoid and epithelial components, the latter with fetal and embryonal cell types. H&E; × 300. (D) Higher magnification (× 600) of the embryonal hepatoblasts showing more irregularity in nuclear configuration and cell size and a more solid growth pattern in comparison to the fetal hepatoblastoma in (C). H&E.

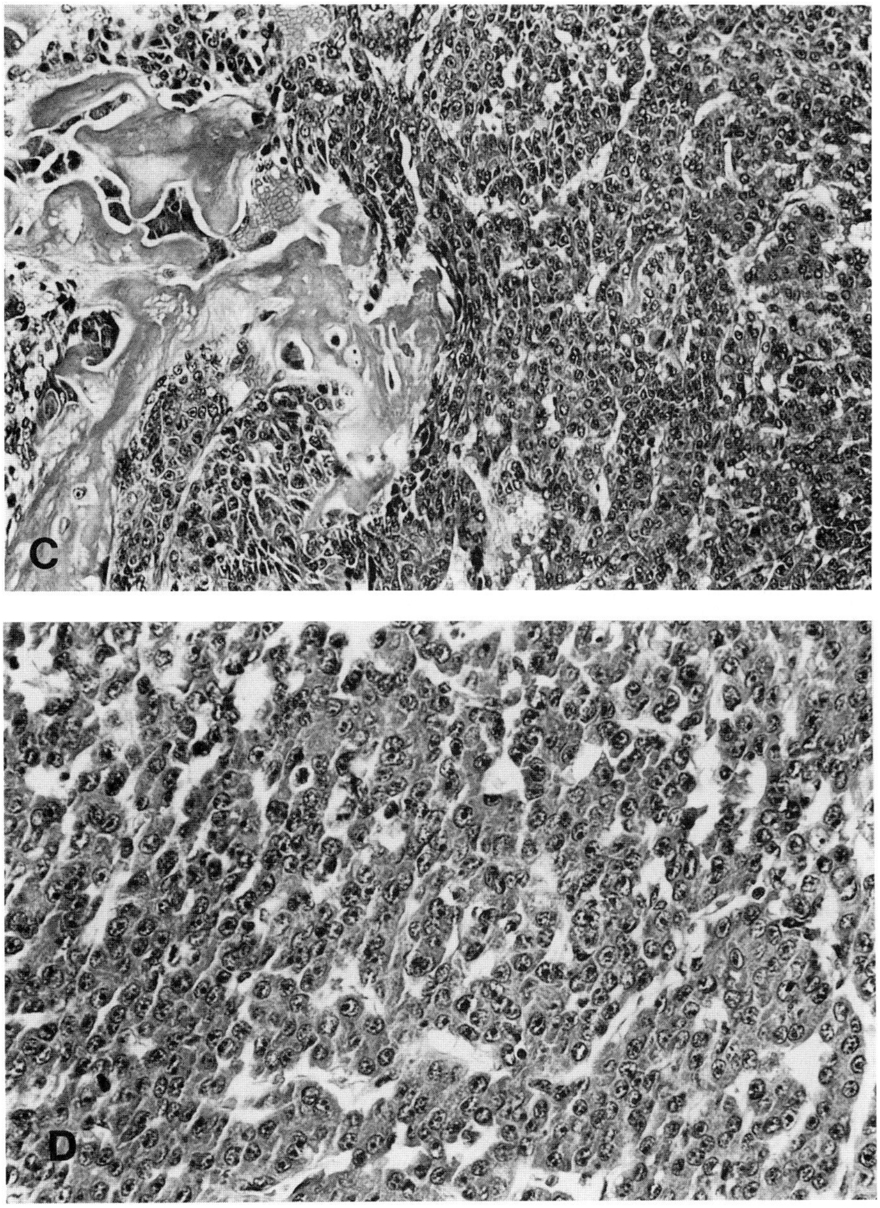

FIGURE 11.4. (*Continued*)

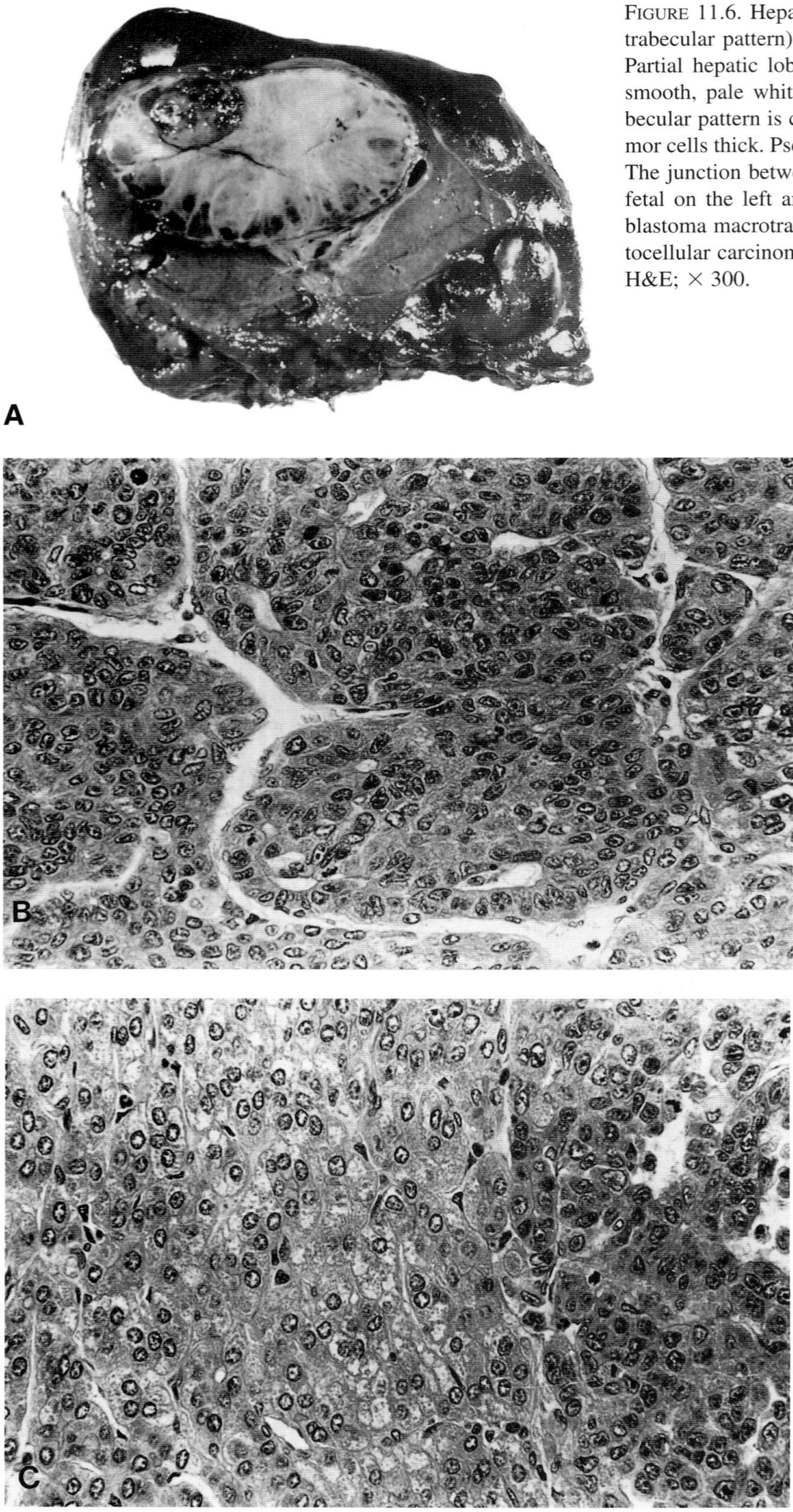

FIGURE 11.6. Hepatoblastoma (epithelial, embryonal, with the macrotrabecular pattern) in a 1-year-old male with an abdominal mass. (A) Partial hepatic lobectomy revealed a 7×5 cm^2 tumor composed of smooth, pale white-gray areas and nodular areas. (B) The macrotrabecular pattern is characterized by trabeculae (cords) more than 10 tumor cells thick. Pseudoacinar formations are present. H&E; $\times$ 300. (C) The junction between the two epithelial components of the tumor, the fetal on the left and embryonal on the right. Sometimes the hepatoblastoma macrotrabecular pattern is difficult to distinguish from hepatocellular carcinoma (cf with Figure 11.17, hepatocellular carcinoma). H&E; $\times$ 300.

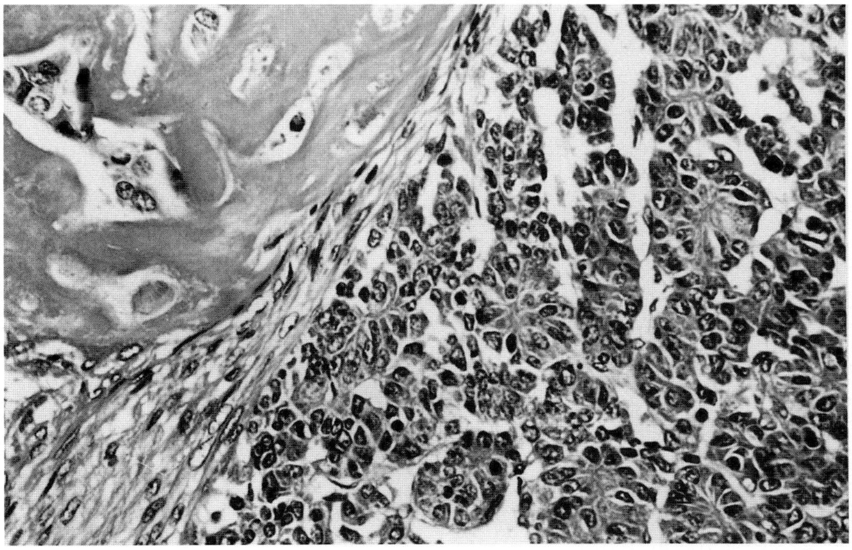

FIGURE 11.7. Hepatoblastoma (mixed; epithelial, embryonal with rosette formations), in a 9-month-old female with an abdominal mass, which had calcifications on imaging studies. The right hepatic lobectomy weighed 993 g and measured 14.5×11 cm^2. Adjacent to the island of osteoid in the upper left corner are clusters of embryonal hepatoblasts arranged in rosette formations. H&E; $\times$ 300.

reticulum, tonofilaments, desmosomes, and canalicular microvilli [25]. The tumor cells are cytokeratin positive but are unreactive with vimentin, desmin, S-100, and neuron-specific enolase [25]. When fetal or embryonal epithelial components are found in areas of the tumor, diagnosis of liver cell origin is established. A chromosomal translocation t(10;22) has been described for the anaplastic variant [27].

In addition to the epithelial components already described, mixed hepatoblastomas contain mesenchymal elements consisting of spindle-shaped fibroblastic cells, a pale-staining myxoid stroma, and small foci of osteoid, bone, and less often cartilage or skeletal muscle located within nests of fetal or embryonic epithelium. Osteoid is the most common differentiated tissue, and it is found in 20 to 35% of hepatoblastomas [12] (Figures 11.4C and 11.7). Osteoid is formed by the mesenchymal cells of the tumor [3,7].

The number of cases of "pure" epithelial versus mixed hepatoblastoma varies, but in most series the epithelial type predominates [3,5,8,14]. The presence of a mesenchymal component apparently has no bearing on prognosis [7]. The term "teratoid" is reserved for mixed hepatoblastomas with tissues from all three germ layers (e.g., cartilage, skeletal muscle, intestinal, and keratinizing stratified squamous epithelium) [21,28]. Deposits of melanin pigment are present in some. Hepatoblastoma with a yolk sac tumor component was described in a 6-month-old male [29].

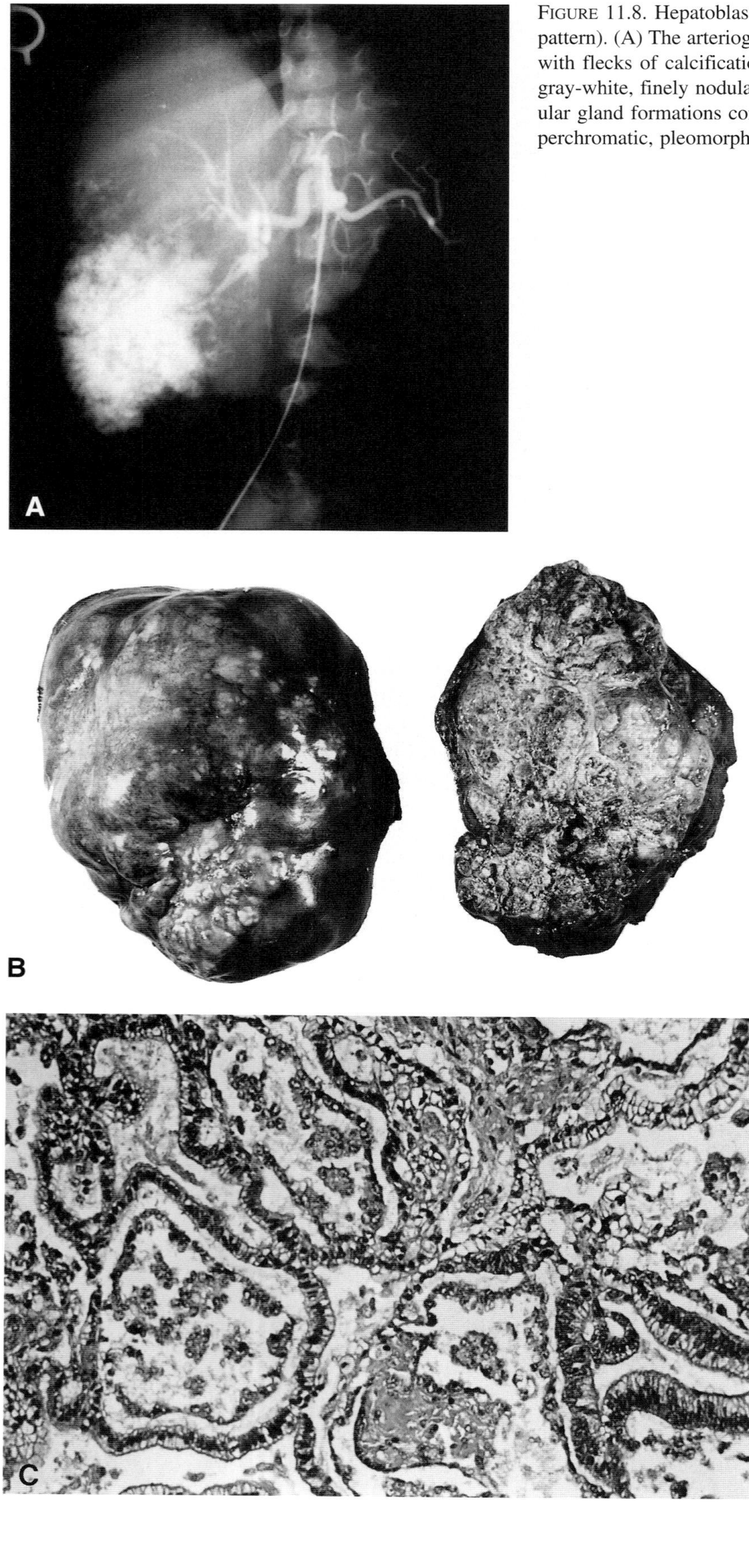

FIGURE 11.8. Hepatoblastoma (mixed; epithelial, embryonal with a glandular pattern). (A) The arteriogram reveals a large mass in the right lobe of the liver with flecks of calcification. (B) Most of the right lobe is replaced by a soft, gray-white, finely nodular tumor mass. (C) Portions of the tumor show irregular gland formations composed of vacuolated columnar epithelium with hyperchromatic, pleomorphic nuclei. H&E; × 300.

TABLE 11.4. Staging classification for hepatoblastoma.

Stage I	Complete resection
Stage II	Microscopic residual, negative nodal involvement, no spilled tumor
Stage III	Gross residual or nodal involvement or spilled tumor
Stage IV	Metastatic disease

Source: Conran RM, et al. [21].

The most common sites of metastases of hepatoblastoma in decreasing order of rank are the lungs, abdomen, brain, lymph nodes, and vena cava [15].

Survival of infants and children with hepatoblastoma depends on three important factors: complete resection of the tumor, absence of metastases, and absence of embryonal or undifferentiated cell types ("unfavorable histology") [7,8]. The patient with the best chance for survival has a small tumor that is completely resected before it has metastasized (i.e., stage I) [9] (Table 11.4). The one exception is the small-cell undifferentiated (anaplastic) variant, which has a particularly poor prognosis regardless of size because it is infiltrative and metastasizes early in the course of the disease [3,25,26]. Some hepatoblastomas are sensitive to chemotherapy, which produces necrosis, fibrosis, and shrinkage, allowing resection of an originally inoperable tumor.

Mesenchymal Hamartoma

Mesenchymal hamartoma is the second most common benign hepatic lesion in the first year of life, following hemangioma [1,3,5,7,12,30,31]. Over half the pediatric cases are observed in infants and about a fourth in newborns [1,7,12]. The chief presenting sign is an expanding abdominal mass. Frequently the diagnosis is suggested preoperatively by imaging studies (Figures 11.9–11.11) [19,32]. Although benign histologically, mesenchymal hamartomas may cause life-threatening complications, particularly in the perinatal period [33,34].

Mesenchymal hamartomas have been detected antenatally by sonography as early as 29 weeks gestation [19,33,34]. Sonograms reveal a complex multicystic, well-circumscribed mass in the upper fetal abdomen, often in the right lobe of the liver [34]. The mass contains areas of echolucency surrounded by multiple echodense membranes (Figure 11.9A). Over time the cystic lesion expands, rapidly displacing adjacent abdominal organs and, if large enough, producing polyhydramnios, placentomegaly, nonimmune hydrops, and stillbirth [33,34].

Most of these lesions arise from the right lobe of the liver, but up to 10% originate from both lobes [1,3,12]. Frequently mesenchymal hamartomas can be diagnosed by their characteristic gross appearance, which consists of a well-circumscribed, demarcated, multiloculated cystic mass with a light gray or pale white bosselated external surface [1,3,12,35] (Figures 11.9–11.11). Usually the adjacent liver is grossly normal. The specimens range considerably in size from 100 g to over a kilogram [1]. The cut surface of the hamartoma reveals multiple cysts varying from a few millimeters to 10 cm or more in diameter. The cysts have gray-white, smooth, and glistening lining and contain clear, pale yellow fluid. Some specimens are composed mostly of multiple cysts (Figures 11.9 and 11.10) while others are more solid consisting of fibrous connective tissue and only one or more small cysts (Figure 11.11 B). Large necrotic centers are noted in a few.

Microscopically the multilocular cysts are lined by endothelium and cuboidal bile duct epithelium or lack a well-defined epithelium (Figures 11.9–11.11). The cysts are surrounded by dense or pale myxoid fibrous connective tissue septa containing blood vessels and small bile ducts [3,19,31,34,35]. The periphery of the lesion is either encircled by a thick fibrous capsule or blends imperceptibly with the adjacent liver parenchyma, which may show compression, fibrosis, and atrophy. Electron microscopy reveals connective tissue, fibroblasts, and bile ducts and normal hepatocytes [3]. Sometimes it may be difficult to distinguish a mesencymal hamartoma from a hemangioma of the liver either grossly or microscopically, particularly if there is extensive hemorrhage into the former.

Mesenchymal hamartomas are cured by surgical resection provided that they are completely removed and there are no associated serious clinical problems such as congestive heart failure, hydrops, pulmonary compromise, or fatal operative complications [15,31].

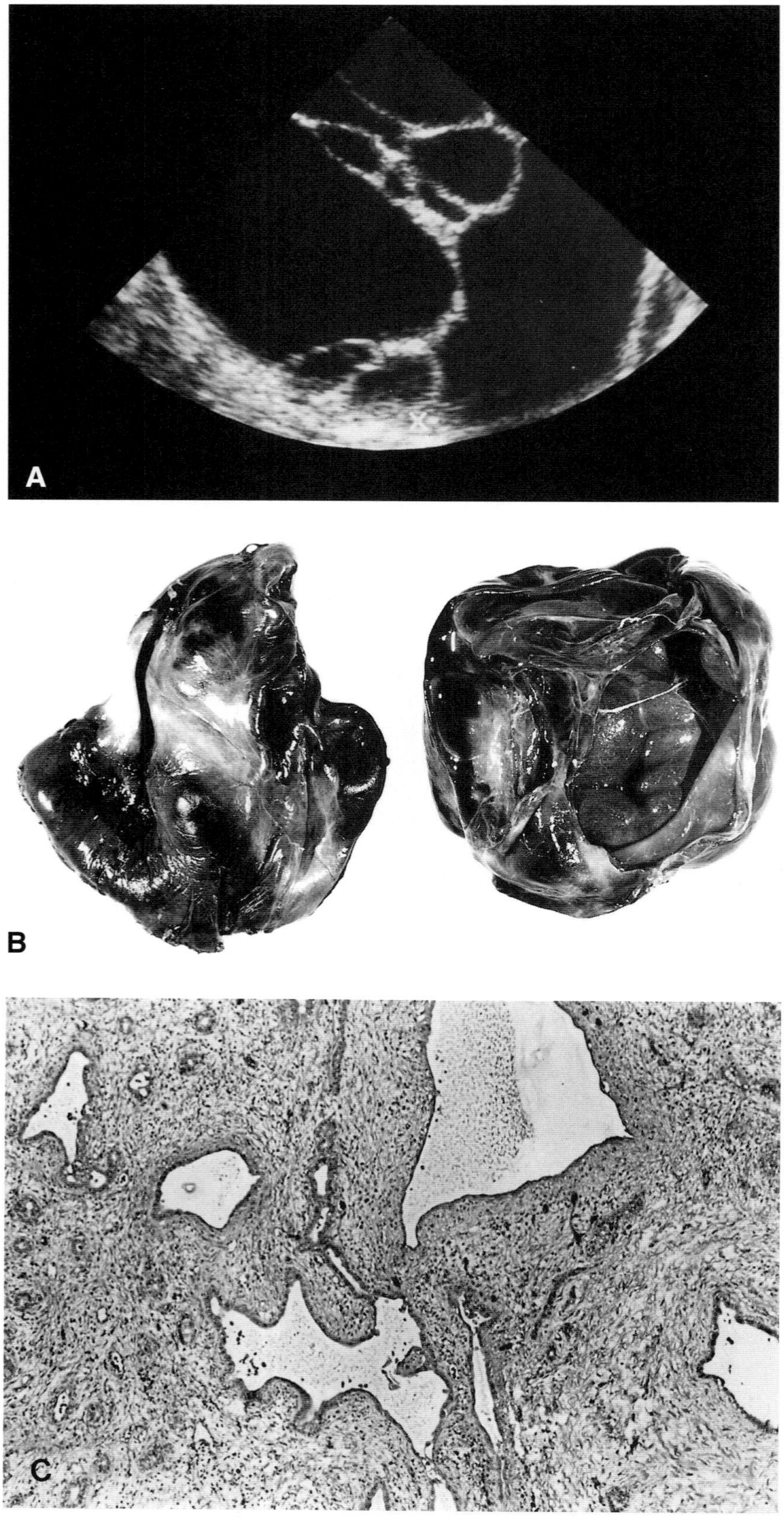

FIGURE 11.9. Mesenchymal hamartoma of the liver in a 1-day-old female, 37 weeks gestation, with abdominal distension noted at birth. (A) Sonogram reveals a large cystic multilocular mass within the right lobe of the liver filled with sonolucent fluid. (B) The specimen consists of a multiloculated hepatic mass, 600 g, $15 \times 12 \times 7$ cm^3, with thick cyst walls. The cut surface is on the right side and the external surface is on the left. (C) Irregular cystic spaces of varying size are surrounded by a loose vascular connective stroma, which is lined with flattened endothelium or low cuboidal epithelium. H&E; $\times$ 48.

FIGURE 11.10. Mesenchymal hamartoma of the liver in a 6-month-male with a rapidly growing abdominal mass, first noticed at 4 months of age. (A) Computerized tomography scan reveals a large, multiloculated, cystic lesion within the liver. (B) The tumor weighed over a kilogram and occupied both the right and left lobes of the liver. It consists of multiple large cysts (filled with clear fluid) separated by white fibrous connective tissue septa. (C) The cysts are lined cuboidal and squamous endothelial cells separated by fibrovascular septa. The interstitial connective tissue contains a few small bile ducts. Mesenchymal hamartoma cysts progressively fill with fluid, presumably lymph and bile, and may expand at an alarming rate, mimicking a malignant tumor. H&E; × 120.

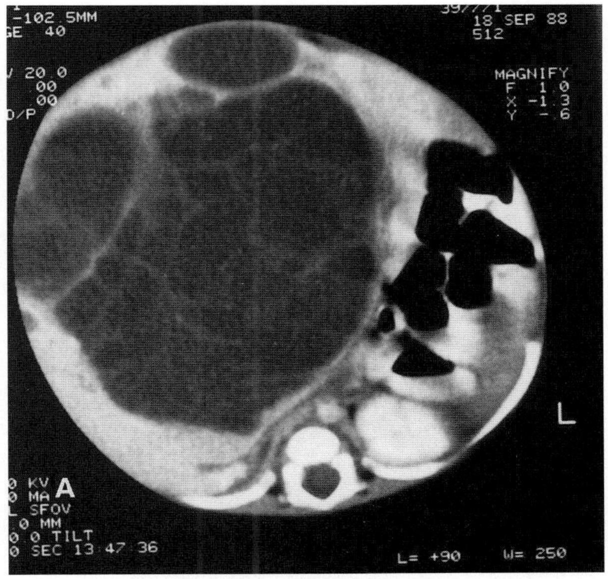

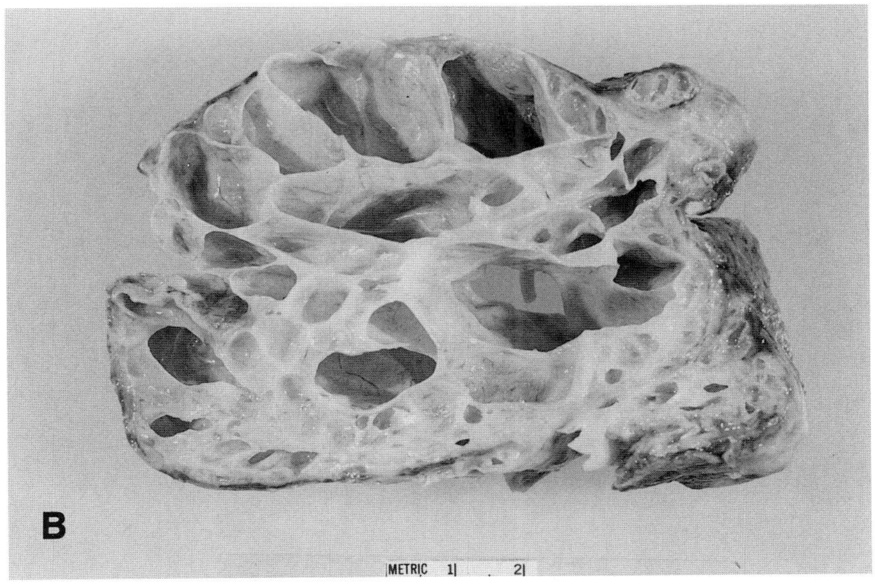

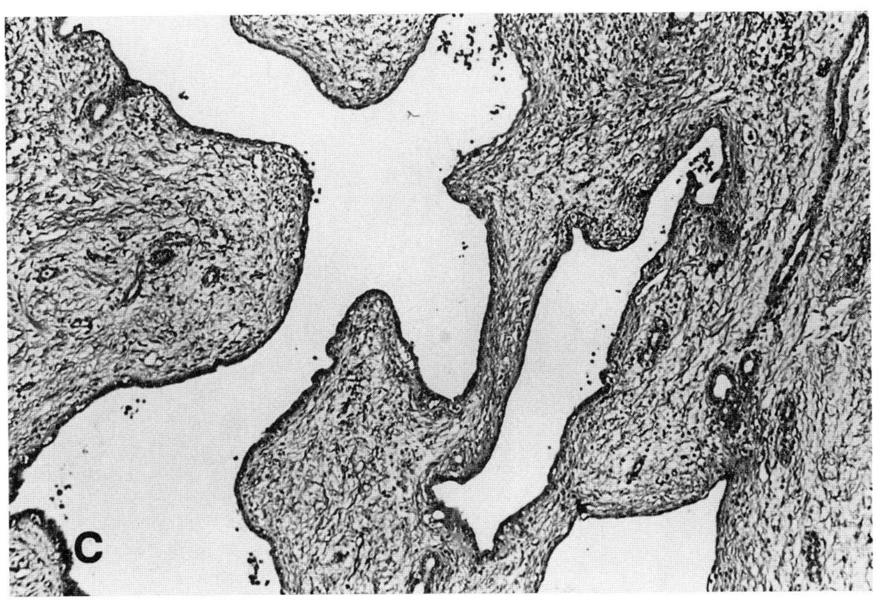

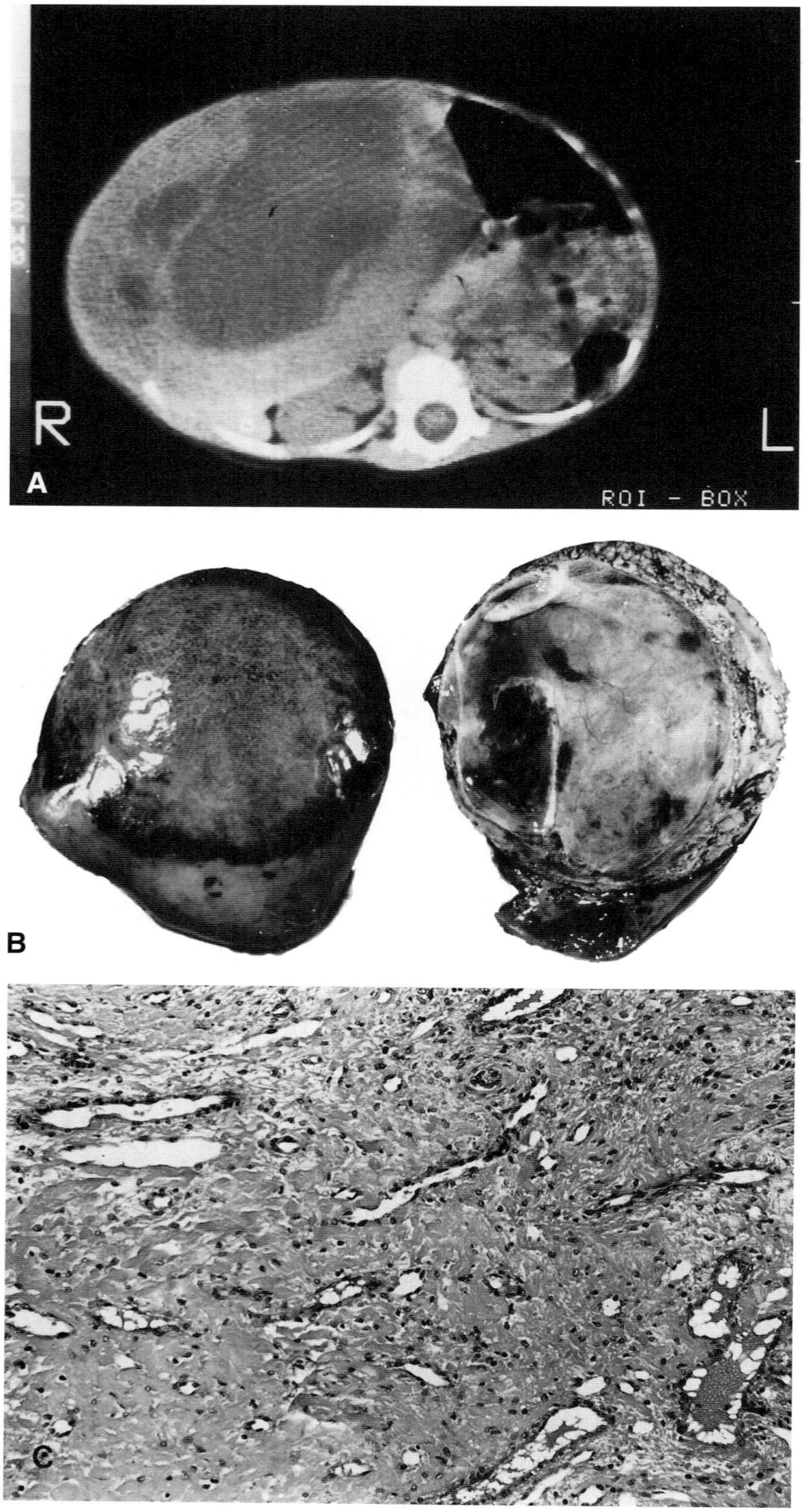

FIGURE 11.11. Mesenchymal hamartoma of the liver in an 11-month-old female with an abdominal mass. (A) Computerized tomography scan reveals a solid and cystic mass within the right lobe of the liver. (B) The partial right hepatectomy specimen consists of solid and cystic myxoid-appearing structures. The capsular surface of the liver is on the left side and the cut surface is on the right. (C) Compact fibrovascular tissue containing slitlike spaces lined with flattened endothelium (mostly) and low cuboidal epithelium. H&E; × 120.

Hemangioma of the Liver

Most hepatic hemangiomas (hemangioendotheliomas) are diagnosed before 6 months of age, and almost half are found within the first week of life [3,7,12,32,36–38] (Figure 11.12). More occur than are actually recorded because many are asymptomatic ("silent") conditions that regress. They are discovered as incidental clinical, prenatal sonogram, or postmortem findings [3,19,38]. Some hemangiomas produce hepatomegaly or a mass lesion, while others cause life-threatening situations in the perinatal period (e.g., rupture, hemoperitoneum during delivery, high output cardiac failure) [7,12,19,36]. Stillbirth and sudden unexpected death are uncommon [38]. Consumptive coagulopathy due to disseminated intravascular coagulation, sequestration of platelets resulting in a bleeding diathesis (Kasabach-Merritt syndrome), and anemia are other findings [12,15]. Polyhydramnios and nonimmune fetal hydrops are poor prognostic signs [37,39]. Hepatic hemangiomas are associated with hemangiomas in the skin and other organs in approximately 40 to 50% of the cases [3,12,36]. Some of these hemangiomas are classified as kaposiform hemangioendotheliomas (see Chapter 3, Soft Tissue Tumors, for further discussion). When a newborn presents with cutaneous hemangiomas, an abdominal mass, and congestive heart failure, a large hepatic hemangioma should be one of the main diagnostic considerations [1] (Figure 11.12).

Grossly hepatic hemangioma appears as a single mass in the right or left lobe or less often as multicentric lesions distributed throughout both lobes. The cut surface of the liver reveals a dark reddish-brown, spongelike mass composed of blood vessels filled with blood with or without areas of fibrosis, calcification, and hemorrhage [1,3] (Figures 11.12–11.14). Hemangiomas are divided histologically into two main types, capillary and cavernous, depending only on the size of the vascular spaces. Cavernous hemangiomas have larger spaces than capillary hemangiomas [15,38] (Figures 11.12E, 11.13C, and 11.14B). The vascular channels are lined by a single layer of flat or plump, regular endothelial cells. The pale-staining connective tissue stroma between the vascular spaces is variable in amount and contains small bile ducts and foci of extramedullary erythropoiesis. It has been shown that with time, capillary hemangiomas become cavernous as the shunting increases. Thrombosis, necrosis, and fibrosis leave large residual vessels [15] (Figure 11.14C,D). Approximately two thirds of hemangiomas undergo spontaneous regression, characterized histologically by thrombosis, fibrosis, and calcification [15] (Figure 11.14C,D). Both types of hemangioma coexist in the same lesion (Figure 11.13 and 11.14); an arteriovenous malformation may be present as well. It was mentioned previously that it may be difficult to distinguish some hemangiomas from a mesenchymal hamartoma because the two share histological features, namely, blood vessels and fibrous connective tissue stroma.

The prognosis of infants with hepatic hemangiomas is favorable provided there are no major clinical complications as described earlier [3,32].

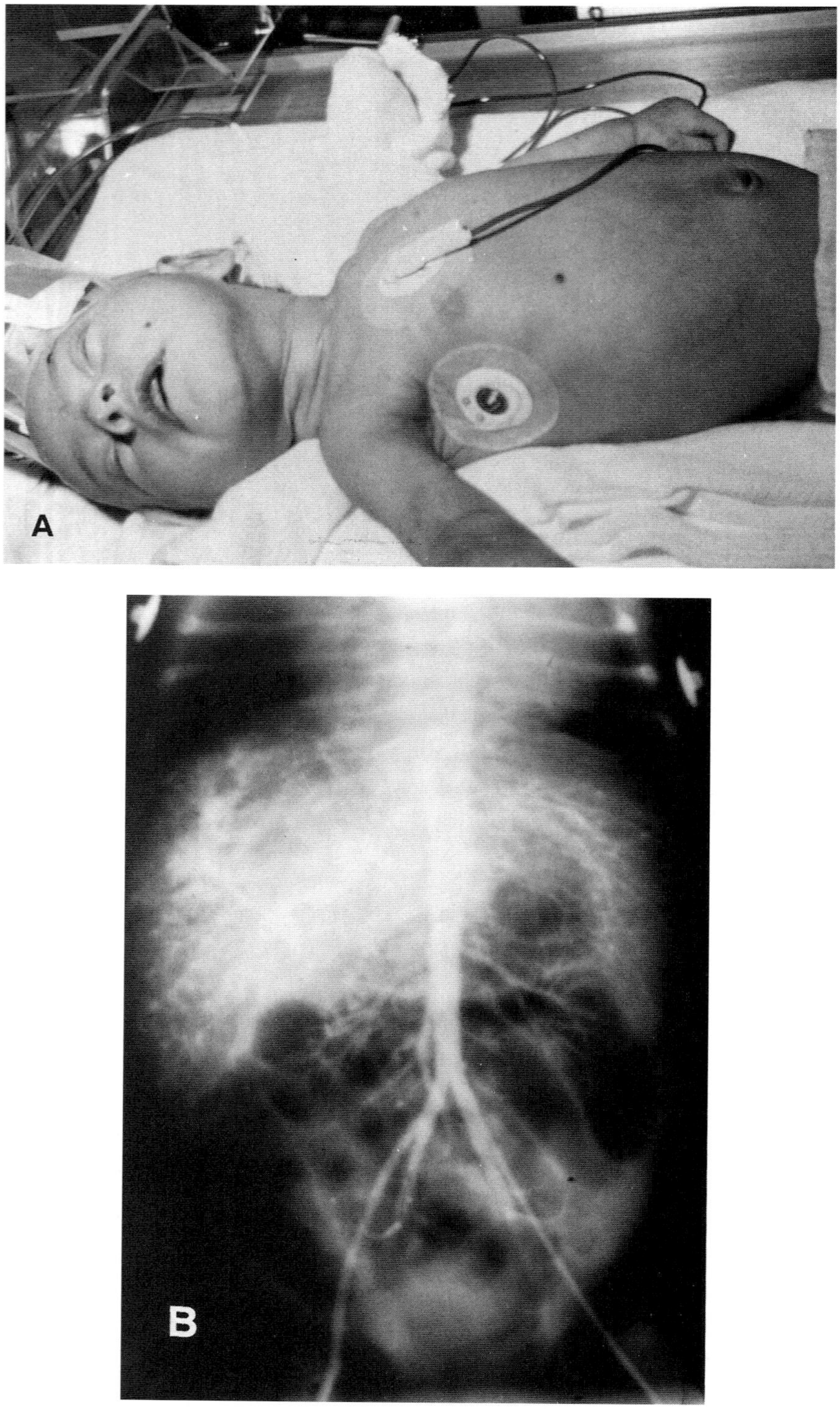

FIGURE 11.12. Hemangioma of the liver. (A) Newborn with congestive heart failure and multiple cutaneous hemangiomas. (B) Arteriogram demonstrates a highly vascular lesion occupying most of the right lobe of the liver. (C) Postmortem photograph showing the hepatic hemangioma in situ arising from the right lobe and occupying much of the abdominal cavity. (D) Cross section of the liver showing the spongelike hemangioma filled with blood. (E) The cavernous hemangioma consists of large vascular channels lined by regular endothelium. The adjacent liver parenchyma exhibits fibrosis and atrophy. H&E; × 32. (From Isaacs H Jr [6], with permission.)

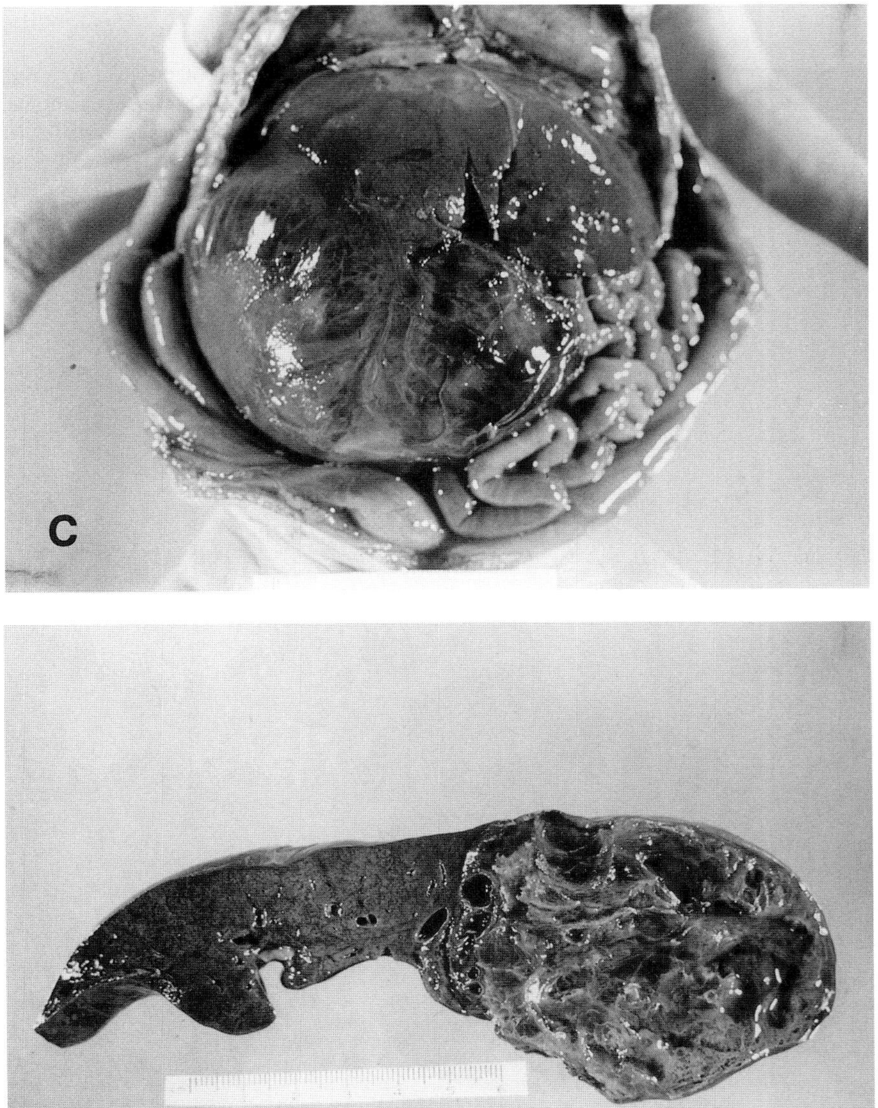

FIGURE 11.12. (*Continued*)

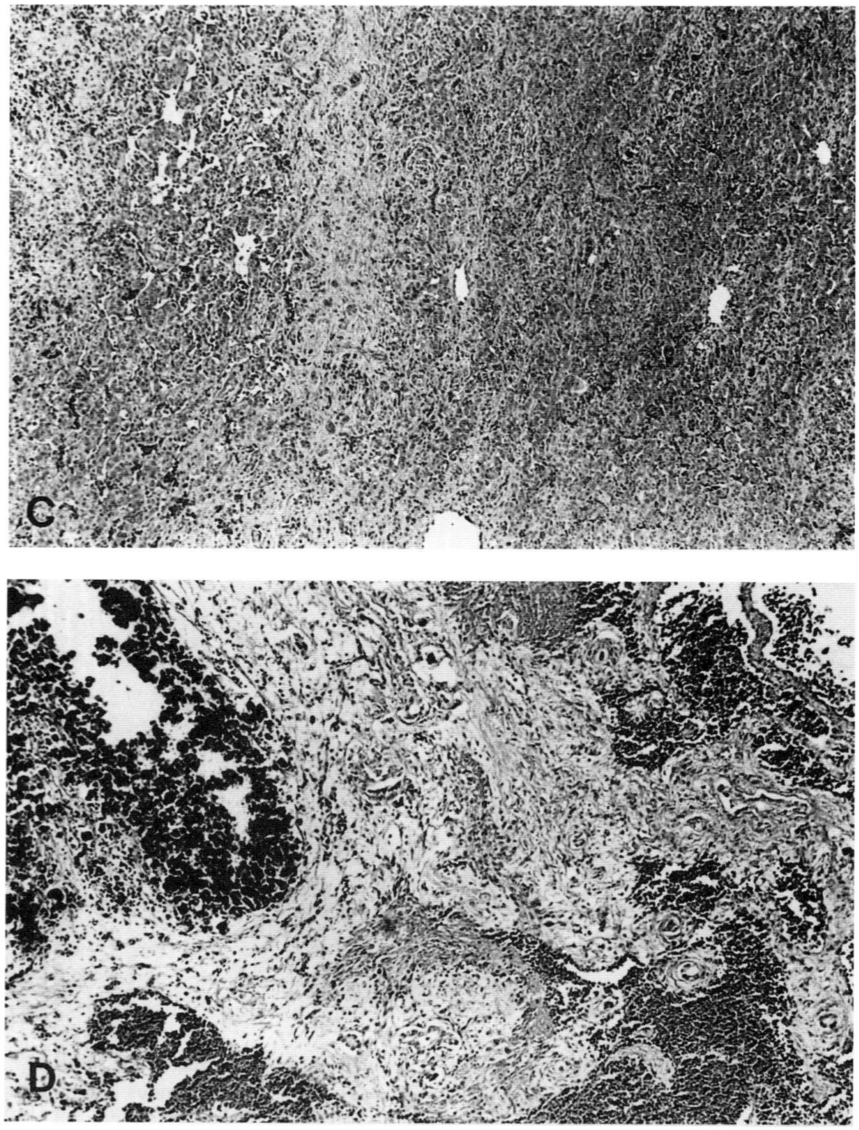

FIGURE 11.14. (*Continued*)

Primary Sarcomas of the Liver

Embryonal rhabdomyosarcoma and embryonal (undifferentiated) sarcoma are the two main primary malignant mesenchymal tumors of the liver in childhood [3,7,40,41]. Hepatic sarcomas are noted infrequently during the first year of life [40]. Embryonal sarcoma affects mostly older children, 6 to 10 years of age, whereas rhabdomyosarcoma is found primarily in younger children between the ages of 2 and 4 years [3,7]. Several case reports have appeared describing instances of embryonal sarcoma arising from a mesenchymal hamartoma and the presence of a common 19q13.4 translocation, suggesting a histogenetic link between the two neoplasms [35,42,43,44]. Angiosarcoma of the liver diagnosed during the first year of life is the subject of an anecdotal case report [45].

An abdominal mass and pain are the main clinical findings. Imaging studies reveal a space-occupying lesion, usually situated in the right lobe of the liver, with solid and cystic areas. Generally the tumors are large, measuring 30 cm or more in diameter, with a fibrous pseudocapsule [3,7,35,40]. The cut surface has a tan to light gray, gelatinous appearance with extensive necrosis and cyst formation (Figure 11.15A). Microscopically there is an abundant pale-staining myxoid stroma with clusters of small dark round, oval, or spindle-shaped cells (Figure 11.15B). Large tumor cells with bizarre nuclei and characteristic eosinophilic-staining cytoplasmic globules are seen (Figures 11.15C,D). Mitoses are not uncommon. Entrapped liver cells and bile ducts are noted about the periphery. Vascular invasion by tumor cells is an important diagnostic finding. Vimentin and the histiocytic markers lysozyme and α_1-antitrypsin are generally immunoreactive, but α-fetoprotein is not [41]. Desmin is variably focally positive. Electron microscopic examination of embryonal sarcoma reveals of a variety of cell types but principally fibroblastic cells with spindle-shaped cytoplasms, containing well-developed rough endoplasmic reticulum. In addition, primitive mesenchymal cells with scanty cytoplasms and few organelles, histiocytoid cells with lysosomal granules and microvillous-like cytoplasmic projections, and myofibroblastic cells with serrated nuclei, abundant rough endoplasmic reticulum, filamentous condensations beneath the cytoplasmic membrane, and focal basal lamina formation are described [41].

The liver is rarely the primary site of an embryonal rhabdomyosarcoma arising from either the intrahepatic bile ducts or from the extrahepatic biliary system [1,3,7,46]. The child presents with an abdominal mass and jaundice if the biliary tract is obstructed. When the sarcoma occurs in one of the major bile ducts, it forms tiny, pale, gray-tan grapelike, polypoid masses filling the ductal lumen, similar to those arising from hollow viscera elsewhere. The histological appearance is that of a botryoid embryonal rhabdomyosarcoma consisting of polypoid projections composed of a pale myxoid stroma with a cellular rind of small, dark, round or spindle-shaped rhabdomyoblastic cells situated beneath the bile duct epithelium. Cytoplasmic cross-striations may be evident with H&E staining. Desmin and other skeletal markers are immunoreactive, and the ultrastructure is diagnostic (see Chapter 3, Soft Tissue Tumors). When the tumor arises from the liver parenchyma, it forms a pale gray-tan, bulging, gelatinous-appearing intrahepatic mass with foci of necrosis and hemorrhage and similar histology.

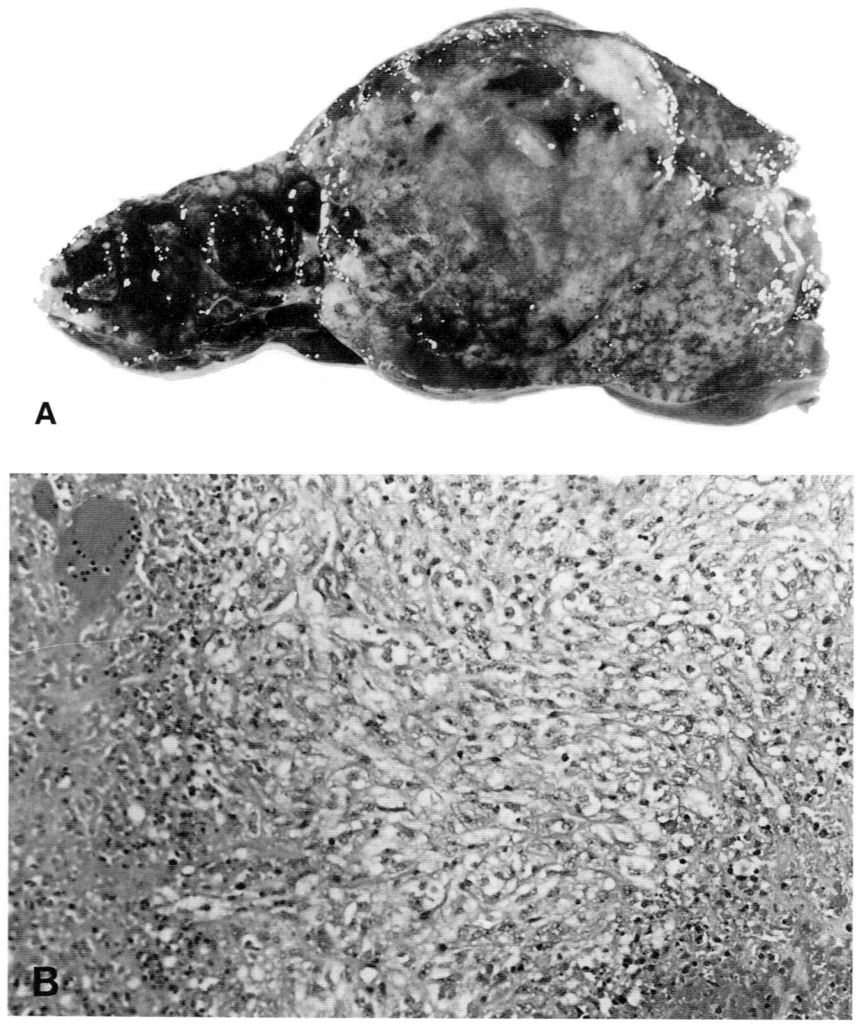

FIGURE 11.15. Embryonal (undifferentiated) sarcoma of the liver in a 2-year-old girl with an abdominal mass. Initial imaging studies revealed a large tumor with solid and cystic areas situated within the right lobe of the liver. Hepatic abscess was one of the diagnostic considerations prior to surgery. (A) Most of the liver is replaced by friable, gelatinous tissue with large areas of cystic necrosis and hemorrhage. (B) The tumor consists of an abundant, pale-staining, myxoid stroma with clusters of small round, oval, or spindle-shaped vacuolated cells. Extensive necrosis is present, which is a characteristic finding. If pieces of tumor showing only necrosis and a few spindle-shaped cells are submitted for frozen section diagnosis, a diagnostic problem may be created. H&E; × 120. (C) Large tumor cells with bizarre nuclei and characteristic eosinophilic staining, PAS positive; are diastase resistant, cytoplasmic globules are present. Mitoses are not uncommon. Vascular invasion by tumor cells, not pictured here, is an important diagnostic finding. Masson trichrome; × 120. (D) Higher power (× 380) view showing the anaplastic giant cells, globules, and vacuolated cells in more detail. Masson trichrome.

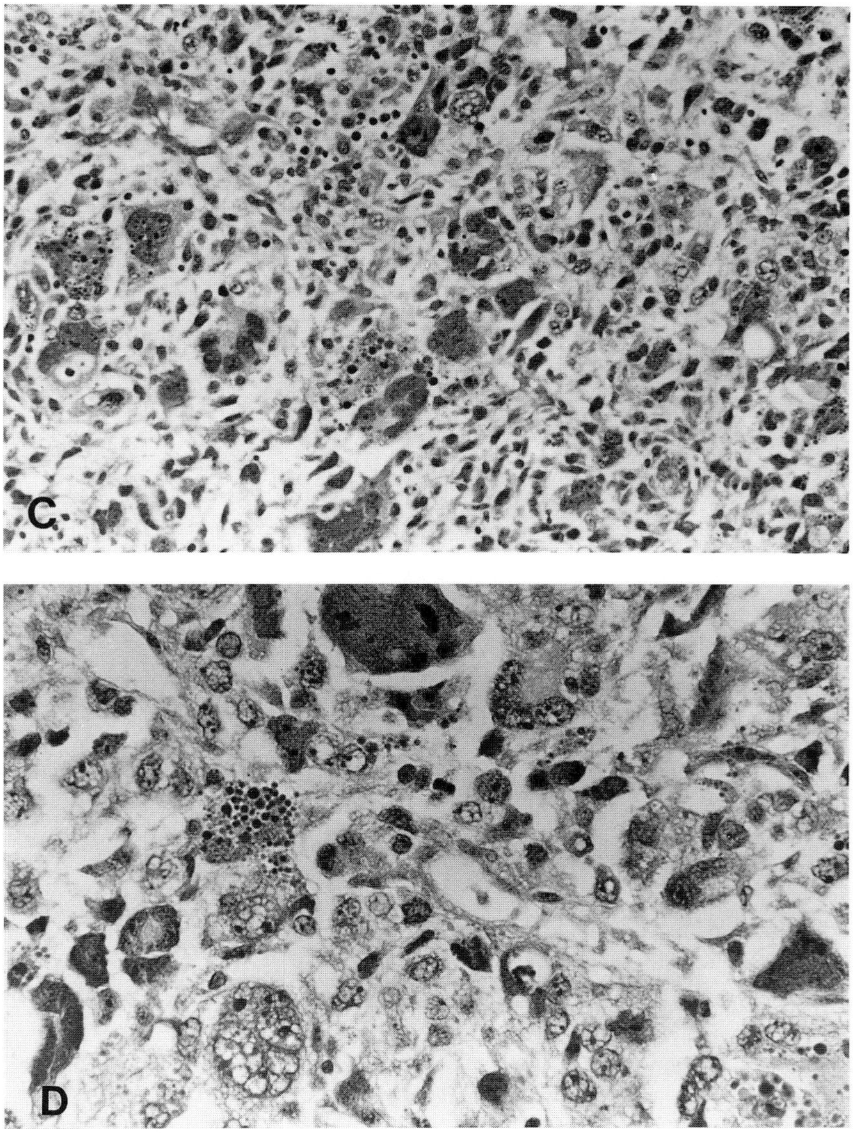

FIGURE 11.15. (*Continued*)

Germ Cell Tumors

Germ cell tumors metastatic to the liver, namely, yolk sac tumors spreading from the sacrococcygeal area or testis, are observed more often than primary ones during the first year of life [3,7,15] (Figure 2.29B). Although few in number, examples of teratoma of the liver are reported more frequently than cases of primary yolk sac tumor [3,7,47–49]. Teratoma of the liver is more common in the newborn and infant than in older children and adults [47]. Intrahepatic fetus-in-fetu detected by antenatal sonography is described in association with separate nodules of mature teratoma in the same liver [49]. Hepatic teratomas typically are cystic, composed of mature and to a lesser extent immature neuroglial elements, and usually lacking in malignant germ cell tumor components [15]. Surgical excision results in cure provided no residual tumor is left behind.

Yolk sac tumor is observed rarely in association with hepatoblastoma [50]. Primary yolk sac tumor of the falciform ligament and left lobe of the liver is described [51] (Figure 11.16).

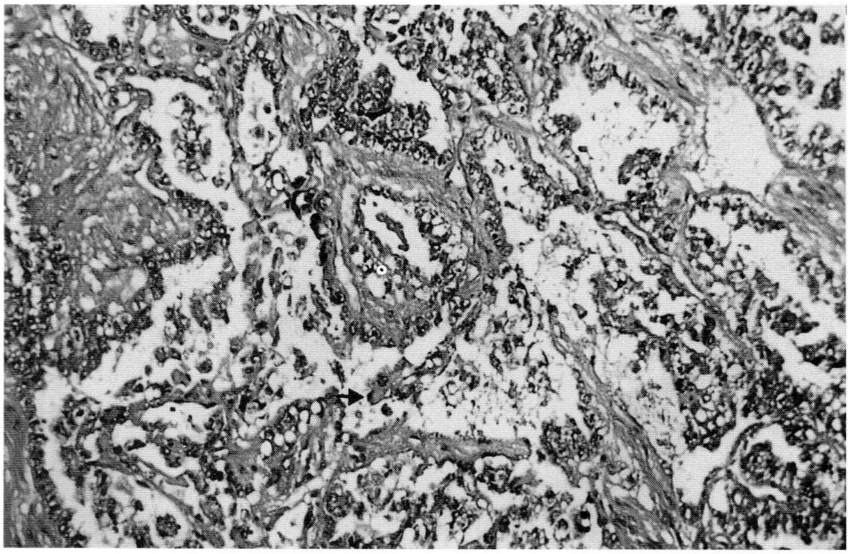

FIGURE 11.16. Yolk sac tumor of the liver of a 1-year-old male with a large, friable, gelatinous mass invading the abdominal wall, the falciform ligament, and the left lobe of the liver. The elevated serum α-fetoprotein was 76,000. The tumor shows papillary and reticular growth patterns consisting of small vacuolated cells with irregular nuclei. Tiny globules of intracellular material (arrows) are reactive with a-fetoprotein antibody. H&E; $\times$ 200. (From Atkinson JB, et al. [51], with permission.)

Miscellaneous Hepatic Tumors and Tumorlike Conditions

Adenoma and focal nodular hyperplasia are regarded as benign proliferations of liver cells that occur more often in adults than in infants and children [3,7,12,52–54]. They are seldom noted in the fetus and infant [53,54].

Hepatocellular carcinoma is unusual in children under 6 years of age [3,4]. Often there is an underlying condition associated with this malignancy (e.g., cirrhosis, biliary atresia, tyrosinemia, prolonged neonatal parenteral malnutrition) [55,56]. Grossly the liver is replaced by multifocal grayish-white tumor nodules with abundant necrosis and hemorrhage (Figure 11.17A). Microscopically the tumor shows broad trabecular and acinar formations. Considerable nuclear pleomorphism, prominent nucleoli, and giant cells are characteristic features (Figure 11.17C,D). The tumor cells are strongly reactive with α-fetoprotein antibody, which is observed also with hepatoblastoma and yolk sac tumor cells.

Hepatic ("nonparasitic") cysts may reach an enormous size in the fetus and neonate and fill the peritoneal cavity, producing an abdominal mass and respiratory distress [1,3,57] (Figure 11.18). *Infantile polycystic disease of the kidneys and liver* may be apparent at birth and is manifested by renohepatomegaly with abdominal masses as the result of expanding cyst formations in the involved organs (Figure 11.19).

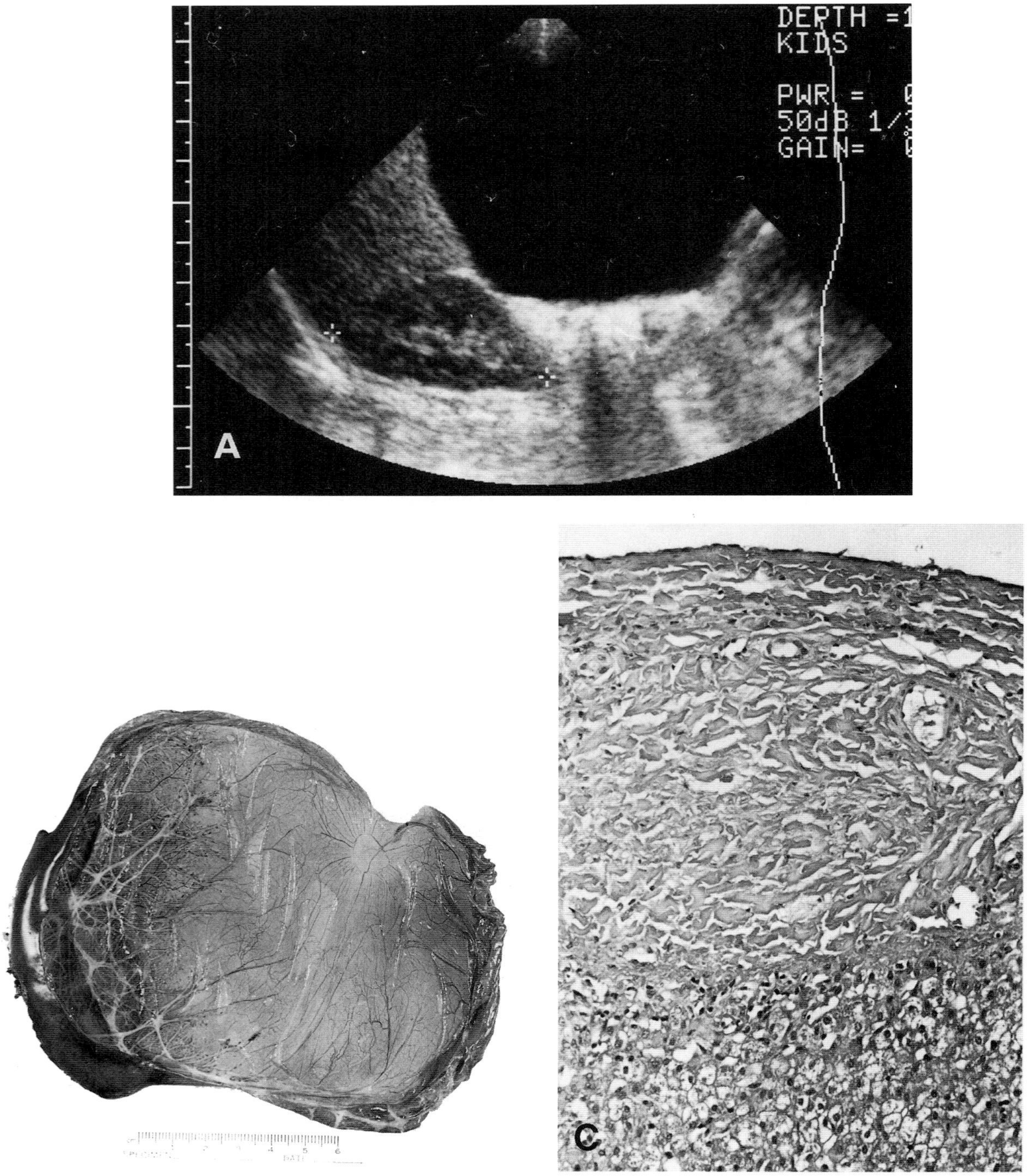

FIGURE 11.18. Unilocular cyst of the liver in an 11-month-old female with a history of progressive abdominal distension over several months. (A) Sonogram reveals a cystic mass occupying the abdominal cavity without a definite site of origin. (B) The hepatic cyst, 1.1 kg and 18 cm in diameter, has a smooth, glistening lining and was filled with clear fluid. (C) The cyst is lined by both cuboidal and flattened squamous epithelium situated on a thick layer of fibrous connective tissue. The underlying liver parenchyma is unremarkable. H&E; × 300.

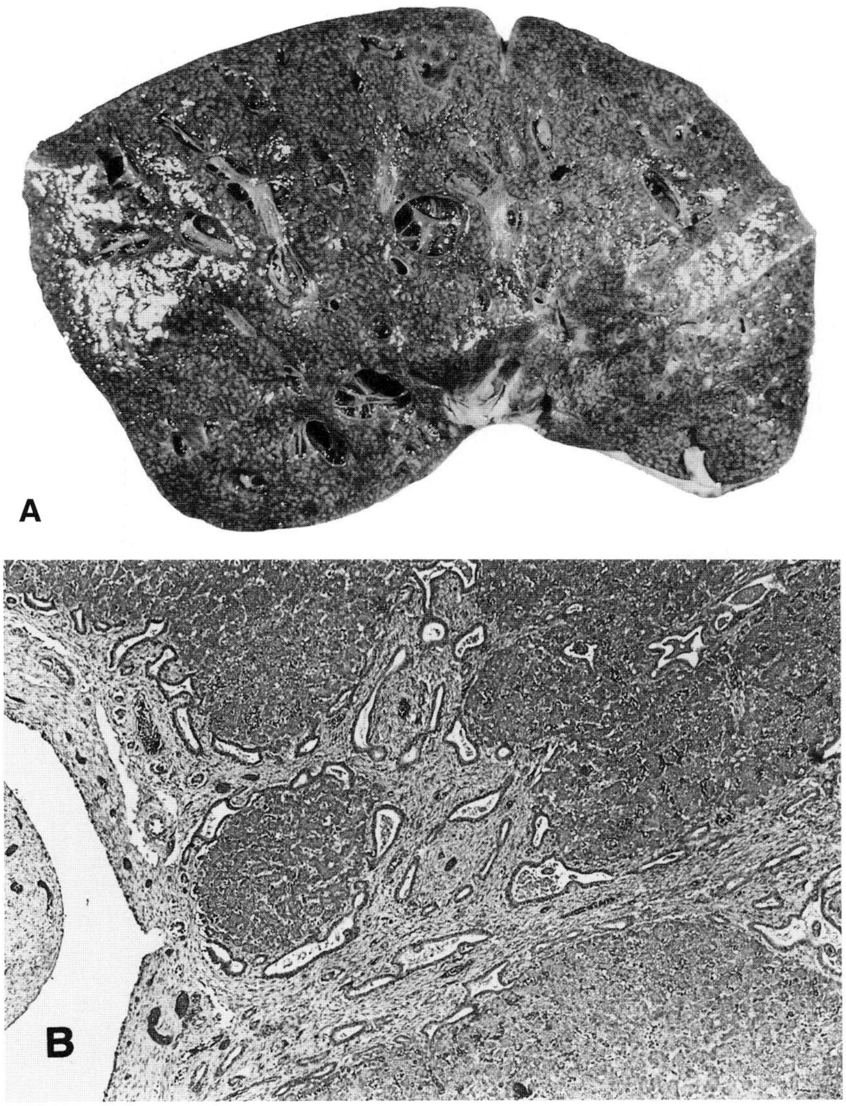

FIGURE 11.19. Infantile polycystic disease of the kidneys and liver (autosomal recessive). The neonate and his cystic kidneys are pictured in Figure 10.3. (A) Gross appearance of the liver with multiple cysts. (B) Angulated branching cystically dilated bile ducts and portal fibrosis are present. H&E; × 48.

References

1. Isaacs H Jr. Liver tumors. In: *Tumors of the Newborn and Infant.* St. Louis, MO: Mosby Year Book; 1991:147.
2. Bader JL, Miller RW. US cancer incidence and mortality in the first year of life. *Am J Dis Child* 1979;133:157.
3. Dehner LP. Hepatic tumors in the pediatric age group: a distinctive clinical pathologic spectrum. *Perspect Pediatr Pathol* 1978;4:217.
4. Ishak KG, Glunz PR. Hepatoblastomas and hepatocarcinoma in infancy and childhood. *Cancer* 1967;20:396.
5. Isaacs H Jr. Neoplasms in infants: a report of 265 cases. *Pathol Annu* 1983;18(2):165.
6. Isaacs H Jr. Perinatal (congenital and neonatal) neoplasms: a report of 110 cases. *Pediatr Pathol* 1985;3:165.
7. Weinberg AG, Finegold MJ. In: Finegold M, ed. Primary hepatic tumors in childhood. In: *Pathology of Neoplasia in Children and Adolescents*, Philadelphia: WB Saunders; 1986. Major Problems in Pathology Series. Vol 18.
8. Lack EE, Neave C, Vawter GF. Hepatoblastoma: a clinical and pathological study of 54 cases. *Am J Surg Pathol* 1982;6:693.
9. Weitzman S, Grant R. Neonatal oncology: diagnostic and therapeutic dilemmas. *Semin Perinatol* 1997;21:102.
10. Bolande RP. Developmental pathology. *Am J Pathol* 1979;94:627.
11. Campbell AN, Chan HSL, O'Brien A, et al. Malignant tumours in the neonate. *Arch Dis Child* 1987;62:19.
12. Dehner LP. Neoplasms of the fetus and neonate. In: Naeye RL, Kissane JM, Kaufman N, eds. *Perinatal Diseases.* Baltimore: Williams & Wilkins; 1981:286. International Academy of Pathology, Monograph No. 22.
13. Stocker JT. Hepatoblastoma. *Semin Diagn Pathol* 1994;11:136.
14. Gonzalez-Crussi F, Upton MP, Mauer HS. Hepatoblastoma: attempt at characterization of histologic subtypes. *Am J Surg Pathol* 1982;6:599.
15. Isaacs H Jr. Tumors. In: Gilbert-Barness E, ed. *Potter's Pathology of the Fetus and Infant.* Vol 2. St. Louis, MO: CV Mosby; 1997:1242.
16. Sotelo-Avila C, Gooch WM. Neoplasms associated with the Beckwith-Wiedemann syndrome. In: Rosenberg HS, Bolande RP, eds. *Perspectives in Pediatric Pathology.* Vol II. Chicago: Year Book Medical Publishers; 1976:255.
17. Giardiello FM, Offerhaus GJA, Krush AJ, et al. Risk of hepatoblastoma in familial adenomatous polyposis. *J Pediatr* 1991;119:766.
18. Orozco-Florian R, McBride JA, Favara BE, et al. Congenital hepatoblastoma and Beckwith-Wiedemann syndrome: a case study including DNA ploidy profiles of tumor and adrenal cytomegaly. *Pediatr Pathol* 1991;11:131.
19. Isaacs H Jr. Liver tumors. In: *Tumors of the Fetus and Newborn.* Philadelphia: WB Saunders; 1997:278. Major Problems in Pathology Series. Vol. 35.
20. Doss BJ, Vicari J, Jacques SM, et al. Placental involvement in congenital hepatoblastoma. *Pediatr Dev Pathol* 1998;1:538.
21. Conran RM, Hitchcock CL, Waclawiw MA, et al. Hepatoblastoma: the prognostic significance of histologic type. *Pediatr Pathol* 1992;12:167.
22. Van Eyken P, Sciot R, Callea F, et al. A cytokeratin-immunohistochemical study of hepatoblastoma. *Hum Pathol* 1990;21:302.
23. Fasano M, Theise ND, Nalesnik M, et al. Immunohistochemical evaluation of hepatoblastomas with use of hepatocyte-specific marker, hepatocyte parrafin 1, and the polyclonal anti-carcinoembryonic antigen. *Mod Pathol* 1998;11:934.
24. Horie A, Kotoo Y, Hayashi I. Ultrastructural comparison of hepatoblastoma and hepatocellular carcinoma. *Cancer* 1979;44:2184.
25. Joshi VV, Kapur P, Ryan B, et al. Mucoid anaplastic hepatoblastoma. A case report. *Cancer* 1984;54:2035.
26. Gonzalez-Crussi F. Case 1: undifferentiated small cell ("anaplastic") hepatoblastoma. *Pediatr Pathol* 1991;11:155.
27. Hansen K, Bagtas J, Mark HF, et al. Undifferentiated small cell hepatoblastoma with a unique chromosomal translocation: a case report. *Pediatr Pathol* 1992;12:457.
28. Manivel C, Wick MR, Abenoza P, et al. Teratoid hepatoblastoma: the nosologic dilemma of solid embryonic neoplasms of childhood. *Cancer* 1986;57:2168.
29. Cross SS, Variend S. Combined hepatoblastoma and yolk sac tumor of the liver. *Cancer* 1992;69:1323.
30. Stocker JT, Ishak KG. Mesenchymal hamartoma of the liver: report of 30 cases and review of the literature. *Pediatr Pathol* 1983;1:245.
31. De Maioribus CA, Lally KP, Sim K, et al. Mesenchymal hamartoma: a 35-year review. *Arch Surg* 1990;125:598.
32. Miller JH, Greenspan BS. Integrated imaging of hepatic tumors in childhood, II: Benign lesions (congenital, reparative, and inflammatory). *Radiology* 1985;154:91.
33. Dickinson JE, Knowles S, Phillips JM. Prenatal diagnosis of hepatic mesenchymal hamartoma. *Pren Diagn* 1999;19:81.
34. Tovbin J, Segal M, Tavori I, et al. Hepatic mesenchymal hamartoma: a pediatric tumor that may be diagnosed prenatally. *Ultrasound Obstet Gynecol* 1997;10:63.
35. Bove KE, Blough RI, Soukup S. Third report of t(19q)(13.4) in mesenchymal hamartoma of the liver with comments on link to embryonal sarcoma. *Pediatr Dev Pathol* 1998;1:438.
36. Stanley P, Geer GD, Miller JH, et al. Infantile hepatic hemangiomas: clinical features, radiologic investigations, and treatment of 20 patients. *Cancer* 1989;64:936.
37. Shturman-Ellstein R, Greco MA, Myrie C, Goldman EK. Hydrops fetalis, hydramnios and hepatic vascular malformation associated with cutaneous hemangioma and chorangioma. *Acta Paediatr Scand* 1978;67:239.
38. Selby DM, Stocker JT, Waclawiw MA, et al. Infantile hemangioendothelioma of the liver. *Hepatology* 1994;20:39.
39. Albano G, Pugliese A, Stabile M, et al. Hydrops foetalis caused by hepatic haemangioma. *Acta Paediatr* 1998;87:1307.
40. Stocker JT, Ishak KG. Undifferentiated (embryonal) sarcoma of the liver. *Cancer* 1978;42:336.
41. Aoyama C, Hachitanda Y, Sato JK, et al. Undifferentiated (embryonal) sarcoma of the liver: a tumor of uncertain histogenesis showing divergent differentiation. *Am J Surg Pathol* 1991;15:615.
42. Lauwers GY, Grant LD, Donnelly DH, et al. Hepatic undifferentiated (embryonal) sarcoma arising from a mesenchymal hamartoma. *Am J Surg Pathol* 1997;21:1248.
43. Mascarello JT, Krous HF. Second report of a translocation involving 19q13.4 in a mesenchymal hamartoma of the liver. *Cancer Genet Cytogenet* 1992;58:141.
44. Ramanujam TM, Ramesh JC, Goh DW, et al. Malignant transformation of mesenchymal hamartoma of the liver: case report and review of the literature. *J Pediatr Surg* 1999;34:1684.

45. Noronha R, Gonzalez-Crussi F. Hepatic angiosarcoma in a child. *Am J Surg Pathol* 1984;8:863.
46. Horowitz ME, Etcubanas E,Webber BL, et al. Hepatic undifferentiated (embryonal) sarcoma and rhabdomyosarcoma in children: results of therapy. *Cancer* 1987;59:396.
47. Dische MR, Gardner HA. Mixed teratoid tumors of the liver and neck in trisomy 13. *Am J Clin Pathol* 1978;69:631.
48. Hart WR. Primary endodermal sinus (yolk sac) tumor of the liver: first reported case. Cancer 1975;35:1453.
49. Magnus KG, Millar AJ, Sinclair-Smith CC, et al. Intrahepatic fetus-in-fetu: a case report and review of the literature. *J Pediatr Surg* 1999;34:1861.
50. Cross SS, Variend S. Combined hepatoblastoma and yolk sac tumor of the liver. *Cancer* 1992;69:1323.
51. Atkinson JB, Foster CE, Lally KP, et al. Primary endodermal sinus (yolk sac) tumor of the falciform ligament. *J Pediatr Surg* 1992;27:105.
52. Dehner LP, Parker ME, Franciosi RA, et al. Focal nodular hyperplasia and adenoma of the liver. A pediatric experience. *Am J Pediatr Hematol Oncol* 1979;1:85.
53. Lack EE, Ornvold K. Focal nodular hyperplasia and hepatic adenoma: a review of eight cases in the pediatric age group. *J Surg Oncol* 1986;33:129.
54. Applegate KE, Ghei M, Perez-Atayde AR. Prenatal detection of a solid liver adenoma. *Pediatr Radiol* 1999;29:92.
55. Patterson K, Kapur SP, Chandra RS. Hepatocellular carcinoma in a noncirrhotic infant after prolonged parenteral nutrition. *J Pediatr* 1985;106:797.
56. McGoldrick JP, Boston VE, Glasgow JFT. Hepatocellular carcinoma associated with macronodular cirrhosis in a neonate. *J Pediatr Surg* 1986;21:277.
57. Takano H, Sato Y, Kao SCS, et al. Enlarging liver cyst in Beckwith-Wiedemann syndrome. *Pediatr Radiol* 1997;27:619.

12 Adrenocortical Tumors

The adrenal cortex is the source of tumors and tumorlike conditions with unusual clinical manifestations in the fetus and infant [1–15] (Table 12.1). Adrenocortical neoplasms are uncommon in this age group and throughout childhood [3,4]. Relatively few examples of adrenal cysts, cortical adenomas, and carcinomas have been described, some in association with the Beckwith-Wiedemann syndrome, hemihypertrophy, cancer family (Li-Fraumeni) syndrome, or other tumors [3,16–18]. The adrenal medulla, with an incidence ratio of 10:1 of neuroblastoma to adrenal cortical tumors in children, is a far more common site of origin of tumors in the young than is the cortex [3,4]. Adrenocortical hyperplasia associated with the adrenogenital and Cushing syndromes occurs more frequently than true adrenal cortical neoplasms in the first year of life [2].

Adrenal Cortical Hyperplasia

Hyperplasia and neoplasms of the adrenal cortex occur in infancy. The excess hormone secretion associated with these conditions results in recognizable clinical syndromes, the manifestations depending on the type(s) of steroid produced [2,6,19]. The hormone excess produces one or more characteristic findings (or a combination): Cushing syndrome from hypersecretion of glucocorticoids; hypertension and hypokalemia from excess mineralocorticoids, or virilization or feminization from increased levels of androgens and estrogens, respectively [2,19]. Inherited enzyme defects in steroid hormone biosynthesis are responsible for most adrenocortical disorders in infants and children [2,6]. Congenital adrenal hyperplasia (adrenogenital syndrome) and Cushing syndrome are the two principal endocrine diseases associated with adrenocortical hyperplasia, and the former is far more common than the latter [1,2,6]. Cushing syndrome in childhood is usually caused by an adrenal cortical tumor rather than by a primary pituitary tumor [1].

Congenital Adrenal Cortical Hyperplasia (Adrenogenital Syndrome)

Congenital adrenal hyperplasia is an autosomal recessive disease caused by one of the five or more known enzymatic defects (blocks) in cortisol biosynthesis from cholesterol (Figure 12.1) [2,6,7,20]. It is the most frequent cause of ambiguous genitalia in the newborn [20] (Figure 12.2). Neonatal screening studies indicate a frequency of 1 in 5000 to 1 in 15,000, which is relatively common for an inborn error of metabolism [7]. Cortisol is synthesized principally in the zona fasciculata of the adrenal cortex. The secretion of cortisol is regulated by a negative feedback mechanism that involves this steroid and adrenocorticotropic hormone (ACTH). Low cortisol levels lead to increased ACTH production by the pituitary gland, which in turn continues to stimulate the adrenal cortex. The end results of this adrenal-pituitary imbalance are bilateral adrenocortical hyperplasia and increased cortical steroid synthesis, with excessive secretion of sex hormones and other steroids depending on the enzymatic block. If not treated promptly, some forms of congenital adrenal hyperplasia, particularly the salt-losing forms, can be fatal in the newborn period [7,20].

Deficiency of the 21-hydroxylase enzyme accounts for more than 90% of cases of congenital adrenal hyperplasia [2,6,7] (Figures 12.2 and 12.3). Next in frequency is 11-hydroxylase deficiency. The genetic defect of congenital adrenal hyperplasia is localized to the 6p chromosome for the 21-hydroxylase deficiency and to the 8q chromosome for the 11-hydroxylase deficiency [2]. The 21-hydroxylase defect is closely linked to the HLA major histocompatibility complex situated on the short arm of of chromosome 6 [7].

Cortisol is not synthesized effectively in patients with the 21-hydroxylase defect, since 17-hydroxyprogesterone is not converted to 11-deoxycortisol, the immediate precursor of cortisol [7] (Figure 12.1). The lack of negative feedback then leads to high levels of ACTH and, in turn, the hypersecretion

TABLE 12.1. Tumors and tumorlike conditiion of the adrenal cortex in fetuses and infants.

Adrenal cortical hyperplasia
Congenital adrenal hyperplasis (adrenogenital syndrome)
Adrenal cortical hyperplasia associated with Cushing syndrome
Adrenocortical adenoma
Adrenocortical carcinoma
Adrenal cytomegaly
Cyst
Adrenal enlargement due to hemorrhage
Adrenal enlargement due to infection or abscess

Source: Isaacs H Jr [12].

of steroid precursors proximal to the 21-hydroxylation step. These precursors are converted to androgens and, in two thirds of the patients with the 21-hydroxylase defect, to the mineralocorticoid aldosterone. Androgens regulate the growth and development of both the male and female external genitalia of the fetus but not the internal female organs—the uterus, fallopian tubes, and ovaries [7].

Virilization is the end result of high levels of circulating androgens and is the most important physical finding in newborns, with both the 11- and 21-hydroxylase enzyme defects. It is manifested in the female by clittoral hypertrophy (a phalluslike clittoris), labial fusion, and sometimes scrotalization of the labia majora [2,6,7,10] (Figure 12.2). The internal female organs are essentially normal. The presence of "ambiguous genitalia" should alert the clinician of the possibility of the diagnosis of the adrenogenital syndrome, the discovery of which may be lifesaving [7]. Virilization in the newborn male is not as obvious as it is in the female and may not become apparent in the former until the baby is several months old. It should be remembered that virilization present at birth due to an adrenal neoplasm can be mistaken for the adrenogenital syndrome [21].

In approximately half to two thirds of patients with the classic 21-hydroxylase defect, aldosterone synthesis is impaired [7]. This form of the disease is called the "salt-wasting type" of congenital adrenal hyperplasia [7,20]. "Simple virilizing disease" is the term used for infants and children without impaired aldosterone synthesis. Excessive salt loss from the renal tubules, secondary to low levels of aldosterone, leads to feeding difficulties, vomiting, hypovolemia, and shock. If left untreated the salt-losing type results in death within the first weeks of life [7].

The adrenal glands of patients with congenital adrenal hyperplasia have a striking, distinctive gross appearance [1,11, 12,22]. First of all the adrenals are enlarged, weighing 15 g or more than the normal expected weight, which is 3 to 10 g [11,12]. Typically the cortex is thrown into numerous folds or convolutions, giving the gland a wrinkled or cerebriform external surface that is even more evident on cross section (Figure 12.3A).

Microscopic examination reveals cortical hyperplasia, often nodular in configuration, with increased numbers of eosinophilic cortical cells extending out toward the cortical surface (Figure 12.3B,C). The cortical cells show less lipid than is found in normal cells [3,11]. Ultrastructurally the cells of the hyperplastic nodules closely resemble those of the zona reticularis [23,24]. They show abundant smooth endoplastic reticulum, spherical and elongated mitochondria with tubulovesicular cristae, in addition to many lysozomes and lipofuscin granules [23,24]. Despite these findings, electron microscopy is not helpful in distinguishing adrenal cortical hyperplasia from either adenoma or carcinoma, nor can it be used to determine whether a cortical lesion is or is not functioning [23,24].

Adrenocortical Hyperplasia Associated with Cushing Syndrome

Adrenal cortical hyperplasia occurs also in infants who do not have a known enzymatic block in steroid metabolism. Cushing syndrome develops when there is excessive cortisol production by either a neoplasm or hyperplasia of the adrenal cortex [1,2,5,25–27].

Clinical Findings

The presenting signs and symptoms of an infant with Cushing syndrome due to either neoplasm or adrenal hyperplasia are variable. They may appear abruptly or gradually, and they may be subtle or obvious [2]. In most series obesity is the most frequent presenting sign in the young [2]. Other clinical findings are "moon facies" with plump cheeks, buffalo hump, and hypertension as observed in older children. Excessive weight gain, although there is failure of normal growth, is characteristic of this condition in the infant [2]. Additional features of Cushing syndrome in this age group are worth mentioning. Virulism is a more common sign in infancy than in older children. Insulin-resistant diabetes is found in all age groups, and although the pathogenesis is not well understood, the diabetes is usually cured by removal of the adrenal lesion. Hypertension accompanies the syndrome, and its pathogenesis also is unknown. The effects of high blood pressure can be serious and if not corrected may result in heart failure.

Pathological Findings

The adrenal glands in infants with Cushing syndrome associated with adrenal cortical nodular hyperplasia are usually but not always enlarged and are characterized by the presence of numerous irregular, ill-defined nodularities, which are seen on both the external and cut surfaces [10,11,25]. Histologically the hyperplastic nodules cover a range of sizes and involve mostly the zona fasciculata. They consist of active-appearing cortical cells varying in size and in degrees of lipid vacuolization. The intervening cortical tissue between the nodules tends to be atrophic. Cells of the fetal cortex may comprise the nodules in the neonate.

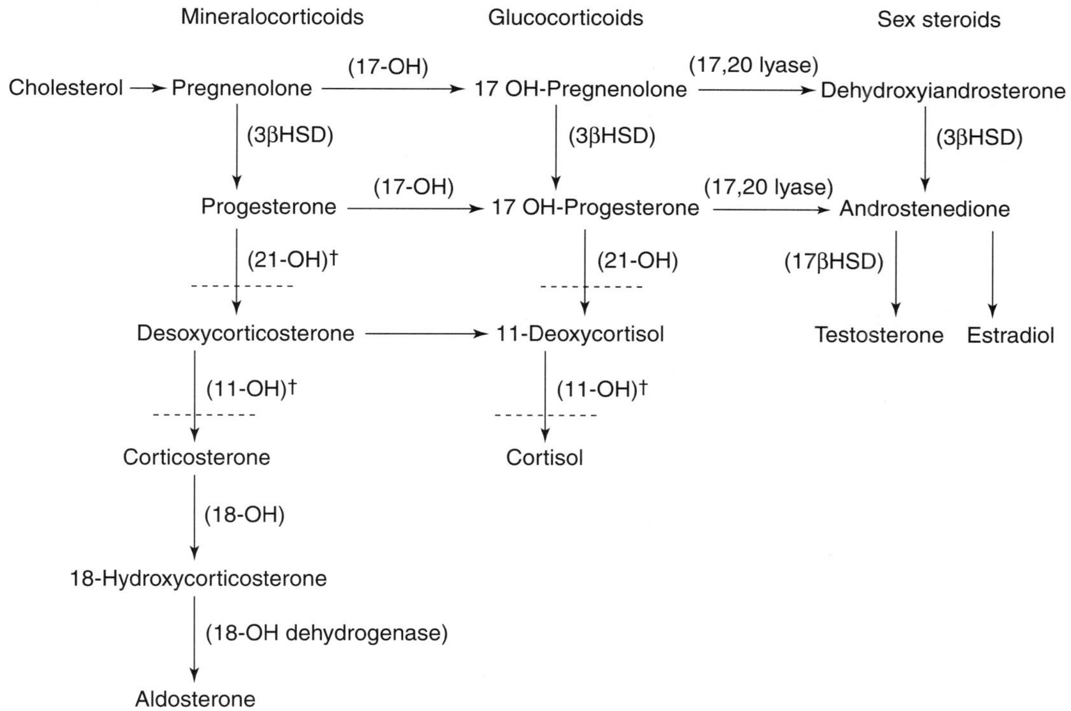

FIGURE 12.1. Adrenocortical steroid biosynthetic pathways. Dashed line = enzyme block. Enzymes: 3βHSD, 3β-hydroxysteroid dehydrogenase; 17βHSD, 17β-hydroxysteroid dehydrogenase; 17-OH, 17-hydroxylase; 11-OH, 11-OH hydroxylase; 18-OH, 18-hydroxylase; 21-OH, 21-hydroxylase. † = most common enzyme defects associated with the adrenogenital syndrome. (From New MI, et al. [2], Hughes IA [6].)

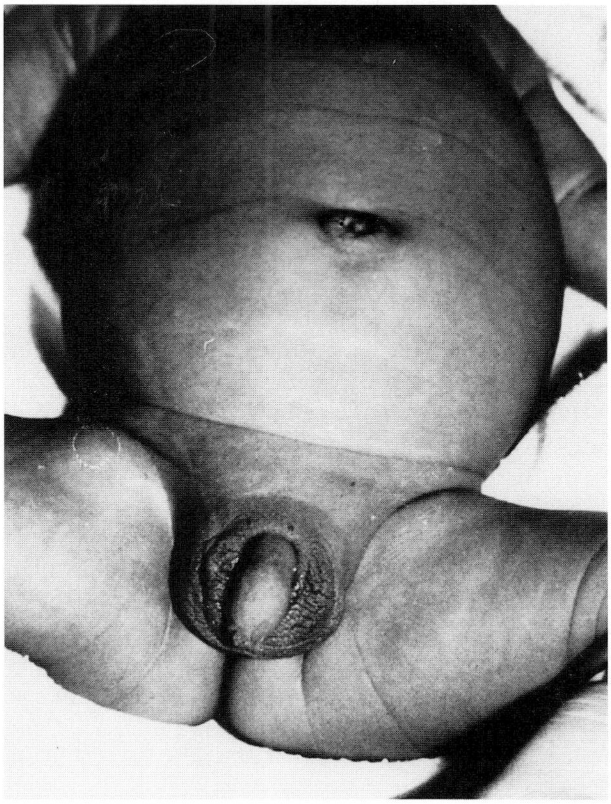

FIGURE 12.2. Congenital adrenal hyperplasia (adrenogenital syndrome) in a newborn female with striking clitoral enlargement, labial fusion ("scrotalization"), and hyperpigmentation. Congenital adrenal hyperplasia is one of the most common causes of ambiguous genitalia in the newborn. (Courtesy of Francine Kaufman, M.D., Division of Endrocrinology and Metabolism, Children's Hospital, Los Angeles.)

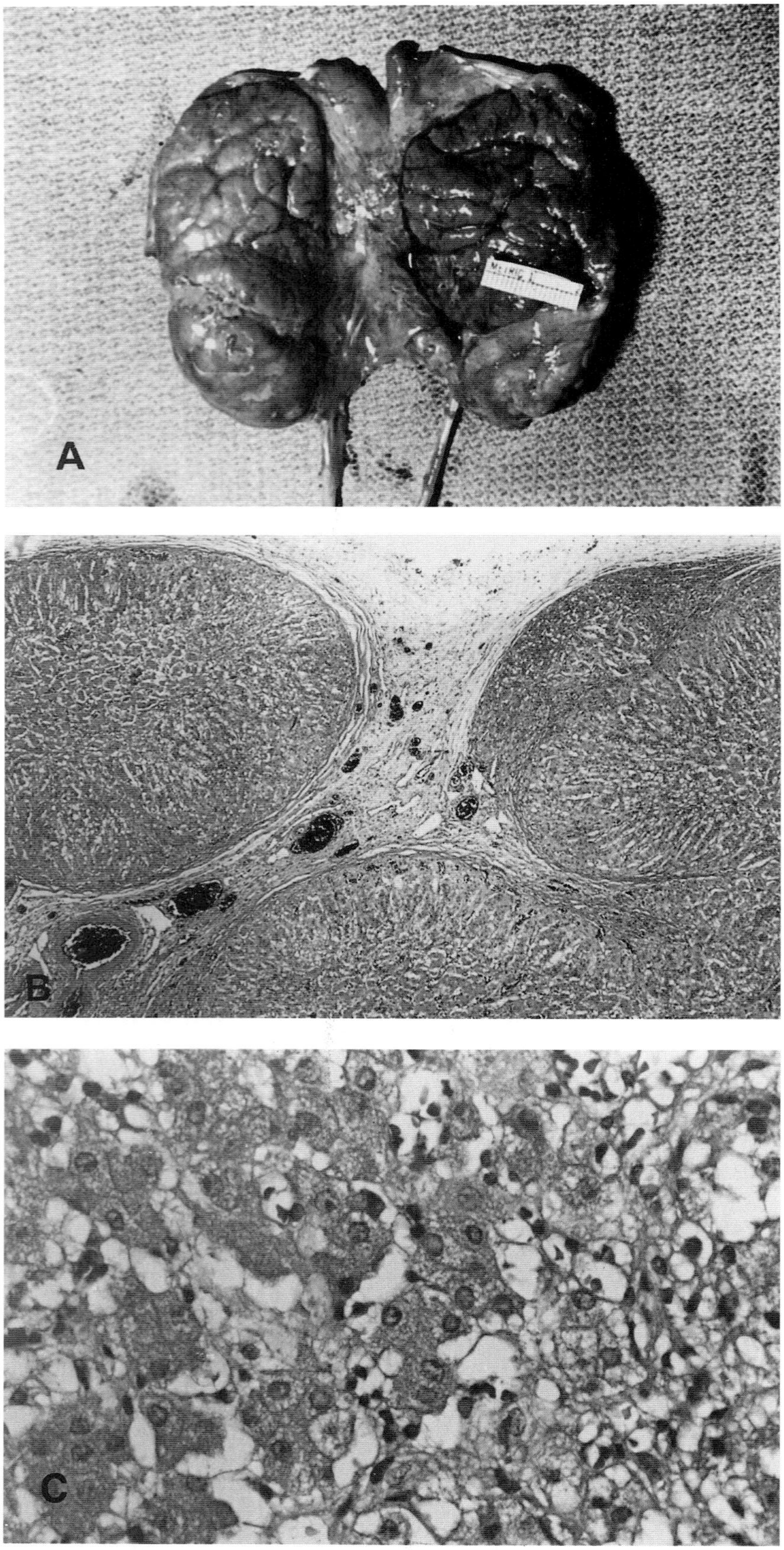

FIGURE 12.3. Congenital adrenal hyperplasia in a 3-year-old male with the salt-losing form of congenital adrenal hyperplasia due to 21-hydroxylase deficiency. (A) The kidneys are dwarfed by large convoluted adrenals (top), weighing 25 g together (normal combined weight is 5.2 g) and having a wrinkled, cerebriform appearance. (B) The cortex is thrown into nodular folds. The hyperplasia involves primarily the zona fasciculata and zona reticularis. The zona glomerulosa is abnormally small and ill defined. H&E; × 30. (C) The adrenal cortical cells are large, vacuolated, and filled with lipid. The darker staining cells in the left half of the field have eosinophilic staining cytoplasms. H&E; × 300.

Adrenocortical Tumors

Clinical Findings

An abdominal mass, virilization, and Cushing syndrome are the main presenting findings in newborns and infants [1–3,26,27]. The clinical findings depend largely upon the type of the hormone(s) secreted by the tumor [2]. In some series carcinomas outnumber adenomas at least 3:1 from birth to 3 years of age and have essentially the same histological appearance seen in tumors of older children and adults [1,3,13,27].

Neoplasms of the adrenal cortex share with Wilms tumor and hepatoblastoma increased associations with both hemihypertrophy and the Beckwith-Wiedemann syndrome, suggesting a common etiology possibly of genetic origin [3, 28–30]. Cytogenetic studies show that a locus exists on chromosome 11p15 (a deletion or loss of heterozygosity) that is found in both adrenocortical carcinoma and the Beckwith-Wiedemann syndrome [29]. Loss of tumor suppressor genes at loci on the short arm of chromosome 11(p) may be involved with the formation of other childhood malignancies as hepatoblastoma, Wilms tumor, and possibly rhabdomyosarcoma [29,30]. Certain families with children who have adrenal cortical tumors exhibit an increased susceptibility to malignancy as part of the cancer family syndrome called SBLA (**s**arcoma, **b**reast and **b**rain tumor, **l**eukemia, **l**aryngeal and **l**ung cancer, and **a**drenal cortical carcinoma), which is probably transmitted in an autosomal dominant fashion [5,16].

Steroid secretion is suppressed in patients with congenital adrenal hyperplasia but not in those with a neoplasm. Urinary 17-ketosteroids are markedly elevated in patients with a functioning adrenocortical carcinoma [9]. Ultimately the diagnosis depends on microscopic examination of the tumor.

Most adrenocortical tumors are encapsulated even when malignant. Both adenomas and carcinomas may consist of large cells with bizarre hyperchromatic nuclei. The histological distinction between benign and malignant adrenal cortical tumors is not always clear, which makes predicting prognosis on the basis of histology alone difficult. Moreover, distinctions between hyperplasia, adenomatous hyperplasia, and adenoma are not always well defined, a problem inherent in dealing with endocrine neoplasms in general [11,27].

Adrenocortical Adenoma

Adrenocortical adenoma consists of a single encapsulated mass surrounded by a thin fibrous capsule. The cut surface is smooth and bulging, varying in color from tannish-gray and tannish-yellow to orange brown (Figures 12.4–12.6). Cyst formations, hemorrhage, and necrosis are uncommon except in some larger specimens. Adjacent to the adenoma, the atrophic adrenal cortex may be compressed into an almost paper-thin, orange-tan shell [11,12,22,27].

The tumor consists of cords and clusters of round, oval, or polygonal cells having the appearance of adrenal cortical cells and surrounded by a prominent sinusoidal network (Figures 12.4–12.6). Cells of two types are present in adenomas from young patients with clinical signs and symptoms of Cushing syndrome and virilization. One is morphologically similar to the eosinophilic compact cells of the zona reticularis, observed commonly in patients with virilization. The other is a larger cell with a vacuolated spongy appearance due to an increase in lipid content that is comparable to that of cells of the zona fasciculata [27]. Not infrequently there is blending of the two types of cell. Moderate variation in cell size and focally giant cells with atypical nuclei may be present. Large areas of necrosis, anaplasia, and significant mitotic activity are characteristically absent. Capsular and vascular invasion, two important criteria of malignancy, are absent, but the latter is sometimes difficult to evaluate particularly in adrenocortical and in other endocrine neoplasms because of the inherent intimate relationship between tumor cells and adjacent vascular sinusoids [11].

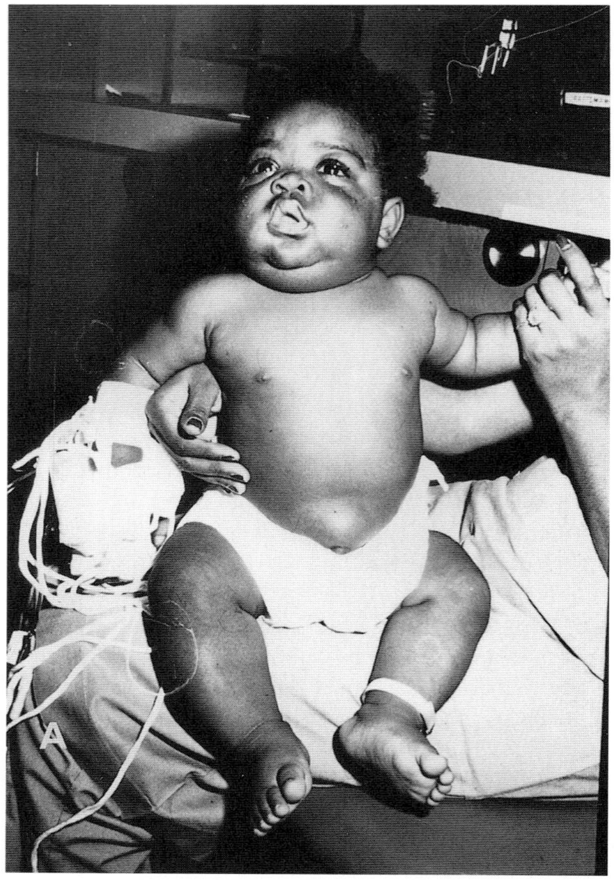

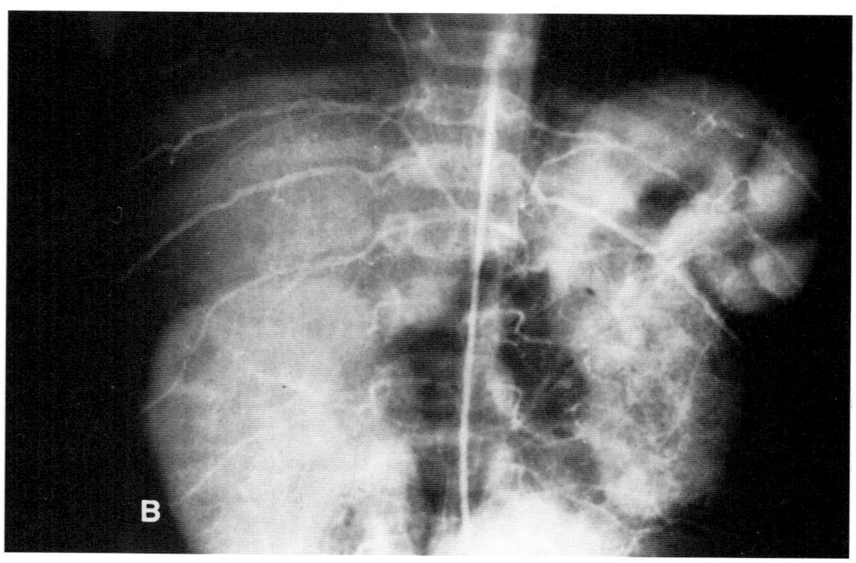

FIGURE 12.4. Adrenocortical adenoma in an 8-month-old female with Cushing syndrome (obesity, hypertension, acne) and a right adrenal mass. (A) Hemihypertrophy involves the right side of the body. The umbilicus is protuberant. (B) Arteriogram displays a right suprarenal mass and a right kidney that is patently larger than the left. (C) The adrenal is replaced by a 21 g, $5.5 \times 3.5\ cm^2$, tumor with a bright orange-tan, smooth bulging cut surface. (D) The tumor consists of nests and cords of clear vaculated cells with regular nuclei and cells with eosinophilic cytoplasms and more variable nuclei. Neither mitoses nor nuclear atypia are present. H&E; × 150. (E) There is a focus of adrenal cytomegaly in the residual fetal cortex, which normally involutes at age 3 months.The patient appears to have two conditions, hemihypertrophy and possibly the incomplete form of the Beckwith-Wiedemann syndrome, both of which are associated with a high incidence of adrenal tumors. H&E; × 300.

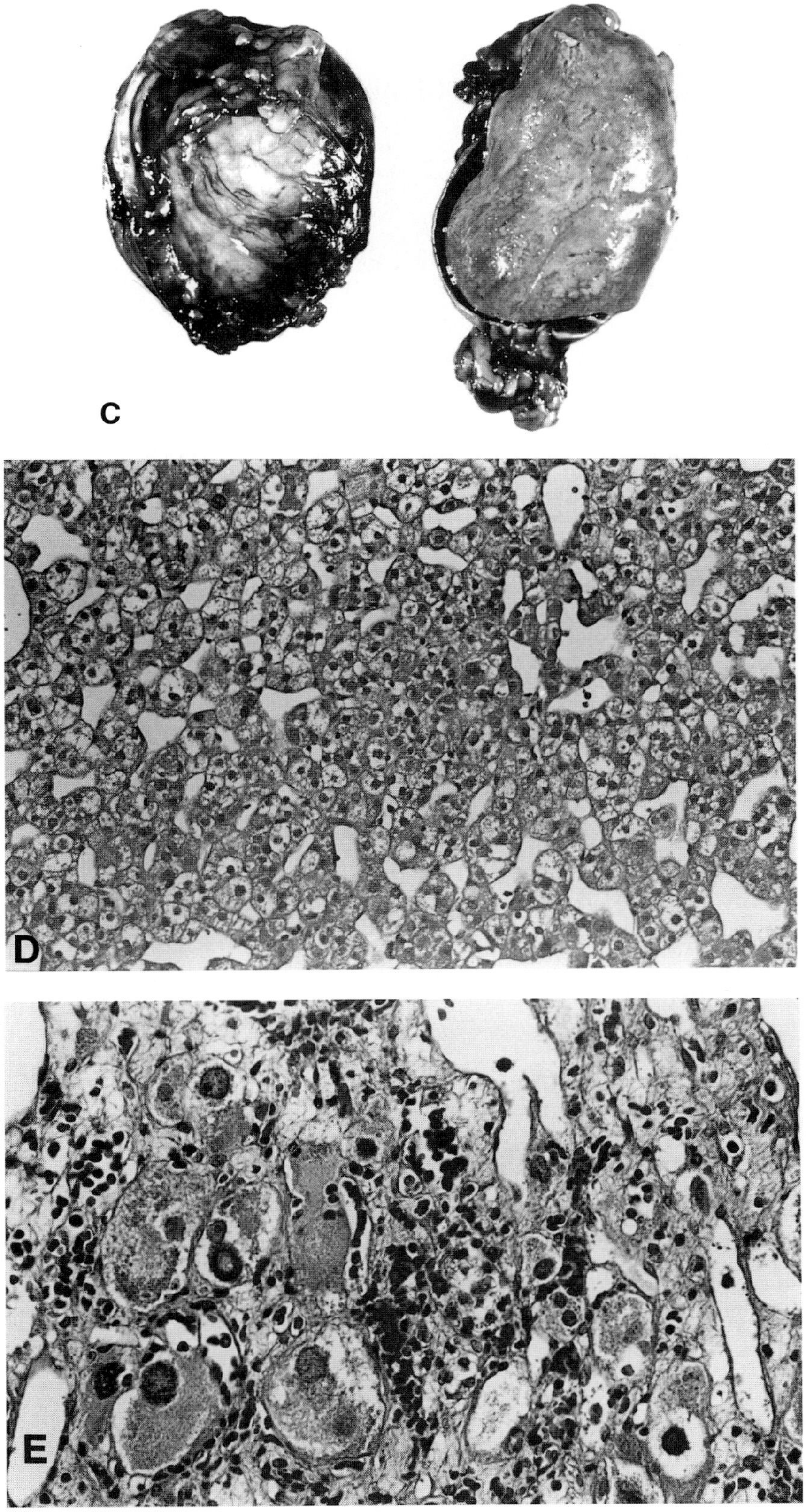

FIGURE 12.4. (*Continued*)

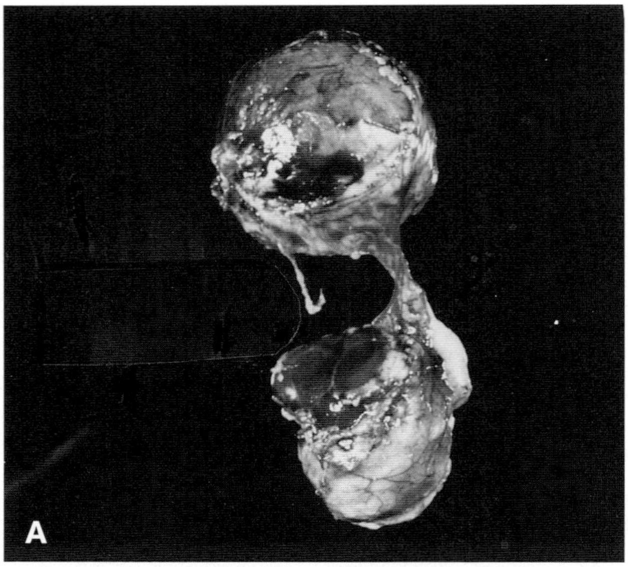

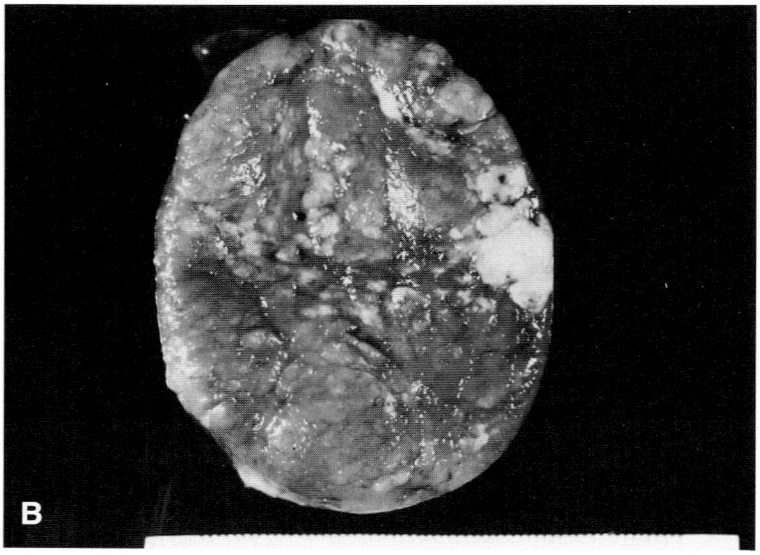

FIGURE 12.5. Adrenocortical adenoma in a 6-month-old female with Cushing syndrome and an abdominal mass. (A) The left adrenal mass (above), 120 g and 6.5×6 cm^2, is attached to the left kidney (below) by fibrofatty adhesions. (B) The cut surface of the tumor is multilobulated, tan to tannish pink with small yellow-white areas of necrosis. (C) The tumor shows lobular and trabeculated growth patterns with a prominent vascular component. The cells are relatively uniform, with round to oval nuclei showing minimal pleomorphism. H&E; × 120. (D) Higher magification (× 300) reveals some large cells with atypical nuclei; 1 to 2 mitoses per 10 high power fields were found. H&E. (Courtesy of Stephen G. Romansky, M.D., Department of Pathology, Long Beach Memorial Medical Center, Long Beach, CA. Used by permission.)

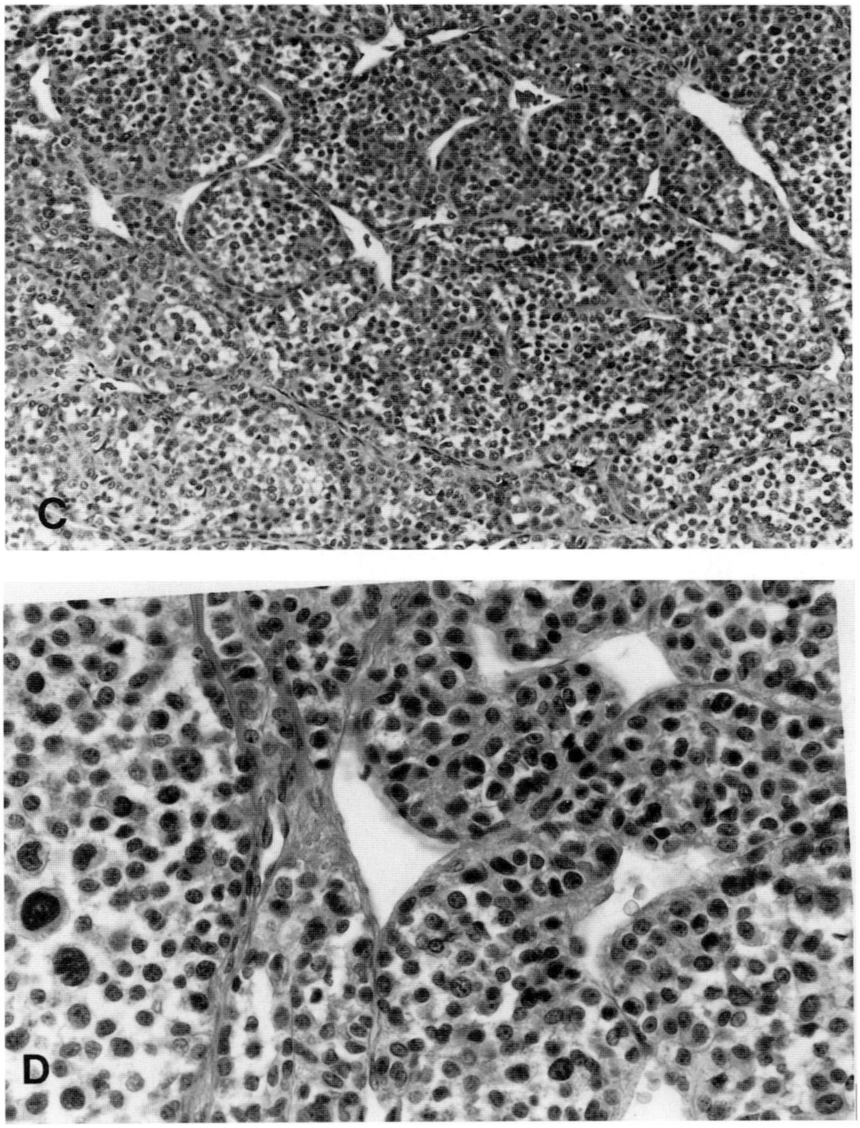

FIGURE 12.5. (*Continued*)

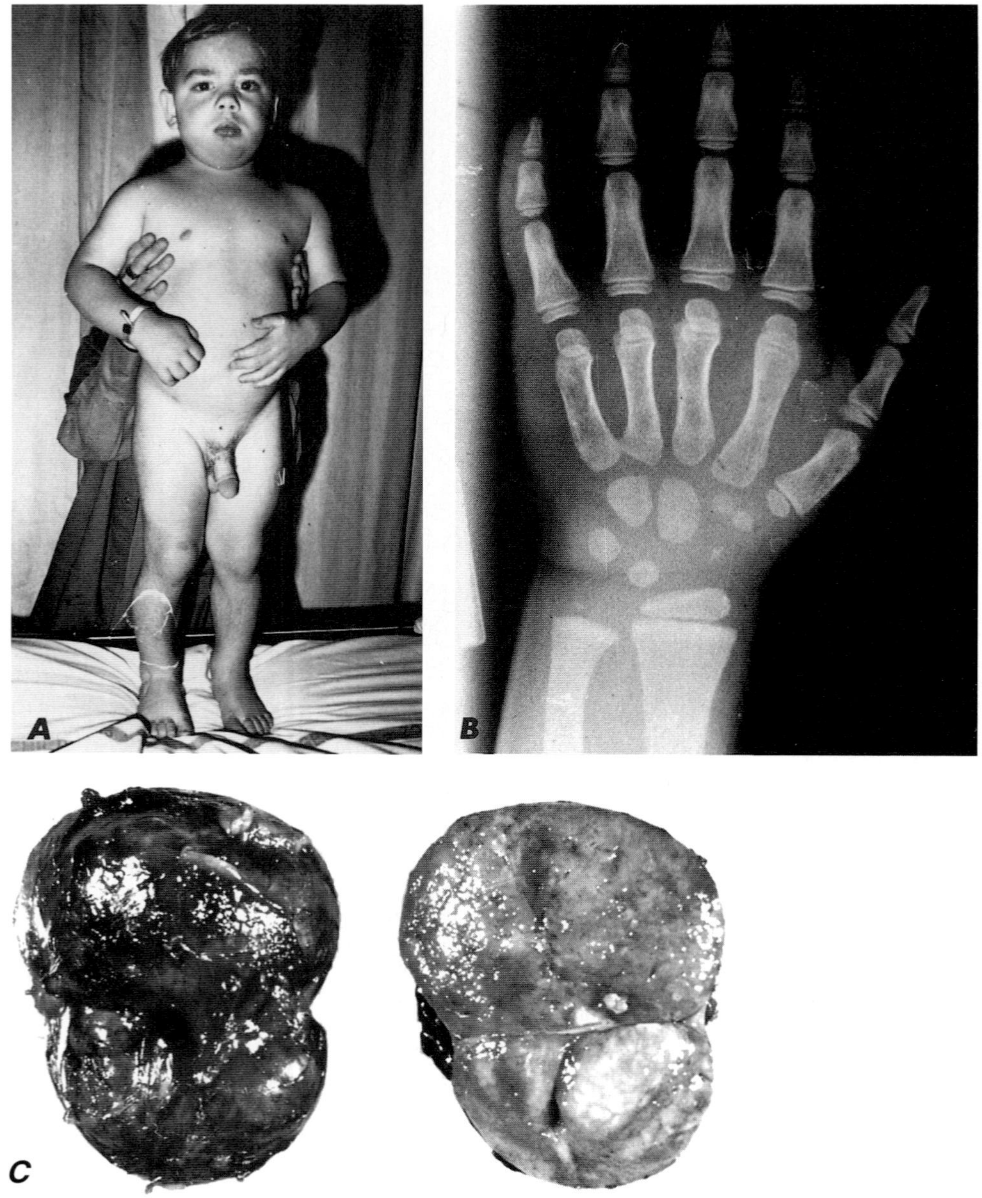

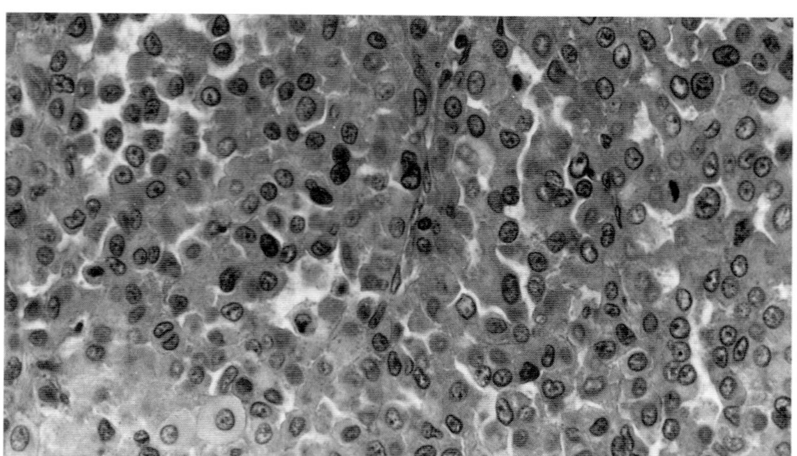

FIGURE 12.6. Adrenocortical adenoma in a 2-year-old male with a masculinizing adrenal cortical tumor. (A) The patient has markedly enlarged genitalia, pubic and body hair, and facial acne. (B) Radiogram of the left hand and wrist showing advanced bone age of a 6-year-old child. (C) The tumor, 17.1 g and 4×3 cm^2, has a slightly roughened, brown capsular surface with yellow highlights (right side of gross photograph). The cut surface (right) is orange-yellow, bulging with small, flat yellow-white areas of necrosis. (D) The tumor forms cords and nests of cells slightly resembling adrenocortical cells. There are cells of two types; one has regular, small, round occasionally vesicular nuclei and prominent pink cytoplasms; the other consists of larger round cells with abundant pale, slightly foamy cytoplasm and regular, round nuclei with prominent nucleoli. Minimal nuclear atypia and rare mitoses were found. H&E; $\times$ 150.

Adrenocortical Carcinoma

Adrenocortical carcinoma is an uncommon neoplasm in infants and children [1,9,31]. Its incidence is estimated at 0.2 to 0.4% of all childhood malignancies [1,5]. In some series adrenocortical carcinoma occurs more often than adenoma in the newborn [3,13,14]. Two age peaks of carcinoma exist, one in infancy and the other in adolescence. Female predominance is at least 2 :1. An abdominal mass, signs and symptoms of Cushing syndrome, and virilization are the presenting findings; most tumors are functional, and virilization is found in over 60% of infants and children [14,26,27,32] (Figure 12.7). Infrequently metastastases, e.g., to the skin, are the initial presenting finding [14].

"Nonfunctioning" adrenal cortical carcinomas sometimes occur in the pediatric age group, including neonates [32]. Some patients present with metastases at the time of diagnosis and tend to have a much worse prognosis [14,26,32].

Adrenocortical carcinomas are usually larger than the adenomas, ranging from 5 to 20 cm in diameter and weighing more than 100 g [1,26,27,33]. Most are encapsulated, but extension through the capsule into adjacent tissues is seen, especially in the larger specimens. Cut surfaces are soft and lobular, varying in color and consistency. Some tumors have a mushy light tan-gray appearance, while others are firmer, yellow-orange to tan. Prominent fibrous bands may be observed in the latter and are considered one of the features of malignancy. Large areas of necrosis, hemorrhage, and cystic degeneration with or without calcification typically are seen in the adrenocortical carcinomas [1,22] (Figure 12.7).

Similar to the adenoma and hyperplasia, adrenocortical carcinoma cells resemble those of the adrenal cortex. Typically the tumor cells are arranged in nests and cords situated adjacent to prominent vascular sinusoids. Some carcinomas consist of cells with compact eosinophilic cytoplasm, while others have larger cells composed of clear or vaculated cytoplasm due to abundant lipid droplets similar to those described for the adenoma (Figures 12.4D and 12.6D). Mixtures of the two cell types are seen. Some tumor cells resemble the peculiar giant cells of adrenal cortical cytomegaly found incidentally in stillborns and neonates with a variety of conditions. These bizarre-appearing tumor giant cells are observed both in adenomas and in carcinomas. The electron microscopic findings are essentially the same as adrenocortical adenoma [23,24]. Carcinomas appear to be less mature than adenomas with less well-defined organelles. Poorly differentiated adrenocortical carcinomas are unusual in the first year of life.

The histological criteria proposed for establishing the diagnosis of adrenocortical carcinoma are capsular and vascular invasion, a diffuse growth pattern, broad fibrous bands, increased mitotic rate with atypical mitotic figures, tumor necrosis, and extreme pleomorphism [1,22,27,31,32,34,35]. Nonhistological factors such as tumor weight exceeding 200 g and weight loss in the patient are used also as indicators of malignancy [36].

Immunohistochemical studies and DNA content (ploidy) do not reliably indicate malignancy [1,35]. DNA ploidy may be of some prognostic value [37]. According to Lack, the approach to the classification of childhood adrenocortical tumors is somewhat different from that used in adults, since some tumors are more likely to be clinically benign than had been previously been thought [22,34]. A high mitotic rate with abnormal mitoses is one of the most reliable indicators of malignancy [1,27,35]. The presence of metastases at the time of diagnosis and tumor weight greater than 200 g and/or size appear to be more reliable than histopathology in the assessment of the potential for malignancy [31,33,36]. Nevertheless small tumors have been known to metastasize, whereas some of the larger ones do not.

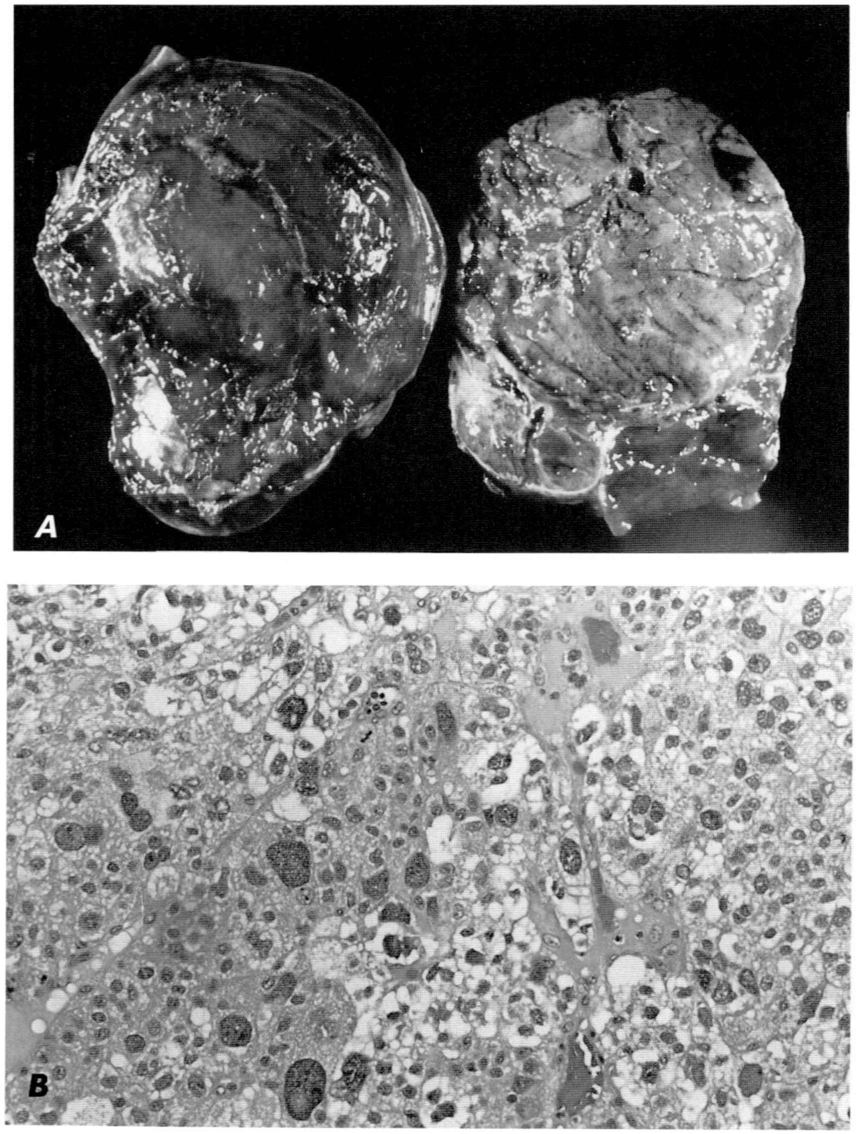

FIGURE 12.7. Adrenocortical carcinoma in an 8-month-old male with a masculinizing adrenal cortical tumor. The boy was big for his age and was noted to have enlarged genitalia and pubic hair. Abdominal CT revealed a large left adrenal tumor. Testosterone levels were markedly elevated. (A) The tumor, 160 g and 9.1×6.5 cm^2, is composed of multiple nodules varying in size, color, and consistency. The largest nodule consists of firm tan tissue with focal areas of hemorrhage and calcification. Nodules composed of bright yellow and brown hemorrhagic tissues are present. (B) The tumor shows a diffuse arrangement of pleomorphic cells. Many cells are large with abundant eosinophilic cytoplasm, big nuclei, prominent nucleoli, and sometimes multiple nuclei, while others are smaller, more regular cells, which have more uniform nuclei with dispersed chromatin and clear to vaculated cytoplasms. H&E; $\times$ 150. (C) Higher magnification ($\times$ 300) showing the two types of tumor and anaplastic giant cells; 15 mitoses per 10 high power fields were found. H&E. (D) Capsular invasion is present. The tumor extends to the line of resection. H&E; $\times$ 150.

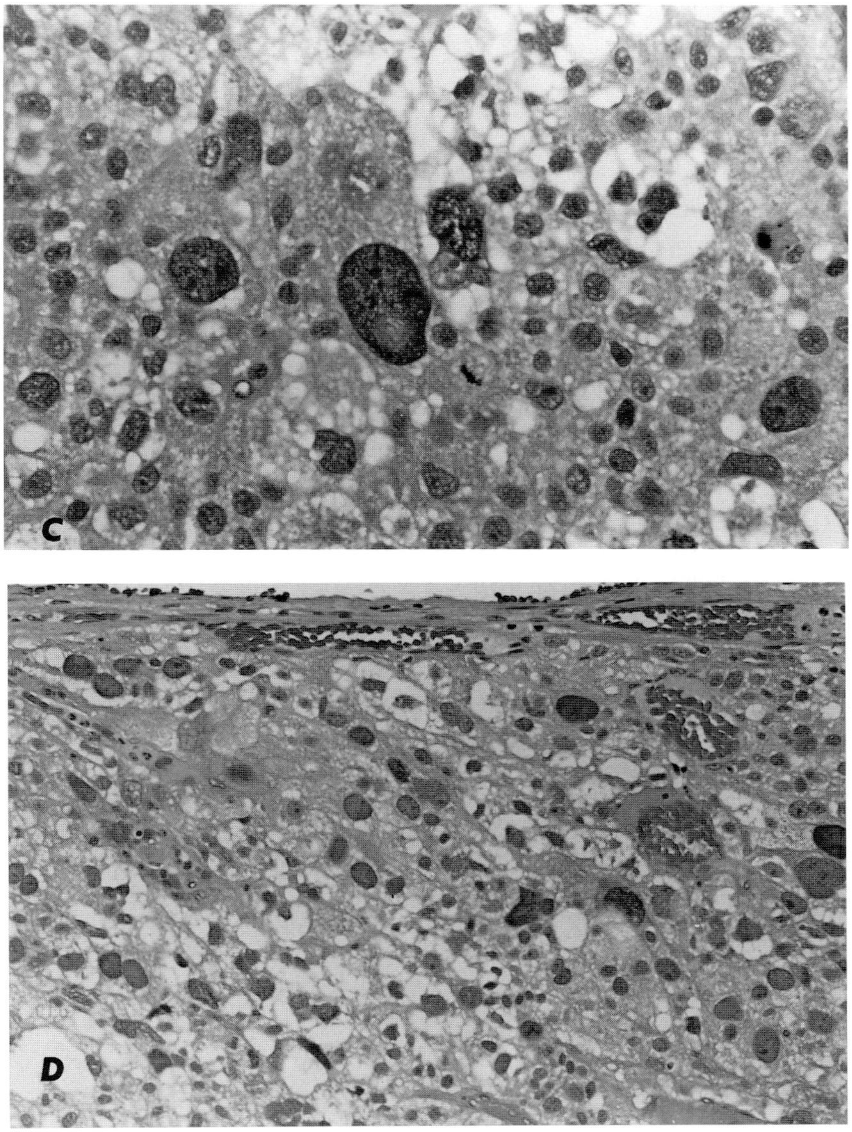

FIGURE 12.7. (*Continued*)

TABLE 12.2. Staging of adrenal cortical carcinoma.

Stage		Staging criteria[a]
I	T1N0M0	T1 Tumor less than or equal to 5 cm, no invasion
		T2 Tumor greater than 5 cm, no invasion
II	T2N0M0	T3 Tumor any size, locally invasive but not involving adjacent organs
		T4 Tumor of any size with invasion of adjacent organs
III	T1N0M0	N0 Negative regional node(s)
	T2N1M0	N1 Positive regional node(s)
	T3N0M0	M0 No distant metastases
		M1 Distant metastases
IV		Any T, any N, M1
		T3, N1
		T4

[a] T, tumor; N, lymph nodes; M, metastases.
Source: Abstracted from Lack EE [22], Table 3.

For some unexplained reason children with adrenocortical carcinoma, particularly those under 1 year of age, have a much better prognosis than adolescents and adults, with survival rates of approximately 53 and 17%, respectively [35].

Local spread within the abdomen portends a poor outcome. The most frequent sites of metastases of adrenocortical carcinoma in decreasing order are the lung, liver, peritoneum, regional lymph nodes, kidney, and brain [1,2,22,31]. Recurrence of tumor or death due to metastases generally occurs within 2 years of diagnosis, but some deaths have been reported as late as 5 years after the original diagnosis [31]. The proposed staging of adrenocortical carcinoma is shown in Table 12.2 [22].

Adrenocortical Cytomegaly

Large, peculiar-appearing adrenal cortical cells were first described by Kampmeier in 1927 [38] and reported later in a detailed study by Craig and Landing, who also noted similar cells in the developing fetal cortex of stillborns and infants up to 2 months of age [39]. Adrenal cytomegaly, a term coined by Potter, is defined as the presence of large, bizarre-looking eosinophilic cells with huge nuclei, distributed focally or diffusely within the fetal cortical zone [12,39,40]. The cytomegalic cells are found only in the fetal cortex and are thought to arise from the large single cells noted early in the developing fetal cortex [12]. They may occur singly or in large numbers as wedge-shaped clusters extending to the surface of the adrenal or as collections of cells distributed diffusely throughout the fetal cortex (Figures 12.4E and 12.8).

An increased incidence of adrenal cytomegaly is noted in fetuses and newborns with a variety of conditions, including erythroblastosis fetalis, maternal toxemia, multiple pregnancies, congenital heart malformations, and hemihypertrophy, and particularly in those with the Beckwith-Wiedemann syndrome [1,12,17,40–43]. The most characteristic and consistent histological finding in the adrenal glands of patients with the Beckwith-Wiedemann syndrome is diffuse, bilateral cytomegaly of the fetal cortex [1,22,44]. Cytomegaly is described essentially in every patient with this syndrome where the adrenals are examined microscopically.

In addition, there is a significant association between adrenal cytomegaly and congenital malformations [40,41]. Cardiac anomalies as truncus arteriosus, atrial and ventricular

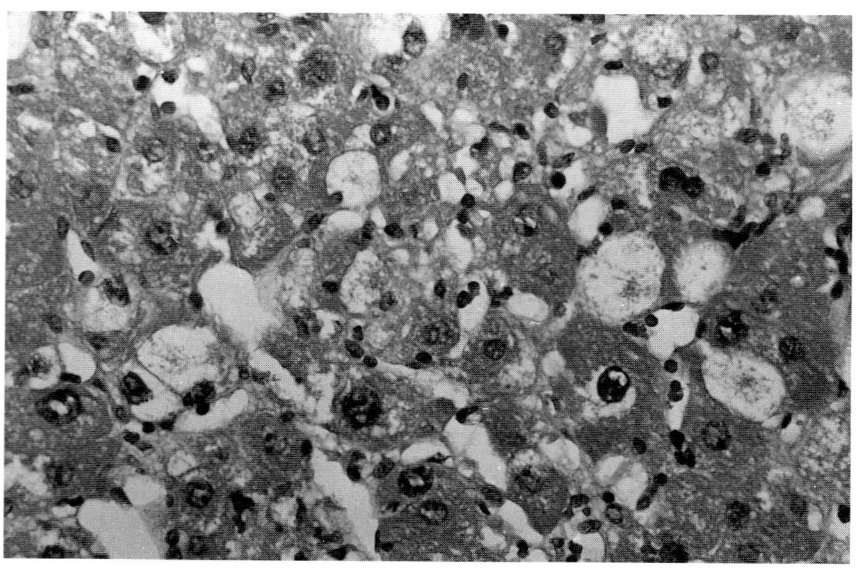

FIGURE 12.8. Adrenal cortical cytomegaly. Incidental necropsy finding in a 4-day-old male with the Pena-Shokeir syndrome (dysmorphic facies, multiple joint contractures, pulmonary hypoplasia). Large, peculiar-appearing cells with huge nuclei are present within the developing fetal adrenal cortex. Focal infiltrates of lymphocytes are present. H&E; × 300.

septal defects, and various types of atresia are the most common ones mentioned [40].

Typically the external appearance of the adrenals is normal. The presence of cytomegaly is an unexpected microscopic finding in postmortem examinations performed on fetuses, stillborns, and newborns [12,39–43].

Microscopically the enlarged nuclei of the cytomegalic cells possess a dumbbell shape and a polylobar configuration. They are irregular, darkly basophilic staining and are composed of spherical acidophilic inclusion-like bodies and vacuoles. The cytoplasm stains uniformly eosinophilic, similar to the rest of the fetal cortical cells, and contains circumscribed masses of finely granular material. There is a peripheral rim of small vacuoles, giving the cell a characteristic foamy appearance. Like the smaller cells of the fetal cortex, the cytomegalic cells have a polygonal outline [12,39,42] (Figures 12.4E and 12.8). Electron microscopic studies show that the nuclear inclusions actually represent invaginations of portions of the cytoplasm into the nucleus surrounded by double membranes [42]. The cytoplasm is packed with numerous mitochondria, vesicles, microsomes, and lysosomes and the cytoplasmic granular material appears to be a lipochrome pigment [42]. Mitoses are found rarely.

The cause of adrenal cytomegaly is unknown. Several etiologies have been suggested [1,17,39,40–44].

Adrenal Cyst

Adrenal cysts are found in both infants and children but are more common in adults [15]. Fewer than 100 cases are reported in the newborn [11,15,45–47]. The cyst may present as an abdominal mass, thus mimicking a tumor. Sometimes it is apparent at birth, or it may be an incidental finding on imaging studies, prenatally or postnatally [11,15,45–47]. Many cysts are asymptomatic and spontaneously resolve [45].

Most adrenal cysts in infants are considered to be the result of perinatal adrenal hemorrhage, which is found in up to 1% of postmortem examinations performed on newborns [11,45]. Trauma associated with delivery is the most important cause, but other less common conditions such as infection, neoplasms (leukemia and neuroblastoma), hemorrhagic disease of the newborn, incompatible blood transfusions, and adrenal vein thrombosis have been implicated [45]. Adrenal cysts seldom arise from a hematoma within a tumor such as neuroblastoma (Figure 5.16). Large adrenocortical cysts occur in fetuses and newborns with the Beckwith-Wiedemann syndrome [47].

Organization of adrenal hemorrhage may lead to a large cyst occasionally, the size of an orange [12,45]. On cross section the cysts are unilocular or multilocular, varying in diameter from 2 to 11 cm. They may contain xanthochromic fluid or purulent material, if infected.

Microscopically the cyst wall consists of fibrous connective tissue containing small clusters of adrenal cortical cells. The cyst lacks an epithelial lining, and sometimes focal calcification is present and can be seen on imaging studies. The cyst contents are variable and are composed of fibrin, necrotic debris, and erythrocytes. If the lesion is infected inflammatory granulation tissue and/or pus may be present. Adrenal cysts occur with in situ neuroblastoma [11,46,47] (Figure 5.16).

References

1. Lack EE, Mulvihill JJ, Travis WD, et al. Adrenal cortical neopasms in the pediatric and adolescent age group: clinicopathologic study of 30 cases with emphasis on epidemiological and prognostic factors. *Pathol Annu* 1992;27(1):1.
2. New MI, del Balzo P, Crawford C, et al. The adrenal cortex. In: Kaplan SA, ed. *Clinical Pediatric Endocrinology.* Philadelphia: WB Saunders; 1990:181.
3. Dehner LP. Neoplasms of the fetus and neonate. In: Naeye RL, Kissane JM, Kaufman N, eds. *Perinatal Diseases.* Baltimore: Williams & Wilkins, 1981:286. International Academy of Pathology, Monograph No. 22.
4. Dehner LP, Franciosi RA. Adrenocortical neoplasia in the pediatric age group. A clinicopathologic study of seven cases. In: Humphrey FB, Grindley LP, Dehner LP, et al, eds. *Adrenal and Endocrine Tumors in Children.* Boston: Martinus Nijhoff; 1984: 291.
5. Hartley AL, Birch JM, Marsden HB, et al. Adrenal cortical tumors: epidemiological and familial aspects. *Arch Dis Child* 1987; 62:683.
6. Hughes IA. Congenital and acquired disorders of the adrenal cortex. *Clin Endocrinol Metabol* 1982;11(1):89.
7. White PC, New MI, Dupont B. Congenital adrenal hyperplasia. *N Engl J Med* 1987;316:1519.
8. Avni EF, Rypens F, Smet MH, et al. Sonographic demonstration of congenital adrenal hyperplasia in the neonate: the cerebriform pattern. *Pediatr Radiol* 1993;23:88.
9. Holcombe JH, Pysher TJ, Kirkland RT. Functioning adrenocortical tumors in childhood. In: Humphrey FB, Grindley LP, Dehner LP, et al, eds. *Adrenal and Endocrine Tumors in Children.* Boston: Martinus Nijhoff; 1984:277.
10. McArthur RG, Bahn RC, Hayles AB. Primary adrenocortical nodular dysplasia as a cause of Cushing's syndrome in infants and children. *Mayo Clin Proc* 1982;57:58.
11. Isaacs H Jr. Tumors. In: Gilbert-Barness E, ed. *Potter's Pathology of the Fetus and Infant.* Vol 2. St. Louis, MO: CV Mosby, 1997:1287.
12. Isaacs H Jr. Adrenocortical tumors. In: *Tumors of the Fetus and Newborn.* Philadelphia: WB Saunders, 1997:298. Major Problems in Pathology Series; Vol 35.
13. Artigas JLR, Niclewicz ED, Silva APG, et al. Congenital adrenal cortical carcinoma. *J Pediatr Surg* 1976;11:247.
14. Saracco S, Abramowsky C, et al. Spontaneously regressing adrenal cortical carcinoma in a newborn: a case report with DNA ploidy analysis. *Cancer* 1988;62:507.
15. Zivkovic SM, Jancic-Zguricas M, Jokanovic R, et al. Adrenal cysts in the newborn. *J Urol* 1983;129:1031.
16. Lynch HT, Katz DA, Bogard PJ, et al. The sarcoma, breast, cancer and adrenocortical carcinoma syndrome revisited. *Am J Dis Child* 1985;139:134.

17. Sotelo-Avila C, Gonzalez-Crussi F, Fowler JW. Complete and incomplete forms of Beckwith-Wiedemann syndrome. Their oncogenic potential. *J Pediatr* 1980;96:47.
18. Tank ES, Kay R. Neoplasms associated with hemihypertrophy, Beckwith-Wiedemann syndrome and aniridia. *J Urol* 1980;124: 266.
19. Telander RL, Wolf SA, Simmons PS, et al. Endocrine disorders of the pancreas and adrenal cortex. *Mayo Clin Proc* 1986;61:459.
20. Mininberg DT, Levine LS, New MI. Current concepts in congenital adrenal hyperplasia. *Pathol Annu* 1982;17(2):179.
21. Pigeon B, Ryckewaert P, Massin B. Corticosurrénalome virilisant de l'enfant à propos d'une observation néo-natale. *Pediatrie* 1985;40:309.
22. Lack EE. Tumors of the adrenal gland and extra-adrenal paraganglia. In: *Atlas of Tumor Pathology*. 3rd ser. Fascicle 19. Washington, DC: Armed Forces Institute of Pathology; 1997.
23. Ghadially FN. *Diagnostic Electron Microscopy of Tumours*. 2nd ed. London: Butterworths, 1985:291.
24. Dickersin GR. Large cell neoplasms. In: *Diagnostic Electron Microscopy: A Text/Atlas*. 2nd ed. New York: Springer-Verlag; 2000:63.
25. Danon M, Robboy SJ, Kim S, et al. Cushing syndrome, sexual precocity, and polyostotic fibrous dysplasia (Albright syndrome) in infancy. *J Pediatr* 1975;87:917.
26. Kakkur N, Vasishta RK, Lamba A, et al. Pathological case of the month: congenital adrenocortical carcinoma. *Arch Pediatr Adolesc Med* 2000;154:1267.
27. Weatherby RP, Carney JA. Pathologic features of childhood adrenocortical tumors. In: Humphrey FB, Grindley LP, Dehner LP, et al, eds. *Adrenal and Endocrine Tumors in Children*. Boston: Martinus Nijhoff; 1984:217.
28. Bolande RP. Neoplasia of early life and its relationships to teratogenesis. In: Rosenberg HS, Bolande RP, eds. *Perspectives in Pediatric Pathology*. Vol 3. Chicago: Year Book Medical Publishers, 1976:145.
29. Henry I, Grandjouan S, Coullin P, et al. Tumor specific loss of 11p15.5 alleles in del 11p13 Wilms' tumor and in familial adrenocortical carcinoma. *Proc Natl Acad Sci USA* 1989;86:3247.
30. Koufos A, Hansen MF, Copeland NG, et al. Loss of heterozygosity in three embryonal tumors suggests a common pathogenetic mechanism. *Nature* 1985;316:330.
31. Lefevre M, Gerard-Marchant R, Gubler JP, et al. Adrenal cortical carcinoma in children: 42 patients treated from 1958 to 1980 at Villejuif. In: Humphrey FB, Grindley LP, Dehner LP, et al, eds. *Adrenal and Endocrine Tumors in Children*. Boston: Martinus Nijhoff; 1984:265.
32. Sasano N, Ojima M, Masuda T. Endocrinologic pathology of functioning adrenocortical tumors. *Pathol Annu* 1980;15(2):105.
33. Wolthers OD, Cameron FJ, Scheimberg I, et al. Androgen secreting adrenocortical tumours. *Arch Dis Child* 1999;80:46.
34. Lack EE. Recommendations for the reporting of tumors of the adrenal cortex and medulla: Association of Directors of Anatomic and Surgical Pathology. *Virchows Arch* 1999;435:87.
35. Medeiros LJ, Weiss LM. New developments in the pathologic diagnosis of adrenal cortical neoplasms: a review. *Am J Clin Pathol* 1992;97:73.
36. Hawkins HP, Cagle PT. Adrenal cortical neoplasms in children [correspondence and corrections]. *Am J Clin Pathol* 1992;98:382.
37. Taylor SR, Roederer BS, Murphy RF. Flow cytometric DNA analysis of adrenocortical tumors in children. *Cancer* 1987;59: 2059.
38. Kampmeier OF. Giant epithelial cells of the human fetal adrenal. *Anat Record* 1927;37:95.
39. Craig JM, Landing BH. Anaplastic cells of fetal adrenal cortex. *Am J Clin Pathol* 1951;21:940.
40. Oppenheimer EH. Adrenal cytomegaly: studies by light and electron microscopy. *Arch Pathol* 1970;90:57.
41. Aterman K, Kerenyi N, Lee M. Adrenal cytomegaly. *Virchows Arch [A] (Pathol Anat)* 1972;355:105.
42. Nakamura Y, Yano H, Nakashima T. False intranuclear inclusions in adrenal cytomegaly. *Arch Pathol Lab Med* 1981;105:358.
43. Favara BE, Steele A, Grant JH, et al. Adrenal cytomegaly: quantitative assessment by image analysis. *Pediatr Pathol* 1991;11: 521.
44. Beckwith JB. Macroglossia, omphalocele, adrenal cytomegaly, gigantism and hyperplastic visceromegaly. In: *Birth Defects: Original Article Series* 1969;5:188.
45. Van De Water JM, Fonkalsrud EW. Adrenal cysts in infancy. *Surgery* 1966;60;1267.
46. Tubergen DG, Heyn RM. In situ neuroblastoma associated with an adrenal cyst. *J Pediatr* 1970;76:451.
47. McCauley RGK, Beckwith JB, Elias ER, et al. Benign hemorrhagic adrenocortical macrocysts in Beckwith-Wiedemann syndrome. *Am J Radiol* 1991;157:549.

13
Pancreatic Tumors

True neoplasms of the pancreas are rare in the newborn and infant [1–6]. Lesions arising from islet cells (e.g., nesidiodysplasia and adenomatous hyperplasia) occur more often than exocrine tumors in this age group [1,5–14]. The "embryonic" tumor of the pancreas, the so-called pancreatoblastoma, a neoplasm unique to infants and children, has been the subject of several case reports [1,5,12,15]. Table 13.1 lists pancreatic tumors and tumorlike conditions found in the fetus and infant.

Nesidioblastosis and Nesidiodysplasia of Infancy

In 1938 Laidlaw created the term "nesidioblastosis" to describe the neoformation of islets of Langerhans from panceatic duct epithelium [16] (Figure 13.1). It is derived from the Greek words *nesidion* for "islet" and *blastos* for "germ." He proposed also the term "nesidioblastoma" for an adenoma composed of islet cells. Yakovac in 1971 was one of the first to describe nesidioblastosis in a group of 12 infants with intractable hypoglycemia [17]. He demonstrated histochemically the presence of islet cells, singly and in clusters of two to six, separate from the islets of Langerhans located in acini and in small pancreatic ducts.

The current view is that nesidioblastosis represents the normal endocrine cell development of the pancreas in the fetus and newborn and thus should not be considered a pathological disorder in this age group [7,18] (Figures 13.1 and 13.2). Since nesidioblastosis can be observed in the pancreas of normoglycemic neonates, it is not pathognomic of hyperinsulinemic hypoglycemia of infancy [8,18].

"Persistent hyperinsulinemic hypoglycemia of infancy" is a term used to designate the condition of patients under a year in age with hyperinsulinism and profound hypoglycemia who require pancreatic resection [7].

Four main histiological findings are associated with persistent hyperinsulinemic hypoglycemia of infancy: nesidiodysplasia (endocrine cell dysplasia), adenomatous hyperplasia, islet cell adenoma (insulinoma), or no recognizable histological abnormality. Nesidiodysplasia is characterized by diffuse disorganization of islet architecture, characterized by loss of discrete centrilobular islet formation and an increase in randomly distributed nests of islet cells. Islet shape, size, and architecture are abnormal [7] (Figure 13.3). Adenomatous hyperplasia (adenomatosis) is distinguished by focal areas of excessive islet cell proliferation, usually greater than 40% of a given low power field, pushing acinar elements aside

TABLE 13.1. Tumors and tumorlike conditions of the pancreas in fetuses and infants.

Endocrine	
	Nesidioblastosis
	Nesidiodysplasia (endocrine cell dysplasia)
	Adenomatous hyperplasia
	Islet cell adenoma (insulinoma)
Exocrine	
	Cystadenoma
	Ductal adenocarcinoma[a]
	Acinar adenocarcinoma[a]
	Hamartoma
Pluripotential	
	Pancreatoblastoma
Mesenchymal	
	Hemangioma
	Lymphangioma
	Fibromatosis
Metastatic neoplasms	
	Neuroblastoma
	Leukemia
	Wilms tumor
	Rhabdoid tumor
	Yolk sac tumor
Miscellaneous	
	Extramedullary hematopoiesis
	Lymphocytic infiltration

[a] Usually does not occur in the infant.
Source: Isaacs H Jr [14].

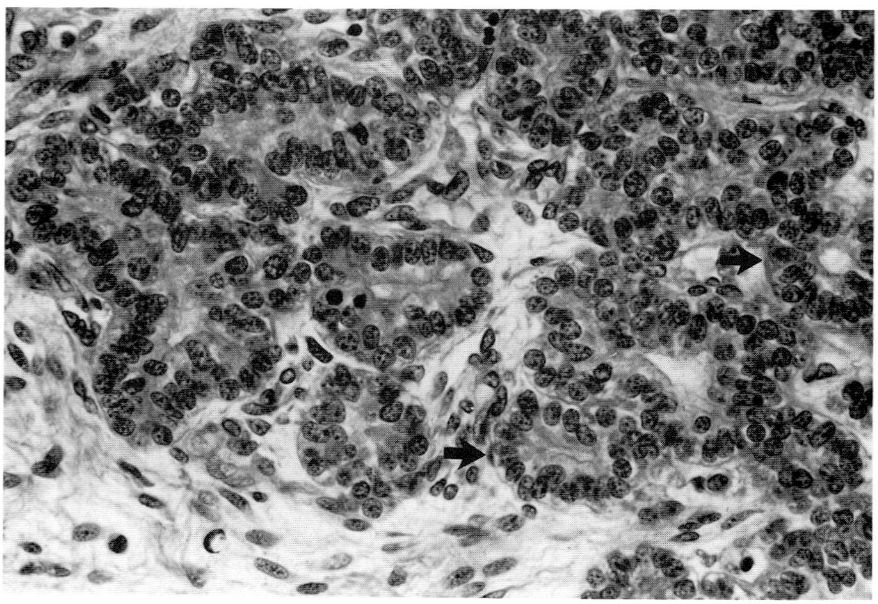

FIGURE 13.1. Developing pancreas: acini (arrow) are budding off from proliferating ducts. From an embryo of estimated 10 to 12 weeks gestation. H&E; × 250.

FIGURE 13.2. Nesidioblastosis: normal pancreas from a 650 g fetus. (A) Clusters of one or more islet cells separate from the islets of Langerhans are situated within acini and small pancreatic ducts. H&E; $\times$ 100. (B) Higher magnification ($\times$ 250) reveals islet cells within acini and in the wall of a small pancreatic duct. H&E. (C) Insulin immunoperoxidase stain showing the presence of β-cells outside of the islets. $\times$ 100. Insulin immunoperoxidase.

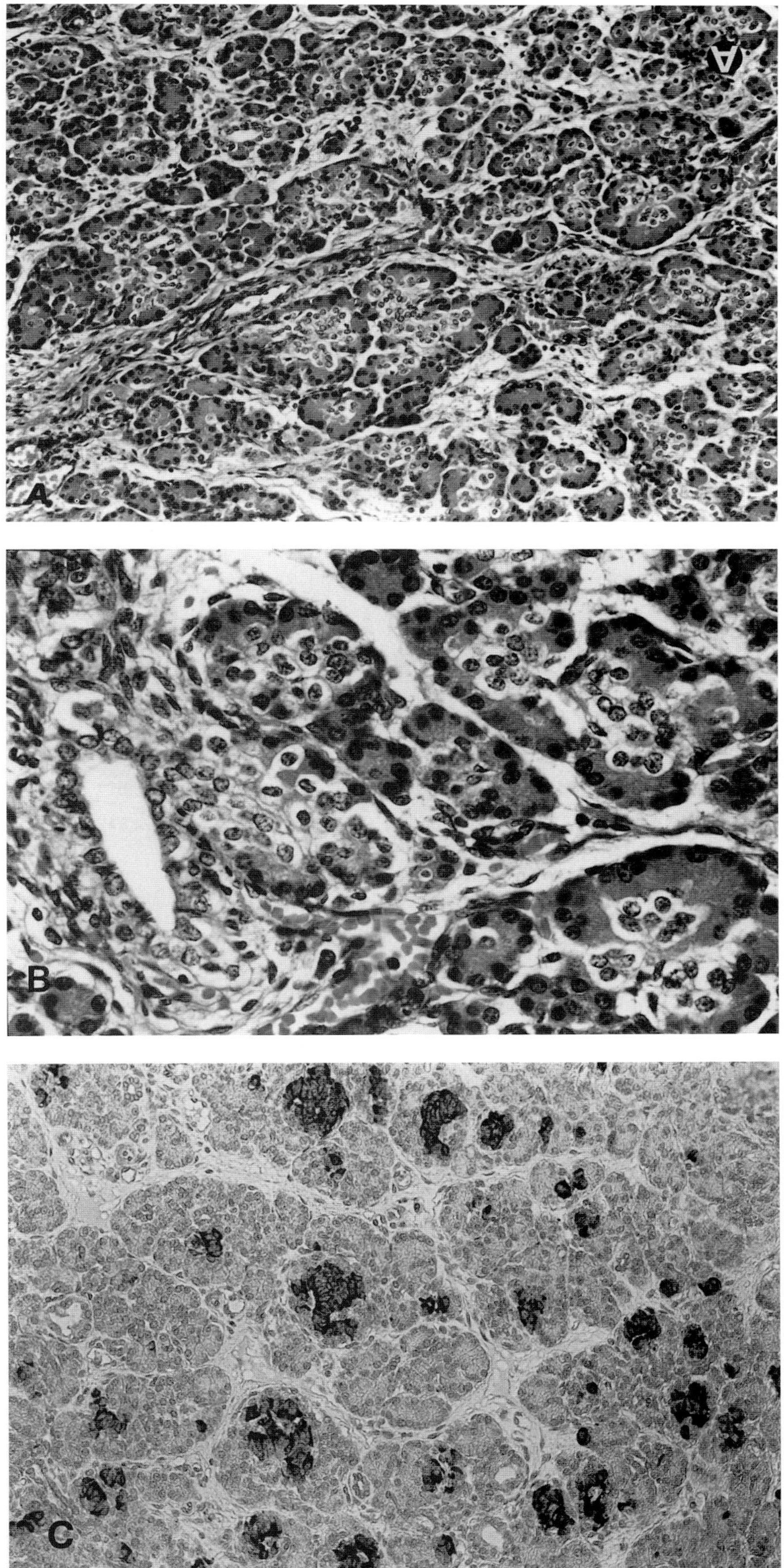

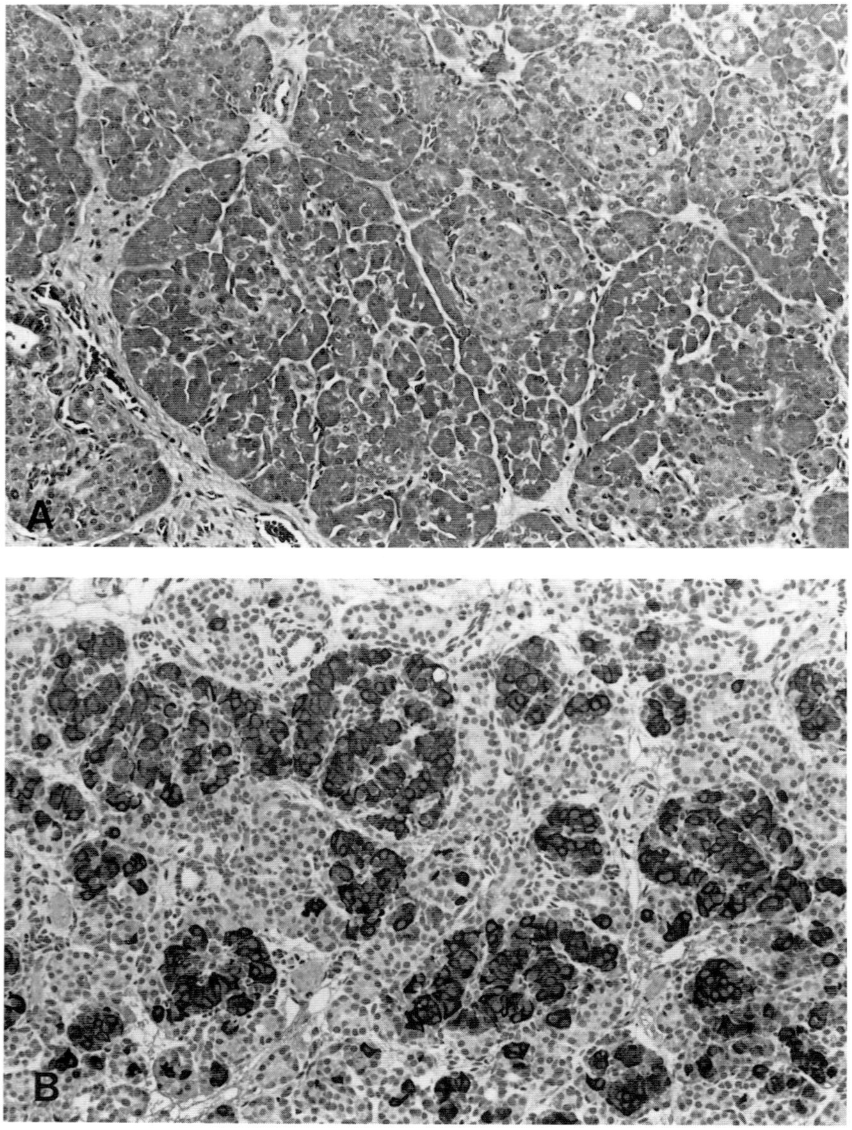

FIGURE 13.3. Nesidiodysplasia (endocrine cell dysplasia) in a 1.5-month-old female with severe, persistent hyperinsulinemic hypoglycemia. (A) Islets of Langerhans are irregularly distributed throughout the pancreas and vary considerably in size; some within the interlobular septa are exceedingly large. H&E; × 120. (B) Immunostain for insulin reveals intense staining of islet cells not only within islets but also as individual cells and clusters within acini. Immunostains for glucagon and somatostatin showed also intense immunoreactivity. Insulin immunoperoxidase; × 120.

or haphazardly incorporating them [7,18–20] (Figure 13.4). Cells comprising the hyperplasia are a mixture of islet cell types, α, β, and δ cells, in contrast to an islet cell adenoma, which contains exclusively β cells [3,10,20] (Figure 13.5). Adenomatous hyperplasia of islet cells with nesidiodysplasia is one of the hallmarks of the Beckwith-Wiedemann syndrome in the pancreas [18–21] (Figure 13.4), and pancreatoblastoma is the exocrine tumor associated with this syndrome [5,22].

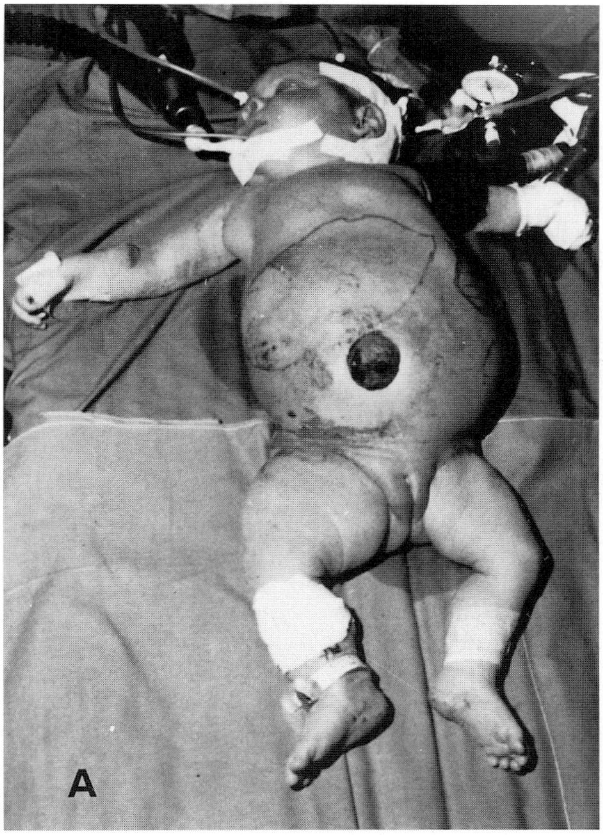

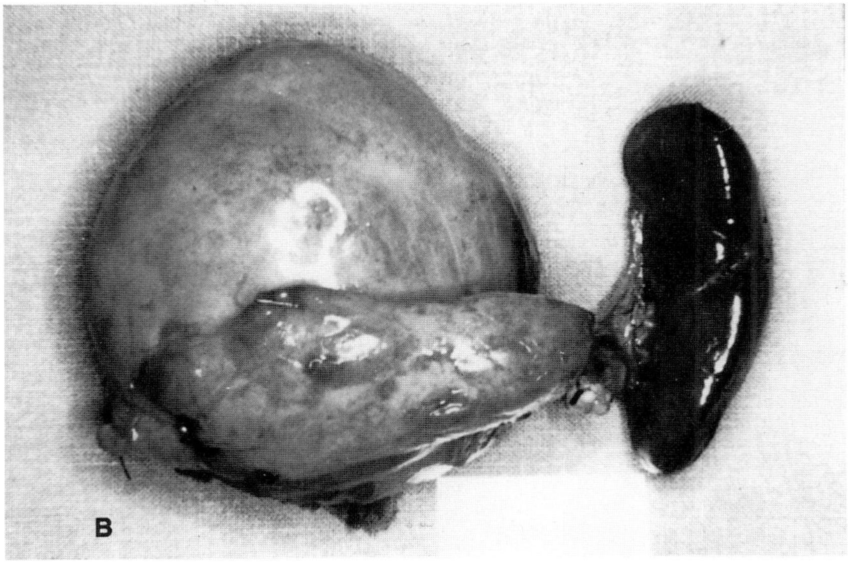

FIGURE 13.4. Pancreas of the Beckwith-Wiedemann syndrome. This 3-week-old female had severe, intractable hyperinsulinemic hypoglycemia, right hemihypertrophy, omphalocele, and abdominal visceromegaly. (A) The enlarged liver and pancreas are outlined on the patient's abdomen. (B) The excised pancreas, 71 g, and spleen. (C) The pancreas reveals diffuse hyperplasia of islet cells with marked reduction in acinar tissue. (D) Higher magnification (× 200) depicting the disorganized proliferation of ductal and islet tissue. "Adenomatous hyperplasia" is a term used to describe this lesion, in which islet tissue is confluent and occupies a much larger area than the acinar pancreas. Other findings in this patient also associated with the Beckwith-Wiedemann syndrome were enlarged adrenal glands with fetal cortical cytomegaly, renomegaly with subcapsular nephrogenic rests, and an umbilical myxoma. H&E; × 100. (A and B, courtesy of Jordon Weitzman, M.D., Department of Surgery, Children's Hospital, Los Angeles. From Roe TF, et al. [19]. Used with permission.)

(*Continued*)

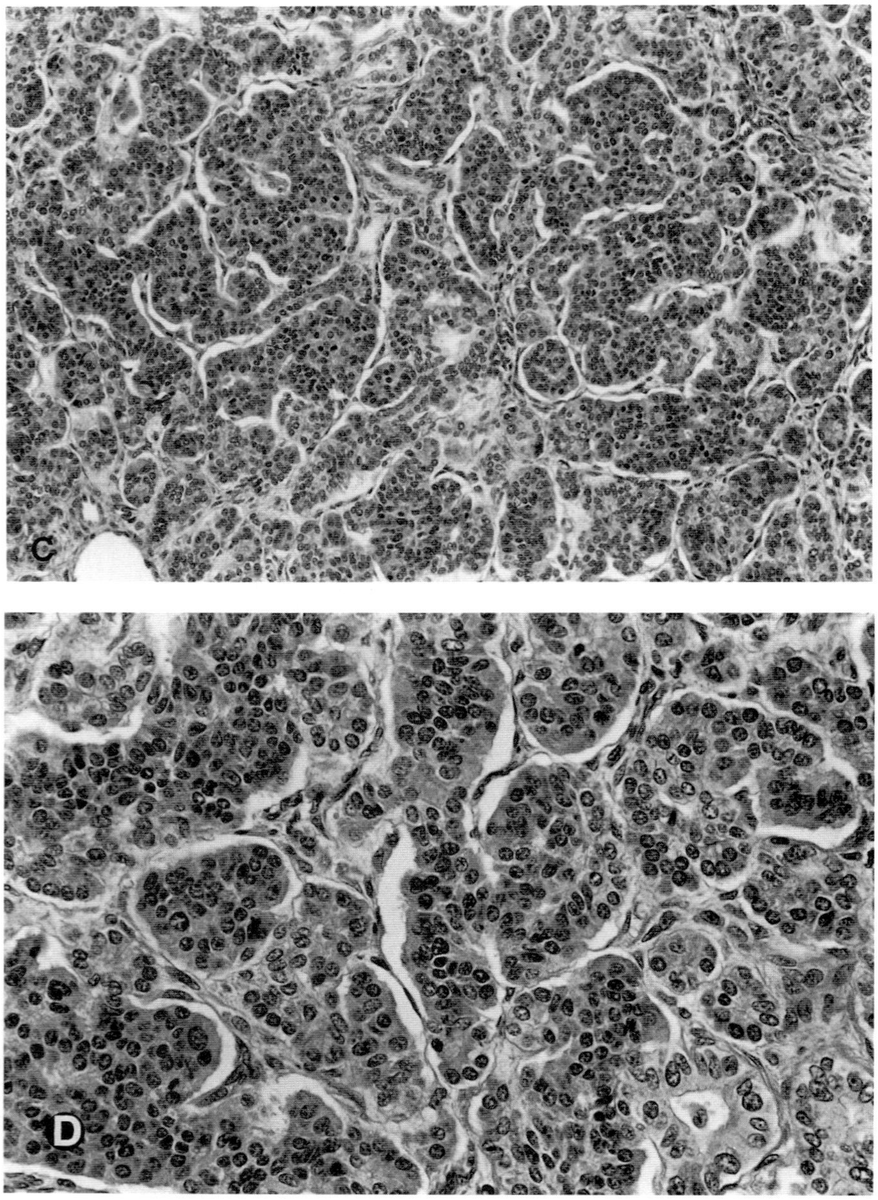

FIGURE 13.4. (*Continued*)

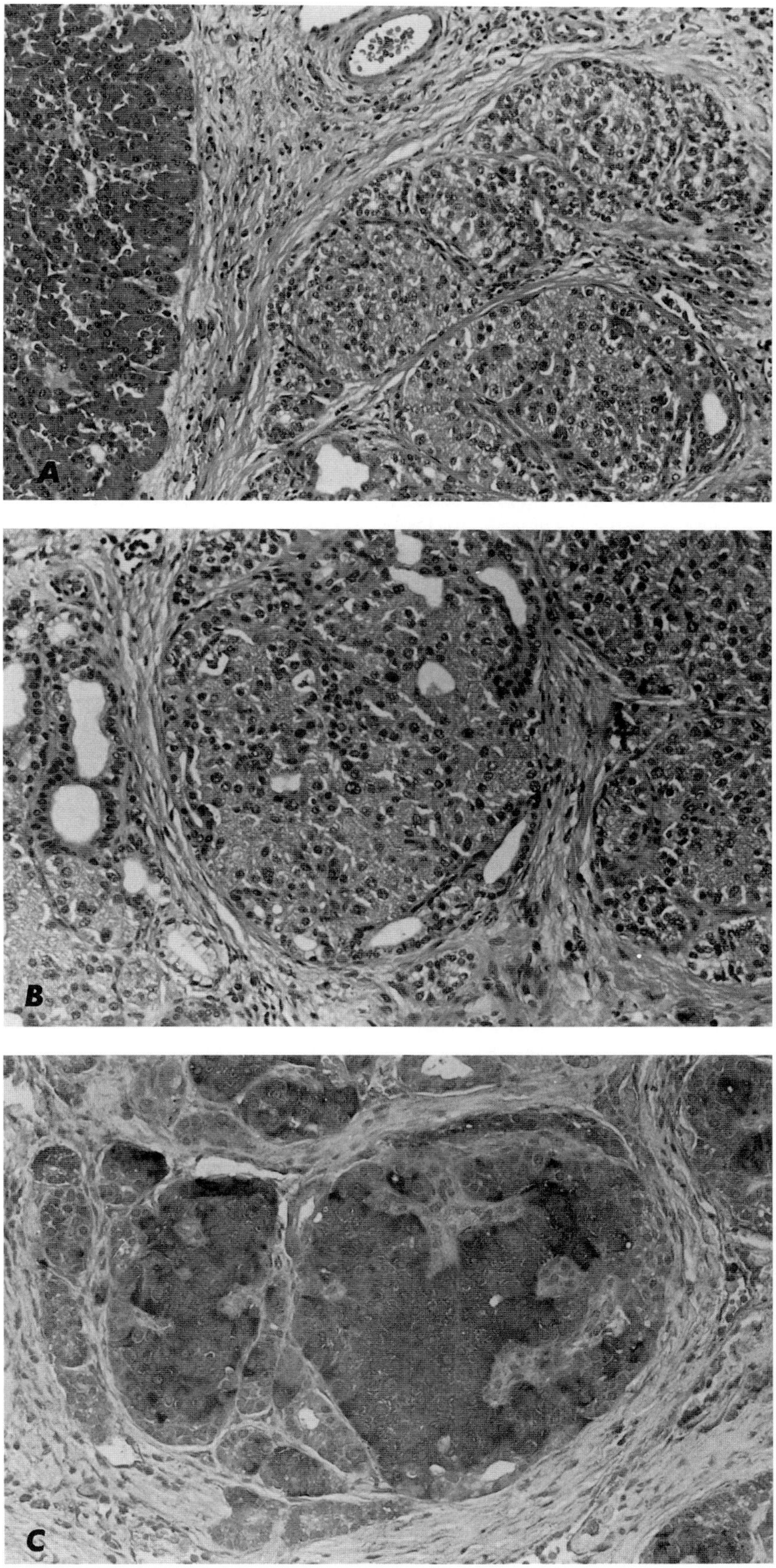

FIGURE 13.5. Islet cell adenoma in a 17-day-old male with severe, persistent hyperinsulinemic hypoglycemia. Within the head of the pancreas there was a firm, reddish-tan tumor nodule, $2 \times 1 \times 1$ cm^3. (A) Nodules of endocrine cells are surrounded by ductules and bands of fibrosis. The tumor is separated from the rest of the pancreas by a fibrous capsule. H&E; $\times$ 120. (B) Higher magnification ($\times$ 150) showing the nodular architecture in more detail. The endocrine cells are small, round, and regular, indistinguishable from islet tissue. H&E. (C) Immunostain for insulin reveals intense staining of the tumor cells. Immunostains for glucagon and somatostatin were nonreactive. Immunostain for insulin in the pancreas outside the tumor revealed intense staining of islet cells not only within islets but also as individual cells and nests within acini similar to that described in Figure 13.2C. Nesidioblastosis and islet cell tumor may occur together in the same pancreas. $\times$ 120.

Islet Cell Adenoma

Islet cell adenoma ("islet cell tumor," "insulinoma," "nesidioblastoma") occurs in the newborn and infant [3,10]. The lesion is found more often in males with a ratio of about 2.7:1. The tumors cause persistent hypoglycemia, and if not treated promptly lead to seizures and permanent brain damage.

Most islet cell adenomas are found in the head of the pancreas, but some occur in the body and tail. Typically islet cell adenomas are small, measuring between 0.1 and 1 cm in diameter. Sometimes the nodules are neither seen nor palpable on the external surface of the pancreas, and multiple cross sections are required to demonstrate the tumor, which has a variable-appearing cut surface described as reddish-tan to white and gritty. Serially sectioning the paraffin block(s) of the excised specimen may be required to find a tiny lesion [13].

Microscopically the adenoma is composed of nodules of islet cells and is separated from acinar tissue by a thin fibrous capsule with small ducts at the periphery of the nests (Figure 13.5). Delicate fibrous connective tissue septa divide the tumor into small lobules. The term "nesidioblastoma" is given to these lesions because of the intimate relationship of the islet cells to small pancreatic ducts [3]. Immunoperoxidase and ultrastructural studies indicate that the tumors are composed primarily of β cells and ductal cells [3]. Islet cell adenomas in the infant may be associated with nesidioblastosis [23].

Exocrine Tumors of the Pancreas

Neoplasms arising from the exocrine pancreas in patients under a year in age are unusual [1,4]. Cystadenoma, hamartoma, pancreatoblastoma, and pancreatic ductal adenocarcinoma are examples of some of the tumors described [1,4,5,9,10,24].

Adenoma and Cystadenoma

Cystadenoma of the pancreas is a benign, well-circumscribed, multilocular cystic neoplasm usually presenting as an abdominal mass [4,9,24,25].The tumor consists of a multiloculated cystic mass occupying the head and/or body of the pancreas and containing serous fluid.

Histologically the cysts are lined by one or more layers of regular cuboidal or columnar epithelium separated by fibrous septa (Figure 13.6).

Pancreatoblastoma

The name pancreatoblastoma ("infantile carcinoma of the pancreas") was coined in 1977 by Horie et al. to describe a unique childhood tumor with distinct histological features and having a better prognosis than adult pancreatic carcinomas [12]. In the first year of life pancreatoblastoma occurs predominantly in males; over half have the Beckwith-Wiedemann syndrome [5,15]. Typically the patients present with an abdominal mass [2,5].

Pancreatoblastomas are encapsulated, are often cystic, and are situated usually in the head but sometimes in the body of the pancreas [2,12] (Figure 13.7). Specimens obtained from neonates consist of cystic, necrotic tumors, 6 to 12 cm in diameter [5,6] (Figure 13.7). Histologically pancreatoblastomas display lobules of regular small cells with vacuolated cytoplasm and uniform round nuclei forming ductular canals and foci of squamous differentiation (Figure 13.7). Some have an organoid growth pattern featuring nodules of squamous cells surrounded by an intermediate medullary zone with small, round, darkly staining cells and a peripheral rim of ductlike tubular structures [2,5,12,26].

The tumors exhibit acinar, endocrine, and ductal differentiation by immunohistochemistry in that they are immunoreactive for pancreatic enzymes, endocrine markers, carcinoembryonic antigen, and keratin [2,26]. α_1-Antitrypsin is immunoreactive, suggesting an acinar cell component [2,5,26,27]. Esterase and esteroprotease are strongly reactive in the squamous cell areas, which is an aid to diagnosis [26]. Electron microscopy reveals zymogen-like granules, luminal microvilli, junctional complexes, and a lamellar arrangement of the granular endoplasmic reticulum within the tumor cells similar to what is observed in normal pancreatic acinar cells [2,5,12,26,27]. Endocrine differentiation has been observed only in a few instances [2].

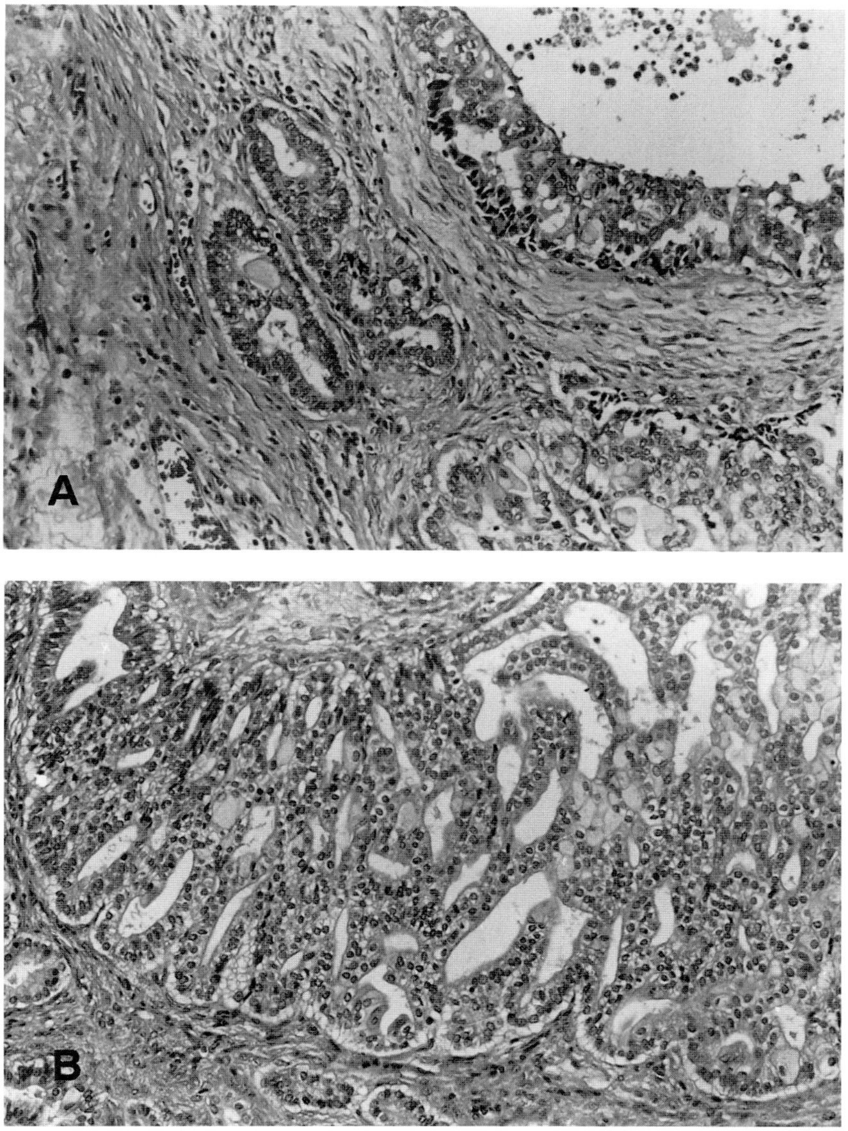

FIGURE 13.6. Cystadenoma of the pancreas in a 3-week-old female with a cystic tumor, 440 g and 13 × 14 cm^2, arising from the body of the pancreas. (A) The wall of the cyst is lined by several rows of vacuolated epithelial cells. Acinar structures are present within the wall. H&E; × 48. (B) The tumor consists of cuboidal to columnar cells with regular round nuclei and vacuolated cytoplasms forming glandular and ductlike structures. H&E; × 150. (Courtesy of Stephen G. Romansky, M.D., Department of Pathology, Long Beach Memorial Medical Center, Long Beach, CA.)

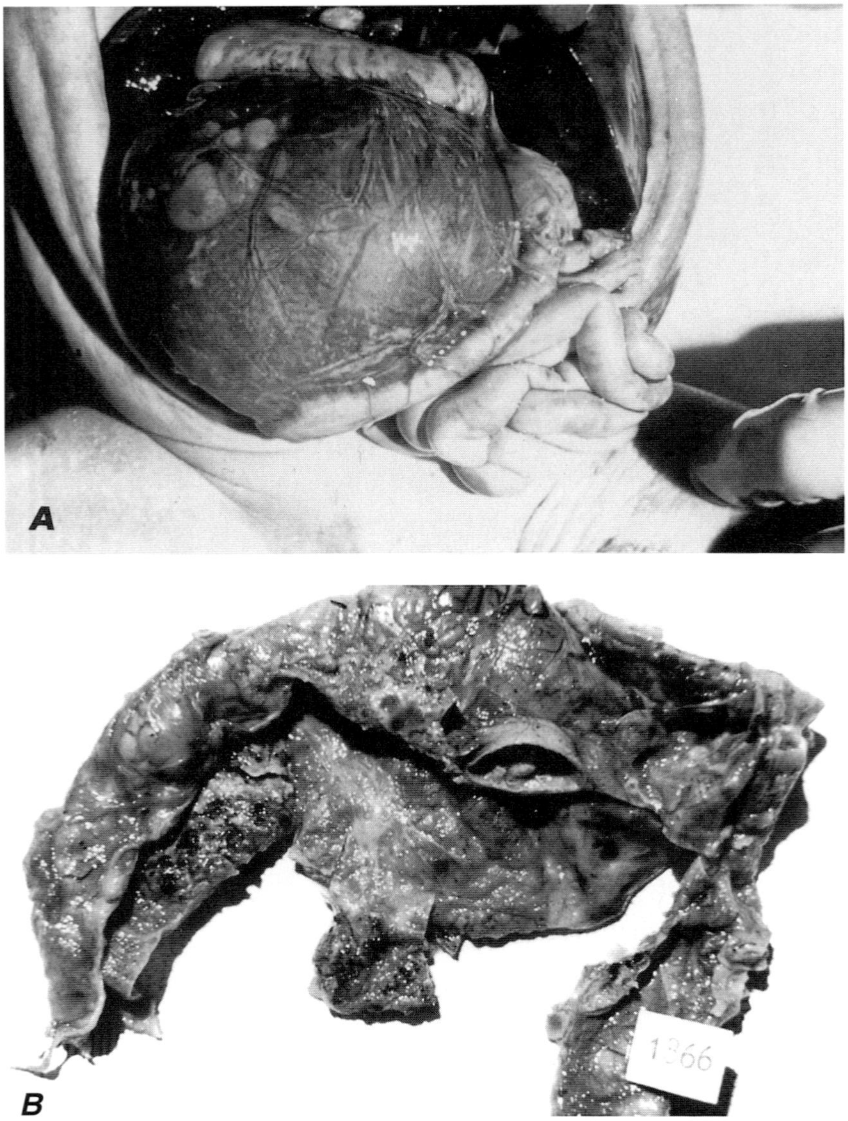

FIGURE 13.7. Pancreatoblastoma. (A) In situ cystic pancreatoblastoma, 12 cm in diameter, in a neonate with the Beckwith-Wiedemann syndrome. (B) The opened specimen reveals a large, cystic, necrotic tumor. (C) Low magnification (× 150) showing the lobular architecture separated by thin fibrovascular septa. H&E. (D) The tumor is composed of uniform cells with round, regular nuclei and vacuolated cytoplasm forming ductular canals. H&E; × 300. (E) Foci of squamous differentiation are noted in some areas. H&E; × 300. (Courtesy of Ricardo Drut, M.D. From Drut R, Jones MC [5]. Used with permission.)

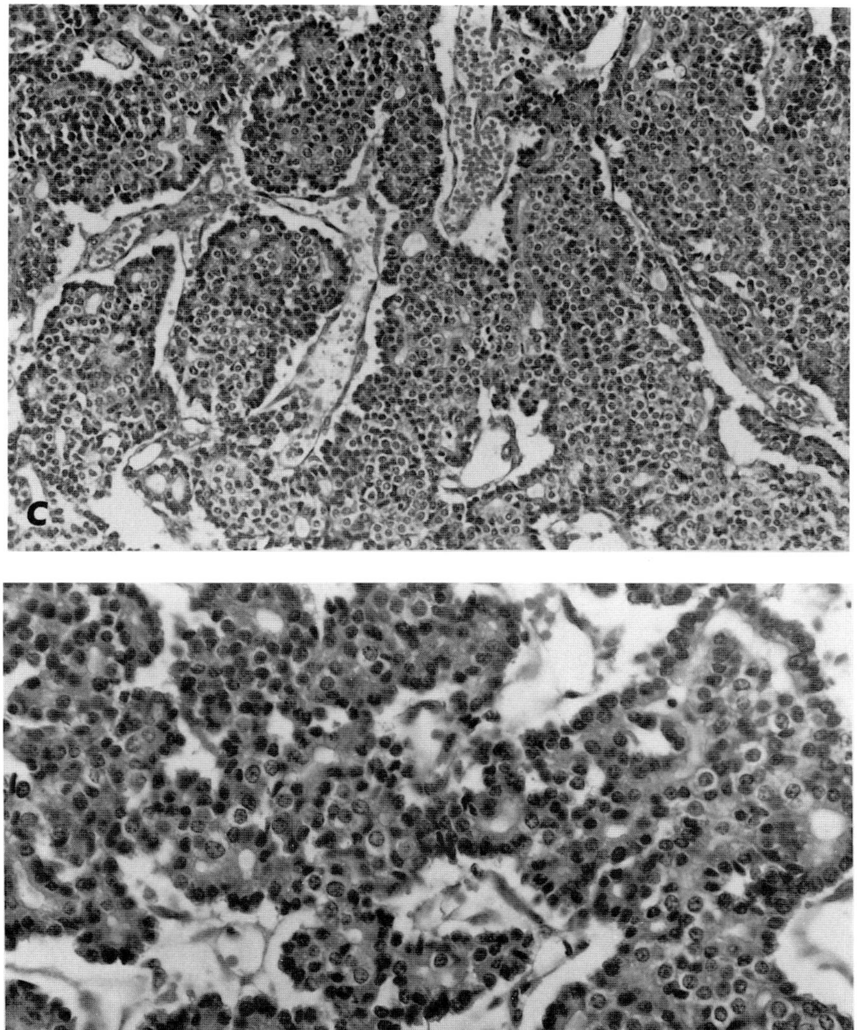

FIGURE 13.7. (*Continued*)

Miscellaneous Tumors and Tumorlike Conditions

Nonepithelial, mesenchymal pancreatic tumors are uncommon in the newborn and infant. Hemangioma and fibromatosis involving the pancreas are described [13]. Extensive leukemic infiltration is seen in some patients with congenital leukemia [8], and the retroperitoneal small-cell malignant tumors such as rhabdomyosarcoma, neuroblastoma, and rhabdoid tumor metastasize to the pancreas [13]. In the fetus and newborn, foci of extramedullary erythro- and myelopoiesis are found frequently in the pancreas and may be seen even at term. Sometimes this finding is mistaken for congenital leukemia. Excessive myelopoiesis is described in patients with a generalized leukemoid reaction associated with infection, trisomies 18 and 21 [8]. Lymphocytic infiltrates normally located in the head of the pancreas are noted often in premature and term infants. Occasionally the infiltrates are so extensive that they may be misinterpreted as lymphoid malignancy rather than as a reaction to some other condition as degenerating islet tissue [8,13] (Figure 13.8).

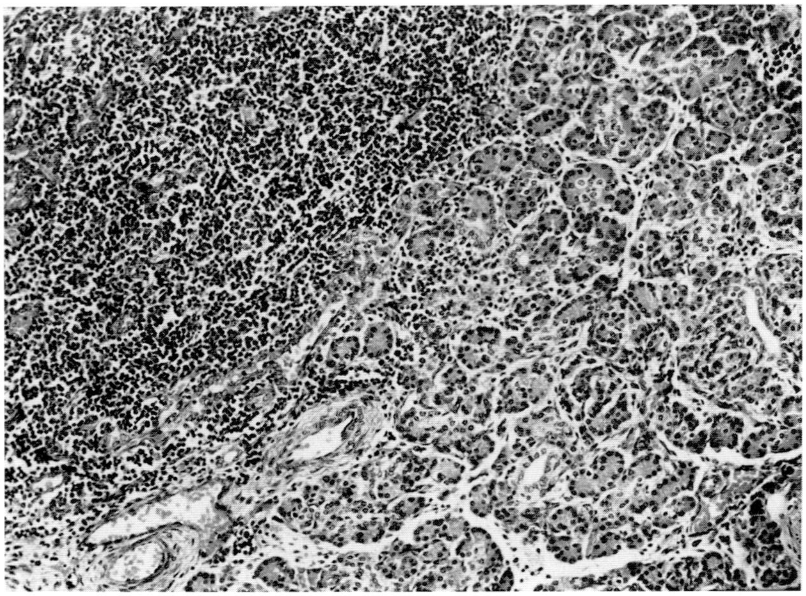

FIGURE 13.8. Lymphocytic infiltration of the pancreas. Photomicrograph was taken from the same 650 g fetus depicted in Figure 13.2. The head of the pancreas reveals extensive infiltration by lymphocytes, which are small, round, and regular, without atypical features. This benign, probably reactive lesion has been misinterpreted as leukemia or lymphoma on occasion. H&E; $\times$ 120.

References

1. Lack EE, Cassady JR, Levey R, et al. Tumors of the exocrine pancreas in children and adolescents. *Am J Surg Pathol* 1983; 7:319.
2. Klimstra DS, Wenig BM, Adair CF, et al. Pancreatoblastoma: a clinicopathologic study and review of the literature. *Am J Surg Pathol* 1995;19(2):1371.
3. Carney CN. Congenital insulinoma (nesidioblastoma): ultrastructural evidence for histogenesis from pancreatic ductal epithelium. *Arch Pathol Lab Med* 1976;100:352.
4. Chang CH, Perrin EV, Hertzler J, et al. Cystadenoma of the pancreas with cytomegalovirus infection in a female infant. *Arch Pathol Lab Med* 1980;104:7.
5. Drut R, Jones MC. Congenital pancreatoblastoma in Beckwith-Wiedemann syndrome: an emerging association. *Pediatr Pathol* 1988;8:331.
6. Robey G, Daneman A, Martin DJ. Pancreatic carcinoma in a neonate. *Pediatr Radiol* 1983;13:284.
7. Jack MM, Walker RM, Thomsett MJ, et al. Histologic findings in persistent hyperinsulinemic hypoglycemia of infancy: Australian experience. *Pediatr Dev Pathol* 2000;3:532.
8. Jaffe R. The pancreas. In: Wigglesworth JS, Singer DB, eds. *Textbook of Fetal and Perinatal Pathology.* 2nd ed. Oxford: Blackwell; 1998:921.
9. Grosfeld JL, Vane DW, Rescoria FJ, et al. Pancreatic tumors in childhood: analysis of 13 cases. *J Pediatr Surg* 1990;25:1057.
10. Rich HR, Weber JL, Shandling B. Adenocarcinoma of the pancreas in the neonate managed by pancreatoduodenectomy. *J Pediatr Surg* 1986;21:806.
11. Burt TB, Condon VR, Matlak ME. Fetal pancreatic hamartoma. *Pediatr Radiol* 1983;13:287.
12. Horie A, Yano Y, Kotoo Y, et al. Morphogenesis of pancreatoblastoma, infantile carcinoma of the pancreas: report of two cases. *Cancer* 1977;39:247.
13. Isaacs H Jr. Tumors. In: Gilbert-Barness E, ed. *Potter's Pathology of the Fetus and Infant.* Vol 2. St. Louis, MO: CV Mosby; 1997:1248.
14. Isaacs H Jr. Pancreatic tumors. In: *Tumors of the Fetus and Newborn.* Philadelphia: WB Saunders; 1997:317. Major Problems in Pathology Series. Vol 35.
15. Potts SR, Brown S, O'Hara MD. Pancreatoblastoma in a neonate associated with Beckwith-Wiedemann syndrome. *Z Kinderchir* 1986;41:56
16. Laidlaw GF. Nesidioblastoma, the islet cell tumor of the pancreas. *Am J Pathol* 1938;14:125.
17. Yakovac WC, Baker L, Hummeler K. Beta cell nesidioblastosis in idiopathic hypoglycemia of infancy. *J Pediatr* 1971;79:226.
18. Dahms BB, Landing BH, Blascovics M, et al. Nesidioblastosis and other islet cell abnormalities in hyperinsulinemic hypoglycemia of childhood. *Hum Pathol* 1980;11:641.
19. Roe TF, Kershnar AK, Weitzman JJ, et al. Beckwith's syndrome with extreme organ hyperplasia. *Pediatr* 1973;52:372.
20. Craver RD, Hill CB. Cure of hypoglycemic hyperinsulinism by enucleation of a focal islet cell adenomatous hyperplasia. *J Pediatr Surg* 1997;32:1526.
21. Sempoux C, Guiot Y, Lefevre A, et al. Neonatal hyperinsulinemic hypoglycemia: heterogeneity of the syndrome and keys for differential diagnosis. *J Clin Endocrinol Metab* 1998;83:1455.
22. Koh THHG, Cooper JE, Newman CL, et al. Pancreatoblastoma in a neonate with Wiedemann-Beckwith syndrome. *Eur J Pediatr* 1986;145:435.
23. Amendt P, Kohnert KD, Kunz J. The hyperinsulinaemic hypoglycaemias in infancy: a study of six cases. *Eur J Pediatr* 1988; 148:107.
24. Jenkins JM, Othersen HB Jr. Cystadenoma of the pancreas in a newborn. *J Pediatr Surg* 1992;27:1569.
25. Tyson RW, Ruyle SZ, Bass K, et al. Serous pancreatic cystadenoma in children under 1 year of age: a report of two cases. *Pediatr Dev Pathol* 1998;1:250. Abstract.
26. Kissane JM. Pancreatoblastoma and solid and cystic papillary tumor: two tumors related to pancreatic ontogeny. *Semin Diagn Pathol* 1994;11:152.
27. Cooper JE, Lake BD. Use of enzyme histochemistry in the diagnosis of pancreatoblastoma. *Histopathology* 1989;15:407.

14 Salivary Gland Tumors

Less than 5% of primary neoplasms of the salivary glands occur in children [1–6], and 8% or less of these are in newborns [4,7–11]. Inflammatory conditions and mucoceles are by far more prevalent than neoplasms [7,13]. Vascular tumor–like lesions, hemangioma and lymphangioma, are more common in the young than epithelial salivary gland tumors [7,9] (Figure 14.1). Capillary hemangioma is the most common parotid tumor in infancy and may present as a facial mass in the neonate [1,7,14,15]. Overall the parotid gland is the primary site more frequently (10:1) than the submandibular or sublingual glands [7]. Plexiform neurofibroma, involving the major salivary glands and producing gland enlargement, occurs in newborns with neurofibromatosis I and even at an early age can cause considerable disfigurement [7] (Figure 14.2). Although pleomorphic adenoma [2,7,16], mucoepidermoid carcinoma [5], undifferentiated carcinoma [1,10,17], and congenital basal cell adenoma [9] are the subjects of sporadic case reports, sialoblastoma ("embryoma") is the major salivary gland tumor of newborns and infants [7,9,11,18–20]. Table 14.1 lists the various salivary gland tumors and tumor-like conditions found in the fetus and infant.

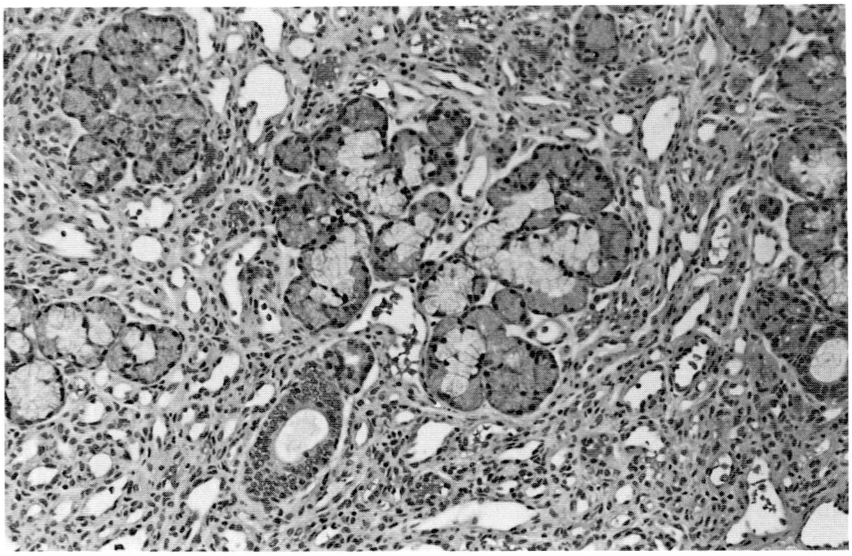

FIGURE 14.1. Capillary hemangioma of the sublingual gland in an 8-month-old male with a sublingual mass, which was biopsied. Clusters of salivary gland tissue composed of both serous and mucinous cells and salivary gland ducts are infiltrated by lobules of proliferating capillaries. The capillary endothelium consists of regular small round to oval cells. H&E; × 120.

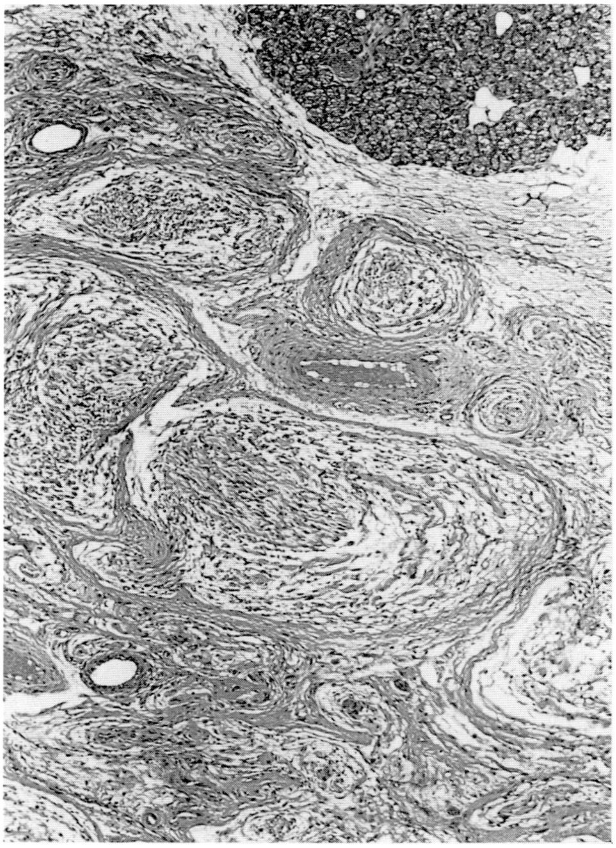

FIGURE 14.2. Plexiform neurofibroma of the parotid gland in a 1-year-old male with neurofibomatosis and a disfiguring facial mass present since birth. The tumor is composed of enlarged, myxoid appearing nerve trunks with proliferations of neurites and Schwann cells. There is a portion of parotid gland within the upper right hand corner. The face and parotid gland are common sites for this lesion in young children with neurofibomatosis. H&E; × 120.

TABLE 14.1. Tumors and tumorlike conditiions of the salivary glands in the fetus and infant.

Epithelial
Sialoblastoma (embryoma)
Mucoepidermoid carcinoma
Undifferentiated carcinoma
Congenital basal cell adenoma
Pleomorphic adenoma of the nasopharynx
Mesenchymal
Hemangioma
Lymphangioma
Neurofibroma
Miscellaneous
Choristoma

Sialoblastoma

The congenital salivary gland tumor originally described and termed "embryoma" by Vawter and Teft [18], the monomorphic adenoma, congenital basal cell adenoma, basaloid adenocarcinoma, congenital hybrid basal cell adenoma–adenoid cystic carcinoma, and congenital carcinoma [3,7,9,15,19–22] all share certain histological features, namely, epithelial and myoepithelial cell components. Currently the preferred term for this puzzling group of neoplasms is "sialoblastoma" [20,21]. This term seems to be appropriate because the histological findings of the tumor resemble the developing salivary gland. To date fewer than 30 cases have been reported.

Sialoblastoma presents as an obvious facial or submandibular mass frequently noticed at birth (Figure 14.3A). When the parotid gland is involved, a seventh nerve facial palsy may be present. Occasionally because of its size the tumor may cause dystocia [22]. Of 16 cases reviewed, 13 arose from the parotid and 2 from the submandibular gland; 1 site was unknown [11].

Sialoblastoma varies considerably in size ranging from 1 to 15 cm in diameter and has a tan-grey to yellow, lobulated, sometimes cystic and focally necrotic cut surface. The tumors tend to be well circumscribed [3,7,18,22].

The characteristic histological features are solid nests of epithelial cells containing small ductlike structures surrounded by narrow bands of fibrous connective tissue [3,18–20,22]. The tumor cells are composed of round to oval nuclei with a fine chromatin pattern, one or more large nucleoli, and a pale- to amphophilic-staining cytoplasm (Figure 14.3B). Small to moderate numbers of mitoses are present. Immunohistochemical studies show that only the ductal cells are immunoreactive with cytokeratin, whereas both ductal cells and epithelial cells express vimentin [3,19,22]. The spindle-shaped myoepithelial cells situated peripherally about the epithelial cell nests stain positively with smooth muscle actin, myosin, and S-100. Electron microscopy confirms the immunohistochemical findings [3,21]. The epithelial nest cells contain smooth oval nuclei, abundant endoplasmic reticulum and numerous ribosomes, and few primitive cell junctions. The ductal epithelial cells are larger, with irregular nuclei and and less prominent cytoplasm containing many lysosomes and free ribosomes. Tight junctions and microvilli are present also. The ductal epithelium is partially surrounded by slender myoepithelial cells with elongated nuclei, intracytoplasmic thin filaments, and replicated basement membranes [3,21].

Of the 16 cases reviewed there was local recurrence in 5, regional lymph node metastases in 1, and death due to local recurrence in 1; there were no instances of distant metastases [11]. These findings indicate that sialoblastoma is locally aggressive and will recur if not adequately excised. The tumor should be considered to be a low grade malignancy and treated as such. Surgery is the treatment of choice, with complete excision at the first surgical procedure if technically possible. Radiation and chemotherapy are definitely contraindicated [3,15]. The long-term prognosis is favorable.

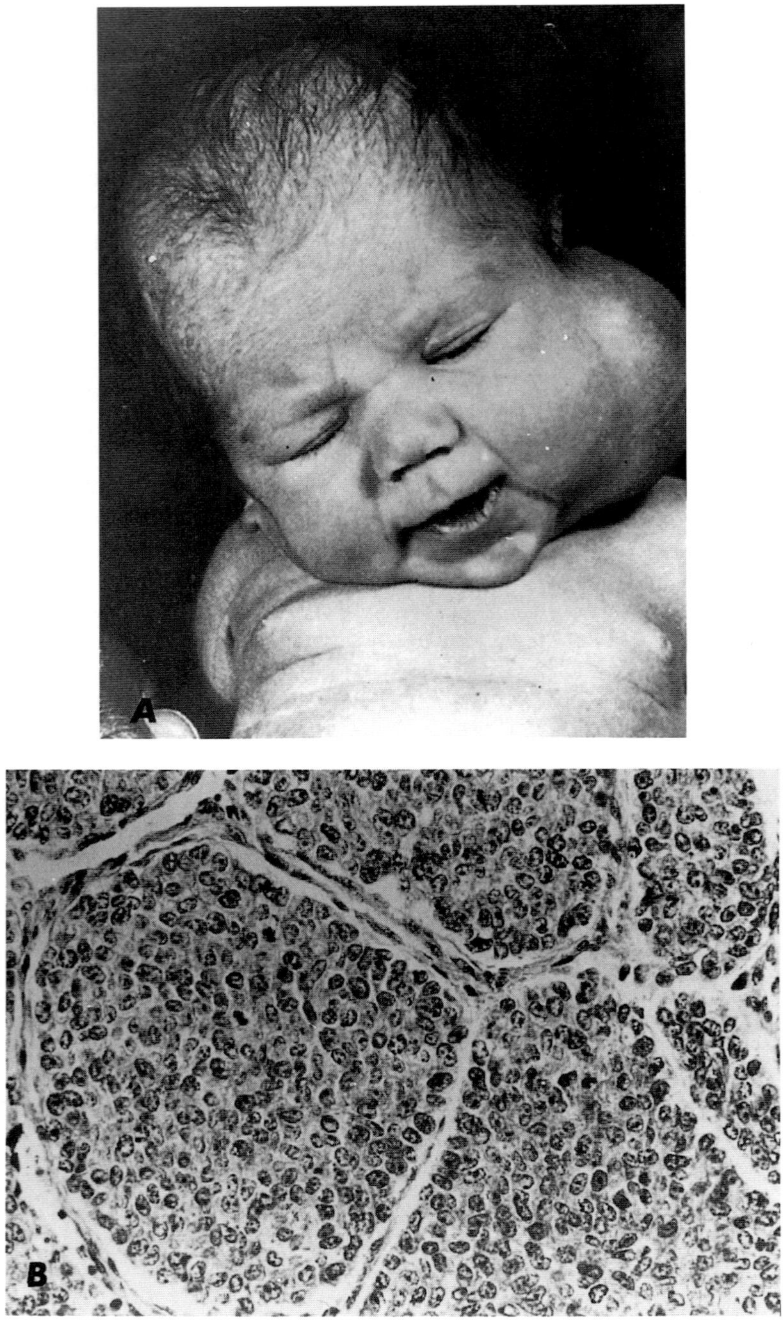

FIGURE 14.3. Sialoblastoma. (A) Newborn with a large tumor of the left parotid gland that overlies the angle of the mandible and distorts the contour of the ear. (B) Irregular lobules and sheets of small cells with pleomorphic nuclei are separated by delicate fibrovascular septa. H&E; × 330. (From Lack EE, Upton MP [7], Copyrighted 1988, American Medical Association, used with permission.)

Congenital Pleomorphic Adenoma (Salivary Gland Anlage Tumor)

The congenital pleomorphic adenoma of the nasopharynx presents in the first 3 months of life with airway obstruction, progressive respiratory distress, and feeding problems [2,11,16] (Figure 14.4). The manifestations of larger lesions may mimic choanal atresia because of the obstructive symptoms.

The midline, pedunculated polypoid tumors of the nasopharynx measure 1 to 3 cm in diameter. Histologically they show an outer surface epithelium composed of nonkeratinized, stratified squamous epithelial cells, a hypocellular submucosa with ductlike structures and nests of squamous epithelium communicating with the surface mucosa, and central (mesenchymal) stroma-like nodules blending with ductules [2,11,16] (Figure 14.4).

Immunohistochemical studies reveal diffuse immunoreactivity for salivary gland amylase, cytokeratin, and epithelial membrane antigen expression in epithelial structures and reactivity with vimentin, cytokeratin, and muscle-specific actin in the stroma-like nodules [2]. Myoepithelial cells characteristically express cytokeratin and actin. Findings from electron microscopy coincide with those of the immunohistochemical studies. Myoepithelial cells with basal lamina, desmosomes, and bundles of microfilaments are noted in the nodules in addition to cells with epithelial features [2].

Salivary gland anlage tumor should be distinguished from sialoblastoma, which occurs predominantly within the parotid, has a different histological appearance, and probably represents a different stage of development of the salivary gland [2]. Nasopharyngeal dermoid ("hairy polyp") is another entity in the differential diagnosis (Figure 2.7). It typically presents as a fingerlike projection protruding from and obstructing the nasopharynx. Histologically the polyp is composed only of skin with hair follicles, fat, and cartilage [11]. Teratoma and encephaloceles are other conditions in the differential.

Although potentially fatal in the neonate if left untreated, surgical excision results in cure [2,16].

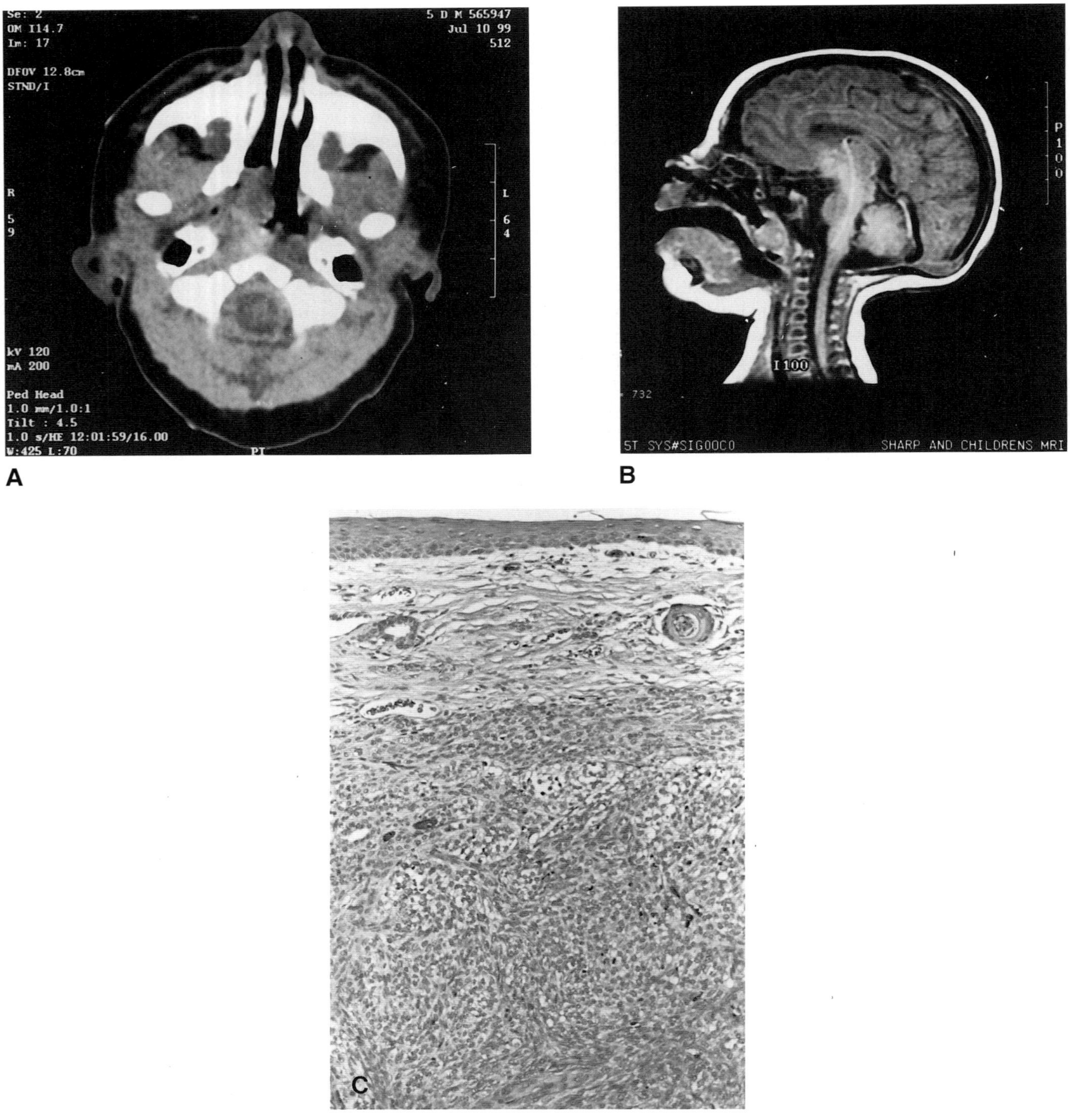

FIGURE 14.4. Pleomorphic adenoma (salivary gland anlage tumor) in a 10-day-old male with respiratory distress and feeding difficulties since birth. (A) Computerized tomography axial scan through the skull base without contrast reveals a nasopharyngeal mass of soft tissue density extending from the nasal choana to the retropharyngeal soft tissues. (B) Magnetic resonance imaging shows a $2 \times 1.4 \times 1.3$ cm^3, pedunculated, enhancing, soft tissue mass within the nasopharynx. (C) Low magnification ($\times$ 120) view, revealing an intact, nonkeratinized, stratified squamous epithelial mucosa with hypocellular submucosa containing a ductlike structure and a round nest of squamous cells. H&E. (D) A cystic squamous nest and ductlike structures are adjacent to a stromal-like nodule (on the left). (E) Higher magnification ($\times$ 150) of a stromal-like nodule displaying the relationship between ducts and stroma. PAS. (F) Squamous nests and cells within a stromal-like nodule showing cytokeratin immunoreactivity. In addition, vimentin and actin were expressed by stromal-like nodule cells. Cytokeratin immunoperoxidase; $\times$ 300. (Louis Dehner, M.D. kindly reviewed the case and provided the diagnosis.)

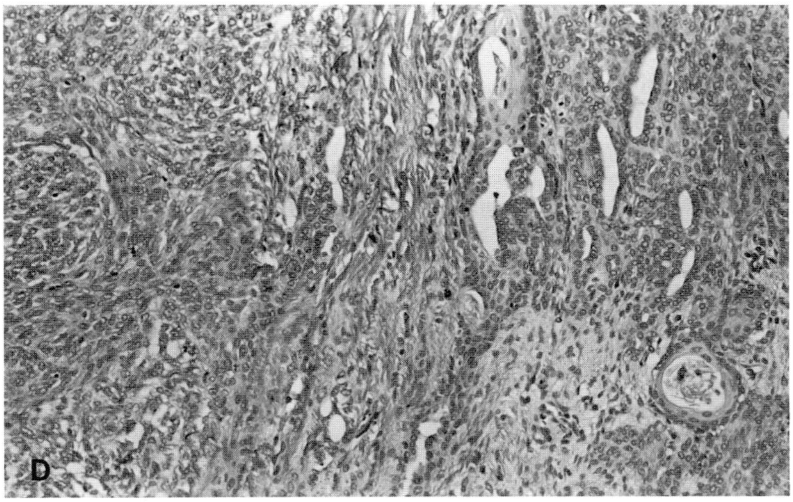

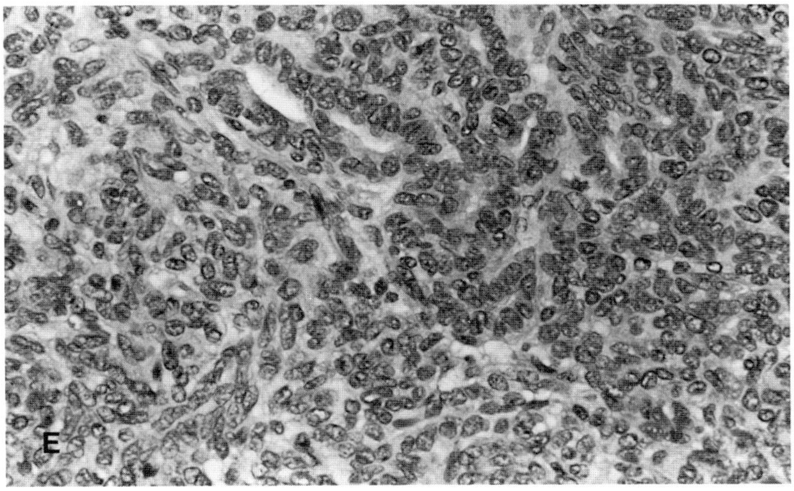

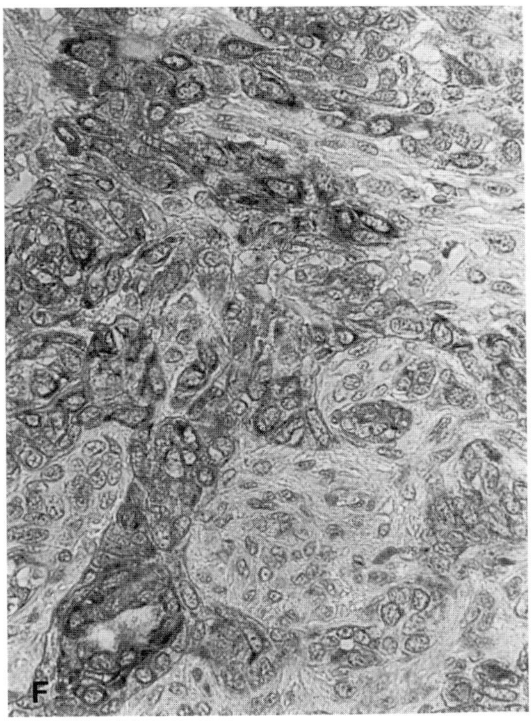

FIGURE 14.4. (*Continued*)

15 Cardiac Tumors

Cardiac tumors are rare in all age groups. Most are histologically benign. Rhabdomyomas, the most common tumor of the heart in the fetus and infant, are probably hamartomas rather than true neoplasms. Because of their location, however, benign tumors and hamartomas of the heart may severely compromise blood flow, interfere with myocardial function, or cause arrhythmias that lead to their clinical discovery or to sudden death [1–5]. Primary tumors are found more often than metastasis to the heart in the fetus and infant [3,6]. Table 15.1 lists the various cardiac tumors and tumorlike conditions found in the fetus and infant.

Primary cardiac tumors are frequently not suspected during life, and death may occur suddenly, with no preceding symptoms of cardiovascular disease [1]. In other instances symptoms may be present and the diagnosis may be established during life through imaging studies both antenatally and postnatally [1,2,7].

Cardiac tumors may imitate such other neonatal conditions as sepsis and some forms of congenital heart disease (e.g., hypoplastic left heart syndrome, subaortic stenosis, mitral stenosis, tricuspid insufficiency) [2,8,9]. In the fetus a cardiac tumor may cause congestive heart failure and hydrops, leading to stillbirth [5,7]. Certain tumors, notably intrapericardial teratomas and fibromas may be successfully removed, as may obstructive intracavitary rhabdomyomas [1,10,11]. Primary tumors in a malformed heart are well documented, but the concurrence is unusual; when present they adversely affect the surgical management and prognosis [8,9]. Cardiac tumors must be considered in the differential diagnosis of any child with unexplained heart murmurs, congestive heart failure, or arrhythmias [10].

Certain tumors, for example, myxoma, are found more often in adults, being seldom observed in the young [1,6,8, 12–14]. On the other hand, neoplasms such as fibromas and intracardiac teratomas are uncommon in adults and more frequent in infants and children [1]. During the first year of life rhabdomyoma and pericardial teratoma comprise over 75% of the primary cardiac tumors [6,13,14].

Finally, cardiac neoplasms may be one of the first manifestations of such genetic disorders as tuberous sclerosis, neurofibromatosis, the Gorlin syndrome (nevoid basal cell carcinoma syndrome), familial myxomas, and rarely the Beckwith-Wiedemann syndrome [5,15,16].

Rhabdomyoma

Rhabdomyoma is the leading cardiac tumor of the fetus and infant, followed by teratoma and fibroma [1–6,7,9,10,13, 17–19]. Cardiac rhabdomyoma(s) are found unexpectedly on routine prenatal sonography and on postmortem examination of a stillborn or a newborn with nonimmune fetal hydrops [1,7,20,21]. In older infants, rhabdomyomas that interfere with the conduction system or obstruct left or right ventricular blood flow result in recurrent arrhythmias or sudden death [1,3]. Cardiomegaly, congestive heart failure, and cardiac murmurs are other clinical findings [1,3]. In the fetus, cardiac arrhythmia is the most frequent sign and is a definite indication for ultrasonography [7]. Half of prenatally diagnosed rhabdomyomas are asymptomatic and are discovered on routine sonography [7] (Figure 15.1). Rhabdomyomas have been detected as early as 22 weeks gestation [19]. The tumor is not often found in association with congenital heart malformations [8,9].

Although they occur independently, rhabdomyomas may be only one manifestation of tuberous sclerosis, a phakomatosis transmitted as an autosomal dominant trait of variable expression. The disease is classically manifested by the triad of mental retardation, epilepsy, and facial angiofibromas. At least one third of infants and children with cardiac rhabdomyomas are stated to have tuberous sclerosis, and the reverse is true also [19].

Grossly rhabdomyomas consist of one or more white to yellow-tan circumscribed nodules in the myocardium, situated most often within the ventricle wall and in the ventricular septum (Figure 15.2). Some tumors occur in the subepicardial

TABLE 15.1. Cardiac tumors and tumorlike conditions of the fetus and infant.

Primary	Hemangiopericytoma
Rhabdomyoma	Lymphangioma[a]
Teratoma	Juvenile xanthogranuloma
Fibroma	Infantile histiocytoid cardiomyopathy
Myxoma	Metastatic
Hemangioma	Neuroblastoma
Rhabdomyosarcoma	Leukemia
Fibrolipoma	Blood cyst of heart valves

[a] Intrapericardial in location.
Source: Isaacs H Jr [4].

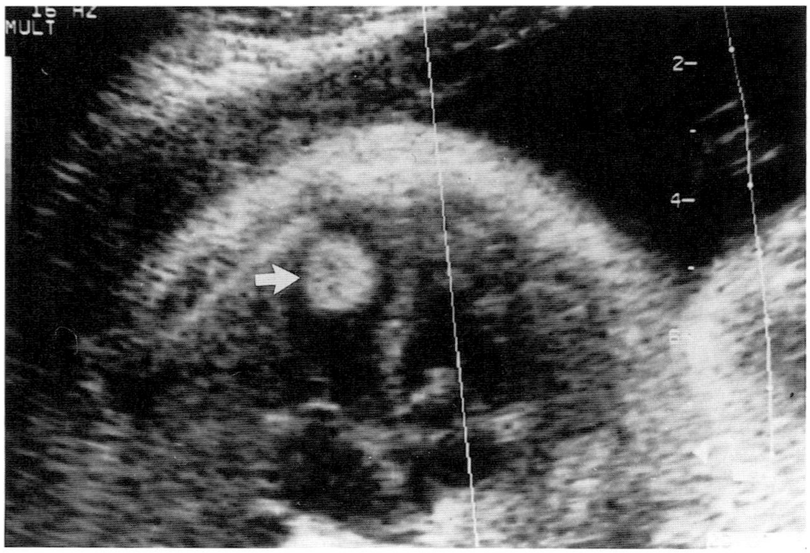

FIGURE 15.1. Cardiac rhabdomyoma. Sonogram of a fetus at 32 weeks gestation demonstrating a round echogenic mass (arrow) bulging into the cavity of the left ventricle. Several additional tumors were visible in other views. The baby proved to have tuberous sclerosis but survived. (Courtesy of Val Catanzarite, M.D., Department of Perinatology, Sharp Memorial Hospital, San Diego.)

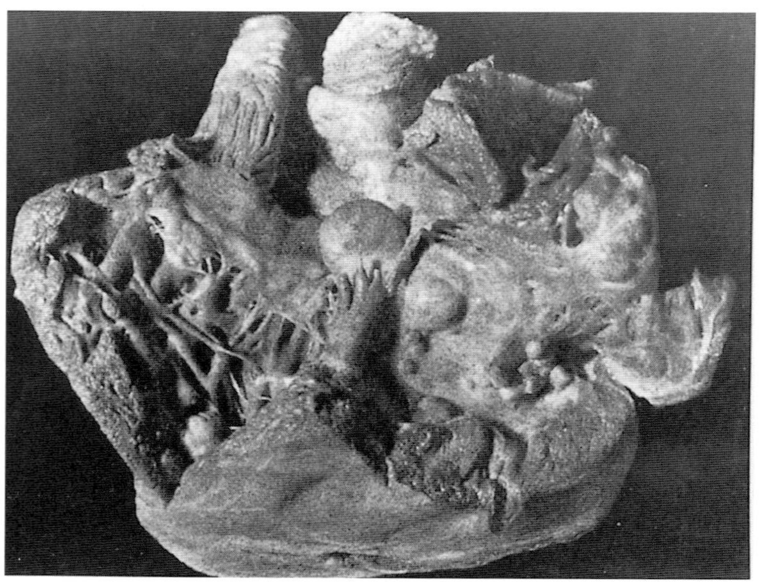

FIGURE 15.2. Cardiac rhabdomyoma. Gross tumors are visible beneath the endocardium. This child with tuberous sclerosis died at age 6 months. (From Isaacs H Jr [5], with permission.)

region and in the atria. In about half the patients the rhabdomyomas are intracavitary, producing obstruction [1,3,5,9,18].

Usually designated rhabdomyomas or hamartomas because of the presence of striated myofibrils, they are considered to be overgrowths of cardiac muscle showing only mildly abnormal changes, not true neoplasia. Their most characteristic components are large round or oval cells with small central nuclei and a thin peripheral rim of cytoplasm. Delicate radiating strands connecting the nucleus with the cell wall are responsible for the designation "spider cells." Cross-striations are noted in these strands and at the periphery of the cells. The intervening vacuoles are filled with glycogen, recognizable with appropriate staining [5,14,18] (Figures 15.3 and 15.4). Such cells are scattered in small groups throughout the musculature and in large aggregates. Immunohistochemical studies show diffuse immunoreactivity with the muscle markers myoglobin, actin, desmin, and vimentin; however S-100 protein is unreactive [20]. Electron microscopy confirms the presence of cross-striations as well as abundant β-glycogen within the cells [18]. Other findings include leptofibrils, a poorly developed sarcoplasmic reticulum, and scattered desmosomes [18].

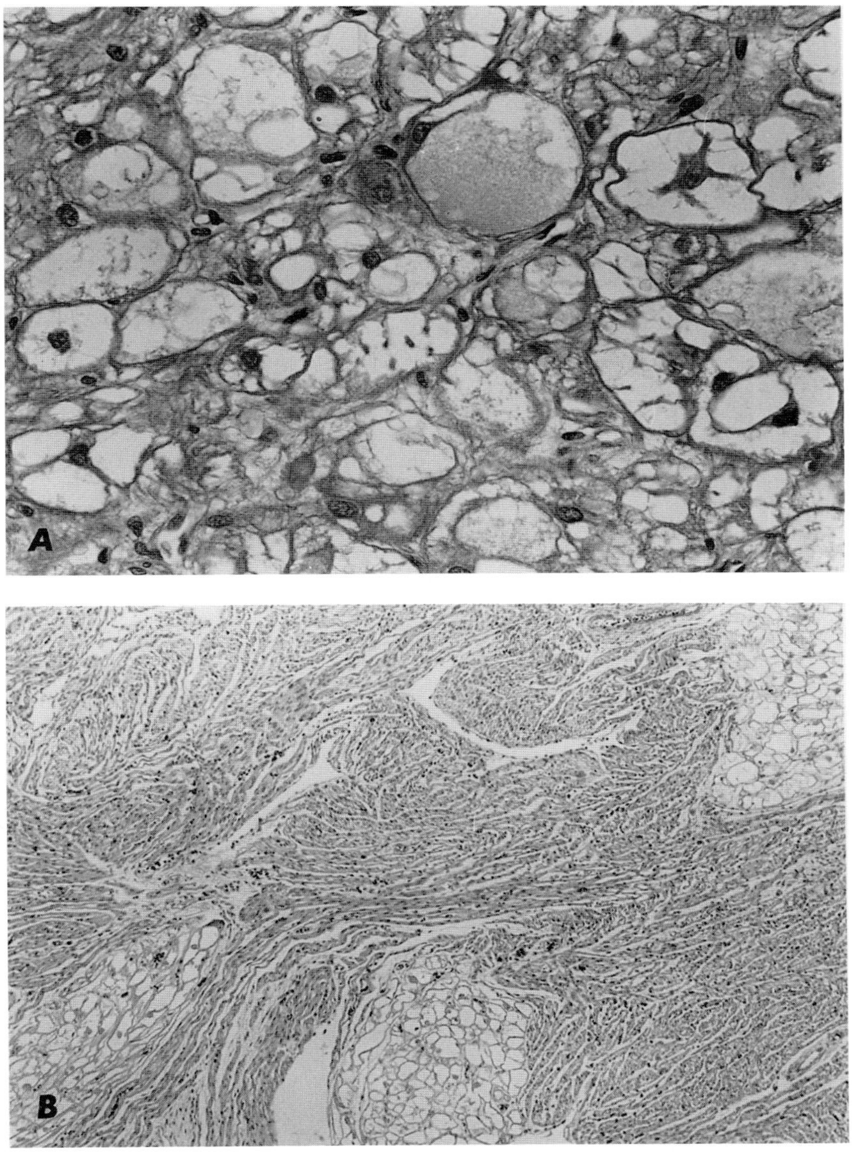

FIGURE 15.3. Cardiac rhabdomyoma in a 6-week-old male who developed an arrhythmia at 1 week of age. Cardiac catherization revealed a tumor in the posterior wall of the left ventricle, and 4 g of yellow-white and lobulated tissue was excised. (A) Large myocardial cells have cytoplasms distended by glycogen-filled vacuoles. Some contain central nuclei with delicate strands extending from the nucleus to the cell wall. These are the so-called spider cells, considered to be pathognomonic for cardiac rhabdomyoma. H&E; $\times 200$. (B) At postmortem examination several additional cardiac rhabdomyomas were found scattered throughout the myocardium. Numerous glial nodules were found in the cerebrum and basal ganglia with imaging studies. H&E; $\times 60$.

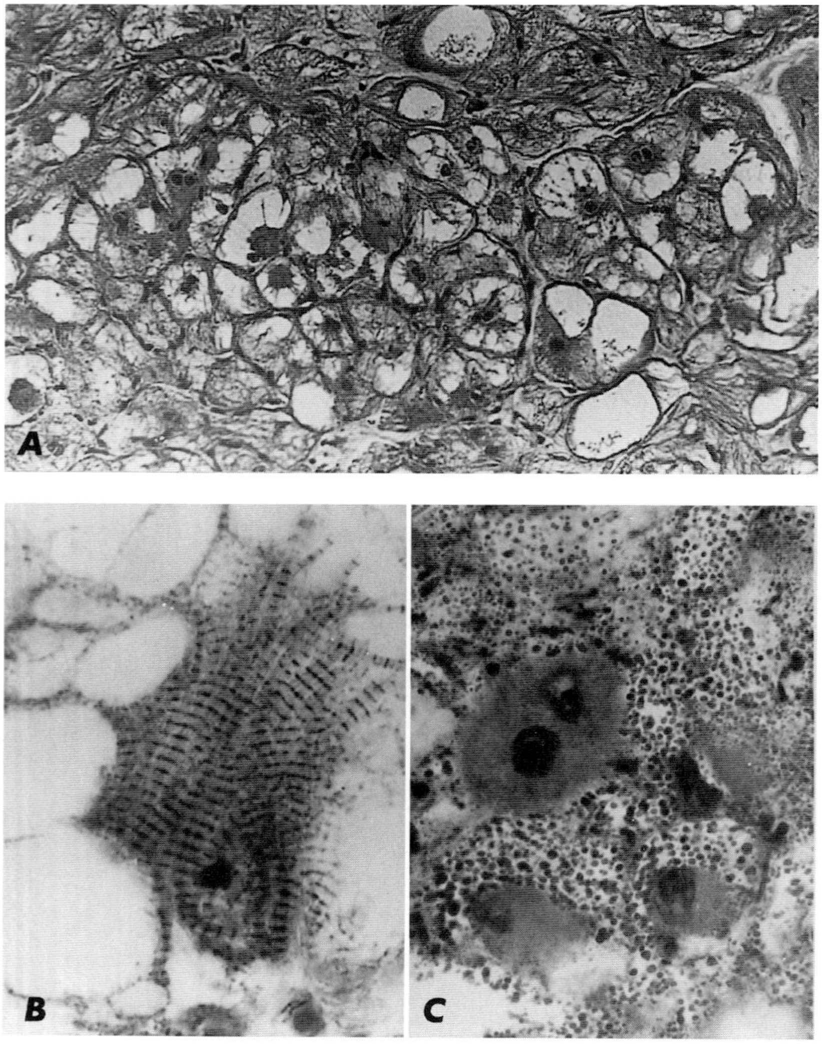

FIGURE 15.4. Cardiac rhabdomyoma. (A) Rhabdomyoma showing typical "spider cells" in a child with tuberous sclerosis who died at 22 days of age (see Figure 8.5 for other lesions found in this infant). H&E; × 300. (B) Cells showing characteristic muscular striations. H&E; × 300. (C) Cells containing large amounts of glycogen. Best carmine. (From Isaacs H Jr [5], with permission.)

Fibroma

Cardiac fibromas are the third most common primary cardiac tumor of the newborn and infant, following rhabdomyoma and teratoma [1–3,6,8,13,22–26]. Cardiac fibromas are connective tissue tumors arising from fibroblasts and myofibroblasts and having the same gross and microscopic appearance and biological behavior as fibromatosis, their soft tissue counterpart (see Chapter 3, Soft Tissue Tumors). Other names applied to this tumor are fibromatosis, myofibromatosis, fibrous hamartoma, and congenital mesoblastic tumor [14,22].

Cardiac fibromas are usually solitary, most often arising from the myocardium of the ventricular septum or free wall of the right or left ventricle [1,14]. The clinical findings depend on the location and extent of the tumor [2,3]. Fibromas arising from the ventricular septum may involve the conduction system, producing ventricular fibrillation. Cardiomegaly and sudden unexpected death occur with tumors situated both in the free wall of the ventricle and in the septum [2]. At either location the tumor obstructs the inflow or outflow tract of the ventricle, leading to progressive congestive heart failure [1,2]. Since most fibromas occur in the left ventricle, the signs and symptoms of obstruction relate mostly to the left side of the heart and are similar to signs and symptoms of the hypoplastic left heart syndrome and of subaortic stenosis [1,3,22].

Extracardiac malformations or syndromes are seldom present, but the tumor has been observed with cleft lip and palate, with the Beckwith-Wiedemann syndrome, and with the basal cell nevus syndrome (Gorlin-Goltz syndrome) [5,15,16]. Moreover, the heart may become involved in congenital generalized fibromatosis [25,26].

Cardiac fibromas have a firm, white, myxoid or trabeculated gross appearance, and they measure 5 cm or more in diameter. The tumor is unencapsulated, blending imperceptibly with the adjacent myocardium. Foci of calcification may be observed within the center [1,14,22]. Histologically the tumor is composed of variable numbers of spindle-shaped fibroblastic and myofibroblastic cells surrounded by abundant collagen fibers and an acid mucopolysaccharide-rich myxoid matrix [26]. Mitoses are noted rarely. Central areas show foci of calcification, elastic fibers, and cystic change. Characteristically entrapped myocardial fibers are present within the tumor [1,14,22]. By electron microscopy the tumor consists of fibroblasts and spindle-shaped cells with abundant myofilaments and dense bodies consistent with myofibroblasts, findings similar to those of the soft tissue fibromatoses [26].

Myxoma

Although myxoma is the leading primary tumor of the heart in adults, it is the subject of a few scattered case reports in the infant and newborn [4,6,8,12,27,28]. The youngest patient recorded was a stillborn infant [28]. Complete surgical excision is the treatment of choice [1]. Myxomas occurring in the newborn and infant arise mostly from the right atrium and often prove fatal [5,8,12].

Teratoma

Teratoma is an unusual primary tumor of the heart or pericardium at any age [4,6,7,10,13,14,29]. Nearly two third of cases are found in infants, and teratoma is the second most common primary cardiac tumor, following rhabdomyoma, in this age group [4,5,14]. Most cardiac teratomas arise from the pericardium; fewer than a half-dozen examples have been reported in the atrium or ventricle [30].

The clinical findings in newborns and infants with intrapericardial teratomas are dyspnea, cardiomegaly, and congestive heart failure, which may be accompanied by cyanosis and a murmur [30]. Pericardial effusion is characteristically present, leading to life-threatening tamponade if not treated. Pericardial teratoma is one of the most common causes of massive pericardial effusion in the neonate [1]. Occasionally the tumor causes sudden unexpected death [1,14].

The diagnosis of cardiac teratomas is established both antenatally and postnatally by their characteristic imaging findings [4,7,31–33] (Figure 15.5A). The antenatal development of fetal hydrops accurately predicts the outcome of fetuses with pericardial teratomas [31]. Fetal hydrops results from obstruction of venous return either by the pericardial effusion or by the mass itself. Unless treatment is begun when hydropic changes appear, these symptoms progress and lead to fetal demise [31].

Typically intrapericardial teratomas arise from the base of the heart and are attached to the root of the pulmonary artery and aorta. The tumor has a smooth lobulated surface (Figure 15.5B). On section it is seen to consist of numerous multiloculated cysts with intervening solid areas. The rare intracardiac tumors arise from the atrial and/or ventricular wall as nodular masses projecting into the cardiac chamber(s) [4,14, 29,30] (Figure 15.6). Histologically pericardial and intracardiac teratomas are identical to those found in other locations with all three germ layers represented [1,5,29]. The cysts are

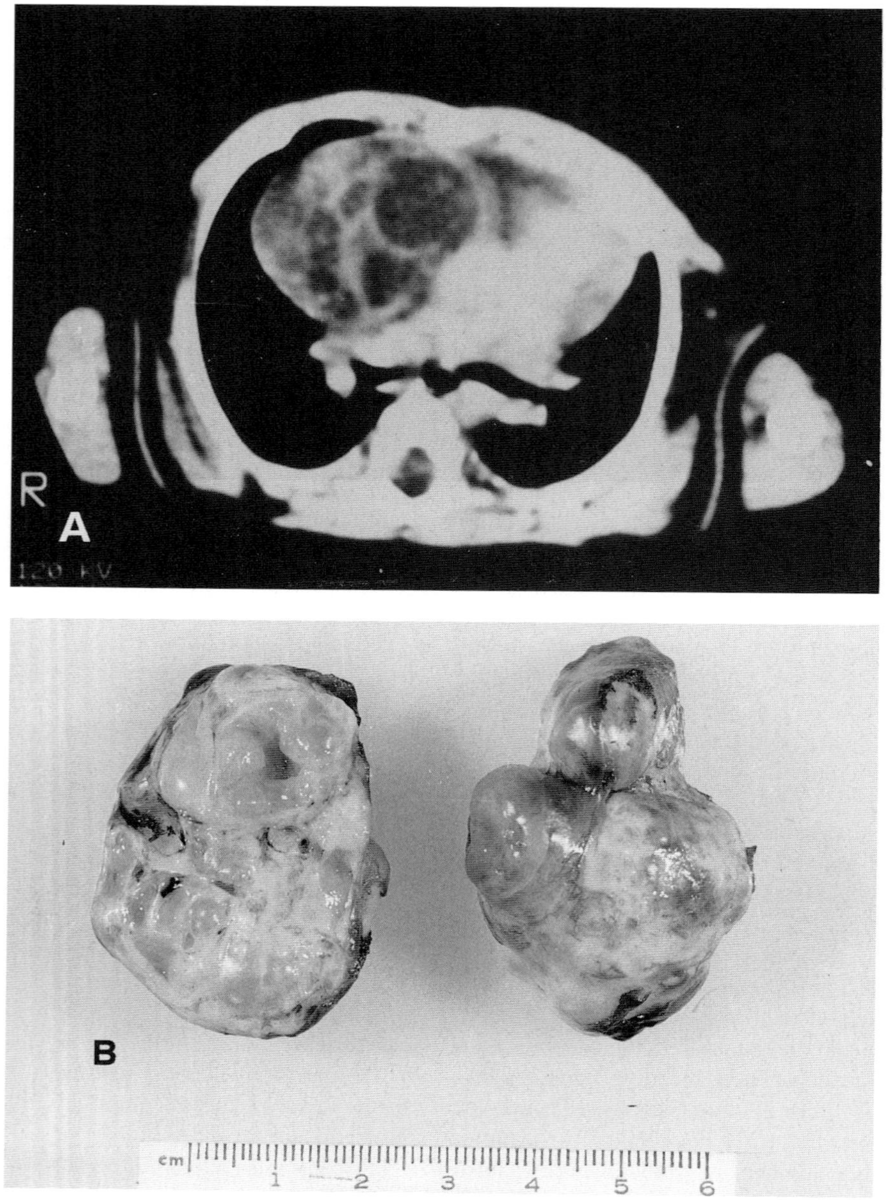

FIGURE 15.5. Intrapericardial teratoma in a 2-month-old male with a history of increasing respiratory distress. (A) Computerized tomography scan of the thorax showing a nodular, cystic density arising from the pericardium and bulging into the right pleural cavity. (B) The tumor, 27.5 g and 5.5×3.6 cm^2, was situated within the pericardial cavity attached to the aorta. It has a lobulated appearance and contains multiple cysts. (C) Mature respiratory epithelium, cartilage, and seromucinous glands are present. H&E; $\times 60$. (D) Immature neuroglial elements are depicted. H&E; $\times 40$.

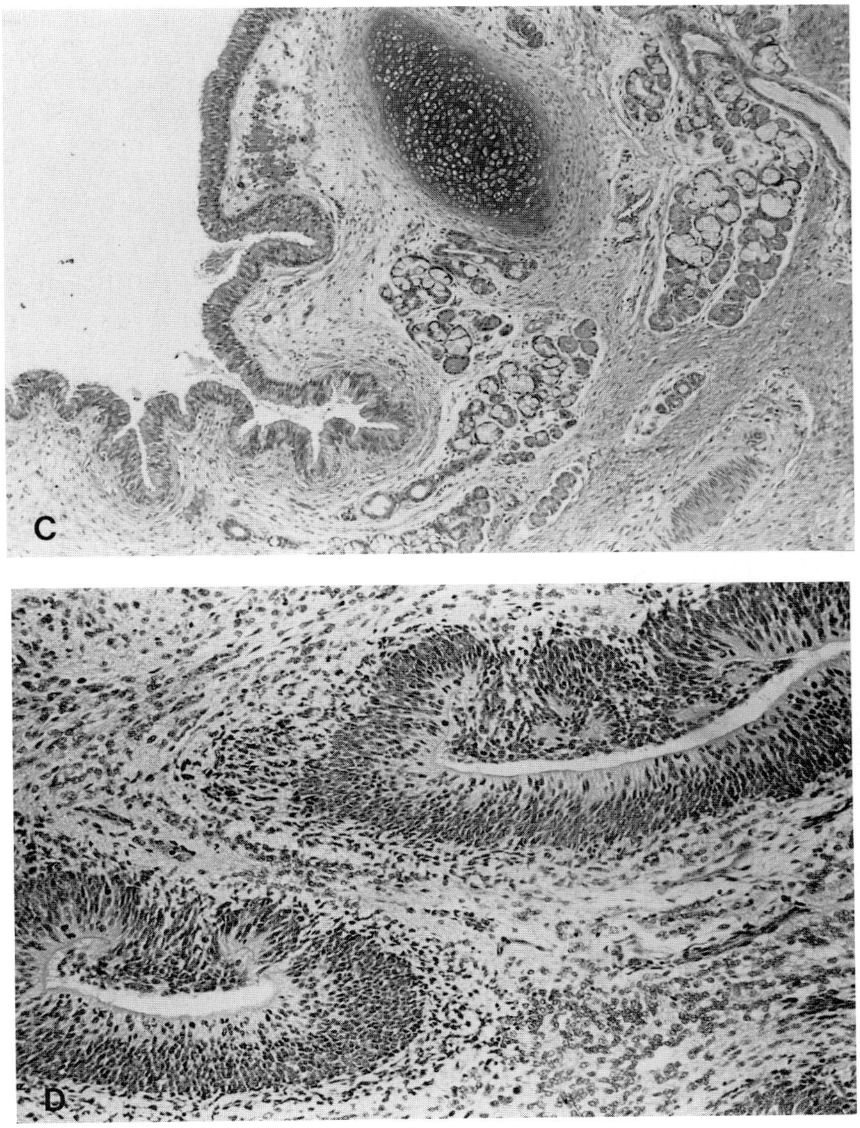

FIGURE 15.5. (*Continued*)

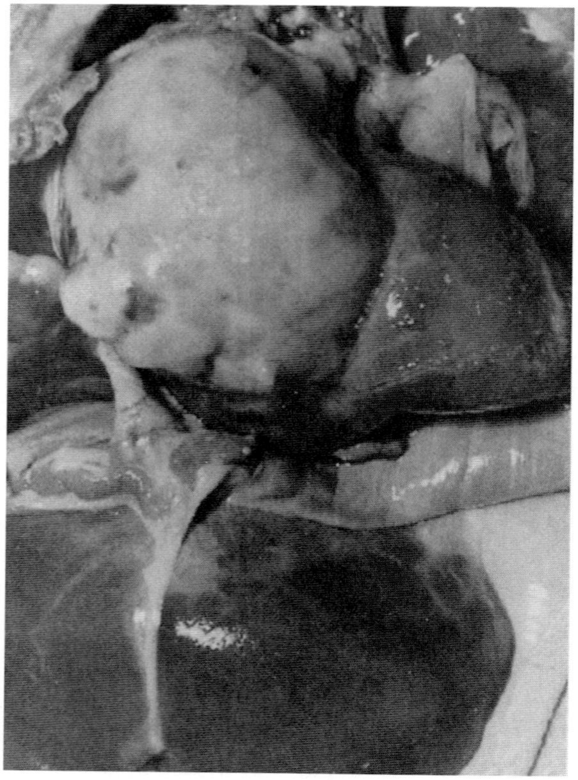

FIGURE 15.6. Intracardiac teratoma of the right atrium in a neonate. Gross appearance. Microscopically the tumor consisted of squamous epithelium, neuroglia, and glands similar to what is shown in Figure 15.5 C, D. (From Isaacs H Jr [5], with permission.)

lined by a variety of epithelia including stratified squamous, cuboidal, mucin-secreting, or respiratory epithelial cells. The more solid areas contain mature or immature neuroglial tissue, thyroid, pancreas, smooth and skeletal muscle, and foci of cartilage and bone. Essentially all pericardial and intracardiac teratomas reported in newborns and infants have been benign, composed of mature tissues with or without immature neuroglial elements [1,5,29] (Figures 15.5 and 15.6).

Hemangioma

Cardiac hemangiomas comprise about 2.8% of all primary cardiac tumors, and most are diagnosed in the neonatal period [11,34–36]. Hemangiomas arise from anywhere in the vicinity of the heart. Pericardial effusion is characteristically present. Other associated findings are cardiomegaly, congestive heart failure, hydrops fetalis, hemopericardium, and thrombocytopenia [1,11,34–36]. Echocardiography shows mixed echogenicity with echogenic and hyperechoic areas. The formation of hyperechoic and cystic spaces is attributed to cavernous lakes, thrombosis and calcification, and changes possibly related to spontaneous regression [34,35].

The tumors appear as sessile, lobulated, and nonencapsulated masses ranging in size from 1.5 to 2.5 cm [34,36]. Thrombosis and calcification may be present. Microscopically the hemangioma consists of capillary and/or cavernous components [34–36]. (See Chapter 3, Soft Tissue Tumors, for further discussion of hemangiomas.)

Miscellaneous Conditions

Myocardial hamartomas, distinctly morphologically different from rhabdomyomas, are described in infants and young children with severe and fatal arrhythmias [5,37–42]. The lesion has been called a variety of names such as histiocytoid cardiomyopathy, Purkinje cell tumor, oncocytic cardiomyopathy, focal lipid cardiomyopathy, and idiopathic infantile cardiomyopathy. There is a high incidence of extracardiac anomalies involving the eyes and brain and of oncocytic cells in various glands.

Grossly yellowish nodules may be found on the endocardium, epicardium, and/or valves. Histologically the hamartoma-like aggregates within the myocardium consist of collections of large round to polygonal cells, resembling histiocytes, and composed of a coarsely granular, pale cytoplasm. Ultrastructurally the granules correspond to numerous mitochondria; some contain inclusions [37,38,41]. The etiology is unknown and the cell of origin (myocyte, Purkinje cell, or histiocyte?) is currently disputed. Some patients with this disease have been cured with electrophysiological mapping and surgical excision [37].

Blood cysts are a common finding in neonates dying of various causes and probably have no clinical significance [43]. Blood cysts may persist and enlarge to form giant cysts of heart valves. They occur most frequently on the tricuspid and mitral valves of fetuses and infants. Congenital blood cysts are believed to form from infoldings of the valvular endothelium that bulges into the atria because of the transvalvular pressure [43].

References

1. Arey JB. Tumors of the heart and pericardium. In: *Cardiovascular Pathology in Infants and Children.* Philadelphia: WB Saunders; 1984:372.
2. Freedom RM, Benson LN. Cardiac neoplasms. In: Freedom RM, Benson LN, Smallhorn JF, eds. *Neonatal Heart Disease.* London: Springer-Verlag; 1992:723.
3. Nadas AS, Ellison RC. Cardiac tumors in infancy. *Am J Cardiol* 1968;21:363.
4. Isaacs H Jr. Cardiac tumors. In: *Tumors of the Fetus and Newborn.* Philadelphia: WB Saunders; 1997:330. Major Problems of Pathology Series; Vol 35.
5. Isaacs H Jr. Tumors. In: Gilbert-Barness E, ed. *Potter's Pathology of the Fetus and Infant.* Vol 2. St. Louis, MO: CV Mosby, 1997:1287.

6. Chan HSL, Sonley MJ, Moes CAF, et al. Primary and secondary tumors of childhood involving the heart, pericardium and great vessels: a report of 75 cases and review of the literature. *Cancer* 1985;56:825.
7. Groves AMM, Fagg NLK, Cook AC, et al. Cardiac tumours in intrauterine life. *Arch Dis Child* 1992;67:1189.
8. Russell GA, Dhasmana JP, Berry PJ, et al. Coexistent cardiac tumors and malformations of the heart. *Int J Cardiol* 1989;22:89.
9. Golding R, Reed G. Rhabdomyoma of the heart. Two unusual clinical presentations. *N Engl J Med* 1967;276:597.
10. Arciniegas E, Hakimi M, Farooke QZ, et al. Primary cardiac tumors in children. *J Thorac Cardiovasc Surg* 1980;79:582.
11. Holley DG, Martin GR, Brenner JI, et al. Diagnosis and management of fetal cardiac tumors: a multicenter experience and review of published reports. *J Am Coll Cardiol* 1995;26:516.
12. Hals J, Ek J, Sandnes K. Cardiac myxoma as the cause of death in an infant. *Acta Paediatr Scand* 1990;79:999.
13. Fyler DC. Report of the New England Regional Infant Cardiac Program. *Pediatrics* 1980;65(suppl):376.
14. McAllister HA Jr, Fenoglio JJ Jr. Tumors of the cardiovascular system. In: *Atlas of Tumor Pathology.* 2nd ser. Fascicle 15. Washington, DC: Armed Forces Institute of Pathology; 1978.
15. Vidaillet HJ. Cardiac tumors associated with hereditary syndromes. *Am J Cardiol* 1988;61:1355.
16. Coffin CM. Case 1. Congenital cardiac fibroma associated with Gorlin syndrome. *Pediatr Pathol* 1992;12:255.
17. Smythe JF, Dyck JD, Smallhorn JF, et al. Natural history of cardiac rhabdomyoma in infancy and childhood. *Am J Cardiol* 1990;66:1247.
18. Fenoglio JJ Jr, McAllister HA, Ferrens VJ. Cardiac rhabdomyoma: a clinicopathologic and electron microscopic study. *Am J Cardiol* 1976;38:241.
19. Webb DW, Thomas RD, Osborne JP. Cardiac rhabdomyomas and their association with tuberous sclerosis. *Arch Dis Child* 1993; 68:367.
20. Burke AP, Virmani R. Cardiac rhabdomyoma: a clinicopathologic study. *Mod Pathol* 1991;4:70.
21. Ostor AG, Fortune DW. Tuberous sclerosis initially seen as hydrops fetalis: report of a case and review of the literature. *Arch Pathol Lab Med* 1978;102:34.
22. Gonzalez-Crussi F, Eberts TJ, Mirkin DL. Congenital fibrous hamartoma of the heart. *Arch Pathol Lab Med* 1978;102:491.
23. Marin-Garcia J, Fitch CW, Shenenfelt RE. Primary right ventricular tumor (fibroma) simulating cyanotic heart disease in a newborn. *J Am Coll Cardiol* 1984;3:868.
24. Beghetti M, Haney I, Williams WG, et al. Massive right ventricular fibroma treated with partial resection and a cavopulmonary shunt. *Ann Thorac Surg* 1996;62:882.
25. Rosenberg HS, Stenback WA, Spjut HJ. The fibromatoses of infancy and childhood. In: Rosenberg HS, Bolande RP, eds. *Perspectives in Pediatric Patholology.* Vol 4. Chicago: Year Book Medical Publishers; 1978:269.
26. Turi GK, Albala A, Fenoglio JJ Jr. Cardiac fibromatosis: an ultrastructural study. *Hum Pathol* 1980;11(suppl):577.
27. Goldberg BE, Romano AA, Amato JJ, et al. Unique association of a rapidly growing right atrial myxoma in a child with double-outlet right ventricle. *Pediatr Cardiol* 1997;18:303.
28. Wold LE, Lie JT. Cardiac myxomas. A clinicopathologic profile. *Am J Pathol* 1980;101:219.
29. Gonzalez-Crussi F. Extragonadal teratomas. In: *Atlas of Tumor Pathology.* Washington, DC: Armed Forces Institute of Pathology; 1982. 2nd ser; Fascicle 18.
30. Costas C, Williams RL, Fortune RL. Intracardiac teratoma in an infant. *Pediatr Cardiol* 1986;7:179.
31. Bruch SW, Adzick NS, Reiss R, et al. Prenatal therapy for pericardial teratomas. *J Pediatr Surg* 1997;32:1113.
32. Sklansky M, Greenberg M, Lucas V, et al. Intrapericardial teratoma in a twin fetus: diagnosis and management. *Obstet Gynecol* 1997;89:807.
33. Catanzarite V, Mehalek K, Maida C, Mendoza A. Early sonographic diagnosis of intrapericardial teratoma. *Ultrasound Obstet Gynecol* 1994;4:505.
34. Leithiser RE, Fyfe D, Weatherby E III, et al. Prenatal sonographic diagnosis of atrial hemangioma. *Am J Radiol* 1986;147:1207.
35. Tseng JJ, Chou MM, Lee YH, et al. In utero diagnosis of cardiac hemangioma. *Ultrasound Obstet Gynecol* 1999;13:363.
36. Platt LD, Geierman CA, Turkel SB, et al. Atrial hemangioma and hydrops fetalis. *Am J Obstet Gynecol* 1981;141:107.
37. Kearney DL, Titus JL, Hawkins EP, et al. Pathologic features of myocardial hamartomas causing childhood tachyarrhythmias. *Circulation* 1987;75:705.
38. Franciosi RA, Singh A. Oncocytic cardiomyopathy syndrome. *Hum Pathol* 1988;19:1361.
39. Gelb AB, Van Meter SH, Billingham ME, et al. Infantile histiocytoid cardiomyopathy-myocardial or conduction system hamartoma: What is the cell type involved? *Hum Pathol* 1993;24:1226.
40. Koponen MA, Siegel RJ. Histiocytoid cardiomyopathy and sudden death. *Hum Pathol* 1996;27:420.
41. Boissy C, Chevallier A, Michiels J-F, et al. Histiocytoid cardiomyopathy: a cause of sudden death in infancy. *Pathol Res Pract* 1997;193:589.
42. Shehata BM, Patterson K, Thomas JE, et al. Histiocytoid cardiomyopathy: three new cases and a review of the literature. *Pediatr Dev Pathol* 1998;1:56.
43. Zimmerman KG, Paplanus SH, Dong S, et al. Congenital blood cysts of the heart valves. *Hum Pathol* 1983;14:699.

16
Tumors of the Lung

Both benign and malignant lung tumors are described in newborns and infants. Metastatic disease is much more prevalent [1–11]. Rhabdomyosarcoma, Wilms tumor, hepatoblastoma, and endodermal sinus tumor are the main malignant neoplasms metastasizing to the lungs during the first year of life. In patients with leukemia the pulmonary vessels may contain leukemic cells, and these cells may infiltrate the alveolar wall. Perivascular collections are found in severe cases [11].

Some nonneoplastic conditions may mimic tumors: for example, abscess and organizing bacterial pneumonia, or congenital anomalies such as bronchogenic cyst, extrapulmonary sequestration, adenomatoid malformation, and pulmonary lymphangiectasis [2,7–9] (Figures 16.1–16.3). Lymphoproliferative disorders may present with pulmonary nodules simulating metastatic tumor [10] (Figure 16.4).

Table 16.1 lists the various tumors and tumorlike conditions of the lung found in the fetus and infant.

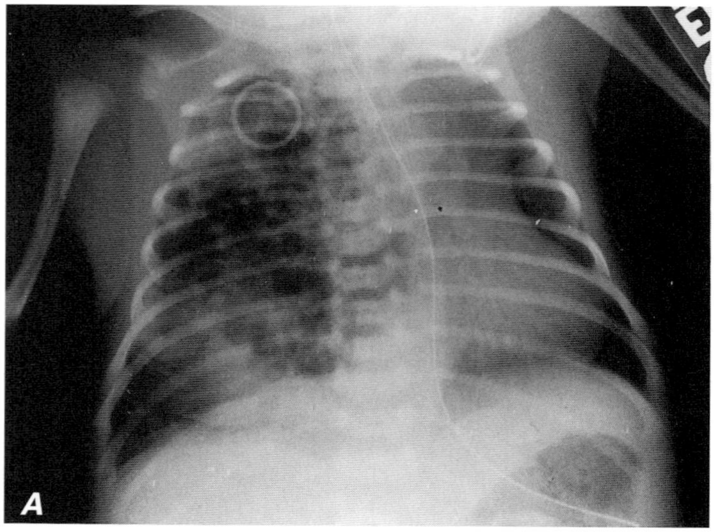

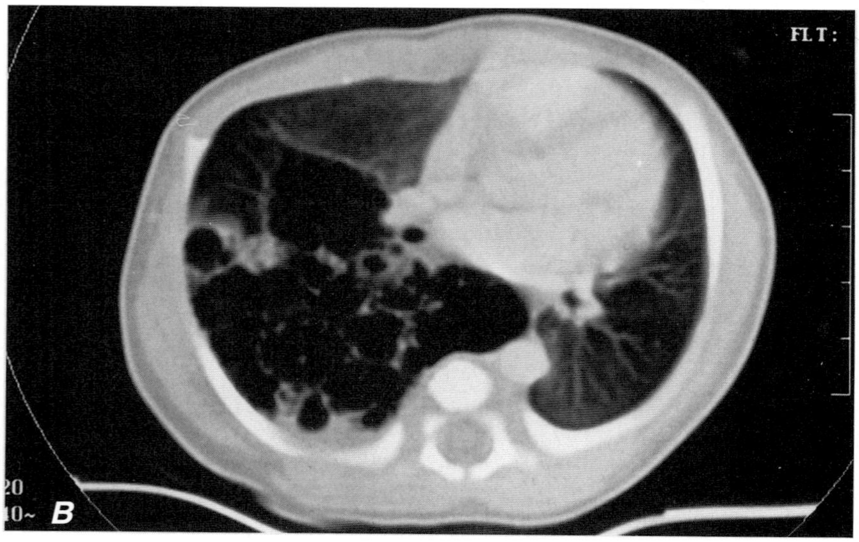

FIGURE 16.1. Cystic adenomatoid malformation, type I, in an 8-day-old male with progressive respiratory distress since birth. (A) Chest x-ray reveals multiple large cysts within the right upper lobe with a shift of mediastinal structures to the left chest. (B) Computerized tomography scan shows the cysts and compression of the right lower and middle lobes in more detail. (C) The specimen consists of the right upper lobe and a portion of the middle lobe measuring $6.0 \times 6.0 \times 1$ cm^3 and weighing 23 g. The upper lobe contains several multilocular cysts, the largest measuring 2.5 cm. (D) Several irregular, small and large cysts are lined by cuboidal epithelium. H&E; $\times 48$.

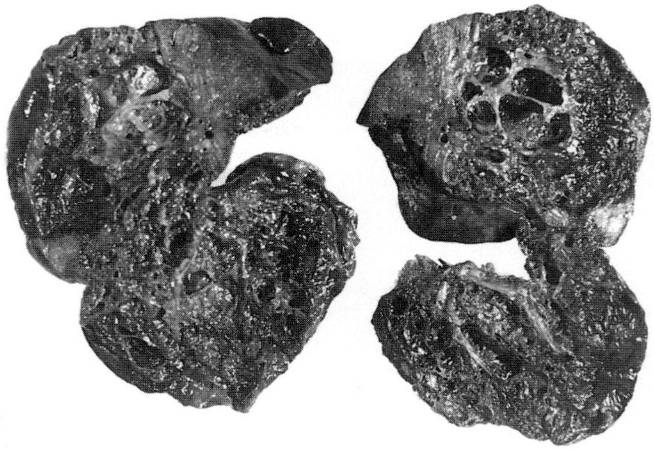

C

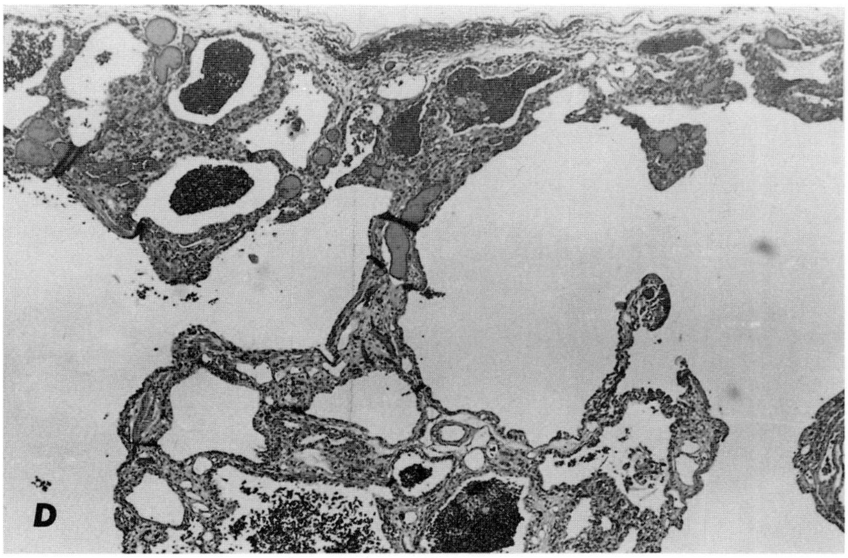

D

FIGURE 16.1. (*Continued*)

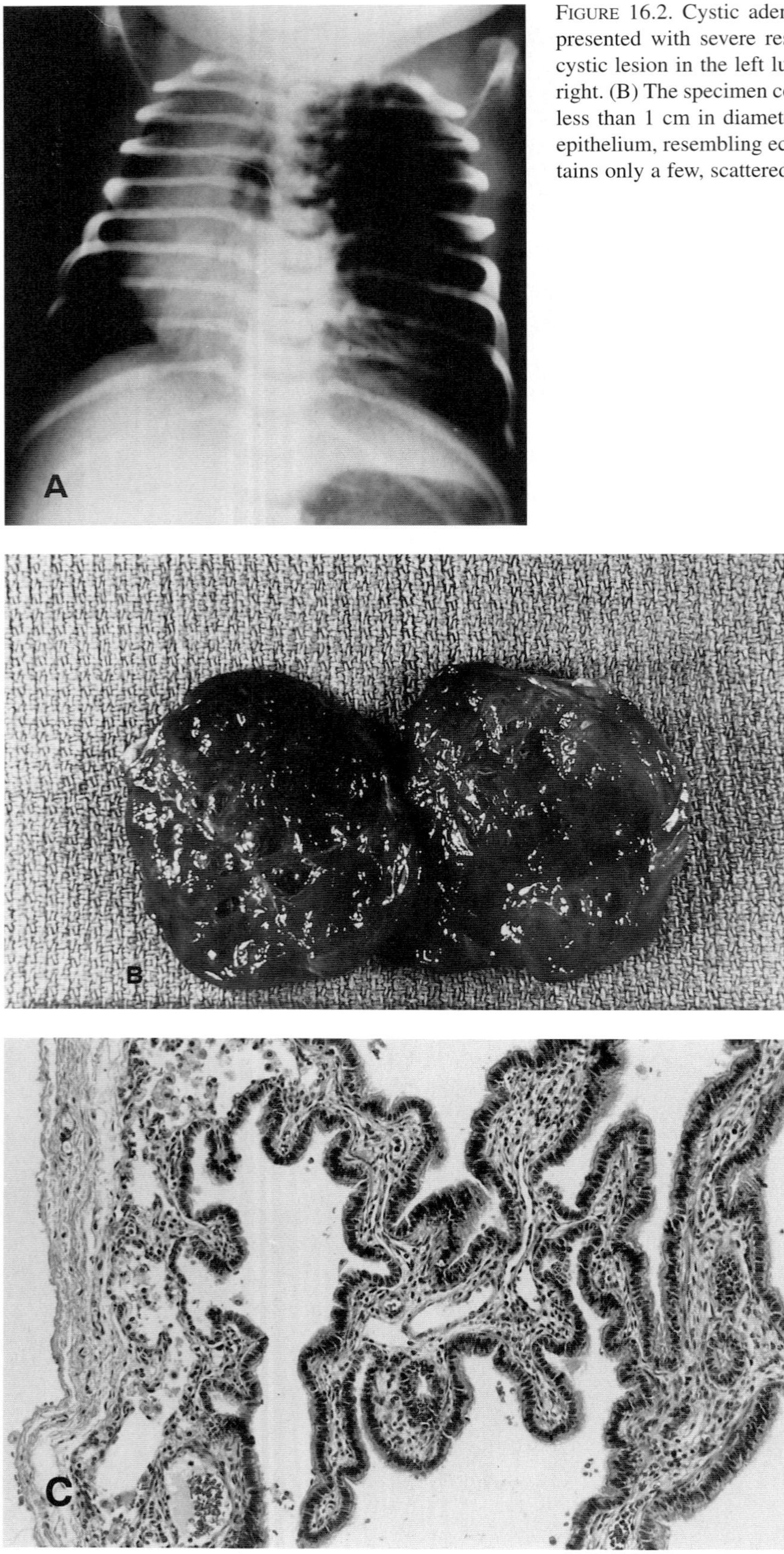

FIGURE 16.2. Cystic adenomatoid malformation, type II, in a newborn who presented with severe respiratory distress. (A) Chest x-ray demonstrating a cystic lesion in the left lung, upper lobe, producing a mediastinal shift to the right. (B) The specimen consists of firm, congested lung with many small cysts less than 1 cm in diameter. (C) The cysts are lined by cuboidal to columnar epithelium, resembling ectatic terminal bronchioles. The wall of the cysts contains only a few, scattered smooth muscle fibers and capillaries. H&E; × 120.

FIGURE 16.3. Congenital pulmonary lymphangiectasis in a 5-month-old female with increasing chronic pulmonary insufficiency. (A) Chest film reveals a coarse reticulonodular pattern within both lungs consistent with pulmonary lymphangiectasis. (B) Severely dilated lymphatics are present within the interlobular septa and beneath the pleura. H&E; $\times 60$.

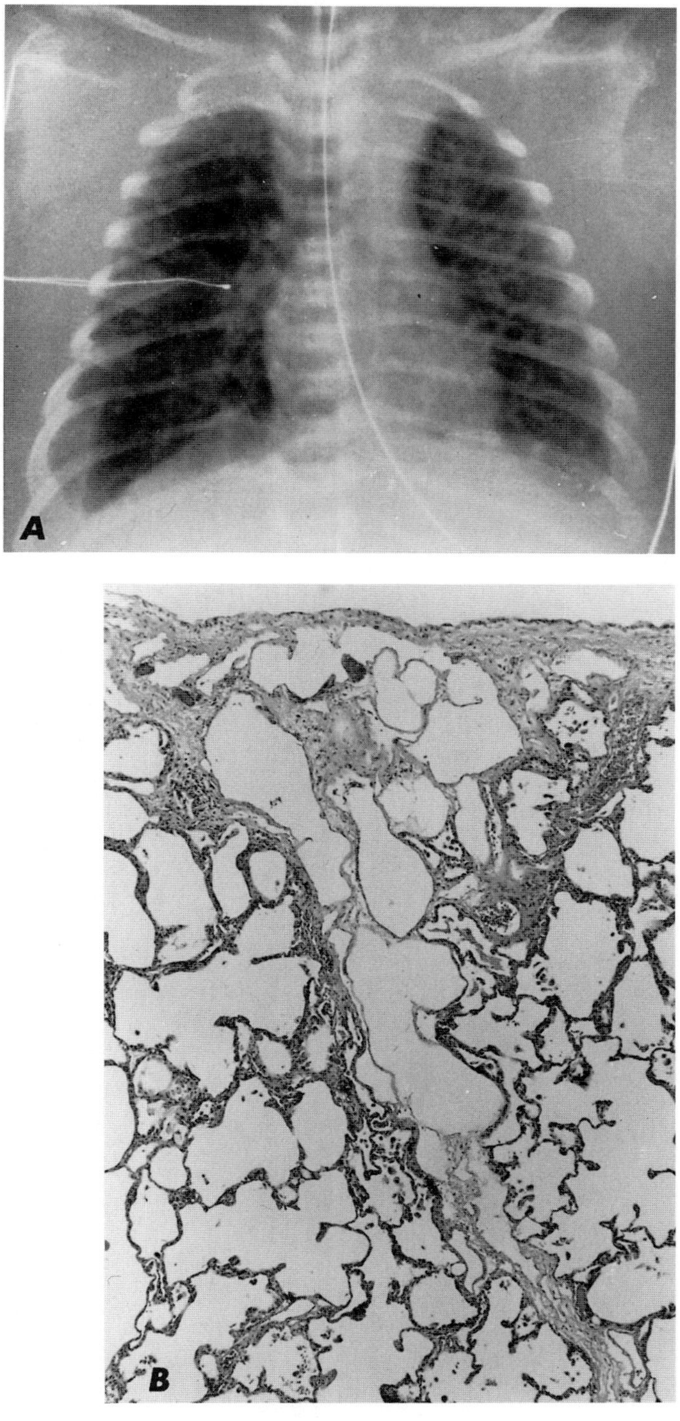

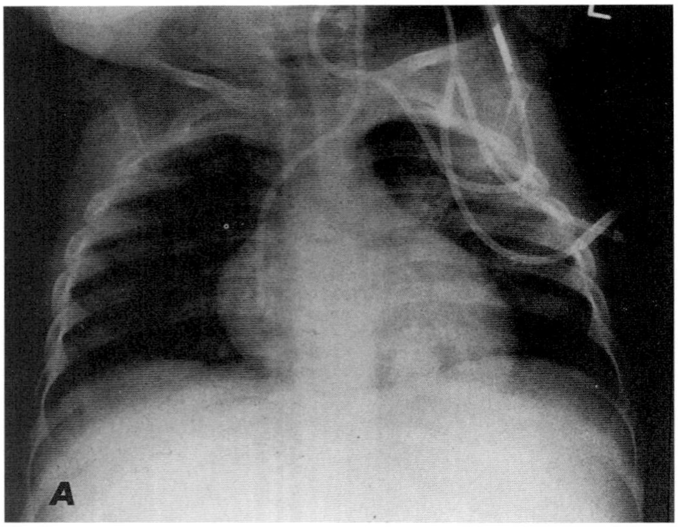

FIGURE 16.4. Lymphomatoid granulomatosis in a 1-year-old male who presented with chronic otitis media and failure to thrive. The baby had hypogammaglobulinema and Epstein-Barr virus infection. (A) Chest radiogram reveals bilateral pulmonary nodules. (B) Lung biopsy material shows a necrotizing vasculitis involving a pulmonary artery (the artery has both an internal and external elastic lamina) and adjacent infarcts. Verhoff–Van Giesen (elastic) stain; × 60. (C) Higher magnification (× 300) displays vascular necrosis, edema, and infiltration by lymphocytes, plasma cells, and atypical lymphoid cells. The vascular lumen is on the left. At postmortem examination similar yellow-tan, necrotic, vasculitic nodules were found in the kidneys, adrenals, liver, and lymph nodes. Giant cells and palisading epithelioid cells characteristic of Wegener's granulomatosis were not found. H&E; × 300. (From Lehman TJA, Church JA, Isaacs H Jr. *J Rheumatol* 1989;16:235. Used with permission.)

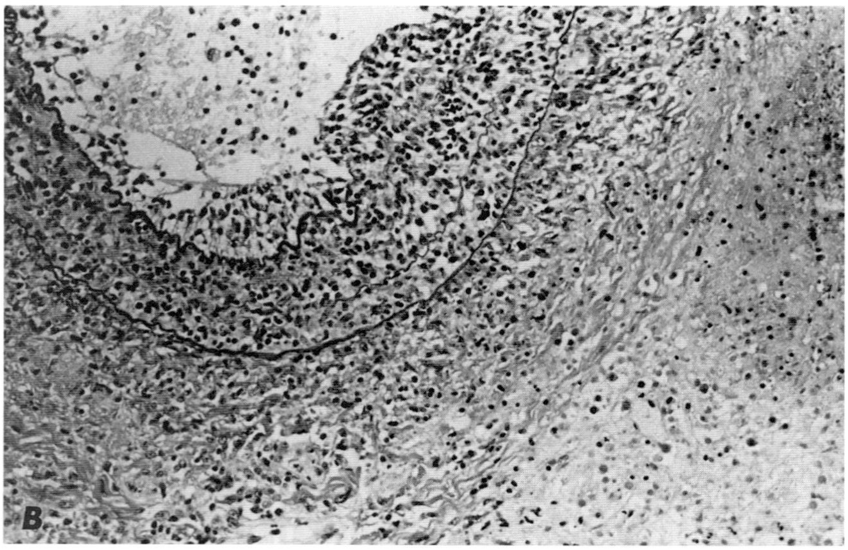

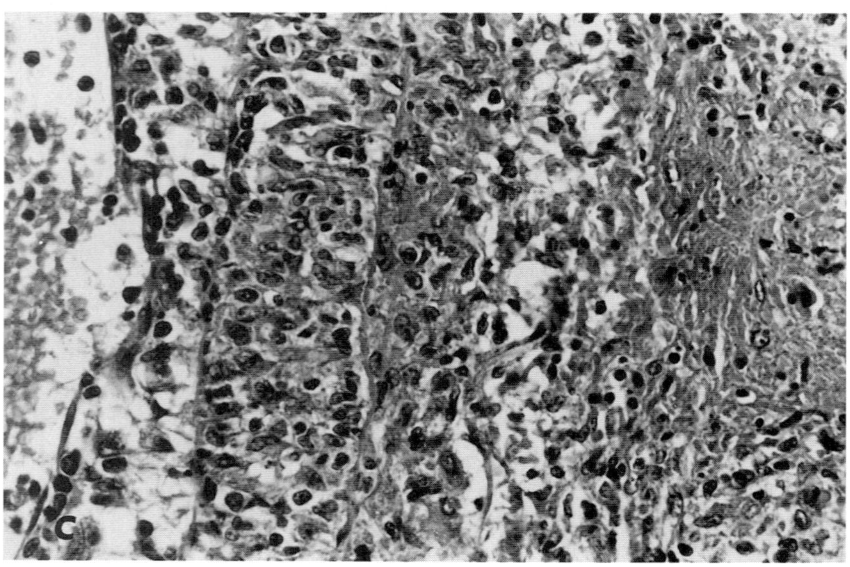

TABLE 16.1. Tumors and tumorlike conditions of the lung in the fetus and infant.

Inflammatory pseudotumor (plasma cell granuloma)
Cystic mesenchymal hamartoma[a]
Rhabdomyomatous dysplasia (rhabdomyomatosis)
Cystic adenomatoid malformation
Extralobar pulmonary sequestration
Congenital pulmonary lymphangiectasis
Lymphangioma
Hemangioma
Myofibromatosis (fibromatosis)
Pulmonary myofibroblastic tumor (bronchopulmonary myofibrosarcoma)
Pleuropulmonary blastoma
Primitive neuroectodermal tumor of the chest wall (Askin tumor)[a]
Teratoma
Langerhans cell histiocytosis (Letterer-Siwe disease)
Laryngotracheal papillomatosis
Juvenile xanthogranuloma
Lymphoproliferative disorders
Tumors metastatic to the lung
Leukemia
Hepatoblastoma
Rhabdomyosarcoma
Neuroblastoma
Wilms tumor
Rhabdoid tumor
Yolk sac tumor

[a] Usually occurs in older children.
Source: Isaacs H Jr [5].

Inflammatory Pseudotumor

Inflammatory pseudotumor is a benign condition, probably a reactive inflammatory process, that can mimic a neoplasm. Although the lung is the most common and well-known site, the lesion occurs also in brain, orbit, small intestine, bladder, mesentery, spleen, retroperitoneum, heart, liver, and skin [2,4,12]. Other names for this condition referred to in the literature include plasma cell granuloma, inflammatory pseudotumor, and inflammatory myofibroblastic tumor [2,4,12]. Inflammatory pseudotumor is characterized by slow growth, instances of spontaneous resolution, and favorable prognosis.

Grossly the inflammatory pseudotumor consists of a firm, circumscribed grayish-white to yellow nodule situated usually at the periphery of the lung. Three different histological patterns are described: organizing pneumonia, fibrous histiocytoma, and lymphoplasmacytic pattern [2,13] (Figure 16.5). Immunohistochemical and ultrastructural studies show that the spindle cells are fibroblasts and myofibroblasts; hence the term "myofibroblastic tumor" has been applied to some [6]. Neither morphological evaluation for cellularity, mitosis, or inflammatory infiltrate nor expression of oncogenes appears to be useful for prediction of clinical outcome, which is generally favorable [12].

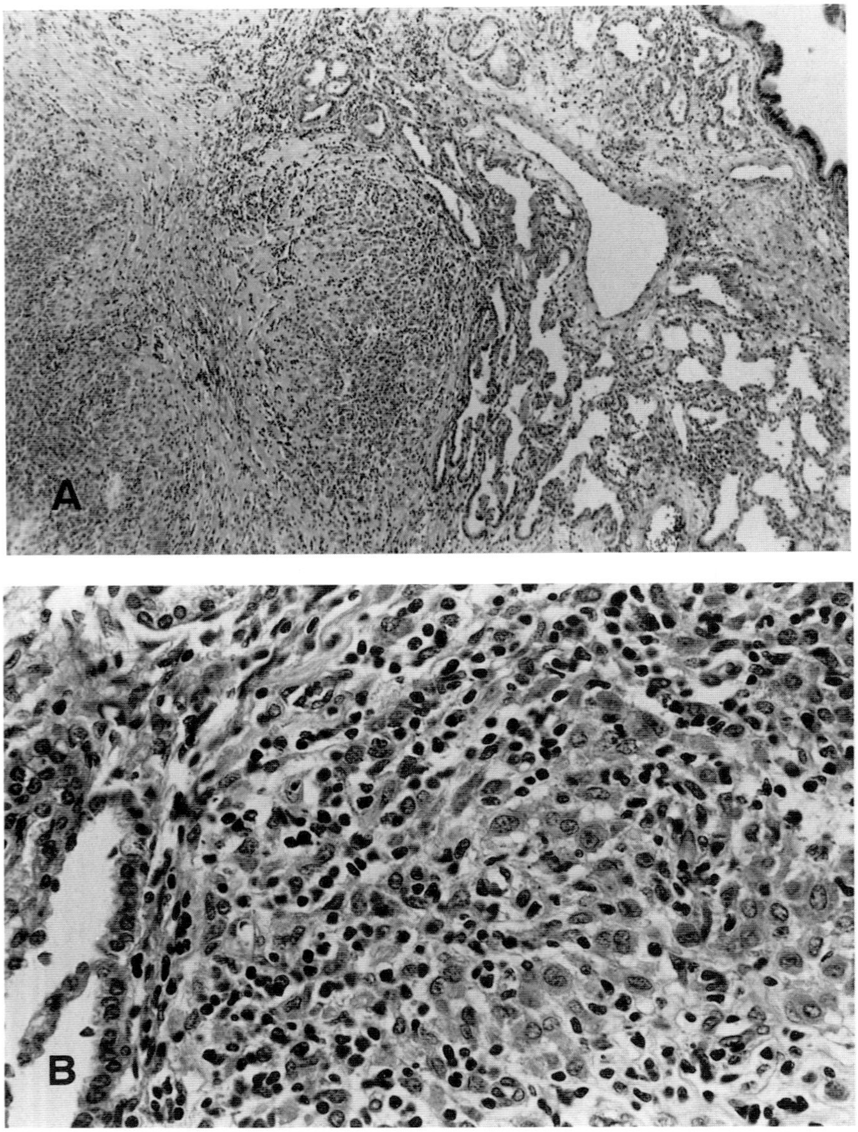

FIGURE 16.5. Inflammatory pseudotumor in a 1-week-old female with a 3.5 × 2.5 cm^2, subpleural mass removed from the left lung, upper lobe. (A) Within the lung there is a nodular inflammatory cell infiltrate with fibrosis. H&E; × 48. (B) The infltrate consists mostly of lymphocytes and histiocytes with a rare plasma cell. H&E; × 300. (Courtesy of Frederick Walker, M.D., Department of Pathology, Kaiser Permanente, San Diego.)

Fibromatoses (Myofibromatoses)

Fibromatosis of the lung(s) usually occurs as part of a congenital disseminated disease termed generalized or visceral fibromatosis (myofibromatosis) [1,4,14]. In this disease multiple fibrous nodules are found in the skin, musculoskeletal system, and viscera. The tumors are composed of myofibroblasts showing storiform, vascular, and hemangiopericytomatous growth patterns. Congenital generalized fibromatosis is classified into two main types: type I, patients with involvement of the skin, subcutaneous tissue, muscle and bone, and type II, those with visceral involvement [14]. Pulmonary involvement in generalized fibromatosis usually portends a poor prognosis [14] (see Chapter 3, Soft Tissue Tumors).

Vascular Conditions

Vascular conditions seldom occur in the lungs of infants and children. Hemangiomas of the lung are uncommon [6]. Most lesions reported as hemangiomas in infants and children are congenital malformations of the pulmonary vasculature, variously called pulmonary arteriovenous fistula or aneurysm, congenital arteriovenous varix, and pulmonary angiomatosis [11]. Patients with these conditions often have associated vascular malformations elsewhere as in the mucous membranes, the skin, or the brain, forming part of a more generalized syndrome of hereditary telangiectasia (e.g., the Rendu-Osler-Weber disease). Familial occurrence is reported [11]. Multiple pulmonary hemangiomas may occur as part of a fatal syndrome, diffuse hemangiomatosis of the newborn, in which cutaneous hemangiomas are associated with similar lesions in the brain, liver, lung, and other organs (see Chapter 3, Soft Tissue Tumors, and Chapter 4, Tumors and Tumorlike Conditions of the Skin).

Similar to the soft tissues, hemangiomas are composed of capillary and/or cavernous components, typically showing a lobulular growth pattern. They are accompanied by prominent "feeding" arterial and venous blood vessels, sometimes forming large arteriovenous malformations. Capillary and cavernous hemangiomas are lined by regular plump or flattened endothelial cells and differ only in their diameter (see Chapter 3, Soft Tissue Tumors).

Congenital Pulmonary Cystic Diseases

Congenital cystic diseases of the lung—for example, cystic adenomatoid malformation—may clinically mimic a neoplasm and are a source of malignancy within the cyst(s) [1,2, 6,15,16] (Table 16.2). Rhabdomyosarcoma and pleuropulmonary blastoma are examples of malignancies arising from lung cysts [16,17].

TABLE 16.2. Pulmonary cystic lesions of the fetus and infant.

Congenital cystic adenomatoid malformation
Extralobar pulmonary sequestration
Bronchogenic cyst
Interstitial emphysema
Infantile lobar emphysema (overinflation)
Congenital pulmonary lymphangiectasis
Lymphangioma
Cystic pleuropulmonary blastoma
Mesenchymal cystic hamartoma
Cystic bronchiectasis
Late stages of bronchopulmonary dysplasia

Source: Colby TV, Koss MN, Travis WD [28]; Katzenstein A-LA, Askin FB [6].

Cystic Adenomatoid Malformation

Cystic adenomatoid malformation is an uncommon congenital pulmonary anomaly with a variable prognosis depending on the size of the lesion, the pathological subtype, and the presence or absence of clinical complications [18–21]. In fetuses with large lesions and associated hydrops, the prognosis for survival is bleak [18,19]. The malformation is almost always unilateral, without predilection for right or left side and usually limted to one lobe. An arrest in lung development has been suggested as the cause [20]. There are a wide variety of congenital anomalies associated with cystic adenomatoid malformation. Bilateral renal agenesis, renal dysplasia, hydrocephalus, extralobar pulmonary sequestration, cardiac defects, and diaphragmatic hernia are a few examples [18]. The lesion is detected antenatally as early as 21 weeks gestation, presenting as a mediastinal shift by sonography [19]. Most affected infants develop respiratory distress on the first day of life, but the malformation may not become clinically evident until later on in childhood [19].

On the basis of its gross and histological findings, cystic adenomatoid malformation is divided into three main types [2,15]. Type I consists of one or more large cysts with fibrous septa lined by columnar or cuboidal epithelial cells (Figure 16.1). Type II contains many smaller cysts (<1.2 cm) lined by ciliated columnar epithelial cells and resembles ectatic terminal bronchioles (Figure 16.2). Skeletal muscle may be found in the septa or adjacent to the cysts. The less common solid type, type III, is composed of a firm, bulky mass that occupies much of a lung lobe. Histologically type III consists of dilated, irregularly branching bronchiole-like structures separated by alveolar-like spaces lined by low cuboidal epithelium [2,6,15]. The more prevalent macrocystic lesion, type I, is less often associated with hydrops fetalis and has a more favorable prognosis than types II and III. Particularly the solid (type III) can mimic a tumor on imaging studies, grossly and microscopically.

Mesenchymal Cystic Hamartoma of the Lung

Pulmonary mesenchymal cystic hamartomas are nodules or cysts composed of both mesenchymal and epithelial elements [22]. The nodules grow slowly, and when they reach approximately 1 cm in diameter they become cystic and visible on imaging studies. The cysts are lined by respiratory epithelium forming papillary projections. Beneath the epithelium is a cambium layer composed of mesenchymal cells resembling those of botyroid embryonal rhabdomyosarcoma [22]. However, the mesenchymal cells of the hamartoma do not show the immaturity, the mitotic activity, and the cellular atypia of rhabdomyosarcoma. Hamartomas usually occur as solitary lesions in the young, but they may be multifocal or bilateral in older individuals. If large enough, the cysts and nodules cause hemoptysis, pneumothorax, hemothorax, chest pain, or mild dyspnea [22]. Infants and children with pulmonary cysts, particularly mesenchymal cystic hamartomas, are at risk for developing malignancies within these cysts and should receive periodic follow-up imaging studies [17]. Pleuropulmonary blastoma and rhabdomyosarcoma arise from within these cysts [17,23,24].

Laryngotracheal Papilloma

Laryngotracheal papilloma (juvenile papilloma, squamous papilloma) is a relatively frequent, benign tumor arising most often in the larynx. The occurrence of many, recurrent papillomas in the larynx and tracheobronchial tree is called papillomatosis. In 5 to 10% of individuals these tumors extend distally into the trachea, and in less than 1% into the adjacent lung [2]. The stated incidence is 1500 to 2000 infants and children in the United States per year. Human papillomavirus is considered as the etiological agent; condylomas situated around the birth canal have been proposed as the source of the virus [2,25,26]. The tumor occurs most often during the first 3 years of life.

Papillomas in the larynx cause hoarseness and inspiratory stridor, eventually leading to severe respiratory distress if not treated [2,25]. Typically the clinical course is characterized by multiple recurrences requiring numerous resections and often a life-saving tracheostomy. Laser therapy has met with some success [27]. Spontaneous regression usually occurs by late adolescence, but sometimes there is extension into the peripheral airway and lungs, with cavity formations filled with numerous papillomas [2,28]. Progression to squamous cell carcinoma has been documented [27,29]. Extensive laryngeal-tracheal papillomatosis can be the cause of sudden unexpected death from suffocation in a young child [5].

Papillomas present as one or more tiny papillary or sessile wartlike growths projecting from the laryngeal and tracheal mucosa. Occasionally the mucosa is covered by numerous lesions that obstruct the airway (Figure 16.6). Microscopically they consist of papillary formations composed of stratified squamous epithelium displaying maturation toward the surface and varying degrees of koilocytosis. Focal atypia and increased mitotic activity are not unusual. The epithelium rests on a slender fibrovascular core. Koilocytosis (i.e., the presence of vacuolated epithelial cells with clear cytoplasm or perinuclear halos and angular nuclear pyknosis) is a prominent feature and is indicative of viral infection. Human papillomavirus, specifically type 11 and less often type 6, has been isolated from these lesions and from maternal condylomas [28].

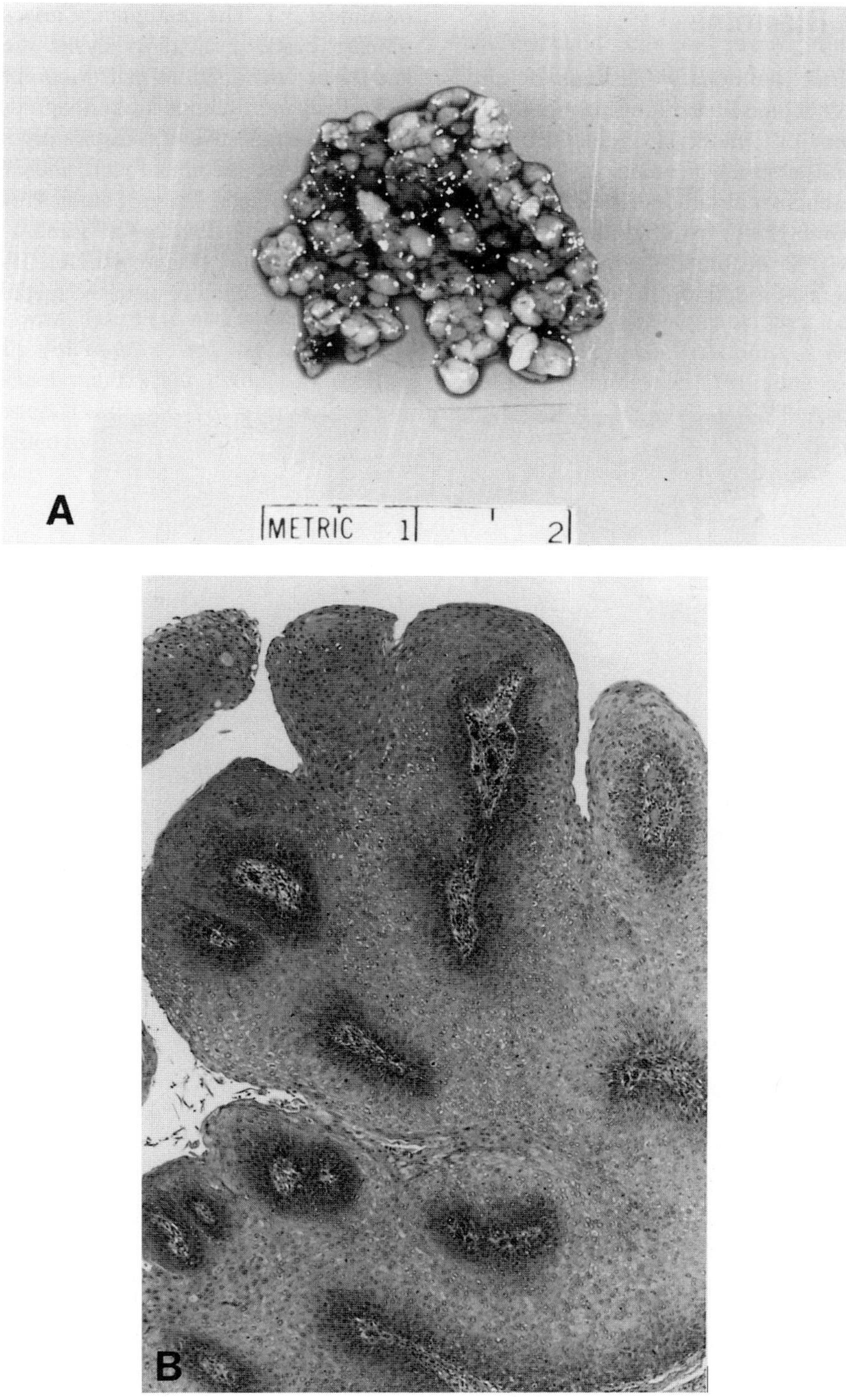

FIGURE 16.6. Laryngotracheal papilloma. (A) Multiple papillomas removed from the larynx of a child with hoarseness and respiratory distress consist of pale gray-white, cauliflower-like growths. (B) Papillary formations composed of nonkeratinized stratified squamous epithelium rest on a fibrovascular core. Minimal koilocytotic atypia is present. H&E; $\times$ 60.

from pulmonary blastoma histologically [28]. The presence of malignant skeletal muscle and/or cartilage rules out the former.

Immunohistochemical studies show that tumor cells are immunoreactive with vimentin, desmin, α_1-antitrypsin, and α_1-chymotrypsin and focally positive for S-100 protein. The entrapped epithelial elements are positive for epithelial membrane antigen and cytokeratin [17,28,31–34]. Ultrastructurally the blastemal cells are primitive, appearing with round to oval nuclei and scant cytoplasm, containing a few organelles. Spindle-shaped fibroblasts and myofibroblasts with a well-formed endoplasmic reticulum, actinlike filaments with focal condensations beneath the cytoplasmic membrane, and rhabdomyoblasts showing thick and thin filaments and Z-band formations are present. There are fibrohistiocytoid and histiocytoid cells with numerous lysosomes [17,28,30–32].

Congenital Pulmonary Myofibroblastic Tumor

Congenital pulmonary myofibroblastic tumor is a spindle cell neoplasm unique to the neonate. The tumor has been variously called bronchopulmonary fibrosarcoma, leiomyosarcoma, hamartoma, mesenchymal tumor, or malformation [2, 28,35]. Although pulmonary myofibroblastic tumor is rare, it is probably the most frequent congenital primary tumor of the lung [36–38]. Newborns with this neoplasm present with severe respiratory distress occurring soon after birth [36]. McGinnis et al. proposed the name for this tumor and suggested that it is histologically analogous to the renal mesoblastic nephroma [38]. Despite its ominous histological appearance, it behaves in a benign fashion if adequately treated [35].

Pulmonary myofibroblastic tumor arises either from the bronchus or from the lung parenchyma. McGinnis et al. suggest that the neoplasm develops from the mesenchyme that surrounds primitive respiratory ducts [38] (Figure 16.8).

The tumors range in size from 3.5 to 6 cm. Grossly they are firm, gray, yellow, or light tan with focal hemorrhage and cyst formation. Histological examination reveals a cellular tumor with spindle-shaped cells forming interdigitating fascicles similar to those observed in the soft tissue congenital fibrosarcoma. Central necrosis and dystrophic calcification may be present [35]. The mitotic rate varies from moderate to high. Benign entrapped bronchoalveolar elements are present.

Immunohistochemical and ultrastructural findings suggest a fibroblastic-myofibroblastic origin. The tumor cells are immunoreactive with vimentin; the muscle markers desmin and actin are focally positive. The tumor cells are nonreactive with neuron-specific enolase, S-100, Leu-7, and α_1-antichymotrypsin [35,37,38]. Ultrastructurally the spindle-shaped tumor

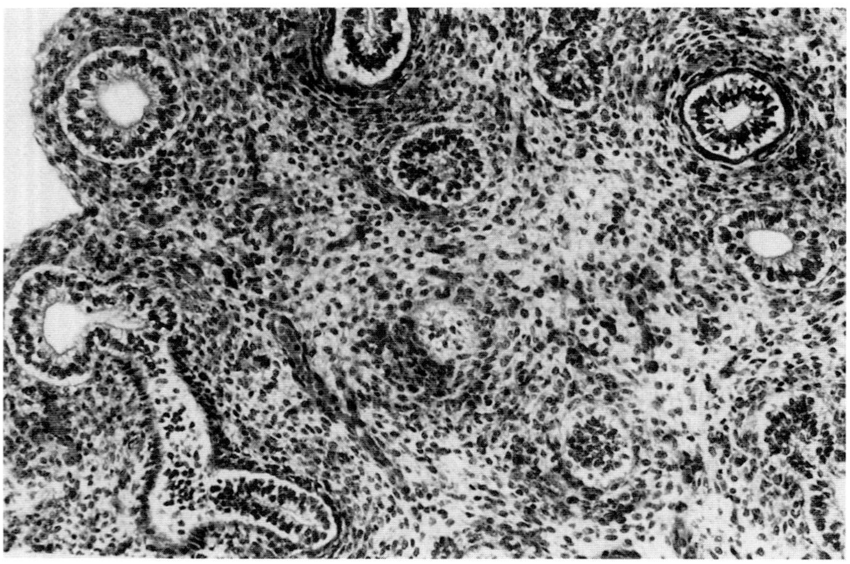

FIGURE 16.8. In the developing lung, branches of the pulmonary tree are uniformly lined by tall columnar cells and widely separated by primitive mesenchyme containing tiny spindle-shaped cells. From a fetus of estimated 10 to 12 weeks gestation. H&E; × 120.

cells display indented nuclei and infrequent nucleoli. The cytoplasm contains a prominent rough endoplasmic reticulum and intermediate filaments. Primitive cell junctions and extracellular deposits of collagen are present [36,37]. Cells are noted with cytoplasmic actin filaments, dense bodies, and attachment plaques having features of myofibroblasts [35,38]. The lesion is benign; a favorable prognosis following complete resection should be anticipated [2,35,36,38].

Skeletal Muscle Tumors

There are well-documented examples of striated muscle occurring in the lungs of fetuses and newborns, in lungs with cystic adenomatoid malformation and extralobar pulmonary sequestration [2,4,39–42]. The presence of striated muscle in the lungs, termed "rhabdomyomatous dysplasia" or "rhabdomyomatosis," is often accompanied by significant malformations of the heart and lungs [40]. It may present as a solid tumorlike mass occupying one or more lobes [41,42]. Microscopically the dysplasia consists of many bundles of striated fibers arranged haphazardly in expanded alveolar septa. The skeletal muscle fibers may show marked variation in differentiation and are desmin and actin immunoreactive [41,42].

Primary pulmonary rhabdomyosarcoma in childhood is rare [3,4,43]. The tumor occurs either as a solid mass in the lung or as part of a congenital cyst such as mesenchymal cystic hamartoma, bronchogenic cyst, or cystic adenomatoid malformation. In some of the older literature examples of pulmonary blastoma were called rhabdomyosarcoma. Immunohistochemistry, namely, the presence of the reactive skeletal muscle markers myoglobin, desmin, and myosin, is helpful in establishing the diagnosis of rhabdomyosarcoma and electron microscopy in doubtful cases [43].

Peripheral Neuroectodermal Tumor of Thoracopulmonary Origin

The small-cell malignant tumor of thoracopulmonary origin (Askin tumor, PNET of chest wall and lung) is one of the small blue-cell tumors of childhood associated with a rapidly progressive, downhill clinical course and a poor outcome. Current thinking based on immunohistochemical, ultrastructural, and cytogenetic studies is that this tumor is essentially the same entity as the extraskeletal Ewing sarcoma and PNET of the soft tissues [44,45]. A neural crest origin is proposed for these malignancies.

The PNET of thoracopulmonary origin occurs in two age peaks, one in late infancy and the other in adolescence [1]. The malignancy presents with fever, chest pain, a palpable mass in the chest wall, and pleural effusion [1]. The tumor arises from the soft tissues of the chest wall; but because of its proximity, the lung and pleura become involved during the course of the disease.

On gross inspection the chest wall and adjacent lung are replaced by one or more pale gray tumor nodules measuring up to 10 cm in greatest diameter. The pale gray gelatinous cut surfaces show foci of necrosis, hemorrhage, and cyst formation [1]. Histologically the tumor is highly cellular consisting of small, round, darkly staining cells with scant cytoplasm occasionally arranged in lobules (Figure 16.9). Pseudorosettes are noted in some. Neural differention is manifested by the presence of neuritic processes, dense core granules, and neurotubules. Glycogen deposits are variably present. Neuron-specific esterase and more specifically the cell surface antigen HBA71 (Ewing antibody 013) are diagnostically immunoreactive. Chromosomal analysis from tissue culture material reveals a characteristic t(11;22) rearrangement, which helps to establish the diagnosis [44,45].

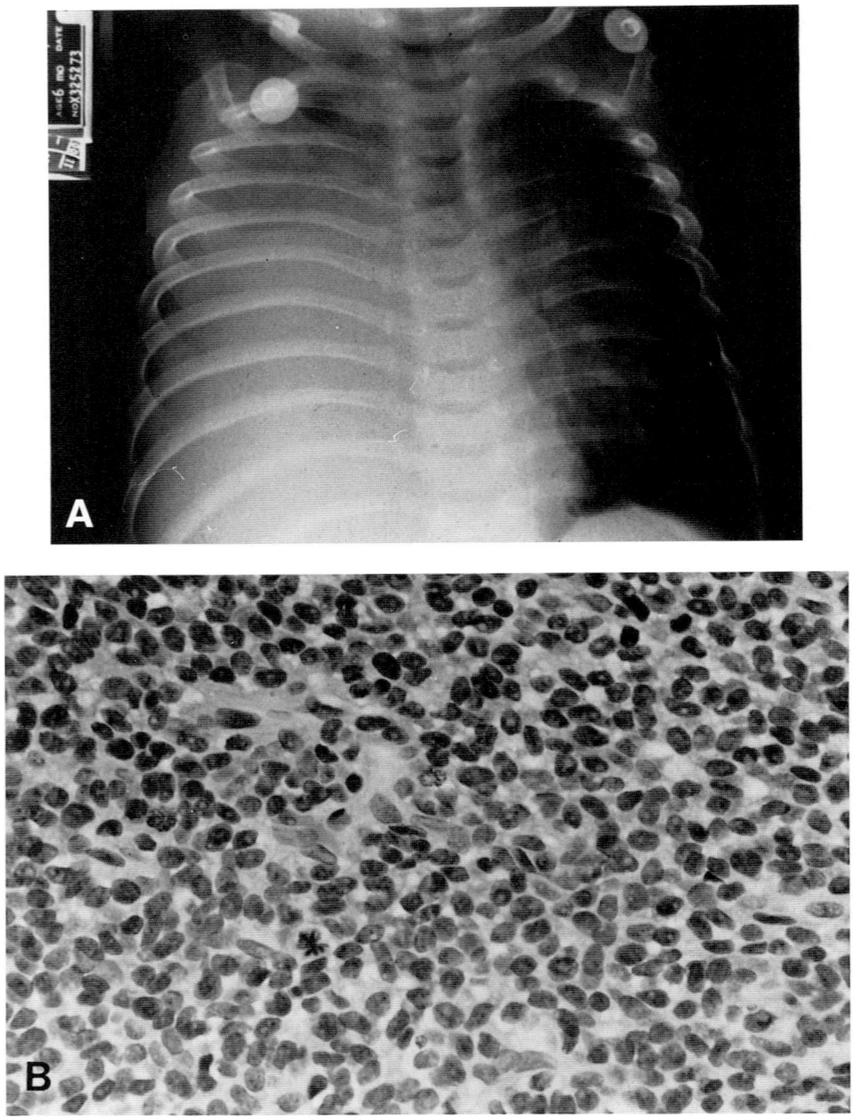

FIGURE 16.9. Primitive neuroectodermal tumor of thoracopulmonary origin in a 6-month-old male with a chest wall mass. (A) Chest radiograph reveals consolidation of the right thorax by an infiltrative mass. (B) Biopsy material shows a small-cell malignant tumor forming nests of cells, which have round or oval nuclei, coarse or clumped chromatin, small nucleoli, and indistinct or scant cytoplasms. Several mitoses are present. H&E; × 380. (C) Tumor cells are diffusely immunoreactive with the Ewing antibody 013, neuron-specific esterase, and neurofilament. They are weakly reactive with synaptophysin but unreactive with actin and desmin. MIC2 (013) immunoperoxidase; × 380. (D) Ultrastructural features consist of small primitive-appearing cells with sparse organelles. Well-defined cell junctions, neurosecretory granules, or neurofilaments are not present in this electron micrograph (× 10,000).

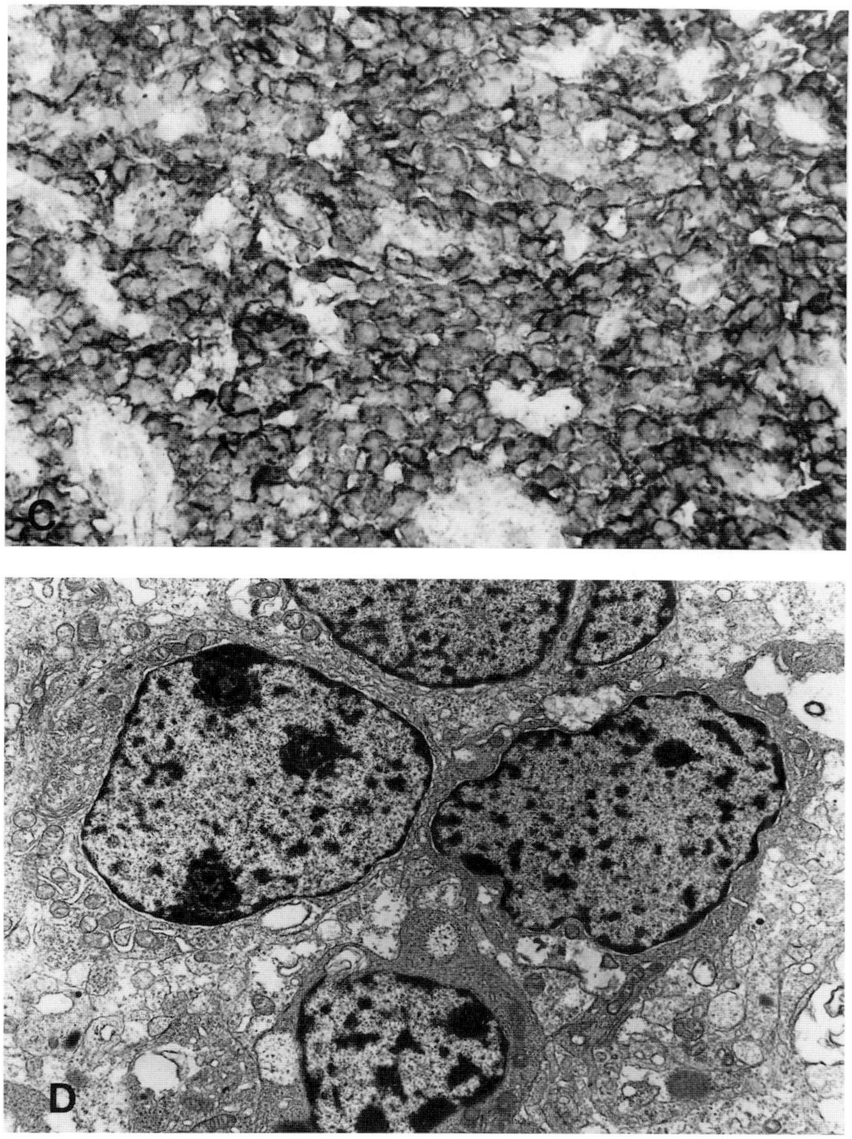

FIGURE 16.9. (*Continued*)

References

1. Dehner LP. Tumors and tumor-like lesions of the lung and chest wall in childhood: clinical and pathologic review. In: Stocker JT, ed. *Pediatric Pulmonary Disease*. Washington, DC: Hemisphere; 1989:207.
2. Stocker JT. The respiratory tract. In: Stocker JT, Dehner LP, eds. *Pediatric Pathology*. Vol I. Philadelphia: JB Lippincott; 1992: 505.
3. Hartman GE, Schochat SJ. Primary pulmonary neoplasms of childhood: a review. *Ann Thorac Surg* 1983;36:108.
4. Isaacs H Jr. Tumors. In: Gilbert-Barness E, ed. *Potter's Pathology of the Fetus and Infant*. Vol 2. St. Louis, MO: CV Mosby; 1997:1287.
5. Isaacs H Jr. Tumors of the lung. In: *Tumors of the Fetus and Newborn*. Philadelphia: WB Saunders; 1997:344. Major Problems in Pathology Series; Vol 35.
6. Katzenstein A-LA, Askin FB. Surgical pathology of non-neoplastic lung disease. In: Bennington JL, ed. Philadelphia: WB Saunders; 1990. *Major Problems in Pathology* Series; 2nd ed. Vol 13.
7. Conran RM, Stocker JT. Extralobar sequestration with frequently associated congenital cystic adenomatoid malformation, type 2: report of 50 cases. *Pediatr Develop Pathol* 1999;2:454.
8. Hunter WS, Becroft DMO. Congenital pulmonary lymphangiectasis associated with pleural effusions. *Arch Dis Child* 1984; 59:278.
9. Wilson RH, Duncan A, Hume R, et al. Prenatal pleural effusion associated with congenital lymphangiectasis. *Prenat Diagn* 1985; 5:73.
10. Lehman TJA, Church JA, Isaacs HJR. Lymphomatoid granulomatosis in a 13-month-old Infant. *J Rheumatol* 1989;16:235
11. Cox JN. Respiratory system. In: Berry CL, ed. *Paediatric Pathology*. 2nd ed. London: Springer-Verlag; 1989:293.
12. Hussong JW, Brown M, Perkins SJ, et al. Comparison of DNA ploidy, histologic, and immunohistochemical findings with clinical outcome in inflammatory myofibroblastic tumors. *Mod Pathol* 1999;12:279.
13. Matsubara O, Tan Liu NS, Kenney RM, et al. Inflammatory pseudotumor of the lung: progression from organizing pneumonia to fibrous histiocytoma or to plasma cell granuloma in 32 cases. *Hum Pathol* 1988;19:807.
14. Rosenberg HS, Stenback WA, Spjut HJ. The fibromatoses of infancy and childhood. In: Rosenberg HS, Bolande RP, eds. *Perspectives in Pediatric Pathology*. Vol 4. Chicago: Year Book Medical Publishers; 1978:269.
15. Stocker JT, Madewell JE, Drake RM. Congenital cystic adenomatoid malformation of the lung. Classification and morphologic spectrum. *Hum Pathol* 1977:8:155.
16. Holland-Moritz RM, Heyn RM. Pulmonary blastoma associated with cystic lesions in children. *Med Pediatr Oncol* 1984;12:85.
17. Manivel JC, Priest JR, Watterson J, et al. Pleuropulmonary blastoma: the so-called pulmonary blastoma of childhood. *Cancer* 1988;62:1516.
18. Kuller JA, Yankowitz J, Goldberg JD, et al. Outcome of antenatally diagnosed cystic adenomatoid malformations. *Am J Obstet Gynecol* 1992;167:1038.
19. Cha I, Adzick NS, Harrison MR, et al. Fetal congenital cystic adenomatoid malformations of the lung: a clinicopathologic study of eleven cases. *Am J Surg Pathol* 1997;21:537.
20. Morotti RA, Cangiarella J, Gutierrez MC, et al. Congenital cystic adenomatoid malformation of the lung (CCAM): evaluation of the cellular components. *Hum Pathol* 1999;30:618.
21. Moerman P, Fryns JP, Vanden Berghe K, et al. Pathogenesis of congenital cystic adenomatoid malformation of the lung. *Histopathology* 1992;21:315.
22. Mark EJ. Mesenchymal cystic hamartoma of the lung. *N Engl J Med* 1986;315:1255.
23. Becroft DMO, Jagusch MF. Pulmonary sarcomas arising in mesenchymal hamartomas. *Pediatr Pathol* 1987;7:478.
24. Bove KE. Sarcoma arising in pulmonary mesenchymal cystic hamartoma. *Pediatr Pathol* 1989;9:785.
25. Cohen SR, Geller KA, Selzer S, et al. Papilloma of the larynx and tracheobronchial bronchial tree in children. A retrospective study. *Ann Otol Rhinol Laryngol* 1980;89:497.
26. Mounts P, Kashima H. Association of human papilloma virus subtype and clinical course in respiratory papillomatosis. *Laryngoscope* 1984;94:28.
27. Kramer SS, Wehunt WD, Stocker JT, et al. Pulmonary manifestations of juvenile laryngotracheal papillomatosis. *Am J Radiol* 1985;144:687.
28. Colby TV, Koss MN, Travis WD. Tumors of the lower respiratory tract. In: *Atlas of Tumor Pathology*. 3rd series. Fascicle 13. Washington, DC: Armed Forces Institute of Pathology; 1995.
29. Siegel SE, Cohen SR, Isaacs H, et al. Malignant transformation of tracheobronchial juvenile papillomatosis without prior radiotherapy. *Ann Otol Rhinol Laryngol* 1979;88:192.
30. Dehner LP, Watterston J, Priest. Pleuropulmonary blastoma: a unique intrathoracic-pulmonary neoplasm of childhood. In: Askin FB, Langston C, Rosenberg HS, Bernstein J, eds. *Pulmonary Disease*. Basel: Karger, 1995:214. Perspectives in Pediatric Pathology Series; Vol 18.
31. Hachitanda Y, Aoyama C, Sato JK, et al. Pleuropulmonary blastoma in childhood: a tumor of divergent differentiation. *Am J Surg Pathol* 1993;17:382.
32. Priest JR, McDermott, Bhatia S, et al. Pleuropulmonary blastoma. A clinicopathologic study of 50 cases. *Cancer* 1997;80: 147.

32a. Barnard WG. Embryoma of the lung. *Thorax* 1952;7:299.

33. Ashworth TG. Pulmonary blastoma, a true congenital neoplasm. *Histopathology* 1983;7:585.
34. Koss MN, Hochholzer L, O'Leary T. Pulmonary blastomas. *Cancer* 1991;67:2368.
35. Alobeid B, Beneck D, Sreekantaiah C, et al. Congenital pulmonary myofibroblastic tumor: a case report with cytogenetic analysis and review of the literature. *Am J Surg Pathol* 1997; 21:610.
36. Jimenez JF, Uthman EO, Townsend JW, et al. Primary bronchopulmonary leiomyosarcoma in childhood. *Arch Pathol Lab Med* 1986;110:348.
37. Pettinato G, Manivel C, Saldana MJ, et al. Primary bronchopulmonary fibrosarcoma of childhood and adolescence. Reassessment of a low grade malignancy—clinicopathologic study of five cases and a review of the literature. *Hum Pathol* 1989;20:463.
38. McGinnis M, Jacobs G, El-Naggar A, et al. Congenital peribronchial myofibroblastic tumor (so-called "congenital leiomyosarcoma"). A distinct neonatal lung lesion associated with nonimmune hydrops fetalis. *Mod Pathol* 1993;6:487.
39. Chen MF, Onerheim R, Wang NS, et al. Rhabdomyomatosis of newborn lung: a case report with immunohistochemical and electron microscopic characterization of striated muscle cells in the lung. *Pediatr Pathol* 1991;11:123.

40. Drut RM, Quijano G, Drut R, et al. Rhabdomyomatous dysplasia of the lung. *Pediatr Pathol* 1988;8:385.
41. Hardisson D, Garcia-Jimenez JA, Jimenez-Heffernan JA, et al. Rhabdomyomatosis of the newborn lung unassociated with other malformations. *Histopathology* 1997;31:474.
42. Ramaswamy A, Weyers I, Duda V, et al. A tumorous type of pulmonary rhabdomyomatous dysplasia. *Pathol Res Pract* 1998; 194:639.
43. Allan BT, Day DL, Dehner LP. Primary pulmonary rhabdomyosarcoma of the lung in children. Report of two cases presenting with spontaneous pneumothorax. *Cancer* 1987;59: 1005.
44. Schmidt D, Harms D, Burdach S. Malignant peripheral neuroectodermal tumours of childhood and adolescence. *Virchows Arch [Pathol Anat]* 1985;406:351.
45. Tsokos M. Peripheral primitive neuroectodermal tumors: diagnosis, classification and prognosis. *Perspect Pediatr Pathol* 1992; 16:27.

17 Tumors and Tumorlike Conditions of Bone

Primary tumors of bone occur infrequently during the first decade of life and are particularly uncommon in infancy. Few cases have been reported [1–15]. In a series of 249 children with bone lesions all subjects were under 10 years of age and 16 were less than a year old [1]. Langerhans cell histiocytosis was the dominant neoplasm (10 cases), and the skull was the site most frequently involved (Figure 17.1). In addition, there were two examples of chest wall hamartoma and one each of Ewing sarcoma of the scapula, fibrous dysplasia of the turbinate, osteochondroma of the rib, and myofibromatosis of the mandible [1].

Symptoms of skeletal neoplasms are usually nonspecific, often leading to an erroneous diagnosis. Skeletal tumors may mimic benign conditions such as congenital malformations, osteomyelitis, or trauma [1].

Tumors and tumorlike conditions of bone in the fetus and infant are described in Table 17.1.

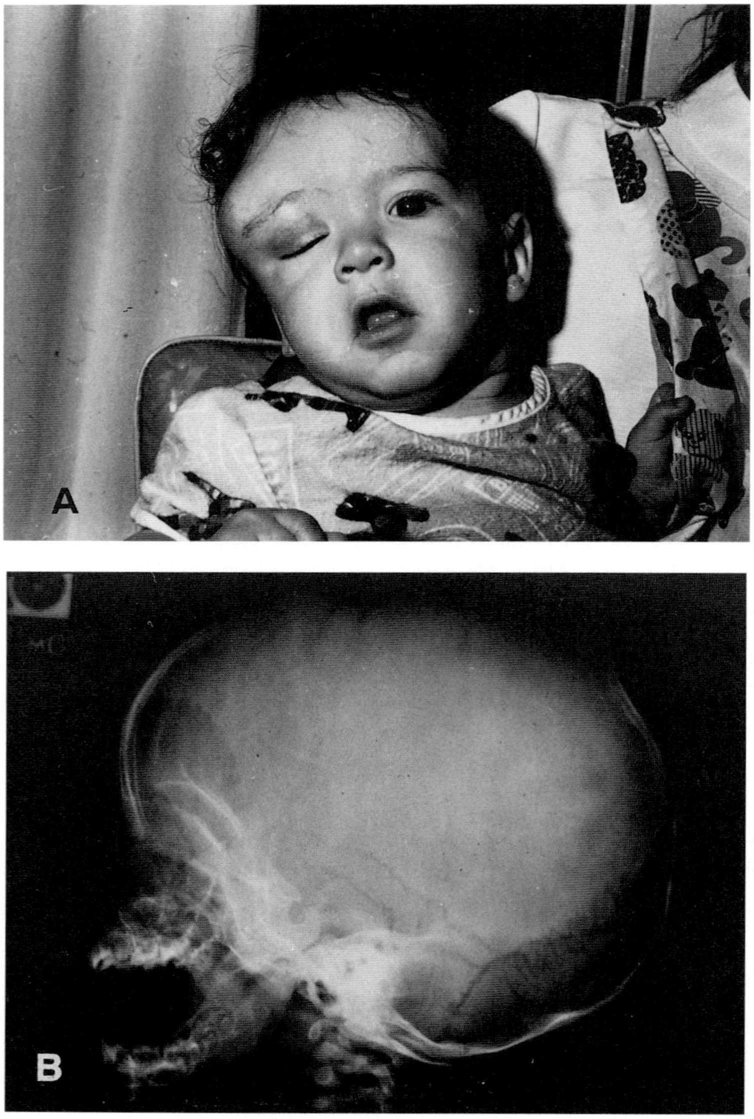

FIGURE 17.1. Langerhans cell histiocytosis. (A) A 7-month-old male with painful swelling, 3×3 cm^2, over the right orbital ridge noted 1 month prior to biopsy. The skull is the bone most commonly affected by this disease. (B) Skull radiograph, lateral view, reveals a lytic defect in the right frontal bone. Characteristically no reaction is present along the advancing margins of the lesion. (C) Cranial bone is replaced by a cellular infiltrate. H&E; $\times 48$. (D) The infiltrate consists mostly of histiocytes with eosinophilic cytoplasm, occasionally vacuolated, and infolded nuclei and one or two small nucleoli. A few lymphocytes and neutrophils are present; S-100 is strongly positive in the histiocytes. H&E; $\times 150$. (From Isaacs H Jr [11], with permission.)

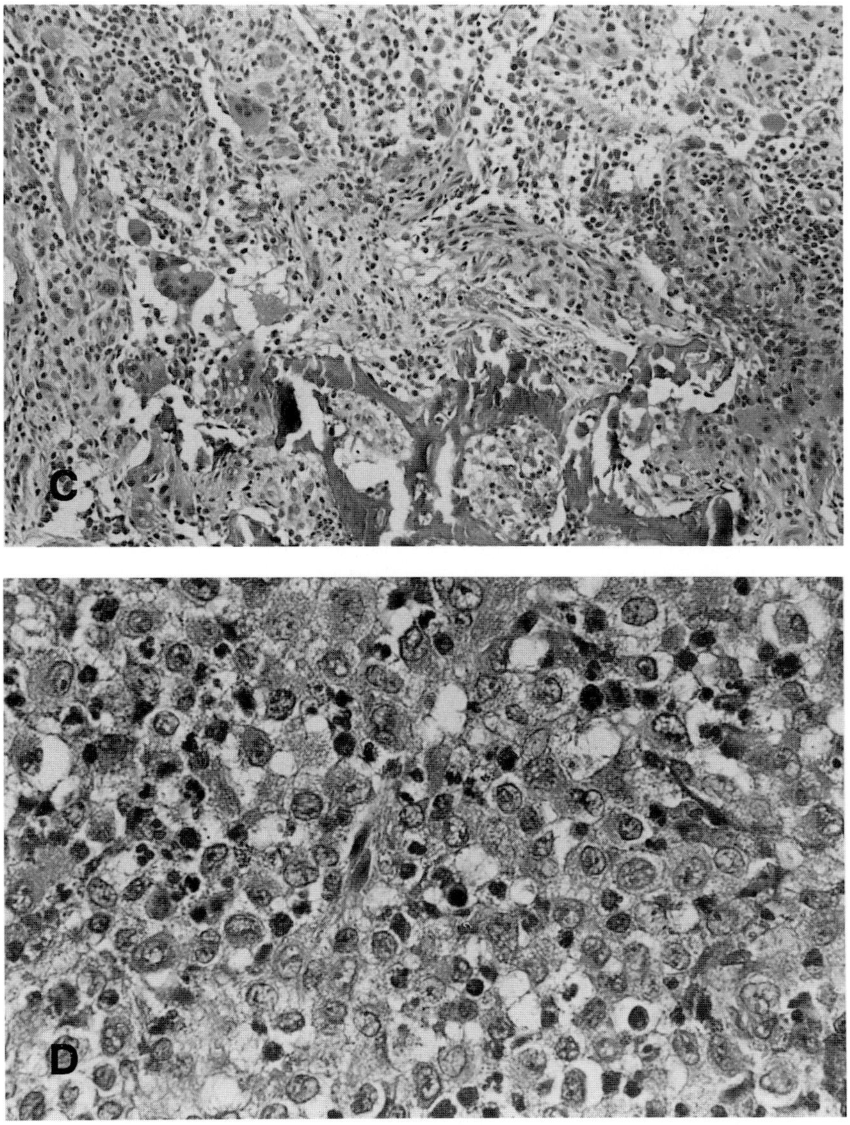

FIGURE 17.1. (*Continued*)

TABLE 17.1. Tumors and tumorlike conditions of bone in the fetus and infant.

Chest wall hamartoma
Nasal chondromesenchymal hamartoma
Langerhans cell histiocytosis
Congenital myofibromatosis
Fibrous dysplasia
Osteofibrous dysplasia
Melanotic neuroectodermal tumor of infancy
Enchondroma (Ollier disease)
Reparative giant cell granuloma
Juvenile xanthogranuloma
Osteochondromyxoma
Congenital fibrosarcoma
Chordoma
Osteolysis ("disappearing bone disease")

Source: Isaacs H Jr [10].

Chest Wall Hamartoma

The chest wall hamartoma has been variously referred to as benign mesenchymoma, aneurysmal bone cyst, and cartilagenous hamartoma [8,11–13,16,17]. Characteristically it is a tumorlike condition of newborns and infants involving one or more adjacent ribs and producing a mass and deformity of the chest wall; occasionally these lesions are multifocal [11, 17]. When the hamartomas are extensive, they compress the lungs, causing respiratory distress. The lesion has been detected antenatally by sonography [16].

The involved rib is expanded, partially or totally replaced by a cystic lesion composed of cartilagenous, osseous, fibrous, and vascular elements (Figures 17.2–17.4). Microscopically

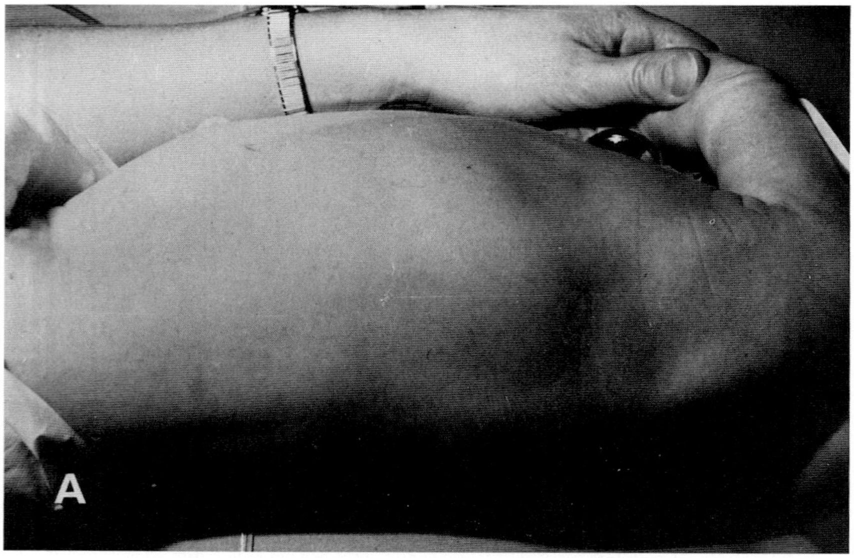

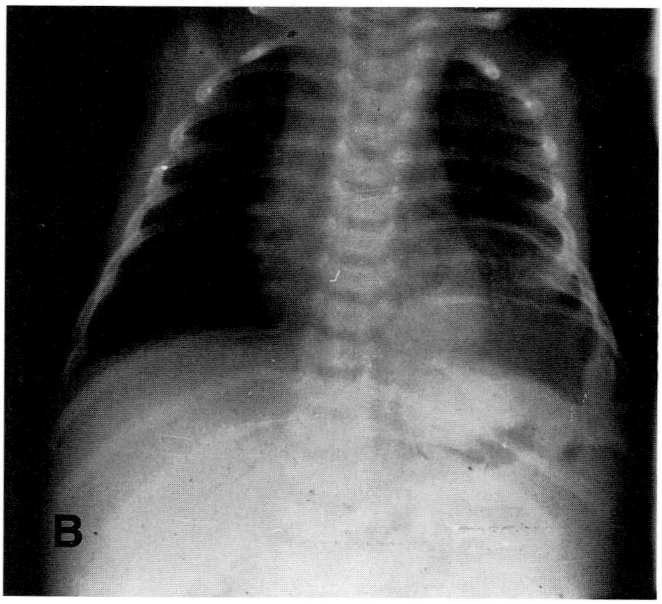

FIGURE 17.2. Chest wall hamartoma. (A) A 1-month-old male with a bulge over the left chest wall since birth. (B) Cystic, expansile lesion involving the left eighth rib. (C) The soft, hemorrhagic contents of the markedly expanded rib are covered by a thin, "eggshell" layer of bone. (D) The vascular component consists of irregular vascular channels filled with blood separated by fibrovascular tissue containing osteoclasts. H&E; × 150. (E) Nodules of cartilage and spicules of partially decalcified bone are surrounded by vascular connective tissue. H&E; × 48. (D and E from Isaacs H Jr [11], with permission.)

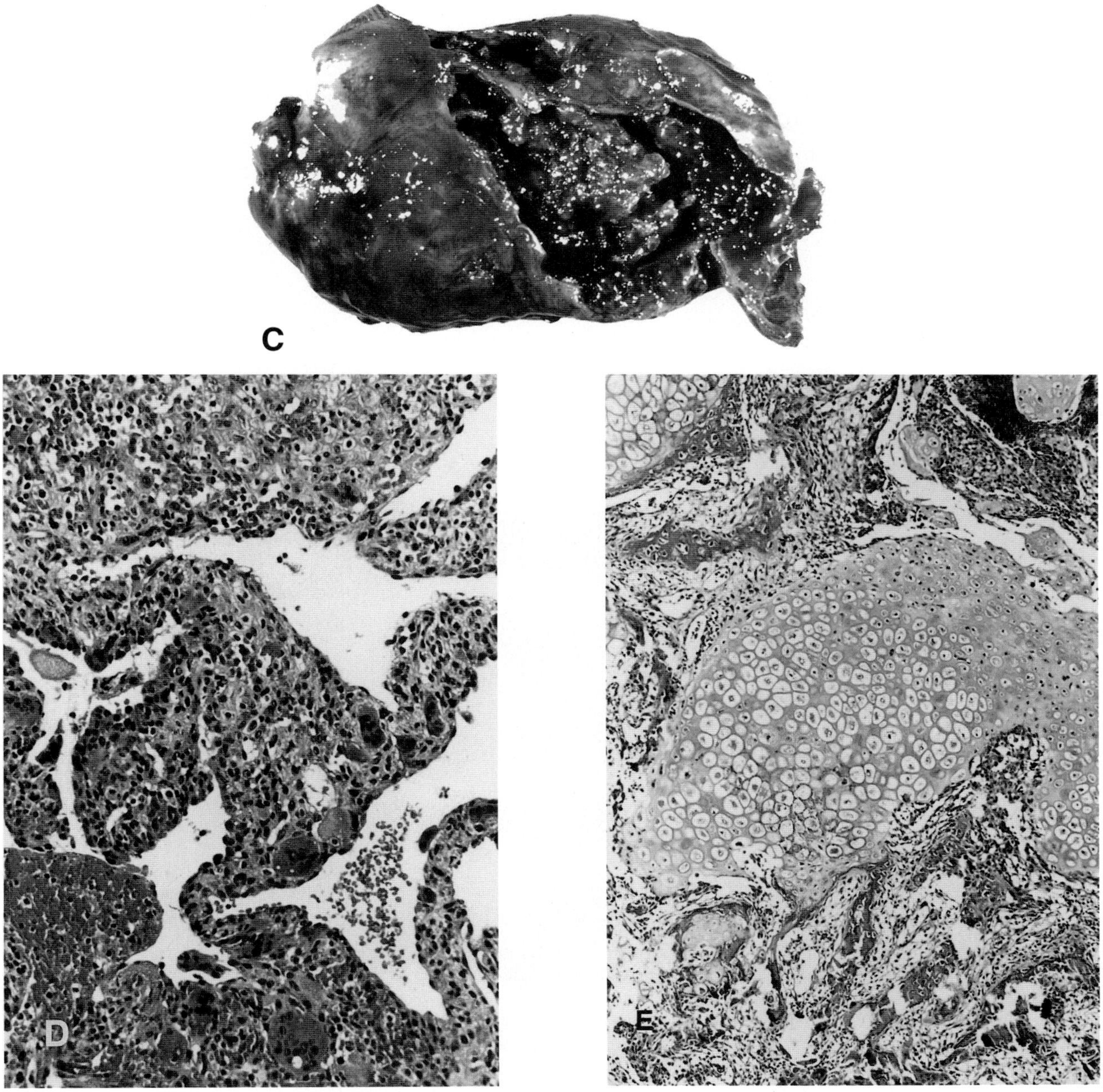

FIGURE 17.2. (*Continued*)

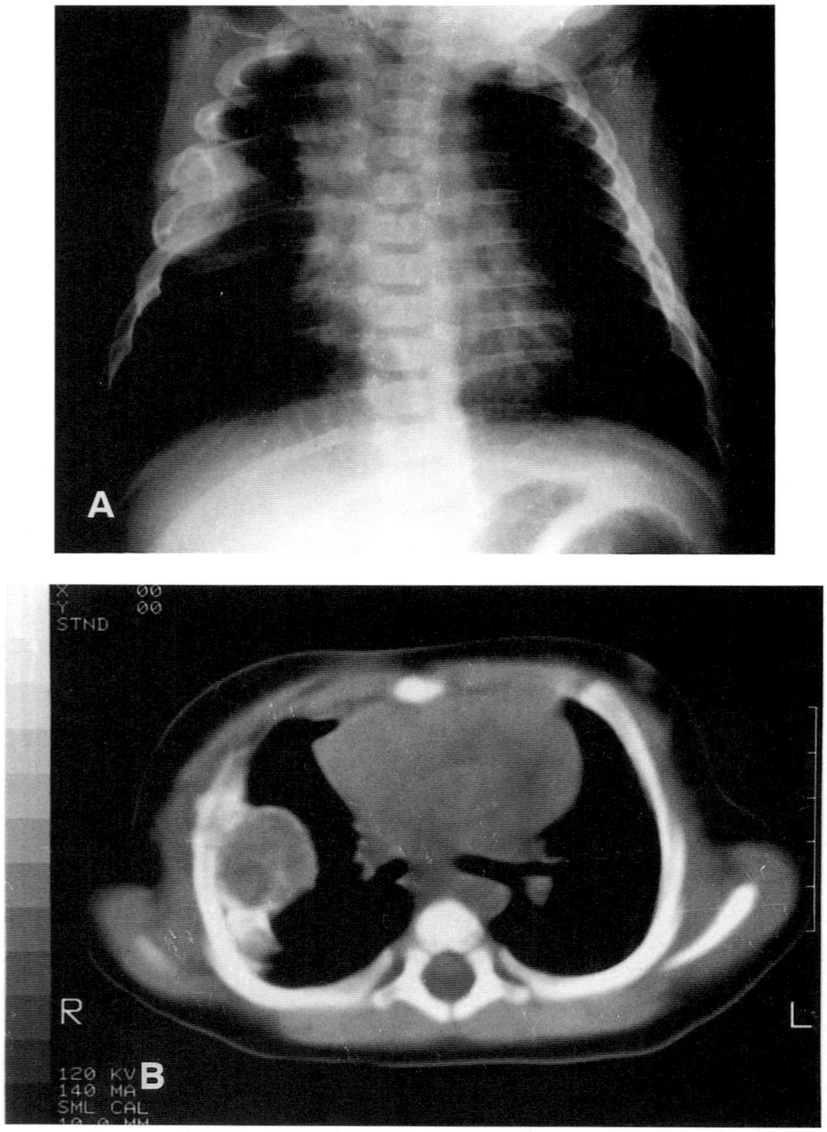

FIGURE 17.3. Chest wall hamartoma in a 5-month-old male with a chest mass. (A) A cystic, expansile lesion involves the right fourth rib. Scoliosis and chest wall deformity, present here, are complications of this entity. (B) Computerized tomography scan reveals a defect in the right chest wall and a cystic lesion bulging into the thoracic cavity. (C) The involved rib is divided in half, revealing an expansile, spongelike mass occupying the rib and reminiscent of an aneurysmal bone cyst. (D) Vascular, cartilaginous, and osseous components are illustrated. H&E; $\times 48$.

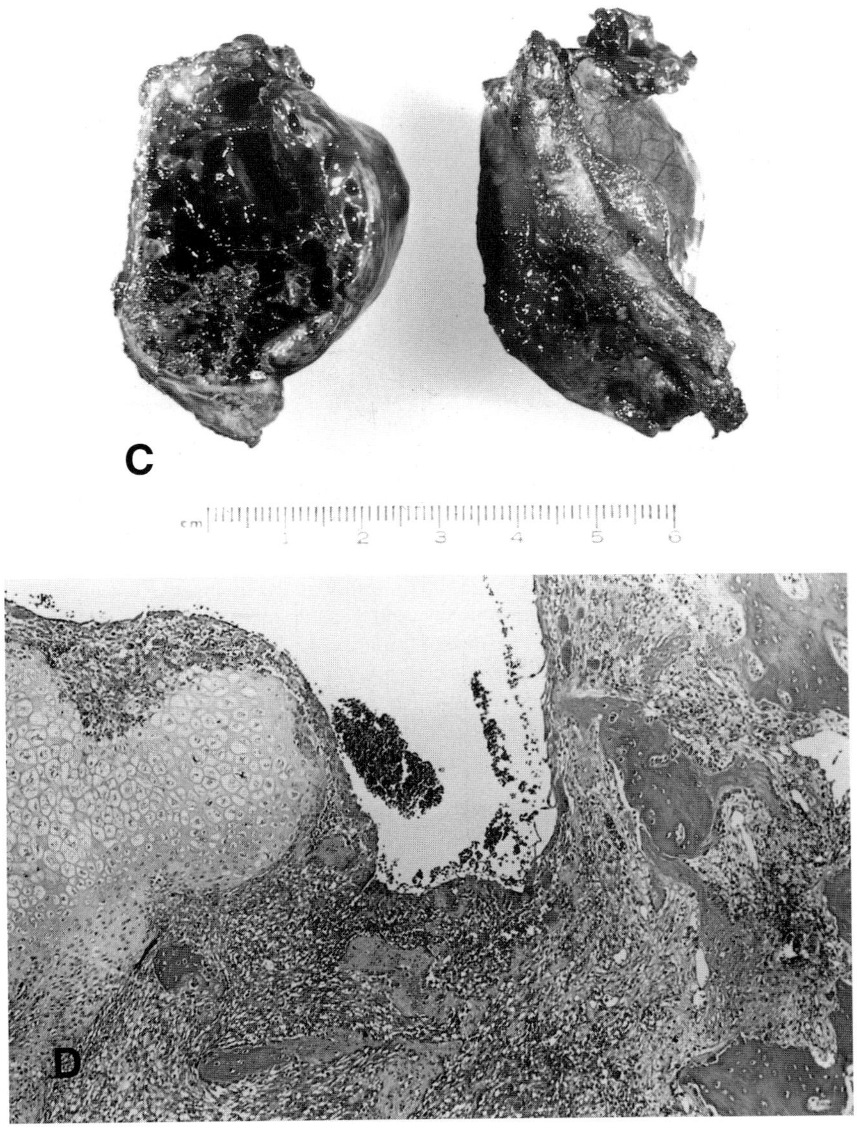

FIGURE 17.3. (*Continued*)

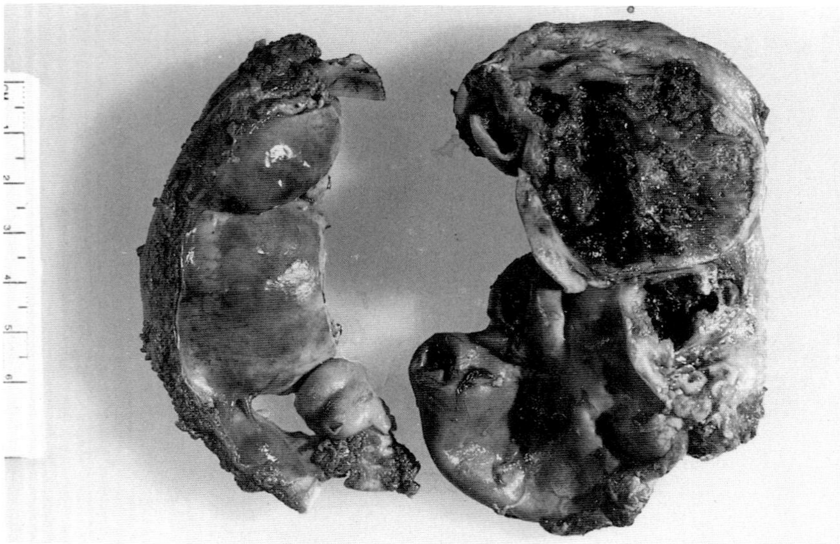

FIGURE 17.4. Chest wall hamartoma in a 4-month-old male with a chest mass. The lesion measured $10 \times 5 \times 5$ cm^2 and occupied two contiguous ribs and the intercostal muscle in between. The external surface is on the left and the cut surface on the right side of the photograph. The ribs are nodular, expanded, and deformed and are replaced vascular connective tissue surrounded by a thin rim of bone and periosteum. Microscopic sections of the hamartoma revealed prominent vascular channels separated by fibrous connective tissue septa with foci of osteoid and proliferating cartilage similar to that depicted in Figures 17.2 & 17.3.

the hamartoma resembles an aneurysmal bone cyst with striking proliferation of hyaline cartilage in addition to bone and fibrous connective tissue [8,11,12,13]. The lesion probably represents a malformation rather than a true neoplasm [13]. Nasal chondromesenchymal hamartoma shares certain histological features with chest wall hamartoma [14,15].

Nasal Chondromesenchymal Hamartoma

Nasal chondromesenchymal hamartoma is the name applied by McDermott, Ponder, and Dehner to describe a tumor of the nasal passages and paranasal sinuses with a detectable mass in the nose [14]. The patients are usually 3 months of age or less at the time of diagnosis and present with a nasal mass. Imaging studies reveal a complex solid and cystic lesion occupying the nasal cavity, with extension into the ethmoid sinus [14,15,18] (Figure 17.5). Occasionally intracranial and orbital involvement occurs [18].

Grossly the hamartoma consists of fragments of white, firm fibrous tissue containing nodules of bone and cartilage [18]. It displays a varied histological appearance consisting of islands of osseous tissue, nodules of cartilage composed of both mature and immature (but not atypical-looking) chondrocytes, a myxoid stroma with spindle-shaped fibroblastic cells, focal osteoclastic giant cells, and a prominent vascular component similar to an aneurysmal bone cyst [14,15,18] (Figure 17.5). Immunohistochemical studies reveal cartilage nodules with S-100 reactivity and mesenchymal spindle cells showing vimentin and variable actin expression [14]. The stromal spindle-shaped cells exhibit fibroblastic and myofibroblastic features by electron microscopy, complementing the immunohistochemical findings [15,18].

Surgical excision is the treatment of choice [14]. Neither recurrence nor instances of metastases have been reported. However, patients with residual, unresectable tumors are described in several reports [14,18].

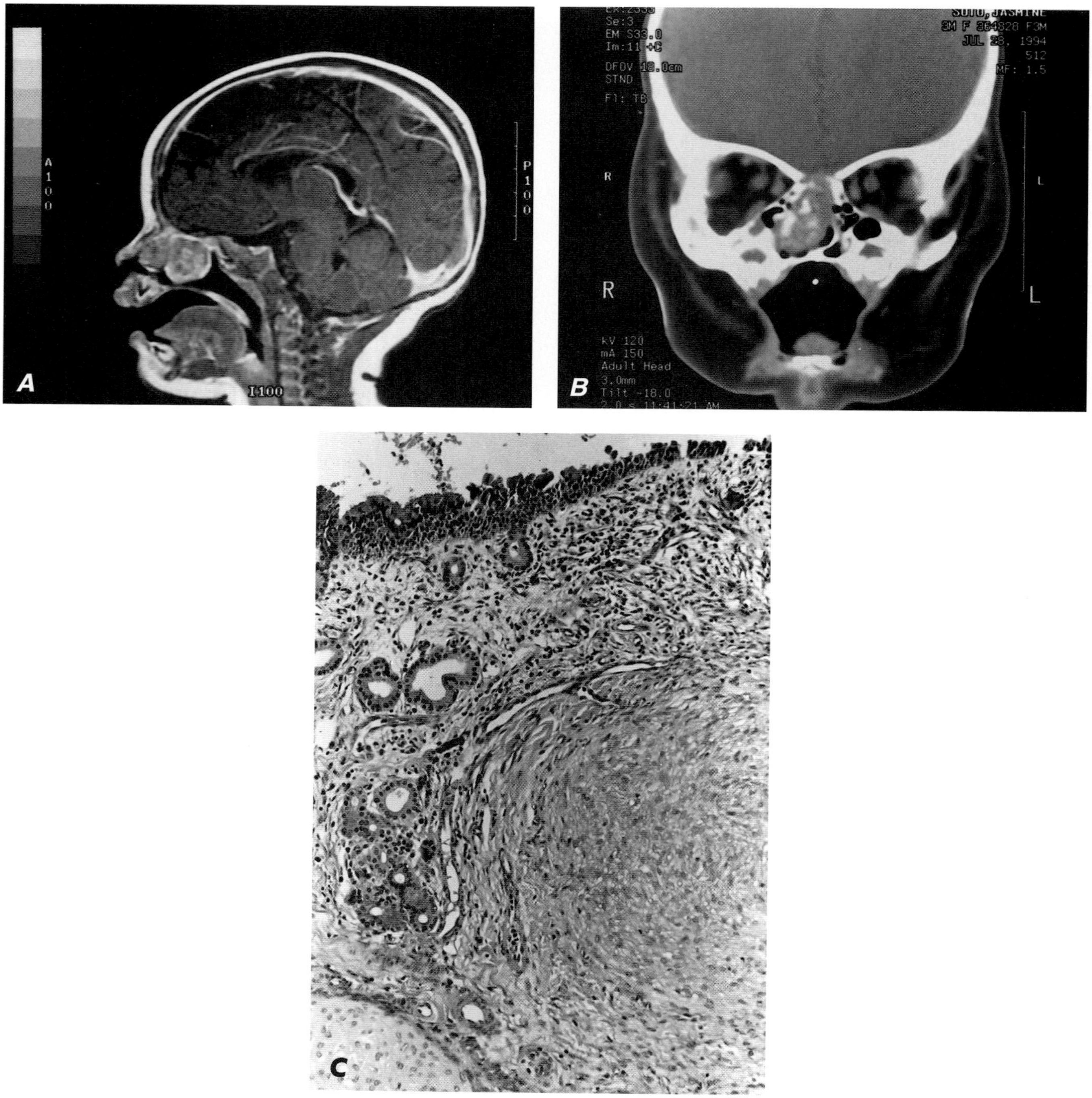

FIGURE 17.5. Nasal chondromesenchymal hamartoma in a 3-month-old female with a nasal mass. (A) Magnetic resonance imaging shows a mass with heterogeneous areas of high density situated within the nasal cavity beneath the cribriform plate. (B) Computerized tomography scan demonstrates the mass within the right nasal cavity and adjacent paranasal sinuses. The lesion extends across the midline, deviating the nasal septum, and perforates the cribriform plate. (C) Chondroid nodules are present beneath metaplastic, chronically inflamed respiratory mucosa. The stroma is composed of tiny round and spindle-shaped cells surrounding hyaline cartilage. H&E; × 120. (D) Reactive new bone formation, vascular channels, and a cellular spindle cell stroma are present. H&E; × 120. (E) Hyalinization about blood vessels is a prominent feature. H&E; × 120. (The author expresses appreciation to Louis Dehner, M.D., Barnes-Jewish Hospital and St. Louis Children's Hospital, St. Louis, Missouri, for reviewing the case and for providing the diagnosis.)

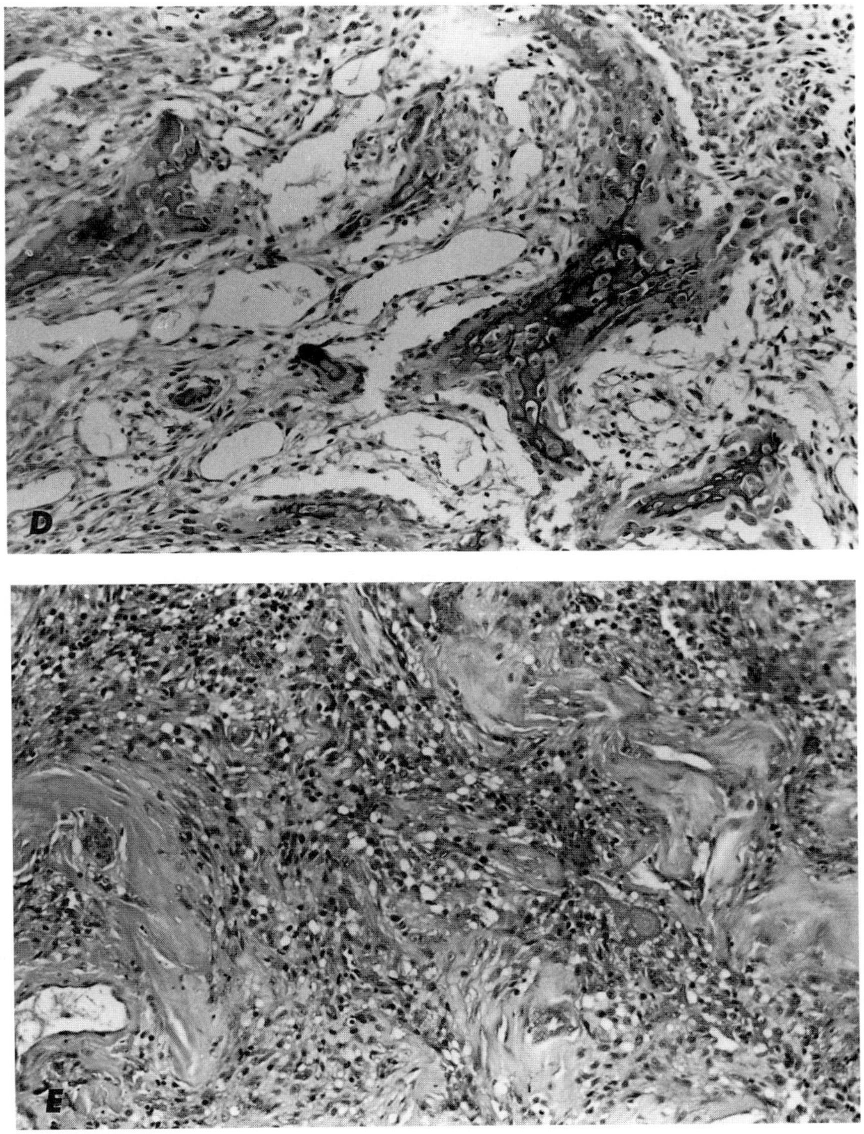

FIGURE 17.5. (*Continued*)

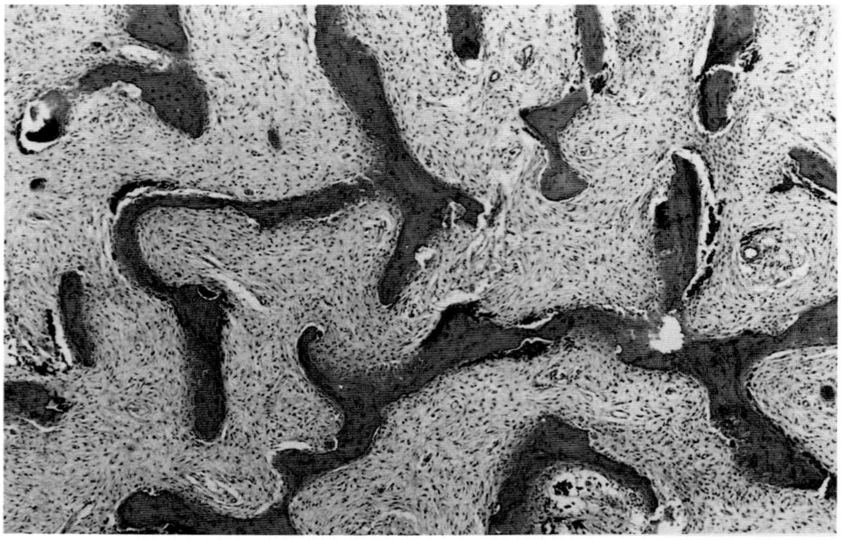

FIGURE 17.6. Fibrous dysplasia: irregularly shaped, narrow, immature trabeculae of bone are situated in a moderately cellular, pale-staining fibrous stroma. There is little or no osteoblastic activity. H&E; ×48.

Fibrous Dysplasia of Bone

This disease has several variations, in all of which the fundamental lesion is a fibrous dysplasia of localized portions of one or several bones. It appears to be a congenital developmental defect, without evidence of a genetic cause. Fibrous dysplasia may be present at birth, or it may not become evident until later in childhood [9].

The monostotic form usually affects children or young adults and is rare in infants. The bones most often involved are the ribs, tibia, femur, mandible, and maxilla. The first sign may be a fracture, or there may be local swelling and tenderness. The child is usually otherwise normal [9].

Histologically the marrow and osseous structures in well-delineated areas are replaced by fibrous connective tissue and poorly formed, haphazardly arranged, immature bone that forms trabeculae lacking normal osteoblastic activity [9] (Figure 17.6).

Polyostotic fibrous dysplasia is characterized by lesions in many bones, cutaneous café-au-lait spots, and sexual precocity, also known as the McCune-Albright syndrome. Early lesions of fibrous dysplasia (e.g., osteoporosis) may appear as early as one month of age [19]. Later on portions of flat bones become distended into tumorlike masses, while the long bones are curved, shortened, and thickened. The serum calcium and phosphorus levels are normal, but serum phosphatase is elevated [9].

The McCune-Albright syndrome when associated with Cushing syndrome and adrenal nodular hyperplasia may become evident shortly after birth, causing life-threatening metabolic problems or even sudden, unexpected death [20] (see Chapter 12, Adrenocortical Tumors).

Enchondromatosis (Ollier Disease)

Although enchondromatosis is rarely recognized at birth, it may be identified as early as 6 months of age [9,10]. The lesion consists of multiple cartilagenous tumors originating in the enchondral growth zone. The tumors grow with the remainder of the shaft and extend in a triangular fashion toward the diaphysis. Radiologically they appear as rarefactions in the metaphysis and the end of the diaphysis, with expansion of the cortex around the tumor. The process is accompanied by shortening and bowing of the involved long bones and by distortion of the nearby joints. The tumors are usually unilateral and irregularly distributed [9].

Histologically enchondromas are lobulated with marked variation in cellularity, organization, calcification, and degeneration of cartilage (Figure 17.7). In about a third of adults with multiple enchondromatosis the enchondromas undergo malignant change to chondrosarcoma [21].

Unless sarcomas develop, the prognosis is relatively good, although fractures and significant joint deformities occur. Some cases are complicated in later childhood by the presence of vascular anomalies, particularly angiomas, phlebectasis, or hemangiomas [9,21]. When all these findings are present, the disease is referred to as the Maffucci-Kast syndrome.

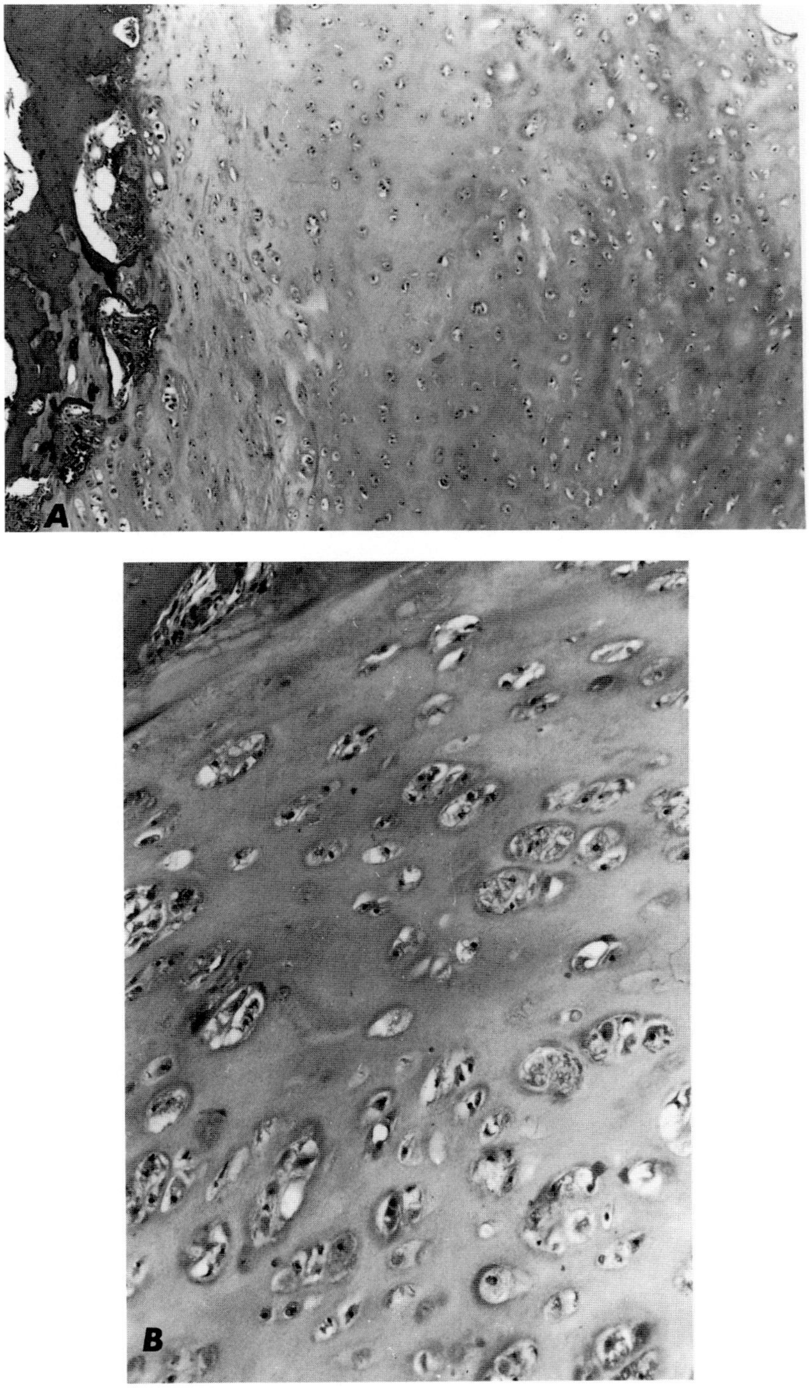

FIGURE 17.7. Enchondromatosis (Ollier disease) in a child. (A) The tumor consists of chondrocytes in lacunae surrounded by abundant hyaline chondroid matrix. Normal bone is to the left. (B) The tumor is hypercellular, with large numbers of multinucleated chondrocytes. (Courtesy of Rob Newbury, M.D., Department of Pathology, Children's Hospital, San Diego.)

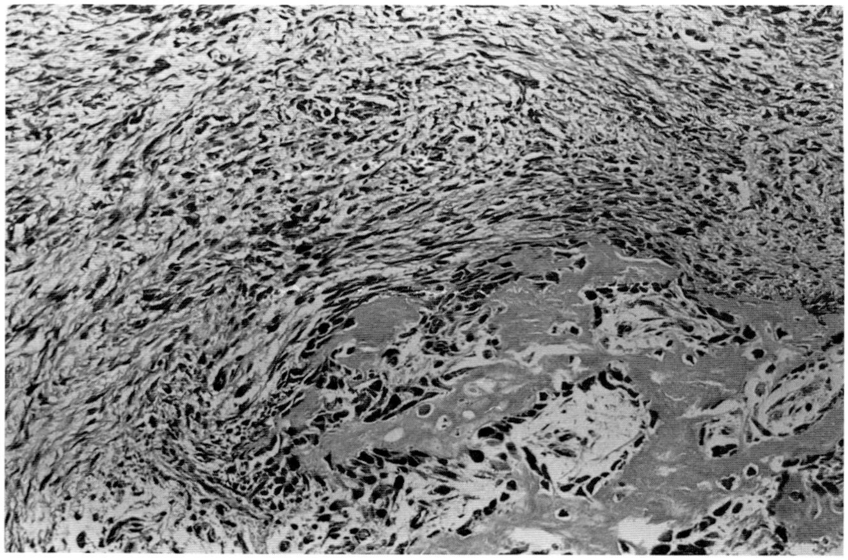

FIGURE 17.8. Myofibromatosis in a 3-week-old male with multiple subcutaneous nodules situated over the face, chest, arms, and in the bones. Biopsy of a jaw lesion shows an island of reactive bone surrounded by spindle-shaped myofibroblastic cells. H&E; × 150.

Congenital Generalized Fibromatosis (Infantile Myofibromatosis)

Generalized myofibromatosis, discussed also in Chapter 3, Soft Tissue Tumors, occcasionally presents predominantly as bone lesions or may be limited entirely to the bones [7,22–25]. The lesions may remain unnoticed until the infant presents with a pathological fracture. Solitary myofibromatosis of bone, typically involving the craniofacial bones, occurs [7,25]. The tumor may cause significant orbital bone destruction and proptosis, mimicking a malignant neoplasm [25]. The radiological features consist of a lytic lesion with a sclerotic rim involving the craniofacial bones, femur, tibia, vertebra, and rib [7,24,25].

Microscopically myofibroma consists of bundles of spindle-shaped fibroblasts and myofibroblasts with central vascular spaces (Figure 17.8). The light microscopic, immunohistochemical, and ultrastructural features are similar to those of myofibromatoses occurring in skin, soft tissue, or viscera [7,23,25].

Ewing Sarcoma

Ewing sarcoma is rare in infants. It is primarily a tumor of older children and adolescents, found more often in girls than in boys during the first 3 years of life [1,6,26]. Ewing sarcoma and primitive neuroectodermal tumor (PNET) are considered histogenetically related, both arising from the neural crest. These highly malignant tumors have a consistent cytogenetic abnormality, that is, a translocation t(11;22)(q24;Q12) and similar histological findings [26,27].

Clinically, young patients with Ewing sarcoma present with pain and swelling of the affected part for several weeks, sometimes accompanied by fever and leukocytosis, mimicking osteomyelitis. Pathological fracture may be the first sign of the tumor [1,26].

On plain films or CT scans the tumor, which frequently affects a rib or flat bone, has a lytic, destructive ("moth-eaten") appearance. The diaphysis is the site of involvement in the long bones (e.g., the humerus, tibia, femur). A soft tissue mass and layers of periosteal new bone formation ("onionskin" appearance) are present about the periphery. The radiographic differential diagnosis of Ewing sarcoma in the infant includes osteomyelitis, Langerhans cell histiocytosis, and metastatic neuroblastoma [1,9].

Tumor currettings or biopsy specimens tend to be soft, mushy, light gray with extensive necrosis. Some are almost liquid in consistency. Because of these features, not infrequently a barely adequate biopsy sample is taken for diagnosis.

Ewing sarcoma is a prime example of one of the small blue-cell malignant tumors of childhood consisting of sheets or nests of small cells containing round or oval nuclei with a fine, diffuse chromatin pattern and tiny nucleoli [9,26–28] (Figure 17.9). Mitotic figures are common. Periodic acid–Schiff stain for glycogen is positive in about half the cases. Immunoperoxidase studies show that Ewing sarcoma cells are immunoreactive with vimentin and the cell surface antigen HBA71 (013, CD99 Ewing antigen) and occasionally are focally reactive with neuron-specific enolase and S-100 protein [27,28] (Table 3.2). The skeletal muscle markers desmin and

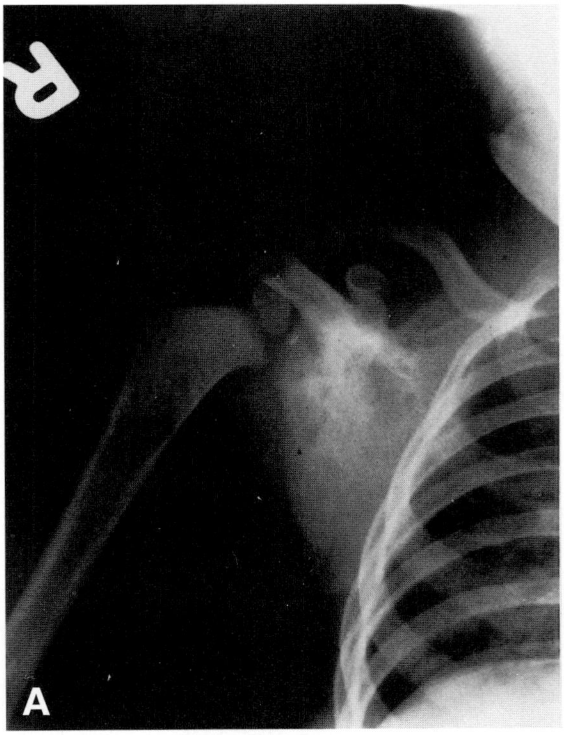

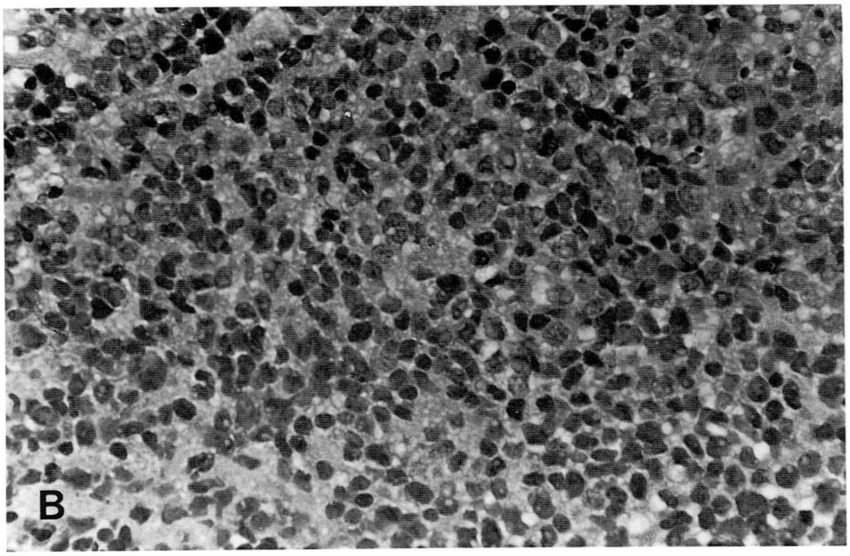

FIGURE 17.9. Ewing sarcoma in a 9-month-old female who presented with pain, a mass in the right shoulder, and refusal to use the right extremity. (A) Radiograph depicts diffuse loss of the normal contour of the right scapula. The tumor shows mixed lytic and sclerotic areas and an adjacent soft tissue mass. (B) The tumor consists of nests of small cells containing round or oval nuclei with a fine, diffuse chromatin pattern and tiny nucleoli. The cytoplasm is scanty and indistinct. Immunohistochemical studies revealed strong immunoreactivity for the Ewing sarcoma antigen 013 and weak reactivity with neuron-specific enolase; vimentin, leukocyte common antigen, desmin, and actin were nonreactive. H&E; × 300. (C) "Touch" preparation imprint shows scanty to practically absent cytoplasm. One or more tiny nucleoli are present. Wright stain; × 480. (D) Ultrastructural findings consist of small primitive cells with sparse organelles and poorly formed intercellular attachments. This electron micrograph (× 6000) shows no evidence of neural differentiation.

(*Continued*)

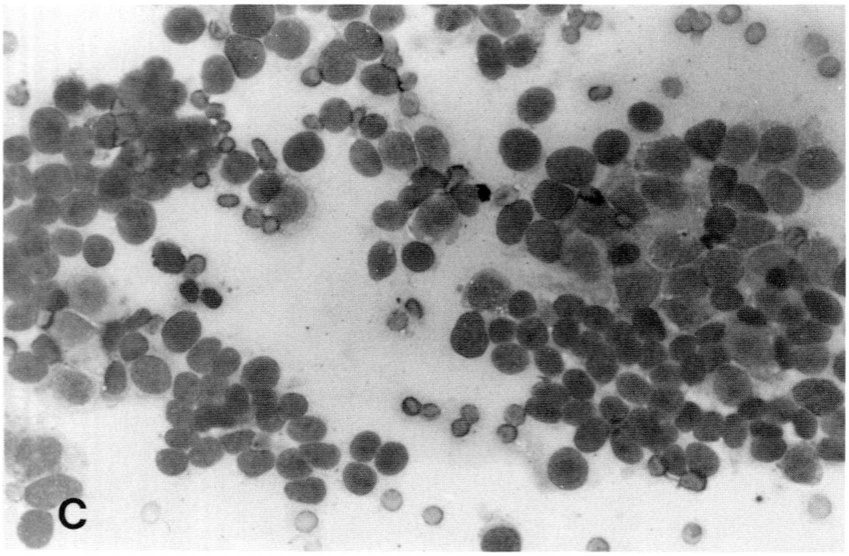

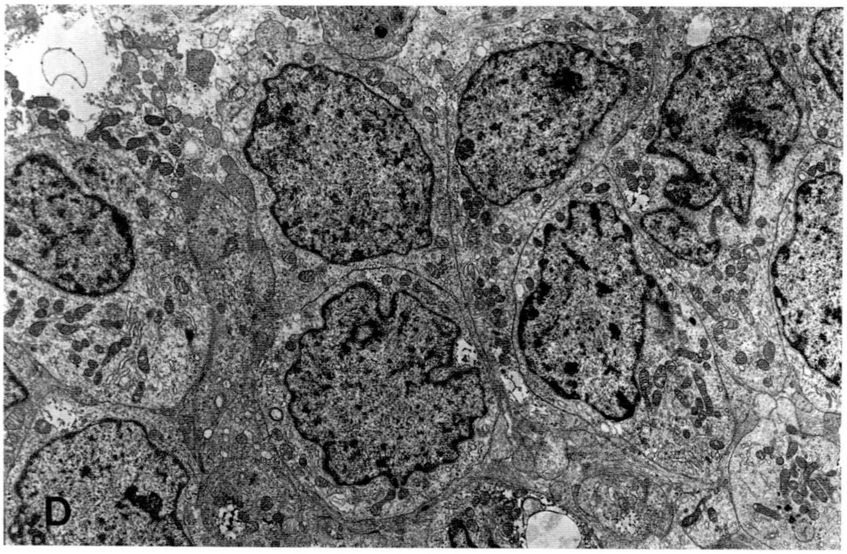

FIGURE 17.9. (*Continued*)

actin are not expressed. Electron microscopy reveals primitive-appearing small cells with a high nuclear-to-cytoplasmic ratio, sparse organelles, variable glycogen deposits, and rudimentary cell junctions (Figure 17.9D). Occasionally neural differentiation in the form of dense core granules and neurofilaments is found (see Chapter 3, Soft Tissue Tumors).

Congenital Fibrosarcoma

Congenital fibrosarcoma of the soft tissues is relatively common, but primary osseous fibrosarcoma is unusual [10,29,30]. Distinguishing between congenital fibrosarcoma and fibromatosis of bone may be as difficult as it is in the soft tissues [30]. Compared with fibrosarcomas, fibromatoses are less cellular and more fibrous in histological appearance [10,23]. Congenital fibrosarcoma is a highly cellular neoplasm composed of small spindle-shaped cells arranged in whorled and herringbone growth patterns with relatively little interstitial collagen. Nuclei are small and slightly irregular, and many mitotic figures are present [10,29,30]. The femur and humerus are the primary sites most often reported in infants [10,29,30]. If the tumor is extensive, limb salvage or amputation may be required for cure.

References

1. Senac MO Jr, Isaacs H Jr, Gwinn JL. Primary lesions of bone in the 1st decade of life: Retrospective survey of biopsy results. *Radiology* 1986;160:491.
2. Carney JA, Boccon-Gibod L. Jarka DE, et al. Osteochondromyxoma of bone. A congenital tumor associated with lentigines and other unusual disorders. *Am J Surg Pathol* 2001;25:164.
3. Anderson MJ, Townsend DR, Johnston JO, et al. Osteofibrous dysplasia in the newborn: report of a case. *J Bone Joint Surg* 1993;75-A:265.
4. Coffin CM, Swanson PE, Wick MR, et al. Chordoma in childhood and adolescence. A clinicopathologic analysis of 12 cases. *Arch Pathol Lab Med* 1993;117:927.
5. Cohen D, Granda-Ricart MC. Giant cell reparative granuloma of the base of the skull in a 4-month-old infant—CT findings. *Pediatr Radiol* 1993;23:319.
6. Hsieth H-Y, Hsiao C-C, Chen W-S, et al. Congenital Ewing's sarcoma of the humerus. *Br J Radiol* 1998;71:1313.
7. Inwards CY, Unni KK, Beabout JW, et al. Solitary congenital fibromatosis (infantile myofibromatosis) of bone. *Am J Surg Pathol* 1991;15:935.
8. Cohen MC, Drut R, Garcia C, et al. Mesenchymal hamartoma of the chest wall: a cooperative study with review of the literature. *Pediatr Pathol* 1992;12:525.
9. Isaacs H Jr. Tumors. In: Gilbert-Barness E, ed. *Potter's Pathology of the Fetus and Infant.* Vol 2. St. Louis, MO: CV Mosby, 1997:1287.
10. Isaacs H Jr. Bone tumors. In: *Tumors of the Fetus and Newborn.* Philadelphia: WB Saunders; 1997:357. Major Problems in Pathology Series; Vol 35.
11. Isaacs H Jr. Neoplasms in infants: a report of 265 cases. *Pathol Annu* 1983;18(2):165.
12. McCarthy EF, Dorfman HD. Vascular and cartilagenous hamartoma of the ribs in infancy with secondary aneurysmal bone cyst formation. *Am J Surg Pathol* 1980;4:247.
13. Odell JM, Benjamin DR. Mesenchymal hamartoma of chest wall in infancy: natural history of two cases. *Pediatr Pathol* 1986; 5:135.
14. McDermott MB, Ponder TB, Dehner LP. Nasal chondromesenchymal hamartoma. An upper respiratory tract analogue of the chest wall mesenchymal hamartoma. *Am J Surg Pathol* 1998; 22:425.
15. Hsueh C, Hsueh S, Gonzalez-Crussi F, et al. Nasal chondromesenchymal hamartoma in children. Report of 2 cases with review of the literature. *Arch Pathol Lab Med* 2001;125:400.
16. Jung AL, Johnson DG, Condon VR, et al. Congenital chest wall mesenchymal hamartoma. *J Perinatol* 1994;149:487.
17. Troum S, Dalton ML, Donner RS, et al. Multifocal mesenchymal hamartoma of the chest wall in infancy. *J Pediatr Surg* 1996; 31:713.
18. Kato K, Ijiri R, Tanaka, et al. Nasal chondromesenchymal hamartoma of infancy: the first Japanese case report. *Pathol Int* 1999; 49:731.
19. Kirk JMW, Brain CE, Carson DJ, et al. Cushing's syndrome caused by nodular adrenal hyperplasia in children with McCune-Albright syndrome. *J Pediatr* 1999;134:789.
20. Shenker A, Weinstein LS, Moran A, et al. Severe endocrine and nonendocrine manifestations of the McCune-Albright syndrome associated with activating mutations of stimulatory G protein Gs. *J Pediatr* 1993;123:509.
21. Fechner RE, Mills SE. Tumors of the bones and joints. In: *Atlas of Tumor Pathology.* 3rd ser. Fascicle 8. Washington, DC: Armed Forces Institute of Pathology; 1991.
22. Chung EB, Enzinger FM. Infantile myofibromatosis. *Cancer* 1981;48:1807.
23. Rosenberg HS, Stenback WA, Spjut HJ. The fibromatoses of infancy and childhood. In: Rosenberg HS, Bolande RP, eds. *Perspectives in Pediatric Pathology.* Vol 4. Chicago: Year Book Medical Publishers; 1978:269.
24. Chan Y-F, Lau JHK, Tong CY. Congenital generalized fibromatosis with predominant osseous involvement in a Chinese newborn. *J Pediatr Orthopaed* 1989;9:64.
25. Shields CL, Husson M, Shields JA, et al. Solitary intraosseous infantile myofibroma of the orbital roof. *Arch Ophthalmol* 1998;116:1528.
26. Maygarden SJ, Askin FB, Siegel GP, et al. Ewing sarcoma of bone in infants and toddlers: a clinicopathologic report from the Intergroup Ewing's Study. *Cancer* 1993;71:2109.
27. Dehner LP. Neuroepithelioma (primitive neuroectodermal tumor) and Ewing's sarcoma. At least a partial consensus. *Arch Pathol Lab Med* 1994;118:606.
28. Navarro S, Cavazzana AO, Llombart-Bosch A, et al. Comparison of Ewing's sarcoma of bone and peripheral neuroepithelioma. An immunocytochemical and ultrastructural analysis of two primitive neuroectodermal neoplasms. *Arch Pathol Lab Med* 1994;118:608.
29. Bernado L, Admella C, Lucaya J, et al. Infantile fibrosarcoma of the femur. *Pediatr Pathol* 1987;7:201.
30. Faure C, Gruner M, Boccon-Gibod L. Case report 149. *Skeletal Radiol* 1981;6:208.

Index

ISBN 0-387-95186-5

EAN

9 780387 951867 >